D0301576

ELEMENTARY REAL ANALYSIS

Brian S. Thomson / Simon Fraser University

Judith B. Bruckner

Andrew M. Bruckner / University of California,
Santa Barbara

PRENTICE-HALL, Upper Saddle River, New Jersey 07458

Library of Congress Cataloging-in-Publication Data

Thomson, Brian S.
Elementary real analysis/ Brian S. Thomson, Judith B. Bruckner, Andrew M. Bruckner.
p. cm.
Includes index.
ISBN: 0-13-019075-6
1. Mathematical analysis. I. Bruckner, Judith B. II. Bruckner, Andrew M. III. Title
QA300. T45 2001 00-041679
515–dc21 CIP

Acquisitions Editor: *George Lobell*
Editor in Chief: *Sally Yagan*
Production Editor: *Lynn Savino Wendel*
Assistant Vice President Production and Manufacturing: *David W. Riccardi*
Executive Managing Editor: *Kathleen Schiaparelli*
Senior Managing Editor: *Linda Mihatov Behrens*
Manufacturing Buyer: *Alan Fischer*
Manufacturing Manager: *Trudy Pisciotti*
Director of Marketing: *John Tweeddale*
Marketing Manager: *Angela Battle*
Marketing Assistant: *Vince Jansen*
Art Director: *Jayne Conte*
Editorial Assistant: *Gale Epps*
Cover Designer: *Kiwi Design*

Prentice Hall

Printed in the United States of America
10 9 8 7 6 5 4 3 2 1

ISBN: 0-13-019075-6

Prentice-Hall International (UK) Limited, *London*
Prentice-Hall of Australia Pty. Limited, *Sydney*
Prentice-Hall Canada, Inc., *Toronto*
Prentice-Hall Hispanoamericana, S.A., *Mexico*
Prentice-Hall of India Private Limited, *New Delhi*
Prentice-Hall of Japan, Inc., *Tokyo*
Pearson Education Asia Pte. Ltd.
Editora Prentice-Hall do Brasil, Ltda., *Rio de Janeriro*

CONTENTS

[1]Optional sections are marked † for "Enrichment" and * for "Advanced."

PREFACE

University mathematics departments have for many years offered courses with titles such as *Advanced Calculus* or *Introductory Real Analysis.* These courses are taken by a variety of students, serve a number of purposes, and are written at various levels of sophistication. The students range from ones who have just completed a course in elementary calculus to beginning graduate students in mathematics. The purposes are multifold:

1. To present familiar concepts from calculus at a more rigorous level.

2. To introduce concepts that are not studied in elementary calculus but that are needed in more advanced undergraduate courses. This would include such topics as point set theory, uniform continuity of functions, and uniform convergence of sequences of functions.

3. To provide students with a level of mathematical sophistication that will prepare them for graduate work in mathematical analysis, or for graduate work in several applied fields such as engineering or economics.

4. To develop many of the topics that the authors feel all students of mathematics should know.

There are now many texts that address some or all of these objectives. These books range from ones that do little more than address objective (1) to ones that try to address all four objectives. The books of the first extreme are generally aimed at one-term courses for students with minimal background. Books at the other extreme often contain substantially more material than can be covered in a one-year course.

The level of rigor varies considerably from one book to another, as does the style of presentation. Some books endeavor to give a very efficient streamlined development; others try to be more user friendly. We have opted for the user-friendly approach. We feel this approach makes the concepts more meaningful to the student.

Our experience with students at various levels has shown that most students have difficulties when topics that are entirely new to them first appear. For some students that might occur almost immediately when rigorous proofs are required, for example, ones needing ε-δ arguments. For others, the difficulties begin with elementary point set theory, compactness arguments, and the like.

To help students with the transition from elementary calculus to a more rigorous course, we have included motivation for concepts most students have not seen before and provided more details in proofs when we introduce new methods. In addition, we have tried to give students ample opportunity to see the new tools in action.

For example, students often feel uneasy when they first encounter the various compactness arguments (Heine-Borel theorem, Bolzano-Weierstrass theorem, Cousin's lemma, introduced in Section 4.5). To help the student see why such theorems are useful, we pose the problem of determining circumstances under which local boundedness of a function f on a set E implies global boundedness of f on E. We show by example that some conditions on E are needed, namely that E be closed and bounded, and then show how each of several theorems could be used to show that closed and boundedness of the set E suffices. Thus we introduce students to the theorems by showing how the theorems can be used in natural ways to solve a problem.

We have also included some optional material, marked as "Advanced" or "Enrichment" and flagged with the symbol ✂.

Enrichment

We have indicated as "Enrichment"' some relatively elementary material that could be added to a longer course to provide enrichment and additional examples. For example, in Chapter 3 we have added to the study of series a section on infinite products. While such a topic plays an important role in the representation of analytic functions, it is presented here to allow the instructor to explore ideas that are closely related to the study of series and that help illustrate and review many of the fundamental ideas that have played a role in the study of series.

Advanced

We have indicated as "Advanced" material of a more mathematically sophisticated nature that can be omitted without loss of continuity. These topics might be needed in more advanced courses in real analysis or in certain of the marked sections or exercises that appear later in this book. For example, in Chapter 2 we have added to the study of sequence limits a section on lim sups and lim infs. For an elementary first course this can be considered somewhat advanced and skipped. Later problems and text material that require these concepts are carefully indicated. Thus, even though the text carries on to relatively advanced undergraduate analysis, a first course can be presented by avoiding these advanced sections.

We apply these markings to some entire chapters as well as to some sections within chapters and even to certain exercises. We do not view these markings as absolute. They can simply be interpreted in the following ways. Any unmarked material will not depend, in any substantial way, on earlier marked sections. In addition, if a section has been flagged and will be used in a much later section of this book, we indicate where it will be required.

The material marked "Advanced" is in line with goals (2) and (3). We resist the temptation to address objective (4). There are simply too many additional topics that one might feel every student should know (e.g., functions of bounded variation, Riemann-Stieltjes and Lebesgue integrals). To cover these topics in the manner we cover other material would render the book more like a reference book than a text that could reasonably be covered in a year. Students who have completed this book will be in a good position to study such topics at rigorous levels.

We include, however, a chapter on metric spaces. We do this for two reasons: to offer a more general framework for viewing concepts treated in earlier chapters, and to illustrate how the abstract viewpoint can be applied to solving concrete problems. The metric space presentation in Chapter 13 can be considered more advanced as the reader would require a reasonable level of preparation. Even so, it is more readable and accessible than many other presentations of metric space theory, as we have prepared it with the assumption that the student has just the minimal background. For example, it is easier than the corresponding chapter in our graduate level text (*Real Analysis*, Prentice Hall, 1997) in which the student is expected to have studied the Lebesgue integral and to be at an appropriately sophisticated level.

The Exercises

The exercises form an integral part of the book. Many of these exercises are routine in nature. Others are more demanding. A few provide examples that are not usually presented in books of this type but that students have found challenging, interesting, and instructive.

Some exercises have been flagged with the ✂ symbol to indicate that they require material from a flagged section. For example, a first course is likely to skip over the section on lim sups and lim infs of sequences. Exercises that require those concepts are flagged so that the instructor can decide whether they can be used or not. Generally, that symbol on an exercise warns that it might not be suitable for routine assignments.

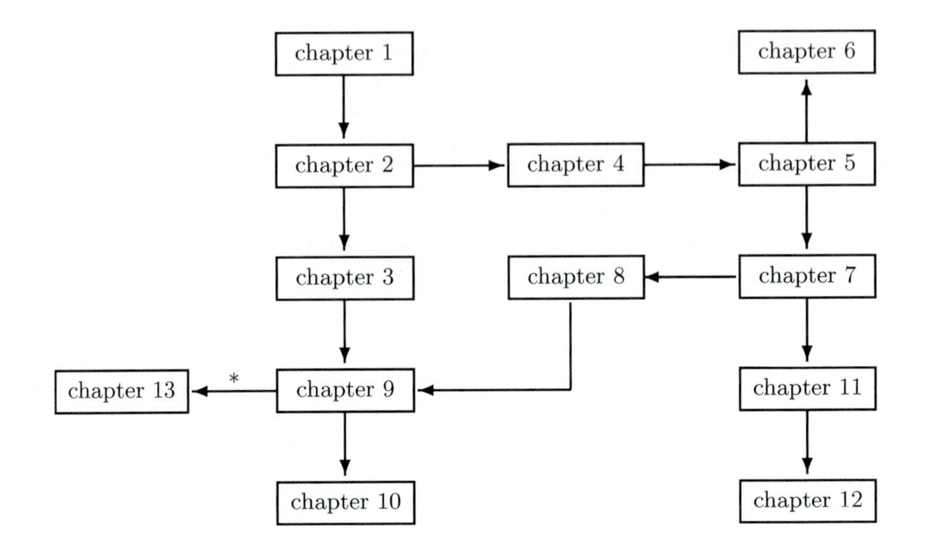

Figure 0.1. Chapter Dependencies (Unmarked Sections).

The exercises at the end of some of the chapters can be considered more challenging. They include some Putnam problems and some problems from the journal *American Mathematical Monthly.* They do not require more knowledge than is in the text material but often need a bit more persistence and some clever ideas. Students should be made aware that solutions to Putnam problems can be found on various Web sites and that solutions to *Monthly* problems are published; even so, the fun in such problems is in the attempt rather than in seeing someone else's solution.

Designing a Course

We have attempted to write this book in a manner sufficiently flexible to make it possible to use the book for courses of various lengths and a variety of levels of mathematical sophistication.

Much of the material in the book involves rigorous development of topics of a relatively elementary nature, topics that most students have studied at a nonrigorous level in a calculus course. A short course of moderate mathematical sophistication intended for students of minimal background can be based entirely on this material. Such a course might meet objective (1).

We have written this book in a leisurely style. This allows us to provide motivational discussions and historical perspective in a number of places.

Even though the book is relatively large (in terms of number of pages), we can comfortably cover most of the the main sections in a full-year course, including many of the interesting exercises.

Instructors teaching a short course have several options. They can base a course entirely on the unmarked material of Chapters 1, 2, 4, 5, and 7. As time permits, they can add the early parts of Chapters 3 and 8 or parts of Chapters 11 and 12 and some of the enrichment material.

Background

We should make one more point about this book. We do assume that students are familiar with nonrigorous calculus. In particular, we assume familiarity with the elementary functions and their elementary properties. We also assume some familiarity with computing derivatives and integrals. This allows us to illustrate various concepts using examples familiar to the students. For example, we begin Chapter 2, on sequences, with a discussion of approximating $\sqrt{2}$ using Newton's method. This is merely a motivational discussion, so we are not bothered by the fact that that we don't treat the derivative formally until Chapter 7 and haven't yet proved that $\frac{d}{dx}(x^2 - 2) = 2x$. For students with minimal background we provide an appendix that informally covers such topics as notation, elementary set theory, functions, and proofs.

Acknowledgments

A number of friends, colleagues, and students have made helpful comments and suggestions while the text was being written. We are grateful to the reviewers of the text: Professors Eugene Allgower (Colorado State University), Stephen Breen (California State University, Northridge), Robert E. Fennell (Clemson University), Jan E. Kucera (Washington State University), and Robert F. Lax (Louisiana State University). The authors are particularly grateful to Professors Steve Agronsky (California Polytechnic State University), Peter Borwein (Simon Fraser University), Paul Humke (St. Olaf College), T. H. Steele (Weber State University), and Clifford Weil (Michigan State University) for using preliminary versions of the book in their classes.

A. M. B.
J. B. B.
B. S. T.

Chapter 1

PROPERTIES OF THE REAL NUMBERS

1.1 Introduction

The goal of any analysis course is to do some analysis. There are some wonderfully important and interesting facts that can be established in a first analysis course.

Unfortunately, all of the material we wish to cover rests on some foundations, foundations that may not have been properly set down in your earlier courses. Calculus courses traditionally avoid any foundational problems by simply not proving the statements that would need them. Here we cannot do this. We must start with the real number system.

Historically much of real analysis was undertaken without any clear understanding of the real numbers. To be sure the mathematicians of the time had a firm intuitive grasp of the nature of the real numbers and often found precisely the right tool to use in their proofs, but in many cases the tools could not be justified by any line of reasoning.

By the 1870s mathematicians such as Georg Cantor (1845–1918) and Richard Dedekind (1831–1916) had found ways to describe the real numbers that seemed rigorous. We could follow their example and find a presentation of the real numbers that starts at the very beginning and leads up slowly (very slowly) to the exact tools that we need to study analysis. This subject is, perhaps, best left to courses in logic, where other important foundation issues can be discussed.

The approach we shall take (and most textbooks take) is simply to list all the tools that are needed in such a study and take them for granted. You may consider that the real number system is exactly as you have always imagined it. You can sketch pictures of the real line and measure distances and consider the order just as before. Nothing is changed from high school algebra or calculus. But when we come to prove assertions about real numbers or real functions or real sets, we must use exactly the tools here and not rely on our intuition.

1.2 The Real Number System

To do real analysis we should know exactly what the real numbers are. Here is a loose exposition, suitable for calculus students but (as we will see) not suitable for us.

The Natural Numbers We start with the natural numbers. These are the counting numbers

$$1, 2, 3, 4, \ldots.$$

The symbol $\mathbb{N}$ is used to indicate this collection. Thus $n \in \mathbb{N}$ means that n is a natural number, one of these numbers $1, 2, 3, 4, \ldots$.

There are two operations on the natural numbers, addition and multiplication:

$$m + n \text{ and } m \cdot n.$$

There is also an order relation

$$m < n.$$

Large amounts of time in elementary school are devoted to an understanding of these operations and the order relation.

(Subtraction and division can also be defined, but not for all pairs in $\mathbb{N}$. While $7 - 5$ and $10/5$ are assigned a meaning [we say $x = 7 - 5$ if $x + 5 = 7$ and we say $x = 10/5$ if $5 \cdot x = 10$] there is no meaning that can be attached to $5 - 7$ and $5/10$ in this number system.)

The Integers For various reasons, usually well motivated in the lower grades, the natural numbers prove to be rather limited in representing problems that arise in applications of mathematics to the real world. Thus they are enlarged by adjoining the negative integers and zero. Thus the collection

$$\ldots, -4, -3, -2, -1, 0, 1, 2, 3, 4, \ldots$$

is denoted $\mathbb{Z}$ and called the integers. (The symbol $\mathbb{N}$ seems obvious enough [N for "natural"] but the symbol $\mathbb{Z}$ for the integers originates in the German word for whole number.)

Once again, there are two operations on $\mathbb{Z}$, addition and multiplication:

$$m + n \text{ and } m \cdot n.$$

Again there is an order relation

$$m < n.$$

Fortunately, the rules of arithmetic and order learned for the simpler system $\mathbb{N}$ continue to hold for $\mathbb{Z}$, and young students extend their abilities perhaps painlessly.

(Subtraction can now be defined in this larger number system, but division still may not be defined. For example, $-9/3$ is defined but $3/(-9)$ is not.)

The Rational Numbers At some point the problem of the failure of division in the sets $\mathbb{N}$ and $\mathbb{Z}$ becomes acute and the student must progress to an understanding of fractions. This larger number system is denoted $\mathbb{Q}$, where the symbol chosen is meant to suggest quotients, which is after all what fractions are.

The collection of all "numbers" of the form

$$\frac{m}{n},$$

where $m \in \mathbb{Z}$ and $n \in \mathbb{N}$ is called the set of *rational numbers* and is denoted by the symbol $\mathbb{Q}$.

A higher level of sophistication is demanded at this stage. Equality has a new meaning. In $\mathbb{N}$ or $\mathbb{Z}$ a statement $m = n$ meant merely that m and n were the same object. Now

$$\frac{m}{n} = \frac{a}{b}$$

for $m, a \in \mathbb{Z}$ and $n, b \in \mathbb{N}$ means that

$$m \cdot b = a \cdot n.$$

Addition and multiplication present major challenges too. Ultimately the students must learn that

$$\frac{m}{n} + \frac{a}{b} = \frac{mb + na}{nb}$$

and

$$\frac{m}{n} \cdot \frac{a}{b} = \frac{ma}{nb}.$$

Subtraction and division are similarly defined. Fortunately, once again the rules of arithmetic are unchanged. The associative rule, distributive rule, etc. all remain true even in this number system.

Again, too, an order relation

$$\frac{m}{n} < \frac{a}{b}$$

is available. It can be defined by requiring, for $m, a \in \mathbb{Z}$ and $n, b \in \mathbb{N}$,

$$mb < na.$$

The same rules for inequalities learned for integers and natural numbers are valid for rationals.

The Real Numbers Up to this point in developing the real numbers we have encountered only arithmetic operations. The progression from $\mathbb{N}$ to $\mathbb{Z}$ to $\mathbb{Q}$ is simply algebraic. All this algebra might have been a burden to the weaker students in the lower grades, but conceptually the steps are easy to grasp with a bit of familiarity.

The next step, needed for all calculus students, is to develop the still larger system of real numbers, denoted as $\mathbb{R}$. We often refer to the real number system as *the real line* and think about it as a geometrical object, even though nothing in our definitions would seem at first sight to allow this.

Most calculus students would be hard pressed to say exactly what these numbers are. They recognize that $\mathbb{R}$ includes all of $\mathbb{N}$, $\mathbb{Z}$, and $\mathbb{Q}$ and also many new numbers, such as $\sqrt{2}$, e, and π. But asked what a real number is, many would return a blank stare. Even just asked what $\sqrt{2}$, e, or π are often produces puzzlement. Well, $\sqrt{2}$ is a number whose square is 2. But is there a number whose square is 2? A calculator might oblige with 1.4142136, but

$$(1.4142136)^2 \neq 2.$$

So what exactly "is" this number $\sqrt{2}$? If we are unable to write down a number whose square is 2, why can we claim that there is a number whose square is 2? And π and e are worse.

Some calculus texts handle this by proclaiming that real numbers are obtained by infinite decimal expansions. Thus while rational numbers have infinite decimal expansions that terminate (e.g., $1/4 = 0.25$) or repeat (e.g., $1/3 = 0.333333\ldots$), the collection of real numbers would include *all* infinite decimal expansions whether repeating, terminating, or not. In that case the claim would be that there is some infinite decimal expansion $1.414213\ldots$ whose square really is 2 and that infinite decimal expansion is the number we mean by the symbol $\sqrt{2}$.

This approach is adequate for applications of calculus and is a useful way to avoid doing any hard mathematics in introductory calculus courses. But you should recall that, at certain stages in the calculus textbook that you used, appeared a phrase such as "the proof of this next theorem is beyond the level of this text." It was beyond the level of the text only because the real numbers had not been properly treated and so there was no way that a proof could have been attempted.

We need to construct such proofs and so we need to abandon this loose, descriptive way of thinking about the real numbers. Instead we will define the real numbers to be a complete, ordered field. In the next sections each of these terms is defined.

1.3 Algebraic Structure

We describe the real numbers by assuming that they have a collection of properties. We do not construct the real numbers, we just announce what properties they are to have. Since the properties that we develop are familiar and acceptable and do in fact describe the real numbers that we are accustomed to using, this approach should not cause any distress. We are just stating rather clearly what it is about the real numbers that we need to use.

We begin with the algebraic structure.

In elementary algebra courses one learns many formulas that are valid for real numbers. For example, the formula

$$(x+y)+z = x+(y+z)$$

called the associative rule is learned. So also is the useful factoring rule

$$x^2 - y^2 = (x-y)(x+y).$$

It is possible to reduce the many rules to one small set of rules that can be used to prove all the other rules.

These rules can be used for other kinds of algebra, algebras where the objects are not real numbers but some other kind of mathematical constructions. This particular structure occurs so frequently, in fact, and in so many different applications that it has its own name. Any set of objects that has these same features is called a *field*. Thus we can say that the first important structure of the real number system is the field structure.

The following nine properties are called the *field axioms*. When we are performing algebraic manipulations in the real number system it is the field axioms that we are really using.

Assume that the set of real numbers $\mathbb{R}$ has two operations, called addition "+" and multiplication "·" and that these operations satisfy the field axioms. The operation $a \cdot b$ (multiplication) is most often written without the dot as ab.

A1 For any a, $b \in \mathbb{R}$ there is a number $a+b \in \mathbb{R}$ and $a+b = b+a$.

A2 For any a, b, $c \in \mathbb{R}$ the identity

$$(a+b)+c = a+(b+c)$$

is true.

A3 There is a unique number $0 \in \mathbb{R}$ so that, for all $a \in \mathbb{R}$,

$$a+0 = 0+a = a.$$

A4 For any number $a \in \mathbb{R}$ there is a corresponding number denoted by $-a$ with the property that

$$a + (-a) = 0.$$

M1 For any a, $b \in \mathbb{R}$ there is a number $ab \in \mathbb{R}$ and $ab = ba$.

M2 For any a, b, $c \in \mathbb{R}$ the identity

$$(ab)c = a(bc)$$

is true.

M3 There is a unique number $1 \in \mathbb{R}$ so that

$$a1 = 1a = a$$

for all $a \in \mathbb{R}$.

M4 For any number $a \in \mathbb{R}$, $a \neq 0$, there is a corresponding number denoted a^{-1} with the property that

$$aa^{-1} = 1.$$

AM1 For any a, b, $c \in \mathbb{R}$ the identity

$$(a + b)c = ac + bc$$

is true.

Note that we have labeled the axioms with letters indicating which operations are affected, thus A for addition and M for multiplication. The distributive rule AM1 connects addition and multiplication.

How are we to use these axioms? The answer likely is that, in an analysis course, you would not. You might try some of the exercises to understand what a field is and why the real numbers form a field. In an algebra course it would be most interesting to consider many other examples of fields and some of their applications. For an analysis course, understand that we are trying to specify exactly what we mean by the real number system, and these axioms are just the beginning of that process. The first step in that process is to declare that the real numbers form a field under the two operations of addition and multiplication.

Exercises

1.3.1 The field axioms include rules known often as associative rules, commutative rules and distributive rules. Which are which and why do they have these names?

1.3.2 To be precise we would have to say what is meant by the operations of addition and multiplication. Let S be a set and let $S \times S$ be the set of all ordered pairs (s_1, s_2) for $s_1,\ s_2 \in S$. A *binary operation* on S is a function $B : S \times S \to S$. Thus the operation takes the pair (s_1, s_2) and outputs the element $B(s_1, s_2)$. For example, addition is a binary operation. We could write $(s_1, s_2) \to A(s_1, s_2)$ rather than the more familiar $(s_1, s_2) \to s_1 + s_2$.

(a) Rewrite axioms A1–A4 using this notation $A(s_1, s_2)$ instead of the sum notation.

(b) Define a binary operation on $\mathbb{R}$ different from addition, subtraction, multiplication, or division and determine some of its properties.

(c) For a binary operation B define what you might mean by the commutative, associative, and distributive rules.

(d) Does the binary operation of subtraction satisfy any one of the commutative, associative, or distributive rules?

1.3.3 If in the field axioms for $\mathbb{R}$ we replace $\mathbb{R}$ by any other set with two operations $+$ and $\cdot$ that satisfy these nine properties, then we say that that structure is a field. For example, $\mathbb{Q}$ is a field. The rules are valid since $\mathbb{Q} \subset \mathbb{R}$. The only thing that needs to be checked is that $a + b$ and $a \cdot b$ are in $\mathbb{Q}$ if both a and b are. For this reason $\mathbb{Q}$ is called a *subfield* of $\mathbb{R}$. Find another subfield.

1.3.4 Let S be a set consisting of two elements labeled as A and B. Define $A+A = A$, $B + B = A$, $A + B = B + A = B$, $A \cdot A = A$, $A \cdot B = B \cdot A = A$, and $B \cdot B = B$. Show that all nine of the axioms of a field hold for this structure.

1.3.5 Using just the field axioms, show that

$$(x + 1)^2 = x^2 + 2x + 1$$

for all $x \in \mathbb{R}$. Would this identity be true in any field?

1.3.6 Define operations of addition and multiplication on $\mathbb{Z}_5 = \{0, 1, 2, 3, 4\}$ as follows:

+	0	1	2	3	4
0	0	1	2	3	4
1	1	2	3	4	0
2	2	3	4	0	1
3	3	4	0	1	2
4	4	0	1	2	3

×	0	1	2	3	4
0	0	0	0	0	0
1	0	1	2	3	4
2	0	2	4	1	3
3	0	3	1	4	2
4	0	4	3	2	1

Show that $\mathbb{Z}_5$ satisfies all the field axioms.

1.3.7 Define operations of addition and multiplication on $\mathbb{Z}_6 = \{0, 1, 2, 3, 4, 5\}$ as follows:

+	0	1	2	3	4	5
0	0	1	2	3	4	5
1	1	2	3	4	5	0
2	2	3	4	5	0	1
3	3	4	5	0	1	2
4	4	5	0	1	2	3
5	5	0	1	2	3	4

×	0	1	2	3	4	5
0	0	0	0	0	0	0
1	0	1	2	3	4	5
2	0	2	4	0	2	4
3	0	3	0	3	0	3
4	0	4	2	0	4	2
5	0	5	4	3	2	1

Which of the field axioms does $\mathbb{Z}_6$ fail to satisfy?

1.4 Order Structure

The real number system also enjoys an order structure. Part of our usual picture of the reals is the sense that some numbers are "bigger" than others or more to the "right" than others. We express this by using inequalities $x < y$ or $x \leq y$. The order structure is closely related to the field structure. For example, when we use inequalities in elementary courses we frequently use the fact that if $x < y$ and $0 < z$, then $xz < yz$ (i.e., that inequalities can be multiplied through by positive numbers).

This structure, too, can be axiomatized and reduced to a small set of rules. Once again, these same rules can be found in other applications of mathematics. When these rules are added to the field axioms the result is called an *ordered field*.

The real number system is an ordered field, satisfying the four additional axioms. Here $a < b$ is now a statement that is either true or false. (Before $a + b$ and $a \cdot b$ were not statements, but elements of $\mathbb{R}$.)

O1 For any a, $b \in \mathbb{R}$ exactly one of the statements $a = b$, $a < b$ or $b < a$ is true.

O2 For any a, b, $c \in \mathbb{R}$ if $a < b$ is true and $b < c$ is true, then $a < c$ is true.

O3 For any a, $b \in \mathbb{R}$ if $a < b$ is true, then $a + c < b + c$ is also true for any $c \in \mathbb{R}$.

O4 For any a, $b \in \mathbb{R}$ if $a < b$ is true, then $a \cdot c < b \cdot c$ is also true for any $c \in \mathbb{R}$ for which $c > 0$.

Exercises

1.4.1 Using just the axioms, prove that $ad + bc < ac + bd$ if $a < b$ and $c < d$.

1.4.2 Show for every $n \in \mathbb{N}$ that $n^2 \geq n$.

1.4.3 Using just the axioms, prove the *arithmetic-geometric mean inequality*:

$$\sqrt{ab} \leq \frac{a+b}{2}$$

for any a, $b \in \mathbb{R}$ with $a > 0$ and $b > 0$. (Assume, for the moment, the existence of square roots.)

1.5 Bounds

Let E be some set of real numbers. There may or may not be a number M that is bigger than every number in the set E. If there is, we say that M is an upper bound for the set. If there is no upper bound, then the set is said to be *unbounded above* or to have no upper bound. This is a simple enough idea, but it is critical to an understanding of the real numbers and so we shall look more closely at it and give some precise definitions.

Definition 1.1 (Upper Bounds) Let E be a set of real numbers. A number M is said to be an *upper bound* for E if $x \leq M$ for all $x \in E$.

Definition 1.2 (Lower Bounds) Let E be a set of real numbers. A number m is said to be a *lower bound* for E if $m \leq x$ for all $x \in E$.

It is often important to note whether a set has bounds or not. A set that has an upper bound and a lower bound is called *bounded.*

A set can have many upper bounds. Indeed every number is an upper bound for the empty set $\emptyset$. A set may have no upper bounds. We can use the phrase "E is unbounded above" if there are no upper bounds. For some sets the most natural upper bound (from among the infinitely many to choose) is just the largest member of the set. This is called the maximum. Similarly, the most natural lower bound for some sets is the smallest member of the set, the minimum.

Definition 1.3 (Maximum) Let E be a set of real numbers. If there is a number M that belongs to E and is larger than every other member of E, then M is called the maximum of the set E and we write $M = \max E$.

Definition 1.4 (Minimum) Let E be a set of real numbers. If there is a number m that belongs to E and is smaller than every other member of E, then m is called the minimum of the set E and we write $m = \min E$.

Example 1.5 The interval

$$[0, 1] = \{x : 0 \leq x \leq 1\}$$

has a maximum and a minimum. The maximum is 1 and 1 is also an upper bound for the set. (If a set has a maximum, then that number must certainly be an upper bound for the set.) Any number larger than 1 is also an upper bound. The number 0 is the minimum and also a lower bound. ◀

Example 1.6 The interval

$$(0, 1) = \{x : 0 < x < 1\}$$

has no maximum and no minimum. At first glance some novices insist that the maximum should be 1 and the minimum 0 as before. But look at the definition. The maximum must be both an upper bound and also a member of the set. Here 1 and 0 are upper and lower bounds, respectively, but do not belong to the set. ◀

Example 1.7 The set $\mathbb{N}$ of natural numbers has a minimum but no maximum and no upper bounds at all. We would say that it is bounded below but not bounded above. ◀

1.6 Sups and Infs

Let us return to the subject of maxima and minima again. If E has a maximum, say M, then that maximum could be described by the statement

> M is the least of all the upper bounds of E,

that is to say, M is the minimum of all the upper bounds. The most frequent language used here is "M is the least upper bound." It is possible for a set to have no maximum and yet be bounded above. In any example that comes to mind you will see that the set appears to have a least upper bound.

Example 1.8 The open interval $(0, 1)$ has no maximum, but many upper bounds. Certainly 2 is an upper bound and so is 1. The least of all the upper bounds is the number 1. Note that 1 cannot be described as a maximum because it fails to be in the set. ◀

Definition 1.9 (Least Upper Bound/Supremum) Let E be a set of real numbers that is bounded above and nonempty. If M is the least of all the upper bounds, then M is said to be *the least upper bound of* E or *the supremum* of E and we write $M = \sup E$.

Definition 1.10 (Greatest Lower Bound/Infimum) Let E be a set of real numbers that is bounded below and nonempty. If m is the greatest of all the lower bounds of E, then m is said to be *the greatest lower bound of* E or *the infimum* of E and we write $M = \inf E$.

To complete the definition of $\inf E$ and $\sup E$ it is most convenient to be able write this expression even for $E = \emptyset$ or for unbounded sets. Thus we write

1. $\inf \emptyset = \infty$ and $\sup \emptyset = -\infty$.
2. If E is unbounded above, then $\sup E = \infty$.
3. If E is unbounded below, then $\inf E = -\infty$.

The Axiom of Completeness Any example of a nonempty set that you are able to visualize that has an upper bound will also have a least upper bound. Pages of examples might convince you that all nonempty sets bounded above must have a least upper bound. Indeed your intuition will forbid you to accept the idea that this could not always be the case. To prove such an assertion is not possible using only the axioms for an ordered field. Thus we shall assume one further axiom, known as the axiom of completeness.

> **Completeness Axiom** A nonempty set of real numbers that is bounded above has a least upper bound (i.e., if E is nonempty and bounded above, then $\sup E$ exists and is a real number).

This now is the totality of all the axioms we need to assume. We have assumed that $\mathbb{R}$ is a field with two operations of addition and multiplication, that $\mathbb{R}$ is an ordered field with an inequality relation "<", and finally that $\mathbb{R}$ is a complete ordered field. This is enough to characterize the real numbers and the phrase "complete ordered field" refers to the system of real numbers and to no other system. (We shall not prove this statement; see Exercise 1.11.3 for a discussion.)

Exercises

1.6.1 Show that a set of real numbers E is bounded if and only if there is a positive number r so that $|x| < r$ for all $x \in E$.

1.6.2 Find $\sup E$ and $\inf E$ and (where possible) $\max E$ and $\min E$ for the following examples of sets:

(a) $E = \mathbb{N}$

(b) $E = \mathbb{Z}$

(c) $E = \mathbb{Q}$
(d) $E = \mathbb{R}$
(e) $E = \{-3, 2, 5, 7\}$
(f) $E = \{x : x^2 < 2\}$
(g) $E = \{x : x^2 - x - 1 < 0\}$
(h) $E = \{1/n : n \in \mathbb{N}\}$
(i) $E = \{\sqrt[n]{n} : n \in \mathbb{N}\}$

1.6.3 Under what conditions does $\sup E = \max E$?

1.6.4 Show for every nonempty, finite set E that $\sup E = \max E$.

1.6.5 For every $x \in \mathbb{R}$ define

$$[x] = \max\{n \in \mathbb{Z} : n \leq x\}$$

called the *greatest integer function.* Show that this is well defined and sketch the graph of the function.

1.6.6 Let A be a set of real numbers and let $B = \{-x : x \in A\}$. Find a relation between $\max A$ and $\min B$ and between $\min A$ and $\max B$.

1.6.7 Let A be a set of real numbers and let $B = \{-x : x \in A\}$. Find a relation between $\sup A$ and $\inf B$ and between $\inf A$ and $\sup B$.

1.6.8 Let A be a set of real numbers and let $B = \{x+r : x \in A\}$ for some number r. Find a relation between $\sup A$ and $\sup B$.

1.6.9 Let A be a set of real numbers and let $B = \{xr : x \in A\}$ for some positive number r. Find a relation between $\sup A$ and $\sup B$. (What happens if r is negative?)

1.6.10 Let A and B be sets of real numbers such that $A \subset B$. Find a relation among $\inf A$, $\inf B$, $\sup A$, and $\sup B$.

1.6.11 Let A and B be sets of real numbers and write $C = A \cup B$. Find a relation among $\sup A$, $\sup B$, and $\sup C$.

1.6.12 Let A and B be sets of real numbers and write $C = A \cap B$. Find a relation among $\sup A$, $\sup B$, and $\sup C$.

1.6.13 Let A and B be sets of real numbers and write

$$C = \{x + y : x \in A,\ y \in B\}.$$

Find a relation among $\sup A$, $\sup B$, and $\sup C$.

1.6.14 Let A and B be sets of real numbers and write

$$C = \{x + y : x \in A,\ y \in B\}.$$

Find a relation among $\inf A$, $\inf B$, and $\inf C$.

1.6.15 Let A be a set of real numbers and write $A^2 = \{x^2 : x \in A\}$. Are there any relations you can find between the infs and sups of the two sets?

1.6.16 Let E be a set of real numbers. Show that x is not an upper bound of E if and only if there exists a number $e \in E$ such that $e > x$.

1.6.17 Let A be a set of real numbers. Show that a real number x is the supremum of A if and only if $a \leq x$ for all $a \in A$ and for every positive number ε there is an element $a' \in A$ such that $x - \varepsilon < a'$.

1.6.18 Formulate a condition analogous to the preceding exercise for an infimum.

1.6.19 Using the completeness axiom, show that every nonempty set E of real numbers that is bounded below has a greatest lower bound (i.e., $\inf E$ exists and is a real number).

1.6.20 A function is said to be *bounded* if its range is a bounded set. Give examples of functions $f : \mathbb{R} \to \mathbb{R}$ that are bounded and examples of such functions that are unbounded. Give an example of one that has the property that

$$\sup\{f(x) : x \in \mathbb{R}\}$$

is finite but $\max\{f(x) : x \in \mathbb{R}\}$ does not exist.

1.6.21 The rational numbers $\mathbb{Q}$ satisfy the axioms for an ordered field. Show that the completeness axiom would not be satisfied. That is show that this statement is false: Every nonempty set E of rational numbers that is bounded above has a least upper bound (i.e., $\sup E$ exists and is a rational number).

1.6.22 Let F be the set of all numbers of the form $x + \sqrt{2}y$, where x and y are rational numbers. Show that F has all the properties of an ordered field but does not have the completeness property.

1.6.23 Let A and B be nonempty sets of real numbers and let

$$\delta(A, B) = \inf\{|a - b| : a \in A,\ b \in B\}.$$

$\delta(A, B)$ is often called the "distance" between the sets A and B.

(a) Let $A = \mathbb{N}$ and $B = \mathbb{R} \setminus \mathbb{N}$. Compute $\delta(A, B)$

(b) If A and B are finite sets, what does $\delta(A, B)$ represent?

(c) Let $B = [0, 1]$. What does the statement $\delta(\{x\}, B) = 0$ mean for the point x?

(d) Let $B = (0, 1)$. What does the statement $\delta(\{x\}, B) = 0$ mean for the point x?

1.7 The Archimedean Property

There is an important relationship holding between the set of natural numbers $\mathbb{N}$ and the larger set of real numbers $\mathbb{R}$. Because we have a well-formed mental image of what the set of reals "looks like," this property is entirely intuitive and natural. It hardly seems that it would require a proof. It says that the set of natural numbers $\mathbb{N}$ has no upper bound (i.e., that there is no real number x so that $n \leq x$ for all $n = 1, 2, 3, \dots$).

At first sight this seems to be a purely algebraic and order property of the reals. In fact it cannot be proved without invoking the completeness property of Section 1.6.

The property is named after the famous Greek mathematician known as Archimedes of Syracuse (287 B.C.–212 B.C.).[1]

Theorem 1.11 (Archimedean Property of $\mathbb{R}$) *The set of natural numbers $\mathbb{N}$ has no upper bound.*

Proof The proof is obtained by contradiction. If the set $\mathbb{N}$ does have an upper bound, then it must have a least upper bound. Let $x = \sup \mathbb{N}$, supposing that such does exist as a finite real number. Then $n \leq x$ for all integers n but $n \leq x - 1$ cannot be true for all integers n. Choose some integer m with $m > x - 1$. Then $m + 1$ is also an integer and $m + 1 > x$. But that cannot be so since we defined x as the supremum. From this contradiction the theorem follows. ■

The archimedean theorem has some consequences that have a great impact on how we must think of the real numbers.

1. No matter how large a real number x is given, there is always an integer n larger.

2. Given any positive number y, no matter how large, and any positive number x, no matter how small, one can add x to itself sufficiently many times so that the result exceeds y (i.e., $nx > y$ for some integer $n \in \mathbb{N}$).

3. Given any positive number x, no matter how small, one can always find a fraction $1/n$ with n a positive integer that is smaller (i.e., so that $1/n < x$).

Each of these is a consequence of the archimedean theorem, and the archimedean theorem in turn can be derived from any one of these.

[1] Archimedes seems to be the archetypical absent-minded mathematician. The historian Plutarch tells of his death at the hand of an invading army: "As fate would have it, Archimedes was intent on working out some problem by a diagram, and having fixed both his mind and eyes upon the subject of his speculation, he did not notice the entry of the Romans nor that the city was taken. In this transport of study a soldier unexpectedly came up to him and commanded that he accompany him. When he declined to do this before he had finished his problem, the enraged soldier drew his sword and ran him through." For this biographical detail and many others on all the mathematicians in this book consult *http://www-history.mcs.st-and.ac.uk/history*.

Exercises

1.7.1 Using the archimedean theorem, prove each of the three statements that follow the proof of the archimedean theorem.

1.7.2 Suppose that it is true that for each $x > 0$ there is an $n \in \mathbb{N}$ so that $1/n < x$. Prove the the archimedean theorem using this assumption.

1.7.3 Without using the archimedean theorem, show that for each $x > 0$ there is an $n \in \mathbb{N}$ so that $1/n < x$.

1.7.4 Let x be any real number. Show that there is an integer $m \in \mathbb{Z}$ so that

$$m \leq x < m + 1.$$

Show that m is unique.

1.7.5 The mathematician Leibniz based his calculus on the assumption that there were "infinitesimals," positive real numbers that are extremely small—smaller than all positive rational numbers certainly. Some calculus students also believe, apparently, in the existence of such numbers since they can imagine a number that is "just next to zero." Is there a positive real number smaller than all positive rational numbers?

1.7.6 The archimedean property asserts that if $x > 0$, then there is an integer N so that $1/N < x$. The proof requires the completeness axiom. Give a proof that does not use the completeness axiom that works for x rational. Find a proof that is valid for $x = \sqrt{y}$, where y is rational.

1.7.7 In Section 1.2 we made much of the fact that there is a number whose square is 2 and so $\sqrt{2}$ does exist as a real number. Show that

$$\alpha = \sup\{x \in \mathbb{R} : x^2 < 2\}$$

exists as a real number and that $\alpha^2 = 2$.

1.8 Inductive Property of $\mathbb{N}$

Since the natural numbers are included in the set of real numbers there are further important properties of $\mathbb{N}$ that can be deduced from the axioms. The most important of these is the principle of induction. This is the basis for the technique of proof known as induction, which is often used in this text. For an elementary account and some practice, see Section A.8 in the appendix.

We first prove a statement that is equivalent.

Theorem 1.12 (Well-Ordering Property) *Every nonempty subset of $\mathbb{N}$ has a smallest element.*

Proof Let $S \subset \mathbb{N}$ and $S \neq \emptyset$. Then $\alpha = \inf S$ must exist and be a real number since S is bounded below. If $\alpha \in S$, then we are done since we have found a minimal element.

Suppose not. Then, while α is the greatest lower bound of S, α is not a minimum. There must be an element of S that is smaller than $\alpha + 1$ since α is the greatest lower bound of S. That element cannot be α since we have assumed that $\alpha \notin S$. Thus we have found $x \in S$ with

$$\alpha < x < \alpha + 1.$$

Now x is not a lower bound of S, since it is greater than the greatest lower bound of S, so there must be yet another element y of S such that

$$\alpha < y < x < \alpha + 1.$$

But now we have reached an impossibility, for x and y are in S and both integers, but $0 < x - y < 1$, which cannot happen for integers. From this contradiction the proof now follows. ■

Now we can state and prove the principle of induction.

Theorem 1.13 (Principle of Induction) *Let $S \subset \mathbb{N}$ so that $1 \in S$ and, for every integer n, if $n \in S$ then so also is $n + 1$. Then $S = \mathbb{N}$.*

Proof Let $E = \mathbb{N} \setminus S$. We claim that $E = \emptyset$ and then it follows that $S = \mathbb{N}$ proving the theorem. Suppose not (i.e., suppose $E \neq \emptyset$). By Theorem 1.12 there is a first element α of E. Can $\alpha = 1$? No, because $1 \in S$ by hypothesis. Thus $\alpha - 1$ is also an integer and, since it cannot be in E it must be in S. By hypothesis it follows that $\alpha = (\alpha - 1) + 1$ must be in S. But it is in E. This is impossible and so we have obtained a contradiction, proving our theorem. ■

Exercises

1.8.1 Show that any bounded, nonempty set of natural numbers has a maximal element.

1.8.2 Show that any bounded, nonempty subset of $\mathbb{Z}$ has a maximum and a minimum.

1.8.3 For further exercises on proving statements using induction as a method, see Section A.8.

1.9 The Rational Numbers Are Dense

There is an important relationship holding between the set of rational numbers $\mathbb{Q}$ and the larger set of real numbers $\mathbb{R}$. The rational numbers are dense. They make an appearance in every interval; there are no gaps, no intervals that miss having rational numbers.

For practical purposes this has great consequences. We need never actually compute with arbitrary real numbers, since close by are rational numbers that can be used. Thus, while π is irrational, in routine computations with a practical view any nearby fraction might do. At various times people have used 3, 22/7, and 3.14159, for example.

For theoretical reasons this fact is of great importance too. It allows many arguments to replace a consideration of the set of real numbers with the smaller set of rationals. Since every real is as close as we please to a rational and since the rationals can be carefully described and easily worked with, many simplifications are allowed.

Definition 1.14 (Dense Sets) A set E of real numbers is said to be *dense* (or *dense in* $\mathbb{R}$) if every interval (a, b) contains a point of E.

Theorem 1.15 *The set* $\mathbb{Q}$ *of rational numbers is dense.*

Proof Let $x < y$ and consider the interval (x, y). We must find a rational number inside this interval.

By the archimedean theorem, Theorem 1.11, there is a positive integer

$$n > \frac{1}{y-x}.$$

This means that $ny > nx + 1$.

Let m be chosen as the integer just less than $nx+1$; more precisely (using Exercise 1.7.4), find an integer $m \in \mathbb{Z}$ so that

$$m \leq nx + 1 < m + 1.$$

Now some arithmetic on these inequalities shows that

$$m - 1 \leq nx < ny$$

and then

$$x < \frac{m}{n} \leq x + \frac{1}{n} < y$$

thus exhibiting a rational number m/n in the interval (x, y). ■

Exercises

1.9.1 Show that the definition of "dense" could be given as

> A set E of real numbers is said to be *dense* if every interval (a, b) contains infinitely many points of E.

1.9.2 Find a rational number between $\sqrt{10}$ and π.

1.9.3 If a set E is dense, what can you conclude about a set $A \supset E$?

1.9.4 If a set E is dense, what can you conclude about the set $\mathbb{R} \setminus E$?

1.9.5 If two sets E_1 and E_2 are dense, what can you conclude about the set $E_1 \cap E_2$?

1.9.6 Show that the *dyadic rationals* (i.e., rational numbers of the form $m/2^n$ for $m \in \mathbb{Z}$, $n \in \mathbb{N}$) are dense.

1.9.7 Are the numbers of the form

$$\pm m/2^{100}$$

for $m \in \mathbb{N}$ dense? What is the length of the largest interval that contains no such number?

1.9.8 Show that the numbers of the form

$$\pm m\sqrt{2}/n$$

for $m, n \in \mathbb{N}$ are dense.

1.10 The Metric Structure of $\mathbb{R}$

In addition to the algebraic and order structure of the real numbers, we need to make measurements. We need to describe distances between points. These are the metric properties of the reals, to borrow a term from the Greek for measure (*metron*).

As usual, the distance between a point x and another point y is either $x - y$ or $y - x$ depending on which is positive. Thus the distance between 3 and -4 is 7. The distance between π and $\sqrt{10}$ is $\sqrt{10} - \pi$. To describe this in general requires the absolute value function which simply makes a choice between positive and negative.

Definition 1.16 (Absolute Value) For any real number x write

$$|x| = x \quad \text{if } x \geq 0$$

and

$$|x| = -x \quad \text{if } x < 0\ .$$

(Beginners tend to think of the absolute value function as "stripping off the negative sign," but the example

$$|\pi - \sqrt{10}| = \sqrt{10} - \pi$$

shows that this is a limited viewpoint.)

Properties of the Absolute Value Since the absolute value is defined directly in terms of inequalities (i.e., the choice $x \geq 0$ or $x < 0$), there are a number of properties that can be proved directly from properties of inequalities. These properties are used routinely and the student will need to have a complete mastery of them.

Theorem 1.17 *The absolute value function has the following properties:*

1. *For any* $x \in \mathbb{R}$, $-|x| \leq x \leq |x|$.
2. *For any* $x, y \in \mathbb{R}$, $|xy| = |x|\,|y|$.
3. *For any* $x, y \in \mathbb{R}$, $|x+y| \leq |x| + |y|$.
4. *For any* $x, y \in \mathbb{R}$, $|x| - |y| \leq |x-y|$ *and* $|y| - |x| \leq |x-y|$.

Distances on the Real Line Using the absolute value function we can define the distance function or metric.

Definition 1.18 (Distance) The distance between two real numbers x and y is

$$d(x,y) = |x-y|.$$

We hardly ever use the notation $d(x,y)$ in elementary analysis, preferring to write $|x-y|$ even while we are thinking of this as the distance between the two points. Thus if a sequence of points x_1, x_2, x_3, ... is growing ever closer to a point c, we should perhaps describe $d(x_n, c)$ as getting smaller and smaller, thus emphasizing that the distances are shrinking; more often we would simply write $|x_n - c|$ and expect you to interpret this as a distance.

Properties of the Distance Function The main properties of the distance function are just interpretations of the absolute value function. Expressed in the language of a distance function, they are geometrically very intuitive:

1. $d(x,y) \geq 0$

   (all distances are positive or zero).

2. $d(x,y) = 0$ if and only if $x = y$

   (different points are at positive distance apart).

3. $d(x,y) = d(y,x)$

   (distance is symmetric, that is the distance from x to y is the same as from y to x)).

4. $d(x, y) \leq d(x, z) + d(z, y)$

   (the triangle inequality, that is it is no longer to go directly from x to y than to go from x to z and then to y).

In Chapter 13 we will study general structures called metric spaces, where exactly such a notion of distance satisfying these four properties is used. For now we prefer to rewrite these properties in the language of the absolute value, where they lose some of their intuitive appeal. But it is in this form that we are likely to use them.

1. $|a| \geq 0$.

2. $|a| = 0$ if and only if $a = 0$.

3. $|a| = |-a|$.

4. $|a + b| \leq |a| + |b|$ (the triangle inequality).

Exercises

1.10.1 Show that $|x| = \max\{x, -x\}$.

1.10.2 Show that $\max\{x, y\} = |x - y|/2 + (x + y)/2$. What expression would give $\min\{x, y\}$?

1.10.3 Show that the inequalities $|x - a| < \varepsilon$ and

$$a - \varepsilon < x < a + \varepsilon$$

are equivalent.

1.10.4 Show that if $\alpha < x < \beta$ and $\alpha < y < \beta$, then $|x - y| < \beta - \alpha$ and interpret this geometrically as a statement about the interval (α, β).

1.10.5 Show that $||x| - |y|| \leq |x - y|$ assuming the triangle inequality (i.e., that $|a + b| \leq |a| + |b|$). This inequality is also called the triangle inequality.

1.10.6 Under what conditions is it true that $|x + y| = |x| + |y|$?

1.10.7 Under what conditions is it true that

$$|x - y| + |y - z| = |x - z|?$$

1.10.8 Show that

$$|x_1 + x_2 + \cdots + x_n| \leq |x_1| + |x_2| + \cdots + |x_n|$$

for any numbers $x_1, x_2, \ldots, x_n$.

1.10.9 Let E be a a set of real numbers and let $A = \{|x| : x \in E\}$. What relations can you find between the infs and sups of the two sets?

1.10.10 Find the inf and sup of the set $\{x : |2x + \pi| < \sqrt{2}\}$.

1.11 Challenging Problems for Chapter 1

1.11.1 The complex numbers $\mathbb{C}$ are defined as equal to the set of all ordered pairs of real numbers subject to these operations:

$$(a_1, b_1) + (a_2, b_2) = (a_1 + a_2, b_1 + b_2)$$

and

$$(a_1, b_1) \cdot (a_2, b_2) = (a_1 a_2 - b_1 b_2, a_1 b_2 + a_2 b_1).$$

(a) Show that $\mathbb{C}$ is a field.

(b) What are the additive and multiplicative identity elements?

(c) What are the additive and multiplicative inverses of an element (a, b)?

(d) Solve $(a, b)^2 = (1, 0)$ in $\mathbb{C}$.

(e) We identify $\mathbb{R}$ with a subset of $\mathbb{C}$ by identifying the elements $x \in \mathbb{R}$ with the element $(x, 0)$ in $\mathbb{C}$. Explain how this can be interpreted as saying that "$\mathbb{R}$ is a subfield of $\mathbb{C}$."

(f) Show that there is an element $i \in \mathbb{C}$ with $i^2 = -1$ so that every element $z \in \mathbb{C}$ can be written as $z = x + iy$ for $x,\ y \in \mathbb{R}$.

(g) Explain why the equation $x^2 + x + 1 = 0$ has no solution in $\mathbb{R}$ but two solutions in $\mathbb{C}$.

1.11.2 Can an order be defined on the field $\mathbb{C}$ of Exercise 1.11.1 in such a way so to make it an ordered field?

1.11.3 The statement that every complete ordered field "is" the real number system means the following. Suppose that F is a nonempty set with operations of addition "+" and multiplication "·" and an order relation "<" that satisfies all the axioms of an ordered field and also the axiom of completeness. Then there is a one-to-one onto function $f : \mathbb{R} \to F$ that has the following properties:

(a) $f(x + y) = f(x) + f(y)$ for all $x,\ y \in \mathbb{R}$.

(b) $f(x \cdot y) = f(x) \cdot f(y)$ for all $x,\ y \in \mathbb{R}$.

(c) $f(x) < f(y)$ if and only if $x < y$ for $x,\ y \in \mathbb{R}$.

Thus, in a certain sense, F and $\mathbb{R}$ are essentially the same object. Attempt a proof of this statement. [Note that $x + y$ for $x,\ y \in \mathbb{R}$ refers to the addition in the reals whereas $f(x) + f(y)$ refers to the addition in the set F.]

1.11.4 We have assumed in the text that the set $\mathbb{N}$ is obviously contained in $\mathbb{R}$. After all, 1 is a real number (it's in the axioms), 2 is just $1 + 1$ and so real, 3 is $2 + 1$ etc. In that way we have been able to prove the material of Section 1.8. But there is a logical flaw here. We would need induction really to define $\mathbb{N}$ in this way (and not just say "etc."). Here is a set of exercises that would remedy that for students with some background in set manipulations.

(a) Define a set $S \subset \mathbb{R}$ to be *inductive* if $1 \in S$ and if $x \in S$ implies that $x + 1 \in S$. Show that $\mathbb{R}$ is inductive.

(b) Show that there is a smallest inductive set by showing that the intersection of the family of all inductive sets is itself inductive.

(c) Define $\mathbb{N}$ to be that smallest inductive set.

(d) Prove Theorem 1.13 now. (That is, show that any set S with the property stated there is inductive and conclude that $S = \mathbb{N}$.)

(e) Prove Theorem 1.12 now. (That is, with this definition of $\mathbb{N}$ prove the well-ordering property.)

1.11.5 Use this definition of "dense in a set" to answer the following questions:

> A set E of real numbers is said to be *dense in a set* A if every interval (a, b) that contains a point of A also contains a point of E.

(a) Show that dense in the set of all reals is the same as dense.

(b) Give an example of a set E dense in $\mathbb{N}$ but with $E \cap \mathbb{N} = \emptyset$.

(c) Show that the irrationals are dense in the rationals. (A real number is *irrational* if it is not rational, that is if it belongs to $\mathbb{R}$ but not to $\mathbb{Q}$.)

(d) Show that the rationals are dense in the irrationals.

(e) What property does a set E have that is equivalent to the assertion that $\mathbb{R} \setminus E$ is dense in E?

1.11.6 Let G be a subgroup of the real numbers under addition (i.e., if x and y are in G, then $x + y \in G$ and $-x \in G$). Show that either G is a dense subset of $\mathbb{R}$ or else there is a real number α so that

$$G = \{n\alpha : n = 1, \pm 1, \pm 2, \pm 3, \dots\}.$$

Chapter 2

SEQUENCES

2.1 Introduction

Let us start our discussion with a method for solving equations that originated with Newton in 1669. To solve an equation $f(x) = 0$ the method proposes the introduction of a new function

$$F(x) = x - \frac{f(x)}{f'(x)}.$$

We begin with a guess at a solution of $f(x) = 0$, say x_1 and compute $x_2 = F(x_1)$ in the hopes that x_2 is closer to a solution than x_1 was. The process is repeated so that $x_3 = F(x_2)$, $x_4 = F(x_3)$, $x_5 = F(x_4)$, ... and so on until the desired accuracy is reached. Processes of this type have been known for at least 3500 years although not in such a modern notation.

We illustrate by finding an approximate value for $\sqrt{2}$ this way. We solve the equation $f(x) = x^2 - 2 = 0$ by computing the function

$$F(x) = x - \frac{f(x)}{f'(x)} = x - \frac{x^2 - 2}{2x}$$

and using it to improve our guess. A first (very crude) guess of $x_1 = 1$ will produce the following list of values for our subsequent steps in the procedure. We have retained 60 digits in the decimal expansions to show how this is working:

$$x_1 = 1.00000000000000000000000000000000000000000000000000000000000$$

$$x_2 = 1.50000000000000000000000000000000000000000000000000000000000$$

$$x_3 = 1.41666666666666666666666666666666666666666666666666666666667$$

$$x_4 = 1.41421568627450980392156862745098039215686274509803921568628$$

$$x_5 = 1.41421356237468991062629557889013491011655962211574404458490$$

$$x_6 = 1.41421356237309504880168962350253024361498192577619742849829$$

$$x_7 = 1.41421356237309504880168872420969807856967187537723400156101.$$

To compare, here is the value of the true solution $\sqrt{2}$, computed in a different fashion to the same number of digits:

$$\sqrt{2} = 1.41421356237309504880168872420969807856967187537694807317668.$$

Note that after only four steps the procedure gives a value differing from the true value only in the sixth decimal place, and all subsequent values remain this close. A convenient way of expressing this is to write that

$$|x_n - \sqrt{2}| < 10^{-5} \text{ for all } n \geq 4.$$

By the seventh step, things are going even better and we can claim that

$$|x_n - \sqrt{2}| < 10^{-47} \text{ for all } n \geq 7.$$

It is inconceivable that anyone would require any further accuracy for any practical considerations. The error after the sixth step cannot exceed 10^{-47}, which is a tiny number. Even so, as mathematicians we can ask what may seem an entirely impractical sort of question. Can this accuracy of approximation continue forever? Is it possible that, if we wait long enough, we can find an approximation to $\sqrt{2}$ with *any* degree of accuracy?

Expressed more formally, if we are given a positive number ε (we call it epsilon to suggest that it measures an *error*) no matter how small, can we find a stage in this procedure so that the value computed and all subsequent values are closer to $\sqrt{2}$ than ε? In symbols, is there an integer n_0 (which will depend on just how small ε is) that is large enough so that

$$|x_n - \sqrt{2}| < \varepsilon \text{ for all } n \geq n_0?$$

If this is true then this sequence has a remarkable property. It is not merely in its first few terms a convenient way of computing $\sqrt{2}$ to some accuracy; the sequence truly represents the number $\sqrt{2}$ itself, and it cannot represent any other number. We shall say that the sequence converges to $\sqrt{2}$ and write

$$\lim_{n \to \infty} x_n = \sqrt{2}.$$

This is the beginning of the theory of convergence that is central to analysis. If mathematicians had never considered the ultimate behavior of such sequences and had contented themselves with using only the first few terms for practical computations, there would have been no subject known as analysis. These ideas lead, as you might imagine, to an ideal world of infinite precision, where sequences are not merely useful gadgets for getting good computations but are precise tools in discussing real numbers. From the theory of sequences and their convergence properties has developed a vast world of beautiful and useful mathematics.

For the student approaching this material for the first time this is a critical test. All of analysis, both pure and applied, rests on an understanding of limits. What you learn in this chapter will offer a foundation for all the rest that you will have to learn later.

2.2 Sequences

A sequence (of real numbers, of sets, of functions, of anything) is simply a list. There is a first element in the list, a second element, a third element, and so on continuing in an order forever. In mathematics a finite list is not called a sequence; a sequence must continue without interruption.

For a more formal definition notice that the natural numbers are playing a key role here. Every item in the sequence (the list) can be labeled by its position; label the first item with a "1," the second with a "2," and so on. Seen this way a sequence is merely then a function mapping the natural numbers $\mathbb{N}$ into some set. We state this as a definition. Since this chapter is exclusively about sequences of real numbers, the definition considers just this situation.

Definition 2.1 By a *sequence of real numbers* we mean a function

$$f : \mathbb{N} \to \mathbb{R}.$$

Thus the sequence *is* the function. Even so, we usually return to the list idea and write out the sequence f as

$$f(1), f(2), f(3), \ldots, f(n), \ldots$$

with the ellipsis (i.e., the three dots) indicating that the list is to continue in this fashion. The function values $f(1)$, $f(2)$, $f(3)$, $\ldots$ are called the *terms* of the sequence.

If we need to return to the formality of functions we do, but try to keep the intuitive notion of a sequence as an unending list in mind. While computer scientists much prefer the function notation, mathematicians have become more accustomed to a subscript notation and would rather have the terms of the preceding sequence rendered as

$$f_1, f_2, f_3, \ldots, f_n, \ldots.$$

In this chapter we study sequences of real numbers. Later on we will encounter the same word applied to other lists of objects (e.g., sequences of intervals, sequences of sets, sequences of functions. In all cases the word sequence simply indicates a list of objects).

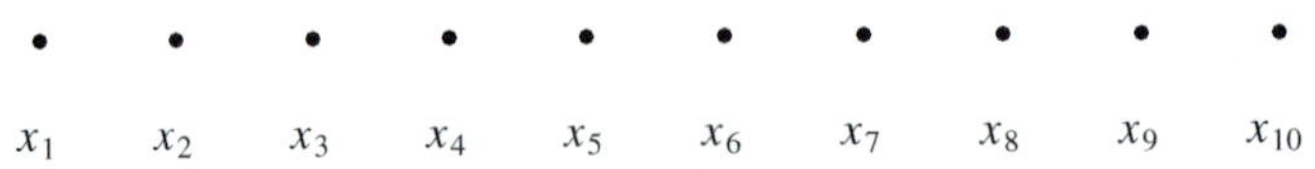

$x_1 \quad x_2 \quad x_3 \quad x_4 \quad x_5 \quad x_6 \quad x_7 \quad x_8 \quad x_9 \quad x_{10}$

Figure 2.1. An arithmetic progression.

2.2.1 Sequence Examples

In order to specify some sequence we need to communicate what every term in the sequence is. For example, the sequence of even integers

$$2, 4, 6, 8, 10, \ldots$$

could be communicated in precisely that way: "Consider the sequence of even integers." Perhaps more direct would be to give a formula for all of the terms in the sequence: "Consider the sequence whose nth term is $x_n = 2n$." Or we could note that the sequence starts with 2 and then all the rest of the terms are obtained by adding 2 to the previous term: "Consider the sequence whose first term is 2 and whose nth term is 2 added to the $(n-1)$st term," that is,

$$x_n = 2 + x_{n-1}.$$

Often an explicit formula is best. Frequently though, a formula relating the nth term to some preceding term is preferable. Such formulas are called *recursion formulas* and would usually be more efficient if a computer is used to generate the terms.

Arithmetic Progressions The simplest types of sequences are those in which each term is obtained from the preceding by adding a fixed amount. These are called *arithmetic progressions.* The sequence

$$c, c+d, c+2d, c+3d, c+4d, \ldots, c+(n-1)d, \ldots$$

is the most general arithmetic progression. The number d is called the *common difference.*

Every arithmetic progression could be given by a formula

$$x_n = c + (n-1)d$$

or a recursion formula

$$x_1 = c \qquad x_n = x_{n-1} + d.$$

Note that the explicit formula is of the form $x_n = f(n)$, where f is a linear function, $f(x) = dx + b$ for some b. Figure 2.1 shows the points of an arithmetic progression plotted on the line. If, instead, you plot the points (n, x_n) you will find that they all lie on a straight line with slope d.

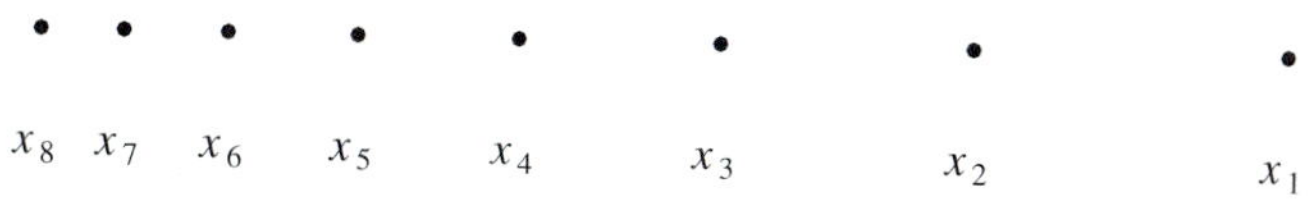

x_8 x_7 x_6 x_5 x_4 x_3 x_2 x_1

Figure 2.2. A geometric progression.

Geometric Progressions. A variant on the arithmetic progression is obtained by replacing the addition of a fixed amount by the multiplication by a fixed amount. These sequences are called *geometric progressions.* The sequence

$$c, cr, cr^2, cr^3, cr^4, \ldots, cr^n, \ldots$$

is the most general geometric progression. The number r is called the *common ratio.*

Every geometric progression could be given by a formula

$$x_n = cr^{n-1}$$

or a recursion formula

$$x_1 = c \qquad x_n = rx_{n-1}.$$

Note that the explicit formula is of the form $x_n = f(n)$, where f is an exponential function $f(x) = br^x$ for some b. Figure 2.2 shows the points of a geometric progression plotted on the line. Alternatively, plot the points (n, x_n) and you will find that they all lie on the graph of an exponential function. If $c > 0$ and the common ratio r is larger than 1, the terms increase in size, becoming extremely large. If $0 < r < 1$, the terms decrease in size, getting smaller and smaller. (See Figure 2.2.)

Iteration The examples of an arithmetic progression and a geometric progression are special cases of a process called *iteration.* So too is the sequence generated by Newton's method in the introduction to this chapter.

Let f be some function. Start the sequence $\{x_n\}$ by assigning some value in the domain of f, say $x_1 = c$. All subsequent values are now obtained by feeding these values through the function repeatedly:

$$c, f(c), f(f(c)), f(f(f(c))), f(f(f(f(c)))), \ldots.$$

As long as all these values remain in the domain of the function f, the process can continue indefinitely and defines a sequence. If f is a function of the form $f(x) = x + b$, then the result is an arithmetic progression. If f is a function of the form $f(x) = ax$, then the result is a geometric progression.

A recursion formula best expresses this process and would offer the best way of writing a computer program to compute the sequence:

$$x_1 = c \qquad x_n = f(x_{n-1}).$$

Sequence of Partial Sums. If a sequence

$$x_1, x_2, x_3, x_4, \ldots$$

is given, we can construct a new sequence by adding the terms of the old one:

$$s_1 = x_1$$

$$s_2 = x_1 + x_2$$

$$s_3 = x_1 + x_2 + x_3$$

$$s_4 = x_1 + x_2 + x_3 + x_4$$

and continuing in this way. The process can also be described by a recursion formula:

$$s_1 = x_1, \qquad s_n = s_{n-1} + x_n.$$

The new sequence is called the *sequence of partial sums* of the old sequence $\{x_n\}$. We shall study such sequences in considerable depth in the next chapter.

For a particular example we could use $x_n = 1/n$ and the sequence of partial sums could be written as

$$s_n = 1 + 1/2 + 1/3 + \cdots + 1/n.$$

Is there a more attractive and simpler formula for s_n? The answer is no.

Example 2.2 The examples, given so far, are of a general nature and describe many sequences that we will encounter in analysis. But a sequence is just a list of numbers and need not be defined in any manner quite so systematic. For example, consider the sequence defined by $a_n = 1$ if n is divisible by three, $a_n = n$ if n is one more than a multiple of three, and $a_n = -2^n$ if n is two more than a multiple of three. The first few terms are evidently

$$1, 2, -8, 1, 5, -64, \ldots.$$

What would be the next three terms? ◀

Exercises

2.2.1 Let a sequence be defined by the phrase "consider the sequence of prime numbers $2, 3, 5, 7, 11, 13 \ldots$". Are you sure that this defines a sequence?

2.2.2 On IQ tests one frequently encounters statements such as "what is the next term in the sequence 3, 1, 4, 1, 5, ... ?". In terms of our definition of a sequence is this correct usage? (By the way, what do you suppose the next term in the sequence might be?)

2.2.3 Give two different formulas (for two different sequences) that generate a sequence whose first four terms are 2, 4, 6, 8.

2.2.4 Give a formula that generates a sequence whose first five terms are 2, 4, 6, 8, π.

2.2.5 The examples listed here are the first few terms of a sequence that is either an arithmetic progression or a geometric progression. What is the next term in the sequence? Give a general formula for the sequence.

(a) 7, 4, 1, ...

(b) .1, .01, .001, ...

(c) 2, $\sqrt{2}$, 1, ...

2.2.6 Consider the sequence defined recursively by

$$x_1 = \sqrt{2}, \quad x_n = \sqrt{2} + x_{n-1}.$$

Find an explicit formula for the nth term.

2.2.7 Consider the sequence defined recursively by

$$x_1 = \sqrt{2}, \quad x_n = \sqrt{2}x_{n-1}.$$

Find an explicit formula for the nth term.

2.2.8 Consider the sequence defined recursively by

$$x_1 = \sqrt{2}, \quad x_n = \sqrt{2 + x_{n-1}}.$$

Show, by induction, that $x_n < 2$ for all n.

2.2.9 Consider the sequence defined recursively by

$$x_1 = \sqrt{2}, \quad x_n = \sqrt{2 + x_{n-1}}.$$

Show, by induction, that $x_n < x_{n+1}$ for all n.

2.2.10 The sequence defined recursively by

$$f_1 = 1, \; f_2 = 1, \quad f_{n+2} = f_n + f_{n+1}$$

is called the *Fibonacci sequence*. It is possible to find an explicit formula for this sequence. Give it a try.

2.3 Countable Sets

✂ *Enrichment*

A sequence of real numbers, formally, is a function whose domain is the set $\mathbb{N}$ of natural numbers and whose range is a subset of the reals $\mathbb{R}$. What sets might be the range of some sequence? To put it another way, what sets can have their elements arranged into an unending list? Are there sets that cannot be arranged into a list?

The arrangement of a collection of objects into a list is sometimes called an *enumeration*. Thus another way of phrasing this question is to ask what sets of real numbers can be *enumerated*?

The set of natural numbers is already arranged into a list in its natural order. The set of integers (including 0 and the negative integers) is not usually presented in the form of a list but can easily be so presented, as the following scheme suggests:

$$0, 1, -1, 2, -2, 3, -3, 4, -4, 5, -5, 6, -6, 7, -7, \ldots.$$

Example 2.3 The rational numbers can also be listed but this is quite remarkable, for at first sight no reasonable way of ordering them into a sequence seems likely to be possible. The usual order of the rationals in the reals is of little help.

To find such a scheme define the "rank" of a rational number m/n in its lowest terms (with $n \geq 1$) to be $|m|+n$. Now begin making a finite list of all the rational numbers at a various rank; list these from smallest to largest. At rank 1 we would have only the rational number $0/1$. At rank 2 we would have only the rational numbers $-1/1$, $1/1$. At rank 3 we would have only the rational numbers $-2/1$, $-1/2$, $1/2$, $2/1$. Carry on in this fashion through all the ranks. Now construct the final list by concatenating these shorter lists in order of the ranks:

$$0/1, -1/1, 1/1, -2/1, -1/2, 1/2, 2/1, \ldots.$$

The range of this sequence is the set of all rational numbers. ◀

Your first impression might be that few sets would be able to be the range of a sequence. But having seen in Example 2.3 that even the set of rational numbers $\mathbb{Q}$ that is seemingly so large can be listed, it might then appear that all sets can be so listed. After all, can you conceive of a set that is "larger" than the rationals in some way that would stop it being listed? The remarkable fact that there are sets that cannot be arranged to form the elements of some sequence was proved by Georg Cantor (1845–1918). This proof is essentially his original proof. (Note that this requires some familiarity with infinite decimal expansions; the exercises review what is needed.)

Theorem 2.4 (Cantor) *No interval (a, b) of real numbers can be the range of some sequence.*

Proof It is enough to prove this for the interval $(0, 1)$ since there is nothing special about it (see Exercise 2.3.1). The proof is a proof by contradiction. We suppose that the theorem is false and that there is a sequence $\{s_n\}$ so that every number in the interval $(0, 1)$ appears at least once in the sequence. We obtain a contradiction by showing that this cannot be so. We shall use the sequence $\{s_n\}$ to find a number c in the interval $(0, 1)$ so that $s_n \neq c$ for all n.

Each of the points s_1, s_2, s_3 ... in our sequence is a number between 0 and 1 and so can be written as a decimal fraction. If we write this sequence out in decimal notation it might look like

$$s_1 = 0.x_{11}x_{12}x_{13}x_{14}x_{15}x_{16}\dots$$

$$s_2 = 0.x_{21}x_{22}x_{23}x_{24}x_{25}x_{26}\dots$$

$$s_3 = 0.x_{31}x_{32}x_{33}x_{34}x_{35}x_{36}\dots$$

etc. Now it is easy to find a number that is not in the list. Construct

$$c = 0.c_1c_2c_3c_4c_5c_6\dots$$

by choosing c_i to be either 5 or 6 whichever is different from x_{ii}. This number cannot be equal to any of the listed numbers s_1, s_2, s_3 ... since c and s_i differ in the ith position of their decimal expansions. This gives us our contradiction and so proves the theorem. ■

Definition 2.5 (Countable) A nonempty set S of real numbers is said to be *countable* if there is a sequence of real numbers whose range is the set S.

In the language of this definition then we can see that (1) any finite set is countable, (2) the natural numbers and the integers are countable, (3) the rational numbers are countable, and (4) no interval of real numbers is countable. By convention we also say that the empty set $\emptyset$ is countable.

Exercises

2.3.1 Show that, once it is known that the interval $(0, 1)$ cannot be expressed as the range of some sequence, it follows that any interval (a, b), $[a, b)$, $(a, b]$, or $[a, b]$ has the same property.

2.3.2 Some novices, on reading the proof of Cantor's theorem, say "Why can't you just put the number c that you found at the front of the list." What is your rejoinder?

2.3.3 A set (any set of objects) is said to be *countable* if it is either finite or there is an enumeration (list) of the set. Show that the following properties hold for arbitrary countable sets:

(a) All subsets of countable sets are countable.

(b) Any union of a pair of countable sets is countable.

(c) All finite sets are countable.

2.3.4 Show that the following property holds for countable sets: If

$$S_1,\ S_2,\ S_3,\ \dots$$

is a sequence of countable sets of real numbers, then the set S formed by taking all elements that belong to at least one of the sets S_i is also a countable set.

2.3.5 Show that if a nonempty set is contained in the range of some sequence of real numbers, then there is a sequence whose range is precisely that set.

2.3.6 In Cantor's proof presented in this section we took for granted material about infinite decimal expansions. This is entirely justified by the theory of sequences studied later. Explain what it is that we need to prove about infinite decimal expansions to be sure that this proof is valid.

2.3.7 Define a relation on the family of subsets of $\mathbb{R}$ as follows. Say that $A \sim B$, where A and B are subsets of $\mathbb{R}$, if there is a function

$$f : A \to B$$

that is one-to-one and onto. (If $A \sim B$ we would say that A and B are "cardinally equivalent.") Show that this is an *equivalence relation*, that is, show that

(a) $A \sim A$ for any set A.

(b) If $A \sim B$ then $B \sim A$.

(c) If $A \sim B$ and $B \sim C$ then $A \sim C$.

2.3.8 Let A and B be finite sets. Under what conditions are these sets cardinally equivalent (in the language of Exercise 2.3.7)?

2.3.9 Show that an infinite set of real numbers that is countable is cardinally equivalent (in the language of Exercise 2.3.7) to the set $\mathbb{N}$. Give an example of an infinite set that is not cardinally equivalent to $\mathbb{N}$.

2.3.10 We define a real number to be *algebraic* if it is a solution of some polynomial equation

$$a_n x^n + a_{n-1} x^{n-1} + \cdots + a_1 x + a_0 = 0,$$

where all the coefficients are integers. Thus $\sqrt{2}$ is algebraic because it is a solution of $x^2 - 2 = 0$. The number π is not algebraic because no such polynomial equation can ever be found (although this is hard to prove). Show that the set of algebraic numbers is countable. A real number that is not algebraic is said to be *transcendental*. For example, it is known that e and π are transcendental. What can you say about the existence of other transcendental numbers?

2.4 Convergence

The sequence

$$1, \frac{1}{2}, \frac{1}{3}, \frac{1}{4}, \frac{1}{5}, \frac{1}{6}, \dots$$

is getting closer and closer to the number 0. We say that this sequence *converges* to 0 or that the limit of the sequence is the number 0. How should this idea be properly defined?

The study of convergent sequences was undertaken and developed in the eighteenth century without any precise definition. The closest one might find to a definition in the early literature would have been something like

> *A sequence $\{s_n\}$ converges to a number L if the terms of the sequence get closer and closer to L.*

Apart from being too vague to be used as anything but a rough guide for the intuition, this is misleading in other respects. What about the sequence

$$.1, .01, .02, .001, .002, .0001, .0002, .00001, .00002, \ldots?$$

Surely this should converge to 0 but the terms do not get steadily "closer and closer" but back off a bit at each second step. Also, the sequence

$$.1, .11, .111, .1111, .11111, .111111, \ldots$$

is getting "closer and closer" to .2, but we would not say the sequence converges to .2. A smaller number (1/9, which it is also getting closer and closer to) is the correct limit. We want not merely "closer and closer" but somehow a notion of "arbitrarily close."

The definition that captured the idea in the best way was given by Augustin Cauchy in the 1820s. He found a formulation that expressed the idea of "arbitrarily close" using inequalities. In this way the notion of limit is defined by a straightforward mathematical statement about inequalities.

Definition 2.6 (Limit of a Sequence) Let $\{s_n\}$ be a sequence of real numbers. We say that $\{s_n\}$ *converges* to a number L and write

$$\lim_{n\to\infty} s_n = L$$

or

$$s_n \to L \text{ as } n \to \infty$$

provided that for every number $\varepsilon > 0$ there is an integer N so that

$$|s_n - L| < \varepsilon$$

whenever $n \geq N$.

A sequence that converges is said to be *convergent*. A sequence that fails to converge is said to *diverge*. We are equally interested in both convergent and divergent sequences.

Note. In the definition the N depends on ε. If ε is particularly small, then N might have to be chosen large. In fact, then N is really a function of ε. Sometimes it is best to emphasize this and write $N(\varepsilon)$ rather than N.

Note, too, that if an N is found, then any larger N would also be able to be used. Thus the definition requires us to find some N but not necessarily the smallest N that would work.

While the definition does not say this, the real force of the definition is that the N can be determined *no matter how small a number ε is chosen.* If ε is given as rather large there may be no trouble finding the N value. If you find an N that works for $\varepsilon = .1$ that same N would work for all larger values of ε.

Example 2.7 Let us use the definition to prove that

$$\lim_{n\to\infty} \frac{n^2}{2n^2+1} = \frac{1}{2}.$$

It is by no means clear from the definition how to obtain that the limit is the number $L = \frac{1}{2}$. Indeed the definition is not intended as a method of finding limits. It assigns a precise meaning to the statement about the limit but offers no way of computing that limit. Fortunately most of us remember some calculus devices that can be used to first obtain the limit before attempting a proof of its validity.

$$\lim_{n\to\infty} \frac{n^2}{2n^2+1} = \lim_{n\to\infty} \frac{1}{2+1/n^2} = \frac{1}{\lim_{n\to\infty}(2+1/n^2)}$$

$$= \frac{1}{2+\lim_{n\to\infty}(1/n^2)} = \frac{1}{2}.$$

Indeed this would be a proof that the limit is 1/2 provided that we could prove the validity of each of these steps. Later on we will prove this and so can avoid the ε, N arguments that we now use.

Let any positive ε be given. We need to find a number N [or $N(\varepsilon)$ if you prefer] so that every term in the sequence on and after the Nth term is closer to 1/2 than ε, that is, so that

$$\left| \frac{n^2}{2n^2+1} - \frac{1}{2} \right| < \varepsilon$$

for $n = N$, $n = N+1$, $n = N+2$, ... It is easiest to work backward and discover just how large n should be for this. A little work shows that this will happen if

$$\frac{1}{2(2n^2+1)} < \varepsilon$$

or

$$4n^2 + 2 > \frac{1}{\varepsilon}.$$

The smallest n for which this statement is true could be our N. Thus we could use any integer N with

$$N^2 > \frac{1}{4}\left(\frac{1}{\varepsilon} - 2\right).$$

There is no obligation to find the smallest N that works and so, perhaps, the most convenient one here might be a bit larger, say take any integer N larger than

$$N > \frac{1}{2\sqrt{\varepsilon}}.$$

◀

The real lesson of the example, perhaps, is that we wish never to have to use the definition to check any limit computation. The definition offers a rigorous way to develop a theory of limits but an impractical method of computation of limits and a clumsy method of verification. Only rarely do we have to do a computation of this sort to verify a limit.

Uniqueness of Sequence Limits Let us take the first step in developing a theory of limits. This is to ensure that our definition has defined *limit* unambiguously. Is it possible that the definition allows for a sequence to converge to two different limits? If we have established that $s_n \to L$ is it possible that $s_n \to L_1$ for a different number L_1?

Theorem 2.8 (Uniqueness of Limits) *Suppose that*

$$\lim_{n\to\infty} s_n = L_1 \text{ and } lim_{n\to\infty} s_n = L_2$$

are both true. Then $L_1 = L_2$.

Proof Let ε be any positive number. Then, by definition, we must be able to find a number N_1 so that

$$|s_n - L_1| < \varepsilon$$

whenever $n \geq N_1$. We must also be able to find a number N_2 so that

$$|s_n - L_2| < \varepsilon$$

whenever $n \geq N_2$. Take m to be the maximum of N_1 and N_2. Then both assertions

$$|s_m - L_1| < \varepsilon \quad \text{and} \quad |s_m - L_2| < \varepsilon$$

are true.

This allows us to conclude that

$$|L_1 - L_2| \leq |L_1 - s_m| + |s_m - L_2| < 2\varepsilon$$

so that

$$|L_1 - L_2| < 2\varepsilon.$$

But ε can be any positive number whatsoever. This could only be true if $L_1 = L_2$, which is what we wished to show. ■

Exercises

2.4.1 Give a precise ε, N argument to prove that $\lim_{n\to\infty} \frac{1}{n} = 0$.

2.4.2 Give a precise ε, N argument to prove the existence of

$$\lim_{n\to\infty} \frac{2n+3}{3n+4}.$$

2.4.3 Show that a sequence $\{s_n\}$ converges to a limit L if and only if the sequence $\{s_n - L\}$ converges to zero.

2.4.4 Show that a sequence $\{s_n\}$ converges to a limit L if and only if the sequence $\{-s_n\}$ converges to $-L$.

2.4.5 Show that Definition 2.6 is equivalent to the following slight modification:

> We write $\lim_{n\to\infty} s_n = L$ provided that for every positive integer m there is a real number N so that $|s_n - L| < 1/m$ whenever $n \geq N$.

2.4.6 Compute the limit

$$\lim_{n\to\infty} \frac{1+2+3+\cdots+n}{n^2}$$

and verify it by the definition.

2.4.7 Compute the limit

$$\lim_{n\to\infty} \frac{1^2+2^2+3^2+\cdots+n^2}{n^3}.$$

2.4.8 Suppose that $\{s_n\}$ is a convergent sequence. Prove that $\lim_{n\to\infty} 2s_n$ exists.

2.4.9 Prove that $\lim_{n\to\infty} n$ does not exist.

2.4.10 Prove that $\lim_{n\to\infty}(-1)^n$ does not exist.

2.4.11 The sequence $s_n = (-1)^n$ does not converge. For what values of $\varepsilon > 0$ is it nonetheless true that there is an integer N so that $|s_n - 1| < \varepsilon$ whenever $n \geq N$? For what values of $\varepsilon > 0$ is it nonetheless true that there is an integer N so that $|s_n - 0| < \varepsilon$ whenever $n \geq N$?

2.4.12 Let $\{s_n\}$ be a sequence that assumes only integer values. Under what conditions can such a sequence converge?

2.4.13 Let $\{s_n\}$ be a sequence and obtain a new sequence (sometimes called the "tail" of the sequence) by writing

$$t_n = s_{M+n} \quad \text{for } n = 1, 2, 3, \ldots$$

where M is some integer (perhaps large). Show that $\{s_n\}$ converges if and only if $\{t_n\}$ converges.

2.4.14 Show that the the statement "$\{s_n\}$ converges to L" is false if and only if there is a positive number c so that the inequality

$$|s_n - L| > c$$

holds for infinitely many values of n.

2.4.15 If $\{s_n\}$ is a sequence of positive numbers converging to 0, show that $\{\sqrt{s_n}\}$ also converges to zero.

2.4.16 If $\{s_n\}$ is a sequence of positive numbers converging to a positive number L, show that $\{\sqrt{s_n}\}$ converges to $\sqrt{L}$.

2.5 Divergence

A sequence that fails to converge is said to *diverge.* Some sequences diverge in a particularly interesting way, and it is worthwhile to have a language for this.

The sequence $s_n = n^2$ diverges because the terms get larger and larger. We are tempted to write

$$n^2 \to \infty \quad \text{or} \quad \lim_{n\to\infty} n^2 = \infty.$$

This conflicts with our definition of limit and so needs its own definition. We do not say that this sequence "converges to ∞" but rather that it "diverges to ∞."

Definition 2.9 (Divergence to ∞) Let $\{s_n\}$ be a sequence of real numbers. We say that $\{s_n\}$ *diverges* to ∞ and write

$$\lim_{n\to\infty} s_n = \infty$$

or

$$s_n \to \infty \text{ as } n \to \infty$$

provided that for every number M there is an integer N so that

$$s_n > M$$

whenever $n \geq N$.

Note. The definition does not announce this, but the force of the definition is that the choice of N is possible *no matter how large M is chosen.* There may be no difficulty in finding an N if the M given is not big.

Example 2.10 Let us prove that

$$\frac{n^2+1}{n+1} \to \infty$$

using the definition. If M is any positive number we need to find some point in the sequence after which all terms exceed M. Thus we need to consider the inequality

$$\frac{n^2+1}{n+1} \geq M.$$

After some arithmetic we see that this is equivalent to

$$n + \frac{1}{n+1} - \frac{n}{n+1} \geq M.$$

Since

$$\frac{n}{n+1} < 1$$

we see that, as long as $n \geq M + 1$ this will be true. Thus take any integer $N \geq M + 1$ and it will be true that

$$\frac{n^2+1}{n+1} \geq M$$

for all $n \geq N$. (Any larger value of N would work too.) ◀

Exercises

2.5.1 Formulate the definition of a sequence diverging to $-\infty$.

2.5.2 Show, using the definition, that $\lim_{n\to\infty} n^2 = \infty$.

2.5.3 Show, using the definition, that $\lim_{n\to\infty} \frac{n^3+1}{n^2+1} = \infty$.

2.5.4 Prove that if $s_n \to \infty$ then $-s_n \to -\infty$.

2.5.5 Prove that if $s_n \to \infty$ then $(s_n)^2 \to \infty$ also.

2.5.6 Prove that if $x_n \to \infty$ then the sequence $s_n = \frac{x_n}{x_n+1}$ is convergent. Is the converse true?

2.5.7 Suppose that $\lim_{n\to\infty} s_n = 0$. Show that $\lim_{n\to\infty} 1/s_n = \infty$. Is the converse true?

2.5.8 Suppose that a sequence $\{s_n\}$ of positive numbers satisfies the condition $s_{n+1} > \alpha s_n$ for all n where $\alpha > 1$. Show that $s_n \to \infty$.

2.5.9 The sequence $s_n = (-1)^n$ does not diverge to ∞. For what values of M is it nonetheless true that there is an integer N so that $s_n > M$ whenever $n \geq N$?

2.5.10 Show that the sequence

$$n^p + \alpha_1 n^{p-1} + \alpha_2 n^{p-2} + \dots \alpha_p$$

diverges to ∞, where here p is a positive integer and $\alpha_1, \alpha_2, \dots, \alpha_p$ are real numbers (positive or negative).

2.6 Boundedness Properties of Limits

A sequence is said to be *bounded* if its range is a bounded set. Thus a sequence $\{s_n\}$ is bounded if there is a number M so that every term in the sequence satisfies

$$|s_n| \leq M.$$

For such a sequence, every term belongs to the interval $[-M, M]$.

It is fairly evident that a sequence that is not bounded could not converge. This is important enough to state and prove as a theorem.

Theorem 2.11 *Every convergent sequence is bounded.*

Proof Suppose that $s_n \to L$. Then for every number $\varepsilon > 0$ there is an integer N so that

$$|s_n - L| < \varepsilon$$

whenever $n \geq N$. In particular we could take just one value of ε, say $\varepsilon = 1$, and find a number N so that

$$|s_n - L| < 1$$

whenever $n \geq N$. From this we see that

$$|s_n| = |s_n - L + L| \leq |s_n - L| + |L| < |L| + 1$$

for all $n \geq N$. This number $|L| + 1$ would be an upper bound for all the numbers $|s_n|$ except that we have no indication of the values for $|s_1|$, $|s_2|$, $\dots$, $|s_{N-1}|$.

Thus if we write

$$M = \max\{|s_1|, |s_2|, \dots, |s_{N-1}|, |L| + 1\}$$

we must have

$$|s_n| \leq M$$

for *every* value of n. This is an upper bound, proving the theorem. ■

As a consequence of this theorem we can conclude that an unbounded sequence must diverge. Thus, even though it is a rather crude test, we can prove the divergence of a sequence if we are able somehow to show that it is unbounded. The next example illustrates this technique.

Example 2.12 We shall show that the sequence

$$s_n = 1 + \frac{1}{2} + \frac{1}{3} + \frac{1}{4} + \cdots + \frac{1}{n}$$

diverges. The easiest proof of this is to show that it is unbounded and hence, by Theorem 2.11, could not converge.

We watch only at the steps 1, 2, 4, 8, ... and make a rough lower estimate of s_1, s_2, s_4, s_8, ... in order to show that there can be no bound on the sequence. After a bit of arithmetic we see that

$$s_1 = 1$$

$$s_2 = 1 + \frac{1}{2}$$

$$s_4 = 1 + \frac{1}{2} + \left(\frac{1}{3} + \frac{1}{4}\right) > 1 + \frac{1}{2} + 2\left(\frac{1}{4}\right)$$

$$s_8 = 1 + \frac{1}{2} + \left(\frac{1}{3} + \frac{1}{4}\right) + \left(\frac{1}{5} + \frac{1}{6} + \frac{1}{7} + \frac{1}{8}\right)$$

$$\geq 1 + \frac{1}{2} + 2\left(\frac{1}{4}\right) + 4\left(\frac{1}{8}\right)$$

and, in general, that

$$s_{2^n} \geq 1 + n/2$$

for all $n = 0, 1, 2, \ldots$. Thus the sequence is not bounded and so must diverge. ◀

Example 2.13 As a variant of the sequence of the preceding example consider the sequence

$$t_n = 1 + \frac{1}{2^p} + \frac{1}{3^p} + \frac{1}{4^p} \cdots + \ldots \frac{1}{n^p}$$

where p is any positive real number. The case $p = 1$ we have just found diverges.

For $p < 1$ the sequence is larger than it is for $p = 1$ and so the case is even stronger for divergence. For $p > 1$ the sequence is smaller and we cannot see immediately whether it is bounded or unbounded; in fact, with some effort we can show that such a sequence is bounded. What can we conclude? Nothing yet. An unbounded sequence diverges. A bounded sequence may converge or diverge. ◀

Exercises

2.6.1 Which statements are true?

(a) If $\{s_n\}$ is unbounded then it is true that either $\lim_{n\to\infty} s_n = \infty$ or else $\lim_{n\to\infty} s_n = -\infty$.

(b) If $\{s_n\}$ is unbounded then $\lim_{n\to\infty} |s_n| = \infty$.

(c) If $\{s_n\}$ and $\{t_n\}$ are both bounded then so is $\{s_n + t_n\}$.

(d) If $\{s_n\}$ and $\{t_n\}$ are both unbounded then so is $\{s_n + t_n\}$.

(e) If $\{s_n\}$ and $\{t_n\}$ are both bounded then so is $\{s_n t_n\}$.

(f) If $\{s_n\}$ and $\{t_n\}$ are both unbounded then so is $\{s_n t_n\}$.

(g) If $\{s_n\}$ is bounded then so is $\{1/s_n\}$.

(h) If $\{s_n\}$ is unbounded then $\{1/s_n\}$ is bounded.

2.6.2 If $\{s_n\}$ is bounded prove that $\{s_n/n\}$ is convergent.

2.6.3 State the converse of Theorem 2.11. Is it true?

2.6.4 State the contrapositive of Theorem 2.11. Is it true?

2.6.5 Suppose that $\{s_n\}$ is a sequence of positive numbers converging to a positive limit. Show that there is a positive number c so that $s_n > c$ for all n.

2.6.6 As a computer experiment compute the values of the sequence

$$s_n = 1 + \frac{1}{2} + \frac{1}{3} + \frac{1}{4} + \cdots + \frac{1}{n}$$

for large values of n. Is there any indication in the numbers that you see that this sequence fails to converge or must be unbounded?

2.7 Algebra of Limits

Sequences can be combined by the usual arithmetic operations (addition, subtraction, multiplication, and division). Indeed most sequences we are likely to encounter can be seen to be composed of simpler sequences combined together in this way.

In Example 2.7 we suggested that the computations

$$\lim_{n\to\infty} \frac{n^2}{2n^2+1} = \lim_{n\to\infty} \frac{1}{2+1/n^2} = \frac{1}{\lim_{n\to\infty}(2+1/n^2)}$$

$$= \frac{1}{2+\lim_{n\to\infty} 1/n^2} = \frac{1}{2}$$

could be justified. Note how this sequence has been obtained from simpler ones by ordinary processes of arithmetic. To justify such a method we need to investigate how the limit operation is influenced by algebraic operations.

Suppose that

$$s_n \to S \text{ and } t_n \to T.$$

Then we would expect

$$Cs_n \to CS$$

$$s_n + t_n \to S + T$$

$$s_n - t_n \to S - T$$

$$s_n t_n \to ST$$

and

$$s_n/t_n \to S/T.$$

Each of these statements must be justified, however, solely on the basis of the definition of convergence, not on intuitive feelings that this should be the case. Thus we need to develop what could be called the "algebra of limits."

Theorem 2.14 (Multiples of Limits) *Suppose that $\{s_n\}$ is a convergent sequence and C a real number. Then*

$$\lim_{n\to\infty} Cs_n = C\left(\lim_{n\to\infty} s_n\right).$$

Proof Let $S = \lim_{n\to\infty} s_n$. In order to prove that $\lim_{n\to\infty} Cs_n = CS$ we need to prove that, no matter what positive number ε is given, we can find an integer N so that, for all $n \geq N$,

$$|Cs_n - CS| < \varepsilon.$$

Note that

$$|Cs_n - CS| = |C|\,|s_n - S|$$

by properties of absolute values. This gives us our clue.

Suppose first that $C \neq 0$ and let $\varepsilon > 0$. Choose N so that

$$|s_n - S| < \varepsilon/|C|$$

if $n \geq N$. Then if $n \geq N$ we must have

$$|Cs_n - CS| = |C|\,|s_n - S| < |C|\,(\varepsilon/|C|) = \varepsilon.$$

This is precisely the statement that

$$\lim_{n\to\infty} Cs_n = CS$$

and the theorem is proved in the case $C \neq 0$. The case $C = 0$ is obvious. (Now we should probably delete our first paragraph since it does not contribute to the proof; it only serves to motivate us in finding the correct proof.) ■

Theorem 2.15 (Sums and Differences of Limits) *Suppose that the sequences $\{s_n\}$ and $\{t_n\}$ are convergent. Then*

$$\lim_{n\to\infty} (s_n + t_n) = \lim_{n\to\infty} s_n + \lim_{n\to\infty} t_n$$

and

$$\lim_{n\to\infty} (s_n - t_n) = \lim_{n\to\infty} s_n - \lim_{n\to\infty} t_n.$$

Proof Let $S = \lim_{n\to\infty} s_n$ and $T = \lim_{n\to\infty} t_n$. In order to prove that

$$\lim_{n\to\infty} (s_n + t_n) = S + T$$

we need to prove that no matter what positive number ε is given we can find an integer N so that

$$|(s_n + t_n) - (S + T)| < \varepsilon$$

if $n \geq N$. Note that

$$|(s_n + t_n) - (S + T)| \leq |s_n - S| + |t_n - T|$$

by the triangle inequality. Thus we can make this expression smaller than ε by making each of the two expressions on the right smaller than $\varepsilon/2$. This provides the method.

Suppose that $\varepsilon > 0$. Choose N_1 so that

$$|s_n - S| < \varepsilon/2$$

if $n \geq N_1$ and also choose N_2 so that

$$|t_n - T| < \varepsilon/2$$

if $n \geq N_2$. Then if n is greater than both N_1 and N_2 both of these inequalities will be true. Set

$$N = \max\{N_1, N_2\}$$

and note that if $n \geq N$ we must have

$$|(s_n + t_n) - (S + T)| \leq |s_n - S| + |t_n - T| < \varepsilon/2 + \varepsilon/2 = \varepsilon.$$

This is precisely the statement that

$$\lim_{n\to\infty} (s_n + t_n) = S + T$$

and the first statement of the theorem is proved. The second statement is similar and is left as an exercise. (Once again, for a more formal presentation, we would delete the first paragraph.) ∎

Theorem 2.16 (Products of Limits) *Suppose that $\{s_n\}$ and $\{t_n\}$ are convergent sequences. Then*

$$\lim_{n\to\infty} (s_n t_n) = \left(\lim_{n\to\infty} s_n\right)\left(\lim_{n\to\infty} t_n\right).$$

Proof Let $S = \lim_{n\to\infty} s_n$ and $T = \lim_{n\to\infty} t_n$. In order to prove that $\lim_{n\to\infty}(s_n t_n) = ST$ we need to prove that no matter what positive number ε is given we can find an integer N so that, for all $n \geq N$,

$$|s_n t_n - ST| < \varepsilon.$$

It takes some experimentation with different ways of writing this to find the most useful version. Here is an inequality that offers the best approach:

$$|s_nt_n - ST| = |s_n(t_n - T) + s_nT - ST|$$
$$\leq |s_n|\,|t_n - T| + |T|\,|s_n - S|. \qquad (1)$$

We can control the size of $|s_n - S|$ and $|t_n - T|$, T is constant, and $|s_n|$ cannot be too big. To control the size of $|s_n|$ we need to recall that convergent sequences are bounded (Theorem 2.11) and get a bound from there. With these preliminaries explained the rest of the proof should seem less mysterious. (Now this paragraph can be deleted for a more formal presentation.)

Suppose that $\varepsilon > 0$. Since $\{s_n\}$ converges it is bounded and hence, by Theorem 2.11, there is a positive number M so that $|s_n| \leq M$ for all n. Choose N_1 so that

$$|s_n - S| < \frac{\varepsilon}{2|T| + 1}$$

if $n \geq N_1$. [We did not use $\varepsilon/(2T)$ since there is a possibility that $T = 0$.] Also, choose N_2 so that

$$|t_n - T| < \frac{\varepsilon}{2M}$$

if $n \geq N_2$. Set $N = \max\{N_1, N_2\}$ and note that if $n \geq N$ we must have

$$|s_nt_n - ST| \leq |s_n|\,|t_n - T| + |T|\,|s_n - S|$$
$$\leq M\left(\frac{\varepsilon}{2M}\right) + |T|\left(\frac{\varepsilon}{2|T| + 1}\right) < \varepsilon.$$

This is precisely the statement that

$$\lim_{n\to\infty} s_nt_n = ST$$

and the theorem is proved. ■

Theorem 2.17 (Quotients of Limits) *Suppose that $\{s_n\}$ and $\{t_n\}$ are convergent sequences. Suppose further that $t_n \neq 0$ for all n and that the limit*

$$\lim_{n\to\infty} t_n \neq 0.$$

Then

$$\lim_{n\to\infty}\left(\frac{s_n}{t_n}\right) = \frac{\lim_{n\to\infty} s_n}{\lim_{n\to\infty} t_n}.$$

Proof Rather than prove the theorem at once as it stands let us prove just a special case of the theorem, namely that

$$\lim_{n\to\infty}\left(\frac{1}{t_n}\right) = \frac{1}{\lim_{n\to\infty} t_n}.$$

Let $T = \lim_{n\to\infty} t_n$. We need to show that no matter what positive number ε is given we can find an integer N so that

$$\left|\frac{1}{t_n} - \frac{1}{T}\right| < \varepsilon$$

if $n \geq N$. To work with this inequality requires us to consider

$$\left|\frac{1}{t_n} - \frac{1}{T}\right| = \frac{|t_n - T|}{|t_n|\,|T|}.$$

It is only the $|t_n|$ in the denominator that offers any trouble since if it is too small we cannot control the size of the fraction. This explains the first step in the proof that we now give, which otherwise might have seemed strange.

Suppose that $\varepsilon > 0$. Choose N_1 so that

$$|t_n - T| < |T|/2$$

if $n \geq N_1$ and also choose N_2 so that

$$|t_n - T| < \varepsilon|T|^2/2$$

if $n \geq N_2$. From the first inequality we see that

$$|T| - |t_n| \leq |T - t_n| < |T|/2$$

and so

$$|t_n| \geq |T|/2$$

if $n \geq N_1$. Set $N = \max\{N_1, N_2\}$ and note that if $n \geq N$ we must have

$$\left|\frac{1}{t_n} - \frac{1}{T}\right| = \frac{|t_n - T|}{|t_n|\,|T|}$$

$$< \frac{\varepsilon|T|^2/2}{|T|^2/2} = \varepsilon.$$

This is precisely the statement that $\lim_{n\to\infty}(1/t_n) = 1/T$.

We now complete the proof of the theorem by applying the product theorem along with what we have just proved to obtain

$$\lim_{n\to\infty}\left(\frac{s_n}{t_n}\right) = \left(\lim_{n\to\infty} s_n\right)\left(\lim_{n\to\infty}\frac{1}{t_n}\right) = \frac{\lim_{n\to\infty} s_n}{\lim_{n\to\infty} t_n}$$

as required. ■

Exercises

2.7.1 By imitating the proof given for the first part of Theorem 2.15 show that $\lim_{n\to\infty}(s_n - t_n) = \lim_{n\to\infty} s_n - \lim_{n\to\infty} t_n$.

2.7.2 Show that $\lim_{n\to\infty}(s_n)^2 = (\lim_{n\to\infty} s_n)^2$ using the theorem on products and also directly from the definition of limit.

2.7.3 Explain which theorems are needed to justify the computation of the limit $\lim_{n\to\infty} \frac{n^2}{2n^2+1}$ that introduced this section.

2.7.4 Prove Theorem 2.16 but verifying and using the inequality

$$|s_n t_n - ST| \le |(s_n - S)(t_n - T)| + |S(t_n - T)| + |T(s_n - S)|$$

in place of the inequality (1). Which proof do you prefer?

2.7.5 Which statements are true?

(a) If $\{s_n\}$ and $\{t_n\}$ are both divergent then so is $\{s_n + t_n\}$.

(b) If $\{s_n\}$ and $\{t_n\}$ are both divergent then so is $\{s_n t_n\}$.

(c) If $\{s_n\}$ and $\{s_n + t_n\}$ are both convergent then so is $\{t_n\}$.

(d) If $\{s_n\}$ and $\{s_n t_n\}$ are both convergent then so is $\{t_n\}$.

(e) If $\{s_n\}$ is convergent so too is $\{1/s_n\}$.

(f) If $\{s_n\}$ is convergent so too is $\{(s_n)^2\}$.

(g) If $\{(s_n)^2\}$ is convergent so too is $\{s_n\}$.

2.7.6 Note that there are extra hypotheses in the quotient theorem (Theorem 2.17) that were not in the product theorem (Theorem 2.16). Explain why both of these hypotheses are needed.

2.7.7 A careless student gives the following as a proof of Theorem 2.16. Find the flaw:

"Suppose that $\varepsilon > 0$. Choose N_1 so that

$$|s_n - S| < \frac{\varepsilon}{2|T| + 1}$$

if $n \ge N_1$ and also choose N_2 so that

$$|t_n - T| < \frac{\varepsilon}{2|s_n| + 1}$$

if $n \ge N_2$. If $n \ge N = \max\{N_1, N_2\}$ then

$$|s_n t_n - ST| \le |s_n|\,|t_n - T| + |T|\,|s_n - S|$$

$$\le |s_n|\left(\frac{\varepsilon}{2|s_n| + 1}\right) + |T|\left(\frac{\varepsilon}{2|T| + 1}\right) < \varepsilon.$$

Well, that works!"

2.7.8 Why are Theorems 2.15 and 2.16 no help in dealing with the limits

$$\lim_{n\to\infty} \left(\sqrt{n+1} - \sqrt{n}\right)$$

and

$$\lim_{n\to\infty} \sqrt{n}\left(\sqrt{n+1} - \sqrt{n}\right)?$$

What else can you do?

2.7.9 In calculus courses one learns that a function $f : \mathbb{R} \to \mathbb{R}$ is continuous at y if for every $\varepsilon > 0$ there is a $\delta > 0$ so that $|f(x) - f(y)| < \varepsilon$ for all $|x - y| < \delta$. Show that if f is continuous at y and $s_n \to y$ then $f(s_n) \to f(y)$. Use this to prove that $\lim_{n\to\infty}(s_n)^2 = (\lim_{n\to\infty} s_n)^2$.

2.8 Order Properties of Limits

In the preceding section we discussed the algebraic structure of limits. It is a natural mathematical question to ask how the algebraic operations are preserved under limits. As it happens, these natural mathematical questions usually are important in applications. We have seen that the algebraic properties of limits can be used to great advantage in computations of limits.

There is another aspect of structure of the real number system that plays an equally important role as the algebraic structure and that is the order structure. Does the limit operation preserve that order structure the same way that it preserves the algebraic structure? For example, if

$$s_n \leq t_n$$

for all n, can we conclude that

$$\lim_{n\to\infty} s_n \leq \lim_{n\to\infty} t_n?$$

In this section we solve this problem and several others related to the order structure. These results, too, will prove to be most useful in handling limits.

Theorem 2.18 *Suppose that $\{s_n\}$ and $\{t_n\}$ are convergent sequences and that*

$$s_n \leq t_n$$

for all n. Then

$$\lim_{n\to\infty} s_n \leq \lim_{n\to\infty} t_n.$$

Proof Let $S = \lim_{n\to\infty} s_n$ and $T = \lim_{n\to\infty} t_n$ and suppose that $\varepsilon > 0$. Choose N_1 so that

$$|s_n - S| < \varepsilon/2$$

if $n \geq N_1$ and also choose N_2 so that

$$|t_n - T| < \varepsilon/2$$

if $n \geq N_2$. Set $N = \max\{N_1, N_2\}$ and note that if $n \geq N$ we must have

$$0 \leq t_n - s_n = T - S + (t_n - T) + (S - s_n) < T - S + \varepsilon/2 + \varepsilon/2.$$

This shows that

$$-\varepsilon < T - S.$$

This statement is true for *any* positive number ε. It would be false if $T - S$ is negative and hence $T - S$ is positive or zero (i.e., $T \geq S$ as required). ■

Note. There is a trap here that many students have fallen into. Since the condition $s_n \leq t_n$ implies

$$\lim_{n\to\infty} s_n \leq \lim_{n\to\infty} t_n$$

would it not follow "similarly" that the condition $s_n < t_n$ implies

$$\lim_{n\to\infty} s_n < \lim_{n\to\infty} t_n?$$

Be careful with this. It is false. See Exercise 2.8.1.

Corollary 2.19 *Suppose that $\{s_n\}$ is a convergent sequence and that*

$$\alpha \leq s_n \leq \beta$$

for all n. Then

$$\alpha \leq \lim_{n\to\infty} s_n \leq \beta.$$

Proof Consider that the assumption here can be read as $\alpha_n \leq s_n \leq \beta_n$ where $\{\alpha_n\}$ and $\{\beta_n\}$ are constant sequences. Now apply the theorem. ■

Note. Again, don't forget the trap. The condition $\alpha < s_n < \beta$ for all n implies that

$$\alpha \leq \lim_{n\to\infty} s_n \leq \beta.$$

It would not imply that

$$\alpha < \lim_{n\to\infty} s_n < \beta.$$

The Squeeze Theorem The next theorem is another useful variant on these themes. Here an unknown sequence is sandwiched between two convergent sequences, allowing us to conclude that that sequence converges. This theorem is often taught as "the squeeze theorem," which seems a convenient label.

Theorem 2.20 (Squeeze Theorem) *Suppose that $\{s_n\}$ and $\{t_n\}$ are convergent sequences, that*

$$\lim_{n\to\infty} s_n = \lim_{n\to\infty} t_n$$

and that

$$s_n \leq x_n \leq t_n$$

for all n. Then $\{x_n\}$ is also convergent and

$$\lim_{n\to\infty} x_n = \lim_{n\to\infty} s_n = \lim_{n\to\infty} t_n.$$

Proof Let L be the limit of the two sequences. Choose N_1 so that

$$|s_n - L| < \varepsilon$$

if $n \geq N_1$ and also choose N_2 so that

$$|t_n - L| < \varepsilon$$

if $n \geq N_2$. Set $N = \max\{N_1, N_2\}$. Note that

$$s_n - L \leq x_n - L \leq t_n - L$$

for all n and so

$$-\varepsilon < s_n - L \leq x_n - L \leq t_n - L < \varepsilon$$

if $n \geq N$. From this we see that

$$-\varepsilon < x_n - L < \varepsilon$$

or, to put it in a more familiar form,

$$|x_n - L| < \varepsilon$$

proving the statement of the theorem. ■

Example 2.21 Let θ be some real number and consider the computation of

$$\lim_{n\to\infty} \frac{\sin n\theta}{n}.$$

While this might seem hopeless at first sight since the values of $\sin n\theta$ are quite unpredictable, we recall that none of these values lies outside the interval $[-1, 1]$. Hence

$$-\frac{1}{n} \leq \frac{\sin n\theta}{n} \leq \frac{1}{n}.$$

The two outer sequences converge to the same value 0 and so the inside sequence (the "squeezed" one) must converge to 0 as well. ◀

Absolute Values A further theorem on the theme of order structure is often needed. The absolute value, we recall, is defined directly in terms of the order structure. Is absolute value preserved by the limit operation?

Theorem 2.22 (Limits of Absolute Values) *Suppose that $\{s_n\}$ is a convergent sequence. Then the sequence $\{|s_n|\}$ is also a convergent sequence and*

$$\lim_{n\to\infty} |s_n| = \left|\lim_{n\to\infty} s_n\right|.$$

Proof Let $S = \lim_{n\to\infty} s_n$ and suppose that $\varepsilon > 0$. Choose N so that

$$|s_n - S| < \varepsilon$$

if $n \geq N$. Observe that, because of the triangle inequality, this means that

$$||s_n| - |S|| \leq |s_n - S| < \varepsilon$$

for all $n \geq N$. By definition

$$\lim_{n\to\infty} |s_n| = |S|$$

as required. ■

Maxima and Minima Since maxima and minima can be expressed in terms of absolute values, there is a corollary that is sometimes useful.

Corollary 2.23 (Max/Min of Limits) *Suppose that $\{s_n\}$ and $\{t_n\}$ are convergent sequences. Then the sequences*

$$\{\max\{s_n, t_n\}\} \quad \text{and} \quad \{\min\{s_n, t_n\}\}$$

are also convergent and

$$\lim_{n\to\infty} \max\{s_n, t_n\} = \max\{\lim_{n\to\infty} s_n, \lim_{n\to\infty} t_n\}$$

and

$$\lim_{n\to\infty} \min\{s_n, t_n\} = \min\{\lim_{n\to\infty} s_n, \lim_{n\to\infty} t_n\}.$$

Proof The first of these follows from the identity

$$\max\{s_n, t_n\} = \frac{s_n + t_n}{2} + \frac{|s_n - t_n|}{2}$$

and the theorem on limits of sums and the theorem on limits of absolute values. In the same way the second assertion follows from

$$\min\{s_n, t_n\} = \frac{s_n + t_n}{2} - \frac{|s_n - t_n|}{2}.$$

■

Exercises

2.8.1 Show that the condition $s_n < t_n$ does not imply that

$$\lim_{n\to\infty} s_n < \lim_{n\to\infty} t_n.$$

(If the proof of Theorem 2.18 were modified in an attempt to prove this false statement, where would the modifications fail?)

2.8.2 If $\{s_n\}$ is a sequence all of whose values lie inside an interval $[a,b]$ prove that $\{s_n/n\}$ is convergent.

2.8.3 A careless student gives the following as a proof of the squeeze theorem. Find the flaw:

"If $\lim_{n\to\infty} s_n = \lim_{n\to\infty} t_n = L$, then take limits in the inequality

$$s_n \leq x_n \leq t_n$$

to get $L \leq \lim_{n\to\infty} x_n \leq L$. This can only be true if $\lim_{n\to\infty} x_n = L$."

2.8.4 Suppose that $s_n \leq t_n$ for all n and that $s_n \to \infty$. What can you conclude?

2.8.5 Suppose that $\lim_{n\to\infty} \frac{s_n}{n} > 0$ Show that $s_n \to \infty$.

2.8.6 Suppose that $\{s_n\}$ and $\{t_n\}$ are sequences of positive numbers, that

$$\lim_{n\to\infty} \frac{s_n}{t_n} = \alpha$$

and that $s_n \to \infty$. What can you conclude?

2.8.7 Suppose that $\{s_n\}$ and $\{t_n\}$ are sequences of positive numbers, that

$$\lim_{n\to\infty} \frac{s_n}{t_n} = \infty$$

and that $t_n \to \infty$. What can you conclude?

2.8.8 Suppose that $\{s_n\}$ and $\{t_n\}$ are sequences of positive numbers, that

$$\lim_{n\to\infty} \frac{s_n}{t_n} = \infty$$

and that $\{s_n\}$ is bounded. What can you conclude?

2.8.9 Let $\{s_n\}$ be a sequence of positive numbers. Show that the condition

$$\lim_{n\to\infty} \frac{s_{n+1}}{s_n} < 1$$

implies that $s_n \to 0$.

2.8.10 Let $\{s_n\}$ be a sequence of positive numbers. Show that the condition

$$\lim_{n\to\infty} \frac{s_{n+1}}{s_n} > 1$$

implies that $s_n \to \infty$.

2.9 Monotone Convergence Criterion

In many applications of sequence theory we find that the sequences that arise are going in one direction: The terms steadily get larger or steadily get smaller. The analysis of such sequences is much easier than for general sequences.

Definition 2.24 (Increasing) We say that a sequence $\{s_n\}$ is *increasing* if

$$s_1 < s_2 < s_3 < \cdots < s_n < s_{n+1} < \ldots.$$

Definition 2.25 (Decreasing) We say that a sequence $\{s_n\}$ is *decreasing* if

$$s_1 > s_2 > s_3 > \cdots > s_n > s_{n+1} > \ldots.$$

Often we encounter sequences that "increase" except perhaps occasionally successive values are equal rather than strictly larger. The following language is usually used in this case.

Definition 2.26 (Nondecreasing) We say that a sequence $\{s_n\}$ is *nondecreasing* if

$$s_1 \leq s_2 \leq s_3 \leq \cdots \leq s_n \leq s_{n+1} \leq \ldots.$$

Definition 2.27 (Nonincreasing) We say that a sequence $\{s_n\}$ is *nonincreasing* if

$$s_1 \geq s_2 \geq s_3 \geq \cdots \geq s_n \geq s_{n+1} \geq \ldots.$$

Thus every increasing sequence is also nondecreasing but not conversely. A sequence that has any one of these four properties (increasing, decreasing, nondecreasing, or nonincreasing) is said to be *monotonic.* Monotonic sequences are often easier to deal with than sequences that can go both up and down.

Note. In some texts you will find that a nondecreasing sequence is said to be increasing and an increasing sequence is said to be *strictly* increasing. The way in which we intend these terms should be clear and intuitive. If your monthly salary occasionally rises but sometimes stays the same you would not likely say that it is increasing. You might, however, say "at least it never decreases" (i.e., it is nondecreasing).

The convergence issue for a monotonic sequence is particularly straightforward. We can imagine that an increasing sequence could increase up to some limit, or we could imagine that it could increase indefinitely and diverge to $+\infty$. It is impossible to imagine a third possibility. We express this as a theorem that will become our primary theoretical tool in investigating convergence of sequences.

Theorem 2.28 (Monotone Convergence Theorem) *Suppose that* $\{s_n\}$ *is a monotonic sequence. Then* $\{s_n\}$ *is convergent if and only if* $\{s_n\}$ *is bounded. More specifically,*

1. *If* $\{s_n\}$ *is nondecreasing then either* $\{s_n\}$ *is bounded and converges to* $\sup\{s_n\}$ *or else* $\{s_n\}$ *is unbounded and* $s_n \to \infty$.

2. *If* $\{s_n\}$ *is nonincreasing then either* $\{s_n\}$ *is bounded and converges to* $\inf\{s_n\}$ *or else* $\{s_n\}$ *is unbounded and* $s_n \to -\infty$.

Proof If the sequence is unbounded then it diverges. This is true for any sequence, not merely monotonic sequences.

Thus the proof is complete if we can show that for any bounded monotonic sequence $\{s_n\}$ the limit is $\sup\{s_n\}$ in case the sequence is nondecreasing, or it is $\inf\{s_n\}$ in case the sequence is nonincreasing. Let us prove the first of these cases.

Let $\{s_n\}$ be assumed to be nondecreasing and bounded, and let

$$L = \sup\{s_n\}.$$

Then $s_n \leq L$ for all n and if $\beta < L$ there must be some term s_m say, with $s_m > \beta$. Let $\varepsilon > 0$. We know that there is an m so that

$$s_n \geq s_m > L - \varepsilon$$

for all $n \geq m$. But we already know that every term $s_n \leq L$. Putting these together we have that

$$L - \varepsilon < s_n \leq L < L + \varepsilon$$

or

$$|s_n - L| < \varepsilon$$

for all $n \geq m$. By definition then $s_n \to L$ as required. ■

How would we normally apply this theorem? Suppose a sequence $\{s_n\}$ were given that we recognize as increasing (or maybe just nondecreasing). Then to establish that $\{s_n\}$ converges we need only show that the sequence is bounded above, that is, we need to find just one number M with

$$s_n \leq M$$

for all n. Any crude upper estimate would verify convergence.

Example 2.29 Let us show that the sequence $s_n = 1/\sqrt{n}$ converges. This sequence is evidently decreasing. Can we find a lower bound? Yes, all of the terms are positive so that 0 is a lower bound. Consequently, the sequence must converge. If we wish to show that

$$\lim_{n\to\infty} \frac{1}{\sqrt{n}} = 0$$

we need to do more. But to conclude convergence we needed only to make a crude estimate on how low the terms might go. ◀

Example 2.30 Let us examine the sequence

$$s_n = 1 + \frac{1}{2} + \frac{1}{3} + \frac{1}{4} + \cdots + \frac{1}{n}.$$

This sequence is evidently increasing. Can we find an upper bound? If we can then the series does converge. If we cannot then the series diverges. We have already (earlier) checked this sequence. It is unbounded and so $\lim_{n\to\infty} s_n = \infty$. ◀

Example 2.31 Let us examine the sequence

$$\sqrt{2},\ \sqrt{2+\sqrt{2}},\ \sqrt{2+\sqrt{2+\sqrt{2}}},\ \sqrt{2+\sqrt{2+\sqrt{2+\sqrt{2}}}}, \ldots.$$

Handling such a sequence directly by the limit definition seems quite impossible. This sequence can be defined recursively by

$$x_1 = \sqrt{2} \quad x_n = \sqrt{2 + x_{n-1}}.$$

The computation of a few terms suggests that the sequence is increasing and so should be accessible by the methods of this section.

We prove this by induction. That $x_1 < x_2$ is just an easy computation (do it). Let us suppose that $x_{n-1} < x_n$ for some n and show that it must follow that $x_n < x_{n+1}$. But

$$x_n = \sqrt{2 + x_{n-1}} < \sqrt{2 + x_n} = x_{n+1}$$

where the middle step is the induction hypothesis (i.e., that $x_{n-1} < x_n$). It follows by induction that the sequence is increasing.

Now we show inductively that the sequence is bounded above. Any crude upper bound will suffice. It is clear that $x_1 < 10$. If $x_{n-1} < 10$ then

$$x_n = \sqrt{2 + x_{n-1}} < \sqrt{2 + 10} < 10$$

and so it follows, again by induction, that all terms of the sequence are smaller than 10. We conclude from the monotone convergence theorem that this sequence is convergent.

But to what? (Certainly it does not converges to 10 since that estimate was extremely crude.) That is not so easy to sort out, it seems. But perhaps it is, since we know that the sequence converges to something, say L. In the equation

$$(x_n)^2 = 2 + x_{n-1},$$

obtained by squaring the recursion formula given to us, we can take limits as $n \to \infty$. Since $x_n \to L$ so too does $x_{n-1} \to L$ and $(x_n)^2 \to L^2$. Hence

$$L^2 = 2 + L.$$

The only possibilities for L in this quadratic equation are $L = -1$ and $L = 2$. We know the limit L exists and we know that it is either -1 or 2. We can clearly rule out -1 as none of the numbers in our sequence were negative. Hence $x_n \to 2$.

◀

Exercises

2.9.1 Define a sequence $\{s_n\}$ recursively by setting $s_1 = \alpha$ and

$$s_n = \frac{(s_{n-1})^2 + \beta}{2s_{n-1}}$$

where $\alpha,\ \beta > 0$.

(a) Show that for $n = 1, 2, 3, \ldots$

$$\frac{(s_n - \sqrt{\beta})^2}{2s_n} = s_{n+1} - \beta.$$

(b) Show that $s_n > \sqrt{\beta}$ for all $n = 2, 3, 4, \ldots$ unless $\alpha = \sqrt{\beta}$. What happens if $\alpha = \sqrt{\beta}$?

(c) Show that $s_2 > s_3 > s_4 > \ldots s_n > \ldots$ except in the case $\alpha = \sqrt{\beta}$.

(d) Does this sequence converge? To what?

(e) What is the relation of this sequence to the one introduced in Section 2.1 as Newton's method?

2.9.2 Define a sequence $\{t_n\}$ recursively by setting $t_1 = 1$ and

$$t_n = \sqrt{t_{n-1} + 1}.$$

Does this sequence converge? To what?

2.9.3 Consider the sequence $s_1 = 1$ and $s_n = \frac{2}{s_{n-1}^2}$. We argue that if $s_n \to L$ then $L = \frac{2}{L^2}$ and so $L^3 = 2$ or $L = \sqrt[3]{2}$. Our conclusion is that $\lim_{n\to\infty} s_n = \sqrt[3]{2}$. Do you have any criticisms of this argument?

2.9.4 Does the sequence

$$\frac{1 \cdot 3 \cdot 5 \cdot \cdots \cdot (2n-1)}{2 \cdot 4 \cdot 6 \cdot \cdots \cdot (2n)}$$

converge?

2.9.5 Does the sequence

$$\frac{2 \cdot 4 \cdot 6 \cdot \cdots \cdot (2n) \cdot 1}{1 \cdot 3 \cdot 5 \cdot \cdots \cdot (2n-1) \cdot n^2}$$

converge?

2.9.6 Several nineteenth-century mathematicians used, without proof, a principle in their proofs that has come to be known as the *nested interval property*:

Given a sequence of closed intervals

$$[a_1, b_1] \supset [a_2, b_2] \supset [a_3, b_3] \supset \dots$$

arranged so that each interval is a subinterval of the one preceding it and so that the lengths of the intervals shrink to zero, then there is exactly one point that belongs to every interval of the sequence.

Prove this statement. Would it be true for a descending sequence of open intervals

$$(a_1, b_1) \supset (a_2, b_2) \supset (a_3, b_3) \supset \dots ?$$

2.10 Examples of Limits

The theory of sequence limits has now been developed far enough that we may investigate some interesting limits. Each of the limits in this section has some cultural interest. Most students would be expected to know and recognize these limits as they arise quite routinely. For us they are also an opportunity to show off our methods. Mostly we need to establish inequalities and use some of our theory. We do not need to use an ε, N argument since we now have more subtle and powerful tools at hand.

Example 2.32 (Geometric Progressions) Let r be a real number. What is the limiting behavior of the sequence

$$1, r, r^2, r^3, r^4, \dots, r^n, \dots$$

forming a geometric progression? If $r > 1$ then it is not hard to show that

$$r^n \to \infty.$$

If $r \leq -1$ the sequence certainly diverges. If $r = 1$ this is just a constant sequence.

The interesting case is

$$\lim_{n\to\infty} r^n = 0 \qquad \text{if } -1 < r < 1.$$

To prove this we shall use an easy inequality. Let $x > 0$ and n an integer. Then, using the binomial theorem (or induction if you prefer), we can show that

$$(1 + x)^n > nx.$$

Case (i): Let $0 < r < 1$. Then

$$r = \frac{1}{1 + x}$$

(where $x = 1/r - 1 > 0$) and so

$$0 < r^n = \frac{1}{(1+x)^n} < \frac{1}{nx} \to 0$$

as $n \to \infty$. By the squeeze theorem we see that $r^n \to 0$ as required.

Case (ii): If $-1 < r < 0$ then $r = -t$ for $0 < t < 1$. Thus

$$-t^n \leq r^n \leq t^n.$$

By case (i) we know that $t^n \to 0$. By the squeeze theorem we see that $r^n \to 0$ again as required. ◀

Example 2.33 (Roots) An interesting and often useful limit is

$$\lim_{n \to \infty} \sqrt[n]{n} = 1.$$

To show this we once again derive an inequality from the binomial theorem. If $n \geq 2$ and $x > 0$ then

$$(1+x)^n > n(n-1)x^2/2.$$

For $n \geq 2$ write

$$\sqrt[n]{n} = 1 + x_n$$

(where $x_n = \sqrt[n]{n} - 1 > 0$) and so

$$n = (1+x_n)^n > n(n-1)x_n^2/2$$

or

$$0 < x_n^2 < \frac{2}{n-1} \to 0$$

as $n \to \infty$. By the squeeze theorem we see that $x_n \to 0$ and it follows that $\sqrt[n]{n} \to 1$ as required.

As a special case of this example note that

$$\sqrt[n]{C} \to 1$$

as $n \to \infty$ for any positive constant C. This is true because if $C > 1$ then

$$1 < \sqrt[n]{C} < \sqrt[n]{n}$$

for large enough n. By the squeeze theorem this shows that $\sqrt[n]{C} \to 1$. If, however, $0 < C < 1$ then

$$\sqrt[n]{C} = \frac{1}{\sqrt[n]{1/C}} \to 1$$

by the first case since $1/C > 1$. ◀

Example 2.34 (Sums of Geometric Progressions) For all values of x in the interval $(-1,1)$ the limit

$$\lim_{n\to\infty}\left(1+x+x^2+x^3+\cdots+x^n\right)=\frac{1}{1-x}.$$

While at first a surprising result, this is quite evident once we check the identity

$$(1-x)\left(1+x+x^2+x^3+\cdots+x^n\right)=1-x^{n+1},$$

which just requires a straightforward multiplication. Thus

$$\lim_{n\to\infty}\left(1+x+x^2+x^3+\cdots+x^n\right)=\lim_{n\to\infty}\frac{1-x^{n+1}}{1-x}=\frac{1}{1-x}$$

where we have used the result we proved previously, namely that

$$x^{n+1}\to 0 \quad \text{if } |x|<1.$$

One special case of this is useful to remember. Set $x=1/2$. Then

$$\lim_{n\to\infty}\left(1+\frac{1}{2}+\frac{1}{2^2}+\frac{1}{2^3}+\cdots+\frac{1}{2^n}\right)=2.$$

◀

Example 2.35 (Decimal Expansions) What meaning is assigned to the infinite decimal expansion

$$x=0.d_1d_2d_3d_4\ldots d_n\ldots$$

where the choices of integers $0\le d_i\le 9$ can be made in any way? Repeating decimals can always be converted into fractions and so the infinite process can be avoided. But if the pattern does not repeat, a different interpretation must be made.

The most obvious interpretation of this number x is to declare that it is the limit of the sequence

$$\lim_{n\to\infty} 0.d_1d_2d_3d_4\ldots d_n.$$

But how do we know that the limit exists? Our theory provides an immediate answer. Since this sequence is nondecreasing and every term is smaller than 1, by the monotone convergence theorem the sequence converges. This is true no matter what the choices of the decimal digits are. ◀

Example 2.36 (Expansion of e^x) Let $x>0$ and consider the two closely related sequences

$$s_n=1+x+\frac{x^2}{2!}+\frac{x^3}{3!}+\cdots+\frac{x^n}{n!}$$

and

$$t_n = \left(1 + \frac{x}{n}\right)^n.$$

The relation between the two sequences becomes more apparent once the binomial theorem is used to expand the latter.

In more advanced mathematics it is shown that both sequences converge to e^x. Let us be content to prove that

$$\lim_{n\to\infty} s_n = \lim_{n\to\infty} t_n.$$

The sequence $\{s_n\}$ is clearly increasing since each new term is the preceding term with a positive number added to it. To show convergence then we need only show that the sequence is bounded. This takes some arithmetic, but not too much.

Choose an integer N larger than $2x$. Note then that

$$\frac{x^{N+1}}{(N+1)!} < \frac{1}{2}\left(\frac{x^N}{(N)!}\right)$$

that

$$\frac{x^{N+2}}{(N+2)!} < \frac{1}{4}\left(\frac{x^N}{(N)!}\right)$$

and that

$$\frac{x^{N+3}}{(N+3)!} < \frac{1}{8}\left(\frac{x^N}{(N)!}\right).$$

Thus

$$s_n \leq \left[1 + x + \frac{x^2}{2!} + \ldots \frac{x^{N-1}}{(N-1)!}\right] + \frac{x^N}{(N)!}\left(1 + \frac{1}{2} + \frac{1}{4}\ldots\right)$$

$$\leq \left[1 + x + \frac{x^2}{2!} + \ldots \frac{x^{N-1}}{(N-1)!}\right] + 2\frac{x^N}{(N)!}.$$

Here we have used the limit for the sum of a geometric progression from Example 2.34 to make an upper estimate on how large this sum can get. Note that the N is fixed and so the number on the right-hand side of this inequality is just a number, and it is larger than every number in the sequence $\{s_n\}$.

It follows now from the monotone convergence theorem that $\{s_n\}$ converges. To handle $\{t_n\}$, first apply the binomial theorem to obtain

$$t_n = 1 + x + \frac{1 - 1/n}{2!}x^2 + \frac{(1 - 1/n)(1 - 2/n)}{3!}x^3 + \cdots \leq s_n.$$

From this we see that $\{t_n\}$ is increasing and that it is smaller than the convergent sequence $\{s_n\}$. It follows, again from the monotone convergence theorem, that $\{t_n\}$ converges. Moreover,

$$\lim_{n\to\infty} t_n \leq \lim_{n\to\infty} s_n.$$

If we can obtain the opposite inequality we will have proved our assertion. Let m be a fixed number and let $n > m$. Then, from the preceding expansion, we note that

$$t_n > 1 + x + \frac{1 - 1/n}{2!}x^2 + \frac{(1 - 1/n)(1 - 2/n)}{3!}x^3$$
$$+ \cdots + \frac{(1 - 1/n)(1 - 2/n)\dots(1 - [m-1]/n)}{m!}x^m.$$

We can hold m fixed and allow $n \to \infty$ in this inequality and obtain that

$$\lim_{n\to\infty} t_n \geq s_m$$

for each m. From this it now follows that

$$\lim_{n\to\infty} t_n \geq \lim_{n\to\infty} s_n$$

and we have completed our task. ◀

Exercises

2.10.1 Since we know that

$$1 + x + x^2 + x^3 + \cdots + x^n \to \frac{1}{1-x}$$

this suggests the formula

$$1 + 2 + 4 + 8 + 16 + \cdots = \frac{1}{1-2} = -1.$$

Do you have any criticisms?

2.10.2 Let α and β be positive numbers. Discuss the convergence behavior of the sequence

$$\frac{\alpha^{\beta n}}{\beta^{\alpha n}}.$$

2.10.3 Define

$$e = \lim_{n\to\infty}\left(1 + \frac{1}{n}\right)^n.$$

Show that $2 < e < 3$.

2.10.4 Show that

$$\lim_{n\to\infty}\left(1 + \frac{1}{2n}\right)^n = \sqrt{e}.$$

2.10.5 Check the simple identity

$$\left(1+\frac{2}{n}\right) = \left(1+\frac{1}{n+1}\right)\left(1+\frac{1}{n}\right)$$

and use it to show that

$$\lim_{n\to\infty}\left(1+\frac{2}{n}\right)^n = e^2.$$

2.11 Subsequences

The sequence

$$1, -1, 2, -2, 3, -3, 4, -4, 5, -5, \dots$$

appears to contain within itself the two sequences

$$1, 2, 3, 4, 5, \dots$$

and

$$-1, -2, -3, -4, -5, \dots.$$

In order to have a language to express this we introduce the term *subsequence.* We would say that the latter two sequences are subsequences of the first sequence. Often a sequence is best studied by looking at some of its subsequences. But what is a proper definition of this term? We need a formal mathematical way of expressing the vague idea that a subsequence is obtained by crossing out some of the terms of the original sequence.

Definition 2.37 (Subsequences) Let

$$s_1, s_2, s_3, s_4, \dots$$

be any sequence. Then by a *subsequence* of this sequence we mean any sequence

$$s_{n_1}, s_{n_2}, s_{n_3}, s_{n_4}, \dots$$

where

$$n_1 < n_2 < n_3 < \dots$$

is an increasing sequence of natural numbers.

Example 2.38 We can consider

$$1, 2, 3, 4, 5, \dots$$

to be a subsequence of sequence

$$1, -1, 2, -2, 3, -3, 4, -4, 5, -5, \dots$$

because it contains just the first, third, fifth, etc. terms of the original sequence. Here $n_1 = 1$, $n_2 = 3$, $n_3 = 5$, $\dots$. ◄

In many applications of sequences it is the subsequences that need to be studied. For example, what can we say about the existence of monotonic subsequences, or bounded subsequences, or divergent subsequences, or convergent subsequences? The answers to these questions have important uses.

Existence of Monotonic Subsequences Our first question is easy to answer for any specific sequence, but harder to settle in general. Given a sequence can we always select a subsequence that is monotonic, either monotonic nondecreasing or monotonic nonincreasing?

Theorem 2.39 *Every sequence contains a monotonic subsequence.*

Proof We construct first a nonincreasing subsequence if possible. We call the mth element x_m of the sequence $\{x_n\}$ a turn-back point if all later elements are less than or equal to it, in symbols if $x_m \geq x_n$ for all $n > m$. If there is an infinite subsequence of turn-back points $x_{m_1}, x_{m_2}, x_{m_3}, x_{m_4}, \dots$ then we have found our nonincreasing subsequence since

$$x_{m_1} \geq x_{m_2} \geq x_{m_3} \geq x_{m_4} \geq \dots .$$

This would not be possible if there are only finitely many turn-back points. Let us suppose that x_M is the last turn-back point so that any element x_n for $n > M$ is not a turn-back point. Since it is not there must be an element further on in the sequence greater than it, in symbols $x_m > x_n$ for some $m > n$. Thus we can choose $x_{m_1} > x_{M+1}$ with $m_1 > M + 1$, then $x_{m_2} > x_{m_1}$ with $m_2 > m_1$, and then $x_{m_3} > x_{m_2}$ with $m_3 > m_2$, and so on to obtain an increasing subsequence

$$x_{M+1} < x_{m_1} < x_{m_2} < x_{m_3} < x_{m_4} < \dots$$

as required. ■

Existence of Convergent Subsequences Having answered this question about the existence of monotonic subsequences, we can also now answer the question about the existence of convergent subsequences. This might, at first sight, seem just a curiosity, but it will give us later one of our most important tools in analysis.

The theorem is traditionally attributed to two major nineteenth-century mathematicians, Karl Theodor Wilhelm Weierstrass (1815-1897) and Bernhard Bolzano (1781–1848). These two mathematicians, the first German and the second Czech, rank with Cauchy among the founders of our subject.

Theorem 2.40 (Bolzano-Weierstrass) *Every bounded sequence contains a convergent subsequence.*

Proof By Theorem 2.39 every sequence contains a monotonic subsequence. Here that subsequence would be both monotonic and bounded, and hence convergent. ■

Other (less important) questions of this type appear in the exercises.

Exercises

2.11.1 Show that, according to our definition, every sequence is a subsequence of itself. How would the definition have to be reworded to avoid this if, for some reason, this possibility were to have been avoided?

2.11.2 Show that every subsequence of a subsequence of a sequence $\{x_n\}$ is itself a subsequence of $\{x_n\}$.

2.11.3 If $\{s_{n_k}\}$ is a subsequence of $\{s_n\}$ and $\{t_{m_k}\}$ is a subsequence of $\{t_n\}$ then is it true that $\{s_{n_k} + t_{m_k}\}$ is a subsequence of $\{s_n + t_n\}$?

2.11.4 If $\{s_{n_k}\}$ is a subsequence of $\{s_n\}$ is $\{(s_{n_k})^2\}$ a subsequence of $\{(s_n)^2\}$?

2.11.5 Describe all sequences that have only finitely many different subsequences.

2.11.6 Establish which of the following statements are true.

(a) A sequence is convergent if and only if all of its subsequences are convergent.

(b) A sequence is bounded if and only if all of its subsequences are bounded.

(c) A sequence is monotonic if and only if all of its subsequences are monotonic.

(d) A sequence is divergent if and only if all of its subsequences are divergent.

2.11.7 Establish which of the following statements are true for an arbitrary sequence $\{s_n\}$.

(a) If all monotone subsequences of a sequence $\{s_n\}$ are convergent, then $\{s_n\}$ is bounded.

(b) If all monotone subsequences of a sequence $\{s_n\}$ are convergent, then $\{s_n\}$ is convergent.

(c) If all convergent subsequences of a sequence $\{s_n\}$ converge to 0, then $\{s_n\}$ converges to 0.

(d) If all convergent subsequences of a sequence $\{s_n\}$ converge to 0 and $\{s_n\}$ is bounded, then $\{s_n\}$ converges to 0.

2.11.8 Where possible find subsequences that are monotonic and subsequences that are convergent for the following sequences

(a) $\{(-1)^n n\}$

(b) $\{\sin(n\pi/8)\}$

(c) $\{n\sin(n\pi/8)\}$

(d) $\{\frac{n+1}{n}\sin(n\pi/8)\}$

(e) $\{1+(-1)^n\}$

(f) $\{r_n\}$ consists of all rational numbers in the interval $(0,1)$ arranged in some order.

2.11.9 Describe all subsequences of the sequence

$$1,0,1,0,1,0,1,0,1,0,1,0,\ldots.$$

Describe all convergent subsequences. Describe all monotonic subsequences.

2.11.10 If $\{s_{n_k}\}$ is a subsequence of $\{s_n\}$ show that $n_k \geq k$ for all $k=1,2,3,\ldots$.

2.11.11 Give an example of a sequence that contains subsequences converging to every natural number (and no other numbers).

2.11.12 Give an example of a sequence that contains subsequences converging to every number in $[0,1]$ (and no other numbers).

2.11.13 Show that there cannot exist a sequence that contains subsequences converging to every number in $(0,1)$ and no other numbers.

2.11.14 Show that if $\{s_n\}$ has no convergent subsequences, then $|s_n| \to \infty$ as $n \to \infty$.

2.11.15 If a sequence $\{x_n\}$ has the property that

$$\lim_{n\to\infty} x_{2n} = \lim_{n\to\infty} x_{2n+1} = L$$

show that the sequence $\{x_n\}$ converges to L.

2.11.16 If a sequence $\{x_n\}$ has the property that

$$\lim_{n\to\infty} x_{2n} = \lim_{n\to\infty} x_{2n+1} = \infty$$

show that the sequence $\{x_n\}$ diverges to ∞.

2.11.17 Let α and β be positive real numbers and define a sequence by setting $s_1=\alpha$, $s_2=\beta$ and $s_{n+2}=\frac{1}{2}(s_n+s_{n+1})$ for all $n=1,2,3,\ldots$. Show that the subsequences $\{s_{2n}\}$ and $\{s_{2n-1}\}$ are monotonic and convergent. Does the sequence $\{s_n\}$ converge? To what?

2.11.18 Without appealing to any of the theory of this section prove that every unbounded sequence has a strictly monotonic subsequence (i.e., either increasing or decreasing).

2.11.19 Show that if a sequence $\{x_n\}$ converges to a finite limit or diverges to $\pm\infty$ then every subsequence has precisely the same behavior.

2.11.20 Suppose a sequence $\{x_n\}$ has the property that every subsequence has a further subsequence convergent to L. Show that $\{x_n\}$ converges to L.

2.11.21 Let $\{x_n\}$ be a bounded sequence and let $x = \sup\{x_n : n \in \mathbb{N}\}$. Suppose that, moreover, $x_n < x$ for all n. Prove that there is a subsequence convergent to x.

2.11.22 Let $\{x_n\}$ be a bounded sequence, let

$$y = \inf\{x_n : n \in \mathbb{N}\} \quad \text{and} \quad x = \sup\{x_n : n \in \mathbb{N}\}.$$

Suppose that, moreover, $y < x_n < x$ for all n. Prove that there is a pair of convergent subsequences $\{x_{n_k}\}$ and $\{x_{m_k}\}$ so that

$$\lim_{k\to\infty} |x_{n_k} - x_{m_k}| = x - y.$$

2.11.23 Does every divergent sequence contain a divergent monotonic sequence?

2.11.24 Does every divergent sequence contain a divergent bounded sequence?

2.11.25 Construct a proof of the Bolzano-Weierstrass theorem for bounded sequences using the nested interval property and not appealing to the existence of monotonic subsequences.

2.11.26 Construct a direct proof of the assertion that every convergent sequence has a convergent, monotonic subsequence (i.e., without appealing to Theorem 2.39).

2.11.27 Let $\{x_n\}$ be a bounded sequence that we do not know converges. Suppose that it has the property that every one of its convergent subsequences converges to the same number L. What can you conclude?

2.11.28 Let $\{x_n\}$ be a bounded sequence that diverges. Show that there is a pair of convergent subsequences $\{x_{n_k}\}$ and $\{x_{m_k}\}$ so that

$$\lim_{k\to\infty} |x_{n_k} - x_{m_k}| > 0.$$

2.11.29 Let $\{x_n\}$ be a sequence. A number z with the property that for all $\varepsilon > 0$ there are infinitely many terms of the sequence in the interval $(z-\varepsilon, z+\varepsilon)$ is said to be a *cluster point* of the sequence. Show that z is a cluster point of a sequence if and only if there is a subsequence $\{x_{n_k}\}$ converging to z.

2.12 Cauchy Convergence Criterion

What property of a sequence characterizes convergence? As a "characterization" we would like some necessary and sufficient condition for a sequence to converge. We could simply write the definition and consider that that is a characterization. Thus the following technical statement would, indeed, be a characterization of the convergence of a sequence $\{s_n\}$.

A sequence $\{s_n\}$ is convergent if and only if $\exists L$ so that $\forall \varepsilon > 0$ $\exists N$ with the property that

$$|s_n - L| < \varepsilon$$

whenever $n \geq N$.

In mathematics when we ask for a characterization of a property we can expect to find many answers, some more useful than others. The limitation of this particular characterization is that it requires us to find the number L which is the limit of the sequence in advance. Compare this with a characterization of convergence of a monotonic sequence $\{s_n\}$.

> *A monotonic sequence $\{s_n\}$ is convergent if and only if it is bounded.*

This is a wonderful and most useful characterization. But it applies only to monotonic sequences.

A correct and useful characterization, applicable to all sequences, was found by Cauchy. This is the content of the next theorem. Note that it has the advantage that it describes a convergent sequence with no reference whatsoever to the actual value of the limit. Loosely it asserts that a sequence converges if and only if the terms of the sequence are eventually arbitrarily close together.

Theorem 2.41 (Cauchy Criterion) *A sequence $\{s_n\}$ is convergent if and only if for each $\varepsilon > 0$ there exists an integer N with the property that*

$$|s_n - s_m| < \varepsilon$$

whenever $n \geq N$ and $m \geq N$.

Proof This property of the theorem is so important that it deserves some terminology. A sequence is said to be a *Cauchy sequence* if it satisfies this property. Thus the theorem states that a sequence is convergent if and only if it is a Cauchy sequence. The terminology is most significant in more advanced situations where being a Cauchy sequence is not necessarily equivalent with being convergent.

Our proof is a bit lengthy and will require an application of the Bolzano-Weierstrass theorem.

The proof in one direction, however, is easy. Suppose that $\{s_n\}$ is convergent to a number L. Let $\varepsilon > 0$. Then there must be an integer N so that

$$|s_k - L| < \frac{\varepsilon}{2}$$

whenever $k \geq N$. Thus if both m and n are larger than N,

$$|s_n - s_m| \leq |s_n - L| + |L - s_m| < \frac{\varepsilon}{2} + \frac{\varepsilon}{2} = \varepsilon$$

which shows that $\{s_n\}$ is a Cauchy sequence.

Now let us prove the opposite (and more difficult) direction.

For the first step we show that every Cauchy sequence is bounded. Since the proof of this can be obtained by copying and modifying the proof of

Theorem 2.11, we have left this as an exercise. (It is not really interesting that Cauchy sequences are bounded since after the proof is completed we know that all Cauchy sequences are convergent and so must, indeed, be bounded.)

For the second step we apply the Bolzano-Weierstrass theorem to the (bounded) sequence $\{s_n\}$ to obtain a convergent subsequence $\{s_{n_k}\}$.

The final step is a feature of Cauchy sequences. Once we know that $s_{n_k} \to L$ and that $\{s_n\}$ is Cauchy, we can show that $s_n \to L$ also. Let $\varepsilon > 0$ and choose N so that

$$|s_n - s_m| < \varepsilon/2$$

for all m, $n \geq N$. Choose K so that

$$|s_{n_k} - L| < \varepsilon/2$$

for all $k \geq K$. Suppose that $n \geq N$. Set m equal to any value of n_k that is larger than N and so that $k \geq K$. For this value $s_m = s_{n_k}$

$$|s_n - L| \leq |s_n - s_{n_k}| + |s_{n_k} - L| < \varepsilon/2 + \varepsilon/2 = \varepsilon.$$

By definition, $\{s_n\}$ converges to L and so the proof is complete. ■

Example 2.42 The Cauchy criterion is most useful in theoretical developments rather than applied to concrete examples. Even so, occasionally it is the fastest route to a proof of convergence. For example, consider the sequence $\{x_n\}$ defined by setting $x_1 = 1$, $x_2 = 2$ and then, recursively,

$$x_n = \frac{x_{n-1} + x_{n-2}}{2}.$$

Each term after the second is the average of the preceding two terms. The distance between x_1 and x_2 is 1, that between x_2 and x_3 is $1/2$, between x_3 and x_4 is $1/4$, and so on. We see then that after the N stage all the distances are smaller than 2^{-N+1}, that is, that for all $n \geq N$ and $m \geq N$

$$|x_n - x_m| \leq \frac{1}{2^{N-1}}.$$

This is exactly the Cauchy criterion and so this sequence converges. Note that the Cauchy criterion offers no information on what the sequence is converging to. You must come up with another method to find out. ◀

Exercises

2.12.1 Show directly that the sequence $s_n = 1/n$ is a Cauchy sequence.

2.12.2 Show directly that any multiple of a Cauchy sequence is again a Cauchy sequence.

2.12.3 Show directly that the sum of two Cauchy sequences is again a Cauchy sequence.

2.12.4 Show directly that any Cauchy sequence is bounded.

2.12.5 The following criterion is weaker than the Cauchy criterion. Show that it is not equivalent:

For all $\varepsilon > 0$ there exists an integer N with the property that
$$|s_{n+1} - s_n| < \varepsilon$$
whenever $n \geq N$.

2.12.6 A careless student believes that the following statement is the Cauchy criterion.

For all $\varepsilon > 0$ and all positive integers p there exists an integer N with the property that
$$|s_{n+p} - s_n| < \varepsilon$$
whenever $n \geq N$.

Is this statement weaker, stronger, or equivalent to the Cauchy criterion?

2.12.7 Show directly that if $\{s_n\}$ is a Cauchy sequence then so too is $\{|s_n|\}$. From this conclude that $\{|s_n|\}$ converges whenever $\{s_n\}$ converges.

2.12.8 Show that every subsequence of a Cauchy sequence is Cauchy. (Do not use the fact that every Cauchy sequence is convergent.)

2.12.9 Show that every bounded monotonic sequence is Cauchy. (Do not use the monotone convergence theorem.)

2.12.10 Show that the sequence in Example 2.42 converges to 5/3.

2.13 Upper and Lower Limits

✂ *Advanced*

If $\lim_{n\to\infty} x_n = L$ then, according to our definition, numbers α and β on either side of L, that is, $\alpha < L < \beta$, have the property that
$$\alpha < x_n \text{ and } x_n < \beta$$
for all sufficiently large n. In many applications only *half* of this information is used.

Example 2.43 Here is an example showing how half a limit is as good as a whole limit. Let $\{x_n\}$ be a sequence of positive numbers with the property that
$$\lim_{n\to\infty} \sqrt[n]{x_n} = L < 1.$$
Then we can prove that $x_n \to 0$. To see this pick numbers α and β so that
$$\alpha < L < \beta < 1.$$

There must be an integer N so that

$$\alpha < \sqrt[n]{x_n} < \beta < 1$$

for all $n \geq N$. Forget half of this and focus on

$$\sqrt[n]{x_n} < \beta < 1.$$

Then we have

$$x_n < \beta^n$$

for all $n \geq N$ and it is clear now why $x_n \to 0$. ◀

This example suggests that the definition of limit might be weakened to handle situations where less is needed. This way we have a tool to discuss the limiting behavior of sequences that may not necessarily converge. Even if the sequence does converge this often offers a tool that can be used without first finding a proof of convergence.

We break the definition of sequence limit into two half-limits as follows.

Definition 2.44 (Lim Sup) A *limit superior* of a sequence $\{x_n\}$, denoted as

$$\limsup_{n\to\infty} x_n,$$

is defined to be the infimum of all numbers β with the following property:

There is an integer N so that $x_n < \beta$ for all $n \geq N$.

Definition 2.45 (Lim Inf) A *limit inferior* of a sequence $\{x_n\}$, denoted as

$$\liminf_{n\to\infty} x_n,$$

is defined to be the supremum of all numbers α with the following property:

There is an integer N so that $\alpha < x_n$ for all $n \geq N$.

Note. In interpreting this definition note that, by our usual rules on infs and sups, the values $-\infty$ and ∞ are allowed. If there are *no* numbers β with the property of the definition, then the sequence is simply unbounded above. The infimum of the empty set is taken as ∞ and so

$$\limsup_{n\to\infty} x_n = \infty \Leftrightarrow \text{the sequence } \{x_n\} \text{ has no upper bound.}$$

On the other hand, if *every* number β has the property of the definition this means exactly that our sequence must be diverging to $-\infty$. The infimum of the set of *all* real numbers is taken as $-\infty$ and so

$$\limsup_{n\to\infty} x_n = -\infty \Leftrightarrow \text{the sequence } \{x_n\} \to -\infty.$$

The same holds in the other direction. A sequence that is unbounded below can be described by saying $\liminf_{n\to\infty} x_n = -\infty$. A sequence that diverges to ∞ can be described by saying $\liminf_{n\to\infty} x_n = \infty$.

We refer to these concepts as "upper limits" and "lower limits" or "extreme limits." They extend our theory describing the limiting behavior of sequences to allow precise descriptions of divergent sequences. Obviously, we should establish very quickly that the upper limit is indeed greater than or equal to the lower limit since our language suggests this.

Theorem 2.46 *Let $\{x_n\}$ be a sequence of real numbers. Then*

$$\liminf_{n\to\infty} x_n \le \limsup_{n\to\infty} x_n.$$

Proof If $\limsup_{n\to\infty} x_n = \infty$ or if $\liminf_{n\to\infty} x_n = -\infty$ we have nothing to prove. If not then take any number β larger than $\limsup_{n\to\infty} x_n$ and any number α smaller than $\liminf_{n\to\infty} x_n$. By definition then there is an integer N so that $x_n < \beta$ for all $n \ge N$ and an integer M so that $\alpha < x_n$ for all $n \ge M$. It must be true that $\alpha < \beta$. But β is *any* number larger than $\limsup_{n\to\infty} x_n$. Hence

$$\alpha \le \limsup_{n\to\infty} x_n.$$

Similarly, α is *any* number smaller than $\liminf_{n\to\infty} x_n$. Hence

$$\liminf_{n\to\infty} x_n \le \limsup_{n\to\infty} x_n$$

as required. ■

How shall we use the limit superior of a sequence $\{x_n\}$? If

$$\limsup_{n\to\infty} x_n = L$$

then every number $\beta > L$ has the property that $x_n < \beta$ for all n large enough. This is because L is the infimum of such numbers β. On the other hand, any number $b < L$ cannot have this property so $x_n \ge b$ for infinitely many indices n. Thus numbers slightly larger than L must be upper bounds for the sequence eventually. Numbers slightly less than L are not upper bounds eventually. To express this a little more precisely, the number L is the limit superior of a sequence $\{x_n\}$ exactly when the following holds:

> For every $\varepsilon > 0$ there is an integer N so that $x_n < L + \varepsilon$ for all $n \ge N$ and $x_n > L - \varepsilon$ for infinitely many $n \ge N$.

The next theorem gives another characterization which is sometimes easier to apply. This version also better explains why we describe this notion as a "lim sup" and "lim inf."

Theorem 2.47 *Let $\{x_n\}$ be a sequence of real numbers. Then*

$$\limsup_{n\to\infty} x_n = \lim_{n\to\infty} \sup\{x_n, x_{n+1}, x_{n+2}, x_{n+3}, \dots\}$$

and

$$\liminf_{n\to\infty} x_n = \lim_{n\to\infty} \inf\{x_n, x_{n+1}, x_{n+2}, x_{n+3}, \dots\}.$$

Proof Let us prove just the statement for lim sups as the lim inf statement can be proved similarly.

Write

$$y_n = \sup\{x_n, x_{n+1}, x_{n+2}, x_{n+3}, \dots\}.$$

Then $x_n \le y_n$ for all n and so, using the inequality promised in Exercise 2.13.5,

$$\limsup_{n\to\infty} x_n \le \limsup_{n\to\infty} y_n.$$

But $\{y_n\}$ is a nonincreasing sequence and so

$$\limsup_{n\to\infty} y_n = \lim_{n\to\infty} y_n.$$

From this it follows that

$$\limsup_{n\to\infty} x_n \le \lim_{n\to\infty} \sup\{x_n, x_{n+1}, x_{n+2}, x_{n+3}, \dots\}.$$

Let us now show the reverse inequality. If $\limsup_{n\to\infty} x_n = \infty$ then the sequence is unbounded above. Thus for all n

$$\sup\{x_n, x_{n+1}, x_{n+2}, x_{n+3}, \dots\} = \infty$$

and so, in this case,

$$\limsup_{n\to\infty} x_n = \lim_{n\to\infty} \sup\{x_n, x_{n+1}, x_{n+2}, x_{n+3}, \dots\}$$

must certainly be true.

If

$$\limsup_{n\to\infty} x_n < \infty$$

then take any number β larger than $\limsup_{n\to\infty} x_n$. By definition then there is an integer N so that $x_n < \beta$ for all $n \ge N$. It follows that

$$\lim_{n\to\infty} \sup\{x_n, x_{n+1}, x_{n+2}, x_{n+3}, \dots\} \le \beta.$$

But β is *any* number larger than $\limsup_{n\to\infty} x_n$. Hence

$$\lim_{n\to\infty} \sup\{x_n, x_{n+1}, x_{n+2}, x_{n+3}, \dots\} \le \limsup_{n\to\infty} x_n.$$

We have proved both inequalities, the equality follows, and the theorem is proved.

■

The connection between limits and extreme limits is close. If a limit exists then the upper and lower limits must be the same.

Theorem 2.48 *Let $\{x_n\}$ be a sequence of real numbers. Then $\{x_n\}$ is convergent if and only if $\limsup_{n\to\infty} x_n = \liminf_{n\to\infty} x_n$ and these are finite. In this case*

$$\limsup_{n\to\infty} x_n = \liminf_{n\to\infty} x_n = \lim_{n\to\infty} x_n.$$

Proof Let $\varepsilon > 0$. If $\limsup_{n\to\infty} x_n = L$ then there is an integer N_1 so that $x_n < L + \varepsilon$ for all $n \geq N_1$. If it is also true that $\liminf_{n\to\infty} x_n = L$ then there is an integer N_2 so that $x_n > L - \varepsilon$ for all $n \geq N_2$. Putting these together we have

$$L - \varepsilon < x_n < L + \varepsilon$$

for all

$$n \geq N = \max\{N_1, N_2\}.$$

By definition then $\lim_{n\to\infty} x_n = L$.

Conversely, if $\lim_{n\to\infty} x_n = L$ then for some N,

$$L - \varepsilon < x_n < L + \varepsilon$$

for all $n \geq N$. Thus

$$L - \varepsilon \leq \liminf_{n\to\infty} x_n \leq \limsup_{n\to\infty} x_n \leq L + \varepsilon.$$

Since ε is an arbitrary positive number we must have

$$L = \liminf_{n\to\infty} x_n = \limsup_{n\to\infty} x_n$$

as required. ■

In the exercises you will be asked to compute several lim sups and lim infs. This is just for familiarity with the concepts. Computations are not so important. What is important is the use of these ideas in theoretical developments. More critical is how these limit operations relate to arithmetic or order properties. The limit of a sum is the sum of the two limits. Is this true for lim sups and lim infs? (See Exercise 2.13.9.) Do not skip these exercises.

Exercises

2.13.1 Complete Example 2.43 by showing that if $\{x_n\}$ is a sequence of positive numbers with the property that $\limsup_{n\to\infty} \sqrt[n]{x_n} < 1$ then $x_n \to 0$. Show that if

$$\liminf_{n\to\infty} \sqrt[n]{x_n} > 1$$

then $x_n \to \infty$. What can you conclude if $\limsup_{n\to\infty} \sqrt[n]{x_n} > 1$ or if $\liminf_{n\to\infty} \sqrt[n]{x_n} < 1$?

2.13.2 Compute lim sups and lim infs for the following sequences

(a) $\{(-1)^n n\}$
(b) $\{\sin(n\pi/8)\}$
(c) $\{n\sin(n\pi/8)\}$
(d) $\{[(n+1)\sin(n\pi/8)]/n\}$
(e) $\{1+(-1)^n\}$
(f) $\{r_n\}$ consists of all rational numbers in the interval $(0,1)$ arranged in some order.

2.13.3 Give examples of sequences of rational numbers $\{a_n\}$ with

(a) upper limit $\sqrt{2}$ and lower limit $-\sqrt{2}$,
(b) upper limit $+\infty$ and lower limit $\sqrt{2}$,
(c) upper limit π and lower limit e.

2.13.4 Show that $\limsup_{n\to\infty}(-x_n) = -(\liminf_{n\to\infty} x_n)$.

2.13.5 If two sequences $\{a_n\}$ and $\{b_n\}$ satisfy the inequality $a_n \leq b_n$ for all sufficiently large n, show that

$$\limsup_{n\to\infty} a_n \leq \limsup_{n\to\infty} b_n \quad \text{and} \quad \liminf_{n\to\infty} a_n \leq \liminf_{n\to\infty} b_n.$$

2.13.6 Show that $\lim_{n\to\infty} x_n = \infty$ if and only if

$$\limsup_{n\to\infty} x_n = \liminf_{n\to\infty} x_n = \infty.$$

2.13.7 Show that if $\limsup_{n\to\infty} a_n = L$ for a finite real number L and $\varepsilon > 0$, then $a_n > L+\varepsilon$ for only finitely many n and $a_n > L-\varepsilon$ for infinitely many n.

2.13.8 Show that for any monotonic sequence $\{x_n\}$

$$\limsup_{n\to\infty} x_n = \liminf_{n\to\infty} x_n = \lim_{n\to\infty} x_n$$

(including the possibility of infinite limits).

2.13.9 Show that for any bounded sequences $\{a_n\}$ and $\{b_n\}$

$$\limsup_{n\to\infty}(a_n+b_n) \leq \limsup_{n\to\infty} a_n + \limsup_{n\to\infty} b_n.$$

Give an example to show that the equality need not occur.

2.13.10 What is the correct version for the lim inf of Exercise 2.13.9?

2.13.11 Show that for any bounded sequences $\{a_n\}$ and $\{b_n\}$ of positive numbers

$$\limsup_{n\to\infty}(a_n b_n) \leq (\limsup_{n\to\infty} a_n)(\limsup_{n\to\infty} b_n).$$

Give an example to show that the equality need not occur.

2.13.12 Correct the careless student proof in Exercise 2.8.3 for the squeeze theorem by replacing lim with limsup and liminf in the argument.

2.13.13 What relation, if any, can you state for the lim sups and lim infs of a sequence $\{a_n\}$ and one of its subsequences $\{a_{n_k}\}$?

2.13.14 If a sequence $\{a_n\}$ has no convergent subsequences, what can you state about the lim sups and lim infs of the sequence?

2.13.15 Let S denote the set of all real numbers t with the property that some subsequence of a given sequence $\{a_n\}$ converges to t. What is the relation between the set S and the lim sups and lim infs of the sequence $\{a_n\}$?

2.13.16 Prove the following assertion about the upper and lower limits for any sequence $\{a_n\}$ of positive real numbers:

$$\liminf_{n\to\infty} \frac{a_{n+1}}{a_n} \le \liminf_{n\to\infty} \sqrt[n]{a_n} \le \limsup_{n\to\infty} \sqrt[n]{a_n} \le \limsup_{n\to\infty} \frac{a_{n+1}}{a_n}.$$

Give an example to show that each of these inequalities may be strict.

2.13.17 For any sequence $\{a_n\}$ write $s_n = (a_1 + a_2 + \dots a_n)/n$. Show that

$$\liminf_{n\to\infty} a_n \le \liminf_{n\to\infty} s_n \le \limsup_{n\to\infty} s_n \le \limsup_{n\to\infty} a_n.$$

Give an example to show that each of these inequalities may be strict.

2.14 Challenging Problems for Chapter 2

2.14.1 Let α and β be positive numbers. Show that

$$\lim_{n\to\infty} \sqrt[n]{\alpha^n + \beta^n} = \max\{\alpha, \beta\}.$$

2.14.2 For any convergent sequence $\{a_n\}$ write $s_n = (a_1 + a_2 + \dots a_n)/n$, the sequence of averages. Show that

$$\lim_{n\to\infty} a_n = \lim_{n\to\infty} s_n.$$

Give an example to show that $\{s_n\}$ could converge even if $\{a_n\}$ diverges.

2.14.3 Let $a_1 = 1$ and define a sequence recursively by

$$a_{n+1} = \sqrt{a_1 + a_2 + \dots + a_n}.$$

Show that $\lim_{n\to\infty} \frac{a_n}{n} = 1/2$.

2.14.4 Let $x_1 = \theta$ and define a sequence recursively by

$$x_{n+1} = \frac{x_n}{1 + x_n/2}.$$

For what values of θ is it true that $x_n \to 0$?

2.14.5 Let $\{a_n\}$ be a sequence of numbers in the interval $(0, 1)$ with the property that

$$a_n < \frac{a_{n-1} + a_{n+1}}{2}$$

for all $n = 2, 3, 4, \dots$. Show that this sequence is convergent.

2.14.6 For any convergent sequence $\{a_n\}$ write

$$s_n = \sqrt[n]{(a_1 a_2 \ldots a_n)},$$

the sequence of geometric averages. Show that $\lim_{n\to\infty} a_n = \lim_{n\to\infty} s_n$. Give an example to show that $\{s_n\}$ could converge even if $\{a_n\}$ diverges.

2.14.7 If

$$\lim_{n\to\infty} \frac{s_n - \alpha}{s_n + \alpha} = 0$$

what can you conclude about the sequence $\{s_n\}$?

2.14.8 A function f is defined by

$$f(x) = \lim_{n\to\infty} \left(\frac{1 - x^2}{1 + x^2}\right)^n$$

at every value x for which this limit exists. What is the domain of the function?

2.14.9 A function f is defined by

$$f(x) = \lim_{n\to\infty} \frac{1}{x^n + x^{-n}}$$

at every value x for which this limit exists. What is the domain of the function?

2.14.10 Suppose that $f : \mathbb{R} \to \mathbb{R}$ is a positive function with a derivative f' that is everywhere continuous and negative. Apply Newton's method to obtain a sequence

$$x_1 = \theta, \quad x_{n+1} = x_n - \frac{f(x_n)}{f'(x_n)}.$$

Show that $x_n \to \infty$ for any starting value θ.

2.14.11 Let $f(x) = x^3 - 3x + 3$. Apply Newton's method to obtain a sequence

$$x_1 = \theta, \quad x_{n+1} = x_n - \frac{f(x_n)}{f'(x_n)}.$$

Show that for any positive integer p there is a starting value θ such that the sequence $\{x_n\}$ is periodic with period p.

2.14.12 Determine all subsequential limit points of the sequence $x_n = \cos n$.

2.14.13 A sequence $\{s_n\}$ is said to be *contractive* if there is a positive number $0 < r < 1$ so that

$$|s_{n+1} - s_n| \le r|s_n - s_{n-1}|$$

for all $n = 2, 3, 4, \ldots$.

(a) Show that the sequence defined by $s_1 = 1$ and $s_n = (4 + s_{n-1})^{-1}$ for $n = 2, 3, \ldots$ is contractive.

(b) Show that every contractive sequence is Cauchy.

(c) Show that a sequence can satisfy the condition

$$|s_{n+1} - s_n| < |s_n - s_{n-1}|$$

for all $n = 2, 3, 4, \ldots$ and not be contractive, nor even convergent.

(d) Is every convergent sequence contractive?

2.14.14 The sequence defined recursively by

$$f_1 = 1, \; f_2 = 1 \quad f_{n+2} = f_n + f_{n+1}$$

is called the *Fibonacci sequence.* Let

$$r_n = f_{n+1}/f_n$$

be the sequence of ratios of successive terms of the Fibonacci sequence.

(a) Show that $r_1 < r_3 < r_5 \cdots < r_6 < r_4 < r_2$.

(b) Show that $r_{2n} - r_{2n-1} \to 0$.

(c) Deduce that the sequence $\{r_n\}$ converges. Can you find a way to determine that limit? (This is related to the roots of the equation $x^2 - x - 1 = 0$.)

2.14.15 A sequence of real numbers $\{x_n\}$ has the property that

$$(2 - x_n)x_{n+1} = 1.$$

Show that $\lim_{n\to\infty} x_n = 1$.

2.14.16 Let $\{a_n\}$ be an arbitrary sequence of positive real numbers. Show that

$$\limsup_{n\to\infty} \left(\frac{a_1 + a_{n+1}}{a_n}\right)^n \geq e.$$

2.14.17 Suppose that the sequence whose nth term is

$$s_n + 2s_{n+1}$$

is convergent. Show that $\{s_n\}$ is also convergent.

2.14.18 Show that the sequence

$$\sqrt{7}, \sqrt{7 - \sqrt{7}}, \sqrt{7 - \sqrt{7 + \sqrt{7}}}, \sqrt{7 - \sqrt{7 + \sqrt{7 - \sqrt{7}}}}, \ldots$$

converges and find its limit.

2.14.19 Let a_1 and a_2 be positive numbers and suppose that the sequence $\{a_n\}$ is defined recursively by

$$a_{n+2} = \sqrt{a_n} + \sqrt{a_{n+1}}.$$

Show that this sequence converges and find its limit.

Chapter 3

INFINITE SUMS

✂ This chapter on infinite sums and series may be skipped over in designing a course or covered later as the need arises. The basic material in Sections 3.4, 3.5, and parts of 3.6 will be needed, but not before the study of series of functions in Chapter 9. All of the enrichment or advanced sections may be omitted and are not needed in the sequel.

3.1 Introduction

The use of infinite sums goes back in time much further, apparently, than the study of sequences. The sum

$$1+\frac{1}{2}+\frac{1}{4}+\frac{1}{8}+\frac{1}{16}+\frac{1}{32}+\frac{1}{64}+\cdots=2$$

has been long known. It is quite easy to convince oneself that this must be valid by arithmetic or geometric "reasoning." After all, just start adding and keeping track of the sum as you progress:

$$1,\ 1\tfrac{1}{2},\ 1\tfrac{3}{4},\ 1\tfrac{7}{8},\ 1\tfrac{15}{16},\ \ldots.$$

Figure 3.1 makes this seem transparent.

But there is a serious problem of meaning here. A finite sum is well defined, an infinite sum is not. Neither humans nor computers can add an infinite column of numbers.

The meaning that is commonly assigned to the preceding sum appears in the following computations:

$$1+\frac{1}{2}+\frac{1}{4}+\frac{1}{8}+\cdots=\lim_{n\to\infty}\left\{1+\frac{1}{2}+\frac{1}{4}+\frac{1}{8}+\cdots+\frac{1}{2^n}\right\}$$

$$=\lim_{n\to\infty}\left\{2-\frac{1}{2^{n+1}}\right\}=2.$$

This reduces the computation of an infinite sum to that of a finite sum followed by a limit operation. Indeed this is exactly what we were doing

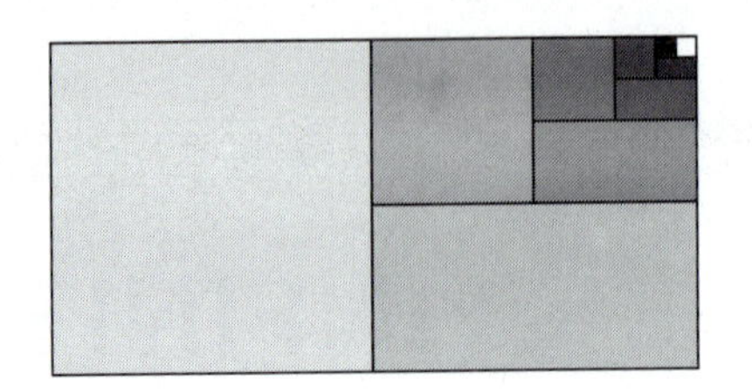

Figure 3.1. $1 + 1/2 + 1/4 + 1/8 + 1/16 + \cdots = 2.$

when we computed $1, 1\frac{1}{2}, 1\frac{3}{4}, 1\frac{7}{8}, 1\frac{15}{16}, \ldots$ and felt that this was a compelling reason for thinking of the sum as 2.

In terms of the development of the theory of this textbook this seems entirely natural and hardly surprising. We have mastered sequences in Chapter 2 and now pass to infinite sums in Chapter 3 using the methods of sequences. Historically this was not the case. Infinite summations appear to have been studied and used long before any development of sequences and sequence limits. Indeed, even to form the notion of an infinite sum as previously, it would seem that we should already have some concept of sequences, but this is not the way things developed.

It was only by the time of Cauchy that the modern theory of infinite summation was developed using sequence limits as a basis for the theory. We can transfer a great deal of our expertise in sequential limits to the problem of infinite sums. Even so, the study in this chapter has its own character and charm. In many ways infinite sums are much more interesting and important to analysis than sequences.

3.2 Finite Sums

We should begin our discussion of infinite sums with finite sums. There is not much to say about finite sums. Any finite collection of real numbers may be summed in any order and any grouping. That is not to say that we shall not encounter *practical* problems in this. For example, what is the sum of the first 10^{100} prime numbers? No computer or human could find this within the time remaining in this universe. But there is no *mathematical* problem in saying that it is defined; it is a sum of a finite number of real numbers.

There are a number of notations and a number of skills that we shall need to develop in order to succeed at the study of infinite sums that is to come. The notation of such summations may be novel. How best to write out a symbol indicating that some set of numbers

$$\{a_1, a_2, a_3, \ldots, a_n\}$$

has been summed? Certainly

$$a_1 + a_2 + a_3 + \cdots + a_n$$

is too cumbersome a way of writing such sums. The following have proved to communicate much better:

$$\sum_{i \in I} a_i$$

where I is the set $\{1, 2, 3, \ldots, n\}$ or

$$\sum_{1 \le i \le n} a_i \quad \text{or} \quad \sum_{i=1}^{n} a_i.$$

Here the Greek letter Σ, corresponding to an uppercase "S," is used to indicate a sum.

It is to Leonhard Euler (1707–1783) that we owe this sigma notation for sums (first used by him in 1755). The notations $f(x)$ for functions, e and π, i for $\sqrt{-1}$ are also his. These alone indicate the level of influence he has left. In his lifetime he wrote 886 papers and books and is considered the most prolific writer of mathematics that has lived.

The usual rules of elementary arithmetic apply to finite sums. The commutative, associative, and distributive rules assume a different look when written in Euler's notation:

$$\sum_{i \in I} a_i + \sum_{i \in I} b_i = \sum_{i \in I} (a_i + b_i),$$

$$\sum_{i \in I} c a_i = c \sum_{i \in I} a_i,$$

and

$$\left(\sum_{i \in I} a_i \right) \times \left(\sum_{i \in J} b_j \right) = \sum_{i \in I} \left(\sum_{j \in J} a_i b_j \right) = \sum_{j \in J} \left(\sum_{i \in I} a_i b_j \right).$$

Each of these can be checked mainly by determining the meanings and seeing that the notation produces the correct result.

Occasionally in applications of these ideas one would like a simplified expression for a summation. The best known example is perhaps

$$\sum_{k=1}^{n} k = 1 + 2 + 3 + \cdots + n = \frac{n(n+1)}{2},$$

which is easily proved. When a sum of n terms for a general n has a simpler expression such as this it is usual to say that it has been expressed in *closed form*. Novices, seeing this, usually assume that any summation with some degree of regularity should allow a closed form expression and that it is always important to get a closed form expression. If not, what can you do with a sum that cannot be simplified?

One of the simplest of sums

$$1 + \tfrac{1}{2} + \tfrac{1}{3} + \cdots + \tfrac{1}{n} = \textstyle\sum_{k=1}^{n} \tfrac{1}{k}$$

does not allow any convenient formula, expressing the sum as some simple function of n. This is typical. It is only the rarest of summations that will allow simple formulas. Our work is mostly in *estimating* such expressions; we hardly ever succeed in computing them exactly.

Even so, there are a few special cases that should be remembered and which make our task in some cases much easier.

Telescoping Sums. If a sum can be rewritten in the special form below, a simple computation (canceling s_1, s_2, etc.) gives the following closed form:

$$(s_1 - s_0) + (s_2 - s_1) + (s_3 - s_2) + (s_4 - s_3) + \cdots + (s_n - s_{n-1}) = s_n - s_0.$$

It is convenient to call such a sum "telescoping" as an indication of the method that can be used to compute it.

Example 3.1 For a specific example of a sum that can be handled by considering it as telescoping, consider the sum

$$\sum_{k=1}^{n} \frac{1}{k(k+1)} = \frac{1}{1\cdot 2} + \frac{1}{2\cdot 3} + \frac{1}{3\cdot 4} + \frac{1}{4\cdot 5} \cdots + \frac{1}{(n-1)\cdot n}.$$

A closed form is available since, using partial fractions, each term can be expressed as

$$\frac{1}{k(k+1)} = \frac{1}{k} - \frac{1}{k+1}.$$

Thus

$$\sum_{k=1}^{n} \frac{1}{k(k+1)} =$$

$$\sum_{k=1}^{n} \left(\frac{1}{k} - \frac{1}{k+1}\right) = 1 - \frac{1}{n+1}.$$

The exercises contain a number of other examples of this type. ◀

Geometric Progressions. If the terms of a sum are in a geometric progression (i.e., if each term is some constant factor times the previous term), then a closed form for any such sum is available:

$$1 + r + r^2 + \cdots + r^{n-1} + r^n = \frac{1 - r^{n+1}}{1 - r}. \tag{1}$$

This assumes that $r \neq 1$; if $r = 1$ the sum is easily seen to be just $n+1$. The formula in (1) can be proved by converting to a telescoping sum. Consider instead $(1 - r)$ times the preceding sum:

$$(1-r)(1 + r + r^2 + \cdots + r^{n-1} + r^n) = (1 - r) + (r - r^2) + \ldots (r^n - r^{n+1}).$$

Now add this up as a telescoping sum to obtain the formula stated in (1).

Any geometric progression assumes the form

$$A + Ar + Ar^2 + \cdots + Ar^n = A(1 + r + r^2 + \cdots + r^n)$$

and formula (1) (which should be memorized) is then applied.

Summation By Parts. Sums are frequently given in a form such as

$$\sum_{k=1}^{n} a_k b_k$$

for sequences $\{a_k\}$ and $\{b_k\}$. If a formula happens to be available for

$$s_n = a_1 + a_2 + \ldots a_n,$$

then there is a frequently useful way of rewriting this sum (using $s_0 = 0$ for convenience):

$$\sum_{k=1}^{n} a_k b_k = \sum_{k=1}^{n} (s_k - s_{k-1}) b_k$$

$$= s_1(b_1 - b_2) + s_2(b_2 - b_3) \cdots + s_{n-1}(b_{n-1} - b_n) + s_n b_n.$$

Usually some extra knowledge about the sequences $\{s_k\}$ and $\{b_k\}$ can then be used to advantage. The computation is trivial (it is all contained in the preceding equation which is easily checked). Sometimes this summation formula is referred to as Abel's transformation after the Norwegian mathematician Niels Henrik Abel (1802–1829), who was one of the founders of the rigorous theory of infinite sums. It is the analog for finite sums of the integration by parts formula of calculus.

Abel's most important contributions are to analysis but he is forever immortalized in group theory (to which he made a small contribution) by the fact that commutative groups are called "Abelian."

Exercises

3.2.1 Prove the formula

$$\sum_{k=1}^{n} k = \frac{n(n+1)}{2}.$$

3.2.2 Give a formal definition of $\sum_{i \in I} a_i$ for any finite set I and any function $a : I \to \mathbb{R}$ that uses induction on the number of elements of I.

Your definition should be able to handle the case $I = \emptyset$.

3.2.3 Check the validity of the formulas given in this section for manipulating finite sums. Are there any other formulas you can propose and verify?

3.2.4 Is the formula

$$\sum_{i \in I \cup J} a_i = \sum_{i \in I} a_i + \sum_{i \in J} a_i$$

valid?

3.2.5 Let $I = \{(i,j) : 1 \le i \le m,\ 1 \le j \le n\}$. Show that

$$\sum_{(i,j) \in I} a_{ij} = \sum_{i=1}^{m} \sum_{j=1}^{n} a_{ij}.$$

3.2.6 Give a formula for the sum of n terms of an arithmetic progression. (An arithmetic progression is a list of numbers, each of which is obtained by adding a fixed constant to the previous one in the list.) For the purposes of infinite sums (our concern in this chapter) such a formula will be of little use. Explain why.

3.2.7 Obtain formulas (or find a source for such formulas) for the sums

$$\sum_{k=1}^{n} k^p = 1^p + 2^p + 3^p + \cdots + n^p$$

of the pth powers of the natural numbers where $p = 1, 2, 3, 4, \ldots$. Again, for the purposes of infinite sums such formulas will be of little use.

3.2.8 Explain the (vague) connection between integration by parts and summation by parts.

3.2.9 Obtain a formula for $\sum_{k=1}^{n} (-1)^k$.

3.2.10 Obtain a formula for

$$2 + 2\sqrt{2} + 4 + 4\sqrt{2} + 8 + 8\sqrt{2} + \cdots + 2^m.$$

3.2.11 Obtain the formula

$$\sin\theta + \sin 2\theta + \sin 3\theta + \sin 4\theta + \cdots + \sin n\theta = \frac{\cos\theta/2 - \cos(2n+1)\theta/2}{2\sin\theta/2}.$$

How should the formula be interpreted if the denominator of the fraction is zero?

3.2.12 Obtain the formula

$$\cos\theta + \cos 3\theta + \cos 5\theta + \cos 7\theta + \cdots + \cos(2n-1)\theta = \frac{\sin 2n\theta}{2\sin\theta}.$$

3.2.13 If

$$s_n = 1 - \frac{1}{2} + \frac{1}{3} - \frac{1}{4} \cdots + \ldots (-1)^{n+1}\frac{1}{n}$$

show that $1/2 \leq s_n \leq 1$ for all n.

3.2.14 If

$$s_n = 1 + \frac{1}{2} + \frac{1}{3} + \frac{1}{4} \cdots + \cdots + \frac{1}{n}$$

show that $s_{2^n} \geq 1 + n/2$ for all n.

3.2.15 Obtain a closed form for

$$\sum_{k=1}^{n} \frac{1}{k(k+2)(k+4)}.$$

3.2.16 Obtain a closed form for

$$\sum_{k=1}^{n} \frac{\alpha r + \beta}{k(k+1)(k+2)}.$$

3.2.17 Let $\{a_k\}$ and $\{b_k\}$ be sequences with $\{b_k\}$ decreasing and

$$|a_1 + a_2 + \cdots + a_k| \leq K$$

for all k. Show that

$$\left|\sum_{k=1}^{n} a_k b_k\right| \leq K b_1$$

for all n.

3.2.18 If r is the interest rate (e.g., $r = .06$) over a period of years, then

$$P(1+r)^{-1} + P(1+r)^{-2} + \ldots P(1+r)^{-n}$$

is the present value of an annuity of P dollars paid every year, starting next year and for n years. Give a shorter formula for this. (A perpetuity has nearly the same formula but the payments continue forever. See Exercise 3.4.12.)

3.2.19 Define a finite product (product of a finite set of real numbers) by writing

$$\prod_{k=1}^{n} a_k = a_1 a_2 a_3 \ldots a_n.$$

What elementary properties can you determine for products?

3.2.20 Find a closed form expression for

$$\prod_{k=1}^{n} \frac{k^3 - 1}{k^3 + 1}.$$

3.3 Infinite Unordered sums

✂ *Advanced*

We now pass to the study of infinite sums. We wish to interpret

$$\sum_{i \in I} a_i$$

for an index set I that is infinite. The study of finite sums involves no analysis, no limits, no ε's, in short none of the processes that are special to analysis. To define and study infinite sums requires many of our skills in analysis.

To begin our study imagine that we are given a collection of numbers a_i indexed over an infinite set I (i.e., there is a function $a : I \to \mathbb{R}$) and we wish the sum of the totality of these numbers. If the set I has some structure, then we can use that structure to decide how to start adding the numbers. For example, if a is a sequence so that $I = \mathbb{N}$, then we should likely start adding at the beginning of the sequence:

$$a_1,\ a_1 + a_2,\ a_1 + a_2 + a_3,\ a_1 + a_2 + a_3 + a_4, \ldots$$

and so defining the sum as the limit of this sequence of partial sums.

Another set I would suggest a different order. For example, if $I = \mathbb{Z}$ (the set of all integers), then a popular method of adding these up would be to start off:

$$a_0,\ a_{-1} + a_0 + a_1,$$

$$a_{-2} + a_{-1} + a_0 + a_1 + a_2,$$

$$a_{-3} + a_{-2} + a_{-1} + a_0 + a_1 + a_2 + a_3,\ \ldots$$

once again defining the sum as the limit of this sequence.

It seems that the method of summation and hence defining the meaning of the expression

$$\sum_{i \in I} a_i$$

for infinite sets I must depend on the nature of the set I and hence on the particular problems of the subject one is studying. This is true to some extent. But it does not stop us from inventing a method that will apply to *all* infinite sets I. We must make a definition that takes account of no extra structure or ordering for the set I and just treats it as a set. This is called the unordered sum and the notation $\sum_{i \in I} a_i$ is always meant to indicate that an unordered sum is being considered. The key is just how to pass from finite sums to infinite sums. Both of the previous examples used the idea of taking some finite sums (in a systematic way) and then passing to a limit.

Definition 3.2 Let I be an infinite set and a a function $a : I \to \mathbb{R}$. Then we write

$$\sum_{i \in I} a_i = c$$

and say that the sum *converges* if for every $\varepsilon > 0$ there is a finite set $I_0 \subset I$ so that, for every finite set J, $I_0 \subset J \subset I$,

$$\left| \sum_{i \in J} a_i - c \right| < \varepsilon.$$

A sum that does not converge is said to *diverge*.

Note that we never form a sum of infinitely many terms. The definition always computes finite sums.

Example 3.3 Let us show, directly from the definition, that

$$\sum_{i \in \mathbb{Z}} 2^{-|i|} = 3.$$

If we first sum

$$\sum_{-N \le i \le N} 2^{-|i|}$$

by rearranging the terms into the sum

$$1 + 2(2^{-1} + 2^{-2} + \dots 2^{-N})$$

we can see why the sum is likely to be 3. Let $\varepsilon > 0$ and choose N so that $2^{-N} < \varepsilon/4$. From the formula for a finite geometric progression we have

$$\left| \sum_{-N \le i \le N} 2^{-|i|} - 3 \right| = 2|(2^{-1} + 2^{-2} + \dots 2^{-N}) - 1| < 2(2^{-N}) < \varepsilon/2.$$

Also, if $J \subset \mathbb{Z}$ with J finite and $j > N$ for all $j \in J$, then

$$\sum_{j \in J} 2^{-|j|} < 2(2^{-N}) < \varepsilon/2$$

again from the formula for a finite geometric progression. Let

$$I_0 = \{i \in \mathbb{Z} : -N \le i \le N\}.$$

If $I_0 \subset J \subset \mathbb{Z}$ with J finite then

$$\left| \sum_{i \in J} 2^{-|i|} - 3 \right| = \left| \sum_{-N \le i \le N} 2^{-|i|} - 3 \right| + \sum_{i \in J \setminus I_0} 2^{-|i|} < \varepsilon$$

as required.

◄

3.3.1 Cauchy Criterion

Advanced

In most theories of convergence one asks for a necessary and sufficient condition for convergence. We saw in studying sequences that the Cauchy criterion provided such a condition for the convergence of a sequence. There is usually in any theory of this kind a type of Cauchy criterion. Here is the Cauchy criterion for sums.

Theorem 3.4 *A necessary and sufficient condition that the sum $\sum_{i\in I} a_i$ converges is that for every $\varepsilon > 0$ there is a finite set I_0 so that*

$$\left|\sum_{i\in J} a_i\right| < \varepsilon$$

for every finite set $J \subset I$ that contains no elements of I_0 (i.e., for all finite sets $J \subset I \setminus I_0$).

Proof As usual in Cauchy criterion proofs, one direction is easy to prove. Suppose that $\sum_{i\in I} a_i = C$ converges. Then for every $\varepsilon > 0$ there is a finite set I_0 so that

$$\left|\sum_{i\in K} a_i - C\right| < \varepsilon/2$$

for every finite set $I_0 \subset K \subset I$. Let $J \subset I \setminus I_0$ and consider taking a sum over $K = I_0 \cup J$. Then

$$\left|\sum_{i\in I_0\cup J} a_i - C\right| < \varepsilon/2$$

and

$$\left|\sum_{i\in I_0} a_i - C\right| < \varepsilon/2.$$

By subtracting these two inequalities and remembering that

$$\sum_{i\in I_0\cup J} a_i = \sum_{i\in J} a_i + \sum_{i\in I_0} a_i$$

(since I_0 and J are disjoint) we obtain

$$\left|\sum_{i\in J} a_i\right| < \varepsilon.$$

This is exactly the Cauchy criterion.

Conversely, suppose that the sum does satisfy the Cauchy criterion. Then, applying that criterion to $\varepsilon = 1,\ 1/2,\ 1/3,\ \dots$ we can choose a sequence of finite sets $\{I_n\}$ so that

$$\left|\sum_{i\in J} a_i\right| < 1/n$$

for every finite set $J \subset I \setminus I_n$. We can arrange our choices to make

$$I_1 \subset I_2 \subset I_3 \subset \dots$$

so that the sequence of sets is increasing.

Let

$$c_n = \sum_{i\in I_n} a_i$$

Then for any $m > n$,

$$|c_n - c_m| = \left|\sum_{i\in I_m\setminus I_n} a_i\right| < 1/n.$$

It follows from this that $\{c_n\}$ is a Cauchy sequence of real numbers and hence converges to some real number c. Let $\varepsilon > 0$ and choose N so that $N > 2/\varepsilon$. Then, for any $n > N$ and any finite set J with $I_N \subset J \subset I$,

$$\left|\sum_{i\in J} a_i - c\right| \leq \left|\sum_{i\in I_N} a_i - c_N\right| + |c_N - c| + \left|\sum_{i\in J\setminus I_N} a_i\right| < 0 + 2/N < \varepsilon.$$

By definition, then,

$$\sum_{i\in I_n} a_i = c$$

and the theorem is proved. ■

All But Countably Many Terms in a Convergent Sum Are Nonzero. Our next theorem shows that having "too many" numbers to add up causes problems. If the set I is not countable then most of the a_i that we are to add up should be zero if the sum is to exist. This shows too that the theory of sums is in an essential way limited to taking sums over countable sets. It is notationally possible to have a sum

$$\sum_{x\in[0,1]} f(x)$$

but that sum cannot be defined unless $f(x)$ is mostly zero with only countably many exceptions.

Theorem 3.5 *Suppose that $\sum_{i \in I} a_i$ converges. Then $a_i = 0$ for all $i \in I$ except for a countable subset of I.*

Proof We shall use Exercise 3.3.2, where it is proved that for any convergent sum there is a positive integer M so that all the sums

$$\left|\sum_{i \in I_0} a_i\right| \leq M$$

for any finite set $I_0 \subset I$. Let m be an integer. We ask how many elements a_i are there such that $a_i > 1/m$? It is easy to see that there are at most Mm of them since if there were any more our sum would exceed M. Similarly, there are at most Mm terms such that $-a_i > 1/m$. Thus each element of $\{a_i : i \in I\}$ that is not zero can be given a "rank" m depending on whether

$$1/m < a_i \leq 1/(m-1) \text{ or } 1/m < -a_i \leq 1/(m-1).$$

As there are only finitely many elements at each rank, this gives us a method for listing all of the nonzero elements in $\{a_i : i \in I\}$ and so this set is countable. ■

The elementary properties of unordered sums are developed in the exercises. These sums play a small role in analysis, a much smaller role than the ordered sums we shall consider in the next sections. The methods of proof, however, are well worth studying since they are used in some form or other in many parts of analysis. These exercises offer an interesting setting in which to test your skills in analysis, skills that will play a role in all of your subsequent study.

Exercises

3.3.1 Show that if $\sum_{i \in I} a_i$ converges, then the sum is unique.

3.3.2 Show that if $\sum_{i \in I} a_i$ converges, then there is a positive number M so that all the sums

$$\left|\sum_{i \in I_0} a_i\right| \leq M$$

for any finite set $I_0 \subset I$.

3.3.3 Suppose that all the terms in the sum $\sum_{i \in I} a_i$ are nonnegative and that there is a positive number M so that all the sums

$$\sum_{i \in I_0} a_i \leq M$$

for any finite set $I_0 \subset I$. Show that $\sum_{i \in I} a_i$ must converge.

3.3.4 Show that if $\sum_{i \in I} a_i$ converges so too does $\sum_{i \in J} a_i$ for every subset $J \subset I$.

3.3.5 Show that if $\sum_{i\in I} a_i$ converges and each $a_i \geq 0$, then

$$\sum_{i\in I} a_i = \sup\left\{\sum_{i\in J} a_i \,:\, J \subset I,\ J \text{ finite}\right\}.$$

3.3.6 Each of the rules for manipulation of the finite sums of Section 3.2 can be considered for infinite unordered sums. Formulate the correct statement and prove what you think to be the analog of these statements that we know hold for finite sums:

$$\sum_{i\in I} a_i + \sum_{i\in I} b_i = \sum_{i\in I} (a_i + b_i)$$

$$\sum_{i\in I} ca_i = c\sum_{i\in I} a_i$$

$$\sum_{i\in I} a_i \times \sum_{i\in J} b_j = \sum_{i\in I}\sum_{j\in J} a_i b_j = \sum_{j\in J}\sum_{i\in I} a_i b_j.$$

3.3.7 Prove that

$$\sum_{i\in I\cup J} a_i + \sum_{i\in I\cap J} a_i = \sum_{i\in I} a_i + \sum_{i\in J} a_i$$

under appropriate convergence assumptions.

3.3.8 Let $\sigma : I \to J$ one-to-one and onto. Establish that

$$\sum_{j\in J} a_j = \sum_{i\in I} a_{\sigma(i)}$$

under appropriate convergence assumptions.

3.3.9 Find the sum

$$\sum_{i\in \mathbb{N}} \frac{1}{2^i}.$$

3.3.10 Show that

$$\sum_{i\in \mathbb{N}} \frac{1}{i}$$

diverges. Are there any infinite subsets $J \subset \mathbb{N}$ such that

$$\sum_{i\in J} \frac{1}{i}$$

converges?

3.3.11 Show that $\sum_{i\in I} a_i$ converges if and only if both $\sum_{i\in I}[a_i]^+$ and $\sum_{i\in I}[a_i]^-$ converge and that

$$\sum_{i\in I} a_i = \sum_{i\in I}[a_i]^+ - \sum_{i\in I}[a_i]^-$$

and

$$\sum_{i\in I} |a_i| = \sum_{i\in I}[a_i]^+ + \sum_{i\in I}[a_i]^-.$$

3.3.12 Compute

$$\sum_{(i,j)\in\mathbb{N}\times\mathbb{N}} 2^{-i-j}.$$

What kind of *ordered* sum would seem natural here (in the way that ordered sums over $\mathbb{N}$ and $\mathbb{Z}$ were considered in this section)?

3.4 Ordered Sums: Series

For the vast majority of applications, one wishes to sum not an arbitrary collection of numbers but most commonly some sequence of numbers:

$$a_1 + a_2 + a_3 + \ldots.$$

The set $\mathbb{N}$ of natural numbers has an order structure, and it is not in our best interests to ignore that order since that is the order in which the sequence is presented to us.

The most compelling way to add up a sequence of numbers is to begin accumulating:

$$a_1,\ a_1 + a_2,\ a_1 + a_2 + a_3,\ a_1 + a_2 + a_3 + a_4, \ldots$$

and to define the sum as the limit of this sequence. This is what we shall do.

If you studied Section 3.3 on unordered summation you should also compare this "ordered" method with the unordered method. The ordered sum of a sequence is called a *series* and the notation

$$\sum_{k=1}^{\infty} a_k$$

is used exclusively for this notion.

Definition 3.6 Let $\{a_k\}$ be a sequence of real numbers. Then we write

$$\sum_{k=1}^{\infty} a_k = c$$

and say that the series *converges* if the sequence

$$s_n = \sum_{k=1}^{n} a_k$$

(called the *sequence of partial sums of the series*) converges to c. If the series does not converge it is said to be *divergent.*

This definition reduces the study of series to the study of sequences. We already have a highly developed theory of convergent sequences in Chapter 2 that we can apply to develop a theory of series. Thus we can rapidly produce a fairly deep theory of series from what we already know. As the theory develops, however, we shall see that it begins to take a character of its own and stops looking like a mere application of sequence ideas.

3.4.1 Properties

The following short harvest of theorems we obtain directly from our sequence theory. The convergence or divergence of a series $\sum_{k=1}^{\infty} a_k$ depends on the convergence or divergence of the sequence of partial sums

$$s_n = \sum_{k=1}^{n} a_k$$

and the value of the series is the limit of the sequence. To prove each of the theorems we now list requires only to find the correct theorem on sequences from Chapter 2. This is left as Exercise 3.4.2.

Theorem 3.7 *If a series $\sum_{k=1}^{\infty} a_k$ converges, then the sum is unique.*

Theorem 3.8 *If both series $\sum_{k=1}^{\infty} a_k$ and $\sum_{k=1}^{\infty} b_k$ converge, then so too does the series*

$$\sum_{k=1}^{\infty}(a_k + b_k)$$

and

$$\sum_{k=1}^{\infty}(a_k + b_k) = \sum_{k=1}^{\infty} a_k + \sum_{k=1}^{\infty} b_k.$$

Theorem 3.9 *If the series $\sum_{k=1}^{\infty} a_k$ converges, then so too does the series $\sum_{k=1}^{\infty} ca_k$ for any real number c and*

$$\sum_{k=1}^{\infty} ca_k = c\sum_{k=1}^{\infty} a_k.$$

Theorem 3.10 *If both series $\sum_{k=1}^{\infty} a_k$ and $\sum_{k=1}^{\infty} b_k$ converge and $a_k \leq b_k$ for each k, then*

$$\sum_{k=1}^{\infty} a_k \leq \sum_{k=1}^{\infty} b_k.$$

Theorem 3.11 *Let $M \geq 1$ be any integer. Then the series*

$$\sum_{k=1}^{\infty} a_k = a_1 + a_2 + a_3 + a_4 + \dots$$

converges if and only if the series

$$\sum_{k=1}^{\infty} a_{M+k} = a_{M+1} + a_{M+2} + a_{M+3} + a_{M+4} + \dots$$

converges.

Note. If we call $\sum_p^\infty a_i$ a "tail" for the series $\sum_1^\infty a_i$, then we can say that this last theorem asserts that it is the behavior of the tail that determines the convergence or divergence of the series. Thus in questions of convergence we can easily ignore the first part of the series—however many terms we like. Naturally, the actual sum of the series will depend on having all the terms.

3.4.2 Special Series

Telescoping Series Any series for which we can find a closed form for the partial sums we should probably be able to handle by sequence methods. Telescoping series are the easiest to deal with.

If the sequence of partial sums of a series can be computed in some closed form $\{s_n\}$, then the series can be rewritten in the telescoping form

$$(s_1) + (s_2 - s_1) + (s_3 - s_2) + (s_4 - s_3) + \dots + (s_n - s_{n-1}) \dots$$

and the series studied by means of the sequence $\{s_n\}$.

Example 3.12 Consider the series

$$\sum_{k=1}^{\infty} \frac{1}{k(k+1)} = \sum_{k=1}^{\infty} \left(\frac{1}{k} - \frac{1}{k+1}\right) = \lim_{n\to\infty} \left(1 - \frac{1}{n+1}\right) = 1$$

with an easily computable sequence of partial sums. ◀

Do not be too encouraged by the apparent ease of the method illustrated by the example. In practice we can hardly ever do anything but make a crude estimate on the size of the partial sums. An exact expression, as we have here, would be rarely available. Even so, it is entertaining and instructive to handle a number of series by such a method (as we do in the exercises).

Geometric Series Geometric series form another convenient class of series that we can handle simply by sequence methods. From the elementary formula

$$1 + r + r^2 + \dots + r^{n-1} + r^n = \frac{1 - r^{n+1}}{1 - r} \quad (r \neq 1)$$

we see immediately that the study of such a series reduces to the computation of the limit

$$\lim_{n\to\infty} \frac{1 - r^{n+1}}{1 - r} = \frac{1}{1 - r}$$

which is valid for $-1 < r < 1$ (which is usually expressed as $|r| < 1$) and invalid for all other values of r. Thus, for $|r| < 1$ the series

$$\sum_{k=1}^{\infty} r^{k-1} = 1 + r + r^2 + \cdots = \frac{1}{1 - r} \tag{2}$$

and is convergent and for $|r| \geq 1$ the series diverges. It is well worthwhile to memorize this fact and formula (2) for the sum of the series.

Harmonic Series As a first taste of an elementary looking series that presents a new challenge to our methods, consider the series

$$\sum_{k=1}^{\infty} \frac{1}{k} = 1 + \frac{1}{2} + \frac{1}{3} + \ldots,$$

which is called the *harmonic series*. Let us show that this series diverges.

This series has no closed form for the sequence of partial sums $\{s_n\}$ and so there seems no hope of merely computing $\lim_{n\to\infty} s_n$ to obtain convergence/divergence of the harmonic series. But we can make estimates on the size of s_n even if we cannot compute it directly. The sequence of partial sums increases at each step, and if we watch only at the steps 1, 2, 4, 8, ... and make a rough lower estimate of s_1, s_2, s_4, s_8, ... we see that $s_{2^n} \geq 1 + n/2$ for all n (see Exercise 3.2.14). From this we see that $\lim_{n\to\infty} s_n = \infty$ and so the series diverges.

Alternating Harmonic Series A variant on the harmonic series presents immediately a new challenge. Consider the series

$$\sum_{k=1}^{\infty} (-1)^{k-1} \frac{1}{k} = 1 - \frac{1}{2} + \frac{1}{3} - \frac{1}{4} \ldots,$$

which is called the *alternating harmonic series.*

The reason why this presents a different challenge is that the sequence of partial sums is no longer increasing. Thus estimates as to how big that sequence get may be of no help. We can see that the sequence is bounded, but that does not imply convergence for a non monotonic sequence. Once again, we have no closed form for the partial sums so that a routine computation of a sequence limit is not available.

By computing the partial sums s_2, s_4, s_6, ... we see that the subsequence $\{s_{2n}\}$ is increasing. By computing the partial sums s_1, s_3, s_5, ... we see that the subsequence $\{s_{2n-1}\}$ is decreasing. A few more observations show us that

$$1/2 = s_2 < s_4 < s_6 < \cdots < s_5 < s_3 < s_1 = 1. \tag{3}$$

Our theory of sequences now allows us to assert that both limits

$$\lim_{n\to\infty} s_{2n} \text{ and } \lim_{n\to\infty} s_{2n-1}$$

exist. Finally, since

$$s_{2n} - s_{2n-1} = \frac{-1}{2n} \to 0$$

we can conclude that $\lim_{n\to\infty} s_n$ exists. [It is somewhere between $\frac{1}{2}$ and 1 because of the inequalities (3) but exactly what it is would take much further analysis.] Thus we have proved that the alternating harmonic series converges (which is in contrast to the divergence of the harmonic series).

p-Harmonic Series The series

$$\sum_{k=1}^{\infty} \frac{1}{k^p} = 1 + \frac{1}{2^p} + \frac{1}{3^p} + \ldots$$

for any parameter $0 < p < \infty$ is called the *p-harmonic series.* The methods we have used in the study of the harmonic series can be easily adapted to handle this series. As a first observation note that if $0 < p < 1$, then

$$\frac{1}{k^p} > \frac{1}{k}.$$

Thus the p-harmonic series for $0 < p < 1$ is larger than the harmonic series itself. Since the latter series has unbounded partial sums it is easy to argue that our series does too and, hence, diverges for all $0 < p \leq 1$.

What about $p > 1$? Now the terms are smaller than the harmonic series, small enough it turns out that the series converges. To show this we can group the terms in the same manner as before for the harmonic series and obtain

$$1 + \left[\frac{1}{2^p} + \frac{1}{3^p}\right] + \left[\frac{1}{4^p} + \frac{1}{5^p} + \frac{1}{6^p} + \frac{1}{7^p}\right] + \left[\frac{1}{8^p} + \cdots + \frac{1}{15^p}\right] + \ldots$$

$$\leq 1 + \frac{2}{2^p} + \frac{4}{4^p} + \frac{8}{8^p} \leq \frac{1}{1 - 2^{1-p}}$$

since we recognize the latter series as a convergent geometric series with ratio 2^{1-p}. In this way we obtain an upper bound for the partial sums of the series

$$\sum_{k=1}^{\infty} \frac{1}{k^p}$$

for all $p > 1$. Since the partial sums are increasing and bounded above, the series must converge.

Size of the Terms It should seem apparent from the examples we have seen that a convergent series must have ultimately small terms. If $\sum_{k=1}^{\infty} a_k$ converges, then it seems that a_k must tend to 0 as k gets large. Certainly for the geometric series that idea precisely described the situation:

$$\sum_{k=1}^{\infty} r^{k-1}$$

converges if $|r| < 1$, which is exactly when the terms tend to zero and diverges when $|r| \geq 1$, which is exactly when the terms do not tend to zero.

A reasonable conjecture might be that this is always the situation: A series $\sum_{k=1}^{\infty} a_k$ converges if and only if $a_k \to 0$ as $k \to \infty$. But we have already seen the harmonic series diverges even though its terms do get small; they simply don't get small fast enough. Thus the correct observation is simple and limited.

If $\sum_{k=1}^{\infty} a_k$ converges, then $a_k \to 0$ as $k \to \infty$.

To check this is easy. If $\{s_n\}$ is the sequence of partial sums of a convergent series $\sum_{k=1}^{\infty} a_k = C$, then

$$\lim_{n\to\infty} a_n = \lim_{n\to\infty} (s_n - s_{n-1}) = \lim_{n\to\infty} s_n - \lim_{n\to\infty} s_{n-1} = C - C = 0.$$

The converse, as we just noted, is false. To obtain convergence of a series it is not enough to know that the terms tend to zero. We shall see, though, that many of the tests that follow discuss the *rate* at which the terms tend to zero.

Exercises

3.4.1 Let $\{s_n\}$ be any sequence of real numbers. Show that this sequence converges to a number S if and only if the series

$$s_1 + \sum_{k=2}^{\infty} (s_n - s_{n-1})$$

converges and has sum S.

3.4.2 State which theorems from Chapter 2 would be used to prove Theorems 3.7–3.11.

3.4.3 If $\sum_{k=1}^{\infty}(a_k + b_k)$ converges, what can you say about the series

$$\sum_{k=1}^{\infty} a_k \text{ and } \sum_{k=1}^{\infty} b_k?$$

3.4.4 If $\sum_{k=1}^{\infty}(a_k + b_k)$ diverges, what can you say about the series

$$\sum_{k=1}^{\infty} a_k \text{ and } \sum_{k=1}^{\infty} b_k?$$

3.4.5 If the series $\sum_{k=1}^{\infty}(a_{2k}+a_{2k-1})$ converges, what can you say about the series $\sum_{k=1}^{\infty} a_k$?

3.4.6 If the series $\sum_{k=1}^{\infty} a_k$ converges, what can you say about the series

$$\sum_{k=1}^{\infty}(a_{2k} + a_{2k-1})?$$

3.4.7 If both series $\sum_{k=1}^{\infty} a_k$ and $\sum_{k=1}^{\infty} b_k$ converge, what can you say about the series $\sum_{k=1}^{\infty} a_k b_k$?

3.4.8 How should we interpret

$$\sum_{k=0}^{\infty} a_{k+1}, \sum_{k=-5}^{\infty} a_{k+6} \text{ and } \sum_{k=5}^{\infty} a_{k-4}?$$

3.4.9 If s_n is a strictly increasing sequence of positive numbers, show that it is the sequence of partial sums of some series with positive terms.

3.4.10 If $\{a_{n_k}\}$ is a subsequence of $\{a_n\}$, is there anything you can say about the relation between the convergence behavior of the series $\sum_{k=1}^{\infty} a_k$ and its "subseries" $\sum_{k=1}^{\infty} a_{n_k}$?

3.4.11 Express the infinite repeating decimal

$$.123451234512345123451234512345\ldots$$

as the sum of a convergent geometric series and compute its sum (as a rational number) in this way.

3.4.12 Using your result from Exercise 3.2.18, obtain a formula for a *perpetuity* of P dollars a year paid every year, starting next year and for every after. You most likely used a geometric series; can you find an argument that avoids this?

3.4.13 Suppose that a bird flying 100 miles per hour (mph) travels back and forth between a train and the railway station, where the train and the bird start off together 1 mile away and the train is approaching the station at a fixed rate of 60 mph. How far has the bird traveled when the train arrives? You most likely did not use a geometric series; can you find an argument that does?

3.4.14 What proportion of the area of the square in Figure 3.2 is black?

3.4.15 Does the series

$$\sum_{k=1}^{\infty} \log\left(\frac{k+1}{k}\right)$$

converge or diverge?

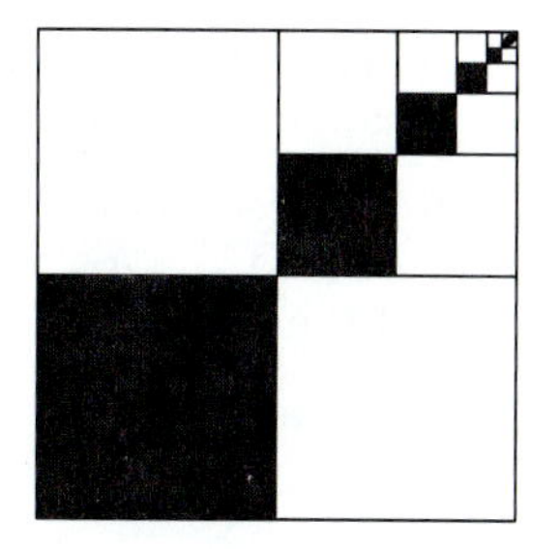

Figure 3.2. What is the area of the black region?

3.4.16 Show that
$$\frac{1}{r-1} = \frac{1}{r+1} + \frac{2}{r^2+1} + \frac{4}{r^4+1} + \frac{8}{r^8+1} + \dots$$
for all $r > 1$.

3.4.17 Obtain a formula for the sum
$$2 + \frac{2}{\sqrt{2}} + 1 + \frac{1}{\sqrt{2}} + \frac{1}{2} + \frac{1}{2\sqrt{2}} + \dots.$$

3.4.18 Obtain a formula for the sum
$$\sum_{k=1}^{\infty} \frac{1}{k(k+2)(k+4)}.$$

3.4.19 Obtain a formula for the sum
$$\sum_{k=1}^{\infty} \frac{\alpha r + \beta}{k(k+1)(k+2)}.$$

3.4.20 Find all values of x for which the the following series converges and determine the sum:
$$x + \frac{x}{1+x} + + \frac{x}{(1+x)^2} + \frac{x}{(1+x)^3} + \frac{x}{(1+x)^4} + \dots.$$

3.4.21 Determine whether the series
$$\sum_{k=1}^{\infty} \frac{1}{a+kb}$$
converges or diverges where a and b are positive real numbers.

3.4.22 We have proved that the harmonic series diverges. A computer experiment seems to show otherwise. Let s_n be the sequence of partial sums and, using a computer and the recursion formula
$$s_{n+1} = s_n + \frac{1}{n+1},$$

compute s_1, s_2, s_3, ... and stop when it appears that the sequence is no longer changing. This does happen! Explain why this is not a contradiction.

3.4.23 Let M be any integer. In Theorem 3.11 we saw that the series $\sum_{k=1}^{\infty} a_k$ converges if and only if the series $\sum_{k=1}^{\infty} a_{M+k}$ converges. What is the exact relation between the sums of the two series?

3.4.24 Write up a formal proof that the p-harmonic series

$$\sum_{k=1}^{\infty} \frac{1}{k^p}$$

converges for $p > 1$ using the method sketched in the text.

3.4.25 With a short argument using what you know about the harmonic series, show that the p-harmonic series for $0 < p \leq 1$ is divergent.

3.4.26 Obtain the divergence of the improper calculus integral

$$\int_0^{\infty} \frac{|\sin x|}{x}\, dx$$

by comparing with the harmonic series.

3.4.27 We have seen that the condition $a_n \to 0$ is a necessary, but not sufficient, condition for convergence of the series $\sum_{k=1}^{\infty} a_k$. Is the condition $na_n \to 0$ either necessary or sufficient for the convergence? This says terms are going to zero *faster* than $1/k$.

3.4.28 Let p be an integer greater or equal to 2 and let x be a real number in the interval $[0,1)$. Construct a sequence of integers $\{k_n\}$ as follows: Divide the interval $[0,1)$ into p intervals of equal length

$$[0,1/p), [1/p, 2/p), \ldots [(p-1)/p, 1)$$

and label them from left to right as 0, 1, ... $p-1$. Then k_1 is chosen so that x belongs to the k_1th interval. Repeat the process applying it now to the interval $[(k_1-1)/p, k_1/p)$ in which x lies, dividing it into p intervals of equal length and choose k_2 so that x belongs to the k_2th interval of the new subintervals. Continue this process inductively to define the sequence $\{k_n\}$. Show that

$$x = \sum_{i=1}^{\infty} \frac{k_i}{p^i}.$$

[This is called the *p-adic representation* of the number x.]

3.5 Criteria for Convergence

How do we determine the convergence or divergence of a series? The meaning of convergence or divergence is directly given in terms of the sequence of partial sums. But usually it is very difficult to say much about that sequence. Certainly we hardly ever get a closed form for the partial sums.

For a successful theory of series we need some criteria that will enable us to assert the convergence or divergence of a series without much bothering with an intimate acquaintance with the sequence of partial sums. The following material begins the development of these criteria.

3.5.1 Boundedness Criterion

If a series $\sum_{k=1}^{\infty} a_k$ consists entirely of nonnegative terms, then it is clear that the sequence of partial sums forms a monotonic sequence. It is strictly increasing if all terms are positive.

We have a well-established fundamental principle for the investigation of all monotonic sequences:

A monotonic sequence is convergent if and only if it is bounded.

Applied to the study of series, this principle says that a series $\sum_{k=1}^{\infty} a_k$ consisting entirely of nonnegative terms will converge if the sequence of partial sums is bounded and will diverge if the sequence of partial sums is unbounded.

This reduces the study of the convergence/divergence behavior of such series to inequality problems:

Is there or is there not a number M so that

$$s_n = \sum_{k=1}^{n} a_k \leq M$$

for all integers n?

This is both good news and bad. Theoretically it means that convergence problems for this special class of series reduce to another problem: one of boundedness. That is good news, reducing an apparently difficult problem to one we already understand. The bad news is that inequality problems may still be difficult.

Note. A word of warning. The boundedness of the partial sums of a series is not of as great an interest for series where the terms can be both positive and negative. For such series the boundedness of the partial sums does not guarantee convergence.

3.5.2 Cauchy Criterion

One of our main theoretical tools in the study of convergent sequences is the Cauchy criterion describing (albeit somewhat technically) a necessary and sufficient condition for a sequence to be convergent.

If we translate that criterion to the language of series we shall then have a necessary and sufficient condition for a series to be convergent. Again it is rather technical and mostly useful in developing a theory rather than in testing specific series. The translation is nearly immediate.

Definition 3.13 The series

$$\sum_{k=1}^{\infty} a_k$$

is said to satisfy the *Cauchy criterion for convergence* provided that for every $\varepsilon > 0$ there is an integer N so that all of the finite sums

$$\left|\sum_{k=n}^{m} a_k\right| < \varepsilon$$

for any $N \leq n < m < \infty$.

Now we have a principle that can be applied in many theoretical situations:

> *A series $\sum_{k=1}^{\infty} a_k$ converges if and only if it satisfies the Cauchy criterion for convergence.*

Note. It may be useful to think of this conceptually. The criterion asserts that convergence is equivalent to the fact that blocks of terms

$$\sum_{k=N}^{M} a_k$$

added up and taken from far on in the series must be small. Loosely we might describe this by saying that a convergent series has a "small tail."

Note too that if the series converges, then this criterion implies that for every $\varepsilon > 0$ there is an integer N so that

$$\left|\sum_{k=n}^{\infty} a_k\right| < \varepsilon$$

for every $n \geq N$.

3.5.3 Absolute Convergence

If a series consists of nonnegative terms only, then we can obtain convergence or divergence by estimating the size of the partial sums. If the partial sums remain bounded, then the series converges; if not, the series diverges.

No such conclusion can be made for a series $\sum_{k=1}^{\infty} a_k$ of positive and negative numbers. Boundedness of the partial sums does not allow us to conclude anything about convergence or divergence since the sequence of partial sums would not be monotonic. What we can do is ask whether there is any relation between the two series

$$\sum_{k=1}^{\infty} a_k \quad \text{and} \quad \sum_{k=1}^{\infty} |a_k|$$

where the latter series has had the negative signs stripped from it. We shall see that convergence of the series of absolute values ensures convergence of the original series. Divergence of the series of absolute values gives, however, no information.

This gives us a useful test that will prove the convergence of a series $\sum_{k=1}^{\infty} a_k$ by investigating instead the related series $\sum_{k=1}^{\infty} |a_k|$ without the negative signs.

Theorem 3.14 *If the series $\sum_{k=1}^{\infty} |a_k|$ converges, then so too does the series $\sum_{k=1}^{\infty} a_k$.*

Proof The proof takes two applications of the Cauchy criterion. If $\sum_{k=1}^{\infty} |a_k|$ converges, then for every $\varepsilon > 0$ there is an integer N so that all of the finite sums

$$\sum_{k=n}^{m} |a_k| < \varepsilon$$

for any $N < n < m < \infty$. But then

$$\left| \sum_{k=n}^{m} a_k \right| \leq \sum_{k=n}^{m} |a_k| < \varepsilon.$$

It follows, by the Cauchy criterion applied to the series $\sum_{k=1}^{\infty} a_k$, that this series is convergent. ■

Note. Note that there is no claim in the statement of this theorem that the two series have the same sum, just that the convergence of one implies the convergence of the other.

For theoretical reasons it is important to know when the series $\sum_{k=1}^{\infty} |a_k|$ of absolute values converges. Such series are "more" than convergent. They are convergent in a way that allows more manipulations than would otherwise be available. They can be thought of as more robust; a series that converges, but whose absolute series does not converge is in some ways fragile. This leads to the following definitions.

Definition 3.15 A series $\sum_{k=1}^{\infty} a_k$ is said to be *absolutely convergent* if the related series $\sum_{k=1}^{\infty} |a_k|$ converges.

Definition 3.16 A series $\sum_{k=1}^{\infty} |a_k|$ is said to be *nonabsolutely convergent* if the series $\sum_{k=1}^{\infty} a_k$ converges but the series $\sum_{k=1}^{\infty} |a_k|$ diverges.

Note that every *absolutely* convergent series is also convergent. We think of it as "more than convergent." Fortunately, the terminology preserves the meaning even though the "absolutely" refers to the absolute value, not to any other implied meaning. This play on words would not be available in all languages.

Example 3.17 Using this terminology, applied to series we have already studied, we can now assert the following:

> Any geometric series $1+r+r^2+r^3+\ldots$ is absolutely convergent if $|r|<1$ and divergent if $|r|\geq 1$.

and

> The alternating harmonic series $1-\frac{1}{2}+\frac{1}{3}-\frac{1}{4}\ldots$ is nonabsolutely convergent.

◄

Exercises

3.5.1 Suppose that $\sum_{k=1}^{\infty} a_k$ is a convergent series of positive terms. Show that $\sum_{k=1}^{\infty} a_k^2$ is convergent. Does the converse hold?

3.5.2 Suppose that $\sum_{k=1}^{\infty} a_k$ is a convergent series of positive terms. Show that $\sum_{k=1}^{\infty} \sqrt{a_k a_{k+1}}$ is convergent. Does the converse hold?

3.5.3 Suppose that both series

$$\sum_{k=1}^{\infty} a_k \text{ and } \sum_{k=1}^{\infty} b_k$$

are absolutely convergent. Show that then so too is the series $\sum_{k=1}^{\infty} a_k b_k$. Does the converse hold?

3.5.4 Suppose that both series

$$\sum_{k=1}^{\infty} a_k \text{ and } \sum_{k=1}^{\infty} b_k$$

are nonabsolutely convergent. Show that it does not follow that the series $\sum_{k=1}^{\infty} a_k b_k$ is convergent.

3.5.5 Alter the harmonic series $\sum_{k=1}^{\infty} 1/k$ by deleting all terms in which the denominator contains a specified digit (say 3). Show that the new series converges.

3.5.6 Show that the geometric series $\sum_{n=1}^{\infty} r^n$ is convergent for $|r| < 1$ by using directly the Cauchy convergence criterion.

3.5.7 Show that the harmonic series is divergent by using directly the Cauchy convergence criterion.

3.5.8 Obtain a proof that every series $\sum_{k=1}^{\infty} a_k$ for which $\sum_{k=1}^{\infty} |a_k|$ converges must itself be convergent without using the Cauchy criterion.

3.5.9 Show that a series $\sum_{k=1}^{\infty} a_k$ is absolutely convergent if and only if two at least of the series

$$\sum_{k=1}^{\infty} a_k \text{ , } \sum_{k=1}^{\infty} [a_k]^+ \text{ and } \sum_{k=1}^{\infty} [a_k]^-$$

converge. (If two converge, then all three converge.)

3.5.10 The sum rule for convergent series

$$\sum_{k=1}^{\infty}(a_k + b_k) = \sum_{k=1}^{\infty} a_k + \sum_{k=1}^{\infty} b_k$$

can be expressed by saying that if any two of these series converges so too does the third. What kind of statements can you make for absolute convergence and for nonabsolute convergence?

3.5.11 Show that a series $\sum_{k=1}^{\infty} a_k$ is absolutely convergent if and only if every subseries $\sum_{k=1}^{\infty} a_{n_k}$ converges.

3.5.12 A sequence $\{x_n\}$ of real numbers is said to be of *bounded variation* if the series

$$\sum_{k=2}^{\infty} |x_k - x_{k-1}|$$

converges.

(a) Show that every sequence of bounded variation is convergent.

(b) Show that not every convergent sequence is of bounded variation.

(c) Show that all monotonic convergent sequences are of bounded variation.

(d) Show that any linear combination of two sequences of bounded variation is of bounded variation.

(e) Is the product of of two sequences of bounded variation also of bounded variation?

3.5.13 Establish the Cauchy-Schwarz inequality: For any finite sequences

$$\{a_1, a_2, \ldots , a_m\} \text{ and } \{b_1, b_2, \ldots , b_m\}$$

the inequality

$$\left|\sum_{k=1}^{n} a_k b_k\right| \leq \left(\sum_{k=1}^{n}(a_k)^2\right)^{\frac{1}{2}} \left(\sum_{k=1}^{n}(b_k)^2\right)^{\frac{1}{2}}$$

must hold.

3.5.14 Using the Cauchy-Schwarz inequality (Exercise 3.5.13), show that if $\{a_n\}$ is a sequence of nonnegative numbers for which $\sum_{n=1}^{\infty} a_n$ converges, then the series

$$\sum_{n=0}^{\infty} \frac{\sqrt{a_n}}{n^p}$$

also converges for any $p > \frac{1}{2}$. Without the Cauchy-Schwarz inequality what is the best you can prove for convergence?

3.5.15 Suppose that $\sum_{n=1}^{\infty} a_n^2$ converges. Show that

$$\limsup_{n\to\infty} \frac{a_1 + \sqrt{2}a_2 + \sqrt{3}a_3 + \sqrt{4}a_4 + \cdots + \sqrt{n}a_n}{n} < \infty.$$

3.5.16 Let x_1, x_2, x_3 be a sequence of positive numbers and write

$$s_n = \frac{x_1 + x_2 + x_3 + \cdots + x_n}{n}$$

and

$$t_n = \frac{\frac{1}{x_1} + \frac{1}{x_2} + \frac{1}{x_3} + \cdots + \frac{1}{x_n}}{n}.$$

If $s_n \to S$ and $t_n \to T$, show that $ST \geq 1$.

3.6 Tests for Convergence

In many investigations and applications of series it is important to recognize that a given series converges, converges absolutely, or diverges. Frequently the sum of the series is not of much interest, just the convergence behavior. Over the years a battery of tests have been developed to make this task easier.

There are only a few basic principles that we can use to check convergence or divergence and we have already discussed these in Section 3.5. One of the most basic is that a series of nonnegative terms is convergent if and only if the sequence of partial sums is bounded. Most of the tests in the sequel are just clever ways of checking that the partial sums are bounded without having to do the computations involved in finding that upper bound.

3.6.1 Trivial Test

The first test is just an observation that we have already made about series: If a series $\sum_{k=1}^{\infty} a_k$ converges, then $a_k \to 0$. We turn this into a divergence test. For example, some novices will worry for a long time over a series such as

$$\sum_{k=1}^{\infty} \frac{1}{\sqrt[k]{k}}$$

applying a battery of tests to it to determine convergence. The simplest way to see that this series diverges is to note that the terms tend to 1 as $k \to \infty$. Perhaps this is the first thing that should be considered for any series. If the terms do not get small there is no point puzzling whether the series converges. It does not.

3.18 (Trivial Test) *If the terms of the series* $\sum_{k=1}^{\infty} a_k$ *do not converge to* 0, *then the series diverges.*

Proof We have already proved this, but let us prove it now as a special case of the Cauchy criterion. For all $\varepsilon > 0$ there is an N so that

$$|a_n| = \left| \sum_{k=n}^{n} a_k \right| < \varepsilon$$

for all $n \geq N$ and so, by definition, $a_k \to 0$. ■

3.6.2 Direct Comparison Tests

A series $\sum_{k=1}^{\infty} a_k$ with all terms nonnegative can be handled by estimating the size of the partial sums. Rather than making a direct estimate it is sometimes easier to find a bigger series that converges. This larger series provides an upper bound for our series without the need to compute one ourselves.

Note. Make sure to apply these tests only for series with nonnegative terms since, for arbitrary series, this information is useless.

3.19 (Direct Comparison Test I) *Suppose that the terms of the series* $\sum_{k=1}^{\infty} a_k$ *are each smaller than the corresponding terms of the series* $\sum_{k=1}^{\infty} b_k$; *that is, that*

$$0 \leq a_k \leq b_k$$

for all k. *If the larger series converges, then so does the smaller series.*

Proof If $0 \leq a_k \leq b_k$ for all k, then

$$\sum_{k=1}^{n} a_k \leq \sum_{k=1}^{n} b_k \leq \sum_{k=1}^{\infty} b_k.$$

Thus the number $B = \sum_{k=1}^{\infty} b_k$ is an upper bound for the sequence of partial sums of the series $\sum_{k=1}^{\infty} a_k$. It follows that $\sum_{k=1}^{\infty} a_k$ must converge. ■

Note. In applying this and subsequent tests that demand that all terms of a series satisfy some requirement, we should remember that convergence and divergence of a series $\sum_{k=1}^{\infty} a_k$ depends only on the behavior of a_k for large values of k. Thus this test (and many others) could be reformulated so as to apply only for k greater than some integer N.

3.20 (Direct Comparison Test II) *Suppose that the terms of the series $\sum_{k=1}^{\infty} a_k$ are each larger than the corresponding terms of the series $\sum_{k=1}^{\infty} c_k$; that is, that*

$$0 \leq c_k \leq a_k$$

for all k. If the smaller series diverges, then so does the larger series.

Proof This follows from Test 3.19 since if the larger series did not diverge, then it must converge and so too must the smaller series. ■

Here are two examples illustrating how these tests may be used.

Example 3.21 Consider the series

$$\sum_{k=1}^{\infty} \frac{k+5}{k^3+k^2+k+1}.$$

While the partial sums might seem hard to estimate at first, a fast glance suggests that the terms (crudely) are similar to $1/k^2$ for large values of k and we know that the series $\sum_{k=1}^{\infty} 1/k^2$ converges. Note that

$$\frac{k+5}{k^3+k^2+k+1} = \frac{1+5/k}{k^2(1+1/k+1/k^2+1/k^3)} \leq \frac{C}{k^2}$$

for some choice of C (e.g., $C = 6$ will work). We now claim that our given series converges by a direct comparison with the convergent series $\sum_{k=1}^{\infty} C/k^2$. (This is a p-harmonic series with $p = 2$.) ◀

Example 3.22 Consider the series

$$\sum_{k=1}^{\infty} \sqrt{\frac{k+5}{k^2+k+1}}.$$

Again, a fast glance suggests that the terms (crudely) are similar to $1/\sqrt{k}$ for large values of k and we know that the series $\sum_{k=1}^{\infty} 1/\sqrt{k}$ diverges. Note that

$$\frac{k+5}{k^2+k+1} = \frac{1+5/k}{k(1+1/k+1/k^2)} \geq \frac{C}{k}$$

for some choice of C (e.g., $C = \frac{1}{4}$ will work). We now claim that our given series diverges by a direct comparison with the divergent series $\sum_{k=1}^{\infty} \sqrt{C}/\sqrt{k}$. (This is a p-harmonic series with $p = 1/2$.) ◀

The examples show both advantages and disadvantages to the method. We must invent the series that is to be compared and we must do some amount of inequality work to show that comparison. The next tests replace the inequality work with a limit operation, which is occasionally easier to perform.

3.6.3 Limit Comparison Tests

We have seen that a series $\sum_{k=1}^{\infty} a_k$ with all terms nonnegative can be handled by comparing with a larger convergent series or a smaller divergent series. Rather than check all the terms of the two series being compared, it is convenient sometimes to have this checked automatically by the computation of a limit. In this section, since the tests involve a fraction, we must be sure not only that all terms are nonnegative, but also that we have not divided by zero.

3.23 (Limit Comparison Test I) *Let each $a_k \geq 0$ and $b_k > 0$. If the terms of the series $\sum_{k=1}^{\infty} a_k$ can be compared to the terms of the series $\sum_{k=1}^{\infty} b_k$ by computing*

$$\lim_{k\to\infty} \frac{a_k}{b_k} < \infty$$

and if the latter series converges, then so does the former series.

Proof The proof is easy. If the stated limit exists and is finite then there are numbers M and N so that

$$\frac{a_k}{b_k} < M$$

for all $k \geq N$. This shows that $a_k \leq Mb_k$ for all $k \geq N$. Consequently, applying the direct comparison test, we find that the series $\sum_{k=N}^{\infty} a_k$ converges by comparison with the convergent series $\sum_{k=N}^{\infty} Mb_k$. ■

3.24 (Limit Comparison Test II) *Let each $a_k > 0$ and $c_k > 0$. If the terms of the series $\sum_{k=1}^{\infty} a_k$ can be compared to the terms of the series $\sum_{k=1}^{\infty} c_k$ by computing*

$$\lim_{k\to\infty} \frac{a_k}{c_k} > 0$$

and if the latter series diverges, then so does the original series.

Proof Since the limit exists and is not zero there are numbers $\varepsilon > 0$ and N so that

$$\frac{a_k}{c_k} > \varepsilon$$

for all $k \geq N$. This shows that, for all $k \geq N$,

$$a_k \geq \varepsilon c_k.$$

Consequently, by the direct comparison test the series $\sum_{k=N}^{\infty} a_k$ diverges by comparison with the divergent series $\sum_{k=N}^{\infty} \varepsilon c_k$. ■

We repeat our two examples, Example 3.21 and 3.22, where we previously used the direct comparison test to check for convergence.

Example 3.25 We look again at the series

$$\sum_{k=1}^{\infty} \frac{k+5}{k^3+k^2+k+1},$$

comparing it, as before, to the convergent series $\sum_{k=1}^{\infty} 1/k^2$. This now requires computing the limit

$$\lim_{k\to\infty} \frac{k^2(k+5)}{k^3+k^2+k+1},$$

which elementary calculus arguments show is 1. Since it is not infinite, the original series can now be claimed to converge by a limit comparison. ◀

Example 3.26 Again, consider the series

$$\sum_{k=1}^{\infty} \sqrt{\frac{k+5}{k^2+k+1}}$$

by comparing with the divergent series $\sum_{k=1}^{\infty} 1/\sqrt{k}$. We are required to compute the limit

$$\lim_{k\to\infty} \sqrt{k}\sqrt{\frac{k+5}{k^2+k+1}},$$

which elementary calculus arguments show is 1. Since it is not zero, the original series can now be claimed to diverge by a limit comparison. ◀

3.6.4 Ratio Comparison Test

Again we wish to compare two series $\sum_{k=1}^{\infty} a_k$ and $\sum_{k=1}^{\infty} b_k$ composed of positive terms. Rather than directly comparing the size of the terms we compare the ratios of the terms. The inspiration for this test rests on attempts to compare directly a series with a convergent geometric series. If $\sum_{k=1}^{\infty} b_k$ is a geometric series with common ratio r, then evidently

$$\frac{b_{k+1}}{b_k} = r.$$

This suggests that perhaps a comparison of ratios of successive terms would indicate how fast a series might be converging.

3.27 (Ratio Comparison Test) *If the ratios satisfy*

$$\frac{a_{k+1}}{a_k} \leq \frac{b_{k+1}}{b_k}$$

for all k (or just for all k sufficiently large) and the series $\sum_{k=1}^{\infty} b_k$, with the larger ratio is convergent, then the series $\sum_{k=1}^{\infty} a_k$ is also convergent.

Proof As usual, we assume all terms are positive in both series. If the ratios satisfy

$$\frac{a_{k+1}}{a_k} \leq \frac{b_{k+1}}{b_k}$$

for $k > N$, then they also satisfy

$$\frac{a_{k+1}}{b_{k+1}} \leq \frac{a_k}{b_k},$$

which means that the sequence $\{a_k/b_k\}$ is decreasing for $k > N$. In particular, that sequence is bounded above, say by C, and so

$$a_k \leq Cb_k.$$

Thus an application of the direct comparison test shows that the series $\sum_{k=1}^{\infty} a_k$ converges. ■

3.6.5 d'Alembert's Ratio Test

The ratio comparison test requires selecting a series for comparison. Often a geometric series $\sum_{k=1}^{\infty} r^k$ for some $0 < r < 1$ may be used. How do we compute a number r that will work? We would wish to use $b_k = r^k$ with a choice of r so that

$$\frac{a_{k+1}}{a_k} \leq \frac{b_{k+1}}{b_k} = \frac{r^{k+1}}{r^k} = r.$$

One useful and easy way to find whether there will be such an r is to compute the limit of the ratios.

3.28 (Ratio Test) *If terms of the series $\sum_{k=1}^{\infty} a_k$ are all positive and the ratios satisfy*

$$\lim_{k\to\infty} \frac{a_{k+1}}{a_k} < 1$$

then the series $\sum_{k=1}^{\infty} a_k$ is convergent.

Proof The proof is easy. If

$$\lim_{k\to\infty} \frac{a_{k+1}}{a_k} < 1,$$

then there is a number $\beta < 1$ so that

$$\frac{a_{k+1}}{a_k} < \beta$$

for all sufficiently large k. Thus the series $\sum_{k=1}^{\infty} a_k$ converges by the ratio comparison test applied to the convergent geometric series $\sum_{k=1}^{\infty} \beta^k$. ■

Note. The ratio test can also be pushed to give a divergence answer: If

$$\lim_{k\to\infty} \frac{a_{k+1}}{a_k} > 1 \tag{4}$$

then the series $\sum_{k=1}^{\infty} a_k$ is divergent. But it is best to downplay this test or you might think it gives an answer as useful as the convergence test. From (4) it follows that there must be an N and β so that

$$\frac{a_{k+1}}{a_k} > \beta > 1$$

for all $k \geq N$. Then

$$a_{N+1} > \beta a_N,$$

$$a_{N+2} > \beta a_{N+1} > \beta^2 a_N,$$

and

$$a_{N+3} > \beta a_{N+2} > \beta^3 a_N.$$

We see that the terms a_k of the series are growing large at a geometric rate. Not only is the series diverging, but it is diverging in a dramatic way.

We can summarize how this test is best applied. If terms of the series $\sum_{k=1}^{\infty} a_k$ are all positive, compute

$$\lim_{k\to\infty} \frac{a_{k+1}}{a_k} = L.$$

1. If $L < 1$, then the series $\sum_{k=1}^{\infty} a_k$ is convergent.
2. If $L > 1$, then the series $\sum_{k=1}^{\infty} a_k$ is divergent; moreover, the terms $a_k \to \infty$.
3. If $L = 1$, then the series $\sum_{k=1}^{\infty} a_k$ may diverge or converge, the test being inconclusive.

Example 3.29 The series

$$\sum_{k=0}^{\infty} \frac{(k!)^2}{(2k)!}$$

is particularly suited for an application of the ratio test since the ratio is easily computed and a limit taken: If we write $a_k = (k!)^2/(2k)!$, then

$$\frac{a_{k+1}}{a_k} = \frac{((k+1)!)^2}{(2k+2)!}\frac{(2k)!}{(k!)^2} = \frac{(k+1)^2}{(2k+2)(2k+1)} \to \frac{1}{4}.$$

Consequently, this is a convergent series. More than that, it is converging faster than any geometric series

$$\sum_{k=0}^{\infty}\left(\frac{1}{4}+\varepsilon\right)^k$$

for any positive ε. (To make this expression "converging faster" more precise, see Exercise 3.12.5.) ◀

3.6.6 Cauchy's Root Test

There is yet another way to achieve a comparison with a convergent geometric series. We suspect that a series $\sum_{k=1}^{\infty} a_k$ can be compared to some geometric series $\sum_{k=1}^{\infty} r^k$ but do not know how to compute the value of r that might work. The limiting values of the ratios

$$\frac{a_{k+1}}{a_k}$$

provide one way of determining what r might work but often are difficult to compute. Instead we recognize that a comparison of the form

$$a_k \le Cr^k$$

would mean that

$$\sqrt[k]{a_k} \le \sqrt[k]{C}r.$$

For large k the term $\sqrt[k]{C}$ is close to 1, and this motivates our next test, usually attributed to Cauchy.

3.30 (Root Test) *If terms of the series $\sum_{k=1}^{\infty} a_k$ are all nonnegative and if the roots satisfy*

$$\lim_{k\to\infty} \sqrt[k]{a_k} < 1,$$

then that series converges.

Proof This is almost trivial. If

$$(a_k)^{1/k} < \beta < 1$$

for all $k \ge N$, then

$$a_k < \beta^k$$

and so $\sum_{k=1}^{\infty} a_k$ converges by direct comparison with the convergent geometric series $\sum_{k=1}^{\infty} \beta^k$. ■

Again we can summarize how this test is best applied. The conclusions are nearly identical with those for the ratio test. Compute

$$\lim_{k \to \infty} (a_k)^{1/k} = L.$$

1. If $L < 1$, then the series $\sum_{k=1}^{\infty} a_k$ is convergent.
2. If $L > 1$, then the series $\sum_{k=1}^{\infty} a_k$ is divergent; moreover, the terms $a_k \to \infty$.
3. If $L = 1$, then the series $\sum_{k=1}^{\infty} a_k$ may diverge or converge, the test being inconclusive.

Example 3.31 In Example 3.29 we found the series

$$\sum_{k=0}^{\infty} \frac{(k!)^2}{(2k)!}$$

to be handled easily by the ratio test. It would be extremely unpleasant to attempt a direct computation using the root test. On the other hand, the series

$$\sum_{k=0}^{\infty} kx^k = x + 2x^2 + 3x^3 + 4x^4 + \dots$$

for $x > 0$ can be handled by either of these tests. You should try the ratio test while we try the root test:

$$\lim_{k \to \infty} \left(kx^k\right)^{1/k} = \lim_{k \to \infty} \sqrt[k]{k}x = x$$

and so convergence can be claimed for all $0 < x < 1$ and divergence for all $x > 1$. The case $x = 1$ is inconclusive for the root test, but the trivial test shows instantly that the series diverges for $x = 1$. ◀

3.6.7 Cauchy's Condensation Test

✂ *Enrichment*

Occasionally a method that is used to study a specific series can be generalized into a useful test. Recall that in studying the sequence of partial sums of the harmonic series it was convenient to watch only at the steps 1, 2, 4, 8, ... and make a rough lower estimate. The reason this worked was simply that the terms in the harmonic series decrease and so estimates of s_1, s_2, s_4, s_8, ... were easy to obtain using just that fact. This turns quickly into a general test.

3.32 (Cauchy's Condensation Test) *If the terms of a series* $\sum_{k=1}^{\infty} a_k$ *are nonnegative and decrease monotonically to zero, then that series converges if and only if the related series*

$$\sum_{j=1}^{\infty} 2^j a_{2^j}$$

converges.

Proof Since all terms are nonnegative, we need only compare the size of the partial sums of the two series. Computing first the sum of $2^{p+1} - 1$ terms of the original series, we have

$$a_1 + (a_2 + a_3) + \cdots + (a_{2^p} + a_{2^p+1} + \cdots + a_{2^{p+1}-1})$$
$$\leq a_1 + 2a_2 + \cdots + 2^p a_{2^p}.$$

And, with the inequality sign in the opposite direction, we compute the sum of 2^p terms of the original series to obtain

$$a_1 + a_2 + (a_3 + a_4) + \cdots + (a_{2^{p-1}+1} + a_{2^{p-1}+2} + \cdots + a_{2^p})$$
$$\geq \frac{1}{2}(a_1 + 2a_2 + \cdots + 2^p a_{2^p}).$$

If either series has a bounded sequence of partial sums so too then does the other series. Thus both converge or else both diverge. ■

Example 3.33 Let us use this test to study the p-harmonic series:

$$\sum_{k=1}^{\infty} \frac{1}{k^p}$$

for $p > 0$. The terms decrease to zero and so the convergence of this series is equivalent to the convergence of the series

$$\sum_{j=1}^{\infty} 2^j \left(\frac{1}{2^j}\right)^p$$

and this series is a geometric series

$$\sum_{j=1}^{\infty} (2^{1-p})^j.$$

This converges precisely when $2^{1-p} < 1$ or $p > 1$ and diverges when $2^{1-p} \geq 1$ or $p \leq 1$. Thus we know exactly the convergence behavior of the p-harmonic series for all values of p. (For $p \leq 0$ we have divergence just by the trivial test.) ◄

It is worth deriving a simple test from the Cauchy condensation test as a corollary. This is an improvement on the trivial test. The trivial test requires that $\lim_{k\to\infty} a_k = 0$ for a convergent series $\sum_{k=1}^{\infty} a_k$. This next test, which is due to Abel, shows that slightly more can be said if the terms form a monotonic sequence. The sequence $\{a_k\}$ must go to zero faster than $\{1/k\}$.

Corollary 3.34 *If the terms of a convergent series $\sum_{k=1}^{\infty} a_k$ decrease monotonically, then*

$$\lim_{k\to\infty} k a_k = 0.$$

Proof By the Cauchy condensation test we know that

$$\lim_{j\to\infty} 2^j a_{2^j} = 0.$$

If $2^j \leq k \leq 2^{j+1}$, then $a_k \leq a_{2^j}$ and so

$$k a_k \leq 2\left(2^j a_{2^j}\right),$$

which is small for large j. Thus $k a_k \to 0$ as required. ■

3.6.8 Integral Test

✂ *Enrichment*

To determine the convergence of a series $\sum_{k=1}^{\infty} a_k$ of nonnegative terms it is often necessary to make some kind of estimate on the size of the sequence of partial sums. Most of our tests have done this automatically, saving us the labor of computing such estimates. Sometimes those estimates can be obtained by calculus methods. The integral test allows us to estimate the partial sums $\sum_{k=1}^{n} f(k)$ by computing instead $\int_1^n f(x)\,dx$ in certain circumstances. This is more than a convenience; it also shows a close relation between series and infinite integrals, which is of much importance in analysis.

3.35 (Integral Test) *Let f be a nonnegative decreasing function on $[1,\infty)$ such that the integral $\int_1^X f(x)\,dx$ can be computed for all $X > 1$. If*

$$\lim_{X\to\infty} \int_1^X f(x)\,dx < \infty$$

exists, then the series $\sum_{k=1}^{\infty} f(k)$ converges. If

$$\lim_{X\to\infty} \int_1^X f(x)\,dx = \infty,$$

then the series $\sum_{k=1}^{\infty} f(k)$ diverges.

Proof Since the function f is decreasing we must have

$$\int_k^{k+1} f(x)\,dx \le f(k) \le \int_{k-1}^{k} f(x)\,dx.$$

Applying these inequalities for $k = 2, 3, 4, \dots$ we obtain

$$\int_1^{n+1} f(x)\,dx \le \sum_{k=1}^{n} f(k) \le f(1) + \int_1^{n} f(x)\,dx. \tag{5}$$

The series converges if and only if the partial sums are bounded. But we see from the inequalities (5) that if the limit of the integral is finite, then these partial sums are bounded. If the limit of the integral is infinite, then these partial sums are unbounded. ■

Note. The convergence of the integral yields the convergence of the series. There is no claim that the sum of the series $\sum_{k=1}^{\infty} f(k)$ and the value of the infinite integral $\int_1^{\infty} f(x)\,dx$ are the same. In this regard, however, see Exercise 3.6.21.

Example 3.36 According to this test the harmonic series $\sum_{k=1}^{\infty} \frac{1}{k}$ can be studied by computing

$$\lim_{X\to\infty} \int_1^X \frac{dx}{x} = \lim_{X\to\infty} \log X = \infty.$$

For the same reasons the p-harmonic series

$$\sum_{k=1}^{\infty} \frac{1}{k^p}$$

for $p > 1$ can be studied by computing

$$\lim_{X\to\infty} \int_1^X \frac{dx}{x^p} = \lim_{X\to\infty} \frac{1}{p-1}\left(1 - \frac{1}{X^{p-1}}\right) = \frac{1}{p-1}.$$

In both cases we obtain the same conclusion as before. The harmonic series diverges and, for $p > 1$, the p-harmonic series converges. ◀

3.6.9 Kummer's Tests

✂ *Advanced*

The ratio test requires merely taking the limit of the ratios

$$\frac{a_{k+1}}{a_k}$$

but often fails. We know that if this tends to 1, then the ratio test says nothing about the convergence or divergence of the series $\sum_{k=1}^{\infty} a_k$.

Kummer's tests provide a collection of ratio tests that can be designed by taking different choices of sequence $\{D_k\}$. The choices $D_k = 1$, $D_k = k$

and $D_k = k \ln k$ are used in the following tests. Ernst Eduard Kummer (1810–1893) is probably most famous for his contributions to the study of Fermat's last theorem; his tests arose in his study of hypergeometric series.

3.37 (Kummer's Tests) *The series $\sum_{k=1}^{\infty} a_k$ can be tested by the following criteria. Let $\{D_k\}$ denote any sequence of positive numbers and compute*

$$L = \liminf_{k\to\infty} \left[D_k \frac{a_k}{a_{k+1}} - D_{k+1} \right].$$

If $L > 0$ the series $\sum_{k=1}^{\infty} a_k$ converges. On the other hand, if

$$\left[D_k \frac{a_k}{a_{k+1}} - D_{k+1} \right] \leq 0$$

for all sufficiently large k and if the series

$$\sum_{k=1}^{\infty} \frac{1}{D_k}$$

diverges, then the series $\sum_{k=1}^{\infty} a_k$ diverges.

Proof If $L > 0$, then we can choose a positive number $\alpha < L$. By the definition of a liminf this means there must exist an integer N so that for all $k \geq N$,

$$\alpha < \left[D_k \frac{a_k}{a_{k+1}} - D_{k+1} \right].$$

Rewriting this, we find that

$$\alpha a_{k+1} < D_k a_k - D_{k+1} a_{k+1}.$$

We can write this inequality for $k = N, N+1, N+2, \dots N+p$ to obtain

$$\alpha a_{N+1} < D_N a_N - D_{N+1} a_{N+1}$$

$$\alpha a_{N+2} < D_{N+1} a_{N+1} - D_{N+2} a_{N+2}$$

and so on. Adding these (note the telescoping sums), we find that

$$\alpha(a_{N+1} + a_{N+2} + \cdots + a_{N+p+1})$$

$$< D_{N+1} a_{N+1} - D_{N+p+1} a_{N+p+1} < D_{N+1} a_{N+1}.$$

(The final inequality just uses the fact that all the terms here are positive.)

From this inequality we can determine that the partial sums of the series $\sum_{k=1}^{\infty} a_k$ are bounded. By our usual criterion, this proves that this series converges.

The second part of the theorem requires us to establish divergence. Suppose now that

$$D_k \frac{a_k}{a_{k+1}} - D_{k+1} \leq 0$$

for all $k \geq N$. Then

$$D_k a_k \leq D_{k+1} a_{k+1}.$$

Thus the sequence $\{D_k a_k\}$ is increasing after $k = N$. In particular,

$$D_k a_k \geq C$$

for some C and all $k \geq N$ and so

$$a_k \geq \frac{C}{D_k}.$$

It follows by a direct comparison with the divergent series $\sum C/D_k$ that our series also diverges. ■

Note. In practice, for the divergence part of the test, it may be easier to compute

$$L = \limsup_{k \to \infty} \left[D_k \frac{a_k}{a_{k+1}} - D_{k+1} \right].$$

If $L < 0$, then we would know that

$$\left[D_k \frac{a_k}{a_{k+1}} - D_{k+1} \right] \leq 0$$

for all sufficiently large k and so, if the series $\sum_{k=1}^{\infty} \frac{1}{D_k}$ diverges, then the series $\sum_{k=1}^{\infty} a_k$ diverges.

Example 3.38 What is Kummer's test if the sequence used is the simplest possible $D_k = 1$ for all k? In this case it is simply the ratio test. For example, suppose that

$$\lim_{k \to \infty} \frac{a_{k+1}}{a_k} = r.$$

Then, replacing $D_k = 1$, we have

$$\lim_{k \to \infty} \left[D_k \frac{a_k}{a_{k+1}} - D_{k+1} \right] = \lim_{k \to \infty} \left[\frac{a_k}{a_{k+1}} - 1 \right] = \frac{1}{r} - 1.$$

Thus, by Kummer's test, if $1/r - 1 < 0$ we have divergence while if $1/r - 1 > 0$ we have convergence. These are just the cases $r > 1$ and $r < 1$ of the ratio test. ◀

3.6.10 Raabe's Ratio Test

Advanced

A simple variant on the ratio test is known as Raabe's test. Suppose that

$$\lim_{k\to\infty} \frac{a_k}{a_{k+1}} = 1$$

so that the ratio test is inconclusive. Then instead compute

$$\lim_{k\to\infty} k\left(\frac{a_k}{a_{k+1}} - 1\right).$$

The series $\sum_{k=1}^{\infty} a_k$ converges or diverges depending on whether this limit is greater than or less than 1.

3.39 (Raabe's Test) *The series $\sum_{k=1}^{\infty} a_k$ can be tested by the following criterion. Compute*

$$L = \lim_{k\to\infty} k\left(\frac{a_k}{a_{k+1}} - 1\right).$$

Then

1. *If $L > 1$, the series $\sum_{k=1}^{\infty} a_k$ converges.*
2. *If $L < 1$, the series $\sum_{k=1}^{\infty} a_k$ diverges.*
3. *If $L = 1$, the test is inconclusive.*

Proof This is Kummer's test using the sequence $D_k = k$. ■

Example 3.40 Consider the series

$$\sum_{k=0}^{\infty} \frac{k^k}{e^k k!}.$$

An attempt to apply the ratio test to this series will fail since the ratio will tend to 1, the inconclusive case. But if instead we consider the limit

$$\lim_{k\to\infty} k\left(\left(\frac{k^k}{e^k k!}\right)\left(\frac{e^{k+1}(k+1)!}{(k+1)^{k+1}}\right) - 1\right)$$

as called for in Raabe's test, we can use calculus methods (L'Hôpital's rule) to obtain a limit of $\frac{1}{2}$. Consequently, this series diverges. ◀

3.6.11 Gauss's Ratio Test

Advanced

Raabe's test can be replaced by a closely related test due to Gauss. We might have discovered while using Raabe's test that

$$\lim_{k\to\infty} k\left(\frac{a_k}{a_{k+1}} - 1\right) = L.$$

This suggests that in any actual computation we will have discovered, perhaps by division, that

$$\frac{a_k}{a_{k+1}} = 1 + \frac{L}{k} + \text{ terms involving } \frac{1}{k^2} \text{ etc.}$$

The case $L > 1$ corresponds to convergence and the case $L < 1$ to divergence, both by Raabe's test. What if $L = 1$, which is considered inconclusive in Raabe's test?

Gauss's test offers a different way to look at Raabe's test and also has an added advantage that it handles this case that was left as inconclusive in Raabe's test.

3.41 (Gauss's Test) *The series $\sum_{k=1}^{\infty} a_k$ can be tested by the following criterion. Suppose that*

$$\frac{a_k}{a_{k+1}} = 1 + \frac{L}{k} + \frac{\phi(k)}{k^2}$$

where $\phi(k)$ $(k = 1, 2, 3, \dots)$ forms a bounded sequence. Then

1. *If $L > 1$ the series $\sum_{k=1}^{\infty} a_k$ converges.*

2. *If $L \leq 1$ the series $\sum_{k=1}^{\infty} a_k$ diverges.*

Proof As we noted, for $L > 1$ and $L < 1$ this is precisely Raabe's test. Only the case $L = 1$ is new! Let us assume that

$$\frac{a_k}{a_{k+1}} = 1 + \frac{1}{k} + \frac{x_k}{k^2}$$

where $\{x_k\}$ is a bounded sequence.

To prove this case (that the series diverges) we shall use Kummer's test with the sequence $D_k = k \log k$. We consider the expression

$$\left[D_k \frac{a_k}{a_{k+1}} - D_{k+1} \right],$$

which now assumes the form

$$k \log k \frac{a_k}{a_{k+1}} - (k+1)\log(k+1)$$

$$= k \log k \left(1 + \frac{1}{k} + \frac{x_k}{k^2} \right) - (k+1)\log(k+1).$$

We need to compute the limit of this expression as $k \to \infty$. It takes only a few manipulations (which you should try) to see that the limit is -1. For this use the facts that

$$(\log k)/k \to 0$$

and

$$(k+1)\log(1+1/k) \to 1$$

as $k \to \infty$.

We are now in a position to claim, by Kummer's test, that our series $\sum_{k=1}^{\infty} a_k$ diverges. To apply this part of the test requires us to check that the series

$$\sum_{k=2}^{\infty} \frac{1}{k \log k}$$

diverges. Several tests would work for this. Perhaps Cauchy's condensation test is the easiest to apply, although the integral test can be used too [see Exercise 3.6.2(c)]. ■

Note. In Gauss's test you may be puzzling over how to obtain the expression

$$\frac{a_k}{a_{k+1}} = 1 + \frac{L}{k} + \frac{\phi(k)}{k^2}.$$

In practice often the fraction a_k/a_{k+1} is a ratio of polynomials and so usual algebraic procedures will supply this. In theory, though, there is no problem. For any L we could simply write

$$\phi(k) = k^2\left(\frac{a_k}{a_{k+1}} - 1 + \frac{L}{k}\right).$$

Thus the real trick is whether it can be done in such a way that the $\phi(k)$ do not grow too large.

Also, in some computations you might prefer to leave the ratio as a_{k+1}/a_k the way it was for the ratio test. In that case Gauss's test would assume the form

$$\frac{a_{k+1}}{a_k} = 1 - \frac{L}{k} + \frac{\phi(k)}{k^2}.$$

(Note the minus sign.) The conclusions are exactly the same.

Example 3.42 The series

$$1 + mx + \frac{m(m-1)}{2!}x^2 +$$

$$\frac{m(m-1)(m-2)}{3!}x^3 + \frac{m(m-1)\dots(m-k+1)}{k!}x^k + \dots$$

is called the *binomial series*. When m is a positive integer all terms for $k > m$ are zero and the series reduces to the binomial formula for $(1+x)^m$. Here now m is any real number and the hope remains that the formula might still be valid, but using a series rather than a finite sum. This series plays an important role in many applications. Let us check for absolute convergence at $x = 1$. We can assume that $m \neq 0$ since that case is trivial.

If we call the absolute value of the $k+1$–st term a_k so

$$a_{k+1} = \left| \frac{m(m-1)\dots(m-k+1)}{k!} \right|,$$

then a simple calculation shows that for large values of k

$$\frac{a_{k+1}}{a_k} = 1 - \frac{m+1}{k}.$$

Here we are using the version a_{k+1}/a_k rather than the reciprocal; see the preceding note.

There are no higher-order terms to worry about in Gauss's test here and so the series $\sum a_k$ converges if $m+1 > 1$ and diverges if $m+1 < 1$. Thus the binomial series converges absolutely for $x = 1$ if $m > 0$. For $m = 0$ the series certainly converges since all terms except for the first one are identically zero. For $m < 0$ we know so far only that it does not converge absolutely. A closer analysis, for those who might care to try, will show that the series is nonabsolutely convergent for $-1 < m < 0$ and divergent for $m \le -1$. ◀

3.6.12 Alternating Series Test

We pass now to a number of tests that are needed for studying series of terms that may change signs. The simplest first step in studying a series $\sum_{i=1}^{\infty} a_i$, where the a_i are both negative and positive, is to apply one from our battery of tests to the series $\sum_{i=1}^{\infty} |a_i|$. If any test shows that this converges, then we know that our original series converges absolutely. This is even better than knowing it converges.

But what shall we do if the series is not absolutely convergent or if such attempts fail? One method applies to special series of positive and negative terms. Recall how we handled the series

$$\sum_{k=1}^{\infty} (-1)^{k-1} \frac{1}{k} = 1 - \frac{1}{2} + \frac{1}{3} - \frac{1}{4} \dots$$

(called the alternating harmonic series). We considered separately the partial sums $s_2, s_4, s_6, \dots$ and $s_1, s_3, s_5, \dots$. The special pattern of $+$ and $-$ signs alternating one after the other allowed us to see that each subsequence $\{s_{2n}\}$ and $\{s_{2n-1}\}$ was monotonic. All the features of this argument can be put into a test that applies to a wide class of series, similar to the alternating harmonic series.

3.43 (Alternating Series Test) *The series*

$$\sum_{k=1}^{\infty} (-1)^{k-1} a_k,$$

whose terms alternate in sign, converges if the sequence $\{a_k\}$ decreases monotonically to zero. Moreover, the value of the sum of such a series lies between the values of the partial sums at any two consecutive stages.

Proof The proof is just exactly the same as for the alternating harmonic series. Since the a_k are nonnegative and decrease, we compute that

$$a_1 - a_2 = s_2 \leq s_4 \leq s_6 \leq \cdots \leq s_5 \leq s_3 \leq s_1 = a_1.$$

These subsequences then form bounded monotonic sequences and so

$$\lim_{n\to\infty} s_{2n} \text{ and } \lim_{n\to\infty} s_{2n-1}$$

exist. Finally, since

$$s_{2n} - s_{2n-1} = -a_{2n} \to 0$$

we can conclude that $\lim_{n\to\infty} s_n = L$ exists. From the proof it is clear that the value L lies in each of the intervals $[s_2, s_1]$, $[s_2, s_3]$, $[s_4, s_3]$, $[s_4, s_5]$, ... and so, as stated, the sum of the series lies between the values of the partial sums at any two consecutive stages. ■

3.6.13 Dirichlet's Test

✂ *Advanced*

Our next test derives from the summation by parts formula

$$\sum_{k=1}^{n} a_k b_k = s_1(b_1 - b_2) + s_2(b_2 - b_3) \cdots + s_{n-1}(b_{n-1} - b_n) + s_n b_n$$

that we discussed in Section 3.2. We can see that if there is some special information available about the sequences $\{s_n\}$ and $\{b_n\}$ here, then the convergence of the series $\sum_{k=1}^{n} a_k b_k$ can be proved. The test gives one possibility for this. The next section gives a different variant.

The test is named after Lejeune Dirichlet[1] (1805–1859) who is most famous for his work on Fourier series, in which this test plays an important role.

3.44 (Dirichlet Test) *If $\{b_n\}$ is a sequence decreasing to zero and the partial sums of the series $\sum_{k=1}^{\infty} a_k$ are bounded, then the series $\sum_{k=1}^{\infty} a_k b_k$ converges.*

---

[1]One of his contemporaries described him thus: "He is a rather tall, lanky-looking man, with moustache and beard about to turn grey with a somewhat harsh voice and rather deaf. He was unwashed, with his cup of coffee and cigar. One of his failings is forgetting time, he pulls his watch out, finds it past three, and runs out without even finishing the sentence." (From *http://www-history.mcs.st-and.ac.uk/history*.

Proof Write $s_n = \sum_{k=1}^{n} a_k$. By our assumptions on the series $\sum_{k=1}^{\infty} a_k$ there is a positive number M so that $|s_n| \leq M$ for all n. Let $\varepsilon > 0$ and choose N so large that $b_n < \varepsilon/(2M)$ if $n \geq N$.

The summation by parts formula shows that for $m > n \geq N$

$$\left|\sum_{k=n}^{m} a_k b_k\right| = |a_n b_n + a_{n+1} b_{n+1} \cdots + a_m b_m|$$

$$= |-s_{n-1} b_n + s_n(b_n - b_{n+1}) + \ldots s_{m-1}(b_{m-1} - b_m) + s_m b_m|$$

$$\leq |-s_{n-1} b_n| + |s_n(b_n - b_{n+1})| + \ldots |s_{m-1}(b_{m-1} - b_m)| + |s_m b_m|$$

$$\leq M(b_n + [b_n - b_m] + b_m) \leq 2M b_n < \varepsilon.$$

Notice that we have needed to use the fact that

$$b_{k-1} - b_k \geq 0$$

for each k. This is precisely the Cauchy criterion for the series $\sum_{k=1}^{\infty} a_k b_k$ and so we have proved convergence. ■

Example 3.45 The series

$$1 - \frac{1}{2} + \frac{1}{3} - \frac{1}{4} + \frac{1}{5} - \frac{1}{6} \cdots$$

converges by the alternating series test. What other pattern of + and − signs could we insert and still have convergence? Let $a_k = \pm 1$. If the partial sums

$$\sum_{k=1}^{n} a_k$$

remain bounded, then, by Dirichlet's test, the series

$$\sum_{k=1}^{n} \frac{a_k}{k}$$

must converge. Thus, for example, the pattern

$$+ - + + - - + - + + - - + - + + - - \ldots$$

would produce a convergent series (that is not alternating). ◀

3.6.14 Abel's Test

✂ *Advanced*

The next test is another variant on the same theme as the Dirichlet test. There the series $\sum_{k=1}^{\infty} a_k b_k$ was proved to be convergent by assuming a fairly weak fact for the series $\sum_{k=1}^{\infty} a_k$ (i.e., bounded partial sums) and a strong fact for $\{b_k\}$ (i.e., monotone convergence to 0). Here we strengthen the first and weaken the second.

3.46 (Abel Test) *If $\{b_n\}$ is a convergent monotone sequence and the series $\sum_{k=1}^{\infty} a_k$ is convergent, then the series $\sum_{k=1}^{\infty} a_k b_k$ converges.*

Proof Suppose first that b_k is decreasing to a limit B. Then $b_k - B$ decreases to zero. We can apply Dirichlet's test to the series

$$\sum_{k=1}^{\infty} a_k(b_k - B)$$

to obtain convergence, since if $\sum_{k=1}^{\infty} a_k$ is convergent, then it has a bounded sequence of partial sums.

But this allows us to express our series as the sum of two convergent series:

$$\sum_{k=1}^{\infty} a_k b_k = \sum_{k=1}^{\infty} a_k(b_k - B) + B \sum_{k=1}^{\infty} a_k.$$

If the sequence b_k is instead increasing to some limit then we can apply the first case proved to the series $-\sum_{k=1}^{\infty} a_k(-b_k)$. ■

Exercises

3.6.1 Let $\{a_n\}$ be a sequence of positive numbers. If $\lim_{n\to\infty} n^2 a_n = 0$, what (if anything) can be said about the series $\sum_{n=1}^{\infty} a_n$. If $\lim_{n\to\infty} n a_n = 0$, what (if anything) can be said about the series $\sum_{n=1}^{\infty} a_n$. (If we drop the assumption about the sequence $\{a_n\}$ being positive does anything change?)

3.6.2 Which of these series converge?

(a) $\displaystyle\sum_{n=1}^{\infty} \frac{n(n+1)}{(n+2)^2}$

(b) $\displaystyle\sum_{n=1}^{\infty} \frac{3n(n+1)(n+2)}{n^3\sqrt{n}}$

(c) $\displaystyle\sum_{n=2}^{\infty} \frac{1}{n^s \log n}$

(d) $\displaystyle\sum_{n=1}^{\infty} a^{1/n} - 1$

(e) $\displaystyle\sum_{n=2}^{\infty} \frac{1}{n(\log n)^t}$

(f) $\displaystyle\sum_{n=2}^{\infty} \frac{1}{n^s(\log n)^t}$

(g) $\displaystyle\sum_{n=1}^{\infty}\left(1-\frac{1}{n}\right)^{n^2}$

3.6.3 For what values of x do the following series converge?

(a) $\sum_{n=2}^{\infty}\frac{x^n}{\log n}$

(b) $\sum_{n=2}^{\infty}(\log n)x^n$

(c) $\sum_{n=1}^{\infty}e^{-nx}$

(d) $1+2x+\frac{3^2x^2}{2!}+\frac{4^3x^3}{3!}+\dots.$

3.6.4 Let a_k be a sequence of positive numbers and suppose that

$$\lim_{k\to\infty} ka_k = L.$$

What can you say about the convergence of the series $\sum_{k=1}^{\infty}a_k$ if $L=0$? What can you say if $L>0$?

3.6.5 Let $\{a_k\}$ be a sequence of nonnegative numbers. Consider the following conditions:

(a) $\limsup_{k\to\infty}\sqrt{k}a_k>0$

(b) $\limsup_{k\to\infty}\sqrt{k}a_k<\infty$

(c) $\liminf_{k\to\infty}\sqrt{k}a_k>0$

(d) $\liminf_{k\to\infty}\sqrt{k}a_k<\infty$

Which condition(s) imply convergence or divergence of the series $\sum_{k=1}^{\infty}a_k$? Supply proofs. Which conditions are inconclusive as to convergence or divergence? Supply examples.

3.6.6 Suppose that $\sum_{n=1}^{\infty}a_n$ is a convergent series of positive terms. Must the series $\sum_{n=1}^{\infty}\sqrt{a_n}$ also be convergent?

3.6.7 Give examples of series both convergent and divergent that illustrate that the ratio test is inconclusive when the limit of the ratios L is equal to 1.

3.6.8 Give examples of series both convergent and divergent that illustrate that the root test is inconclusive when the limit of the roots L is equal to 1.

3.6.9 Apply both the root test and the ratio test to the series

$$\alpha+\alpha\beta+\alpha^2\beta+\alpha^2\beta^2+\alpha^3\beta^2+\alpha^3\beta^3\dots$$

where α, β are positive real numbers.

3.6.10 Show that the limit comparison test applied to series with positive terms can be replaced by the following version. If

$$\limsup_{k\to\infty} \frac{a_k}{b_k} < \infty$$

and if $\sum_{k=1}^{\infty} b_k$ converges, then so does $\sum_{k=1}^{\infty} a_k$. If

$$\liminf_{k\to\infty} \frac{a_k}{c_k} > 0$$

and if $\sum_{k=1}^{\infty} c_k$ diverges, then so does $\sum_{k=1}^{\infty} a_k$.

3.6.11 Show that the ratio test can be replaced by the following version. Compute

$$\liminf_{k\to\infty} \frac{a_{k+1}}{a_k} = L \quad \text{and} \limsup_{k\to\infty} \frac{a_{k+1}}{a_k} = M.$$

(a) If $M < 1$, then the series $\sum_{k=1}^{\infty} a_k$ is convergent.

(b) If $L > 1$, then the series $\sum_{k=1}^{\infty} a_k$ is divergent; moreover, the terms $a_k \to \infty$.

(c) If $L \leq 1 \leq M$, then the series $\sum_{k=1}^{\infty} a_k$ may diverge or converge, the test being inconclusive.

3.6.12 Show that the root test can be replaced by the following version. Compute

$$\limsup_{k\to\infty} \sqrt[k]{a_k} = L.$$

(a) If $L < 1$, then the series $\sum_{k=1}^{\infty} a_k$ is convergent.

(b) If $L > 1$, then the series $\sum_{k=1}^{\infty} a_k$ is divergent; moreover, some subsequence of the terms $a_{k_j} \to \infty$.

(c) If $L = 1$, then the series $\sum_{k=1}^{\infty} a_k$ may diverge or converge, the test being inconclusive.

3.6.13 Show that for any sequence of positive numbers $\{a_k\}$

$$\liminf_{k\to\infty} \frac{a_{k+1}}{a_k} \leq \liminf_{k\to\infty} \sqrt[k]{a_k} \leq \limsup_{k\to\infty} \sqrt[k]{a_k} \leq \limsup_{k\to\infty} \frac{a_{k+1}}{a_k}.$$

What can you conclude about the relative effectiveness of the root and ratio tests?

3.6.14 Give examples of series for which one would clearly prefer to apply the root (ratio) test in preference to the ratio (root) test. How would you answer someone who claims that "Exercise 3.6.13 shows clearly that the ratio test is inferior and should be abandoned in favor of the root test?"

3.6.15 Let $\{a_n\}$ be a sequence of positive numbers and write

$$L_n = \frac{\log\left(\frac{1}{a_n}\right)}{\log n}.$$

Show that if $\liminf L_n > 1$, then $\sum a_n$ converges. Show that if $L_n \leq 1$ for all sufficiently large n, then $\sum a_n$ diverges.

3.6.16 Apply the test in Exercise 3.6.15 to obtain convergence or divergence of the following series (x is positive):

(a) $\sum_{n=2}^{\infty} x^{\log n}$

(b) $\sum_{n=2}^{\infty} x^{\log \log n}$

(c) $\sum_{n=2}^{\infty} (\log n)^{-\log n}$

3.6.17 Prove the alternating series test directly from the Cauchy criterion.

3.6.18 Determine for what values of p the series

$$\sum_{k=1}^{\infty} (-1)^{k-1} \frac{1}{k^p} = 1 - \frac{1}{2^p} + \frac{1}{3^p} - \frac{1}{4^p} \ldots$$

is absolutely convergent and for what values it is nonabsolutely convergent.

3.6.19 How many terms of the series

$$\sum_{k=1}^{\infty} \frac{(-1)^{k-1}}{k^2}$$

must be taken to obtain a value differing from the sum of the series by less than 10^{-10}?

3.6.20 If the sequence $\{x_n\}$ is monotonically decreasing to zero then prove that the series

$$x_1 - \frac{1}{2}(x_1 + x_2) + \frac{1}{3}(x_1 + x_2 + x_3) - \frac{1}{4}(x_1 + x_2 + x_3 + x_4) \ldots$$

converges.

3.6.21 This exercise attempts to squeeze a little more information out of the integral test. In the notation of that test consider the sequence

$$e_n = \sum_{k=1}^{n} f(k) - \int_1^{n+1} f(x)\, dx$$

Show that the sequence $\{e_n\}$ is increasing and that $0 \leq e_n \leq f(1)$. What is the exact relation between $\sum_{k=1}^{\infty} f(k)$ and $\int_1^{\infty} f(x)\, dx$?

3.6.22 Show that

$$\lim_{n \to \infty} \left(\sum_{k=1}^{n} \frac{1}{k} - \int_1^{n+1} \frac{1}{x}\, dx \right) = \gamma$$

for some number γ, $.5 < \gamma < 1$.

3.6.23 Show that

$$\lim_{n \to \infty} \sum_{k=n+1}^{2n} \frac{1}{k} = \log 2.$$

3.6.24 Let F be a positive function on $[1, \infty)$ with a positive, decreasing and continuous derivative F'.

(a) Show that $\sum_{k=1}^{\infty} F'(k)$ converges if and only if

$$\sum_{k=1}^{\infty} \frac{F'(k)}{F(k)}$$

converges.

(b) Suppose that $\sum_{k=1}^{\infty} F'(k)$ diverges. Show that

$$\sum_{k=1}^{\infty} \frac{F'(k)}{[F(k)]^p}$$

converges if and only if $p > 1$.

3.6.25 This collection of exercises develops some convergence properties of *power series*; that is, series of the form

$$\sum_{k=0}^{\infty} a_k x^k = a_0 + a_1 x + a_2 x^2 + a_3 x^3 + \dots .$$

A full treatment of power series appears in Chapter 10.

(a) Show that if a power series converges absolutely for some value $x = x_0$ then the series converges absolutely for all $|x| \leq |x_0|$.

(b) Show that if a power series converges for some value $x = x_0$ then the series converges absolutely for all $|x| < |x_0|$.

(c) Let

$$R = \sup\{t : \sum_{k=0}^{\infty} a_k t^k \text{ converges }\}.$$

Show that the power series $\sum_{k=0}^{\infty} a_k x^k$ must converge absolutely for all $|x| < R$ and diverge for all $|x| > R$. [The number R is called the *radius of convergence* of the series. The explanation for the word "radius" (which conjures up images of circles) is that for complex series the set of convergence is a disk.]

(d) Give examples of power series with radius of convergence 0, ∞, 1, 2, and $\sqrt{2}$.

(e) Explain how the radius of convergence of a power series may be computed with the help of the ratio test.

(f) Explain how the radius of convergence of a power series may be computed with the help of the root test.

✂ (g) Establish the formula

$$R = \frac{1}{\limsup_{k \to \infty} \sqrt[k]{|a_k|}}$$

for the radius of convergence of the power series $\sum_{k=0}^{\infty} a_k x^k$.

(h) Give examples of power series $\sum_{k=0}^{\infty} a_k x^k$ with radius of convergence R so that the series converges absolutely at both endpoints of the interval $[-R, R]$. Give another example so that the series converges at the right-hand endpoint but diverges at the left-hand endpoint of $[-R, R]$. What other possibilities are there?

3.6.26 The series

$$1 + mx + \frac{m(m-1)}{2!}x^2 +$$
$$\frac{m(m-1)(m-2)}{3!}x^3 + \frac{m(m-1)\dots(m-k+1)}{k!}x^k + \dots$$

is called the *binomial series.* Here m is any real number. (See Example 3.42.)

(a) Show that if m is a positive integer then this is precisely the expansion of $(1+x)^m$ by the binomial theorem.

(b) Show that this series converges absolutely for any m and for all $|x| < 1$.

(c) Obtain convergence for $x = 1$ if $m > -1$.

(d) Obtain convergence for $x = -1$ if $m > 0$.

3.7 Rearrangements

✂ *Enrichment*

Any finite sum may be rearranged and summed in any order. This is because addition is commutative. We might expect the same to occur for series. We add up a series $\sum_{k=1}^{\infty} a_k$ by starting at the first term and adding in the order presented to us. If the terms are rearranged into a different order do we get the same result?

Example 3.47 The most famous example of a series that cannot be freely rearranged without changing the sum is the alternating harmonic series. We know that the series

$$1 - \frac{1}{2} + \frac{1}{3} - \frac{1}{4} \dots$$

is convergent (actually nonabsolutely convergent) with a sum somewhere between 1/2 and 1. If we rearrange this so that every positive term is followed by two negative terms, thus,

$$1 - \frac{1}{2} - \frac{1}{4} + \frac{1}{3} - \frac{1}{6} - \frac{1}{8} + \frac{1}{5} - \frac{1}{10} - \frac{1}{12} \dots$$

we shall arrive at a different sum. Grouping these and adding, we obtain

$$\left(1 - \frac{1}{2}\right) - \frac{1}{4} + \left(\frac{1}{3} - \frac{1}{6}\right) - \frac{1}{8} + \left(\frac{1}{5} - \frac{1}{10}\right) - \frac{1}{12} \dots$$

$$= \frac{1}{2}\left(1 - \frac{1}{2} + \frac{1}{3} - \frac{1}{4} \dots\right)$$

whose sum is half the original series. Rearranging the series has changed the sum! ◀

For the theory of unordered sums there is no such problem. If an unordered sum $\sum_{j\in J} a_j$ converges to a number c, then so too does any rearrangement. Exercise 3.3.8 shows that if $\sigma : I \to I$ is one-to-one and onto, then

$$\sum_{i\in I} a_j = \sum_{i\in I} a_{\sigma(i)}.$$

We had hoped for the same situation for series. If $\sigma : \mathbb{N} \to \mathbb{N}$ is one-to-one and onto, then

$$\sum_{k=1}^{\infty} a_k = \sum_{k=1}^{\infty} a_{\sigma(k)}$$

may or may not hold. We call $\sum_{k=1}^{\infty} a_{\sigma(k)}$ a *rearrangement* of the series $\sum_{k=1}^{\infty} a_k$.

We propose now to characterize those series that allow unlimited rearrangements, and those that are more fragile (as is the alternating harmonic series) and cannot permit rearrangement.

3.7.1 Unconditional Convergence

A series is said to be *unconditionally convergent* if all rearrangements of that series converge and have the same sum. Those series that do not allow this but do converge are called *conditionally convergent.* Here the "conditional" means that the series converges in the arrangement given, but may diverge in another arrangement or may converge to a different sum in another arrangement. We shall see that conditionally convergent series are extremely fragile; there are rearrangements that exhibit any behavior desired. There are rearrangements that diverge and there are rearrangements that converge to any desired number.

Our first theorem asserts that any absolutely convergent series may be freely rearranged. All absolutely convergent series are unconditionally convergent. In fact, the two terms are equivalent

unconditionally convergent $\Leftrightarrow$ *absolutely convergent*

although we must wait until the next section to prove that.

Theorem 3.48 (Dirichlet) *Every absolutely convergent series is unconditionally convergent.*

Proof Let us prove this first for series $\sum_{k=1}^{\infty} a_k$ whose terms are all nonnegative. For such series convergence and absolute convergence mean the same thing.

Let $\sum_{k=1}^{\infty} a_{\sigma(k)}$ be any rearrangement. Then for any M

$$\sum_{k=1}^{M} a_{\sigma(k)} \leq \sum_{k=1}^{N} a_k \leq \sum_{k=1}^{\infty} a_k$$

by choosing an N large enough so that $\{1, 2, 3, \dots, N\}$ includes all the integers $\{\sigma(1), \sigma(2), \sigma(3), \dots, \sigma(M)\}$. By the bounded partial sums criterion this shows that $\sum_{k=1}^{\infty} a_{\sigma(k)}$ is convergent and to a sum smaller than $\sum_{k=1}^{\infty} a_k$. But this same argument would show that $\sum_{k=1}^{\infty} a_k$ is convergent and to a sum smaller than $\sum_{k=1}^{\infty} a_{\sigma(k)}$ and consequently all rearrangements converge to the same sum.

We now allow the series $\sum_{k=1}^{\infty} a_k$ to have positive and negative values. Write

$$\sum_{k=1}^{\infty} a_k = \sum_{k=1}^{\infty} [a_k]^+ - \sum_{k=1}^{\infty} [a_k]^-$$

(cf. Exercise 3.5.8) where we are using the notation

$$[X]^+ = \max\{X, 0\} \text{ and } [X]^- = \max\{-X, 0\}$$

and remembering that

$$X = [X]^+ - [X]^- \text{ and } |X| = [X]^+ + [X]^-.$$

Any rearrangement of the series on the left-hand side of this identity just results in a rearrangement in the two series of nonnegative terms on the right. We have just seen that this does nothing to alter the convergence or the sum. Consequently, any rearrangement of our series will have the same sum as required to prove the assertion of the theorem. ■

3.7.2 Conditional Convergence

A convergent series is said to be *conditionally convergent* if it is not unconditionally convergent. Thus such a series converges in the arrangement given, but either there is some rearrangement that diverges or else there is some rearrangement that has a different sum. In fact, both situations always occur.

We have already seen (Example 3.47) how the alternating harmonic series can be rearranged to have a different sum. We shall show that any nonabsolutely convergent series has this property. Our previous rearrangement took advantage of the special nature of the series; here our proof must be completely general and so the method is different.

The following theorem completes Theorem 3.48 and provides the connections:

conditionally convergent $\Leftrightarrow$ *nonabsolutely convergent*

and

unconditionally convergent $\Leftrightarrow$ *absolutely convergent*

Note. You may wonder why we have needed this extra terminology if these concepts are identical. One reason is to emphasize that this is part of the theory. Conditional convergence and nonabsolutely convergence may be equivalent, but they have different underlying meanings. Also, this terminology is used for series of other objects than real numbers and for series of this more general type the terms are not equivalent.

Theorem 3.49 (Riemann) *Every nonabsolutely convergent series is conditionally convergent. In fact, every nonabsolutely convergent series has a divergent rearrangement and can also be rearranged to sum to any preassigned value.*

Proof Let $\sum_{k=1}^{\infty} a_k$ be an arbitrary nonabsolutely convergent series. To prove the first statement it is enough if we observe that both series

$$\sum_{k=1}^{\infty}[a_k]^+ \text{ and } \sum_{k=1}^{\infty}[a_k]^-$$

must diverge in order for $\sum_{k=1}^{\infty} a_k$ to be nonabsolutely convergent. We need to observe as well that $a_k \to 0$ since the series is assumed to be convergent.

Write p_1, p_2, p_3, for the sequence of positive numbers in the sequence $\{a_k\}$ (skipping any zero or negative ones) and write q_1, q_2, q_3, $\dots$ for the sequence of terms that we have skipped. We construct a new series

$$p_1 + p_2 + \cdots + p_{n_1} + q_1 + p_{n_1+1} + p_{n_1+2} + \cdots + p_{n_2} + q_2 + p_{n_2+1} \cdots$$

where we have chosen $0 = n_0 < n_1 < n_2 < n_3 \dots$ so that

$$p_{n_k+1} + p_{n_1+2} + \cdots + p_{n_{k+1}} > 2^k$$

for each $k = 0, 1, 2, \dots$. Since $\sum_{k=1}^{\infty} p_k$ diverges, this is possible. The new series so constructed contains all the terms of our original series and so is a rearrangement. Since the terms $q_k \to 0$, they will not interfere with the goal of producing ever larger partial sums for the new series and so, evidently, this new series diverges to ∞.

The second requirement of the theorem is to produce a convergent rearrangement, convergent to a given number α. We proceed in much the same way but with rather more caution. We leave this to the exercises. ■

3.7.3 Comparison of $\sum_{i=1}^{\infty} a_i$ and $\sum_{i\in\mathbb{N}} a_i$

✂ *Advanced*

The unordered sum of a sequence of real numbers, written as,

$$\sum_{i\in\mathbb{N}} a_i,$$

has an apparent connection with the ordered sum

$$\sum_{i=1}^{\infty} a_i.$$

We should expect the two to be the same when both converge, but is it possible that one converges and not the other?

The answer is that the convergence of $\sum_{i\in\mathbb{N}} a_i$ is equivalent to the *absolute* convergence of $\sum_{i=1}^{\infty} a_i$.

Theorem 3.50 *A necessary and sufficient condition for $\sum_{i\in\mathbb{N}} a_i$ to converge is that the series $\sum_{i=1}^{\infty} a_i$ is absolutely convergent and in this case*

$$\sum_{i\in\mathbb{N}} a_i = \sum_{i=1}^{\infty} a_i.$$

Proof We shall use a device we have seen before a few times: For any real number X write

$$[X]^+ = \max\{X, 0\} \text{ and } [X]^- = \max\{-X, 0\}$$

and note that

$$X = [X]^+ - [X]^- \text{ and } |X| = [X]^+ + [X]^-.$$

The absolute convergence of the series and the convergence of the sum in the statement in the theorem now reduce to considering the equality of the right-hand sides of

$$\sum_{i\in\mathbb{N}} a_i = \sum_{i\in\mathbb{N}} [a_i]^+ - \sum_{i\in\mathbb{N}} [a_i]^-$$

and

$$\sum_{i=1}^{\infty} a_i = \sum_{i=1}^{\infty} [a_i]^+ - \sum_{i=1}^{\infty} [a_i]^-.$$

This reduces our problem to considering just nonnegative series (sums).

Thus we may assume that each $a_i \geq 0$. For any finite set $I \subset \mathbb{N}$ it is clear that

$$\sum_{i\in I} a_i \leq \sum_{i=1}^{\infty} a_i.$$

It follows that if $\sum_{i=1}^{\infty} a_i$ converges, then (by Exercise 3.3.3) so too does $\sum_{i\in\mathbb{N}} a_i$ and

$$\sum_{i\in\mathbb{N}} a_i \leq \sum_{i=1}^{\infty} a_i. \tag{6}$$

Similarly, if N is finite,

$$\sum_{i=1}^{N} a_i \leq \sum_{i\in\mathbb{N}} a_i.$$

It follows that if $\sum_{i\in\mathbb{N}} a_i$ converges, then, by the boundedness criterion, so too does $\sum_{i=1}^{\infty} a_i$ and

$$\sum_{i=1}^{\infty} a_i \leq \sum_{i\in\mathbb{N}} a_i. \tag{7}$$

Together these two assertions and the equations (6) and (7) prove the theorem for the case of nonnegative series (sums). ■

Exercises

3.7.1 Let

$$s = 1 - \frac{1}{2} + \frac{1}{3} - \frac{1}{4} \dots.$$

Show that

$$\frac{3s}{2} = 1 + \frac{1}{3} - \frac{1}{2} + \frac{1}{5} + \frac{1}{7} - \frac{1}{4} \dots.$$

3.7.2 For what values of x does the following series converge and what is the sum?

$$1 + x^2 + x + x^4 + x^6 + x^3 + x^8 + x^{10} + x^5 + \dots$$

3.7.3 For what series is the computation

$$\sum_{k=1}^{\infty} a_k = \sum_{k=1}^{\infty} a_{2k} + \sum_{k=1}^{\infty} a_{2k-1}$$

valid? Is this a rearrangement?

3.7.4 For what series is the computation

$$\sum_{k=1}^{\infty} a_k = \sum_{k=1}^{\infty} (a_{2k} + a_{2k-1})$$

valid? Is this a rearrangement?

3.7.5 For what series is the computation

$$\sum_{k=1}^{\infty} a_k = a_2 + a_1 + a_4 + a_3 + a_6 + a_5 + \dots$$

valid? Is this a rearrangement?

3.7.6 Give an example of an absolutely convergent series for which is it much easier to compute the sum by rearrangement than otherwise.

3.7.7 For what values of α and β does the series

$$\frac{\alpha}{1} - \frac{\beta}{2} + \frac{\alpha}{3} - \frac{\beta}{4} \dots$$

converge?

3.7.8 Let a series be altered by the insertion of zero terms in a completely arbitrary manner. Does this alter the convergence of the series?

3.7.9 Suppose that a convergent series contains only finitely many negative terms. Can it be safely rearranged?

3.7.10 Suppose that a nonabsolutely convergent series has been rearranged and that this rearrangement converges. Does this rearranged series converge absolutely or nonabsolutely?

3.7.11 Is there a divergent series that can be rearranged so as to converge? Can *every* divergent series be rearranged so as to converge? If $\sum_{k=1}^{\infty} a_k$ diverges, but does not diverge to ∞ or $-\infty$, can it be rearranged to diverge to ∞?

3.7.12 How many rearrangements of a nonabsolutely convergent series are there that do not alter the sum?

3.7.13 Complete the proof of Theorem 3.49 by showing that for any nonabsolutely convergent series series $\sum_{k=1}^{\infty} a_k$ and any α there is a rearrangement of the series so that

$$\sum_{k=1}^{\infty} a_{\sigma(k)} = \alpha.$$

3.7.14 Improve Theorem 3.49 by showing that for any nonabsolutely convergent series series $\sum_{k=1}^{\infty} a_k$ and any

$$-\infty \leq \alpha \leq \beta \leq \infty$$

there is a rearrangement of the series so that

$$\alpha = \liminf_{n\to\infty} \sum_{k=1}^{n} a_{\sigma(k)} \leq \limsup_{n\to\infty} \sum_{k=1}^{n} a_{\sigma(k)} = \beta.$$

3.8 Products of Series

✂ *Enrichment*

The rule for the sum of two convergent series[2] in Theorem 3.8

$$\sum_{k=0}^{\infty}(a_k + b_k) = \sum_{k=0}^{\infty} a_k + \sum_{k=0}^{\infty} b_k$$

[2]In the formula for a product of series in this section we prefer to label the series starting with 0. This does not change the series in any way.

is entirely elementary to prove and comes directly from the rule for limits of sums of sequences. If A_n and B_n represent the sum of $n+1$ terms of the two series, then

$$\lim_{n\to\infty} \sum_{k=0}^{\infty}(a_k + b_k) = \lim_{n\to\infty}(A_n + B_n) = \lim_{n\to\infty} A_n + \lim_{n\to\infty} B_n$$

$$= \sum_{k=0}^{\infty} a_k + \sum_{k=0}^{\infty} b_k.$$

At first glance we might expect to have a similar rule for products of series, since

$$\lim_{n\to\infty}(A_n \times B_n) = \lim_{n\to\infty} A_n \times \lim_{n\to\infty} B_n$$

$$= \sum_{k=0}^{\infty} a_k \times \sum_{k=0}^{\infty} b_k.$$

But what is A_nB_n? If we write out this product we obtain

$$A_nB_n = (a_0 + a_1 + a_2 + \cdots + a_n)(b_0 + b_1 + b_2 + \cdots + b_n)$$

$$= \sum_{i=0}^{n}\sum_{j=1}^{n} a_i b_j.$$

From this all we can show is the curious observation that

$$\lim_{n\to\infty} \sum_{i=0}^{n}\sum_{j=1}^{n} a_i b_j = \sum_{k=0}^{\infty} a_k \times \sum_{k=0}^{\infty} b_k.$$

What we would rather see here is a result similar to the rule for sums:

"series + series = series."

Can this result be interpreted as

"series × series = series?"

We need a systematic way of adding up the terms a_ib_j in the double sum so as to form a series. The terms are displayed in a rectangular array in Figure 3.3.

If we replace the series here by a power series, this systematic way will become much clearer. How should we add up

$$\left(a_0 + a_1x + a_2x^2 + \cdots + a_nx^n\right)\left(b_0 + b_1x + b_2x^2 + \cdots + b_nx^n\right)$$

$\times$	a_0	a_1	a_2	a_3	a_4	a_5	$\ldots$
b_0	a_0b_0	a_1b_0	a_2b_0	a_3b_0	a_4b_0	a_5b_0	$\ldots$
b_1	a_0b_1	a_1b_1	a_2b_1	a_3b_1	a_4b_1	a_5b_1	$\ldots$
b_2	a_0b_2	a_1b_2	a_2b_2	a_3b_2	a_4b_2	a_5b_2	$\ldots$
b_3	a_0b_3	a_1b_3	a_2b_3	a_3b_3	a_4b_3	a_5b_3	$\ldots$
b_4	a_0b_4	a_1b_4	a_2b_4	a_3b_4	a_4b_4	a_5b_4	$\ldots$
b_5	a_0b_5	a_1b_5	a_2b_5	a_3b_5	a_4b_5	a_5b_5	$\ldots$
$\ldots$	$\ldots$	$\ldots$	$\ldots$	$\ldots$	$\ldots$	$\ldots$	

Figure 3.3. The product of the two series $\sum_0^\infty a_k$ and $\sum_0^\infty b_k$.

(which with $x = 1$ is the same question we just asked)? The now obvious answer is

$$a_0b_0 + (a_0b_1 + a_1b_0)x + (a_0b_2 + a_1b_1 + a_2b_0)x^2$$
$$+(a_0b_3 + a_1b_2 + a_2b_1 + a_3b_0)x^3 + \ldots.$$

Notice that this method of grouping the terms corresponds to summing along diagonals of the rectangle in Figure 3.3.

This is the source of the following definition.

Definition 3.51 The series

$$\sum_{k=0}^\infty c_k$$

is called the *formal product* of the two series

$$\sum_{k=0}^\infty a_k \text{ and } \sum_{k=0}^\infty b_k$$

provided that

$$c_k = \sum_{i=0}^k a_i b_{k-i}.$$

Our main goal now is to determine if this "formal" product is in any way a genuine product; that is, if

$$\sum_{k=0}^\infty c_k = \sum_{k=0}^\infty a_k \times \sum_{k=0}^\infty b_k.$$

The reason we expect this might be the case is that the series $\sum_{k=0}^\infty c_k$ contains all the terms in the expansion of

$$(a_0 + a_1 + a_2 + a_3 + \ldots)(b_0 + b_1 + b_2 + b_3 + \ldots).$$

A good reason for caution, however, is that the series $\sum_{k=0}^{\infty} c_k$ contains these terms only in a particular arrangement and we know that series can be sensitive to rearrangement.

3.8.1 Products of Absolutely Convergent Series

It is a general rule in the study of series that absolutely convergent series permit the best theorems. We can rearrange such series freely as we have seen already in Section 3.7.1. Now we show that we can form products of such series. We shall have to be much more cautious about forming products of nonabsolutely convergent series.

Theorem 3.52 (Cauchy) *Suppose that $\sum_{k=0}^{\infty} c_k$ is the formal product of two absolutely convergent series*

$$\sum_{k=0}^{\infty} a_k \text{ and } \sum_{k=0}^{\infty} b_k.$$

Then $\sum_{k=0}^{\infty} c_k$ converges absolutely too and

$$\sum_{k=0}^{\infty} c_k = \sum_{k=0}^{\infty} a_k \times \sum_{k=0}^{\infty} b_k.$$

Proof We write

$$A = \sum_{k=0}^{\infty} a_k,\ A' = \sum_{k=0}^{\infty} |a_k|,\ A_n = \sum_{k=0}^{n} a_k,$$

$$B = \sum_{k=0}^{\infty} b_k,\ B' = \sum_{k=0}^{\infty} |b_k|, \text{ and } B_n = \sum_{k=0}^{n} b_k.$$

By definition

$$c_k = \sum_{i=0}^{k} a_i b_{k-i}$$

and so

$$\sum_{k=0}^{N} |c_k| \le \sum_{k=0}^{N} \sum_{i=0}^{k} |a_i| \cdot |b_{k-i}| \le \left(\sum_{i=0}^{N} |a_i| \right) \left(\sum_{i=0}^{N} |b_i| \right) \le A'B'.$$

Since the latter two series converge, this provides an upper bound $A'B'$ for the sequence of partial sums $\sum_{k=1}^{N} |c_k|$ and hence the series $\sum_{k=0}^{\infty} c_k$ converges absolutely.

Let us recall that the formal product of the two series is just a particular rearrangement of the terms $a_i b_j$ taken over all $i \ge 0$, $j \ge 0$. Consider any arrangement of these terms. This must form an absolutely convergent series by the same argument as before since $A'B'$ will be an upper bound for the partial sums of the absolute values $|a_i b_j|$. Thus all rearrangements will converge to the same value by Theorem 3.48.

We can rearrange the terms $a_i b_j$ taken over all $i \geq 0$, $j \geq 0$ in the following convenient way "by squares." Arrange always so that the first $(m+1)^2$ $(m = 0, 1, 2, \dots)$ terms add up to $A_m B_m$. For example, one such arrangement starts off

$$a_0 b_0 + a_1 b_0 + a_0 b_1 + a_1 b_1 + a_2 b_0 + a_2 b_1 + a_0 b_2 + a_1 b_2 + a_2 b_2 + \dots.$$

(A picture helps considerably to see the pattern needed.) We know this arrangement converges and we know it must converge to

$$\lim_{m \to \infty} A_m B_m = AB.$$

In particular, the series $\sum_{k=0}^{\infty} c_k$ which is just another arrangement, converges to the same number AB as required. ■

It is possible to improve this theorem to allow one (but not both) of the series to converge nonabsolutely. The conclusion is that the product then converges (perhaps nonabsolutely), but different methods of proof will be needed. As usual, nonabsolutely convergent series are much more fragile, and the free and easy moving about of the terms in this proof is not allowed.

3.8.2 Products of Nonabsolutely Convergent Series

Let us give a famous example, due to Cauchy, of a pair of convergent series whose product diverges. We know that the alternating series

$$\sum_{k=0}^{\infty} (-1)^k \frac{1}{\sqrt{k+1}}$$

is convergent, but not absolutely convergent since the related absolute series is a p-harmonic series with $p = \frac{1}{2}$.

Let

$$\sum_{k=0}^{\infty} c_k$$

be the formal product of this series with itself. By definition the term c_k is given by

$$(-1)^k \left[\frac{1}{\sqrt{1 \cdot (k+1)}} + \frac{1}{\sqrt{2 \cdot (k)}} + \frac{1}{\sqrt{3 \cdot (k-1)}} \dots + \frac{1}{\sqrt{(k+1) \cdot 1}} \right].$$

There are $k+1$ terms in the sum for c_k and each term is larger than $1/(k+1)$ so we see that $|c_k| \geq 1$. Since the terms of the product series $\sum_{k=0}^{\infty} c_k$ do not tend to zero, this is a divergent series.

This example supplies our observation: The formal product of two nonabsolutely convergent series need not converge. In particular, there may be no convergent series to represent the product

$$\sum_{k=0}^{\infty} a_k \times \sum_{k=0}^{\infty} b_k$$

for a pair of nonabsolutely convergent series. For absolutely convergent series the product always converges.

We should not be too surprised at this result. The theory begins to paint the following picture: Absolutely convergent series can be freely manipulated in most ways and nonabsolutely convergent series can hardly be manipulated in general in any serious manner. Interestingly, the following theorem can be proved that shows that even though, in general, the product might diverge, in cases where it does converge it converges to the "correct" value.

Theorem 3.53 (Abel) *Suppose that $\sum_{k=0}^{\infty} c_k$ is the formal product of two nonabsolutely convergent series $\sum_{k=0}^{\infty} a_k$ and $\sum_{k=0}^{\infty} b_k$ and suppose that this product $\sum_{k=0}^{\infty} c_k$ is known to converge. Then*

$$\sum_{k=0}^{\infty} c_k = \sum_{k=0}^{\infty} a_k \times \sum_{k=0}^{\infty} b_k.$$

Proof The proof requires more technical apparatus and will not be given until Section 3.9.2. ■

Exercises

3.8.1 Form the product of the series $\sum_{k=0}^{\infty} a_k x^k$ with the geometric series

$$\frac{1}{1-x} = 1 + x + x^2 + x^3 + \dots$$

and obtain the formula

$$\frac{1}{1-x} \sum_{k=0}^{\infty} a_k x^k = \sum_{k=0}^{\infty} (a_0 + a_1 + a_2 + \cdots + a_k) x^k.$$

For what values of x would this be valid?

3.8.2 Show that

$$(1-x)^2 = \sum_{k=0}^{\infty} (k+1) x^k$$

for appropriate values of x.

3.8.3 Using the fact that

$$\sum_{k=0}^{\infty} \frac{(-1)^k}{k+1} = \log 2,$$

show that

$$\sum_{k=0}^{\infty} \frac{(-1)^k \sigma_k}{k+2} = \frac{(\log 2)^2}{2}$$

where $\sigma_k = 1 + 1/2 + 1/3 + \ldots 1/(k+1)$.

3.8.4 Verify that $e^{x+y} = e^x e^y$ by proving that

$$\sum_{k=0}^{\infty} \frac{(x+y)^k}{k!} = \sum_{k=0}^{\infty} \frac{x^k}{k!} \sum_{k=0}^{\infty} \frac{y^k}{k!}.$$

3.8.5 For what values of p and q are you able to establish the convergence of the product of the two series

$$\sum_{k=0}^{\infty} \frac{(-1)^k}{(k+1)^p} \quad \text{and} \quad \sum_{k=0}^{\infty} \frac{(-1)^k}{(k+1)^q}?$$

3.9 Summability Methods

✂ *Advanced*

A first course in series methods often gives the impression of being obsessed with the issue of convergence or divergence of a series. The huge battery of tests in Section 3.6 devoted to determining the behavior of series might lead one to this conclusion. Accordingly, you may have decided that convergent series are useful and proper tools of analysis while divergent series are useless and without merit.

In fact divergent series are, in many instances, as important or more important than convergent ones. Many eighteenth century mathematicians achieved spectacular results with divergent series but without a proper understanding of what they were doing. The initial reaction of our founders of nineteenth-century analysis (Cauchy, Abel, and others) was that valid arguments could be based only on convergent series. Divergent series should be shunned. They were appalled at reasoning such as the following: The series

$$s = 1 - 1 + 1 - 1 \ldots$$

can be summed by noting that

$$s = 1 - (1 - 1 + 1 - \ldots) = 1 - s$$

and so $2s = 1$ or $s = \frac{1}{2}$. But the sum $\frac{1}{2}$ proves to be a useful value for the "sum" of this series even though the series is clearly divergent.

There are many useful ways of doing rigorous work with divergent series. One way, which we now study, is the development of *summability methods*.

Suppose that a series $\sum_{k=0}^{\infty} a_k$ diverges and yet we wish to assign a "sum" to it by some method. Our standard method thus far is to take the limit of the sequence of partial sums. We write

$$s_n = \sum_{k=0}^{n} a_k$$

and the sum of the series is $\lim_{n\to\infty} s_n$. If the series diverges, this means precisely that this sequence does not have a limit. How can we use that sequence or that series nonetheless to assign a different meaning to the sum?

3.9.1 Cesàro's Method

Advanced

An infinite series $\sum_{k=0}^{\infty} a_k$ has a sum S if the sequence of partial sums

$$s_n = \sum_{k=0}^{n} a_k$$

converges to S. If the sequence of partial sums diverges, then we must assign a sum by a different method. We will still say that the series diverges but, nonetheless, we will be able to find a number that can be considered the sum.

We can replace $\lim_{n\to\infty} s_n$, which perhaps does not exist, by

$$\lim_{n\to\infty} \frac{s_0 + s_1 + s_2 + \cdots + s_n}{n+1} = C$$

if this exists and use this value for the sum of the series. This is an entirely natural method since it merely takes averages and settles for computing a kind of "average" limit where an actual limit might fail to exist.

For a series $\sum_{k=0}^{\infty} a_k$ often we can use this method to obtain a sum even when the series diverges.

Definition 3.54 If $\{s_n\}$ is the sequence of partial sums of the series $\sum_{k=0}^{\infty} a_k$ and

$$\lim_{n\to\infty} \frac{s_0 + s_1 + s_2 + \cdots + s_n}{n+1} = C$$

then the new sequence

$$\sigma_n = \frac{s_0 + s_1 + s_2 + \cdots + s_n}{n+1}$$

is called the sequence of *averages* or *Cesàro means* and we write

$$\sum_{k=0}^{\infty} a_k = C \quad \text{[Cesàro]}.$$

Thus the symbol [Cesàro] indicates that the value is obtained by this method rather than by the usual method of summation (taking limits of partial sums). The method is named after Ernesto Cesàro (1859–1906).

Our first concern in studying a summability method is to determine whether it assigns the "correct" value to a series that already converges. Does

$$\sum_{k=0}^{\infty} a_k = A \Rightarrow \sum_{k=0}^{\infty} a_k = A \text{ [Cesàro]?}$$

Any method of summing a series is said to be *regular* or a *regular summability method* if this is the case.

Theorem 3.55 *Suppose that a series $\sum_{k=0}^{\infty} a_k$ converges to a value A. Then $\sum_{k=0}^{\infty} a_k = A$ [Cesàro] is also true.*

Proof This is an immediate consequence of Exercise 2.13.17. For any sequence $\{s_n\}$ write

$$\sigma_n = \frac{s_1 + s_2 + \dots s_n}{n}.$$

In that exercise we showed that

$$\liminf_{n\to\infty} s_n \leq \liminf_{n\to\infty} \sigma_n \leq \limsup_{n\to\infty} \sigma_n \leq \limsup_{n\to\infty} s_n.$$

If you skipped that exercise, here is how to prove it. Let

$$\beta > \limsup_{n\to\infty} s_n.$$

(If there is no such β, then $\limsup_{n\to\infty} s_n = \infty$ and there is nothing to prove.) Then $s_n < \beta$ for all $n \geq N$ for some N. Thus

$$\sigma_n \leq \frac{1}{n}(s_1 + s_2 + \dots s_{N-1}) + \frac{(n-N+1)\beta}{n}$$

for all $n \geq N$. Fix N, allow $n \to \infty$, and take limit superiors of each side to obtain

$$\limsup_{n\to\infty} \sigma_n \leq \beta.$$

It follows that

$$\limsup_{n\to\infty} \sigma_n \leq \limsup_{n\to\infty} s_n.$$

The other inequality is similar. In particular, if $\lim_{n\to\infty} s_n$ exists so too does $\lim_{n\to\infty} \sigma_n$ and they are equal, proving the theorem. ∎

Example 3.56 As an example let us sum the series

$$1 - 1 + 1 - 1 + 1 - 1 \dots.$$

The partial sums form the sequence 1, 0, 1, 0, ... , which evidently diverges. Indeed the series diverges merely by the trivial test: The terms do not tend to zero. Can we sum this series by the Cesàro summability method? The averages of the sequence of partial sums is clearly tending to $\frac{1}{2}$. Thus we can write

$$\sum_{k=0}^{\infty}(-1)^k = \frac{1}{2} \text{ [Cesàro]}$$

even though the series is divergent. ◀

3.9.2 Abel's Method

✂ *Advanced*

We require in this section that you recall some calculus limits. We shall need to compute a limit

$$\lim_{x\to 1-} F(x)$$

for a function F defined on $(0, 1)$ where the expression $x \to 1-$ indicates a left-hand limit. In Chapter 5 we present a full account of such limits; here we need remember only what this means and how it is computed.

Suppose that a series $\sum_{k=0}^{\infty} a_k$ diverges and yet we wish to assign a "sum" to it by some other method. If the terms of the series do not get too large, then the series

$$F(x) = \sum_{k=0}^{\infty} a_k x^k$$

will converge (by the ratio test) for all $0 \leq x < 1$. The value we wish for the sum of the series would appear to be $F(1)$, but for a divergent series inserting the value 1 for x gives us nothing we can use. Instead we compute

$$\lim_{x\to 1-} F(x) = \lim_{x\to 1-} \sum_{k=0}^{\infty} a_k x^k = A$$

and use this value for the sum of the series.

Definition 3.57 We write

$$\sum_{k=0}^{\infty} a_k x^k = A \text{ [Abel]}$$

if

$$\lim_{x\to 1-} \sum_{k=0}^{\infty} a_k x^k = A.$$

Here the symbol [Abel] indicates that the value is obtained by this method rather than by the usual method of summation (taking limits of partial sums).

As before, our first concern in studying a summability method is to determine whether it assigns the "correct" value to a series that already converges. Does

$$\sum_{k=0}^{\infty} a_k = A \Rightarrow \sum_{k=0}^{\infty} a_k = A \text{ [Abel]?}$$

We are asking, in more correct language, whether Abel's method of summability of series is *regular.*

Theorem 3.58 (Abel) *Suppose that a series $\sum_{k=0}^{\infty} a_k$ converges to a value A. Then*

$$\lim_{x \to 1-} \sum_{k=0}^{\infty} a_k x^k = A.$$

Proof Our first step is to note that the convergence of the series $\sum_{k=0}^{\infty} a_k$ requires that the terms $a_k \to 0$. In particular, the terms are bounded and so the root test will prove that the series $\sum_{k=0}^{\infty} a_k x^k$ converges absolutely for all $|x| < 1$ at least. Thus we can define

$$F(x) = \sum_{k=0}^{\infty} a_k x^k$$

for $0 \leq x < 1$.

Let us form the product of the series for $F(x)$ with the geometric series

$$\frac{1}{1-x} = 1 + x + x^2 + x^3 + \dots$$

(cf. Exercise 3.8.1). Since both series are absolutely convergent for any $0 \leq x < 1$, we obtain

$$\frac{F(x)}{1-x} = \sum_{k=0}^{\infty} (a_0 + a_1 + a_2 + \cdots + a_k) x^k.$$

Writing

$$s_k = (a_0 + a_1 + a_2 + \cdots + a_k)$$

and using the fact that

$$s_k \to A = \sum_{k=0}^{\infty} a_k,$$

we obtain

$$F(x) = (1-x)\sum_{k=0}^{\infty} s_k x^k = A - (1-x)\sum_{k=0}^{\infty}(s_k - A)x^k.$$

Let $\varepsilon > 0$ and choose N so large that

$$|s_k - A| < \varepsilon/2$$

for $k > N$. Then the inequality

$$|F(x) - A| \leq (1-x)\sum_{k=0}^{N} |s_k - A|x^k + \varepsilon/2$$

holds for all $0 \leq x < 1$. The sum here is just a finite sum, and taking limits in finite sums is routine:

$$\lim_{x\to 1-} (1-x)\sum_{k=0}^{N}(s_k - A)x^k = 0.$$

Thus for $x < 1$ but sufficiently close to 1 we can make this smaller than $\varepsilon/2$ and conclude that

$$|F(x) - A| < \varepsilon.$$

We have proved that

$$\lim_{x\to 1-} F(x) = A$$

and the theorem is proved. ■

Example 3.59 Let us sum the series

$$\sum_{k=0}^{\infty}(-1)^k = 1 - 1 + 1 - 1 + 1 - 1 \ldots$$

by Abel's method. We form

$$F(x) = \sum_{k=0}^{\infty}(-1)^k x^k = \frac{1}{1+x}$$

obtaining the formula by recognizing this as a geometric series. Since

$$\lim_{x\to 1-} F(x) = \frac{1}{2}$$

we have proved that

$$\sum_{k=0}^{\infty}(-1)^k = \frac{1}{2} \quad \text{[Abel]}.$$

Recall that we have already obtained in Example 3.56 that

$$\sum_{k=0}^{\infty}(-1)^k = \frac{1}{2} \quad \text{[Cesàro]}$$

so these two different methods have assigned the same sum to this divergent series. You might wish to explore whether the same thing will happen with all series. ◀

As an interesting application we are now in a position to prove Theorem 3.53 on the product of series.

Theorem 3.60 (Abel) *Suppose that $\sum_{k=0}^{\infty} c_k$ is the formal product of two convergent series $\sum_{k=0}^{\infty} a_k$ and $\sum_{k=0}^{\infty} b_k$ and suppose that $\sum_{k=0}^{\infty} c_k$ is known to converge. Then*

$$\sum_{k=0}^{\infty} c_k = \sum_{k=0}^{\infty} a_k \times \sum_{k=0}^{\infty} b_k.$$

Proof The proof just follows on taking limits as $x \to 1-$ in the expression

$$\sum_{k=0}^{\infty} c_k x^k = \sum_{k=0}^{\infty} a_k x^k \times \sum_{k=0}^{\infty} b_k x^k.$$

Abel's theorem, Theorem 3.58, allows us to do this. How do we know, however, that this identity is true for all $0 \leq x < 1$? All three of these series are absolutely convergent for $|x| < 1$ and, by Theorem 3.52, absolutely convergent series can be multiplied in this way. ■

Exercises

3.9.1 Is the series

$$1 + 1 - 1 + 1 + 1 - 1 + 1 + 1 - 1 + \cdots$$

Cesàro summable?

3.9.2 Is the series

$$1 - 2 + 3 - 4 + 5 - 6 + 7 \cdots$$

Cesàro summable?

3.9.3 Is the series

$$1 - 2 + 3 - 4 + 5 - 6 + 7 \cdots$$

Abel summable?

3.9.4 Show that a divergent series of positive numbers cannot be Cesàro summable or Abel summable.

3.9.5 Find a proof from an appropriate source that demonstrates the exact relation between Cesàro summability and Abel summability.

3.9.6 In an appropriate source find out what is meant by a *Tauberian theorem* and present one such theorem appropriate to our studies in this section.

3.10 More on Infinite Sums

✂ *Advanced*

How should we form the sum of a double sequence $\{a_{jk}\}$ where both j and k can range over all natural numbers? In many applications of analysis such sums are needed. A variety of methods come to mind:

1. We might simply form the unordered sum
$$\sum_{(j,k)\in\mathbb{N}\times\mathbb{N}} a_{jk}.$$

2. We could construct "partial sums" in some systematic method and take limits just as we do for ordinary series:
$$\lim_{N\to\infty}\sum_{j=1}^{N}\sum_{k=1}^{N} a_{jk}.$$
These are called *square sums* and are quite popular. If you sketch a picture of the set of points
$$\{(j,k) : 1 \leq j \leq N,\ 1 \leq k \leq N\}$$
in the plane the square will be plainly visible.

3. We could construct partial sums using *rectangular sums*:
$$\lim_{M,N\to\infty}\sum_{j=1}^{M}\sum_{k=1}^{N} a_{jk}.$$
Here the limit is a double limit, requiring both M and N to get large. If you sketch a picture of the set of points
$$\{(j,k) : 1 \leq j \leq M,\ 1 \leq k \leq N\}$$
in the plane you will see the rectangle.

4. We could construct partial sums using *circular sums*:
$$\lim_{R\to\infty}\sum_{j^2+k^2\leq R^2} a_{jk}.$$
Once again, a sketch would show the circles.

5. We could "iterate" the sums, by summing first over j and then over k:

$$\sum_{j=1}^{\infty}\sum_{k=1}^{\infty} a_{jk}$$

or, in the reverse order,

$$\sum_{k=1}^{\infty}\sum_{j=1}^{\infty} a_{jk}.$$

Our experience in the study of ordinary series suggests that all these methods should produce the same sum if the numbers summed are all nonnegative, but that subtle differences are likely to emerge if we are required to add numbers both positive and negative.

In the exercises there are a number of problems that can be pursued to give a flavor for this kind of theory. At this stage in your studies it is important to grasp the fact that such questions arise. Later, when you have found a need to use these kinds of sums, you can develop the needed theory. The tools for developing that theory are just those that we have studied so far in this chapter.

Exercises

3.10.1 Decide on a meaning for the notion of a double series

$$\sum_{j,k=1}^{\infty} a_{jk} \tag{8}$$

and prove that if all the numbers a_{jk} are nonnegative then this converges if and only if

$$\sum_{(j,k)\in\mathbb{N}\times\mathbb{N}} a_{jk} \tag{9}$$

converges and that the values assigned to (8) and (9) are the same.

3.10.2 Decide on a meaning for the notion of an absolutely convergent double series

$$\sum_{j,k=1}^{\infty} a_{jk}$$

and prove that such a series is absolutely convergent if and only if

$$\sum_{(j,k)\in\mathbb{N}\times\mathbb{N}} a_{jk}$$

converges.

3.10.3 Show that the methods given in the text for forming a sum of a double sequence $\{a_{jk}\}$ are equivalent if all the numbers are nonnegative.

3.10.4 Show that the methods given in the text for forming a sum of a double sequence $\{a_{jk}\}$ are not equivalent in general.

3.10.5 What can you assert about the convergence or divergence of the double series

$$\sum_{j,k=1}^{\infty} \frac{1}{j\,k^4}?$$

3.10.6 What is the sum of the double series

$$\sum_{j,k=0}^{\infty} \frac{x^j y^k}{j!\,k!}?$$

3.11 Infinite Products

Enrichment

In this chapter we studied, quite extensively, infinite sums. There is a similar theory for infinite products, a theory that has much in common with the theory of infinite sums. In this section we shall briefly give an account of this theory, partly to give a contrast and partly to introduce this important topic.

Similar to the notion of an infinite sum

$$\sum_{n=1}^{\infty} a_n = a_1 + a_2 + a_3 + a_4 + \dots$$

is the notion of an infinite product

$$\prod_{n=1}^{\infty} p_n = p_1 \times p_2 \times p_3 \times p_4 \times \dots$$

with a nearly identical definition. Corresponding to the concept of "partial sums" for the former will be the notion of "partial products" for the latter.

The main application of infinite series is that of series representations of functions. The main application of infinite products is exactly the same. Thus, for example, in more advanced material we will find a representation of the sin function as an infinite series

$$\sin x = x - \frac{1}{3!}x^3 + \frac{1}{5!}x^5 - \frac{1}{7!}x^7 \dots$$

and also as an infinite product

$$\sin x = \left(1 - \frac{x^2}{\pi^2}\right)\left(1 - \frac{x^2}{4\pi^2}\right)\left(1 - \frac{x^2}{9\pi^2}\right)\left(1 - \frac{x^2}{16\pi^2}\right)\dots.$$

The most obvious starting point for our theory would be to define an infinite product as the limit of the sequence of partial products in exactly the same way that an infinite sum is defined as the limit of the sequence of partial sums. But products behave differently from sums in one important regard: The number zero plays a peculiar role. This is why the definition we now give is slightly different than a first guess might suggest. Our goal is to define an infinite product in such a way that a product can be zero only if one of the factors is zero (just like the situation for finite products).

Definition 3.61 Let $\{b_k\}$ be a sequence of real numbers. We say that the infinite product

$$\prod_{k=1}^{\infty} b_k$$

converges if there is an integer N so that all $b_k \neq 0$ for $k > N$ and if

$$\lim_{M\to\infty} \prod_{k=N+1}^{M} b_k$$

exists *and is not zero.* For the value of the infinite product we take

$$\prod_{k=1}^{\infty} b_k = b_1 \times b_2 \times \ldots b_N \times \lim_{M\to\infty} \prod_{k=N+1}^{M} b_k.$$

This definition guarantees us that a product of factors can be zero if and only if one of the factors is zero. This is the case for finite products, and we are reluctant to lose this.

Theorem 3.62 *A convergent product*

$$\prod_{k=1}^{\infty} b_k = 0$$

if and only if one of the factors is zero.

Proof This is built into the definition and is one of its features. ■

We expect the theory of infinite products to evolve much like the theory of infinite series. We recall that a series $\sum_{k=1}^{n} a_k$ could converge only if $a_k \to 0$. Naturally, the product analog requires the terms to tend to 1.

Theorem 3.63 *A product*

$$\prod_{k=1}^{\infty} b_k$$

that converges necessarily has $b_k \to 1$ *as* $k \to \infty$.

Proof This again is a feature of the definition, which would not be possible if we had not handled the zeros in this way. Choose N so that none of the factors b_k is zero for $k > N$. Then

$$b_n = \lim_{n\to\infty} \frac{\prod_{k=N+1}^{n} b_k}{\prod_{k=N+1}^{n-1} b_k} = 1$$

as required. ■

As a result of this theorem it is conventional to write all infinite products in the special form

$$\prod_{k=1}^{\infty}(1 + a_k)$$

and remember that the terms $a_k \to 0$ as $k \to \infty$ in a convergent product. Also, our assumption about the zeros allows for $a_k = -1$ only for finitely many values of k. The expressions $(1 + a_k)$ are called the "factors" of the product and the a_k themselves are called the "terms."

A close linkage with series arises because the two objects

$$\sum_{k=1}^{\infty} a_k \quad \text{and} \quad \prod_{k=1}^{\infty}(1 + a_k),$$

the series and the product, have much the same kind of behavior.

Theorem 3.64 *A product*

$$\prod_{k=1}^{\infty}(1 + a_k)$$

where all the terms a_k are positive is convergent if and only if the series $\sum_{k=1}^{\infty} a_k$ converges.

Proof Here we use our usual criterion that has served us through most of this chapter: A sequence that is monotonic is convergent if and only if it is bounded.

Note that

$$a_1 + a_2 + a_3 + \cdots + a_n \leq (1 + a_1)(1 + a_2)(1 + a_3) \times \cdots \times (1 + a_n)$$

so that the convergence of the product gives an upper bound for the partial sums of the series. It follows that if the product converges so must the series.

In the other direction we have

$$(1 + a_1)(1 + a_2)(1 + a_3) \times \cdots \times (1 + a_n) \leq e^{a_1 + a_2 + a_3 + \cdots + a_n}$$

and so the convergence of the series gives an upper bound for the partial products of the infinite product. It follows that if the series converges, so must the product. ■

Exercises

3.11.1 Give an example of a sequence of positive numbers $\{b_k\}$ so that

$$\lim_{n\to\infty} b_1 b_2 b_3 \dots b_n$$

exists, but so that the infinite product

$$\prod_{n=1}^{\infty} b_k$$

nonetheless diverges.

3.11.2 Compute

$$\prod_{k=1}^{\infty} \left(1 - \frac{1}{k^2}\right).$$

3.11.3 In Theorem 3.64 we gave no relation between the value of the product $\prod_{k=1}^{\infty}(1+a_k)$ and the value of the series $\sum_{k=1}^{\infty} a_k$ where all the terms a_k are positive. What is the best you can state?

3.11.4 For what values of p does the product

$$\prod_{n=1}^{\infty} \left(1 + \frac{1}{k^p}\right)$$

converge?

3.11.5 Show that

$$\prod_{k=1}^{\infty}(1+x^{2^k}) = (1+x^2)\times(1+x^4)\times(1+x^8)\times(1+x^{16})\dots$$

converges to $1/(1-x^2)$ for all $-1 < x < 1$ and diverges otherwise.

3.11.6 Find a Cauchy criterion for the convergence of infinite products.

3.11.7 A product

$$\prod_{k=1}^{\infty}(1+a_k)$$

is said to *converge absolutely* if the related product

$$\prod_{k=1}^{\infty}(1+|a_k|)$$

converges.

(a) Show that an absolutely convergent product is convergent.

(b) Show that an infinite product

$$\prod_{k=1}^{\infty}(1+a_k)$$

converges absolutely if and only if the series of its terms $\sum_{k=1}^{\infty} a_k$ converges absolutely.

(c) For what values of x does the product

$$\prod_{k=1}^{\infty}\left(1+\frac{x}{k}\right)$$

converge absolutely?

(d) For what values of x does the product

$$\prod_{k=1}^{\infty}\left(1+\frac{x}{k^2}\right)$$

converge absolutely?

(e) For what values of x does the product

$$\prod_{k=1}^{\infty}\left(1+x^k\right)$$

converge absolutely?

(f) Show that

$$\prod_{k=1}^{\infty}\left(1+\frac{(-1)^k}{k}\right)$$

converges but not absolutely.

3.11.8 Develop a theory that allows for the order of the factors in a product to be rearranged.

3.12 Challenging Problems for Chapter 3

3.12.1 If a_n is a sequence of positive numbers such that $\sum_{n=1}^{\infty} a_n$ diverges what (if anything) can you say about the following three series?

(a) $\sum_{n=1}^{\infty} \frac{a_n}{1+a_n}$

(b) $\sum_{n=1}^{\infty} \frac{a_n}{1+na_n}$

(c) $\sum_{n=1}^{\infty} \frac{a_n}{1+n^2a_n}$

3.12.2 Prove the following variant on the Dirichlet Test 3.44: If $\{b_n\}$ is a sequence of bounded variation (cf. Exercise 3.5.12) that converges to zero and the partial sums of the series $\sum_{k=1}^{\infty} a_k$ are bounded, then the series $\sum_{k=1}^{\infty} a_k b_k$ converges.

3.12.3 Prove this variant on the Cauchy condensation test: If the terms of a series $\sum_{k=1}^{\infty} a_k$ are nonnegative and decrease monotonically to zero, then that series converges if and only if the series

$$\sum_{j=1}^{\infty}(2j+1)a_{j^2}$$

converges.

3.12.4 Prove this more general version of the Cauchy condensation test: If the terms of a series $\sum_{k=1}^{\infty} a_k$ are nonnegative and decrease monotonically to zero, then that series converges if and only if the related series

$$\sum_{j=1}^{\infty}(m_{j+1} - m_j)a_{m_j}$$

converges. Here $m_1 < m_2 < m_3 < m_4 \dots$ is assumed to be an increasing sequence of integers and

$$m_{j+1} - m_j \leq C\,(m_j - m_{j-1})$$

for some positive constant and all j.

3.12.5 For any two series of positive terms write

$$\sum_{k=1}^{\infty} a_k \preceq \sum_{k=1}^{\infty} b_k$$

if $a_k/b_k \to 0$ as $k \to \infty$.

(a) If both series converge, explain why this might be interpreted by saying that $\sum_{k=1}^{\infty} a_k$ is converging faster than $\sum_{k=1}^{\infty} b_k$.

(b) If both series diverge, explain why this might be interpreted by saying that $\sum_{k=1}^{\infty} a_k$ is diverging more slowly than $\sum_{k=1}^{\infty} b_k$.

(c) For convergent series is there any connection between

$$\sum_{k=1}^{\infty} a_k \preceq \sum_{k=1}^{\infty} b_k$$

and

$$\sum_{k=1}^{\infty} a_k \leq \sum_{k=1}^{\infty} b_k?$$

(d) For what values of p, q is

$$\sum_{k=1}^{\infty} \frac{1}{k^p} \preceq \sum_{k=1}^{\infty} \frac{1}{k^q}?$$

(e) For what values of r, s is

$$\sum_{k=1}^{\infty} r^k \preceq \sum_{k=1}^{\infty} s^k?$$

(f) Arrange the divergent series

$$\sum_{k=2}^{\infty} \frac{1}{k}, \sum_{k=2}^{\infty} \frac{1}{k \log k}, \sum_{k=2}^{\infty} \frac{1}{k \log(\log k)}, \sum_{k=2}^{\infty} \frac{1}{k \log(\log(\log k))} \dots$$

into the correct order.

(g) Arrange the convergent series

$$\sum_{k=2}^{\infty} \frac{1}{k^p}, \sum_{k=2}^{\infty} \frac{1}{k(\log k)^p}, \sum_{k=2}^{\infty} \frac{1}{k \log k(\log(\log k))^p},$$

$$\sum_{k=2}^{\infty} \frac{1}{k \log k(\log(\log k))(\log(\log(\log k)))^p} \cdots$$

into the correct order. Here $p > 1$.

(h) Suppose that $\sum_{k=1}^{\infty} b_k$ is a divergent series of positive numbers. Show that there is a series

$$\sum_{k=1}^{\infty} a_k \preceq \sum_{k=1}^{\infty} b_k$$

that also diverges (but more slowly).

(i) Suppose that $\sum_{k=1}^{\infty} a_k$ is a convergent series of positive numbers. Show that there is a series

$$\sum_{k=1}^{\infty} a_k \preceq \sum_{k=1}^{\infty} b_k$$

that also converges (but more slowly).

(j) How would you answer this question? Is there a "mother" of all divergent series diverging so slowly that all other divergent series can be proved to be divergent by a comparison test with that series?

3.12.6 This collection of exercises develops some convergence properties of *trigonometric series*; that is, series of the form

$$a_0/2 + \sum_{k=1}^{\infty} (a_k \cos kx + b_k \sin kx)\,. \tag{10}$$

(a) For what values of x does $\sum_{k=1}^{\infty} \frac{\sin kx}{k^2}$ converge?

(b) For what values of x does $\sum_{k=1}^{\infty} \frac{\sin kx}{k}$ converge?

(c) Show that the condition $\sum_{k=1}^{\infty} (|a_k| + |b_k|) < \infty$ ensures the absolute convergence of the trigonometric series (10) for all values of x.

3.12.7 Let $\{a_k\}$ be a decreasing sequence of positive real numbers with limit 0 such that

$$b_k = a_k - 2a_{k+1} + a_{k+2} \geq 0.$$

Prove that $\sum_{k=1}^{\infty} kb_k = a_1$.

3.12.8 Let $\{a_k\}$ be a monotonic sequence of real numbers such that $\sum_{k=1}^{\infty} a_k$ converges. Show that

$$\sum_{k=1}^{\infty} k(a_k - a_{k+1})$$

converges.

3.12.9 Show that every positive rational number can be obtained as the sum of a finite number of distinct terms of the harmonic series

$$1 + \frac{1}{2} + \frac{1}{3} + \frac{1}{4} + \frac{1}{5} + \dots.$$

3.12.10 Let $\sum_{k=1}^{\infty} x_k$ be a convergent series of positive numbers that is monotonically nonincreasing; that is, $x_1 \geq x_2 \geq x_3 \geq \dots$. Let P denote the set of all real numbers that are sums of finitely or infinitely many terms of the series. Show that P is an interval if and only if

$$x_n \leq \sum_{k=n+1}^{\infty} x_k$$

for every integer n.

3.12.11 Let p_1, p_2, p_3, be a sequence of distinct points that is dense in the interval $(0,1)$. The points p_1, p_2, p_3, ... , p_{n-1} decompose the interval $[0,1]$ into n closed subintervals. The point p_n is an interior point of one of those intervals and decomposes that interval into two closed subintervals. Let a_n and b_n be the lengths of those two intervals. Prove that

$$\sum_{k=1}^{\infty} a_k b_k (a_k + b_k) = 3.$$

3.12.12 Let $\{a_n\}$ be a sequence of positive number such that the series $\sum_{k=1}^{\infty} a_k$ converges. Show that

$$\sum_{k=1}^{\infty} (a_k)^{n/(n+1)}$$

also converges.

3.12.13 Let $\{a_k\}$ be a sequence of positive numbers and suppose that

$$a_k \leq a_{2k} + a_{2k+1}$$

for all $k = 1, 2, 3, 4, \dots$. Show that $\sum_{k=1}^{\infty} a_k$ diverges.

3.12.14 If $\{a_k\}$ is a sequence of positive numbers for which $\sum_{k=1}^{\infty} a_k$ diverges, determine all values of p for which

$$\sum_{k=1}^{\infty} \frac{a_k}{(a_1 + a_2 + \dots + a_k)^p}$$

converges.

3.12.15 Let $\{a_n\}$ be a sequence of real numbers converging to zero. Show that there must exist a monotonic sequence $\{b_n\}$ such that the series $\sum_{k=1}^{\infty} b_k$ diverges and the series $\sum_{k=1}^{\infty} a_k b_k$ is absolutely convergent.

Chapter 4

SETS OF REAL NUMBERS

4.1 Introduction

Modern set theory and the world it has opened to mathematics has its origins in a problem in analysis. A young Georg Cantor in 1870 began to attack a problem given to him by his senior colleague Edward Heine, who worked at the same university. (We shall see Heine playing a key role in some ideas of this chapter too.)

The problem was to determine if the equation

$$\tfrac{1}{2}a_0 + \sum_{k=1}^{\infty} (a_k \cos kx + b_k \sin kx) = 0 \tag{1}$$

must imply that all the coefficients of the series, the $\{a_k\}$ and the $\{b_k\}$ are zero. Cantor solved this using the methods of his time. It was a good achievement, but not the one that was to make him famous. What he did next was to ask, as any good mathematician would, whether his result could be generalized. Suppose that the series (1) converges to zero for all x except possibly for those in a given set E. If this set E is very small, then perhaps, the coefficients of the series should also have to be all zero.

The nature of these exceptional sets (nowadays called sets of uniqueness) required a language and techniques that were entirely new. Previously a number of authors had needed a language to describe sets that arose in various problems. What was used at the time was limited, and few interesting examples of sets were available. Cantor went beyond these, introducing a new collection of ideas that are now indispensable to analysis. We shall encounter in this chapter many of the notions that arose then: accumulation points, derived sets, countable sets, dense sets, nowhere dense sets.

Incidentally, Cantor never did finish his problem of describing the sets of uniqueness, as the development of the new set theory was more important and consumed his energies. In fact, the problem remains unsolved, although much interesting information about the nature of sets of uniqueness has been discovered.

The theory of sets that Cantor initiated has proved to be fundamental to all of mathematics. Very quickly the most talented analysts of that time began applying his ideas to the theory of functions, and by now this material is essential to an understanding of the subject. This chapter contains the most basic material. In Chapter 6 we will need some further concepts.

4.2 Points

In our studies of analysis we shall often need to have a language that describes sets of points and the points that belong to them. That language did not develop until late in the nineteenth century, which is why the early mathematicians had difficulty understanding some problems.

For example, consider the set of solutions to an equation

$$f(x) = 0$$

where f is some well-behaved function. In the simplest cases (e.g., if f is a polynomial function) the solution set could be empty or a finite number of points. There is no difficulty there. But in more general settings the solution set could be very complicated indeed. It may have points that are "isolated," points appearing in clusters, or it may contain intervals or merely fragments of intervals. You can see that we even lack the words to describe the possibilities.

The ideas in this section are all very geometric. Try to draw mental images that depict all of these ideas to get a feel for the definitions. The definitions themselves should be remembered but may prove hard to remember without some associated picture.

The simplest types of sets are intervals. We call

$$[a, b] = \{x : a \leq x \leq b\}$$

a closed interval, and

$$(a, b) = \{x : a < x < b\}$$

an open interval. The other sets that we often consider are the sets $\mathbb{N}$ of natural numbers, $\mathbb{Q}$ of rational numbers, and $\mathbb{R}$ of all real numbers. Use these in your pictures, as well as sets obtained by combining them in many ways.

4.2.1 Interior Points

Every point inside an open interval $I = (a, b)$ has the feature that there is a smaller open interval centered at that point that is also inside I. Thus if $x \in (a, b)$ then for any positive number c that is small enough

$$(x - c, x + c) \subset (a, b).$$

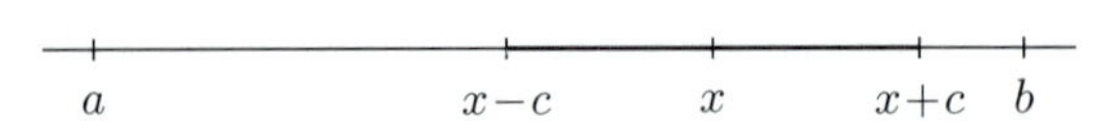

Figure 4.1. Every point in (a, b) is an interior point.

Indeed the arithmetic to show this is easy (and a picture makes it transparent). Let c be any positive number that is smaller than the shortest distance from x to either a or b. Then $(x - c, x + c) \subset (a, b)$. (See Figure 4.1.)

Note. Often we use the following suggestive language. An open interval that contains a point x is said to be a *neighborhood* of x. Thus each point in (a, b) possesses a neighborhood, indeed many neighborhoods, that lie entirely inside the set I. On occasion the point x itself is excluded from the neighborhood: We say that an interval (c, d) is a neighborhood of x if x belongs to the interval and we say that the set $(c, d) \setminus \{x\}$ is a *deleted neighborhood.* This is just the interval with the point x removed.

We can distinguish between points that are merely in a set and points that are more deeply inside the set. The word chosen to convey this image of "inside" is *interior.*

Definition 4.1 (Interior Point) Let E be a set of real numbers. Any point x that belongs to E is said to be *an interior point of E* provided that some interval

$$(x - c, x + c) \subset E.$$

Thus an interior point of the set E is not merely *in the set E*; it is, so to speak, deep inside the set, at a positive distance at least c away from every point that does not belong to E.

Example 4.2 The following examples are immediate if a picture is sketched.

1. Every point x of an open interval (a, b) is an interior point.

2. Every point x of a closed interval $[a, b]$, except the two endpoints a and b, is an interior point.

3. The set of natural numbers $\mathbb{N}$ has no interior points whatsoever.

4. Every point of $\mathbb{R}$ is an interior point.

5. No point of the set of rational numbers $\mathbb{Q}$ is an interior point. [This is because any interval $(x - c, x + c)$ must contain both rational numbers and irrational numbers and, hence, can never be a subset of $\mathbb{Q}$.]

In each case, we should try to find the interval $(x-c, x+c)$ inside the set or explain why there can be no such interval. ◀

4.2.2 Isolated Points

Most sets that we consider will have infinitely many points. Certainly any interval (a, b) or $[a, b]$ has infinitely many points. The set $\mathbb{N}$ of natural numbers also has infinitely many points, but as we look closely at any one of these points we see that each point is all alone, at a certain distance away from every other point in the set. We call these points *isolated points* of the set.

Definition 4.3 (Isolated Point) Let E be a set of real numbers. Any point x that belongs to E is said to be *an isolated point of* E provided that for some interval $(x-c, x+c)$

$$(x-c, x+c) \cap E = \{x\}.$$

Thus an isolated point of the set E is in the set E but has no close neighbors who are also in E. It is at some positive distance at least c away from every other point that belongs to E.

Example 4.4 As before, the examples are immediate if a picture is sketched.

1. No point x of an open interval (a, b) is an isolated point.

2. No point x of a closed interval $[a, b]$ is an isolated point.

3. Every point belonging to the set of natural numbers $\mathbb{N}$ is an isolated point.

4. No point of $\mathbb{R}$ is isolated.

5. No point of $\mathbb{Q}$ is isolated.

In each case, we should try to find the interval $(x-c, x+c)$ that meets the set at no other point or show that there is none. ◀

4.2.3 Points of Accumulation

Most sets that we consider will have infinitely many points. While the isolated points are of interest on occasion, more than likely we would be interested in points that are not isolated. These points have the property that every containing interval contains many points of the set. Indeed we are interested in any point x with the property that the intervals $(x-c, x+c)$ meet the set E at infinitely many points. This could happen even if x itself does not belong to E. We call these points *accumulation points* of the set. An accumulation point need not itself belong to the set.

Definition 4.5 (Accumulation Point) Let E be a set of real numbers. Any point x (not necessarily in E) is said to be *an accumulation point of* E provided that for every $c > 0$ the intersection

$$(x - c, x + c) \cap E$$

contains infinitely many points.

Thus an accumulation point of E is a point that may or may not itself belong to E and that has many close neighbors who are in E.

Note. The definition requires that for all $c > 0$ the intersection

$$(x - c, x + c) \cap E$$

contains infinitely many points of E. In checking for an accumulation point it may be preferable merely to check that there is at least one point in this intersection (other than possibly x itself). If there is always at least one point, then there must in fact be infinitely many (Exercise 4.2.18).

Example 4.6 Yet again, the examples are immediate if a picture is sketched.

1. Every point of an open interval (a, b) is a an accumulation point of (a, b). Moreover, the two endpoints a and b are also accumulation points of (a, b) [although they do not belong themselves to (a, b))].

2. Every point of a closed interval $[a, b]$ is an accumulation point of (a, b). No point outside can be.

3. No point at all is an accumulation point of the set of natural numbers $\mathbb{N}$.

4. Every point of $\mathbb{R}$ is an accumulation point.

5. Every point on the real line, both rational and irrational, is an accumulation point of the set $\mathbb{Q}$.

◀

4.2.4 Boundary Points

The intervals (a, b) and $[a, b]$ have what appears to be an "edge". The points a and b mark the boundaries between the inside of the set (i.e., the interior points) and the "outside" of the set. This inside/outside language with an idea of a boundary between them is most useful but needs a precise definition.

Definition 4.7 (Boundary Point) Let E be a set of real numbers. Any point x (not necessarily in E) is said to be *a boundary point of E* provided that every interval $(x - c, x + c)$ contains at least one point of E and also at least one point that does not belong to E.

This definition is easy to apply to the intervals (a, b) and $[a, b]$ but harder to imagine for general sets. For these intervals the only points that are immediately seen to satisfy the definition are the two endpoints that we would have naturally said to be at the boundary.

Example 4.8 The examples are not all transparent but require careful thinking about the definition.

1. The two endpoints a and b are the only boundary points of an open interval (a, b).
2. The two endpoints a and b are the only boundary points of a closed interval $[a, b]$.
3. Every point in the set $\mathbb{N}$ of natural numbers is a boundary point.
4. No point at all is boundary point of the set $\mathbb{R}$.
5. Every point on the real line, both rational and irrational, is a boundary point of the set $\mathbb{Q}$. (Think for a while about this one!)

◀

Exercises

4.2.1 Determine the set of interior points, accumulation points, isolated points, and boundary points for each of the following sets:

(a) $\{1, 1/2, 1/3, 1/4, 1/5, \dots\}$
(b) $\{0\} \cup \{1, 1/2, 1/3, 1/4, 1/5, \dots\}$
(c) $(0, 1) \cup (1, 2) \cup (2, 3) \cup (3, 4) \cdots \cup (n, n + 1) \cup \dots$
(d) $(1/2, 1) \cup (1/4, 1/2) \cup (1/8, 1/4) \cup (1/16, 1/8) \cup \dots$
(e) $\{x : |x - \pi| < 1\}$
(f) $\{x : x^2 < 2\}$
(g) $\mathbb{R} \setminus \mathbb{N}$
(h) $\mathbb{R} \setminus \mathbb{Q}$

4.2.2 Give an example of each of the following or explain why you think such a set could not exist.

(a) A nonempty set with no accumulation points and no isolated points

(b) A nonempty set with no interior points and no isolated points

(c) A nonempty set with no boundary points and no isolated points

4.2.3 Show that every interior point of a set must also be an accumulation point of that set, but not conversely.

4.2.4 Show that no interior point of a set can be a boundary point, that it is possible for an accumulation point to be a boundary point, and that every isolated point must be a boundary point.

4.2.5 Let E be a nonempty set of real numbers that is bounded above but has no maximum. Let $x = \sup E$. Show that x is a point of accumulation of E. Is it possible for x to also be an interior point of E? Is x a boundary point of E?

4.2.6 State and solve the version of Exercise 4.2.5 that would use the infimum in place of the supremum.

4.2.7 Let A be a set and $B = \mathbb{R} \setminus A$. Show that every boundary point of A is also a boundary point of B.

4.2.8 Let A be a set and $B = \mathbb{R} \setminus A$. Show that every boundary point of A is a point of accumulation of A or else a point of accumulation of B, perhaps both.

4.2.9 Must every boundary point of a set be also an accumulation point of that set?

4.2.10 Show that every accumulation point of a set that does not itself belong to the set must be a boundary point of that set.

4.2.11 Show that a point x is not an interior point of a set E if and only if there is a sequence of points $\{x_n\}$ converging to x and no point $x_n \in E$.

4.2.12 Let A be a set and $B = \mathbb{R} \setminus A$. Show that every interior point of A is not an accumulation point of B.

4.2.13 Let A be a set and $B = \mathbb{R} \setminus A$. Show that every accumulation point of A is not an interior point of of B.

4.2.14 Give an example of a set that has the set $\mathbb{N}$ as its set of accumulation points.

4.2.15 Show that there is no set which has the interval $(0, 1)$ as its set of accumulation points.

4.2.16 Show that there is no set which has the set $\mathbb{Q}$ as its set of accumulation points.

4.2.17 Give an example of a set that has the set

$$E = \{0\} \cup \{1, 1/2, 1/3, 1/4, 1/5, \dots\}$$

as its set of accumulation points.

4.2.18 Show that a point x is an accumulation point of a set E if and only if for every $\varepsilon > 0$ there are at least two points belonging to the set $E \cap (x-\varepsilon, x+\varepsilon)$.

4.2.19 Suppose that $\{x_n\}$ is a convergent sequence converging to a number L and that $x_n \neq L$ for all n. Show that the set

$$\{x : x = x_n \quad \text{for some } n\}$$

has exactly one point of accumulation, namely L. Of what importance was the assumption that $x_n \neq L$ for all n for this exercise?

4.2.20 Let E be a set and $\{x_n\}$ a sequence of distinct elements of E. Suppose that $\lim_{n\to\infty} x_n = x$. Show that x is a point of accumulation of E.

4.2.21 Let E be a set and $\{x_n\}$ a sequence of points, not necessarily elements of E. Suppose that $\lim_{n\to\infty} x_n = x$ and that x is an interior point of E. Show that there is an integer N so that $x_n \in E$ for all $n \geq N$.

4.2.22 Let E be a set and $\{x_n\}$ a sequence of elements of E. Suppose that $\lim_{n\to\infty} x_n = x$ and that x is an isolated point of E. Show that there is an integer N so that $x_n = x$ for all $n \geq N$.

4.2.23 Let E be a set and $\{x_n\}$ a sequence of distinct points, not necessarily elements of E. Suppose that $\lim_{n\to\infty} x_n = x$ and that $x_{2n} \in E$ and $x_{2n+1} \notin E$ for all n. Show that x is a boundary point of E.

4.2.24 If E is a set of real numbers, then E', called the *derived set* of E, denotes the set of all points of accumulation of E. Give an example of each of the following or explain why you think such a set could not exist.

(a) A nonempty set E such that $E' = E$

(b) A nonempty set E such that $E' = \emptyset$

(c) A nonempty set E such that $E' \neq \emptyset$ but $E'' = \emptyset$

(d) A nonempty set E such that E', $E'' \neq \emptyset$ but $E''' = \emptyset$

(e) A nonempty set E such that E', E'', E''', ... are all different

(f) A nonempty set E such that $(E \cup E')' \neq (E \cup E')$

4.2.25 Show that there is no set with uncountably many isolated points.

4.3 Sets

We now begin a classification of sets of real numbers. Almost all of the concepts of analysis (limits, derivatives, integrals, etc.) can be better understood if a classification scheme for sets is in place. By far the most important notions are those of closed sets and open sets. This is the basis for much advanced mathematics and leads to the subject known as topology, which is fundamental to an understanding of many areas of mathematics. On the real line we can master open and closed sets and describe precisely what they are.

4.3.1 Closed Sets

In many parts of mathematics the word "closed" is used to indicate that some operation stays within a system. For example, the set of natural numbers $\mathbb{N}$ is closed under addition and multiplication (any sum or product of two of them is yet another) but not closed under subtraction or division (2 and 3 are natural numbers, but $2-3$ and $3/2$ are not). This same word was employed originally to indicate sets of real numbers that are "closed" under the operation of taking points of accumulation. If all points of accumulation turn out to be in the set, then the set is said to be closed. This terminology has survived and become, perhaps, the best known usage of the word "closed."

Definition 4.9 (Closed) Let E be a set of real numbers. The set E is said to be *closed* provided that every accumulation point of E belongs to the set E.

Thus a set E is not closed if there is some accumulation point of E that does not belong to E. In particular, a set with no accumulation points would have to be closed since there is no point that needs to be checked.

Example 4.10 The examples are immediate since we have previously described all of the accumulation points of these sets.

1. The empty set $\emptyset$ is closed since it contains all of its accumulation points (there are none).

2. The open interval (a, b) is not closed because the two endpoints a and b are accumulation points of (a, b) and yet they do not belong to the set.

3. The closed interval $[a, b]$ is closed since only points that are already in the set are accumulation points.

4. The set of natural numbers $\mathbb{N}$ is closed because it has no points of accumulation.

5. The real line $\mathbb{R}$ is closed since it contains all of its accumulation points, namely every point.

6. The set of rational numbers $\mathbb{Q}$ is not closed. Every point on the real line, both rational and irrational, is an accumulation point of $\mathbb{Q}$, but the set fails to contain any irrationals.

◀

The Closure of a Set If a set is not closed it is because it neglects to contain points that "should" be there since they are accumulation points but not in the set. On occasions it is best to throw them in and consider a larger set composed of the original set together with the offending accumulation points that may not have belonged originally to the set.

Definition 4.11 (Closure) Let E be any set of real numbers and let E' denote the set of all accumulation points of E. Then the set

$$\overline{E} = E \cup E'$$

is called the *closure* of the set E.

For example, $\overline{(a,b)} = [a,b]$, $\overline{[a,b]} = [a,b]$, $\overline{\mathbb{N}} = \mathbb{N}$, and $\overline{\mathbb{Q}} = \mathbb{R}$. Each of these is an easy observation since we know what the points of accumulation of these sets are.

4.3.2 Open Sets

Originally, the word "open" was used to indicate a set that was not closed. In time it was realized that this is a waste of terminology, since the class of "not closed sets" is not of much general interest. Instead the word is now used to indicate a contrasting idea, an idea that is not quite an opposite—just at a different extreme. This may be a bit unfortunate since now a set that is not open need not be closed. Indeed some sets can be both open and closed, and some sets can be both not open and not closed.

Definition 4.12 (Open) Let E be a set of real numbers. Then E is said to be *open* if every point of E is also an interior point of E.

Thus every point of E is not merely a point *in the set* E; it is, so to speak, deep inside the set. For each point x_0 of E there is some positive number δ and all points outside E are at least a distance δ away from x_0. Note that this means that an open set cannot contain any of its boundary points.

Example 4.13 These examples are immediate since we have seen them before in the context of interior points in Section 4.2.1.

1. The empty set $\emptyset$ is open since it contains no points that are not interior points of the set. (This is the first example of a set that is both open and closed.)

2. The open interval (a,b) is open since every point x of an open interval (a,b) is an interior point.

3. The closed interval $[a, b]$ is not open since there are points in the set (namely the two endpoints a and b) that are in the set and yet are not interior points.

4. The set of natural numbers $\mathbb{N}$ has no interior points and so this set is not open; all of its points fail to be interior points.

5. Every point of $\mathbb{R}$ is an interior point and so $\mathbb{R}$ is open. (Remember, $\mathbb{R}$ is also closed so it is both open and closed. Note that $\mathbb{R}$ and $\emptyset$ are the only examples of sets that are both open and closed.)

6. No point of the set of rational numbers $\mathbb{Q}$ is an interior point and so $\mathbb{Q}$ definitely fails to be open.

◀

The Interior of a Set If a set is not open it is because it contains points that "shouldn't" be there since they are not interior. On occasions it is best to throw them away and consider a smaller set composed entirely of the interior points.

Definition 4.14 (Interior) Let E be any set of real numbers. Then the set

$$\text{int}(E)$$

denotes the set of all interior points of E and is called the *interior* of the set E.

For example, $\text{int}((a, b)) = (a, b)$, $\text{int}([a, b]) = (a, b)$, $\text{int}(\mathbb{N}) = \emptyset$, and $\text{int}(\mathbb{Q}) = \emptyset$. Each of these is an easy observation since we know what the interior points of these sets are.

Component Intervals of Open Sets Think of the most general open set G that you can. A first feeble suggestion might be any open interval $G = (a, b)$. We can do a little better. How about the union of two of these

$$G = (a, b) \cup (c, d)?$$

If these are disjoint, then we would tend to think of G as having two "components." It is easy to see that every point is an interior point. We need not stop at two component intervals; any number would work:

$$G = (a_1, b_1) \cup (a_2, b_2) \cup (a_3, b_3) \cup \cdots \cup (a_n, b_n).$$

The argument is the same and elementary. If x is a point in this set, then x is an interior point. Indeed we can form the union of a sequence of such open intervals and it is clear that we shall obtain an open set. For a specific example consider

$$(-\infty, -3) \cup (1/2, 1) \cup (1/8, 1/4) \cup (1/32, 1/16) \cup (1/128, 1/64) \cup \ldots.$$

At this point our imagination stalls and it is hard to come up with any more examples that are not obtained by stringing together open intervals in exactly this way. This suggests that, perhaps, all open sets have this structure. They are either open intervals or else a union of a sequence of open intervals. This theorem characterizes all open sets of real numbers and reveals their exact structure.

Theorem 4.15 *Let G be a nonempty open set of real numbers. Then there is a unique sequence (finite or infinite) of disjoint, open intervals*

$$(a_1, b_1), (a_2, b_2), (a_3, b_3), \ldots, (a_n, b_n), \ldots$$

called the component intervals of G *such that*

$$G = (a_1, b_1) \cup (a_2, b_2) \cup (a_3, b_3) \cup \cdots \cup (a_n, b_n) \cup \ldots.$$

Proof Take any point $x \in G$. We know that there must be some interval (a, b) containing the point x and contained in the set G. This is because G is open and so every point in G is an interior point. We need to take the largest such interval. The easiest way to describe this is to write

$$\alpha = \inf\{t : (t, x) \subset G\}$$

and

$$\beta = \sup\{t : (x, t) \subset G\}.$$

Note that $\alpha < x < \beta$. Then

$$I_x = (\alpha, \beta)$$

is called the *component* of G containing the point x. (It is possible here for $\alpha = -\infty$ or $\beta = \infty$.)

One feature of components that we require is this: If x and y belong to the same component, then

$$I_x = I_y$$

If x and y do not belong to the same component, then I_x and I_y have no points in common. This is easily checked (Exercise 4.3.21).

There remains the task of listing the components as the theorem requires. If the collection

$$\{I_x : x \in G\}$$

is finite, then this presents no difficulties. If it is infinite we need a clever strategy.

Let r_1, r_2, r_3, ... be a listing of all the rational numbers contained in the set G. We construct our list of components of G by writing for the first step

$$(a_1, b_1) = I_{r_1}.$$

The second component must be disjoint from this first component. We cannot simply choose I_{r_2} since if r_2 belongs to (a_1, b_1), then in fact

$$(a_1, b_1) = I_{r_1} = I_{r_2}.$$

Instead we travel along the sequence r_1, r_2, r_3, ... until we reach the first one, say r_{m_2}, that does not already belong to the interval (a_1, b_1). This then serves to define our next interval:

$$(a_2, b_2) = I_{r_{m_2}}.$$

If there is no such point, then the process stops. This process is continued inductively resulting in a sequence of open intervals:

$$(a_1, b_1) \cup (a_2, b_2) \cup (a_3, b_3) \cup \cdots \cup (a_n, b_n) \cup \ldots,$$

which may be infinite or finite. At the kth stage a point r_{m_k} is selected so that r_{m_k} does not belong to any component thus far selected. If this cannot be done, then the process stops and produces only a finite list of components.

The proof is completed by checking that (i) every point of G is in one of these intervals, (ii) every point in one of these interval belongs to G, and (iii) the intervals in the sequence must be disjoint.

For (i) note that if $x \in G$, then there must be rational numbers in the component I_x. Indeed there is a first number r_k in the list that belongs to this component. But then $x \in I_{r_k}$ and so we must have chosen this interval I_{r_k} at some stage. Thus x does belong to one of these intervals.

For (ii) note that if x is in G, then $I_x \subset G$. Thus every point in one of the intervals belongs to G.

For (iii) consider some pair of intervals in the sequence we have constructed. The later one chosen was required to have a point r_{m_k} that did not belong to any of the preceding choices. But that means then that the new component chosen is disjoint from all the previous ones.

This completes the checking of the details and so the proof is done. ■

Exercises

4.3.1 Is it true that a set, all of whose points are isolated, must be closed?

4.3.2 If a set has no isolated points must it be closed? Must it be open?

4.3.3 A careless student, when asked, incorrectly remembers that a set is closed "if all its points are points of accumulation." Must such a set be closed?

4.3.4 A careless student, when asked, incorrectly remembers that a set is open "if it contains all of its interior points." Is there an example of a set that fails to have this property? Is there an example of a nonopen set that has this property?

4.3.5 Determine which of the following sets are open, which are closed, and which are neither open nor closed.

(a) $(-\infty, 0) \cup (0, \infty)$

(b) $\{1, 1/2, 1/3, 1/4, 1/5, \dots\}$

(c) $\{0\} \cup \{1, 1/2, 1/3, 1/4, 1/5, \dots\}$

(d) $(0,1) \cup (1,2) \cup (2,3) \cup (3,4) \cdots \cup (n, n+1) \cup \dots$

(e) $(1/2, 1) \cup (1/4, 1/2) \cup (1/8, 1/4) \cup (1/16, 1/8) \cup \dots$

(f) $\{x : |x - \pi| < 1\}$

(g) $\{x : x^2 < 2\}$

(h) $\mathbb{R} \setminus \mathbb{N}$

(i) $\mathbb{R} \setminus \mathbb{Q}$

4.3.6 Show that the closure operation has the following properties:

(a) If $E_1 \subset E_2$, then $\overline{E_1} \subset \overline{E_2}$.

(b) $\overline{E_1 \cup E_2} = \overline{E_1} \cup \overline{E_2}$.

(c) $\overline{E_1 \cap E_2} \subset \overline{E_1} \cap \overline{E_2}$.

(d) Give an example of two sets E_1 and E_2 such that

$$\overline{E_1 \cap E_2} \neq \overline{E_1} \cap \overline{E_2}.$$

(e) $\overline{\overline{E}} = \overline{E}$.

4.3.7 Show that the interior operation has the following properties:

(a) If $E_1 \subset E_2$, then $\text{int}(E_1) \subset \text{int}(E_2)$.

(b) $\text{int}(E_1 \cap E_2) = \text{int}(E_1) \cap \text{int}(E_2)$.

(c) $\text{int}(E_1 \cup E_2) \supset \text{int}(E_1) \cup \text{int}(E_2)$.

(d) Give an example of two sets E_1 and E_2 such that

$$\text{int}(E_1 \cup E_2) \neq \text{int}(E_1) \cup \text{int}(E_2).$$

(e) $\text{int}(\text{int}(E)) = \text{int}(E)$.

4.3.8 Show that if the set E' of points of accumulation of E is empty, then the set E must be closed.

4.3.9 Show that the set E' of points of accumulation of any set E must be closed.

4.3.10 Show that the set $\text{int}(E)$ of interior points of any set E must be open.

4.3.11 Show that a set E is closed if and only if $\overline{E} = E$.

4.3.12 Show that a set E is open if and only if $\operatorname{int}(E) = E$.

4.3.13 If A is open and B is closed, what can you say about the sets $A \setminus B$ and $B \setminus A$?

4.3.14 If A and B are both open or both closed, what can you say about the sets $A \setminus B$ and $B \setminus A$?

4.3.15 If E is a nonempty bounded, closed set, show that $\max\{E\}$ and $\min\{E\}$ both exist. If E is a bounded, open set, show that neither $\max\{E\}$ nor $\min\{E\}$ exist (although $\sup\{E\}$ and $\inf\{E\}$ do).

4.3.16 Show that if a set of real numbers E has at least one point of accumulation, then for every $\varepsilon > 0$ there exist points x, $y \in E$ so that $0 < |x - y| < \varepsilon$.

4.3.17 Construct an example of a set of real numbers E that has no points of accumulation and yet has the property that for every $\varepsilon > 0$ there exist points x, $y \in E$ so that $0 < |x - y| < \varepsilon$.

4.3.18 Let $\{x_n\}$ be a sequence of real numbers. Let E denote the set of all numbers z that have the property that there exists a subsequence $\{x_{n_k}\}$ convergent to z. Show that E is closed.

4.3.19 Determine the components of the open set $\mathbb{R} \setminus \mathbb{N}$.

4.3.20 Let $F = \{0\} \cup \{1, 1/2, 1/3, 1/4, 1/5, \dots\}$. Show that F is closed and determine the components of the open set $\mathbb{R} \setminus F$.

4.3.21 In the proof of Theorem 4.15 show that if x and y belong to the same component, then $I_x = I_y$, while if x and y do not belong to the same component, then I_x and I_y have no points in common.

4.3.22 In the proof of Theorem 4.15, after obtaining the collection of components $\{I_x : x \in G\}$, there remained the task of listing them. In classroom discussions the following suggestions were made as to how the components might be listed:

(a) List the components from largest to smallest.

(b) List the components from smallest to largest.

(c) List the components from left to right.

(d) List the components from right to left.

For each of these give an example of an open set with infinitely many components for which this strategy would work and also an example where it would fail.

4.3.23 In searching for interesting examples of open sets, you may have run out of ideas. Here is an example of a construction due to Cantor that has become the source for many important examples in analysis. We describe the component intervals of an open set G inside the interval $(0, 1)$. At each "stage" n we shall describe 2^{n-1} components.

At the first stage, stage 1, take $(1/3, 2/3)$ and at stage 2 take $(1/9, 2/9)$ and $(7/9, 8/9)$ and so on so that at each stage we take all the middle third intervals of the intervals remaining inside $(0, 1)$. The set G is the open subset of $(0, 1)$ having these intervals as components.

(a) Describe exactly the collection of intervals forming the components of G.

(b) What are the endpoints of the components. How do they relate to ternary expansions of numbers in $[0, 1]$?

(c) What is the sum of the lengths of all components?

(d) Sketch a picture of the set G by illustrating the components at the first three stages.

(e) Show that if x, $y \in G$, $x < y$, but x and y are not in the same component, then there are infinitely many components of G in the interval (x, y).

4.4 Elementary Topology

The study of open and closed sets in any space is called *topology.* Our goal now is to find relations between these ideas and examine the properties of these sets. Much of this is a useful introduction to topology in any space; some is very specific to the real line, where the topological ideas are easier to sort out.

The first theorem establishes the connection between the open sets and the closed sets. They are not quite opposites. They are better described as "complementary."

Theorem 4.16 (Open vs. Closed) *Let A be a set of real numbers and $B = \mathbb{R} \setminus A$ its complement. Then A is open if and only if B is closed.*

Proof If A is open and B fails to be closed then there is a point z that is a point of accumulation of B and yet is not in B. Thus z must be in A. But if z is a point in an open set it must be an interior point. Hence there is an interval $(z - \delta, z + \delta)$ contained entirely in A; such an interval contains no points of B. Hence z cannot be a point of accumulation of B. This is a contradiction and so we have proved that B must be closed if A is open.

Conversely, if B is closed and A fails to be open, then there is a point $z \in A$ that is not an interior point of A. Hence every interval $(z - \delta, z + \delta)$ must contain points outside of A, namely points in B. By definition this means that z is a point of accumulation of B. But B is closed and so z, which is a point in A, should really belong to B. This is a contradiction and so we have proved that A must be open if B is closed. ■

Theorem 4.17 (Properties of Open Sets) *Open sets of real numbers have the following properties:*

1. *The sets $\emptyset$ and $\mathbb{R}$ are open.*

2. *Any intersection of a finite number of open sets is open.*

3. *Any union of an arbitrary collection of open sets is open.*

4. *The complement of an open set is closed.*

Proof The first assertion is immediate and the last we have already proved. The third is easy. Thus it is enough for us to prove the second assertion. Let us suppose that E_1 and E_2 are open. To show that $E_1 \cap E_2$ is also open we need to show that every point is an interior point. Let $z \in E_1 \cap E_2$. Then, since z is in both of the sets E_1 and E_2 and both are open there are intervals

$$(z - \delta_1, z + \delta_1) \subset E_1$$

and

$$(z - \delta_2, z + \delta_2) \subset E_2.$$

Let $\delta = \min\{\delta_1, \delta_2\}$. We must then have

$$(z - \delta, z + \delta) \subset E_1 \cap E_2,$$

which shows that z is an interior point of $E_1 \cap E_2$. Since z is any point, this proves that $E_1 \cap E_2$ is open.

Having proved the theorem for two open sets, it now follows for three open sets since

$$E_1 \cap E_2 \cap E_3 = (E_1 \cap E_2) \cap E_3.$$

That any intersection of an arbitrary finite number of open sets is open now follows by induction. ■

Theorem 4.18 (Properties of Closed Sets) *Closed sets of real numbers have the following properties:*

1. *The sets $\emptyset$ and $\mathbb{R}$ are closed.*

2. *Any union of a finite number of closed sets is closed.*

3. *Any intersection of an arbitrary collection of closed sets is closed.*

4. *The complement of a closed set is open.*

Proof Except for the second assertion these are easy or have already been proved. Let us prove the second one. Let us suppose that E_1 and E_2 are closed. To show that $E_1 \cup E_2$ is also closed we need to show that every accumulation point belongs to that set. Let z be an accumulation point of $E_1 \cup E_2$ that does not belong to the set. Since z is in neither of the closed sets E_1 and E_2, this point z cannot be a an accumulation point of either. Thus some interval $(z-\delta, z+\delta)$ contains no points of either E_1 or E_2. Consequently, that interval contains no points of $E_1 \cup E_2$ and is not an accumulation point after all, contradicting our assumption. Since z is any accumulation point, this proves that $E_1 \cup E_2$ is closed.

Having proved the theorem for two closed sets, it now follows for three closed sets since

$$E_1 \cup E_2 \cup E_3 = (E_1 \cup E_2) \cup E_3.$$

That any union of an arbitrary finite number of closed sets is closed now follows by induction. ■

Exercises

4.4.1 Explain why it is that the sets $\emptyset$ and $\mathbb{R}$ are open and also closed.

4.4.2 Show that a union of an arbitrary collection of open sets is open.

4.4.3 Show that an intersection of an arbitrary collection of closed sets is closed.

4.4.4 Give an example of a sequence of open sets G_1, G_2, G_3, ... whose intersection is neither open nor closed. Why does this not contradict Theorem 4.17?

4.4.5 Give an example of a sequence of closed sets F_1, F_2, F_3, ... whose union is neither open nor closed. Why does this not contradict Theorem 4.18?

4.4.6 Show that the set $\overline{E}$ can be described as the *smallest closed set that contains every point of E.*

4.4.7 Show that the set $\text{int}(E)$ can be described as the *largest open set that is contained inside E.*

4.4.8 A function $f : \mathbb{R} \to \mathbb{R}$ is said to be *bounded at a point* x_0 provided that there are positive numbers ε and M so that $|f(x)| < M$ for all $x \in (x_0-\varepsilon, x_0+\varepsilon)$. Show that the set of points at which a function is bounded is open. Let E be an arbitrary closed set. Is it possible to construct a function $f : \mathbb{R} \to \mathbb{R}$ so that the set of points at which f is not bounded is precisely the set E?

4.4.9 This exercise continues Exercise 4.3.23. Define the *Cantor ternary set* K to be the complement of the open set G of Exercise 4.3.23 in the interval $[0,1]$.

(a) If all the open intervals up to the nth stage in the construction of G are removed from the interval $[0,1]$, there remains a closed set K_n that is the union of a finite number of closed intervals. How many intervals?

(b) What is the sum of the lengths of these closed intervals that make up K_n?

(c) Show that $K = \bigcap_{n=1}^{\infty} K_n$.

(d) Sketch a picture of the set K by illustrating the sets K_1, K_2, and K_3.

(e) Show that if $x, y \in K$, $x < y$, then there is an open subinterval $I \subset (x, y)$ containing no points of K.

(f) Give an example of a number $z \in K \cap (0,1)$ that is not an endpoint of a component of G.

4.4.10 Express the closed interval $[0,1]$ as an intersection of a sequence of open sets. Can it also be expressed as a union of a sequence of open sets?

4.4.11 Express the open interval $(0,1)$ as a union of a sequence of closed sets. Can it also be expressed as an intersection of a sequence of closed sets?

4.5 Compactness Arguments

✂ Parts of this section could be cut in a short course. For a minimal approach to compactness arguments, you may wish to skip over all but the Bolzano-Weierstrass property. For all purposes of elementary real analysis this is sufficient. Proofs in the sequel that require a compactness argument will be supplied with one that uses the Bolzano-Weierstrass property and, perhaps, another that can be omitted.

In analysis we frequently encounter the problem of arguing from a set of "local" assumptions to a "global" conclusion. Let us focus on just one problem of this type and see the kind of arguments that can be used.

Local Boundedness of a Function Suppose that a function f is *locally bounded* at each point of a set E. By this we mean that for every point $x \in E$ there is an interval $(x-\delta, x+\delta)$ and f is bounded on the points in E that belong to that interval. Can we conclude that f is bounded on the whole of the set E?

Thus we have been given a local condition at each point x in the set E. There must be numbers δ_x and M_x so that

$$|f(t)| \le M_x \text{ for all } t \in E \text{ in the interval } (x-\delta_x, x+\delta_x).$$

The global condition we want, if possible, is to have some single number M that works for all $t \in E$; that is,

$$|f(t)| \le M \text{ for all } t \in E.$$

Two examples show that this depends on the nature of the set E.

Example 4.19 The function $f(x) = 1/x$ is locally bounded at each point x in the set $(0,1)$ but is not bounded on the set $(0,1)$. It is clear that f cannot be bounded on $(0,1)$ since the statement

$$\frac{1}{t} \le M \text{ for all } t \in (0,1)$$

cannot be true for any M. But this function is locally bounded at each point x here. Let $x \in (0,1)$. Take $\delta_x = x/2$ and $M_x = 2/x$. Then

$$f(t) = \frac{1}{t} \leq \frac{2}{x} = M_x$$

if

$$x/2 = x - \delta_x < t < x + \delta_x.$$

What is wrong here? What is there about this set $E = (0,1)$ that does not allow the conclusion? The point 0 is a point of accumulation of $(0,1)$ that does not belong to $(0,1)$, and so there is no assumption that f is bounded at that point. We avoid this difficulty if we assume that E is closed. ◀

Example 4.20 The function $f(x) = x$ is locally bounded at each point x in the set $[0,\infty)$ but is not bounded on the set $[0,\infty)$. It is clear that f cannot be bounded on $[0,\infty)$ since the statement

$$f(t) = t \leq M \text{ for all } t \in [0,\infty)$$

cannot be true for any M. But this function is locally bounded at each point x here. Let $x \in [0,\infty)$. Take $\delta_x = 1$ and $M_x = x+1$. Then

$$f(t) = t \leq x + 1 = M_x$$

if $x - 1 < t < x + 1$.

What is wrong here? What is there about this set $E = [0,\infty)$ that does not allow the conclusion. This set is closed and so contains all of its accumulation points so that the difficulty we saw in the preceding example does not arise. The difficulty is that the set is too big, allowing larger and larger bounds as we move to the right. We could avoid this difficulty if we assume that E is bounded. ◀

Indeed, as we shall see, we have reached the correct hypotheses now for solving our problem. The version of the theorem we were searching for is this:

> **Theorem** *Suppose that a function f is locally bounded at each point of a closed and bounded set E. Then f is bounded on the whole of the set E.*

Arguments that exploit the special features of closed and bounded sets of real numbers are called *compactness arguments.* Most often they are used to prove that some local property has global implications, which is precisely the nature of our boundedness theorem. We now solve our problem using various different compactness arguments. Each of these arguments will become a formidable tool in proving theorems in analysis. Many situations will arise in which some local property must be proved to hold globally, and compactness will play a huge role in these.

4.5.1 Bolzano-Weierstrass Property

A closed and bounded set has a special feature that can be used to design compactness arguments. This property is essentially a repeat of a property about convergent subsequences that we saw in Section 2.11.

Theorem 4.21 (Bolzano-Weierstrass Property) *A set of real numbers E is closed and bounded if and only if every sequence of points chosen from the set has a subsequence that converges to a point that belongs to E.*

Proof Suppose that E is both closed and bounded and let $\{x_n\}$ be a sequence of points chosen from E. Since E is bounded this sequence $\{x_n\}$ must be bounded too. We apply the Bolzano-Weierstrass theorem for sequences (Theorem 2.40) to obtain a subsequence $\{x_{n_k}\}$ that converges. If $x_{n_k} \to z$ then since all the points of the subsequence belong to E either the sequence is constant after some term or else z is a point of accumulation of E. In either case we see that $z \in E$. This proves the theorem in one direction.

In the opposite direction we suppose that a set E, which we do not know in advance to be either closed or bounded, has the Bolzano-Weierstrass property. Then E cannot be unbounded. For example, if E is unbounded above then there is a sequence of points $\{x_n\}$ of E with $x_n \to \infty$ or $-\infty$ and no subsequence of that sequence converges, contradicting the assumption.

Also, E must be closed. If not, there is a point of accumulation z that is not in E. This means that there is a sequence of points $\{x_n\}$ in E converging to z. But any subsequence of $\{x_n\}$ would also converge to z and, since $z \notin E$, we again have a contradiction. ■

This theorem can also be interpreted as a statement about accumulation points.

Corollary 4.22 *A set of real numbers E is closed and bounded if and only if every infinite subset of E has a point of accumulation that belongs to E.*

Let us use the Bolzano-Weierstrass property to prove our theorem about local boundedness.

> **Theorem** *Suppose that a function f is locally bounded at each point of a closed and bounded set E. Then f is bounded on the whole of the set E.*

Proof **(Bolzano-Weierstrass compactness argument)** To use this argument we will need to construct a sequence of points in E that we can use. Our proof is a proof by contradiction. If f is not bounded on E there must be a sequence of points $\{x_n\}$ chosen from E so that

$$|f(x_n)| > n.$$

If such a sequence could not be chosen, then at some stage, N say, there are no more points with $|f(x_N)| > N$ and N is an upper bound.

By compactness (i.e., by Theorem 4.21) there is a convergent subsequence $\{x_{n_k}\}$ converging to a point $z \in E$. By the local boundedness assumption there is an open interval $(z-\delta, z+\delta)$ and a number M_z so that

$$|f(t)| \leq M_z$$

whenever t is in E and inside that interval. But for all sufficiently large values of k, the point x_{n_k} must belong to the interval $(z-\delta, z+\delta)$. The two statements

$$|f(x_{n_k})| > n_k \text{ and } |f(x_{n_k})| \leq M_z$$

cannot both be true for all large k and so we have reached a contradiction, proving the theorem. ■

4.5.2 Cantor's Intersection Property

✂ *Enrichment*

A famous compactness argument, one that is used often in analysis, involves the intersection of a *descending* sequence of sets; that is, a sequence with

$$E_1 \supset E_2 \supset E_3 \supset E_4 \supset \ldots.$$

What conditions on the sequence will imply that

$$\bigcap_{n=1}^{\infty} E_n \neq \emptyset?$$

Example 4.23 An example shows that some conditions are needed. Suppose that for each $n \in \mathbb{N}$ we let $E_n = (0, 1/n)$. Then

$$E_1 \supset E_2 \supset E_3 \supset \ldots,$$

so $\{E_n\}$ is a descending sequence of sets with empty intersection. The same is true of the sequence $F_n = [n, \infty)$. Observe that the sets in the sequence $\{E_n\}$ are bounded (but not closed) while the sets in the sequence $\{F_n\}$ are closed (but not bounded). ◀

In a paper in 1879 Cantor described the following theorem and the role it plays in analysis. He pointed out that variants on this idea had been already used throughout most of that century, notably by Lagrange, Legendre, Dirichlet, Cauchy, Bolzano, and Weierstrass.

Theorem 4.24 *Let $\{E_n\}$ be a sequence of nonempty closed and bounded subsets of real numbers such that $E_1 \supset E_2 \supset E_3 \supset \ldots$. Let $E = \bigcap_{n=1}^{\infty} E_n$. Then E is not empty.*

Proof For each $i \in \mathbb{N}$ choose $x_i \in E_i$. The sequence $\{x_i\}$ is bounded since every point lies inside the bounded set E_1. Therefore, because of Theorem 4.21, $\{x_i\}$ has a convergent subsequence $\{x_{i_k}\}$. Let z denote that limit. Fix an integer m. Because the sets are descending, $x_{i_k} \in E_m$ for all sufficiently large $k \in \mathbb{N}$. But E_m is closed, from which it follows that $z \in E_m$. This is true for all $m \in \mathbb{N}$, so $z \in E$. ■

Corollary 4.25 (Cantor Intersection Theorem) *Suppose that $\{E_n\}$ is a sequence of nonempty closed subsets of real numbers such that*

$$E_1 \supset E_2 \supset E_3 \supset \dots .$$

If

$$\text{diameter } E_n \to 0,$$

then the intersection

$$E = \bigcap_{n=1}^{\infty} E_n$$

consists of a single point.

Proof Here the diameter of a nonempty, closed bounded set E would just be $\max E - \min E$, which exists and is finite for such a set (see Exercise 4.3.15). Since we are assuming that the diameters shrink to zero it follows that, at least for all sufficiently large n, E_n must be bounded.

That $E \neq \emptyset$ follows from Theorem 4.24. It remains to show that E contains only one point. Let $x \in E$ and $y \in \mathbb{R}$, $y \neq x$. Since diameter $E_n \to 0$, there exists $i \in \mathbb{N}$ such that diameter $E_i < |x - y|$. Since $x \in E_i$, y cannot be in E_i. Thus $y \notin E$ and $E = \{x\}$ as required. ■

Now we prove our theorem about local boundedness by using the Cantor intersection property to frame an argument.

> **Theorem** *Suppose that a function f is locally bounded at each point of a closed and bounded set E. Then f is bounded on the whole of the set E.*

Proof **(Cantor intersection compactness argument).** To use this argument we will need to construct a sequence of closed and bounded sets shrinking to a point. Our proof is again a proof by contradiction. Suppose that f is not bounded on E.

Since E is bounded we may assume that E is contained in some interval $[a, b]$. Divide that interval in half, forming two subintervals of the same length, namely $(b - a)/2$. At least one of these intervals contains points of E and f is unbounded on that interval. Call it $[a_1, b_1]$.

Now do the same to the interval $[a_1, b_1]$. Divide that interval in half, forming two subintervals of the same length, namely $(b - a)/4$. At least one of these intervals contains points of E and f is unbounded on that interval. Call it $[a_2, b_2]$. Continue this process inductively, producing a descending sequence of intervals $\{[a_n, b_n]\}$ so that the nth interval $[a_n, b_n]$ has length $(b - a)/2^n$, contains points of E, and f is unbounded on $E \cap [a_n, b_n]$.

By the Cantor intersection property there is a single point $z \in E$ contained in all of these intervals. But by our local boundedness assumption there is an interval $(z - c, z + c)$ so that f is bounded on the points of E in that interval. For any large enough value of n, though, the interval $[a_n, b_n]$ would be contained inside the interval $(z-c, z+c)$. This would be impossible and so we have reached a contradiction, proving the theorem. ■

4.5.3 Cousin's Property

✂ *Enrichment*

Another compactness argument dates back to Pierre Cousin in the last years of the nineteenth century. This exploits the order of the real line and considers how small intervals may be pieced together to give larger intervals. The larger interval $[a, b]$ is subdivided

$$a = x_0 < x_1 < \cdots < x_n = b$$

and then expressed as a finite union of nonoverlapping subintervals said to form a *partition*:

$$[a, b] = \bigcup_{i=1}^{n} [x_{i-1}, x_i].$$

This again provides us with a compactness argument since it allows a way to argue from the local to the global.

Lemma 4.26 (Cousin) *Let $\mathcal{C}$ be a collection of closed subintervals of $[a, b]$ with the property that for each $x \in [a, b]$ there exists $\delta = \delta(x) > 0$ such that $\mathcal{C}$ contains all intervals $[c, d] \subset [a, b]$ that contain x and have length smaller than δ. Then there exists a partition*

$$a = x_0 < x_1 < \cdots < x_n = b$$

of $[a, b]$ such that $[x_{i-1}, x_i] \in \mathcal{C}$ for $i = 1, \ldots, n$.

This lemma makes precise the statement that if a collection of closed intervals contains all "sufficiently small" ones for $[a, b]$, then it contains a partition of $[a, b]$. We shall frequently see the usefulness of such a partition. This is the most elementary of a collection of tools called *covering theorems.* Roughly, a *cover* of a set is a family of intervals covering the set in the sense that each point in the set is contained in one or more of the intervals.

We formalize the assumption in Cousin's lemma in this language:

Definition 4.27 (Full Cover) A collection $\mathcal{C}$ of closed intervals satisfying the hypothesis of Cousin's lemma is called a *full cover* of $[a, b]$.

Proof **(Proof of Cousin's lemma)** Let us, in order to obtain a contradiction, suppose that $\mathcal{C}$ does not contain a partition of the interval $[a, b]$. Let c be the midpoint of that interval and consider the two subintervals $[a, c]$ and $[c, b]$. If $\mathcal{C}$ contains a partition of both intervals $[a, c]$ and $[c, b]$, then by putting those partitions together we can obtain a partition of $[a, b]$, which we have supposed is impossible.

Let $I_1 = [a, b]$ and let I_2 be either $[a, c]$ or $[c, b]$ chosen so that $\mathcal{C}$ contains no partition of I_2. Inductively we can continue in this fashion, obtaining a shrinking sequence of intervals $I_1 \supset I_2 \supset I_3 \supset \dots$ so that the length of I_n is $(b-a)/2^{n-1}$ and $\mathcal{C}$ contains no partition of I_n.

By the Cantor intersection theorem (Theorem 4.25) there is a single point z in all of these intervals. The interval $(z - \delta(z), z + \delta(z))$ contains I_n for all sufficiently large n and so, by definition, $I_n \in \mathcal{C}$. In particular, $\mathcal{C}$ does indeed contain a partition of that interval I_n since the single interval $\{I_n\}$ is itself a partition. But this contradicts the way in which the sequence was chosen and this contradiction completes our proof. ■

Now we reprove our theorem about local boundedness by using Cousin's property to frame an argument.

> **Theorem** *Suppose that a function f is locally bounded at each point of a closed and bounded set E. Then f is bounded on the whole of the set E.*

Proof **(Cousin compactness argument)** The set E is bounded and so is contained in some interval $[a, b]$. Let us say that an interval $[c, d] \subset [a, b]$ is "black" if the following statement is true:

> There is a number M (which may depend on $[c, d]$) so that $|f(t)| \leq M$ for all $t \in E$ that are in the interval $[c, d]$.

The collection of all black intervals is a full cover of $[a, b]$. This is because of the local boundedness assumption on f. Consequently, by Cousin's lemma, there is a partition of the interval $[a, b]$ consisting of black intervals. The function f is bounded in E on each of these finitely many black intervals and so, since there are only finitely many of them, f must be bounded on E in $[a, b]$. But $[a, b]$ includes all of E and so the proof is complete. ■

4.5.4 Heine-Borel Property

✂ *Advanced*

Another famous compactness property involves covers too, as in the Cousin lemma, but this time covers consisting of open intervals. This theorem has wide applications, including again extensions of local properties to global ones. You may find this compactness argument more difficult to work with than the others. On the real line all of the arguments here are equivalent and, in most cases, any one will do the job. Why not use the simpler ones then? The answer is that in more general spaces than the real line these other versions may be more useful. Time spent learning them now will pay off in later courses.

The property we investigate is named after two mathematicians, Émile Borel (1871–1956) and Heinrich Eduard Heine (1821–1881), whose names have become closely attached to these ideas.

We begin with some definitions.

Definition 4.28 (Open Cover) Let $A \subset \mathbb{R}$ and let $\mathcal{U}$ be a family of open intervals. If for every $x \in A$ there exists at least one interval $U \in \mathcal{U}$ such that $x \in U$, then $\mathcal{U}$ is called an *open cover* of A.

Definition 4.29 (Heine-Borel Property) A set $A \subset \mathbb{R}$ is said to have the *Heine-Borel property* if every open cover of A can be reduced to a finite subcover. That is, if $\mathcal{U}$ is an open cover of A, then there exists a finite subset of $\mathcal{U}$, $\{U_1, U_2, \dots, U_n\}$ such that

$$A \subset U_1 \cup U_2 \cup \cdots \cup U_n.$$

Example 4.30 Any finite set has the Heine-Borel property. Just take one interval from the cover for each element in the finite set. ◀

Example 4.31 The set $\mathbb{N}$ does not have the Heine-Borel property. Take, for example, the collection of open intervals

$$\{(0, n) : n = 1, 2, 3, \dots\}.$$

While this forms an open cover of $\mathbb{N}$, no finite subcollection could also be an open cover. ◀

Example 4.32 The set $A = \{1/n : n \in \mathbb{N}\}$ does not have the Heine-Borel property. Take, for example, the collection of open intervals

$$\{(1/n, 2) : n = 1, 2, 3, \dots\}.$$

While this forms an open cover of A, no finite subcollection could also be an open cover. ◀

Observe in these examples that $\mathbb{N}$ is closed (but not bounded) while A is bounded (but not closed). We shall prove, in Theorem 4.33, that a set A has the Heine-Borel property if and only if that set is both closed and bounded.

Theorem 4.33 (Heine-Borel) *A set $A \subset \mathbb{R}$ has the Heine-Borel property if and only if A is both closed and bounded.*

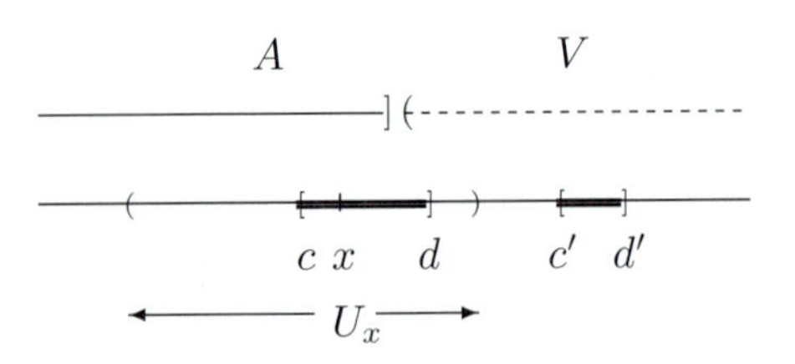

Figure 4.2. The two types of intervals in the proof of Theorem 4.33.

Proof Suppose $A \subset \mathbb{R}$ is both closed and bounded, and $\mathcal{U}$ is an open cover for A. We may assume $A \neq \emptyset$, otherwise there is nothing to prove. Let $[a, b]$ be the smallest closed interval containing A; that is,

$$a = \inf\{x : x \in A\} \text{ and } b = \sup\{x : x \in A\}.$$

Observe that $a \in A$ and $b \in A$. We shall apply Cousin's lemma to the interval $[a, b]$, so we need to first define an appropriate full cover of $[a, b]$.

For each $x \in A$, since $\mathcal{U}$ is an open cover of A, there exists an open interval $U_x \in \mathcal{U}$ such that $x \in U_x$. Since U_x is open, there exists $\delta(x) > 0$ for which $(x - t, x + t) \subset U_x$ for all $t \in (0, \delta(x))$. This defines $\delta(x)$ for points in A. Now consider points in $V = [a, b] \setminus A$. We must define $\delta(x)$ for points of V. Since A is closed and $\{a, b\} \subset A$, V is open (why?); thus for each $x \in V$ there exists $\delta(x) > 0$ such that $(x - t, x + t) \subset V$ for all $t \in (0, \delta(x))$. We can therefore obtain a full cover $\mathcal{C}$ of $[a, b]$ as follows: An interval $[c, d]$ is a member of $\mathcal{C}$ if there exists $x \in [a, b]$ such that either (i) $x \in A$ and $x \in [c, d] \subset U_x$ or (ii) $x \in V$ and $x \in [c, d] \subset V$.

Observe that an interval of type (i) can contain points of V, but an interval of type (ii) cannot contain points of A. Figure 4.2 illustrates examples of both types of intervals. In that figure $[c, d] \subset U_x$ is an interval of type (i) in $\mathcal{C}$; $[c', d'] \subset V$ is an interval of type (ii) in $\mathcal{C}$.

It is clear that $\mathcal{C}$ forms a full cover of $[a, b]$. From Cousin's lemma we infer the existence of a partition $a = x_0 < x_1 < \cdots < x_n = b$ with $[x_{i-1}, x_i] \in \mathcal{C}$ for $i = 1, \ldots, n$. Each of the intervals $[x_{i-1}, x_i]$ is either contained in V (in which case it is disjoint from A) or is contained in some member $U_i \in \mathcal{U}$. We now "throw away" from the partition those intervals that contain only points of V, and the union of the remaining closed intervals covers all of A. Each interval of this finite collection is contained in some open interval U from the cover $\mathcal{U}$. More precisely, let

$$S = \{i : 1 \leq i \leq n \text{ and } [x_{i-1}, x_i] \subset U_i\}.$$

Then

$$A \subset \bigcup_{i \in S} [x_{i-1}, x_i] \subset \bigcup_{i \in S} U_i,$$

so $\{U_i : i \in S\}$ is the required subcover of A.

To prove the converse, we must show that if A is not bounded or is not closed, then there exists an open cover of A with no finite subcover. Suppose first that A is not bounded. Then there must exist either an increasing sequence of points $\{x_n\}$ contained in A so that $x_n \to \infty$ or a decreasing sequence of points $\{x_n\}$ contained in A so that $x_n \to -\infty$. Let us suppose the former. For each $i \in \mathbb{N}$ let $U_1 = (-\infty, x_1)$, $U_{i+1} = (x_i, x_{i+1})$ and $V_i = (x_i - 1, x_{i+1})$. Finally, let $\mathcal{U}$ be the collection of all the intervals U_i, V_i for $i = 1, 2, 3, \ldots$. Then $\mathcal{U}$ is an open cover of A. (Indeed it is an open cover of all of $\mathbb{R}$.) But it is clear that $\mathcal{U}$ contains no finite subcover of A since, for any integer N, the totality of all the sets U_i, V_i for $i = 1, 2, 3, \ldots N$ cannot cover all of A since no point x_n with $n > N$ can belong to any of these intervals.

Now suppose A is not closed. Then there is a point of accumulation z of A that does not belong to A. Then there must exist either an increasing sequence of points $\{x_n\}$ contained in A so that $x_n \to z$ or a decreasing sequence of points $\{x_n\}$ contained in A so that $x_n \to z$. Suppose the former. For each $i \in \mathbb{N}$ let $U_1 = (-\infty, x_1)$, $V_1 = (z, \infty)$, $U_{i+1} = (x_i, x_{i+1})$ and $V_i = (x_i - 1, x_{i+1})$. Then $\mathcal{U}$ is an open cover of A. (Indeed, as before, it is an open cover of all of $\mathbb{R}$.) But it is clear that $\mathcal{U}$ contains no finite subcover of A since, for any integer N, the totality of all the sets U_i, V_i for $i = 1, 2, 3, \ldots N$ cannot cover all of A since no point x_n with $n > N$ can belong to any of these intervals. ■

Once again, we return to our sample theorem, which shows how a local property can be used to prove a global condition, this time using a Heine-Borel compactness argument.

> **Theorem** *Suppose that a function f is locally bounded at each point of a closed and bounded set E. Then f is bounded on the whole of the set E.*

Proof **(Heine-Borel compactness argument).** As f is locally bounded at each point of E, for every $x \in E$ there exists an open interval U_x containing x and a positive number M_x such that $|f(t)| < M_x$ for all $t \in U_x \cap E$. Let $\mathcal{U} = \{U_x : x \in E\}$. Then $\mathcal{U}$ is an open cover of E. By the Heine-Borel theorem there exists

$$\{U_{x_1}, U_{x_2}, \ldots, U_{x_n}\}$$

such that

$$E \subset U_{x_1} \cup U_{x_2} \cup \cdots \cup U_{x_n}.$$

Let

$$M = \max\{M_{x1}, M_{x2}, \ldots, M_{xn}\}.$$

Let $x \in E$. Then there exists i, $1 \leq i \leq n$, for which $x \in U_i$. Since

$$|f(x)| \leq M_{x_i} \leq M$$

we conclude that f is bounded on E. ■

Our ability to reduce $\mathcal{U}$ to a *finite* subcover in the proof of this theorem was crucial. You may wish to use the function $f(x) = 1/x$ on $(0, 1]$ to appreciate this statement.

4.5.5 Compact Sets

We have seen now a wide range of techniques called compactness arguments that can be applied to a set that is closed and bounded. We now introduce the modern terminology for such sets.

Definition 4.34 A set of real numbers E is said to be *compact* if it has any of the following equivalent properties:

1. E is closed and bounded.
2. E has the Bolzano-Weierstrass property.
3. E has the Heine-Borel property.

In spaces more general than the real line there may be analogues of the notions of closed, bounded, convergent sequences, and open covers. Thus there can also be analogues of closed and bounded sets, the Bolzano-Weierstrass property, and the Heine-Borel property. In these more general spaces the three properties are not always equivalent and it is the Heine-Borel property that is normally chosen as the definition of compact sets there. Even so, a thorough understanding of compactness arguments on the real line is an excellent introduction to these advanced and important ideas in other settings.

If we return to our sample theorem we see that now, perhaps, it should best be described in the language of compact sets:

> **Theorem** *Suppose that E is compact. Then every function $f : E \to \mathbb{R}$ that is locally bounded on E is bounded on the whole of the set E. Conversely, if every function $f : E \to \mathbb{R}$ that is locally bounded on E is bounded on the whole of the set E, then E must be compact.*

In real analysis there are many theorems of this type. The concept of compact set captures exactly when many local conditions can have global implications.

Exercises

4.5.1 Give an example of a function $f : \mathbb{R} \to \mathbb{R}$ that is not locally bounded at any point.

4.5.2 Show directly that the interval $[0, \infty)$ does not have the Bolzano-Weierstrass property.

4.5.3 Show directly that the interval $[0, \infty)$ does not have the Heine-Borel property.

4.5.4 Show directly that the set $[0, 1] \cap \mathbb{Q}$ does not have the Heine-Borel property.

4.5.5 Develop the properties of compact sets. For example, is the union of a pair of compact sets compact? The intersection? The union of a family of compact sets?

4.5.6 Show directly that the union of two sets with the Bolzano-Weierstrass property must have the Bolzano-Weierstrass property.

4.5.7 Show directly that the union of two sets with the Heine-Borel property must have the Heine-Borel property.

4.5.8 We defined an open cover of a set E to consist of open *intervals* covering E. Let us change that definition to allow an open cover to consist of any family of open *sets* covering E. What changes are needed in the proof of Theorem 4.33 so that it remains valid in this greater generality?

4.5.9 A function $f : \mathbb{R} \to \mathbb{R}$ is said to be *locally increasing* at a point x_0 if there is a $\delta > 0$ so that

$$f(x) < f(x_0) < f(y)$$

whenever

$$x_0 - \delta < x < x_0 < y < x_0 + \delta.$$

Show that a function that is locally increasing at every point in $\mathbb{R}$ must be increasing; that is, that $f(x) < f(y)$ for all $x < y$.

4.5.10 Let $f : E \to \mathbb{R}$ have this property: For every $e \in E$ there is an $\varepsilon > 0$ so that

$$f(x) > \varepsilon \text{ if } x \in E \cap (e - \varepsilon, e + \varepsilon).$$

Show that if the set E is compact then there is some positive number c so that

$$f(e) > c$$

for all $e \in E$. Show that if E is not closed or is not bounded, then this conclusion may not be valid.

4.5.11 Prove the following variant of Lemma 4.26:

> Let $\mathcal{C}$ be a collection of closed subintervals of $[a, b]$ with the property that for each $x \in [a, b]$ there exists $\delta = \delta(x) > 0$ such that $\mathcal{C}$ contains all intervals $[c, d] \subset [a, b]$ that contain x and have

length smaller than δ. Suppose that $\mathcal{C}$ has the property that if $[\alpha, \beta]$ and $[\beta, \gamma]$ both belong to $\mathcal{C}$ then so too does $[\alpha, \gamma]$. Then $[a, b]$ belongs to $\mathcal{C}$.

4.5.12 Use the version of Cousin's lemma given in Exercise 4.5.11 to give a simpler proof of the sample theorem on local boundedness.

✂ **4.5.13** Give an example of an open covering of the set $\mathbb{Q}$ of rational numbers that does not reduce to a finite subcover.

4.5.14 Suppose that E is closed and K is compact. Show that $E \cap K$ is compact. Do this in two ways (using the definition and using the Bolzano-Weierstrass property).

4.5.15 Prove that every function $f : E \to \mathbb{R}$ that is locally bounded on E is bounded on the whole of the set E only if the set E is compact, by supplying the following two constructions:

(a) Show that if the set E is not bounded, then there is an unbounded function $f : E \to \mathbb{R}$ so that f is locally bounded on E.

(b) Show that if the set E is not closed, then there is an unbounded function $f : E \to \mathbb{R}$ so that f is locally bounded on E.

✂ **4.5.16** Suppose that E is closed and K is compact. Show that $E \cap K$ is compact using the Heine-Borel property.

4.5.17 Suppose that E is compact. Is the set of boundary points of E also compact?

✂ **4.5.18** Prove Lindelöff's covering theorem:

Let $\mathcal{C}$ be a collection of open intervals such that every point of a set E belongs to at least one of the intervals. Then there is a sequence of intervals I_1, I_2, I_3, ... chosen from $\mathcal{C}$ that also covers E.

4.5.19 Describe briefly the distinction between the covering theorem of Lindelöff (Exercise 4.5.18) and that of Heine-Borel.

✂ **4.5.20** We have seen that the following four conditions on a set $A \subset \mathbb{R}$ are equivalent:

(a) A is closed and bounded.

(b) Every infinite subset of A has a limit point in A.

(c) Every sequence of points from A has a subsequence converging to a point in A.

(d) Every open cover of A has a finite subcover.

Prove directly that (b)⇒(c), (b)⇒(d) and (c)⇒(d).

4.5.21 Let f be a function that is locally bounded on a compact interval $[a, b]$. Let

$$S = \{a < x \leq b : \ f \text{ is bounded on } [a, x]\}.$$

(a) Show that $S \neq \emptyset$.

(b) Show that if $z = \sup S$, then $a < z \leq b$.

(c) Show that $z \in S$.

(d) Show that $z = b$ by showing that $z < b$ is impossible.

Using these steps, construct a proof of the sample theorem on local boundedness.

4.6 Countable Sets

As part of our discussion of properties of sets in this chapter let us review a special property of sets that relates, not to their topological properties, but to their size. We can divide sets into finite sets and infinite sets. How do we divide infinite sets into "large" and "larger" infinite sets?

We did this in our discussion of sequences in Section 2.3. (If you skipped over that section now is a good time to go back.) If an infinite set E has the property that the elements of E can be written as a list (i.e., as a sequence)

$$e_1, e_2, e_3, \ldots, e_n \ldots,$$

then that set is said to be *countable.* Note that this property has nothing particularly to do with the other properties of sets encountered in this chapter. It is yet another and different way of classifying sets.

The following properties review our understanding of countable sets. Remember that the empty set, any finite set, and any infinite set that can be listed are all said to be countable. An infinite set that cannot be listed is said to be *uncountable.*

Theorem 4.35 *Countable sets have the following properties:*

1. *Any subset of a countable set is countable.*

2. *Any union of a sequence of countable sets is countable.*

3. *No interval is countable.*

Exercises

4.6.1 Give examples of closed sets that are countable and closed sets that are uncountable.

4.6.2 Is there a nonempty open set that is countable?

4.6.3 If a set is countable, what can you say about its complement?

4.6.4 Is the intersection of two uncountable sets uncountable?

4.6.5 Show that the Cantor set of Exercise 4.3.23 is infinite and uncountable.

4.6.6 Give (if possible) an example of a set with

(a) Countably many points of accumulation
(b) Uncountably many points of accumulation
(c) Countably many boundary points
(d) Uncountably many boundary points
(e) Countably many interior points
(f) Uncountably many interior points

4.6.7 A set is said to be *co-countable* if it has a countable complement. Show that the intersection of finitely many co-countable sets is itself co-countable.

4.6.8 Let E be a set and $f : \mathbb{R} \to \mathbb{R}$ be an *increasing* function [i.e., if $x < y$, then $f(x) < f(y)$]. Show that E is countable if and only if the image set $f(E)$ is countable. (What property other than "increasing" would work here?)

4.6.9 Show that every uncountable set of real numbers has a point of accumulation.

4.6.10 Let $\mathcal{F}$ be a family of (nondegenerate) intervals; that is, each member of $\mathcal{F}$ is an interval (open, closed or neither) but is not a single point. Suppose that any two intervals I and J in the family have no point in common. Show that the family $\mathcal{F}$ can be arranged in a sequence $I_1, I_2, \ldots$.

4.7 Challenging Problems for Chapter 4

4.7.1 Cantor, in 1885, defined a set E to be *dense-in-itself* if $E \subset E'$. Develop some facts about such sets. Include illustrative examples.

4.7.2 One of Cantor's early results in set theory is that for every closed set E there is a set S with $E = S'$. Attempt a proof.

4.7.3 Can the closed interval $[0, 1]$ be expressed as the union of a sequence of disjoint closed subintervals each of length smaller than 1?

4.7.4 In many applications of open sets and closed sets we wish to work just inside some other set A. It is convenient to have a language for this. A set $E \subset A$ is said to be *open relative*

to A if $E = A \cap G$ for some set $G \subset \mathbb{R}$ that is open. A set $E \subset A$ is said to be *closed relative* to A if $E = A \cap F$ for some set $F \subset \mathbb{R}$ that is closed. Answer the following questions.

(a) Let $A = [0, 1]$ describe, if possible, sets that are open relative to A but not open as subsets of $\mathbb{R}$.
(b) Let $A = [0, 1]$ describe, if possible, sets that are closed relative to A but not closed as subsets of $\mathbb{R}$.
(c) Let $A = (0, 1)$ describe, if possible, sets that are open relative to A but not open as subsets of $\mathbb{R}$.

(d) Let $A = (0, 1)$ describe, if possible, sets that are closed relative to A but not closed as subsets of $\mathbb{R}$.

4.7.5 Let $A = \mathbb{Q}$. Give examples of sets that are neither open nor closed but are both relative to $\mathbb{Q}$.

4.7.6 Show that all the subsets of $\mathbb{N}$ are both open and closed relative to $\mathbb{N}$.

4.7.7 Introduce for any set $E \subset \mathbb{R}$ the notation

$$\partial E = \{x : \ x \text{ is a boundary point of } E\}.$$

(a) Show for any set E that $\partial E = \overline{E} \cap \overline{(\mathbb{R} \setminus E)}$.

(b) Show that for any set E the set ∂E is closed.

(c) For what sets E is it true that $\partial E = \emptyset$?

(d) Show that $\partial E \subset E$ for any closed set E.

(e) If E is closed, show that $\partial E = E$ if and only if E has no interior points.

(f) If E is open, show that ∂E can contain no interval.

4.7.8 Let E be a nonempty set of real numbers and define the function

$$f(x) = \inf\{|x - e| : e \in E\}.$$

(a) Show that $f(x) = 0$ for all $x \in E$.

(b) Show that $f(x) = 0$ if and only if $x \in \overline{E}$.

(c) Show for any nonempty closed set E that

$$\{x \in \mathbb{R} : f(x) > 0\} = (\mathbb{R} \setminus E).$$

4.7.9 Let $f : \mathbb{R} \to \mathbb{R}$ have this property: For every $x_0 \in \mathbb{R}$ there is a $\delta > 0$ so that

$$|f(x) - f(x_0)| < |x - x_0|$$

whenever $0 < |x - x_0| < \delta$. Show that for all $x, \ y \in \mathbb{R}, \ x \neq y$,

$$|f(x) - f(y)| < |x - y|.$$

4.7.10 Let $f : E \to \mathbb{R}$ have this property: For every $e \in E$ there is an $\varepsilon > 0$ so that

$$f(x) > \varepsilon \text{ if } x \in E \cap (e - \varepsilon, e + \varepsilon).$$

Show that if the set E is compact, then there is some positive number c so that

$$f(e) > c$$

for all $e \in E$. Show that if E is not closed or is not bounded, then this conclusion may not be valid.

4.7.11 **(Separation of Compact Sets)** Let A and B be nonempty sets of real numbers and let

$$\delta(A, B) = \inf\{|a - b| : a \in A, b \in B\}.$$

$\delta(A, B)$ is often called the "distance" between the sets A and B.

(a) Prove $\delta(A, B) = 0$ if $A \cap B \neq \emptyset$.

(b) Give an example of two closed, disjoint sets in $\mathbb{R}$ for which $\delta(A, B) = 0$.

(c) Prove that if A is compact, B is closed, and $A \cap B = \emptyset$, then $\delta(A, B) > 0$.

4.7.12 Show that every closed set can be expressed as the intersection of a sequence of open sets.

4.7.13 Show that every open set can be expressed as the union of a sequence of closed sets.

4.7.14 A collection of sets $\{S_\alpha : \alpha \in A\}$ is said to have the *finite intersection property* if every finite subfamily has a nonempty intersection.

(a) Show that if $\{S_\alpha : \alpha \in A\}$ is a family of compact sets that has the finite intersection property, then

$$\bigcap_{\alpha \in A} S_\alpha \neq \emptyset.$$

(b) Give an example of a collection of closed sets $\{S_\alpha : \alpha \in A\}$ that has the finite intersection property and yet

$$\bigcap_{\alpha \in A} S_\alpha = \emptyset.$$

4.7.15 A set $S \subset \mathbb{R}$ is said to be *disconnected* if there exist two disjoint open sets U and V each containing a point of S so that $S \subset U \cup V$. A set that is not disconnected is said to be *connected.*

(a) Give an example of a disconnected set.

(b) Show that every compact interval $[a, b]$ is connected.

(c) Show that $\mathbb{R}$ is connected.

(d) Show that every nonempty connected set is an interval.

4.7.16 Show that the only subsets of $\mathbb{R}$ that are both open and closed are $\emptyset$ and $\mathbb{R}$.

4.7.17 Given any uncountable set of real numbers E show that it is possible to extract a sequence $\{a_k\}$ of distinct terms of E so that the series $\sum_{k=1}^{\infty} a_k/k$ diverges.

Chapter 5

CONTINUOUS FUNCTIONS

5.1 Introduction to Limits

The definition of the limit of a function

$$\lim_{x \to x_0} f(x)$$

is given in calculus courses, but in many classes it is not explored to any great depth. Computation of limits is interesting and offers its challenges, but for a course in real analysis we must master the definition itself and derive its consequences.

Our viewpoint is larger than that in most calculus treatments. There it is common to insist, in order for a limit to be defined, that the function f must be defined at least in some interval $(x_0 - \delta, x_0 + \delta)$ that contains the point x_0 (with the possible exception of x_0 itself). Here we must allow a function f that is defined only on some set E and study limits for points x_0 that are not too remote from E. We do not insist that x_0 be in the domain of f but we do require that it be "close." This requirement is expressed using our language from Chapter 4. We must have x_0 a point of accumulation of E.

Except for this detail about the domain of the function the definition we use is the usual ε-δ definition from calculus. Readers familiar with the sequence limit definitions of Chapter 2 will have no trouble handling this definition. It is nearly the same in general form as the ε-N definition for sequences, and many of the proofs use similar ideas.

5.1.1 Limits (ε-δ Definition)

The definition of a sequence limit, $\lim_{n\to\infty} s_n$, made precise the statement that s_n is arbitrarily close to L if n is sufficiently large. The definition of a function limit

$$\lim_{x \to x_0} f(x)$$

is intended, in much the same way, to make precise the statement that $f(x)$ is arbitrarily close to L if x is sufficiently close to x_0. One feature of the definition must be to exclude the value at the point x_0 from consideration; it should be irrelevant to the value of the limit. It is possible (likely even) that $f(x_0) = L$, but whether this is true or false should not be any influence on the existence of the limit.

Thus the definition assumes the following form. The requirement that x_0 be a point of accumulation of E may seem strange at first sight, but we will see that it is needed in order for the definition to have some meaning. Without it any number would be the limit and the theory of limits would be useless.

Definition 5.1 (Limit) Let $f : E \to \mathbb{R}$ be a function with domain E and suppose that x_0 is a point of accumulation of E. Then we write

$$\lim_{x \to x_0} f(x) = L$$

if for every $\varepsilon > 0$ there is a $\delta > 0$ so that

$$|f(x) - L| < \varepsilon$$

whenever x is a point of E differing from x_0 and satisfying $|x - x_0| < \delta$.

Note. The condition on x can be written as

$$0 < |x - x_0| < \delta$$

or as

$$x \in (x_0 - \delta, x_0 + \delta), \quad x \neq x_0$$

or, yet again, as

$$x_0 - \delta < x < x_0 + \delta, \quad x \neq x_0.$$

The exclusion of $x = x_0$ should be seen as an advantage here. An inequality is required to be true for all x satisfying some condition, and we are allowed *not* to have to check $x = x_0$. It may happen to be true that $|f(x) - L| < \varepsilon$ when $x = x_0$ but it is irrelevant to the definition. For example, you will recall that the limit used to define a derivative

$$f'(x_0) = \lim_{x \to x_0} \frac{f(x) - f(x_0)}{x - x_0}$$

must require that the value for $x = x_0$ be excluded; the expression is not defined when $x = x_0$.

See Figure 5.1 for a graphical interpretation of the definition. In the picture a particular value of ε is illustrated and for that value the figure shows a choice of δ that works. Every smaller value of δ would have worked, too. The definition requires doing this, however, for *every* positive ε, and the figure cannot convey that.

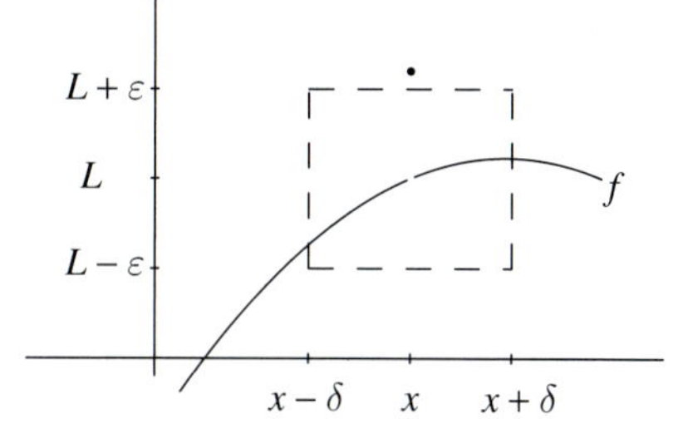

Figure 5.1. Graphical interpretation of the ε-δ limit definition.

We now present some examples illustrating how to prove the existence of a limit directly from the definition. These are to be considered as exercises in understanding the definition. We would rarely use the definition to compute a limit, and we hope seldom to use the definition to verify one; we will use the definition to develop a theory that will verify limits for us.

Example 5.2 Any function $f(x) = ax + b$ will have the easily predicted limit

$$\lim_{x \to x_0} f(x) = \lim_{x \to x_0} (ax + b) = ax_0 + b.$$

If you sketch a picture similar to that of Figure 5.1 you see easily that the choice of δ is monitored by the slope of the line $y = ax + b$. The steeper the slope, the smaller the δ has to be taken in comparison with ε.

Let us do this for the linear function $f(x) = 10x - 11$. We expect that

$$\lim_{x \to 5}(10x - 11) = 10(5) - 11 = 39.$$

Let us prove this. We need a condition ensuring that the expression

$$|(10x - 11) - 39|$$

is smaller than ε. Some arithmetic converts this to

$$|(10x - 11) - 39| = |10x - 50| = |10|\ |x - 5|\,.$$

Now it is clear that if we can insist that $|x - 5| < \varepsilon/10$ we will have $|(10x - 11) - 39| < \varepsilon$. That completes the proof. Better, though, would be to write it in a more straightforward manner that obscures how we did it but gets to the point of the proof more simply:

Let $\varepsilon > 0$. Let $\delta = \varepsilon/10$. Then for all x with $|x - 5| < \delta$ we have

$$|(10x - 11) - 39| = |10|\ |x - 5| < 10\delta = \varepsilon.$$

By definition, $\lim_{x \to 5} 10x - 11 = 39$ as required.

An alert reader of our short proof will know that the choice of δ as $\varepsilon/10$ took some time to compute and is not just an inspired second sentence of the proof. ◀

Example 5.3 Let us use the definition to verify the existence of

$$\lim_{x \to x_0} x^2.$$

Again the definition gives no hints as how to compute the limit; it can be used only to verify the correctness of a limit statement. To keep it simple let us show that $\lim_{x \to 3} x^2 = 9$. We need a condition ensuring that the expression

$$\left|x^2 - 9\right|$$

is smaller than ε. Some arithmetic converts this to

$$\left|x^2 - 9\right| = |x - 3|\ |x + 3|\,.$$

If we insist that

$$|x - 3| < \varepsilon/M,$$

where M is bigger than any value of $|x + 3|$, then we will have $\left|x^2 - 9\right| < \varepsilon$ exactly as we need. But just how big might $|x + 3|$ be? If we remember that we are interested only in values of x close to 3 (not huge values of x), then this is not too big. For example, if x stays inside $(2, 4)$, then $|x + 3| < 7$. These are enough computations to allow us to write up a proof.

> Let $\varepsilon > 0$. Let $\delta = \varepsilon/7$ or $\delta = 1$, whichever is smaller (i.e., $\delta = \min\{\varepsilon/7, 1\}$). Then if $|x - 3| < \delta$ it follows that
>
> $$|x + 3| = |x - 3 + 6| \leq |x - 3| + 6 < 7$$
>
> and hence that
>
> $$\left|x^2 - 9\right| = |x - 3|\ |x + 3| < 7\,|x - 3| < 7(\varepsilon/7) = \varepsilon.$$
>
> By definition, $\lim_{x \to 3} x^2 = 9$ as required.

The finished proof is shorter and lacks all the motivating steps that we just went through. ◀

In spite of these examples and the necessity in elementary courses such as this to work through similar examples, the main goal of our definition is to build up a theory of limits that can then be used to justify other methods of computation and lead to new discoveries. On occasions we must, however, return to the definition to handle an unusual case.

Exercises

5.1.1 Prove the existence of the limit $\lim_{x\to x_0}(4-12x)$.

5.1.2 Prove the validity of the limit $\lim_{x\to x_0}(ax+b)=ax_0+b$.

5.1.3 Prove the existence of the limit $\lim_{x\to -4} x^2$.

5.1.4 Prove the validity of the limit $\lim_{x\to x_0} x^2 = x_0^2$.

5.1.5 Suppose in the definition of the limit that the phrase "x_0 be a point of accumulation of the domain of f" is deleted. Show that then the limit statement $\lim_{x\to -2}\sqrt{x}=L$ would be true for every number L.

5.1.6 Recall that in the definition of $\lim_{x\to x_0} f(x)$ there is a requirement that x_0 be a point of accumulation of the domain of f. Which values of x_0 would be excluded from consideration in the limit

$$\lim_{x\to x_0}\sqrt{x^2-2}?$$

5.1.7 Which values of x_0 would be excluded from consideration in the limit

$$\lim_{x\to x_0}\arcsin|x+2|?$$

5.1.8 Prove the validity of the limit $\lim_{x\to x_0}\sqrt{x}=\sqrt{x_0}$.

5.1.9 Prove that the limit $\lim_{x\to 0}\frac{1}{x}$ fails to exist.

5.1.10 Prove that the limit $\lim_{x\to 0}\sin(1/x)$ fails to exist.

5.1.11 Using the definition, show that if $\lim_{x\to x_0} f(x)=L$, then

$$\lim_{x\to x_0}|f(x)|=|L|.$$

5.1.12 Suppose that x_0 is a point of accumulation of both A and B and that $f: A\to\mathbb{R}$ and $g: B\to\mathbb{R}$. We insist that f and g must agree in the sense that $f(x)=g(x)$ if x is in both A and B.

(a) What conditions on A and B ensure that if $\lim_{x\to x_0} f(x)$ exists so too must $\lim_{x\to x_0} g(x)$?

(b) What conditions on A and B ensure that if

$$\lim_{x\to x_0} f(x) \text{ and } \lim_{x\to x_0} g(x)$$

both exist then they must be equal.

5.1.2 Limits (Sequential Definition)

The theory of function limits can be reduced to the theory of sequence limits. This is a popular device in mathematics. Some new theory turns out to be contained in an old theory. This allows easy proofs of results since the old theory has all the tools needed for constructing proofs in the new subject. If our goal were merely to prove all the properties of limits, this would allow us to skip over ε-δ proofs. But since we are trying in this elementary course to learn many methods of analysis, we shall not escape from learning to use ε-δ arguments. Even so, this is an interesting tool for us to use. We can call upon our sequence experience to discover new facts about function limits.

Definition 5.4 (Limit) Let $f : E \to \mathbb{R}$ be a function with domain E and suppose that x_0 is a point of accumulation of E. Then we write

$$\lim_{x \to x_0} f(x) = L$$

if for every sequence $\{e_n\}$ of points of E with $e_n \neq x_0$ and $e_n \to x_0$ as $n \to \infty$,

$$\lim_{n \to \infty} f(e_n) = L.$$

Note. If x_0 is not a point of accumulation of E, then there would be no sequence $\{e_n\}$ of points of E with $e_n \neq x_0$ for all n and $e_n \to x_0$ as $n \to \infty$. Thus once again this is an essential ingredient of our limit definition.

Before we can use this definition we need to establish that it is equivalent to the ε-δ definition. We prove that now.

Proof (**Definitions 5.1 and 5.4 are equivalent**) Suppose first that $\lim_{x \to x_0} f(x) = L$ according to Definition 5.1 and that $\{e_n\}$, $e_n \neq x_0$, is a sequence of points in the domain of f converging to x_0. Let $\varepsilon > 0$. There must be a positive number δ so that

$$|f(x) - L| < \varepsilon$$

if $0 < |x - x_0| < \delta$. But $e_n \to x_0$ and $e_n \neq x_0$ so there is number N such that $0 < |e_n - x_0| < \delta$ for all $n \geq N$. Putting these together, we find that

$$|f(e_n) - L| < \varepsilon$$

if $n \geq N$. This proves that $\{f(e_n)\}$ converges to L. This verifies that Definition 5.1 implies Definition 5.4.

Conversely, suppose that L is not the limit of $f(x)$ as $x \to x_0$ according to Definition 5.1. We must find a sequence of points $\{e_n\}$ in the domain of f and converging to x_0 such that $f(e_n)$ does not converge to L. Because L is not the limit, there must be some $\varepsilon_0 > 0$ so that for any $\delta > 0$ there will be points x in the domain of f with $0 < |x - x_0| < \delta$ and yet the inequality

$$|f(x) - L| < \varepsilon_0$$

fails. Applying this to $\delta = 1,\ 1/2,\ 1/3,\ 1/4 \ldots$ we obtain a sequence of points x_n with x_n in the domain of f and

$$0 < |x_n - x_0| < 1/n$$

and yet

$$|f(x_n) - L| \geq \varepsilon_0.$$

This is precisely the sequence we wanted since $\{f(x_n)\}$ cannot converge to L. Thus we have shown that Definition 5.4 implies Definition 5.1. ■

Since the two definitions are equivalent, we can use either a sequential argument or an ε-δ argument in our discussions of limits.

Example 5.5 Suppose we wish to prove that $\lim_{x\to x_0} f(x) = L$ implies that $\lim_{x\to x_0} \sqrt{f(x)} = \sqrt{L}$. We could convert this into an ε-δ statement, which will involve us in some unpleasant inequality work. Or we can see that, alternatively, we need to prove that if we know $f(x_n) \to L$, then we can conclude $\sqrt{f(x_n)} \to \sqrt{L}$. But we did study just such problems in our investigation of sequence limits (Exercise 2.4.16). ◀

Exercises

5.1.13 Prove the existence of the limit $\lim_{x\to x_0}(4 - 12x)$ by converting to a statement about sequences.

5.1.14 Prove the validity of the limit

$$\lim_{x\to x_0} (ax + b) = ax_0 + b$$

by converting to a statement about sequences.

5.1.15 Prove the validity of the limit $\lim_{x\to x_0} x^2 = x_0^2$ by converting to a statement about sequences.

5.1.16 Show that $\lim_{x\to 0} |x|/x$ does not exist by using the sequential definition of limit.

5.1.17 Prove that the limit $\lim_{x\to 0} \frac{1}{x}$ fails to exist by converting to a statement about sequences.

5.1.18 Prove that the limit $\lim_{x\to 0} \sin(1/x)$ fails to exist by converting to a statement about sequences.

5.1.19 Let x_0 be an accumulation point of the domain E of a function f. Prove that the limit $\lim_{x\to x_0} f(x)$ fails to exist if and only if there is a sequence of distinct points $\{e_n\}$ of E converging to x_0 but with $\{f(e_n)\}$ divergent.

5.1.20 Let f be the *characteristic function* of the rational numbers; that is, f is defined for all real numbers by setting $f(x) = 1$ if x is a rational number and $f(x) = 0$ if x is not a rational number. Determine where, if possible, the limit $\lim_{x\to 0} f(x)$ exists.

5.1.21 Using the sequential definition, show that if $\lim_{x\to x_0} f(x) = L$, then

$$\lim_{x\to x_0} |f(x)| = |L|.$$

5.1.22 Find hypotheses under which you can prove that if $\lim_{x\to x_0} f(x) = L$, then $\lim_{x\to x_0} \sqrt{f(x)} = \sqrt{L}$.

5.1.3 Limits (Mapping Definition)

✂ *Enrichment*

The essential idea behind a limit

$$\lim_{x \to x_0} f(x) = L$$

is that values of x close to x_0 get mapped by f into values close to L. We have been able to express this idea by using inequalities that express this closeness: δ-close for the x values and ε-close for the $f(x)$ values. This is essentially a mapping property that can be expressed by arbitrary open sets.

The following definition is equivalent to both Definitions 5.1 and 5.4.

Definition 5.6 (Limit) Let $f : E \to \mathbb{R}$ be a function with domain E and suppose that x_0 is a point of accumulation of E. Then we write

$$\lim_{x \to x_0} f(x) = L$$

if for every open set V containing the point L there is an open set U containing the point x_0 and every point $x \neq x_0$ of U that is in the domain of f is mapped into a point in V; that is,

$$f : E \cap U \setminus \{x_0\} \to V.$$

Once again, we must show that this definition is equivalent to the ε-δ definition. We prove that now.

Proof (**Definitions 5.1 and 5.6 are equivalent**) Suppose first that $\lim_{x \to x_0} f(x) = L$ according to Definition 5.1. Let V be an open set containing the point L. Then, since L is an interior point of V there is a positive number ε with

$$(L - \varepsilon, L + \varepsilon) \subset V.$$

Choose $\delta > 0$ so that

$$|f(x) - L| < \varepsilon$$

if $0 < |x - x_0| < \delta$ whenever x is a point in the domain of f. Let U be the open set $(x_0 - \delta, x_0 + \delta)$. Then the inequality we have shows that every point $x \neq x_0$ of U that is in the domain of f is mapped into a point in V. This is precisely Definition 5.6.

Conversely, suppose that $\lim_{x \to x_0} f(x) = L$ according to Definition 5.6. Let $\varepsilon > 0$. Choose $V = (L - \varepsilon, L + \varepsilon)$. By our definition there must be an open set U containing the point x_0 and every point $x \neq x_0$ of U that is in the domain of f is mapped into a point in V. Since x_0 is an interior point of U there must be a positive number δ so that

$$(x_0 - \delta, x_0 + \delta) \subset U.$$

This mapping property implies that

$$|f(x) - L| < \varepsilon$$

if $0 < |x - x_0| < \delta$. This is exactly our ε-δ definition of Definition 5.1. ■

Since all three of our definitions are equivalent we can use either a sequential argument, a mapping argument, or an ε-δ argument in our discussions of limits.

Exercises

5.1.23 Show that $\lim_{x\to 0} |x|/x$ does not exist using the mapping definition of limit.

5.1.24 Prove directly that the sequential definition of limit is equivalent to the mapping definition.

5.1.4 One-Sided Limits

It is possible for a function to fail to have a limit at a point and yet appear to have limits on one side. If we ignore what is happening on the right for a function, perhaps it will have a "left-hand limit." This is easy to achieve. Let f be defined everywhere near a point x_0 and define a new function

$$g(x) = f(x) \text{ for all } x < x_0.$$

This new function g is defined on a set to the left of x_0 and knows nothing of the values of f on the right. Thus the limit

$$\lim_{x\to x_0} g(x)$$

can be thought of as a left-hand limit for f. It would be written as

$$\lim_{x\to x_0-} f(x)$$

where the "x_0-" is the indication that a left-hand limit is used, not an ordinary limit. Similarly, the notation

$$\lim_{x\to x_0+} f(x)$$

denotes a right-hand limit with the "x_0+" indicating the limit on the positive or right side of x_0.

Since these one-sided limits are really just ordinary limits for a different function, they must satisfy all the theory of ordinary limits with no further fuss. We can use them quite freely without worrying that they need a different definition or a different theory. Even so, it is convenient to translate our usual definitions into one-sided limits just to have an expression for them. We give the right-hand version. You can supply a left-hand version.

Definition 5.7 (Right-Hand Limit) Let $f : E \to \mathbb{R}$ be a function with domain E and suppose that x_0 is a point of accumulation of $E \cap (x_0, \infty)$. Then we write

$$\lim_{x \to x_0+} f(x) = L$$

if for every $\varepsilon > 0$ there is a $\delta > 0$ so that

$$|f(x) - L| < \varepsilon$$

whenever $x_0 < x < x_0 + \delta$ and $x \in E$.

An equivalent sequential version can be established.

Definition 5.8 (Right-Hand Limit) Let $f : E \to \mathbb{R}$ be a function with domain E and suppose that x_0 is a point of accumulation of $E \cap (x_0, \infty)$. Then we write

$$\lim_{x \to x_0+} f(x) = L$$

if for every decreasing sequence $\{e_n\}$ of points of E with $e_n > x_0$ and $e_n \to x_0$ as $n \to \infty$,

$$\lim_{n \to \infty} f(e_n) = L.$$

Exercises

5.1.25 Show directly that Definitions 5.7 and 5.8 are equivalent.

5.1.26 Under appropriate additional assumptions about the domain of the function f show that $\lim_{x \to x_0} f(x) = L$ if and only if both

$$\lim_{x \to x_0+} f(x) = L \text{ and } \lim_{x \to x_0-} f(x) = L$$

are valid.

5.1.27 If the two limits

$$\lim_{x \to x_0+} f(x) = L_1 \text{ and } \lim_{x \to x_0-} f(x) = L_2$$

exist and are different, then the function is said to have a *jump discontinuity* at that point. The value $L_1 - L_2$ is called the *magnitude of the jump.* Give an example of a function with a jump of magnitude 3 at the value $x_0 = 2$. Give an example with a jump of magnitude -3.

5.1.28 Compute the one-sided limits of the function

$$f(x) = \frac{x}{|x|}$$

at any point x_0.

5.1.29 Compute, if possible, the one-sided limits of the function

$$f(x) = e^{1/x}$$

at 0.

5.1.30 According to our definitions, is there any distinction between the assertions

$$\lim_{x\to 0} \sqrt{x} = 0 \text{ and } \lim_{x\to 0+} \sqrt{x} = 0?$$

What is the meaning of $\lim_{x\to 0-} \sqrt{x} = 0$?

5.1.5 Infinite Limits

We can easily check that the limits

$$\lim_{x\to 0+} \frac{1}{x} \text{ and } \lim_{x\to 0-} \frac{1}{x}$$

fail to exist. A glance at the graph of the function $f(x) = 1/x$ suggests that we should write instead

$$\lim_{x\to 0+} \frac{1}{x} = \infty \text{ and } \lim_{x\to 0-} \frac{1}{x} = -\infty$$

as a way of conveying more information about what is happening rather than saying merely that the limits do not exist.

In this we are following our custom in the study of divergent sequences. Some sequences merely diverge, some diverge to ∞ or to $-\infty$. If we look back at the definition for sequences and compare it with our function limit definition, we should arrive at the following definition.

Definition 5.9 (Infinite Limit) Let $f : E \to \mathbb{R}$ be a function with domain E and suppose that x_0 is a point of accumulation of $E \cap (x_0, \infty)$. Then we write

$$\lim_{x\to x_0+} f(x) = \infty$$

if for every $M > 0$ there is a $\delta > 0$ so that $f(x) \geq M$ whenever

$$x_0 < x < x_0 + \delta \text{ and } x \in E.$$

Similarly, we can define

$$\lim_{x\to x_0+} f(x) = -\infty$$

if for every $m < 0$ there is a $\delta > 0$ so that $f(x) \leq m$ whenever $x_0 < x < x_0+\delta$ and $x \in E$. The infinite limits on the left are similarly defined and denoted $\lim_{x\to x_0-} f(x) = \infty$ and $\lim_{x\to x_0-} f(x) = -\infty$. Also, two-sided limits are defined in the same manner, but with a two-sided condition.

Note. Just as for sequences, we do not say that the limit of a function *exists* unless that limit is finite. Thus, for example, we would say that the limit $\lim_{x\to 0+} 1/x$ does not exist, and that in fact $\lim_{x\to 0+} 1/x = \infty$. A limit is a real number. The symbols ∞ and $-\infty$ are used to describe certain situations, but they are not interpreted as numbers themselves.

Exercises

5.1.31 Give an equivalent formulation for infinite limits using a sequential version.

5.1.32 Formulate a definition for the statement that $\lim_{x \to x_0-} f(x) = \infty$. Show that $\lim_{x \to x_0-} f(x) = \infty$ if and only if $\lim_{x \to (-x_0)+} f(-x) = \infty$.

5.1.33 Where does the function

$$f(x) = \frac{1}{\sqrt{x^2 - 1}}$$

have infinite limits? Give proofs using the definition.

5.1.34 Formulate a definition for the statements

$$\lim_{x \to \infty} f(x) = L \text{ and } \lim_{x \to -\infty} f(x) = L.$$

5.1.35 Formulate a definition for the statements

$$\lim_{x \to \infty} f(x) = \infty \text{ and } \lim_{x \to -\infty} f(x) = \infty.$$

5.1.36 Let $f : (0, \infty) \to \mathbb{R}$. Show that

$$\lim_{x \to \infty} f(x) = L \text{ if and only if } \lim_{x \to 0+} f(1/x) = L.$$

5.1.37 What are the limits $\lim_{x \to \infty} x^p$ for various real numbers p?

5.1.38 Show that one of the limits $\lim_{x \to 0+} f(x)$ and $\lim_{x \to 0-} f(x)$ of the function

$$f(x) = e^{1/x}$$

at 0 is infinite and one is finite. What can you say about the limits

$$\lim_{x \to \infty} f(x) \quad \text{and} \quad \lim_{x \to -\infty} f(x)?$$

5.2 Properties of Limits

The computation of limits in calculus courses depended on a theory of limits. For most simple computations it was enough to know how to handle functions that were put together by adding, subtracting, multiplying, or dividing other functions. Later, more subtle problems required advanced techniques (e.g., L'Hôpital's rule). Here we develop the rudiments of a theory of function limits.

We start with the uniqueness property, the boundedness property and continue to the algebraic properties. In this we are following much the same path we did when we began our study of sequential limits. Indeed the definitions of sequential limits and function limits are so similar that the theories are necessarily themselves quite similar.

5.2.1 Uniqueness of Limits

When we write the statement

$$\lim_{x \to x_0} f(x) = L$$

we wish to be assured that it is not also true for some other numbers different from L.

Theorem 5.10 (Uniqueness of Limits) *Suppose that*

$$\lim_{x \to x_0} f(x) = L.$$

Then the number L is unique: No other number has this same property.

Proof We suppose that

$$\lim_{x \to x_0} f(x) = L$$

and

$$\lim_{x \to x_0} f(x) = L_1$$

are both true. To prove the theorem we must show that $L = L_1$. If we convert this to a statement about sequences this asserts that any sequence $x_n \to x_0$ with $x_n \neq x_0$ and all points in the domain of f must have

$$f(x_n) \to L$$

and also must have

$$f(x_n) \to L_1.$$

For these limits to exist the point x_0 must be a point of accumulation for the domain of f and so there exists at least one such sequence. But we have already established for sequence limits that this is impossible (Theorem 2.8) unless $L = L_1$. ■

Exercises

5.2.1 Give an ε-δ proof of Theorem 5.10.

5.2.2 Explain why the proof fails if the part of the limit definition that asserts x_0 is to be a point of accumulation of the domain of f were omitted.

5.2.2 Boundedness of Limits

We recall that convergent sequences are bounded. There is a similar statement for functions. If a function limit exists the function cannot be too large; the statement must be made precise, however, since it is really only valid close to the point where the limit is taken.

For example, you will recall from our discussion of local boundedness in Section 4.5 that the function $f(x) = 1/x$ is unbounded and yet locally bounded at each point other than at 0. In the same way we will see that the existence of the limit

$$\lim_{x \to x_0} \frac{1}{x} = \frac{1}{x_0}$$

for every value of $x_0 \neq 0$ also requires that local boundedness property.

Theorem 5.11 (Boundedness of Limits) *Suppose that the limit*

$$\lim_{x \to x_0} f(x) = L$$

exists. Then there is an interval $(x_0 - c, x_0 + c)$ *and a number* M *such that*

$$|f(x)| \leq M$$

for every value of x *in that interval that is in the domain of* f.

Proof There is a $\delta > 0$ so that

$$|f(x) - L| < 1$$

whenever x is a point of E differing from x_0 and satisfying $|x - x_0| < \delta$. If x_0 is not in the domain of f, then this means that

$$|f(x)| = |f(x) - L + L| \leq |f(x) - L| + |L| < |L| + 1$$

for all x in $(x_0 - \delta, x_0 + \delta)$ that are in the domain of f. This would complete the proof since we can take $M = |L| + 1$.

If x_0 is in the domain of f, then take instead

$$M = |L| + 1 + |f(x_0)|.$$

Then

$$|f(x)| \leq M$$

for all x in $(x_0 - \delta, x_0 + \delta)$ that are in the domain of f. ■

A similar statement can be made about boundedness away from zero. This shows that if a function has a nonzero limit, then close by to the point the function stays away from zero. The proof uses similar ideas and is left for the exercises.

Theorem 5.12 (Boundedness Away from Zero) *If the limit*

$$\lim_{x \to x_0} f(x)$$

exists and is not zero, then there is an interval $(x_0 - c, x_0 + c)$ *and a positive number* m *such that*

$$|f(x)| \geq m > 0$$

for every value of x *in that interval that is in the domain of* f.

Exercises

5.2.3 Prove Theorem 5.11 using the sequential definition of limit instead.

5.2.4 Use Theorem 5.11 to show that $\lim_{x\to 0}\frac{1}{x}$ cannot exist.

5.2.5 Prove Theorem 5.12 using an ε-δ argument.

5.2.6 Prove Theorem 5.12 using a sequential argument.

5.2.7 Prove Theorem 5.12 by deriving it from Theorem 5.11 and the fact (proved later) that if

$$\lim_{x\to x_0} f(x) = L \neq 0$$

then

$$\lim_{x\to x_0} \frac{1}{f(x)} = \frac{1}{L}.$$

5.2.3 Algebra of Limits

Functions can be combined by the usual arithmetic operations (addition, subtraction, multiplication and division). Indeed most functions we are likely to have encountered in a calculus course can be seen to be composed of simpler functions combined together in this way.

Example 5.13 The computations

$$\lim_{x\to 3}\frac{2x^3+4}{3x^2+1} = \frac{\lim_{x\to 3}(2x^3+4)}{\lim_{x\to 3}(3x^2+1)}$$

$$= \frac{2(\lim_{x\to 3}x^3)+4}{3(\lim_{x\to 3}x^2)+1} = \frac{2\times 3^3+4}{3\times 3^2+1}$$

should return fond memories of calculus homework assignments. But how are these computations properly justified? ◀

Because of our experience with sequence limits, we can anticipate that there should be an "algebra of function limits" just as there was an algebra of sequence limits. The proofs can be obtained either by imitating the proofs we constructed earlier for sequences or by using the fact that function limits can be reduced to sequential limits.

There is an extra caution here. An example illustrates.

Example 5.14 We know that $\lim_{x\to 0}\sqrt{-x} = 0$ and $\lim_{x\to 0}\sqrt{x} = 0$. Does it follow that

$$\lim_{x\to 0}\left(\sqrt{x}+\sqrt{-x}\right) = 0?$$

There is only one point in the domain of the function

$$f(x) = \sqrt{x}+\sqrt{-x}$$

and so no limit statement is possible. ◀

The extra hypothesis throughout the following theorems appears in order to avoid examples like this. We must assume that the domain of f, call it dom(f), and the domain of g, call it dom(g), must have enough points in common to define the limit at the point x_0 being considered. In most simple applications the domains of the functions do not cause any troubles.

For proofs we have a number of strategies available. We can reduce these limit theorems to statements about sequences and then appeal to the theory of sequential limits that we developed in Chapter 2. Alternatively, we can construct ε-δ proofs by modeling them after the similar statements that we proved for sequences. We do not need any really new ideas. The proofs have, accordingly, been left to the exercises.

Theorem 5.15 (Multiples of Limits) *Suppose that the limit*

$$\lim_{x \to x_0} f(x)$$

exists and that C is a real number. Then

$$\lim_{x \to x_0} Cf(x) = C\left(\lim_{x \to x_0} f(x)\right).$$

Theorem 5.16 (Sums and Differences) *Suppose that the limits*

$$\lim_{x \to x_0} f(x) \quad \textit{and} \quad \lim_{x \to x_0} g(x)$$

exist and that x_0 is a point of accumulation of dom$(f) \cap$ dom(g). Then

$$\lim_{x \to x_0} (f(x) + g(x)) = \lim_{x \to x_0} f(x) + \lim_{x \to x_0} g(x)$$

and

$$\lim_{x \to x_0} (f(x) - g(x)) = \lim_{x \to x_0} f(x) - \lim_{x \to x_0} g(x).$$

Theorem 5.17 (Products of Limits) *Suppose that the limits*

$$\lim_{x \to x_0} f(x) \quad \textit{and} \quad \lim_{x \to x_0} g(x)$$

exist and that x_0 is a point of accumulation of dom$(f) \cap$ dom(g). Then

$$\lim_{x \to x_0} f(x)g(x) = \left(\lim_{x \to x_0} f(x)\right)\left(\lim_{x \to x_0} g(x)\right).$$

Theorem 5.18 (Quotients of Limits) *Suppose that the limits*

$$\lim_{x \to x_0} f(x) \quad \textit{and} \quad \lim_{x \to x_0} g(x)$$

exist and that the latter is not zero and that x_0 is a point of accumulation of dom(f) $\cap$ dom(g). Then

$$\lim_{x \to x_0} \frac{f(x)}{g(x)} = \frac{\lim_{x \to x_0} f(x)}{\lim_{x \to x_0} g(x)}.$$

Exercises

5.2.8 Let f and g be functions with domains dom(f) and dom(g). What are the domains of the functions listed below obtained by combining these functions algebraically or by a composition?

(a) $f + g$
(b) $f - g$
(c) fg
(d) f/g
(e) $f \circ g$
(f) $\sqrt{f+g}$
(g) $\sqrt{fg}$

5.2.9 What exactly is the trouble that arises in the theorems of this section that required us to assume "that x_0 is a point of accumulation of dom(f) $\cap$ dom(g)?"

5.2.10 Is it true that if both $\lim_{x \to x_0} f(x)$ and $\lim_{x \to x_0} g(x)$ fail to exist, then $\lim_{x \to x_0} (f(x) + g(x))$ must also fail to exist?

5.2.11 In the statement of Theorem 5.18 don't we also have to assume that $g(x)$ is never zero?

5.2.12 A careless student gives the following as a proof of Theorem 5.17. Find the flaw: "Suppose that $\varepsilon > 0$. Choose δ_1 so that

$$|f(x) - L| < \frac{\varepsilon}{2|M| + 1}$$

if $0 < |x - x_0| < \delta_1$ and also choose δ_2 so that

$$|g(x) - M| < \frac{\varepsilon}{2|f(x)| + 1}$$

if $0 < |x - x_0| < \delta_2$ Define $\delta = \min\{\delta_1, \delta_2\}$. If $0 < |x - x_0| < \delta$, then we have

$$|f(x)g(x) - LM| \leq |f(x)|\,|g(x) - M| + |M|\,|f(x) - L|$$

$$\leq |f(x)| \left(\frac{\varepsilon}{2|f(x)| + 1} \right) + |M| \left(\frac{\varepsilon}{2|M| + 1} \right) < \varepsilon.$$

Well, that shows $f(x)g(x) \to LM$ if $f(x) \to L$ and $g(x) \to M$."

5.2.13 Prove Theorem 5.15 by using an ε-δ proof and by using the sequential definition of limit.

5.2.14 Prove Theorem 5.16 by using an ε-δ proof and by using the sequential definition of limit.

5.2.15 Prove Theorem 5.18 by using the sequential definition of limit.

5.2.16 Prove Theorem 5.17 by correcting the flawed ε-δ proof in Exercise 5.2.12 and by using the sequential definition of limit. Which method is easier?

5.2.4 Order Properties

Just as we saw that sequence limits preserve both the algebraic structure and the order structure, so we will find that function limits have the same properties. We have just completed the algebraic properties. We turn now to the order properties.

If $f(x) \leq g(x)$ for all x, then we expect to conclude that

$$\lim_{x \to x_0} f(x) \leq \lim_{x \to x_0} g(x).$$

We now prove this and several other properties that relate directly to the order structure of the real numbers.

Theorem 5.19 *Suppose that the limits*

$$\lim_{x \to x_0} f(x) \quad \textit{and} \quad \lim_{x \to x_0} g(x)$$

exist and that x_0 is a point of accumulation of dom$(f) \cap$ dom(g). If

$$f(x) \leq g(x)$$

for all $x \in$ dom$(f) \cap$ dom(g), then

$$\lim_{x \to x_0} f(x) \leq \lim_{x \to x_0} g(x).$$

Proof Let us give an indirect proof. Let

$$L = \lim_{x \to x_0} f(x) \text{ and } M = \lim_{x \to x_0} g(x)$$

and suppose, contrary to the theorem, that $L > M$. Choose ε so small that $M + \varepsilon < L - \varepsilon$; that is, choose

$$\varepsilon < (L - M)/2.$$

By the definition of limits there are numbers δ_1 and δ_2 so that

$$f(x) > L - \varepsilon$$

if $x \neq x_0$ is within δ_1 of x_0 and in the domain of f and

$$g(x) < M + \varepsilon$$

if $x \neq x_0$ is within δ_2 of x_0 and is in the domain of g. But the conditions in the theorem assure us that there must be at least one point, $x = z$ say, that satisfies both conditions. That would mean

$$g(z) < M + \varepsilon < L - \varepsilon < f(z).$$

This is impossible as it contradicts the fact that all the values of $f(x)$ are less than the values $g(x)$. This contradiction completes the proof. ■

Note. There is a trap here that we encountered in our discussions of sequence limits. We remember that the condition $s_n < t_n$ does not imply that

$$\lim_{n\to\infty} s_n < \lim_{n\to\infty} t_n.$$

In the same way the condition $f(x) < g(x)$ does not imply

$$\lim_{x\to x_0} f(x) < \lim_{x\to x_0} g(x).$$

Be careful with this, too.

Corollary 5.20 *Suppose that the limit*

$$\lim_{x\to x_0} f(x)$$

exists and that $\alpha \leq f(x) \leq \beta$ for all x in the domain of f. Then

$$\alpha \leq \lim_{x\to x_0} f(x) \leq \beta.$$

Note. Again, don't forget the trap. The condition $\alpha < f(x) < \beta$ for all x implies at best that

$$\alpha \leq \lim_{x\to x_0} f(x) \leq \beta.$$

It would not imply that

$$\alpha < \lim_{x\to x_0} f(x) < \beta.$$

The next theorem is another useful variant on these themes. Here an unknown function is sandwiched between two functions whose limit behavior is known, allowing us to conclude that a limit exists. This theorem is often taught as "the squeeze theorem" just as the version for sequences in Theorem 2.20 was labeled. Here we need the functions to have the same domain.

Theorem 5.21 (Squeeze Theorem) *Suppose that $f,\ g,\ h : E \to \mathbb{R}$ and that x_0 is a point of accumulation of the common domain E. Suppose that the limits*

$$\lim_{x\to x_0} f(x) = L \quad \text{and} \quad \lim_{x\to x_0} g(x) = L$$

exist and that

$$f(x) \le h(x) \le g(x)$$

for all $x \in E$ except perhaps at $x = x_0$. Then $\lim_{x \to x_0} h(x) = L$.

Proof The easiest proof is to use a sequential argument. This is left as Exercise 5.2.19. ■

Example 5.22 Let us prove that the limit

$$\lim_{x \to 0} x \sin(1/x) = 0$$

is valid. Certainly the expression $\sin(1/x)$ seems troublesome at first. But we notice that the inequalities

$$-|x| \le x \sin(1/x) \le |x|$$

are valid for all x (except $x = 0$ where the function is undefined). Since

$$\lim_{x \to 0} |x| = \lim_{x \to 0} -|x| = 0$$

Theorem 5.21 supplies our result. ◄

A final theorem on the theme of order structure is often needed. The absolute value, we recall, is defined directly in terms of the order structure. Is absolute value preserved by the limit operation? As the proof does not require any new ideas, it is left as Exercise 5.2.21.

Theorem 5.23 (Limits of Absolute Values) *Suppose that the limit*

$$\lim_{x \to x_0} f(x) = L$$

exists. Then

$$\lim_{x \to x_0} |f(x)| = |L|.$$

Since maxima and minima can be expressed in terms of absolute values, there is a corollary that is sometimes useful.

Corollary 5.24 (Max/Min of Limits) *Suppose that the limits*

$$\lim_{x \to x_0} f(x) = L \quad \text{and} \quad \lim_{x \to x_0} g(x) = M$$

exist and that x_0 is a point of accumulation of $dom(f) \cap dom(g)$. Then

$$\lim_{x \to x_0} \max\{f(x), g(x)\} = \max\{L, M\}$$

and

$$\lim_{x \to x_0} \min\{f(x), g(x)\} = \min\{L, M\}.$$

Proof The first of the these follows from the identity

$$\max\{f(x), g(x)\} = \frac{f(x)+g(x)}{2} + \frac{|f(x)-g(x)|}{2}$$

and the theorem on limits of sums and the theorem on limits of absolute values. In the same way the second assertion follows from

$$\min\{f(x), g(x)\} = \frac{f(x)+g(x)}{2} - \frac{|f(x)-g(x)|}{2}.$$

■

Exercises

5.2.17 Show that the condition $f(x) < g(x)$ does not imply that

$$\lim_{x\to x_0} f(x) < \lim_{x\to x_0} g(x).$$

5.2.18 Give a sequential type proof for Theorem 5.19.

5.2.19 Give a sequential type proof for Theorem 5.21.

5.2.20 Give an ε-δ proof of Theorem 5.23.

5.2.21 Give a proof of Theorem 5.23 by converting it to a statement about sequences.

5.2.22 Extend Corollary 5.24 to the case of more than two functions; that is, determine

$$\lim_{x\to x_0} \max\{f_1(x), f_2(x), \ldots, f_n(x)\}.$$

5.2.5 Composition of Functions

You will have observed a pattern that is attractive in the study of limits. These examples suggest the pattern:

$$\lim_{x\to x_0} [f(x)]^2 = \left(\lim_{x\to x_0} f(x)\right)^2,$$

$$\lim_{x\to x_0} \sqrt{f(x)} = \sqrt{\lim_{x\to x_0} f(x)},$$

$$\lim_{x\to x_0} e^{f(x)} = e^{\lim_{x\to x_0} f(x)}.$$

The first is easy to prove since $[f(x)]^2 = f(x)f(x)$ and we can use the product rule. The square root example is harder but could be proved using an ε-δ argument and requires only the assumption that $\lim_{x\to x_0} f(x)$ is positive. It could be false if $\lim_{x\to x_0} f(x) = 0$ and definitely is false if $\lim_{x\to x_0} f(x) < 0$.

The third will require some familiarity with the exponential function and is harder still, though always true.

The general pattern is the following. Some function F is composed with f, and the limit computation we wish to use is

$$\lim_{x \to x_0} F(f(x)) = F\left(\lim_{x \to x_0} f(x)\right).$$

Can this be justified? More correctly, what are the conditions under which it can be justified?

Let us analyze this using a sequence argument since that often simplifies function limits. We suppose $x_n \to x_0$. We have then our supposition that $f(x_n) \to L$. Can we conclude

$$F(f(x_n)) \to F(L)?$$

This is exactly what we are doing when we try to use

$$\lim_{x \to x_0} F(f(x)) = F\left(\lim_{x \to x_0} f(x)\right).$$

The property of the function F that we desire is simple:

$$\textit{If } z_n \to z_0 \textit{ then } F(z_n) \to F(z_0).$$

Think of $z_n = f(x_n)$; then $z_n \to L$ and the required property is

$$F(z_n) \to F(L) \text{ whenever } z_n \to L.$$

This is the same as requiring that

$$\lim_{z \to L} F(z) = F(L).$$

Thus we have proved the following theorem, which completely answers our question about justifying the preceding operations.

Theorem 5.25 *Let F be a function defined in a neighborhood of the point L and such that*

$$\lim_{z \to L} F(z) = F(L).$$

If

$$\lim_{x \to x_0} f(x) = L$$

then

$$\lim_{x \to x_0} F(f(x)) = F\left(\lim_{x \to x_0} f(x)\right) = F(L).$$

The condition on the function F that

$$\lim_{z \to L} F(z) = F(L)$$

is called *continuity at the point L* and is the subject of Section 5.4.

Exercises

5.2.23 Show that

$$\lim_{x \to x_0} \sqrt{f(x)} = \sqrt{\lim_{x \to x_0} f(x)}$$

could be false if $\lim_{x \to x_0} f(x) = 0$ and definitely is false if $\lim_{x \to x_0} f(x) < 0$.

5.2.24 Give a formal proof of Theorem 5.25 using the sequential method sketched in the text.

5.2.25 Give a formal proof of Theorem 5.25 using an ε-δ method.

5.2.26 Give a formal proof of Theorem 5.25 using the mapping idea.

5.2.27 Give an example of a limit for which

$$\lim_{x \to x_0} F(f(x)) \neq F\left(\lim_{x \to x_0} f(x)\right)$$

even though both of the limits in the statement do exist.

5.2.28 Show that

$$\lim_{x \to x_0} |f(x)| = \left|\lim_{x \to x_0} f(x)\right|$$

under some appropriate assumption by applying Theorem 5.25.

5.2.29 Show that

$$\lim_{x \to x_0} \sqrt{|f(x)|} = \sqrt{\left|\lim_{x \to x_0} f(x)\right|}$$

under some appropriate assumption by applying Theorem 5.25.

5.2.30 Obtain Corollary 5.24 as an application of Theorem 5.25.

5.2.6 Examples

There are a number of well-known examples of limits that every student should know. Partly this is because there will be an expectation in later courses that these should have been seen. But, more important, an abundance of examples is needed to gain some insight into when limits exist and when they do not and how they behave.

For any function f defined near a point x_0 there are several possibilities we should look for.

1. Does the limit $\lim_{x \to x_0} f(x)$ exist?

2. If the limit does exist, is the limit the most likely value, namely

$$\lim_{x \to x_0} f(x) = f(x_0)?$$

(Such functions are said to be *continuous* at the point x_0.)

3. If the limit fails to exist, then could it be that the one-sided limits do exist but happen to be unequal; that is,

$$\lim_{x \to x_0+} f(x) \neq \lim_{x \to x_0-} f(x)?$$

(Such a function is said to have a *jump discontinuity* at the point x_0.)

The case that is most familiar, namely where

$$\lim_{x \to x_0} f(x) = f(x_0),$$

is described by the language of continuity. Our study of continuity comes in the next section. But let us be aware now of when a function has this property.

Polynomials All polynomial functions have entirely predictable limits. If

$$p(x) = a_0 + a_1 x + a_2 x^2 + \dots a_n x^n,$$

then

$$\lim_{x \to x_0} p(x) = p(x_0)$$

at every value. (In the language we shall use, these functions are continuous.) To prove this we can use the fact that $\lim_{x \to x_0} a_0 = a_0$ and the fact that $\lim_{x \to x_0} x = x_0$. These are trivial to prove. Then the polynomial is built up from this by additions and multiplications. The theorems of Section 5.2.3 can be used to complete the verification [e.g., $\lim_{x \to x_0} x^2 = x_0^2$ by the product rule, $\lim_{x \to x_0} x^3 = \lim_{x \to x_0} (x)(x^2) = x_0^3$ by the product rule applied again].

Rational Functions A rational function is a function of the form

$$R(x) = \frac{p(x)}{q(x)}$$

where p and q are polynomials (i.e., a ratio of polynomials and hence the name). Since we can take limits

$$\lim_{x \to x_0} R(x) = \frac{\lim_{x \to x_0} p(x)}{\lim_{x \to x_0} q(x)}$$

freely, excepting only the case where the denominator is zero, we have found that

$$\lim_{x \to x_0} R(x) = R(x_0)$$

except at those points where $q(x_0) = 0$. At those points, it is possible that the limit exists. Note, however, that it cannot equal $R(x_0)$ since $R(x_0)$ is not defined. It is also possible that the right-hand and left-hand limits are infinite. There are some examples in the exercises to illustrate these possibilities.

Exponential Functions The exponential function e^x can be proved to have the limiting value that we would expect, namely

$$\lim_{x\to x_0} e^x = e^{x_0}.$$

To prove this depends on how we have defined the exponential function in the first place. There are many ways in which we can develop such a theory. It is usual to wait for more theoretical apparatus and then define the exponential function in an appropriate way that allows that to be exploited. Recall that we mentioned in Example 2.36 that

$$e^x = 1 + x + \frac{x^2}{2!} + \frac{x^3}{3!} + \cdots + \frac{x^n}{n!} + \ldots.$$

Sums like this are called power series. As part of the theory of power series we will discover precisely when they are continuous. Then it is possible to define the exponential function as a power series and claim continuity immediately.

Most of the elementary functions of the calculus (trigonometric functions, inverse trigonometric functions, etc.) can be handled in this way. We do not pause here to worry about limits of such functions.

Characteristic Function of the Rationals The *characteristic function* of a set E of real numbers is the function that assigns value 1 at points in E and value 0 at points outside E. Some authors call it an *indicator function* since it does, indeed, indicate when points are or are not in the set. For an interesting example of a function that would have been considered bizarre in the early days of calculus, consider the characteristic function of the rationals:

$$\chi_{\mathbb{Q}}(x) = 1 \text{ if } x \in \mathbb{Q}$$

and

$$\chi_{\mathbb{Q}}(x) = 0 \text{ if } x \notin \mathbb{Q}.$$

It is an easy exercise to check that

$$\lim_{x\to x_0+} \chi_{\mathbb{Q}}(x) \text{ and } \lim_{x\to x_0-} \chi_{\mathbb{Q}}(x)$$

both fail to exist.

Dirichlet Function The Dirichlet function is defined on $[0,1]$ by

$$f(x) = \begin{cases} 0, & \text{if } x \text{ is irrational or } x = 0 \\ 1/q, & \text{if } x = p/q,\ \ p,q \in \mathbb{N},\ \ p/q \text{ in lowest terms.} \end{cases}$$

To examine the limiting behavior of this function, we need to observe that while there are many points where this function is positive (all rationals) there are not many points where it assumes a value greater than some positive number ε. Indeed if we count them we will see that for any positive integer q the set of points

$$S_q = \{x \in [0,1] : f(x) \geq 1/q\}$$

contains at most $q(q-1)/2$ points. The exact number is not important; all we need to observe is that there are only finitely many such points.

Thus let $\varepsilon > 0$ and choose any integer q large enough so that $1/q < \varepsilon$. For any point x_0 we can choose $\delta > 0$ in such a way that both intervals $(x_0 - \delta, x_0)$ and $(x_0, x_0 + \delta)$ contain no points of the finite set S_q. That must mean that every point x in $(x_0 - \delta, x_0)$ or $(x_0, x_0 + \delta)$ satisfies

$$0 \leq f(x) < 1/q < \varepsilon.$$

Thus it follows that

$$\lim_{x \to x_0} f(x) = 0$$

at every point x_0. In particular, the equation

$$\lim_{x \to x_0} f(x) = f(x_0)$$

will hold at every irrational point x_0 but must fail at every rational point. In the language of continuity we have proved that this function is continuous at every irrational point but discontinuous at every rational point. A curious function: It appears to be continuous at nearly every point and discontinuous at nearly every point. Nineteenth-century mathematicians were quite intrigued by such functions and called them pointwise discontinuous, a term that seems not to have survived.

(We shall return to this example occasionally. For example, Exercise 7.5.4 asks for an account of the local extrema of this function.)

Nondecreasing Functions with Jumps The simplest example of a function with a discontinuity is perhaps

$$H(x) = \begin{cases} 0 & \text{if } x < 0 \\ 1 & \text{if } x \geq 0 \end{cases}$$

This function fails to have a limit at $x = 0$ since $\lim_{x \to 0+} H(x) = 1$ and $\lim_{x \to 0-} H(x) = 0$. In the language introduced earlier in this section we would say that H has a jump (or a jump discontinuity) at the point 0.

The discontinuity can be placed at any point. The function $H(x-c)$ has a jump at $x = c$. Moreover, if $c_1 < c_2 < c_3 < \cdots < c_k$ is a finite sequence of distinct points, then the function

$$F(x) = \sum_{i=1}^{k} H(x - c_i)$$

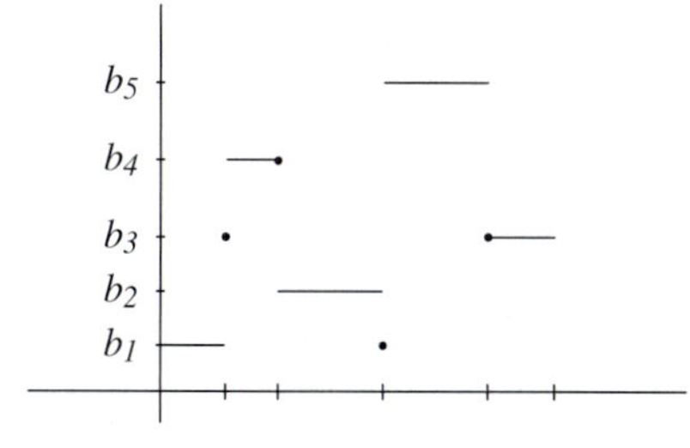

Figure 5.2. Graph of a step function.

is a nondecreasing function with jumps at each of the points $c_1, c_2, c_3, \ldots,$ c_k. At every other point x_0 it is the case that $\lim_{x \to x_0} F(x) = F(x_0)$.

An interesting question now occurs. We have succeeded in constructing a function that is nondecreasing and has jumps at a prescribed finite set of points. Can we construct such a function if we wish to have jumps at a given infinite set of points? This is a question to which we will return.

Step Functions A function f is a *step function* if it assumes finitely many values, say $b_1, b_2, \ldots, b_N$ and for each $1 \leq i \leq N$ the set

$$f^{-1}(b_i) = \{x : f(x) = b_i\},$$

which represents the set of points at which f assumes the value b_i, is a finite union of intervals and singleton point sets. Another way to think about this is that a function of the form

$$f(x) = \sum_{i=1}^{M} a_i \chi_{A_i}(x)$$

is a step function if all the A_i are intervals or singleton sets. (See Figure 5.2 for an illustration.)

Step functions play an important role in integration theory. They offer a crude way of approximating functions. The function

$$H(x) = \begin{cases} 0 & \text{if } x < 0 \\ 1 & \text{if } x \geq 0 \end{cases}$$

that we have just seen is a simple step function since it assumes just two values, 0 and 1, where 0 is assumed on the interval $(-\infty, 0)$ and 1 is assumed on $[0, \infty)$.

The discontinuities of a step function are easy to detect.

Distance of a Closed Set to a Point Let C be a closed set and define a function by writing

$$d(x, C) = \inf\{|x - y| : y \in C\}.$$

This function gives a meaning to the distance between a set C and a point x. If $x_0 \in C$, then $d(x_0, C) = 0$, and if $x_0 \notin C$, then $d(x_0, C) > 0$.

This function is continuous at every point; that is, this function has the property that

$$\lim_{x \to x_0} d(x, C) = d(x_0, C). \tag{1}$$

We can interpret (1) geometrically: If two points x_1 and x_2 are close together, then they are at roughly the same distance from the closed set C.

The Characteristic Function of the Cantor Set Let K be the Cantor set and let χ_K be its characteristic function; that is, let $\chi_K = 1$ if $x \in K$ and $\chi_K(x) = 0$ otherwise. This function has the property that

$$\lim_{x \to x_0} \chi_K(x) = 0$$

if x_0 is not in the Cantor set and the limit exists at no point in the Cantor set. For an easy proof of this you will have to review the properties of the Cantor set and its complement in Exercises 4.3.23 and 4.4.9.

Exercises

5.2.31 Give a proof that includes all necessary details that the limit

$$\lim_{x \to x_0} p(x) = p(x_0)$$

for all polynomials p.

5.2.32 Suppose that you know that

$$\lim_{x \to 2} e^x = e^2.$$

Prove that $\lim_{x \to x_0} e^x = e^{x_0}$ for all x_0.

5.2.33 Suppose that you know that

$$\lim_{x \to 0} \cos x = 1 \text{ and } \lim_{x \to 0} \sin x = 0.$$

Prove that $\lim_{x \to x_0} \sin x = \sin x_0$ for all x_0.

5.2.34 In the text we constructed a nondecreasing function with jumps at each of the points $c_1, c_2, c_3, \dots, c_k$ and continuous everywhere else. Construct an *increasing* function with this property.

5.2.35 Let $f : [a, b] \to \mathbb{R}$ be a step function. Show that there is a partition

$$a = x_0 < x_1 < x_2 < \cdots < x_{n-1} < x_n = b$$

so that f is constant on each interval (x_{i-1}, x_i), $i = 1, 2, \dots n$.

5.2.36 Suppose that

$$f(x) = \sum_{i=1}^{M} a_i \chi_{A_i}$$

where the A_i are intervals. Show that f is a step function; that is, that f assumes finitely many values, and for each b in the range of f the set $f^{-1}(b)$ is a finite union of intervals or singleton sets. Where are the discontinuities of such a function?

5.2.37 Show that the characteristic function of the rationals can also be defined by the formula

$$\chi_{\mathbb{Q}}(x) = \lim_{m\to\infty} \lim_{n\to\infty} |\cos(m!\pi x)|^n.$$

5.2.38 Show that

$$\lim_{x\to x_0+} \chi_{\mathbb{Q}}(x) \text{ and } \lim_{x\to x_0-} \chi_{\mathbb{Q}}(x)$$

both fail to exist, where $\chi_{\mathbb{Q}}$ is the characteristic function of the rationals. What would be the answer to the corresponding question for the characteristic function of the irrationals?

5.2.39 Describe the graph of the function $\chi_{\mathbb{Q}}$. What kind of a sketch would convey this set?

5.2.40 Give an example of a set E such that the characteristic function χ_E of E has limits at every point. Can you describe the most general set E with this property?

5.2.41 Give an example of a set E such that the characteristic function χ_E of E has one-sided limits at every point. Can you describe the most general set E with this property?

5.2.42 Show that

$$\lim_{x\to x_0} d(x, C) = d(x_0, C)$$

at every point x_0 where $d(x, C)$ is the distance from x to the closed set C as defined in this section.

5.2.43 Sketch the graph of the function $d(x, C)$ for several closed sets C (e.g., $\{0\}$, $\mathbb{N}$, $[0,1]$, $\{0\} \cup \{1, 1/2, 1/3, 1/4, \dots\}$, and $[0,1] \cup [2,3]$).

5.2.44 Sketch the graph of the characteristic function χ_K of the Cantor set (Exercises 4.3.23 and 4.4.9) and show that

$$\lim_{x\to x_0} \chi_K(x) = 0$$

at all points x not in the Cantor set and that this limit fails to exist at all points in Cantor set.

5.3 Limits Superior and Inferior

Advanced

If limits fail to exist we need not abandon all hope of discussing the limiting behavior. We saw this situation in our study of sequence limits in Section 2.13. Even if $\{s_n\}$ diverges so that $\lim_{n\to\infty} s_n$ fails to exist, it is possible that the two extreme limits

$$\liminf_{n\to\infty} s_n \text{ and } \limsup_{n\to\infty} s_n$$

provide some meaningful information. These two concepts always exist (possibly as ∞ or $-\infty$). A similar situation occurs for functions. The theory is nearly identical in many respects.

Definition 5.26 (Lim Sup) Let $f : E \to \mathbb{R}$ be a function with domain E and suppose that x_0 is a point of accumulation of E. Then we write

$$\limsup_{x\to x_0} f(x) = \inf_{\delta>0} \sup\{f(x) : x \in (x_0 - \delta, x_0 + \delta) \cap E,\ x \neq x_0\}$$

and

$$\liminf_{x\to x_0} f(x) = \sup_{\delta>0} \inf\{f(x) : x \in (x_0 - \delta, x_0 + \delta) \cap E,\ x \neq x_0\}$$

As this section is for more advanced readers we have left the development of this concept to the exercises.

Exercises

5.3.1 Show from the definition that

$$\limsup_{x\to x_0} f(x) \geq \liminf_{x\to x_0} f(x).$$

5.3.2 Compute each of the following.

(a) $\limsup_{x\to 0} \sin x^{-1}$

(b) $\limsup_{x\to 0} x \sin x^{-1}$

(c) $\limsup_{x\to 0} x^{-1} \sin x^{-1}$

5.3.3 Formulate an equivalent definition for $\limsup_{x\to 0} f(x)$ expressed in terms of sequential limits; that is, in terms of limits of $f(x_n)$ for $x_n \to x_0$. Show that your definition is equivalent to that in the text.

5.3.4 Give an example of a function f so that

$$\liminf_{x\to 0} f(x) = 0 \text{ and } \limsup_{x\to 0} f(x) = 1.$$

5.3.5 What changes, if any, are there if the definition of lim sup had been written as

$$\limsup_{x\to x_0} f(x) = \inf_{\delta>0} \sup\{f(x) : x \in (x_0 - \delta, x_0 + \delta) \cap E\}?$$

5.3.6 Formulate a definition for the one-sided concepts $\limsup_{x\to x_0+} f(x)$ and $\limsup_{x\to x_0-} f(x)$.

5.3.7 Give an example of a function f with the properties $\liminf_{x\to 0+} f(x) = 0$, $\limsup_{x\to 0+} f(x) = \infty$, $\liminf_{x\to 0-} f(x) = -\infty$, and $\limsup_{x\to 0-} f(x) = 1$.

5.3.8 Show that $\lim_{x\to 0} f(x)$ exists if and only if all four of $\liminf_{x\to 0+} f(x)$, $\limsup_{x\to 0+} f(x)$, $\liminf_{x\to 0-} f(x)$, and $\limsup_{x\to 0-} f(x)$ are equal and finite.

5.3.9 Show that $\lim_{x\to 0} f(x) = \infty$ if and only if all four of $\liminf_{x\to 0+} f(x)$, $\limsup_{x\to 0+} f(x)$, $\liminf_{x\to 0-} f(x)$, and $\limsup_{x\to 0-} f(x)$ are ∞.

5.3.10 For the function $\chi_{\mathbb{Q}}$, the characteristic function of the rationals, determine the values of each of the limits $\liminf_{x\to x_0+} \chi_{\mathbb{Q}}(x)$, $\limsup_{x\to x_0+} \chi_{\mathbb{Q}}(x)$, $\liminf_{x\to x_0-} \chi_{\mathbb{Q}}(x)$, and $\limsup_{x\to x_0-} \chi_{\mathbb{Q}}(x)$ at any point x_0.

5.3.11 Give an example of a function f such that

$$\{x_0 : \limsup_{x\to x_0-} f(x) > \limsup_{x\to x_0+} f(x)\}$$

is infinite.

5.4 Continuity

The earliest use of the term "continuity" is somewhat clouded by misconceptions of the nature of a function. If a function was given by a single formula then it was considered in the eighteenth century to be "continuous." If, however, the function had a "break" in the formula—defined differently in one interval than in another—it was considered as "discontinuous." As the subject developed these notions continued to obscure the really important ideas. Augustin Cauchy (1789–1857) was the first to give the modern definition and to focus attention on the concept that has now assumed such an important role in analysis.

5.4.1 How to Define Continuity

Enrichment

Before we proceed to the present day definition, let us consider another notion. Even as late as the middle of the nineteenth century, some mathematicians believed this notion should form the basis for a definition of continuity. This concept is suggested by the phrase "the graph has no jumps." While some instructors of calculus courses might use such phrases to convey a sense of continuity to students, the phrase is not a precise one, nor does it fully convey all we wish a continuous function to be.

This notion *is* related to continuity, however, and has some importance in its own right. We'll begin with a brief discussion of it. Here is one attempt at making our phrase precise. (See Figure 5.3.)

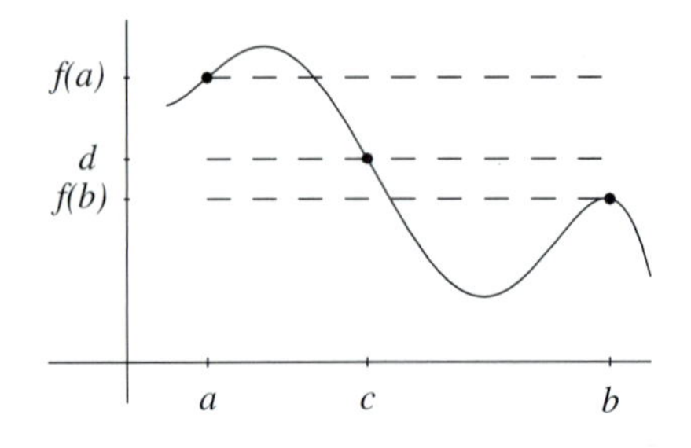

Figure 5.3. At some point between a and b the function assumes any given value d between $f(a)$ and $f(b)$.

Definition 5.27 (Intermediate Value Property) Let f be defined on an interval I. Suppose that for each $a, b \in I$ with $f(a) \neq f(b)$, and for each d between $f(a)$ and $f(b)$, there exists c between a and b for which $f(c) = d$. We then say that f has the *intermediate value property* (IVP) on I.

Functions with this property are called *Darboux* functions after Jean Gaston Darboux (1842–1917), who showed in 1875 that for every differentiable function F on an interval I, the derivative F' has the IVP on I. He is also particularly famous for his 1875 account of the Riemann integral using upper and lower sums; often reference is made to the "Darboux integral," meaning this version of the classical Riemann integral.

Example 5.28 Let

$$F(x) = \begin{cases} \sin x^{-1} & \text{if } x \neq 0 \\ 0 & \text{if } x = 0. \end{cases}$$

The graph of F is shown in Figure 5.4. You may wish to verify that F has the IVP. In particular, F assumes every value in the interval $[-1, 1]$ infinitely often in every neighborhood of $x = 0$. ◀

We haven't yet made precise the phrase "the graph has no jumps," but the IVP seems to convey that idea well enough. Since this property is so easy to describe and appears to have content that is easy to visualize, why *not* take it as the definition of continuity?

Before attempting to answer that question, let us offer a competing phrase to capture the idea of continuity: "If x is near x_0, then $f(x)$ is near $f(x_0)$." As stated, this phrase is not precise, but we can make it precise using the limit concept. This phrase could be interpreted really as asserting

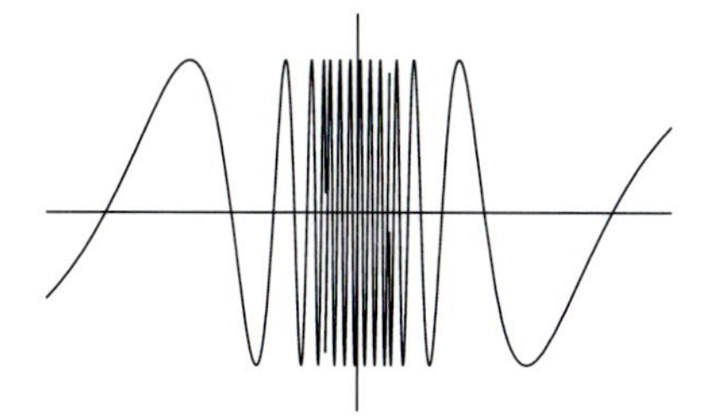

Figure 5.4. Graph of the function $F(x) = \sin x^{-1}$ on $[-\pi/8, \pi/8]$.

that

$$f(x_0) = \lim_{x \to x_0} f(x). \tag{2}$$

According to this criterion our function F of Example 5.28 would *not* be continuous at $x_0 = 0$, because $F(0) = 0$, but $\lim_{x\to 0} F(x)$ does not exist.

We shall see presently that the definition based on limits allows the development of a useful theory. We'll see that the class of continuous functions [as defined using equation (2)] is closed under addition and multiplication, and that such functions have many other desirable properties. For example, the class is closed under certain kinds of limits of sequences, and every continuous function on $[a, b]$ is integrable. On the other hand (as in shown in the exercises), none of the analogous statements is valid for the class of functions defined by IVP.

Thus a theory of continuity based on the limit concept allows a rich structure and enjoys wide applicability, whereas one based on the IVP is rather limited. In addition, the fundamental notion of limit extends to much more general settings than $\mathbb{R}$. In contrast, extensions of IVP, while possible, are peripheral to mathematical analysis.

Exercises

5.4.1 Refer to Example 5.28. Let

$$G(x) = \begin{cases} -F(x) & \text{if } x \neq 0 \\ 1 & \text{if } x = 0. \end{cases}$$

Show that G has the IVP, yet $F + G$ does not. Thus the class of functions with IVP is not closed under addition.

5.4.2 Give an example to show that the class of functions with IVP is not closed under multiplication.

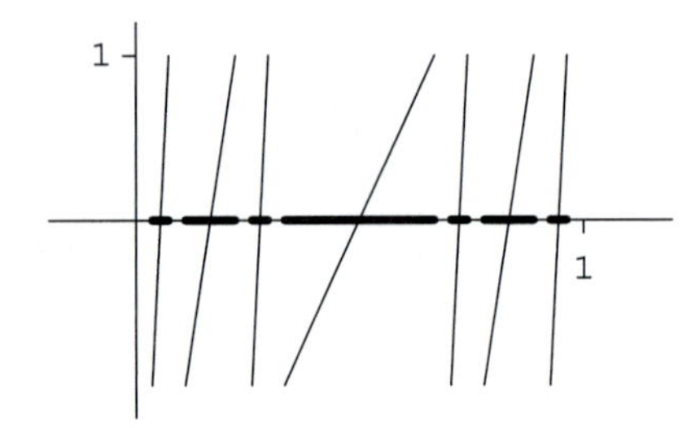

Figure 5.5. Function defined on the complement of the Cantor set, as described in Exercise 5.4.5. The first three stages are shown.

5.4.3 Let

$$H(x) = \begin{cases} x^{-1} \sin x^{-1} & \text{if } x \neq 0 \\ 0 & \text{if } x = 0. \end{cases}$$

Show that H has IVP on $[0, 1]$ but is not bounded there. (This shows that the IVP does not imply boundedness; we shall see that, in contrast, any continuous function on $[0, 1]$ would have to be bounded.)

5.4.4 Give an example of a function f with IVP on $[0, 1]$ that is bounded but achieves no absolute maximum on [0,1].

5.4.5 Let K be the Cantor set of Exercises 4.3.23 and 4.4.9 and let $\{(a_k, b_k)\}$ be the sequence of intervals complementary to K in (0,1).

(a) Write down a set of equations defining a function f that vanishes at every two-sided point of accumulation of K, is continuous on each interval $[a_k, b_k]$, and for which

$$\lim_{x \to a_k+} f(x) = -1 \text{ and } \lim_{x \to b_k-} f(x) = 1.$$

(See Figure 5.5 for an illustration of one possible choice.)

(b) Verify that f has the intermediate value property.

(c) Verify that f is not continuous in the sense that $f(x_0) = \lim_{x \to x_0} f(x)$ fails at certain points. (Which points?)

✂ **5.4.6** We construct a function with IVP whose graph may be more difficult to visualize. Let $I_0 = (0, 1)$. Each $x \in I_0$ has a unique decimal expansion not ending in a string of 9's. For each $n \in \mathbb{N}$ and $x = .a_1 a_2 \dots$ in I_0, let

$$f_n(x) = \frac{a_1(x) + a_2(x) + \cdots + a_n(x)}{n}.$$

Thus $f_n(x)$ represents the average of the first n digits of x. For each $x \in I_0$, let $f(x) = \limsup_n f_n(x)$.

(a) Show that $f : I_0 \to [0, 10]$.

(b) Describe how to construct $x \in I_0$ such that $f(x) = \pi$.

(c) Describe how to construct $x \in (.01, .02)$ such that $f(x) = \pi$.

(d) Show that for each interval $(a, b) \subset I_0$ and each $d \in [0, 10]$ there exists $c \in (a, b)$ such that $f(c) = d$. Thus, f assumes every value in [0,10] in every interval in I_0. In particular, f has IVP.

(e) Let $A = \{x : f(x) = x\}$. Let $g(x) = 0$ if $x \in A$, $g(x) = f(x)$ for $x \notin A$. Show that $g(x)$ has IVP.

(f) Show that $-g(x) + x$ does not have IVP. Thus the sum of a function with IVP with the identity function need not have IVP.

5.4.2 Continuity at a Point

Let us look at Cauchy's concept of *continuous function.* We begin by defining continuity at a point, more specifically continuity at an interior point of the domain of a function f. This way we are assured that if we are interested in what is happening at the point x_0 then f is defined in a neighborhood of x_0; that is, that f is defined in some interval $(x_0 - c, x_0 + c)$ for a positive number c. This simplifies some of the computations.

Definition 5.29 (Continuous) Let f be defined in a neighborhood of x_0. The function f is *continuous* at x_0 provided $\lim_{x \to x_0} f(x) = f(x_0)$.

This means that for each neighborhood V of $f(x_0)$ there is a neighborhood U of x_0 such that $f(U) \subset V$: that is, if $x \in U$, then $f(x) \in V$. We can, of course, state the definition in terms of δ's and ε's: f is continuous at x_0 if for each $\varepsilon > 0$ there exists $\delta > 0$ such that $|f(x) - f(x_0)| < \varepsilon$ whenever $|x - x_0| < \delta$. In the exercises we ask you to verify that the three formulations, involving the language of limits, of neighborhoods, and of δ's and ε's, are equivalent. We believe that this is an important exercise for readers who do not yet feel comfortable with the limit concept. Feeling comfortable with the various forms that continuity takes is essential to feeling comfortable with many of the arguments that appear in the sections and chapters that follow.

Observe that a function f can fail to be continuous at x_0 in three ways:

1. f is not defined at x_0.

2. $\lim_{x \to x_0} f(x)$ fails to exist.

3. f is defined at x_0 and $\lim_{x \to x_0} f(x)$ exists, but
$$f(x_0) \neq \lim_{x \to x_0} f(x).$$

We leave it to you to provide simple examples of each of these possibilities.

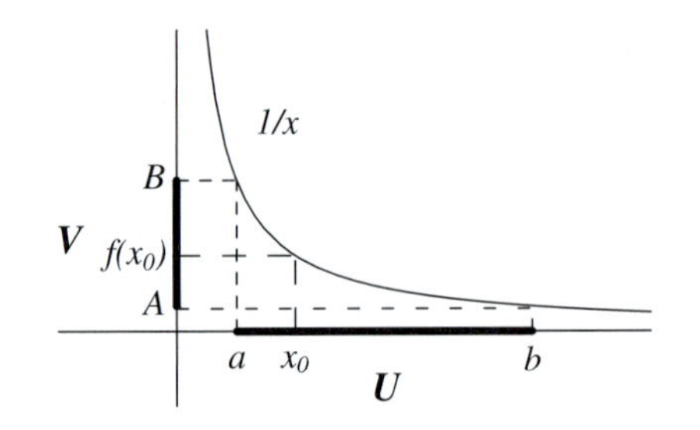

Figure 5.6. Graphical interpretation of the neighborhood definition of continuity for the function $f(x) = x^{-1}$. Note that $a = 1/B$ and $b = 1/A$.

Example 5.30 Let $f : (0, \infty) \to \mathbb{R}$ be defined by $f(x) = 1/x$. We show that if $x_0 \in (0, \infty)$, then f is continuous at x_0.

Let's first try the "neighborhood" definition of continuity. Let V be a neighborhood of $f(x_0)$, say $V = (A, B)$. Thus $A < f(x_0) < B$. We must find a neighborhood $U = (a, b)$ of x_0 such that $f(U) \subset V$. A picture suggests what to do: Let $a = 1/B$, $b = 1/A$. (See Figure 5.6.) But we must be a bit careful here. Nothing in our neighborhood definition of continuity allows us to assume $A > 0$, so b might not be defined (if $A = 0$), or might not be in the domain of f (if $A < 0$). This presents, however, only a minor nuisance. Thus we assume $A > 0$ in our proof.

So, let us assume $A > 0$, $a = 1/B$, $b = 1/A$. Then, since $A < f(x_0) < B$, we have

$$b = \frac{1}{A} > x_0 = \frac{1}{f(x_0)} > \frac{1}{B} = a,$$

so $x_0 \in (a, b) = U$. Furthermore, if $c \in U$, then $a < c < b$ and

$$B > 1/c = f(c) > A,$$

so $f(c) \in V$. This shows that $f(U) \subset V$ as was required. ◀

Let's see how a proof based on the δ-ε definition might look. As with our first proof, we shall provide many details of the proof. After you feel conversant with limits and continuity, you may wish to streamline the proofs somewhat by leaving out details that "any reader finds obvious."

Let $x_0 \in (0, \infty)$ and let $\varepsilon > 0$. We wish to find $\delta > 0$ such that if $|x - x_0| < \delta$ and $x > 0$, then $|1/x - 1/x_0| < \varepsilon$. Rewriting this last inequality as

$$|x - x_0| < \varepsilon x x_0 \tag{3}$$

suggests we try $\delta = \varepsilon x x_0$. But δ should depend only on ε and x_0, not on x. There is no $\delta > 0$ for which the inequality $|x - x_0| < \delta$ implies the inequality

$$|x - x_0| < \varepsilon x x_0$$

for all $x \in (0, \infty)$. We can remove this problem by first requiring x to stay away from 0.

For example, we first require that

$$|x - x_0| < \frac{1}{2} x_0. \tag{4}$$

Then

$$\frac{1}{2} x_0 < x \quad \text{and} \tag{5}$$

$$\frac{1}{2} \varepsilon x_0^2 < \varepsilon x x_0. \tag{6}$$

The inequalities (3), (4), and (6) suggest taking

$$\delta = \min\left(\frac{1}{2} x_0, \frac{1}{2} x_0^2 \varepsilon\right).$$

For this δ , we compute easily that if $|x - x_0| < \delta$, then

$$\left|\frac{1}{x} - \frac{1}{x_0}\right| < \frac{|x - x_0|}{|x x_0|} < \frac{\frac{1}{2} x_0^2 \varepsilon}{\frac{1}{2} x_0^2} = \varepsilon,$$

the last inequality being obtained by using (6) on the numerator $|x - x_0|$ and (5) on the denominator $|x x_0|$.

Exercises

5.4.7 Prove that the function $f(x) = x^2$ is continuous at every point of $\mathbb{R}$ using the δ-ε form of continuity,

5.4.8 Prove that the function $f(x) = |x|$ is continuous at every point of $\mathbb{R}$ using the δ-ε form of continuity,

5.4.9 Show that the three formulations of continuity appearing at the beginning of this section are equivalent.

5.4.10 In the δ-ε verification of continuity of the function $1/x$ we obtained a δ that did the job. We made no claim that this δ is the largest possible δ we could have chosen. Show that for $\varepsilon = 1$ and $x_0 \in (0, 1)$ any δ that works must satisfy $\delta < x_0^2/(1 + x_0)$.

5.4.11 **(Sequence Definition of Continuity)** Prove that f is continuous at x_0 if and only if $\lim_{n\to\infty} f(x_n) = f(x_0)$ for every sequence $\{x_n\} \to x_0$. How would you expect this characterization of continuity at x_0 to be modified if x_0 is not an interior point of its domain.

5.4.12 Give three examples of a function f that fails to be continuous at a point x_0. The first should be discontinuous merely because f is not defined at x_0. The second should be discontinuous because $\lim_{x \to x_0} f(x)$ fails to exist. The third should have neither of these defects but should nonetheless be discontinuous.

5.4.13 A function f is said to be *symmetrically continuous* at a point x if

$$\lim_{h \to 0} [f(x+h) - f(x-h)] = 0.$$

Show that if f is continuous at a point, then it must be symmetrically continuous there and that the converse does not hold.

5.4.3 Continuity at an Arbitrary Point

To this point we have discussed continuity of a function at an interior point of its domain. How should we modify our notions if x_0 is not an interior point?

Continuity at Endpoints For example, if $f : [a, b] \to \mathbb{R}$, how can we define continuity of f at a or at b? Since the function is defined only on the interval $[a, b]$ and we have defined continuity in terms of limits, it seems that we should require, as before, that for any interior point $x_0 \in (a, b)$

$$\lim_{x \to x_0} f(x) = f(x_0)$$

while at the endpoints continuity would be defined by a one-sided limit,

$$\lim_{x \to a+} f(x) = f(a) \quad \text{and} \quad \lim_{x \to b-} f(x) = f(b).$$

We can also reformulate our definition in a way that recognizes that f is defined only on $[a, b]$. In our neighborhood definition we interpret U as a *relative* neighborhood of x_0: We require that $f(U \cap [a, b]) \subset V$. Here by a *relative neighborhood* we mean that part of an ordinary neighborhood that is inside the set where f is defined.

Similarly, for the δ-ε definition, our requirement becomes that

$$|f(x) - f(x_0)| < \varepsilon$$

whenever $|x - x_0| < \delta$ and $x \in [a, b]$. Again we are merely restricting our attention to the set where f is defined.

Continuity on an Arbitrary Set These reformulations would work for any set A, not just an interval. Thus we assume that

$$f : A \to \mathbb{R}$$

so that f is a function with domain A and x_0 is an arbitrary point of A, which need not be an interior point nor even a point of accumulation (it might be isolated in A).

There are four versions of the definition. As before, you should check to see that they are indeed equivalent. Some extra care is needed because x_0 could be any point of A and may be isolated in A.

Definition 5.31 (ε-δ Version) Let f be defined on a set A and let x_0 be any point of A. The function f is *continuous* at x_0 provided for every $\varepsilon > 0$ there is a $\delta > 0$ so that

$$|f(x) - f(x_0)| < \varepsilon$$

for every $x \in A$ for which $|x - x_0| < \delta$.

Definition 5.32 (Limit Version) Let f be defined on a set A and let x_0 be any point of A. The function f is *continuous* at x_0 provided either that x_0 is isolated in A or else that x_0 is a point of accumulation of that set and

$$\lim_{x \to x_0} f(x) = f(x_0).$$

Definition 5.33 (Neighborhood Version) Let f be defined on a set A and let x_0 be any point of A. The function f is *continuous* at x_0 provided that for every open set V containing $f(x_0)$ there is an open set U containing x_0 so that $f(U \cap A) \subset V$.

In other words, the neighborhood version asserts that there is a set $U \cap A$ open relative to A that f maps into V. We recall that a set $B \subset A$ is relatively open relative to A if B is the intersection of some open set (here U) with A.

Definition 5.34 (Sequential Version) Let f be defined on a set A and let x_0 be any point of A. The function f is *continuous* at x_0 provided that for every sequence of points $\{x_n\}$ belonging to A and converging to x_0, it follows that $f(x_n) \to f(x_0)$.

Exercises

5.4.14 Prove the equivalence of the four definitions for the continuity of a function defined on an arbitrary set A.

5.4.15 Let $f : \mathbb{N} \to \mathbb{R}$ by writing $f(n) = 1/n^2$. Is f continuous at any point in its domain?

5.4.16 Using each of the four versions of continuity, show that any function is automatically continuous at any point of its domain that is isolated.

5.4.17 Let f be defined on the set containing the points

$$0,\ 1,\ 1/2,\ 1/4,\ 1/8, \dots\ 1/2^n$$

only. What values can you assign at these points that will make this function continuous everywhere where it is defined?

5.4.18 Let f be defined on the set containing the points 0, ± 1, $\pm 1/2$, $\pm 1/4$, $\pm 1/8$, $\ldots$ $\pm 1/2^n$, $\ldots$ only. What values can you assign at these points that will make this function continuous everywhere where it is defined?

5.4.19 If f is continuous at a point x_0 then is it necessarily true that

$$\lim_{x \to x_0} f(x) = f(x_0)?$$

At what points in the domain of f can you say this?

5.4.20 A function $f : [a, b] \to \mathbb{R}$ is said to be *Lipschitz* if there is a positive number M so that $|f(x) - f(y)| \leq M|x - y|$ for all x, $y \in [a, b]$. Show that a Lipschitz function must be continuous. Is the converse true? [Rudolf Otto Sigismund Lipschitz (1832–1903) is probably best remembered for this condition, now forever attached to his name, which he used in formulating an existence theorem for differential equations of the form $y' = f(x, y)$.]

5.4.4 Continuity on a Set

✂ *Enrichment*

Continuity is defined at points. A function such as $f(x) = x^2$ could be said to be continuous at every real number x_0, meaning only that $\lim_{x \to x_0} x^2 = x_0^2$ for every real number. In many cases the function considered is continuous at every point in its domain. We say simply that f is continuous. But we must remember this is an assertion about every single point where f is defined.

Definition 5.35 Let $f : A \to \mathbb{R}$. Then f is *continuous* (or *continuous on* A) if f is continuous at each point of A.

If we wish to prove directly from this definition that f is continuous, we must show that f is continuous at every $x_0 \in A$. It is sometimes easier to use the *global* characterization of continuity that follows.

Theorem 5.36 *Let $f : A \to \mathbb{R}$. Then f is continuous if and only if for every open set $V \subset \mathbb{R}$, the set $f^{-1}(V) = \{x \in A : f(x) \in V\}$ is open (relative to A).*

Proof Suppose first that f is continuous. Let V be open, let $x_0 \in f^{-1}(V)$ and choose $\alpha < \beta$ so that $(\alpha, \beta) \subset V$ and so that $x_0 \in f^{-1}((\alpha, \beta))$. Then $\alpha < f(x_0) < \beta$. We will find a neighborhood U of x_0 such that $\alpha < f(x) < \beta$ for all $x \in U \cap A$. Let $\varepsilon = \min(\beta - f(x_0), f(x_0) - \alpha)$. Since f is continuous at x_0, there exists $\delta > 0$ such that if

$$x \in A \cap (x_0 - \delta, x_0 + \delta),$$

then

$$|f(x) - f(x_0)| < \varepsilon.$$

Thus

$$f(x) - f(x_0) < \beta - f(x_0),$$

and so $f(x) < \beta$. Similarly,

$$f(x) - f(x_0) > \alpha - f(x_0),$$

and so $f(x) > \alpha$. Thus the relative neighborhood $U = (x_0 - \delta, x_0 + \delta) \cap A$ is a subset of $f^{-1}((\alpha, \beta))$ and hence also a subset of $f^{-1}(V)$. We have shown that each member of $f^{-1}(V)$ has a relative neighborhood in $f^{-1}(V)$. That is, $f^{-1}(V))$ is open relative to A.

To prove the converse, suppose f satisfies the condition that for each open interval (α, β) with $\alpha < \beta$, the set $f^{-1}((\alpha, \beta))$ is open relative to A. Let $x_0 \in A$. We must show that f is continuous at x_0. Let $\varepsilon > 0$, $\beta = f(x_0) + \varepsilon$, $\alpha = f(x_0) - \varepsilon$. Our hypothesis implies that $f^{-1}((\alpha, \beta))$ is open relative to A. Thus

$$f^{-1}((\alpha, \beta)) = \bigcup (a_i, b_i) \cap A,$$

the union being a finite or countable union of pairwise disjoint open intervals. One of these intervals, say (a_j, b_j), contains x_0. Let

$$\delta = \min(x_0 - a_j, b_j - x_0).$$

For $|x - x_0| < \delta$ and $x \in A$ we find

$$\alpha < f(x) < \beta.$$

Because $\beta = f(x_0) + \varepsilon$ and $\alpha = f(x_0) - \varepsilon$ we must have

$$|f(x) - f(x_0)| < \varepsilon.$$

This shows that f is continuous at x_0. ■

We spelled out the details of the proof of Theorem 5.36. This may have caused it to appear rather lengthy. But the proof is nothing more than writing down in a rigorous way what some intuitive pictures indicate. You might find that the neighborhood notion of continuity is a more natural one to use for proving the theorem. We leave this as Exercise 5.4.23.

As a corollary let us point out that we can replace open sets by open intervals; thus to check continuity of a function f it is enough to show that $f^{-1}((\alpha, \beta))$ is open for every interval (α, β).

Corollary 5.37 *Let $f : A \to \mathbb{R}$. Then f is continuous if and only if for every interval (α, β), $f^{-1}((\alpha, \beta))$ is open (relative to A).*

Proof We verify that the conditions (i) $f^{-1}(V)$ is relatively open for all open $V \subset \mathbb{R}$ and (ii) $f^{-1}((\alpha, \beta))$ is relatively open for all $\alpha < \beta$ are equivalent. But this is immediate. If (i) is satisfied, then (ii) is also, since the requirement (ii) is just a special case of (i). On the other hand, if (ii) is satisfied and

$$V = \bigcup (\alpha_i, \beta_i),$$

then

$$f^{-1}(V) = \bigcup f^{-1}((\alpha_i, \beta_i)).$$

Each of the sets $f^{-1}((\alpha_i, \beta_i))$ is open by hypothesis, so $f^{-1}(V)$ is also open because it is a union of a family of open sets. ■

Example 5.38 Let $f(x) = 1/x$ $(x > 0)$. We find

$$f^{-1}((\alpha, \beta)) = \left(\frac{1}{\beta}, \frac{1}{\alpha}\right).$$

Since $(1/\beta, 1/\alpha)$ is open it would follow that f is continuous on $(0, \infty)$. ◀

Exercises

5.4.21 Prove that the function $f(x) = x^2$ is continuous on $\mathbb{R}$ by using Theorem 5.36.

5.4.22 Prove that the function $f(x) = |x|$ is continuous on $\mathbb{R}$ by using Theorem 5.36.

5.4.23 Prove Theorem 5.36 using the neighborhood definition of continuity.

5.4.24 Let f be continuous in a neighborhood U of the point x_0. If $f(x) < \beta$ for all $x \in U \setminus \{x_0\}$, prove that $f(x_0) \leq \beta$. Show by example that we cannot conclude $f(x_0) < \beta$.

5.4.25 Let f, g be defined on $\mathbb{R}$. Suppose $f(0) = 0$ and f is continuous at $x = 0$. Suppose g is bounded in some neighborhood of zero. Prove that fg is continuous at $x = 0$. Apply this to the function $f(x) = x \sin(1/x)$ $(f(0) = 0)$ at $x = 0$.

5.4.26 Let $x_0 \in \mathbb{R}$. Following are four δ-ε conditions on a function $f : \mathbb{R} \to \mathbb{R}$. Which, if any, of these conditions imply continuity of f at x_0? Which, if any, are implied by continuity at x_0?

(a) For every $\varepsilon > 0$ there exists $\delta > 0$ such that if $|x - x_0| < \delta$, then $|f(x) - f(x_0)| < \varepsilon$.

(b) For every $\varepsilon > 0$ there exists $\delta > 0$ such that if $|f(x) - f(x_0)| < \delta$, then $|x - x_0| < \varepsilon$.

(c) For every $\varepsilon > 0$ there exists $\delta > 0$ such that if $|x - x_0| < \varepsilon$, then $|f(x) - f(x_0)| < \delta$.

(d) For every $\varepsilon > 0$ there exists $\delta > 0$ such that if $|f(x) - f(x_0)| < \varepsilon$, then $|x - x_0| < \delta$.

5.4.27 Let $x_0 \in \mathbb{R}$. Following are four δ-ε conditions on a function $f : \mathbb{R} \to \mathbb{R}$. Which, if any, of these conditions imply continuity of f at x_0? Which, if any, are implied by continuity at x_0?

(a) There exists $\varepsilon > 0$ such that for each $\delta > 0$, if $|x - x_0| < \delta$, then $|f(x) - f(x_0)| < \varepsilon$.

(b) There exists $\varepsilon > 0$ such that for each $\delta > 0$, if $|f(x) - f(x_0)| < \delta$, then $|x - x_0| < \varepsilon$.

(c) There exists $\varepsilon > 0$ such that for each $\delta > 0$, if $|x - x_0| < \varepsilon$, then $|f(x) - f(x_0)| < \delta$.

(d) There exists $\varepsilon > 0$ such that for each $\delta > 0$, if $|f(x) - f(x_0)| < \varepsilon$, then $|x - x_0| < \delta$.

5.4.28 For each of the eight conditions of Exercises 5.4.26 and 5.4.27, describe in words which functions satisfy the condition. (Some of these conditions characterize familiar classes of functions, including the empty class.)

5.4.29 Let $A \subset \mathbb{R}$, $f : A \to \mathbb{R}$, $g : f(A) \to \mathbb{R}$. Prove that if f is continuous at $x_0 \in A$ and g is continuous at $f(x_0)$, then $g \circ f$ is continuous at x_0. Apply this to prove that if f is continuous at x_0, then $|f|$ is continuous at x_0.

5.4.30 Using the notions of unilateral or one-sided limits, define *left continuity* of a function f at a point x_0. Do the same for *right continuity*. If f is defined in a neighborhood of x_0, prove that f is continuous at x_0 if and only if f is both left continuous and right continuous at x_0.

5.4.31 Let $f : \mathbb{R} \to \mathbb{R}$. Prove that f is continuous if and only if for every closed set $K \subset \mathbb{R}$, the set $f^{-1}(K)$ is closed in $\mathbb{R}$. State carefully and prove the analogous result if $f : A \to \mathbb{R}$, where A is an arbitrary nonempty subset of $\mathbb{R}$.

5.4.32 Suppose f has the IVP on (a, b) and is discontinuous at $x_0 \in (a, b)$. Prove that there exists $y \in \mathbb{R}$ such that $\{x : f(x) = y\}$ is infinite.

5.5 Properties of Continuous Functions

We now present some of the most basic of the properties of continuous functions. The first theorem is an algebraic one; it asserts that the family of continuous functions defined on a set has many of the properties of an *algebra*: elements may be added, subtracted, multiplied, and (under some conditions) divided.

Theorem 5.39 *Let $f, g : A \to \mathbb{R}$ and let $c \in \mathbb{R}$. Suppose f and g are continuous at $x_0 \in A$. Then cf, $f + g$ and fg are continuous at x_0. Furthermore, if $g(x_0) \neq 0$, then f/g is continuous at x_0.*

Proof The results follow immediately from the limit definition of continuity and the usual algebraic properties of limits. ■

Corollary 5.40 *Every polynomial is continuous on* $\mathbb{R}$.

Proof The functions $f(x) = 1$ and $g(x) = x$ are continuous on $\mathbb{R}$. The corollary follows from Theorem 5.39. ■

Corollary 5.41 *Every rational function is continuous at each point in its domain (i.e., at each* $x \in \mathbb{R}$ *at which the denominator does not vanish).*

One of our most important properties allows us to compose two continuous functions. Be careful, though, with the conditions on the domains as they cannot be overlooked.

Theorem 5.42 *Let* $f : A \to \mathbb{R}$, $g : B \to \mathbb{R}$ *and suppose that* $f(A) \subset B$. *Suppose that* f *is continuous at a point* $x_0 \in A$ *and that* g *is continuous at the point* $y_0 = f(x_0) \in B$. *Then the composition function*

$$g \circ f : A \to \mathbb{R}$$

is continuous at x_0.

Proof This follows from Theorem 5.25. ■

A global version follows as a corollary.

Corollary 5.43 *Let* $f : A \to \mathbb{R}$, $g : B \to \mathbb{R}$ *and suppose that* $f(A) \subset B$. *If* f *is continuous on* A *and* g *is continuous on* B, *then the composition function*

$$g \circ f : A \to \mathbb{R}$$

is continuous on A.

Exercises

5.5.1 If f and g are functions such that $f + g$ is continuous, does it follow that at least one of f or g must be continuous?

5.5.2 If $|f|$ is continuous, does it follow that f is continuous?

5.5.3 If $e^{f(x)}$ is continuous, does it follow that f is continuous?

5.5.4 If $f(f(x))$ is continuous, does it follow that f is continuous?

5.6 Uniform Continuity

Let us take a closer look at the meaning of continuity of a function f on an interval I. The definition asserts that for each $x_0 \in I$ and for every $\varepsilon > 0$, there exists $\delta > 0$ such that if $x \in I$ and $|x - x_0| < \delta$, then

$$|f(x) - f(x_0)| < \varepsilon.$$

Now carefully consider the following statement:

> For every $\varepsilon > 0$, there exists $\delta > 0$ such that if x, $x_0 \in I$ and $|x - x_0| < \delta$, then $|f(x) - f(x_0)| < \varepsilon$.

This may appear at first sight to be just a restatement of the meaning of continuity expressed in the first paragraph. If you cannot detect the difference, then you are in good company: Cauchy did not see any difference and used the property just quoted incorrectly to prove that a continuous function on an interval $[a, b]$ must be integrable.

We need to focus on the fact that the number δ depends not only on f and on ε, but also on x_0; that is, $\delta = \delta(f, \varepsilon, x_0)$.

Example 5.44 Consider the function $f(x) = 1/x$ on the interval $I = (0, 1)$. We found in Exercise 5.4.10 that if we take $\varepsilon = 1$, we can choose

$$\delta(f, 1, x_0) = \frac{x_0^2}{1 + x_0},$$

but we cannot choose a larger value. Thus if $x_0 \to 0$, then $\delta(f, 1, x_0) \to 0$. No number δ is sufficiently small to "work" for *all* $x_0 \in I$. ◀

It is often important to be able to select δ independently of x_0. When this is possible, we say that f is uniformly continuous on I.

Definition 5.45 (Uniformly Continuous) Let f be defined on a set $A \subset \mathbb{R}$. We say that f is *uniformly continuous* (on A) if for every $\varepsilon > 0$ there exists $\delta > 0$ such that if $x, y \in A$ and $|x - y| < \delta$, then $|f(x) - f(y)| < \varepsilon$.

As an illustration of the usefulness of uniform continuity, we note that if f is uniformly continuous on a bounded interval I, then f is bounded on I.

Theorem 5.46 *If a function f is uniformly continuous on a bounded interval I, then f is bounded on I.*

Proof Here we suppose that I is one of (a, b), $[a, b]$, $[a, b)$, or $(a, b]$. To check that f is bounded, choose δ so that $|f(x) - f(y)| < 1$ whenever $x, y \in I$ and $|x - y| < \delta$. There is a finite set $a = x_0 < x_1 < \cdots < x_n = b$ such that $|x_i - x_{i-1}| < \delta$ for $i = 1, \ldots, n$. Our definition of δ implies that f is bounded on each of the intervals $[x_{i-1}, x_i] \cap I$. Let

$$\begin{aligned} m_i &= \inf\{f(x) : x_{i-1} \le x \le x_i \ , \ x \in I\}, \\ M_i &= \sup\{f(x) : x_{i-1} \le x \le x_i \ , \ x \in I\}, \\ m &= \min\{m_1, \ldots, m_n\} \\ M &= \max\{M_1, \ldots, M_n\}. \end{aligned}$$

Then, for every $x \in I$, $m \le f(x) \le M$, so f is bounded on I. ■

Observe that if we tried to present a similar argument for the function $f(x) = 1/x$ on the interval $I = (0, 1)$, the continuity of f would allow us to conclude that every $x \in I$ is in an interval on which f is bounded, but we would be unable to obtain a finite number of such intervals that cover I.

In our illustration that uniform continuity on I implies boundedness, we did not specify whether I contained one or more of its endpoints. Our next objective is to show that when $I = [a, b]$ is a *closed* interval, then every function f that is continuous on I is uniformly continuous on I. (Note also the more general version given in Exercise 5.6.14.)

This result will be of importance in many places. In particular, the important result we will later prove, that a continuous function f on $[a, b]$ is integrable, depends on the uniform continuity of f. Cauchy certainly recognized this fact but failed to distinguish between continuity and uniform continuity.

Theorem 5.47 *Let f be continuous on $[a, b]$. Then f is uniformly continuous.*

Proof Our proof invokes a compactness argument. We recall from our investigations of compactness in Section 4.5 that there are several equivalent formulations possible. We shall use the Bolzano-Weierstrass property. (Exercise 5.6.2 asks for another proof of this same theorem using Cousin's lemma. In Exercise 5.6.13 you are asked to prove it using the Heine-Borel property.)

We use an indirect proof. If f is not uniformly continuous, then there are sequences $\{x_n\}$ and $\{y_n\}$ so that $x_n - y_n \to 0$ but

$$|f(x_n) - f(y_n)| > c$$

for some positive c. (The verification of this step is left as Exercise 5.6.12.)

Now apply the Bolzano-Weierstrass property to obtain a convergent subsequence $\{x_{n_k}\}$. But observe that this requires that $\{x_{n_k}\}$ and $\{y_{n_k}\}$ both converge to the same limit z, which must be a point in the interval $[a, b]$. By the continuity of f, $f(x_{n_k}) \to f(z)$ and $f(y_{n_k}) \to f(z)$. Since $|f(x_n) - f(y_n)| > c$ for all n, this means from our study of sequence limits that

$$|f(z) - f(z)| \geq c > 0$$

and this is impossible. This contradiction proves the theorem. ■

Boundedness of Continuous Functions As an application of Theorem 5.47 we can now prove that any continuous function on a closed bounded interval $[a, b]$ is bounded. Indeed such a function must be uniformly continuous there, and we have already seen in Theorem 5.46 that a uniformly continuous function on a bounded interval is bounded. Thus we have the following useful theorem.

Theorem 5.48 *Let f be continuous on $[a,b]$. Then f is bounded.*

Exercises

5.6.1 Adjust the proof of Theorem 5.47 to show that if f is continuous on a compact set K, then f is uniformly continuous on K.

5.6.2 Give another proof of Theorem 5.47 but this time using Cousin's lemma.

5.6.3 Because of Theorem 5.46 any function that is continuous on $(0,1)$ but unbounded cannot be uniformly continuous there. Give an example of a continuous function on $(0,1)$ that is bounded, but not uniformly continuous.

5.6.4 Let $x_1, x_2, \dots, x_n$ be real numbers, each in the domain of some function f. Show that f is uniformly continuous on the set $X = \{x_1, x_2, \dots, x_n\}$.

5.6.5 Let $X = \{x_1, x_2, \dots, x_n, \dots\}$. What property must X have so that every function continuous on X is uniformly continuous on X?

5.6.6 Suppose f is uniformly continuous on each of the sets X_1, X_2, ..., X_n and also continuous on the union $X = \bigcup_{i=1}^{n} X_i$. Prove that f is uniformly continuous on X.

5.6.7 Suppose f is uniformly continuous on each of the compact sets

$$X_1, X_2, \dots, X_n.$$

Prove that f is uniformly continuous on the set $X = \bigcup_{i=1}^{n} X_i$. Show that this need not be the case if the sets X_k are not closed and need not be the case if the sets X_k are not bounded.

5.6.8 Let f be a uniformly continuous function on a set E. Show that if $\{x_n\}$ is a Cauchy sequence in E then $\{f(x_n)\}$ is a Cauchy sequence in $f(E)$. Show that this need not be true if f is continuous but not uniformly continuous.

5.6.9 A function $f : E \to \mathbb{R}$ is said to be Lipschitz if there is a positive number M so that $|f(x) - f(y)| \leq M|x - y|$ for all x, $y \in E$. Show that such a function must be uniformly continuous on E. Is the converse true?

5.6.10 Explain how Exercise 5.6.4 can be deduced from Exercise 5.6.6 or from Exercise 5.6.7.

5.6.11 Give an example of a function f that is continuous on $\mathbb{R}$ and a sequence of compact intervals X_1, X_2, ..., X_n, ... on each of which f is uniformly continuous, but for which f is not uniformly continuous on $X = \bigcup_{i=1}^{\infty} X_i$.

5.6.12 Show that if f is not uniformly continuous on an interval $[a,b]$ then there are sequences $\{x_n\}$ and $\{y_n\}$ chosen from that interval so that $x_n - y_n \to 0$ but $|f(x_n) - f(y_n)| > c$ for some positive c.

5.6.13 Prove Theorem 5.47 using the Heine-Borel property. ✂

5.6.14 Prove the following more general and complete version of Theorem 5.47.

> Suppose that $f : E \to \mathbb{R}$ is continuous. If E is compact, then f must be uniformly continuous on E. Conversely, if every continuous function $f : E \to \mathbb{R}$ is uniformly continuous, then E must be compact.

5.6.15 Prove Theorem 5.48 without using the fact that such a function is uniformly continuous. Use Cousin's lemma.

5.6.16 Prove Theorem 5.48 without using the fact that such a function is uniformly continuous. Use the Bolzano-Weierstrass property.

✂ **5.6.17** Prove Theorem 5.48 without using the fact that such a function is uniformly continuous. Use the Heine-Borel property.

5.7 Extremal Properties

A familiar kind of problem that we study in elementary calculus involves locating extrema of continuous functions defined on an interval $[a, b]$. The technique entails checking values of the function at points where its derivative is zero, at the endpoints of the interval, and at any points of nondifferentiability. For such a process to work, we must be sure the function *has* a maximum (or minimum) on the interval. We verify this now.

Theorem 5.49 *Let f be continuous on $[a, b]$. Then f possesses both an absolute maximum and an absolute minimum.*

Proof Let $M = \sup\{f(x) : a \leq x \leq b\}$. By Theorem 5.47, f is uniformly continuous on $[a, b]$. Thus, by Theorem 5.48, $M < \infty$. If there exists x_0 such that $f(x_0) = M$, then f achieves a maximum value M. Suppose, then, that $f(x) < M$ for all $x \in [a, b]$. We show this is impossible.

Let $g(x) = 1/(M - f(x))$. For each $x \in [a, b]$, $f(x) \neq M$; as a consequence, g is continuous and $g(x) > 0$ for all $x \in [a, b]$. From the definition of M we see that

$$\inf\{M - f(x) : x \in [a, b]\} = 0,$$

so

$$\sup\left\{\frac{1}{M - f(x)} : x \in [a, b]\right\} = \infty.$$

This means that g is not bounded on $[a, b]$. This is impossible because, as we saw in Section 5.6, a continuous function defined on a closed interval must be bounded. A similar proof would show that f has an absolute minimum on A. ∎

Example 5.50 Does this theorem extend to more general situations? If we replace the interval $[a, b]$ by some other set does the conclusion remain true? The example

$$f(x) = \frac{1}{x} \text{ for } x \in (0, 1)$$

shows that the closed interval cannot be replaced by an open one. On the other hand, the example

$$f(x) = x \text{ for } x \in [0, \infty)$$

shows that the bounded closed interval $[a, b]$ cannot be replaced by an unbounded closed one. ◀

From this example the suggestion that we need a closed and bounded set (i.e., a compact set) seems to offer itself. Indeed that is the correct generalization of Theorem 5.49.

Theorem 5.51 *Let f be continuous on a closed and bounded set A. Then f possesses an absolute maximum and an absolute minimum on A.*

Exercises

5.7.1 Give an example of an everywhere discontinuous function that possesses a unique point at which there is an absolute maximum and a unique point at which there is an absolute minimum.

5.7.2 Show that a continuous function maps compact sets to compact sets.

5.7.3 Prove Theorem 5.49 using a Bolzano-Weierstrass argument.

5.7.4 Give an example of a function defined only on the rationals and continuous at each point in its domain and yet does not have an absolute maximum.

5.7.5 Let $f : \mathbb{R} \to \mathbb{R}$ be a continuous function with the property that

$$\lim_{x\to\infty} f(x) = \lim_{x\to-\infty} f(x) = 0.$$

Show that f has either an absolute maximum or an absolute minimum but not necessarily both.

5.7.6 Let $f : \mathbb{R} \to \mathbb{R}$ be a continuous function that is periodic in the sense that for some number p, $f(x + p) = f(x)$ for all $x \in \mathbb{R}$. Show that f has an absolute maximum and an absolute minimum.

5.8 Darboux Property

We have already observed that the IVP (Darboux property) is not the same as continuity. It is true, however, that if f is continuous on $[a, b]$, then f has the Darboux property. We state Theorem 5.52 in a form that suggests use of Cousin's lemma. (Readers that prefer to use the Bolzano-Weierstrass theorem should see the hint for Exercise 5.8.3.) Expressed this way the theorem asserts that if the graph has no point on some horizontal line $y = c$, then the graph must be entirely above or below that line. Another way to say this (see Exercise 5.8.8) is that the function must assume every value between any two of its values.

Theorem 5.52 *Let f be continuous on $[a,b]$ and let $c \in \mathbb{R}$. If for every $x \in [a,b]$, $f(x) \neq c$, then either $f(x) > c$ for all $x \in [a,b]$ or $f(x) < c$ for all $x \in [a,b]$.*

Proof Again, as in the proof of Theorem 5.47, we must invoke a compactness argument. We shall use Cousin's lemma (Lemma 4.26). In the exercises you are asked to prove this same theorem using the Bolzano-Weierstrass property and the Heine-Borel property.

Let $\mathcal{C}$ denote the collection of closed intervals J such that $f(x) < c$ for all $x \in J$ or $f(x) > c$ for all $x \in J$. We verify that $\mathcal{C}$ forms a full cover of $[a,b]$.

If $x \in [a,b]$, then $|f(x) - c| = \varepsilon > 0$, so there exists $\delta > 0$ such that $|f(t) - f(x)| < \varepsilon$ whenever $|t - x| < \delta$ and $t \in [a,b]$. Thus, if $f(x) < c$, then $f(t) < c$ for all $t \in [x - \delta/2, x + \delta/2]$, while if $f(x) > c$, then $f(t) > c$ for all $t \in [x - \delta/2, x + \delta/2]$. By Cousin's lemma there exists a partition of $[a,b]$, $a = x_0 < x_1 < \cdots < x_n = b$ such that for $i = 1, \ldots, n$, $[x_{i-1}, x_i] \in \mathcal{C}$.

Suppose now that $f(a) < c$. The argument is similar if $f(a) > c$. Since $[a, x_1] = [x_0, x_1] \in \mathcal{C}$, $f(x) < c$ for all $x \in [x_0, x_1]$. Analogously, since $[x_1, x_2] \in \mathcal{C}$, and $f(x_1) < c$, $f(x) < c$ for for $x \in [x_1, x_2]$. Proceeding in this way, we see that $f(x) < c$ for all $x \in [a,b]$. ■

You may wish to look at Exercise 5.8.8 for other wordings of this theorem that suggest IVP as "connectedness."

Exercises

5.8.1 Show that a nondecreasing function with the Darboux property must be continuous.

5.8.2 Show that a continuous function maps compact intervals to compact intervals. Is it true that all continuous functions map closed (open) sets to closed (open) sets?

5.8.3 Prove Theorem 5.52 using the Bolzano-Weierstrass property of sequences rather than Cousin's lemma.

✂ **5.8.4** Prove Theorem 5.52 using the Heine-Borel property.

5.8.5 Prove Theorem 5.52 using the following "last point" argument: suppose that $f(a) < c < f(b)$ and let z be the last point in $[a,b]$ where $f(z) \leq c$, that is, let

$$z = \sup\{x \in [a, b] : f(x) \leq c\}.$$

Show that $f(z) = c$.

5.8.6 A function $f : [a, b] \to [a, b]$ is said to have a *fixed point* $c \in [a, b]$ if $f(c) = c$. Show that every continuous function f mapping $[a, b]$ onto itself has at least one fixed point.

5.8.7 Let $f : [a, b] \to [a, b]$ be continuous. Define a sequence recursively by $z_1 = x_1$, $z_n = f(z_{n-1})$ where $x_1 \in [a, b]$. Show that if the sequence $\{z_n\}$ is convergent, then it must converge to a fixed point of f.

5.8.8 Show that Theorem 5.52 can be reworded in the following ways:

(a) Let f be defined and continuous on an interval I, let $a, b \in I$ with $f(a) \neq f(b)$. Let d lie between $f(a)$ and $f(b)$. Then there exists c between a and b such that $f(c) = d$.

(b) A continuous function defined on an interval I maps subintervals of I onto either single points or else subintervals of $\mathbb{R}$. [Singleton points are often considered to be (degenerate) intervals.]

5.8.9 Show that a continuous function maps compact intervals to compact intervals.

5.8.10 State forms of Theorem 5.52 and its rewordings in Exercise 5.8.8 for continuous functions defined on intervals that need not be closed and/or bounded.

5.9 Points of Discontinuity

In our discussion of continuous functions we have mentioned discontinuities only as a contrast to the notion of continuity. In many applications of mathematics the functions that arise will have discontinuities and it is well to study such functions. We first ask for a language of discontinuity points. Then we investigate an important class of functions, the monotonic functions, and determine just how badly discontinuous they could be.

5.9.1 Types of Discontinuity

Let x_0 be a point of the domain of some function f. If x_0 is a point of discontinuity, then this means that either the limit $\lim_{x \to x_0} f(x)$ fails to exist or else that limit does exist but

$$f(x_0) \neq \lim_{x \to x_0} f(x).$$

Note that when we discuss discontinuity points we are discussing only points at which the function is defined. (Some calculus texts might call x_0 a point of discontinuity even if $f(x_0)$ fails to be defined. This is not our usage here.)

Note, too, that a discontinuity point cannot occur at an isolated point of the domain of the function.

Removable Discontinuities We can separate these cases into situations of increasing severity. The weakest possibility is that $\lim_{x\to x_0} f(x)$ does indeed exist but fails to equal $f(x_0)$. We call this a *removable discontinuity* of f. The word "removable" suggests that were we merely to assign a new value to $f(x_0)$ we would no longer have a discontinuity.

Jump Discontinuities A little more serious case of discontinuity occurs if $\lim_{x\to x_0} f(x)$ does not exist, but it fails to exist only because

$$\lim_{x\to x_0+} f(x) \text{ and } \lim_{x\to x_0-} f(x),$$

the two one-sided limits, exist but disagree. In that case, no matter what value $f(x_0$ assumes, this is a point of discontinuity.

We call this a *jump discontinuity* of f. The difference between the two limits

$$\lim_{x\to x_0+} f(x) - \lim_{x\to x_0-} f(x)$$

is a measure of the "size" of the discontinuity and is called the *jump*.

Essential Discontinuities Finally, the most intractable kind of discontinuity would be the situation in which $\lim_{x\to x_0} f(x)$ does not exist, and at least one of the two right-hand and left-hand limits (perhaps both)

$$\lim_{x\to x_0+} f(x) \text{ and } \lim_{x\to x_0-} f(x)$$

also does not exist. Again, no matter what value $f(x_0$ assumes, this is a point of discontinuity. We call this an *essential discontinuity* of f.

Example 5.53 Let $f(x) = 0$ for all $x \neq 0$ and let $f(0) = 2$. It is clear that 0 is a removable discontinuity of f. Perhaps this example seems entirely artificial. A more natural example would be the function given by the following formula:

$$f(x) = \frac{x+1}{x^2-1}\ (x \neq \pm 1),\quad f(1) = c_1,\quad f(-1) = c_2.$$

This function is clearly continuous at every point other than $x = \pm 1$ but may have two discontinuities, one at -1 and one at 1. One of these is not, however, a serious discontinuity since it is removable. You should try to determine which one is removable and which one is essential. ◄

Example 5.54 Let $f(x)$ be defined as the linear function $x+1$ for $x < 0$ and a different linear function $2x-1$ for $x \geq 0$. Then there is a discontinuity at 0 since

$$\lim_{x \to 0+} f(x) = \lim_{x \to 0+} (2x - 1) = -1$$

but

$$\lim_{x \to 0-} f(x) = \lim_{x \to 0-} (x + 1) = 1.$$

In this case the size of the jump is 2. A picture would show exactly what this jump represents. ◀

Exercises

5.9.1 Show that a function that has the Darboux property cannot have either removable or jump discontinuities.

5.9.2 What kind of discontinuities does the Dirichlet function (see Section 5.2.6) have?

5.9.3 What kind of discontinuities does the characteristic function of the Cantor set (see Section 5.2.6) have?

5.9.4 Let the function $f : \mathbb{R} \to \mathbb{R}$ have just one point of discontinuity and assume only rational values. What kind of discontinuity point must that be?

5.9.5 Classify the discontinuities of the rational function

$$f(x) = \frac{x+1}{x^2 - 1} \ (x \neq \pm 1), \quad f(1) = c_1, \quad f(-1) = c_2.$$

5.9.6 Give an example of a function continuous at 0 but with an essential discontinuity at each other point.

5.9.7 Give an example of a function f with a jump discontinuity and yet $(f)^2$ is continuous everywhere.

5.9.8 Give an example of a function f with an essential discontinuity everywhere and yet $(f)^2$ is continuous everywhere.

5.9.9 Define a function F by the formula

$$F(x) = \lim_{n \to \infty} \frac{x^n}{1 + x^n}.$$

What is the domain of this function? Classify all discontinuities.

5.9.2 Monotonic Functions

In general, there is not too much to say about the continuity of an arbitrary function. It is possible for a function to be discontinuous everywhere. But if the function is monotonic this is not possible. We start with some definitions, needed here and again later in many places.

Definition 5.55 (Nondecreasing) Let f be real valued on an interval I. If $f(x_1) \le f(x_2)$ whenever x_1 and x_2 are points in I with $x_1 < x_2$, we say f is *nondecreasing* on I.

Definition 5.56 (Increasing) Let f be real valued on an interval I. If the strict inequality $f(x_1) < f(x_2)$ holds whenever x_1 and x_2 are points in I with $x_1 < x_2$, we say f is *increasing.*

In the opposite direction we define "nonincreasing" and "decreasing."

Definition 5.57 (Nonincreasing) Let f be real valued on an interval I. If $f(x_1) \ge f(x_2)$ whenever x_1 and x_2 are points in I with $x_1 < x_2$, we say f is *nonincreasing* on I.

Definition 5.58 (Decreasing) Let f be real valued on an interval I. If the strict inequality $f(x_1) > f(x_2)$ holds whenever x_1 and x_2 are points in I with $x_1 < x_2$, we say f is *decreasing.*

A function that is either nonincreasing or nondecreasing is said to be *monotonic.* Sometimes, to emphasize that there is a strict inequality, we say that a function that is increasing or decreasing is *strictly monotonic.*

The class of monotonic functions has a particularly interesting structure as regards continuity. Such functions can never have essential discontinuities. This is because if f is monotonic nondecreasing or monotonic nonincreasing, then at any point both one-sided limits $\lim_{x\to x_0+} f(x)$ and $\lim_{x\to x_0-} f(x)$ exist.

Theorem 5.59 *Let f be monotonic on an interval I. If x_0 is interior to I, then the one-sided limits $\lim_{x\to x_0-} f(x)$ and $\lim_{x\to x_0+} f(x)$ both exist.*

Proof Suppose f is nondecreasing on I; the proof for the case that f is nonincreasing will then follow by noting that in this case $-f$ is nondecreasing. To prove Theorem 5.59 let x_0 be interior to I and let $\{x_k\}$ be an increasing sequence of points in I such that $\lim_{k\to\infty} x_k = x_0$. Then the sequence $\{f(x_k)\}$ is a nondecreasing sequence of numbers bounded from above by $f(x_0)$. Thus by the monotone convergence principle $\{f(x_k)\}$ approaches a limit L.

For $x_k < x < x_0$,

$$f(x_k) \le f(x) \le L.$$

Let $\varepsilon > 0$. Since $f(x_k) \to L$, there exists $N \in \mathbb{N}$ such that

$$L - f(x_k) < \varepsilon$$

whenever $k \ge N$. For all x satisfying $x_N \le x \le x_0$ we thus have

$$L - f(x) \le L - f(x_k) < \varepsilon.$$

It follows that

$$\lim_{x \to x_0-} f(x) = L,$$

so f has a left-sided limit at x_0. A similar argument shows that f also has a right-sided limit at x_0. ■

Monotonic Functions Have Jump Discontinuities Recall that a function f is said to have a *jump* at x_0 if f has limits from the left and from the right at x_0, but these limits are different. Thus, if f is monotonic nondecreasing, say, then clearly

$$\lim_{x \to x_0-} f(x) \leq f(x_0) \leq \lim_{x \to x_0+} f(x).$$

Thus the only possibility of a discontinuity at the point x_0 is if the jump

$$J(x_0) = \lim_{x \to x_0+} f(x) - \lim_{x \to x_0-} f(x)$$

is positive. Thus monotonic functions do not have removable discontinuities nor do they have essential discontinuities. They have only jump discontinuities.

Monotonic Functions Have Countably Many Discontinuities We can go further than this. We can ask about the set of points at which there can be a discontinuity point. We ask how large this set can be. The answer is "not very."

Theorem 5.60 *Let f be monotonic on an interval $[a, b]$. Then the set of points of discontinuity of f in that interval is countable. In particular, f must be continuous at the points of set dense in $[a, b]$.*

Proof We consider again the case that f is nondecreasing since the case that f is nonincreasing follows by considering the function $-f$. If f is nondecreasing and discontinuous at a point x_0 in the interior of I, then the open interval

$$I(x_0) = \left(\lim_{x \to x_0-} f(x), \lim_{x \to x_0+} f(x) \right)$$

either contains no points in the range of f or contains only the single point $f(x_0)$ in the range. (To check this statement, see Exercise 5.9.12.) Thus, each point of discontinuity x_0 of f in I corresponds to an interval $I(x_0)$. For two different points of discontinuity x_1 and x_2, the intervals $I(x_1)$ and $I(x_2)$ are disjoint (because f is nondecreasing). But any collection of disjoint intervals in $\mathbb{R}$ can be arranged into a sequence (Exercise 4.6.10) and so there can be only countably many points of discontinuity of f . ■

It is easy to construct monotonic functions with infinitely many points of discontinuity. For example, if $f(x) = n$ on $[n, n+1)$, then f has jumps at all the integers.

It is natural to ask which countable sets can be the set of discontinuities for some monotonic f. For example, does there exist an increasing function that is discontinuous at every rational number in $\mathbb{R}$? (Exercise 5.9.14 provides an answer.)

Example 5.61 Our theorem shows that a monotonic function has a countable set of points at most where it can be discontinuous. It is easy to find examples of monotonic functions with a prescribed set of discontinuities if the set given to us is finite. Could any countable set be given and we then find a monotonic function that has exactly that set as its points of discontinuity?

The answer, remarkably, is yes. Let C be a countable subset of (a,b). List the elements as c_1, c_2, c_3, $\dots$. Define the function for $a \le x \le b$ as

$$f(x) = \sum_{c_n < x} \frac{1}{2^n}.$$

This function is hard to visualize since it depends on the order of the terms. Clearly, $f(a) = 0$ and $f(b) = 1$. The other values are much less clear. But we can see that there is a jump of magnitude $1/2$ at the point c_1, a jump of magnitude $1/4$ at the point c_2, a jump of magnitude $1/8$ at the point c_3, and so on. The function is strictly increasing on any subinterval in which C is dense and would be constant in any interval that contains no points of C. It can be shown that the only discontinuities occur at the points of C. ◄

Exercises

5.9.10 Construct a function with a jump discontinuity of magnitude -5 at the point $x = 1$ and continuous everywhere else.

5.9.11 Find a monotonic function on $[0,1]$ with discontinuities at $1/3$, $2/3$, and $3/4$ only.

5.9.12 Suppose f is increasing on an interval I. Let x_0 be an interior point of I. Prove that $\lim_{x\to x_0-} f(x) \le f(x_0) \le \lim_{x\to x_0+} f(x)$.

5.9.13 Verify the claims made in Example 5.61 about the function f there.

5.9.14 Using Example 5.61, show that there is a (strictly) increasing function on $[0,1]$ that is discontinuous at each rational number in $(0,1)$ and continuous at each irrational number.

5.9.15 Show that there is no monotonic function on $[0,1]$ that is discontinuous precisely at each irrational number in $(0,1)$.

5.9.16 Show that if $f : [a,b] \to \mathbb{R}$ is continuous and increasing, then the inverse function f^{-1} exists and is also continuous and increasing on the interval on which it is defined.

5.9.17 Let f be a continuous function on an open interval (a, b). Suppose that f has no local maximum or local minimum at any point. Show that f must be monotonic.

5.9.18 Suppose that $f : \mathbb{R} \to \mathbb{R}$ and that $f(x) + \alpha x$ is monotonic for every $\alpha \in \mathbb{R}$. Show that $f(x) = ax + b$ for some a, b.

5.9.19 Let $\{f_n\}$ be a sequence of monotonic functions defined on the interval $[0, 1]$. Suppose that

$$f(x) = \lim_{n \to \infty} f_n(x)$$

exists for each $0 \leq x \leq 1$. Show that f is monotonic. (If the word "monotonic" is replaced throughout this problem by "continuous," the exercise would be invalid: show this, too.)

5.9.20 Can the range of an increasing function on the interval $[0, 1]$ consist only of rational numbers? Can it consist only of irrational numbers?

5.9.3 How Many Points of Discontinuity?

✂ *Advanced*

We have already answered the question as to how many points of discontinuity a monotonic function may have. The set of such points must be countable. We know too that all of these are jump discontinuities; a monotonic function has no removable discontinuities and no essential discontinuities.

What is the situation for an arbitrary function? There are three questions. How many removable discontinuities are possible? How many jump discontinuities are possible? How many essential discontinuities are possible?

Example 5.62 One example that we have seen before shows that there can be a great many essential discontinuities. Let f be the characteristic function of the rational numbers; that is, $f(x)$ is 1 if x is a rational number and is 0 if x is irrational. Clearly,

$$\limsup_{x \to x_0} f(x) = 1$$

and

$$\liminf_{x \to x_0} f(x) = 0$$

at every point x_0. In particular, the limit does not exist anywhere and so every point is an essential discontinuity. ◀

Surprisingly, though, this is not the case for the removable discontinuities or the jump discontinuities. No function can have an uncountable number of such discontinuities.

Theorem 5.63 *Let f be a real function defined on an interval $[a, b]$. The sets of points in $[a, b]$ at which f has a removable discontinuity and at which f has a jump discontinuity are both countable.*

Proof Let J be the set of points at which there is a jump discontinuity. Every point of J is in one of the two sets:

$$J_+ = \{x \in (a,b) : \lim_{y \to x+} f(x) > \lim_{y \to x-} f(x)\}$$

or

$$J_- = \{x \in (a,b) : \lim_{y \to x+} f(x) < \lim_{y \to x-} f(x)\}.$$

We shall show that J_+ is countable.

If $x \in J_+$, then

$$\lim_{y \to x+} f(x) > \lim_{y \to x-} f(x)$$

and so there is for any such x at least one rational number r so that

$$\lim_{y \to x+} f(x) > r > \lim_{y \to x-} f(x).$$

Moreover, there then must exist some integer m (depending on x and r) so that

$$f(z) > r > f(y)$$

whenever $x - 1/m < y < x < z < x + 1/m$.

Let J_{rn}, where r is a rational and n a positive integer, denote the set of all points x with the property that $f(y) < r < f(z)$ whenever

$$x - 1/n < y < x < z < x + 1/n.$$

We claim that this set is countable. If not, then it must have a point of accumulation and, in particular, there would have to be at least three points $a < b < c$, with $c - a < 1/n$, all belonging to J_{rn}. But by the way that J_{rn} was defined this means, since a and $c \in J_{rn}$, that $f(b) < r$ and $r < f(b)$ are both true. Since this is impossible, all points in J_{rn} are isolated and hence J_{rn} is countable. The union

$$\bigcup_{r \in \mathbb{Q}} \bigcup_{n=1}^{\infty} J_{rn}$$

is a countable union of countable sets and is thus also countable. But this set contains every point of J_+ and so that set is also countable. Similarly, it is true that J_- is countable and hence the set of points with jump discontinuities is countable.

That the set of points at which the function has a removable discontinuity is also countable is left as an exercise. The ideas of the proof here can be used to prove it in a similar fashion. Notice especially this technique of inserting a rational number between two unequal numbers. ■

Incidentally, this theorem throws a new light on the theorem about the discontinuity points of monotonic functions. In that proof we used the properties of monotonic functions to show that the collection of discontinuity points was countable. But we know easily that the only such points are the jump discontinuities and any function, monotonic or not, has only countably many of these points by our theorem here. Thus we have another way of looking at Theorem 5.60.

Exercises

5.9.20 Give an example of a function with a dense set of removable discontinuities.

5.9.21 Give an example of a function with a dense set of jump discontinuities.

5.9.22 Prove the remaining statement of Theorem 5.63 that is not proved in the text.

5.10 Challenging Problems for Chapter 5

5.10.1 Suppose that f is a function defined on the real line with the property that $f(x+y) = f(x) + f(y)$ for all x, y. Suppose that f is continuous at 0. Show that f must be continuous everywhere.

5.10.2 Suppose that f is a function defined on the real line with the property that $f(x+y) = f(x) + f(y)$ for all x, y. Suppose that f is continuous at 0. Show that $f(x) = Cx$ for all x and some number C.

5.10.3 Suppose that f is a function defined on the real line with the property that $f(x+y) = f(x)f(y)$ for all x, y. Suppose that f is continuous at 0. Show that f must be continuous everywhere.

5.10.4 Generalize Theorem 5.60 to prove that if a function f (not necessarily monotonic) has left-sided limits and right-sided limits at every point of an open interval I, then f must be continuous except on a countable set.

5.10.5 Determine necessary and sufficient conditions on a pair of sets A and B so that they will have the property that there exists a continuous function $f : \mathbb{R} \to \mathbb{R}$ such that $f(x) = 0$ for all $x \in A$ and $f(x) = 1$ for all $x \in B$.

5.10.6 Let $f : [1, \infty)$ be continuous, positive and increasing with $f(x) \to \infty$ as $x \to \infty$. Show that

$$\sum_{k=1}^{\infty} \frac{1}{f(k)}$$

is convergent if and only if the series

$$\sum_{k=1}^{\infty} \frac{f^{-1}(k)}{k^2}$$

converges (where f^{-1} denotes the inverse function).

5.10.7 **(Extensions of continuous functions)** If $f : A \to \mathbb{R}$, $g : B \to \mathbb{R}$, $A \subset B$, and $f(x) = g(x)$ for all $x \in A$, then the function g is said to be an *extension* of the function f. Prove each of the following:

(a) A function that is continuous on a closed set A can be extended to a function that is continuous on $\mathbb{R}$.

(b) A function that is uniformly continuous on a set A can be extended to a function that is uniformly continuous on $\overline{A}$.

(c) A function that is uniformly continuous on an arbitrary nonempty subset of $\mathbb{R}$ can be extended to a function that is uniformly continuous on all of $\mathbb{R}$.

(d) Give an example of a function f that is continuous on (0,1) but that cannot be extended to a function continuous on [0,1].

✂ **5.10.8** For an arbitrary function $f : \mathbb{R} \to \mathbb{R}$ show that

$$\{x_0 : \limsup_{x \to x_0-} f(x) > \limsup_{x \to x_0+} f(x)\}$$

is countable.

✂ **5.10.9** Give an example of a function $f : \mathbb{R} \to \mathbb{R}$ such that there are infinitely many points x_0 at which either

$$f(x_0) > \limsup_{x \to x_0} f(x) \text{ or } f(x_0) < \liminf_{x \to x_0} f(x).$$

✂ **5.10.10** For an arbitrary function $f : \mathbb{R} \to \mathbb{R}$ show that the set of points x_0 at which $f(x_0)$ does not lie between

$$\liminf_{x \to x_0} f(x) \quad \text{and} \quad \limsup_{x \to x_0} f(x)$$

is countable.

✂ **5.10.11** Let y be a real number or $\pm\infty$ and let $f : E \to \mathbb{R}$ be a function. If there is a sequence $\{x_n\}$ of numbers in E and converging to a point c with $x_n \neq c$ and with $f(x_n) \to y$ then y is called *a cluster value* of f at c. Show that every cluster value at c lies between $\liminf_{x \to c} f(x)$ and $\limsup_{x \to c} f(x)$. Show that both $\liminf_{x \to c} f(x)$ and $\limsup_{x \to c} f(x)$ are themselves cluster values of f at c.

5.10.12 Is there a continuous function $f : \mathbb{R} \to \mathbb{R}$ such that for every real y there are precisely two solutions to the equation $f(x) = y$?

5.10.13 Is there a continuous function $f : \mathbb{R} \to \mathbb{R}$ such that for every real y there are precisely three solutions to the equation $f(x) = y$?

5.10.14 Suppose f has the IVP on (a, b) and is discontinuous at $x_0 \in (a, b)$. Prove that there exists $y \in \mathbb{R}$ such that $\{x : f(x) = y\}$ is infinite.

5.10.15 Prove that if $f : \mathbb{R} \to \mathbb{R}$, then the set

$$\{x : f \text{ is right continuous at } x \text{ but not left continuous at } x\}$$

is countable.

Chapter 6

MORE ON CONTINUOUS FUNCTIONS AND SETS

✂ This chapter can be considered enrichment material containing also several more advanced topics and may be skipped in its entirety. You can proceed directly to the study of derivatives and integrals in Chapters 7 and 8 with no loss in the continuity of the material.

6.1 Introduction

In this chapter we go much more deeply into the analysis of continuous functions. For this we need some new set theoretic ideas and methods.

6.2 Dense Sets

[This section reviews material from Section 1.9.]

Consider the set $\mathbb{Q}$ of rational numbers and let (a, b) be an open interval in $\mathbb{R}$. How do we show that there is a member of $\mathbb{Q}$ in the interval (a, b); that is, that $(a, b) \cap \mathbb{Q} \neq \emptyset$?

Suppose first that $0 < a$. Since $b - a > 0$, the archimedean property (Theorem 1.11) implies that there is a positive integer q such that

$$q(b - a) > 1.$$

Thus

$$qb > 1 + qa.$$

The archimedean property also implies that the set of integers

$$\{m \in \mathbb{N} : m > qa\}$$

is nonempty. Thus, according to the well-ordering principle, there is a smallest integer p in this set and for this p, it is true that $p - 1 \leq qa < p$. It follows that

$$qa < p \leq 1 + qa < qb,$$

which implies $a < \frac{p}{q} < b$. We have shown that, under the assumption $a > 0$, there exists a rational number $r = p/q$ in the interval (a, b).

The same is true under the assumption $a < 0$. To see this observe first that if $a < 0 < b$, we can take $r = 0$. If $a < b < 0$, then $0 < -b < -a$, so the argument of the previous paragraph shows that there exists $r \in \mathbb{Q}$ such that $-b < r < -a$. In this case $a < -r < b$.

The preceding discussion proves that every open interval contains a rational number. We often express this fact by saying that the set of rational numbers is a *dense* set.

Definition 6.1 A set of real numbers A is said to be *dense* (in $\mathbb{R}$) if for each open interval (a, b) the set $A \cap (a, b)$ is nonempty.

It is important to have a more general concept, that of a set A being dense in a set B.

Definition 6.2 Let A and B be subsets of $\mathbb{R}$. If every open interval that intersects B also intersects A, we say that A *is dense in* B.

Thus Definition 6.1 states the special case of Definition 6.2 that occurs when $B = \mathbb{R}$. We should note that some authors require that $A \subset B$ in their version of Definition 6.2. We find it more convenient not to impose this restriction. Thus, for example, in *our* language $\mathbb{Q}$ is dense in $\mathbb{R} \setminus \mathbb{Q}$.

It is easy to verify that A is dense in B if and only if $\overline{A} \supset B$ (Exercise 6.2.1).

Exercises

6.2.1 Verify that A is dense in B if and only if $\overline{A} \supset B$.

6.2.2 Prove that every set A is dense in its closure $\overline{A}$.

6.2.3 Prove that if A is dense in B and $C \subset B$, then A is dense in C.

6.2.4 Prove that if $A \subset B$ and A is dense in B, then $\overline{A} = \overline{B}$. Is the statement correct without the assumption that $A \subset B$?

6.2.5 Is $\mathbb{R} \setminus \mathbb{Q}$ dense in $\mathbb{Q}$?

6.2.6 The following are several pairs (A, B) of sets. In each case determine whether A is dense in B.

(a) $A = \mathbb{N}, B = \mathbb{N}$

(b) $A = \mathbb{N}, B = \mathbb{Z}$

(c) $A = \mathbb{N}, B = \mathbb{Q}$

(d) $A = \left\{x : x = \frac{m}{2^n},\ m \in \mathbb{Z}, n \in \mathbb{N}\right\}, B = \mathbb{Q}$

6.2.7 Let A and B be subsets of $\mathbb{R}$. Prove that A is dense in B if and only if for every $b \in B$ there exists a sequence $\{a_n\}$ of points from A such that $\lim_{n\to\infty} a_n = b$.

6.2.8 Let B be the set of all irrational numbers. Prove that the set

$$A = \{q + \sqrt{2} : q \in \mathbb{Q}\}$$

is a countable subset of B that is dense in B.

6.2.9 Let $f : \mathbb{R} \to \mathbb{R}$ be a strictly increasing continuous function. Does f map dense sets to dense sets; that is, is it true that

$$f(E) = \{f(x) : x \in E\}$$

is dense if E is dense?

6.2.10 Prove that every set $B \subset \mathbb{R}$ contains a countable set A that is dense in B.

6.3 Nowhere Dense Sets

We might view a set A that is dense in $\mathbb{R}$ as being somehow large: Inside every interval, no matter how small, we find points of A. There is an opposite extreme to this situation: A set is said to be *nowhere dense*, and hence is in some sense small, if it is not dense in any interval at all. The precise definition of this important concept of smallness follows.

Definition 6.3 The set $A \subset \mathbb{R}$ is said to be *nowhere dense* in $\mathbb{R}$ provided every open interval I contains an open subinterval J such that $A \cap J = \emptyset$.

We can state this another way: A is nowhere dense provided $\overline{A}$ contains no open intervals. (See Exercise 6.3.4.)

Example 6.4 It is easy to construct examples of nowhere dense sets.

1. Any finite set

2. $\mathbb{N}$

3. $\{1/n : n \in \mathbb{N}\}$

Each of these sets is nowhere dense, as you can verify. ◀

Each of the sets in Example 6.4 is countable and hence also small in the sense of cardinality. It is hard to imagine an uncountable set that is nowhere dense but, as we shall see in Section 6.5, such sets do exist.

We establish a simple result showing that any finite union of nowhere dense sets is again nowhere dense. It is not true that a countable union of nowhere dense sets is again nowhere dense. Indeed countable unions of nowhere dense sets will be important in our subsequent study.

Theorem 6.5 *Let $A_1, A_2, \ldots, A_n$ be nowhere dense in $\mathbb{R}$. Then $A_1 \cup \cdots \cup A_n$ is also nowhere dense in $\mathbb{R}$.*

Proof Let I be any open interval in $\mathbb{R}$. We seek an open interval $J \subset I$ such that $J \cap A_i = \emptyset$ for $i = 1, 2, \ldots, n$.

Since A_1 is nowhere dense, there exists an open interval $I_1 \subset I$ such that $I_1 \cap A_1 = \emptyset$. Now A_2 is also nowhere dense in $\mathbb{R}$, so there exists an open interval $I_2 \subset I_1$ such that $A_2 \cap I_2 = \emptyset$. Proceeding in this way we obtain open intervals

$$I_1 \supset I_2 \supset I_3 \cdots \supset I_n$$

such that for $i = 1, ..., n$, $A_i \cap I_i = \emptyset$. It follows from the fact that $I_n \subset I_i$ for $i = 1, \ldots, n$ that $A_i \cap I_n = \emptyset$ for $i = 1, \ldots, n$. Thus

$$\left(\bigcup_{i=1}^{n} A_i\right) \cap I_n = \bigcup_{i=1}^{n} (A_i \cap I_n) = \bigcup_{i=1}^{n} \emptyset = \emptyset,$$

as was to be proved. ■

Exercises

6.3.1 Give an example of a sequence of nowhere dense sets whose union is not nowhere dense.

6.3.2 Which of the following statements are true?

(a) Every subset of a nowhere dense set is nowhere dense.

(b) If A is nowhere dense, then so too is $A + c = \{t + c : t \in A\}$ for every number c.

(c) If A is nowhere dense, then so too is $cA = \{ct : t \in A\}$ for every positive number c.

(d) If A is nowhere dense, then so too is A', the set of derived points of A.

(e) A nowhere dense set can have no interior points.

(f) A set that has no interior points must be nowhere dense.

(g) Every point in a nowhere dense set must be isolated.

(h) If every point in a set is isolated, then that set must be nowhere dense.

6.3.3 If A is nowhere dense, what can you say about $\mathbb{R} \setminus A$? If A is dense, what can you say about $\mathbb{R} \setminus A$?

6.3.4 Prove that a set $A \subset \mathbb{R}$ is nowhere dense if and only if $\overline{A}$ contains no intervals; equivalently, the interior of $\overline{A}$ is empty.

6.3.5 What should the statement "A is nowhere dense in the interval I" mean? Give an example of a set that is nowhere dense in $[0, 1]$ but is not nowhere dense in $\mathbb{R}$.

6.3.6 Let A and B be subsets of $\mathbb{R}$. What should the statement "A is nowhere dense in the B" mean? Is $\mathbb{N}$ nowhere dense in $[0, 10]$? Is $\mathbb{N}$ nowhere dense in $\mathbb{Z}$? Is $\{4\}$ nowhere dense in $\mathbb{N}$?

6.3.7 Prove that the complement of a dense open subset of $\mathbb{R}$ is nowhere dense in $\mathbb{R}$.

6.3.8 Let $f : \mathbb{R} \to \mathbb{R}$ be a strictly increasing continuous function. Show that f maps nowhere dense sets to nowhere dense sets; that is,

$$f(E) = \{f(x) : x \in E\}$$

is nowhere dense if E is nowhere dense.

6.4 The Baire Category Theorem

✂ *Advanced*

In this section we shall establish the Baire category theorem, which gives a sense in which nowhere dense sets can be viewed as "small:" A union of a sequence of nowhere dense sets cannot fill up an interval. If we interpret Cantor's theorem (Theorem 2.4) as asserting that a union of a sequence of finite sets cannot fill up an interval, then we see the Baire category theorem as a far-reaching generalization.

We motivate this important theorem by way of a game idea that is due to Stefan Banach (1892–1945) and Stanislaw Mazur (1905–1981). Although the origins of the theorem are due to René Baire, after whom the theorem is named, the game approach helps us see why the Baire category theorem might be true. This Banach-Mazur game is just one of many mathematical games that are used throughout mathematics to develop interesting concepts.

6.4.1 A Two-Player Game

✂ *Advanced*

We introduce the Baire category theorem via a game between two players (A) and (B).

Player (A) is given a subset A of $\mathbb{R}$, and player (B) is given the complementary set $B = \mathbb{R} \setminus A$. Player (A) first selects a closed interval $I_1 \subset \mathbb{R}$; then player (B) chooses a closed interval $I_2 \subset I_1$. The players alternate moves, a move consisting of selecting a closed interval inside the previously chosen interval.

The play of the game thus determines a descending sequence of closed intervals

$$I_1 \supset I_2 \supset I_3 \supset \cdots \supset I_n \supset \ldots$$

where player (A) chooses those with odd index and player (B) those with even index. If

$$A \cap \bigcap_{n=1}^{\infty} I_n \neq \emptyset,$$

then player (A) wins; otherwise player (B) wins.

The goal of player (A) is evidently to make sure that the intersection contains a point of A; the goal of player (B) is to ensure that the intersection is empty or contains only points of B. We expect that player (A) should win if his set A is large while player (B) should win if his set is large. It is not, however, immediately clear what "large" might mean for this game.

Example 6.6 If the set A given to player (A) contains an open interval J, then (A) should choose any interval $I_1 \subset J$. No matter how the game continues, player (A) wins. Another way to say this: If the set given to player (B) is not dense, he loses. ◀

Example 6.7 For a more interesting example, let player (A) be dealt the "large" set of all irrational numbers, so that player (B) is dealt the rationals. (Both players have been dealt dense sets now.) Let A consist of the irrational numbers. Player (A) can win by following the strategy we now describe. Let q_1, q_2, q_3, ... be a listing of all of the rational numbers; that is,

$$\mathbb{Q} = \{q_1, q_2, q_3, \dots\}.$$

Player (A) chooses the first interval I_1 as any closed interval such that $\{q_1\} \notin I_1$. Inductively, suppose $I_1, I_2, \dots, I_{2n}$ have been chosen according to the rules of the game so that it is now time for player (A) to choose I_{2n+1}. The set $\{q_1, q_2, \dots, q_n\}$ is finite, so there exists a closed interval $I_{2n+1} \subset I_{2n}$ such that

$$I_{2n+1} \cap \{q_1, q_2, \dots, q_n\}$$

is empty. Player (A) chooses such an interval.

Since for each $n \in \mathbb{N}$, $q_n \notin I_{2n+1}$, the set $\bigcap_{n=1}^{\infty} I_n$ contains no rational numbers, but, as a descending sequence of closed intervals, $\bigcap_{n=1}^{\infty} I_n \neq \emptyset$. Thus $A \cap \bigcap_{n=1}^{\infty} I_n \neq \emptyset$, and (A) wins. ◀

In these two examples, using informal language, we can say that player (A) has a *strategy* to win: No matter how player (B) proceeds, player (A) can "answer" each move and win the game.

In both examples player (A) had a clear advantage: The set A was larger than the set B. But in what sense is it larger? It is not the fact that A is uncountable while B is countable that matters here. It is something else: The fact that given an interval I_{2n}, player (A) can choose I_{2n+1} inside I_{2n} in such a way that I_{2n+1} misses the set $\{q_1, q_2, \dots, q_n\}$.

Let us try to see in the second example a general strategy that should work for player (A) in some cases. The set B was the union of the singleton sets $\{q_n\}$. Suppose instead that B is the union of a sequence of "small" sets Q_n. Then the same "strategy" will prevail if given any interval J and given any $n \in \mathbb{N}$, there exists an interval $I \subset J$ such that

$$I \cap (Q_1 \cup Q_2 \cup \cdots \cup Q_n) = \emptyset.$$

The set $\bigcap_{n=1}^{\infty} I_n$ will be nonempty, and will miss the set $\bigcup_{n=1}^{\infty} Q_n$. Thus, if $B = \bigcup_{n=1}^{\infty} Q_n$, player (A) has a winning strategy. It is in this sense that the set B is "small." The set A is "large" because the set B is "small". If we look carefully at the requirement on the sets Q_k, we see it is just that each of these sets is nowhere dense in $\mathbb{R}$.

Thus the key to player (A) winning rests on the concept of a nowhere dense set. But note that it rests on the set B being the union of a sequence of nowhere dense sets.

6.4.2 The Baire Category Theorem

✂ *Advanced*

We can formulate our result from our discussion of the game in several ways:

1. $\mathbb{R}$ cannot be expressed as a countable union of nowhere dense sets.

2. The complement of a countable union of nowhere dense sets is dense.

The second of these provides a sense in which countable unions of nowhere dense sets are "small:" No matter which countable collection of nowhere dense sets we choose, their union leaves a dense set uncovered.

To formulate the Baire category theorem we need some definitions. This is the original language of Baire and it has survived; he simply places sets in two types or categories. Into the first category he places the sets that are to be considered small and into the second category he puts the remaining (not small) sets.

Definition 6.8 Let A be a set of real numbers.

1. A is said to be of the *first category* if it can be expressed as a countable union of nowhere dense sets.

2. A is said to be of the *second category* if it is not of the first category.

3. A is said to be *residual* in $\mathbb{R}$ if the complement $\mathbb{R} \setminus A$ is of the first category.

The following properties of first category sets and their complements, the residual sets, are easily proved and left as exercises.

Lemma 6.9 *A union of any sequence of first category sets is again a first category set.*

Lemma 6.10 *An intersection of any sequence of residual sets is again a residual set.*

Theorem 6.11 (Baire Category Theorem) *Every residual subset of $\mathbb{R}$ is dense in $\mathbb{R}$.*

Proof The discussion in Section 6.4.1 constitutes a proof. Suppose that player (A) is dealt a set $A = X \cap [a,b]$ where X is residual. Then there is a sequence of nowhere dense sets $\{Q_n\}$ so that

$$X = \mathbb{R} \setminus \bigcup_{n=1}^{\infty} Q_n.$$

Then player (A) wins by choosing any interval $I_1 \subset [a,b]$ that avoids Q_1 and continues following the strategy of Section 6.4.1. In particular, X must contain a point of the interval $[a,b]$, and hence a point of any interval. ■

Theorem 6.11 provides a sense of largeness of sets that is not shared by dense sets in general. The intersection of two dense sets might be empty, but the intersection of two, or even countably many, residual sets must still be dense.

Exercises

6.4.1 Show that the union of any sequence of first category sets is again a first category set.

6.4.2 Show that the intersection of any sequence of residual sets is again a residual set.

6.4.3 Rewrite the proof of Theorem 6.11 without using the games language.

6.4.4 Give an example of two dense sets whose intersection is not dense. Does this contradict Theorem 6.11?

6.4.5 Suppose that $\bigcup_{n=1}^{\infty} A_n$ contains some interval (c,d). Show that there is a set, say A_{n_0}, and a subinterval $(c'd') \subset (c,d)$ so that A_{n_0} is dense in $(c'd')$.

6.4.3 Uniform Boundedness

✂ *Advanced*

There are many applications of the Baire category Theorem in analysis. For now, we present just one application, dealing with the concept of *uniform boundedness.* Suppose we have a collection $\mathcal{F}$ of functions defined on $\mathbb{R}$ with the property that for each $x \in \mathbb{R}$, $\{|f(x)| : f \in \mathcal{F}\}$ is bounded. This means that for each $x \in \mathbb{R}$ there exists a number $M_x \geq 0$ such that $|f(x)| \leq M_x$

for all $f \in \mathcal{F}$. We can describe this situation by saying that $\mathcal{F}$ is *pointwise bounded.* Does this imply that the collection is *uniformly bounded*; that is, that there is a single number M so that $|f(x)| \le M$ for all $f \in \mathcal{F}$ and every $x \in \mathbb{R}$?

Example 6.12 Let $q_1, q_2, q_3, \ldots$ be an enumeration of $\mathbb{Q}$. For each $n \in \mathbb{N}$ we define a function f_n by $f_n(q_k) = k$ if $n \le k$, $f_n(x) = 0$ for all other values x. Let $\mathcal{F} = \{f_n : n \in \mathbb{N}\}$. Then if $x \in \mathbb{R} \setminus \mathbb{Q}$, $f(x) = 0$ for all $f \in \mathcal{F}$, and if $x = q_k$, $|f(x)| \le k$ for all $f \in \mathcal{F}$. Thus, for each $x \in \mathbb{R}$, the set $\{|f(x)| : f \in \mathcal{F}\}$ is bounded. The bounds can be taken to be 0 if $x \in \mathbb{R} \setminus \mathbb{Q}$ ($M_x = 0$ if $x \in \mathbb{R} \setminus \mathbb{Q}$) and we can take $M_{q_k} = k$. But since $\mathbb{Q}$ is dense in $\mathbb{R}$, none of the functions f_n is bounded on any interval. (Verify this.) Thus a collection of functions may be pointwise bounded but not uniformly bounded on any interval. ◀

The functions f_n in Example 6.12 are everywhere discontinuous. Our next theorem shows that if we had taken a collection $\mathcal{F}$ of *continuous* functions, then not only would each $f \in \mathcal{F}$ be bounded on closed intervals (as Theorem 5.48 guarantees), but there would be an interval I on which the entire collection is *uniformly bounded* ; that is, there exists a constant M such that $|f(x)| \le M$ for all $f \in \mathcal{F}$ and each $x \in I$.

Theorem 6.13 *Let $\mathcal{F}$ be a collection of continuous functions on $\mathbb{R}$ such that for each $x \in \mathbb{R}$ there exists a constant $M_x > 0$ such that $|f(x)| \le M_x$ for each $f \in \mathcal{F}$. Then there exists an open interval I and a constant $M > 0$ such that $|f(x)| \le M$ for each $f \in \mathcal{F}$ and $x \in I$.*

Proof For each $n \in \mathbb{N}$, let $A_n = \{x : |f(x)| \le n \text{ for all } f \in \mathcal{F}\}$. By hypothesis, $\mathbb{R} = \bigcup_{n=1}^{\infty} A_n$. Also, by hypothesis, each $f \in \mathcal{F}$ is continuous and so it is easy to check that each of the sets

$$\{x : |f(x)| \le n\}$$

must be closed (e.g., Exercise 5.4.31). Thus

$$A_n = \bigcap_{f \in \mathcal{F}} \{x : |f(x)| \le n\}$$

is an intersection of closed sets and is therefore itself closed. This expresses the real line $\mathbb{R}$ as a union of the sequence of closed sets $\{A_n\}$.

It now follows from the Baire category theorem that at least one of the sets, say A_{n_0}, must be dense in some open interval I. Since A_{n_0} is closed and dense in the interval I, A_{n_0} must contain I. This means that $|f(x)| \le n_0$ for each $f \in \mathcal{F}$ and all $x \in I$. ■

Exercises

6.4.6 Let $\{f_n\}$ be a sequence of continuous functions on an interval $[a,b]$ such that $\lim_{n\to\infty} f_n(x) = f(x)$ exists at every point $x \in [a,b]$. Show that f need not be continuous nor even bounded, but that f must be bounded on some subinterval of $[a,b]$.

6.4.7 Let $\{f_n\}$ be a sequence of continuous functions on $[0,1]$ and suppose that $\lim_{n\to\infty} f_n(x) = 0$ for all $0 \le x \le 1$. Show that there must be an interval $[c,d] \subset [0,1]$ so that, for all sufficiently large n, $|f_n(x)| \le 1$ for all $x \in [c,d]$.

6.4.8 Give an example of a sequence of functions on $[0,1]$ with the property that $\lim_{n\to\infty} f_n(x) = 0$ for all $0 \le x \le 1$ and yet for every interval $[c,d] \subset [0,1]$ and every N there is some $x \in [c,d]$ and $n > N$ with $f_n(x) > 1$.

6.5 Cantor Sets

✂ *Advanced*

We say that a set is *perfect* if it is a nonempty closed set with no isolated points. The only examples that might come to mind are sets that are finite unions of intervals. It might be difficult to imagine a perfect subset of $\mathbb{R}$ that is also nowhere dense. In this section we obtain such a set, the very important classical Cantor set. We also discuss some of its variants. Such sets have historical significance and are of importance in a number of areas of mathematical analysis.

6.5.1 Construction of the Cantor Ternary Set

✂ *Advanced*

We begin with the closed interval $[0,1]$. From this interval we shall remove a dense open set G. The remaining set $K = [0,1] \setminus G$ will then be closed and nowhere dense in [0,1]. We construct G in such a way that K has no isolated points and is nonempty. Thus K will be a nonempty, nowhere dense perfect subset of [0,1].

It is easiest to understand the set G if we construct it in stages. Let $G_1 = \left(\frac{1}{3}, \frac{2}{3}\right)$, and let $K_1 = [0,1] \setminus G_1$. Thus $K_1 = \left[0, \frac{1}{3}\right] \cup \left[\frac{2}{3}, 1\right]$ is what remains when the middle third of the interval [0,1] is removed. This is the first stage of our construction.

We repeat this construction on each of the two component intervals of K_1. Let $G_2 = \left(\frac{1}{9}, \frac{2}{9}\right) \cup \left(\frac{7}{9}, \frac{8}{9}\right)$ and let $K_2 = [0,1] \setminus (G_1 \cup G_2)$. Thus

$$K_2 = \left[0, \frac{1}{9}\right] \cup \left[\frac{2}{9}, \frac{1}{3}\right] \cup \left[\frac{2}{3}, \frac{7}{9}\right] \cup \left[\frac{8}{9}, 1\right].$$

This completes the second stage.

We continue inductively, obtaining two sequences of sets, $\{K_n\}$ and $\{G_n\}$ with the following properties: For each $n \in \mathbb{N}$

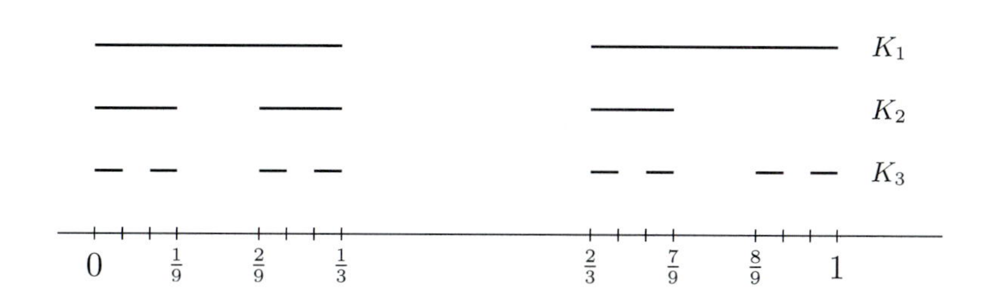

Figure 6.1. The third stage in the construction of the Cantor ternary set.

1. G_n is a union of 2^{n-1} pairwise disjoint open intervals.
2. K_n is a union of 2^n pairwise disjoint closed intervals.
3. $K_n = [0,1] \setminus (G_1 \cup G_2 \cup \cdots \cup G_n)$.
4. Each component of G_{n+1} is the "middle third" of some component of K_n.
5. The length of each component of K_n is $1/3^n$.

Figure 6.1 shows K_1, K_2, and K_3.

Now let

$$G = \bigcup_{n=1}^{\infty} G_n$$

and let

$$K = [0,1] \setminus G = \bigcap_{n=1}^{\infty} K_n.$$

Then G is open and the set K (our Cantor set) is closed.

To see that K is nowhere dense, it is enough, since K is closed, to show that K contains no open intervals (Exercise 6.3.4). Let J be an open interval in $[0,1]$ and let λ be its length. Choose $n \in \mathbb{N}$ such that $1/3^n < \lambda$. By property 5, each component of K_n has length $1/3^n < \lambda$, and by property 2 the components of K_n are pairwise disjoint. Thus K_n cannot contain J, so neither can $K = \bigcap_1^{\infty} K_n$. We have shown that the closed set K contains no intervals and is therefore nowhere dense.

It remains to show that K has no isolated points. Let $x_0 \in K$. We show that x_0 is a limit point of K. To do this we show that for every $\varepsilon > 0$ there exists $x_1 \in K$ such that $0 < |x_1 - x_0| < \varepsilon$. Choose n such that $1/3^n < \varepsilon$. There is a component L of K_n that contains x_0. This component is a closed interval of length $1/3^n < \varepsilon$. The set $K_{n+1} \cap L$ has two components L_0 and L_1, each of which contains points of K. The point x_0 is in one of the components, say L_0. Let x_1 be any point of $K \cap L_1$. Then $0 < |x_0 - x_1| < \varepsilon$. This verifies that x_0 is a limit point of K. Thus K has no isolated points.

The set K is called the *Cantor set.* Because of its construction, it is often called the Cantor middle third set. In a moment we shall present a purely arithmetic description of the Cantor set that suggests another common name for K, the "Cantor ternary set". But first, we mention a few properties of K and of its complement G that may help you visualize these sets.

First note that G is an open dense set in $[0, 1]$. Write $G = \bigcup_{k=1}^{\infty}(a_k, b_k)$. (The component intervals (a_k, b_k) of G can be called the intervals *complementary* to K in $(0, 1)$. Each is a middle third of a component interval of some K_n.) Observe that no two of these component intervals can have a common endpoint. If, for example, $b_m = a_n$, then this point would be an isolated point of K, and K has no isolated points.

Next observe that for each $k \in \mathbb{N}$, the points a_k and b_k are points of K. But there are other points of K as well. In fact, we shall see presently that K is uncountable. These other points are all limit points of the endpoints of the complementary intervals. The set of endpoints is countable, but the closure of this set is uncountable as we shall see. Thus, in the sense of cardinality, "most" points of the Cantor set are *not* endpoints of intervals complementary to K.

Each component interval of the set G_n has length $1/3^n$; thus the sum of the lengths of these component intervals is

$$\frac{2^{n-1}}{3^n} = \frac{1}{2}\left(\frac{2}{3}\right)^n.$$

It follows that the lengths of all component intervals of G forms a geometric series with sum

$$\sum_{n=1}^{\infty} \frac{1}{2}\left(\frac{2}{3}\right)^n = 1.$$

(This also gives us a clue as to why K cannot contain an interval: After removing from the unit interval a sequence of pairwise disjoint intervals with length-sum one, no room exists for any intervals in the set K that remains.)

Exercises

6.5.1 Let E be the set of endpoints of intervals complementary to the Cantor set K. Prove that $\overline{E} = K$.

6.5.2 Let G be a dense open subset of $\mathbb{R}$ and let $\{(a_k, b_k)\}$ be its set of component intervals. Prove that $H = \mathbb{R} \setminus G$ is perfect if and only if no two of these intervals have common endpoints.

6.5.3 Let K be the Cantor set and let $\{(a_k, b_k)\}$ be the sequence of intervals complementary to K in $[0,1]$. For each $k \in \mathbb{N}$, let $c_k = (a_k + b_k)/2$ (the midpoint of the interval (a_k, b_k)) and let $N = \{c_k : k \in \mathbb{N}\}$. Prove each of the following:

(a) Every point of N is isolated.

(b) If $c_i \neq c_j$, there exists $k \in \mathbb{N}$ such that c_k is between c_i and c_j (i.e., no point in N has an immediate "neighbor" in N).

(c) Show that there is an *order-preserving mapping* $\phi : \mathbb{Q} \cap (0,1) \to N$ [i.e., if $x < y \in \mathbb{Q} \cap (0,1)$, then $\phi(x) < \phi(y) \in N$]. This may seem surprising since $\mathbb{Q} \cap (0,1)$ has *no* isolated points while N has *only* isolated points.

6.5.4 It is common now to say that a set E of real numbers is a *Cantor set* if it is nonempty, bounded, perfect, and nowhere dense. Show that the union of a finite number of Cantor sets is also a Cantor set.

6.5.5 Show that every Cantor set is uncountable.

6.5.6 Let A and B be subsets of $\mathbb{R}$. A function h that maps A onto B, is one-to-one, and with both h and h^{-1} continuous is called a *homeomorphism* between A and B. The sets A and B are said to be *homeomorphic.* Prove that a set C is a Cantor set if and only if it is homeomorphic to the Cantor ternary set K.

6.5.2 An Arithmetic Construction of K

Enrichment

We turn now to a purely arithmetical construction for the Cantor set. You will need some familiarity with ternary (base 3) arithmetic here.

Each $x \in [0,1]$ can be expressed in base 3 as

$$x = .a_1 a_2 a_3 \ldots,$$

where $a_i = 0$, 1 or 2, $i = 1, 2, 3, \ldots$. Certain points have two representations, one ending with a string of zeros, the other in a string of twos. For example, $.1000\cdots = .0222\ldots$ both represent the number 1/3 (base ten). Now, if $x \in (1/3, 2/3)$, $a_1 = 1$, thus each $x \in G_1$ must have '1' in the first position of its ternary expansion. Similarly, if

$$x \in G_2 = \left(\frac{1}{9}, \frac{2}{9}\right) \cup \left(\frac{7}{9}, \frac{8}{9}\right),$$

it must have a 1 in the second position of its ternary expansion (i.e., $a_2 = 1$). In general, each point in G_n must have $a_n = 1$. It follows that every point of $G = \bigcup_1^\infty G_n$ must have a 1 someplace in its ternary expansion.

Now endpoints of intervals complementary to K have two representations, one of which involves no 1's. The remaining points of K never fall in the middle third of a component of one of the sets K_n, and so have ternary expansions of the form

$$x = .a_1a_2\ldots \quad a_i = 0 \text{ or } 2.$$

We can therefore describe K arithmetically as the set

$$\{x = .a_1a_2a_3\ldots \text{ (base three) } : a_i = 0 \text{ or } 2 \text{ for each } i \in \mathbb{N}\}.$$

As an immediate result, we see that K is uncountable. In fact, K can be put into 1-1 correspondence with [0,1]: For each

$$x = .a_1a_2a_3\ldots \text{ (base 3)}, \ a_i = 0, 2,$$

in the set K, let there correspond the number

$$y = .b_1b_2b_3\ldots \text{ (base 2)}, \ b_i = a_i/2.$$

This provides a 1-1 correspondence between K (minus endpoints of complementary intervals) and $[0,1]$ (minus the countable set of numbers with two base 2 representations). By allowing these two countable sets to correspond to each other, we obtain a 1-1 correspondence between K and $[0,1]$.

Note. We end this section by mentioning that variations in the constructions of K can lead to interesting situations. For example, by changing the construction slightly, we can remove intervals in such a way that

$$G' = \bigcup_{k=1}^{\infty} (a'_k, b'_k)$$

with

$$\sum_{k=1}^{\infty} (b'_k - a'_k) = 1/2$$

(instead of 1), while still keeping $K' = [0,1] \setminus G'$ nowhere dense and perfect. The resulting set K' created problems for late nineteenth-century mathematicians trying to develop a theory of measure. The "measure" of G' should be 1/2; the "measure" of [0,1] should be 1. Intuition requires that the measure of the nowhere dense set K' should be $1 - \frac{1}{2} = \frac{1}{2}$. How can this be when K' is so "small?"

Exercises

6.5.7 Find a specific irrational number in the Cantor ternary set.

6.5.8 Show that the Cantor ternary set can be defined as

$$K = \left\{ x \in [0,1] : x = \sum_{n=1}^{\infty} \frac{i_n}{3^n} \text{ for } i_n = 0 \text{ or } 2 \right\}.$$

6.5.9 Let

$$D = \left\{ x \in [0,1] : x = \sum_{n=1}^{\infty} \frac{j_n}{3^n} \text{ for } j_n = 0 \text{ or } 1 \right\}.$$

Show that $D + D = \{x + y : x, y \in D\} = [0,1]$. From this deduce, for the Cantor ternary set K, that $K + K = [0,2]$.

6.5.10 A careless student makes the following argument. Explain the error.

> "If $G = (a,b)$, then $\overline{G} = [a,b]$. Similarly, if $G = \bigcup_{i=1}^{\infty} (a_i, b_i)$ is an open set, then $\overline{G} = \bigcup_{i=1}^{\infty} [a_i, b_i]$. It follows that an open set G and its closure $\overline{G}$ differ by at most a countable set."

6.5.3 The Cantor Function

✂ *Advanced*

The Cantor set allows the construction of a rather bizarre function that is continuous and nondecreasing on the interval $[0,1]$. It has the property that it is constant on every interval complementary to the Cantor set and yet manages to increase from $f(0) = 0$ to $f(1) = 1$ by doing all of its increasing on the Cantor set itself. It has sometimes been called "the devil's staircase."

Define the function f in the following way. On $(1/3, 2/3)$, let $f = 1/2$; on $(1/9, 2/9)$, let $f = 1/4$; on $(7/9, 8/9)$, let $f = 3/4$. Proceed inductively. On the $2^{n-1} - 1$ open intervals appearing at the nth stage, define f to satisfy the following conditions:

1. f is constant on each of these intervals.
2. f takes the values
$$\frac{1}{2^n}, \frac{3}{2^n}, \ldots, \frac{2^n - 1}{2^n}$$
on these intervals.
3. If x and y are members of different nth-stage intervals with $x < y$, then $f(x) < f(y)$.

This description defines f on $G = [0,1] \setminus K$. Extend f to all of $[0,1]$ by defining $f(0) = 0$ and, for $0 < x \leq 1$,

$$f(x) = \sup\{f(t) : t \in G, t < x\}.$$

In order to check that this defines the function that we want, we need to check each of the following.

1. $f(G)$ is dense in $[0,1]$.
2. f is nondecreasing on $[0,1]$.

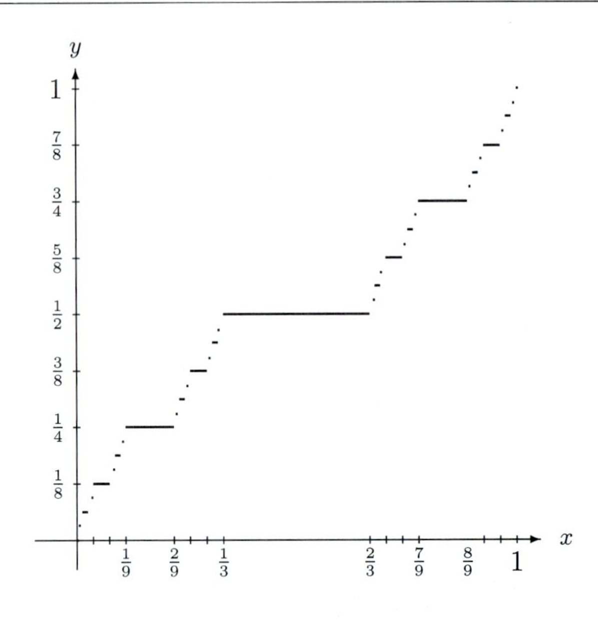

Figure 6.2. The third stage in the construction of the Cantor function.

3. f is continuous on $[0, 1]$.

4. $f(K) = [0, 1]$.

These have been left as exercises.

Figure 6.2 illustrates the construction. The function f is called the *Cantor function*. Observe that f "does all its rising" on the set K.

The Cantor function allows a negative answer to many questions that might be asked about functions and derivatives and, hence, has become a popular counterexample. For example, let us follow this kind of reasoning. If f is a continuous function on $[0, 1]$ and $f'(x) = 0$ for every $x \in (0, 1)$ then f is constant. (This is proved in most calculus courses by using the mean value theorem.) Now suppose that we know less, that $f'(x) = 0$ for every $x \in (0, 1)$ excepting a "small" set E of points at which we know nothing. If E is finite it is still easy to show that f must be constant. If E is countable it is possible, but a bit more difficult, to show that it is still true that f must be constant. The question then arises, just how small a set E can appear here; that is, what would we have to know about a set E so that we could say $f'(x) = 0$ for every $x \in (0, 1) \setminus E$ implies that f is constant?

The Cantor function is an example of a function constant on every interval complementary to the Cantor set K (and so with a zero derivative at those points) and yet is not constant. The Cantor set, since it is nowhere dense, might be viewed as extremely small, but even so it is not insignificant for this problem.

Exercises

6.5.11 In the construction of the Cantor function complete the verification of details.

(a) Show that $f(G)$ is dense in $[0,1]$.

(b) Show that f is nondecreasing on $[0,1]$.

(c) Infer from (a) and (b) that f is continuous on $[0,1]$.

(d) Show that $f(K) = [0,1]$ and thus (again) conclude that K is uncountable.

6.5.12 Find a calculus textbook proof for the statement that a continuous function f on an interval $[a,b]$ that has a zero derivative on (a,b) must be constant. Improve the proof to allow a finite set of points on which f is not known to have a zero derivative.

6.6 Borel Sets

✂ *Advanced*

In our study of continuous functions we have seen that the classes of open sets and closed sets play a significant role. But the class of sets that are of importance in analysis goes beyond merely the open and closed sets. E. Borel (1871–1956) recognized that for many operations of analysis we need to form countable intersections and countable unions of classes of sets. The collection of Borel sets was introduced exactly to allow these operations. We recall that a countable union of closed sets may not be closed (or open) and that a countable intersection of open sets, also, may not be open (or closed).

In this section we introduce two additional types of sets of importance in analysis, sets of type $\mathcal{G}_\delta$ and sets of type $\mathcal{F}_\sigma$. These classes form just the beginning of the large class of Borel sets. We shall find that they are precisely the right classes of sets to solve some fundamental questions about real functions.

6.6.1 Sets of Type G_δ

✂ *Advanced*

Recall that the union of a collection of open sets is open (regardless of how many sets are in the collection), but the intersection of a collection of open sets need not be open if the collection has infinitely many sets. For example,

$$\bigcap_{n=1}^{\infty}\left(-\frac{1}{n},\frac{1}{n}\right) = \{0\}.$$

Similarly, if $q_1, q_2, q_3, \ldots$ is an enumeration of $\mathbb{Q}$, then

$$\bigcap_{k=1}^{\infty} (\mathbb{R} \setminus \{q_k\}) = \mathbb{R} \setminus \mathbb{Q},$$

the set of irrational numbers. The set $\{0\}$ is closed (not open), and $\mathbb{R} \setminus \mathbb{Q}$ is neither open nor closed. The set $\mathbb{R} \setminus \mathbb{Q}$ is a countable intersection of open sets. Such sets are of sufficient importance to give them a name.

Definition 6.14 A subset H of $\mathbb{R}$ is said to be of *type* $\mathcal{G}_\delta$ (or a $\mathcal{G}_\delta$ *set*) if it can be expressed as a countable intersection of open sets, that is, if there exist open sets $G_1, G_2, G_3, \ldots$ such that $H = \bigcap_{k=1}^{\infty} G_k$.

Example 6.15 A closed interval $[a, b]$ or a half-open interval $(a, b]$ is of type $\mathcal{G}_\delta$ since

$$[a, b] = \bigcap_{n=1}^{\infty} \left(a - \frac{1}{n}, b + \frac{1}{n}\right)$$

and

$$(a, b] = \bigcap_{n=1}^{\infty} \left(a, b + \frac{1}{n}\right).$$

◀

Theorem 6.16 *Every open set and every closed set in* $\mathbb{R}$ *is of type* $\mathcal{G}_\delta$.

Proof Let G be an open set in $\mathbb{R}$. It is clear that G is of type $\mathcal{G}_\delta$. We also show that G can be expressed as a countable union of closed sets. Express G in the form

$$G = \bigcup_{k=1}^{\infty} (a_k, b_k)$$

where the intervals (a_k, b_k) are pairwise disjoint. Now for each $k \in \mathbb{N}$ there exist sequences $\{c_{k_j}\}$ and $\{d_{k_j}\}$ such that the sequence $\{c_{k_j}\}$ decreases to a_k, the sequence $\{d_{k_j}\}$ increases to b_k and $c_{k_j} < d_{k_j}$ for each $j \in \mathbb{N}$. Thus

$$(a_k, b_k) = \bigcup_{j=1}^{\infty} [c_{k_j}, d_{k_j}].$$

We have expressed each component interval of G as a countable union of closed sets. It follows that

$$G = \bigcup_{k=1}^{\infty} \bigcup_{j=1}^{\infty} [c_{k_j}, d_{k_j}] = \bigcup_{j,k=1}^{\infty} [c_{k_j}, d_{k_j}]$$

is also a countable union of closed sets. Now take complements. This shows that $\mathbb{R}\setminus G$ can be expressed as a countable intersection of open sets (by using the de Morgan laws). Since every closed set F can be written

$$F = \mathbb{R} \setminus G$$

for some open set G, we have shown that any closed set is of type $\mathcal{G}_\delta$. ■

We observed in Section 6.4 that a dense set can be small in the sense of category. For example, $\mathbb{Q}$ is a first category set. Our next result shows that a dense set *of type* $\mathcal{G}_\delta$ must be large in the sense of category.

Theorem 6.17 *Let H be of type $\mathcal{G}_\delta$ and be dense in $\mathbb{R}$. Then H is residual.*

Proof Write

$$H = \bigcap_{k=1}^{\infty} G_k$$

with each of the sets G_k open. Since H is dense by hypothesis and $H \subset G_k$ for each $k \in \mathbb{N}$, each of the open sets G_k is also dense. Thus $\mathbb{R} \setminus G_k$ is nowhere dense for every $k \in \mathbb{N}$, and so each G_k is residual. The result now follows from Lemma 6.10. ■

Exercises

6.6.1 Which of the following sets are of type $\mathcal{G}_\delta$?

(a) $\mathbb{N}$

(b) $\left\{\frac{1}{n} : n \in \mathbb{N}\right\}$

(c) The set $\{C_n : n \in \mathbb{N}\}$ of midpoints of intervals complementary to the Cantor set

(d) A finite union of intervals (that need not be open or closed)

6.6.2 Prove Theorem 6.17 for the interval $[a, b]$ in place of $\mathbb{R}$.

6.6.3 Prove that a set E of type $\mathcal{G}_\delta$ in $\mathbb{R}$ is either residual or else there is an interval containing no points of E.

6.6.2 Sets of Type F_σ

✂ *Advanced*

Just as the countable intersections of open sets form a larger class of sets, the $\mathcal{G}_\delta$ sets, so also the countable unions of closed sets form a larger class of sets.

The complements of open sets are closed. By dealing with complements of $\mathcal{G}_\delta$ sets we arrive at the dual notion of a set of type $\mathcal{F}_\sigma$.

Definition 6.18 A subset E of $\mathbb{R}$ is said to be of *type* $\mathcal{F}_\sigma$ (or an $\mathcal{F}_\sigma$ set) if it can be expressed as a countable union of closed sets; that is, if there exist closed sets $F_1, F_2, F_3, \ldots$ such that $E = \bigcup_{k=1}^{\infty} F_k$.

Using the de Morgan laws, we verify easily that the complement of a $\mathcal{G}_\delta$ set is an $\mathcal{F}_\sigma$ and vice versa (Exercise 6.6.4). This is closely related to the fact that a set is open if and only if its complement is closed.

Example 6.19 The set of rational numbers, $\mathbb{Q}$ is a set of type $\mathcal{F}_\sigma$. This is clear since it can be expressed as

$$\mathbb{Q} = \bigcup_{n=1}^{\infty} \{r_n\}$$

where $\{r_n\}$ is any enumeration of the rationals. The singleton sets $\{r_n\}$ are clearly closed. But note that $\mathbb{Q}$ is *not* of type $\mathcal{G}_\delta$ also. It follows from Theorem 6.17 that a dense set of type $\mathcal{G}_\delta$ must be uncountable (because a countable set is first category). In particular, $\mathbb{Q}$ is not of type $\mathcal{G}_\delta$ (and therefore $\mathbb{R} \setminus \mathbb{Q}$ is not of type $\mathcal{F}_\sigma$). ◀

Theorem 6.20 *A set is of type $\mathcal{G}_\delta$ if and only if its complement is of type $\mathcal{F}_\sigma$.*

Example 6.21 A half-open interval $(a, b]$ is both of type $\mathcal{G}_\delta$ and of type $\mathcal{F}_\sigma$:

$$(a, b] = \bigcap_{n=1}^{\infty} \left(a, b + \frac{1}{n}\right) = \bigcup_{n=1}^{\infty} \left[a + \frac{b-a}{n}, b\right].$$

◀

Note. The only subsets of $\mathbb{R}$ that are both open and closed are the empty set and $\mathbb{R}$ itself. There are, however, many sets that are of type $\mathcal{G}_\delta$ and also of type $\mathcal{F}_\sigma$. See Exercise 6.6.1.

We can now enlarge on Theorem 6.16. There we showed that all open sets and all closed sets are in the class $\mathcal{G}_\delta$. We now show they are also in the class $\mathcal{F}_\sigma$.

Theorem 6.22 *Every open set and every closed set in $\mathbb{R}$ is both of type $\mathcal{F}_\sigma$ and $\mathcal{G}_\delta$.*

Proof In the proof of Theorem 6.16 we showed explicitly how to express any open set as an $\mathcal{F}_\sigma$. Thus open sets are of type $\mathcal{F}_\sigma$ as well as of type $\mathcal{G}_\delta$ (the latter being trivial). The part pertaining to closed sets now follows by considering complements and using the de Morgan laws. The complement of a closed set is open and therefore the complement of an $\mathcal{F}_\sigma$ set is a $\mathcal{G}_\delta$ set. ■

Exercises

6.6.4 Verify that a subset A of $\mathbb{R}$ is an $\mathcal{F}_\sigma$ ($\mathcal{G}_\delta$) if and only if $\mathbb{R} \setminus A$ is a $\mathcal{G}_\delta$ ($\mathcal{F}_\sigma$).

6.6.5 Which of the following sets are of type $\mathcal{F}_\sigma$?

(a) $\mathbb{N}$

(b) $\left\{\dfrac{1}{n} : n \in \mathbb{N}\right\}$

(c) The set $\{C_n : n \in \mathbb{N}\}$ of midpoints of intervals complementary to the Cantor set

(d) A finite union of intervals (that need not be open or closed)

6.6.6 Prove that a set of type $\mathcal{F}_\sigma$ in $\mathbb{R}$ is either first category or contains an open interval. ✂

6.6.7 Let $\{f_n\}$ be a sequence of real functions defined on $\mathbb{R}$ and suppose that $f_n(x) \to f(x)$ at every point x. Show that ✂

$$\{x : f(x) > \alpha\} = \bigcup_{m=1}^{\infty} \bigcup_{r=1}^{\infty} \bigcap_{n=r}^{\infty} \{x : f_n(x) \geq \alpha + 1/m\}.$$

If each function f_n is continuous, what can you assert about the set

$$\{x : f(x) > \alpha\}?$$

6.7 Oscillation and Continuity

✂ *Advanced*

In this section we return to a problem that we began investigating in Section 5.9 about the nature of the set of discontinuity points of a function. To discuss this set we shall need the notions of $\mathcal{F}_\sigma$ and $\mathcal{G}_\delta$ sets and we need to introduce a new tool, the oscillation of a function.

We begin with an example of a function f that is discontinuous at every rational number and continuous at every irrational number.

Example 6.23 Let $q_1, q_2, q_3, \ldots$ be an enumeration of $\mathbb{Q}$. Define a function f by

$$f(x) = \begin{cases} \frac{1}{k}, & \text{if } x = q_k \\ 0, & \text{if } x \in \mathbb{R} \setminus \mathbb{Q}. \end{cases}$$

Since $\mathbb{R} \setminus \mathbb{Q}$ is dense in $\mathbb{R}$, f can be continuous at a point x only if $f(x) = 0$; that is, only if $x \in \mathbb{R} \setminus \mathbb{Q}$. Thus f is discontinuous at every $x \in \mathbb{Q}$. To check that f is continuous at each point of $\mathbb{R} \setminus \mathbb{Q}$, let $x_0 \in \mathbb{R} \setminus \mathbb{Q}$ and let $\varepsilon > 0$. Choose $k \in \mathbb{N}$ such that $1/k < \varepsilon$. Since the set $q_1, q_2, \ldots, q_k$ is a finite set not containing x_0, there exists $\delta > 0$ such that $|q_i - x_0| \geq \delta$ for each $i = 1, \ldots, k$. Thus if $x \in \mathbb{R}$ and $|x - x_0| < \delta$, then either $x \in \mathbb{R} \setminus \mathbb{Q}$ or $x = q_j$ for some $j > k$. In either case $|f(x) - f(x_0)| \leq \frac{1}{k} < \varepsilon$. This verifies the continuity of f at x_0. Since x_0 was an arbitrary irrational point, we see that f is continuous at every irrational. ◀

Our example shows that it is possible for a function to be continuous at every irrational number and discontinuous at every rational number. Is it possible for the opposite to occur? Does there exist a function f continuous on $\mathbb{Q}$ and discontinuous on $\mathbb{R} \setminus \mathbb{Q}$? More generally, what sets can be the set of points of continuity of some function f defined on an interval.

We answer this question in this section. The principal tool is that of *oscillation* of a function at a point.

6.7.1 Oscillation of a Function

✂ *Advanced*

In order to describe a point of discontinuity we need a way of measuring that discontinuity. For monotonic functions the jump was used previously for such a measure. For general, nonmonotonic, functions a different tool is used.

Definition 6.24 Let f be defined on a nondegenerate interval I. We define the *oscillation of f on I* as the quantity

$$\omega f(I) = \sup_{x,y \in I} |f(x) - f(y)|.$$

Let's see how oscillation relates to continuity. Suppose f is defined in a neighborhood of x_0, and f is continuous at x_0. Then

$$\inf_{\delta > 0} \omega f((x_0 - \delta, x_0 + \delta)) = 0. \tag{1}$$

To see this, let $\varepsilon > 0$. Since f is continuous at x_0, there exists $\delta_0 > 0$ such that

$$|f(x) - f(x_0)| < \varepsilon/2$$

if $|x - x_0| < \delta_0$. If

$$x_0 - \delta_0 < x_1 \leq x_2 < x_0 + \delta_0,$$

then

$$|f(x_1) - f(x_2)| \leq |f(x_1) - f(x_0)| + |f(x_0) - f(x_2)| < \frac{\varepsilon}{2} + \frac{\varepsilon}{2} = \varepsilon. \tag{2}$$

Since (2) is valid for all $x_1, x_2 \in (x_0 - \delta_0, x_0 + \delta_0)$, we have

$$\sup \{|f(x_1) - f(x_2)| : x_0 - \delta_0 < x_1 \leq x_2 < x_0 + \delta_0\} \leq \varepsilon. \tag{3}$$

But (3) implies that if $0 < \delta < \delta_0$, then

$$\omega f([x_0 - \delta, x_0 + \delta]) \leq \varepsilon.$$

Since ε was arbitrary, the result follows.

The converse is also valid. Suppose (1) holds. Let $\varepsilon > 0$. Choose $\delta > 0$ such that

$$\omega f(x_0 - \delta, x_0 + \delta) < \varepsilon.$$

Then

$$\sup \{|f(x) - f(x_0)| : x \in (x_0 - \delta, x_0 + \delta)\} < \varepsilon,$$

so $|f(x) - f(x_0)| < \varepsilon$ whenever $|x - x_0| < \delta$. This implies continuity of f at x_0.

We summarize the preceding as a theorem.

Theorem 6.25 *Let f be defined on an interval I and let $x_0 \in I$. Then f is continuous at x_0 if and only if*

$$\inf_{\delta>0} \omega f((x_0 - \delta, x_0 + \delta)) = 0.$$

The quantity in the statement of the theorem is sufficiently important to have a name.

Definition 6.26 Let f be defined in a neighborhood of x_0. The quantity

$$\omega_f(x_0) = \inf_{\delta>0} \omega f((x_0 - \delta, x_0 + \delta))$$

is called the *oscillation of f at x_0*.

Theorem 6.25 thus states that a function f is continuous at a point x_0 if and only if $\omega_f(x_0) = 0$. Returning to the function that introduced this section, we see that

$$\omega_f(x) = \begin{cases} 1/k, & \text{if } x = q_k \\ 0, & \text{if } x \in \mathbb{R} \setminus \mathbb{Q}. \end{cases}$$

Let's now see how the concept of oscillation relates to the set of points of continuity of a function.

Theorem 6.27 *Let f be defined on a closed interval I (which may be all of $\mathbb{R}$). Let $\gamma > 0$. Then the set*

$$\{x : \omega_f(x) < \gamma\}$$

is open and the set

$$\{x : \omega_f(x) \geq \gamma\}$$

is closed.

Proof Let $A = \{x : \omega_f(x) < \gamma\}$ and let $x_0 \in A$. We wish to find a neighborhood U of x_0 such that $U \subset A$; that is, such that $\omega_f(x) < \gamma$ for all $x \in U$.

Let $\omega_f(x_0) = \alpha < \gamma$ and let $\beta \in (\alpha, \gamma)$. From Definition 6.26 we infer the existence of a number $\delta > 0$ such that

$$|f(u) - f(v)| \leq \beta$$

for $u, v \in (x_0 - \delta, x_0 + \delta)$. Let

$$U = (x_0 - \delta, x_0 + \delta)$$

and let $x \in U$. Since U is open, there exists $\delta_1 < \delta$ such that

$$(x - \delta_1, x + \delta_1) \subset U.$$

Then

$$\begin{aligned} \omega_f(x_0) &\leq \sup \{|f(t) - f(s)| : t, s \in (x - \delta_1, x + \delta_1)\} \\ &\leq \sup \{|f(u) - f(v)| : u, v \in U\} \leq \beta < \gamma, \end{aligned}$$

so $x \in A$. This proves A is open. It follows then that the complement of A in I, the set

$$\{x : \omega_f(x) \geq \gamma\},$$

must be closed. ■

We use the oscillation in the next subsection to answer a question about the nature of the set of points of continuity of a function. We shall encounter the oscillation concept again in Chapter 8 when we study the integrability of functions.

Exercises

6.7.1 Suppose that f is bounded on an interval I. Prove that

$$\omega f(I) = \sup_{x \in I} f(x) - \inf_{x \in I} f(x).$$

6.7.2 A careless student believes that the oscillation can be written as

$$\omega_f(x_0) = \limsup_{x \to x_0} f(x) - \liminf_{x \to x_0} f(x).$$

Show that this is not true, even for bounded functions.

6.7.3 Prove that

$$\omega_f(x_0) = \lim_{\delta \to 0+} \omega f((x_0 - \delta, x_0 + \delta)).$$

6.7.4 Calculate $\omega_f(0)$ for each of the following functions.

(a) $f(x) = \begin{cases} x, & \text{if } x \neq 0 \\ 4, & \text{if } x = 0 \end{cases}$

(b) $f(x) = \begin{cases} 0, & \text{if } x \in \mathbb{Q} \\ 1, & \text{if } x \notin \mathbb{Q} \end{cases}$

(c) $f(x) = \begin{cases} n, & \text{if } x = \frac{1}{n} \\ 0, & \text{otherwise} \end{cases}$

(d) $f(x) = \begin{cases} \sin \frac{1}{x}, & \text{if } x \neq 0 \\ 0, & \text{if } x = 0 \end{cases}$

(e) $f(x) = \begin{cases} \sin \frac{1}{x}, & \text{if } x \neq 0 \\ 7, & \text{if } x = 0 \end{cases}$

(f) $f(x) = \begin{cases} \frac{1}{x} \sin \frac{1}{x}, & \text{if } x \neq 0 \\ 0, & \text{if } x = 0 \end{cases}$

6.7.5 In the proof of Theorem 6.27 we let $\omega_f(x_0) = \alpha < \gamma$ and let $\beta \in (\alpha, \gamma)$. Why was the β introduced? Would the proof have worked if we had used $\beta = \gamma$?

6.7.2 The Set of Continuity Points

Advanced

Given an arbitrary function, how can we describe the nature of the set of points where f is continuous? Can it be any set? Given a set E, how can we know whether there is a function that is continuous at every point of E and discontinuous at every point not in E?

We saw in Example 6.23 that a function exists whose set of continuity points is exactly the irrationals. Can a function exist whose set of continuity points is exactly the rationals? By characterizing the set of such points we can answer this and other questions about the structure of functions.

We now prove the main result of this section using primarily the notion of oscillation introduced in Section 6.7.1.

Theorem 6.28 *Let f be defined on a closed interval I (which may be all of $\mathbb{R}$). Then the set C_f of points of continuity of f is of type $\mathcal{G}_\delta$, and the set D_f of points of discontinuity of f is of type $\mathcal{F}_\sigma$. Conversely, if H is a set of type $\mathcal{G}_\delta$, then there exists a function f defined on $\mathbb{R}$ such that $C_f = H$.*

Proof To prove the first part, let $f : I \to \mathbb{R}$. We show that the set

$$C_f = \{x : \omega_f(x) = 0\}$$

is of type $\mathcal{G}_\delta$. For each $k \in \mathbb{N}$, let

$$B_k = \left\{ x : \omega_f(x) \geq \frac{1}{k} \right\}.$$

By Theorem 6.27, each of the sets B_k is closed. Thus the set

$$B = \bigcup_{k=1}^{\infty} B_k$$

is of type $\mathcal{F}_\sigma$. By Theorem 6.25, $D_f = B$. Therefore, $C_f = I \setminus B$. Since the complement of an $\mathcal{F}_\sigma$ is a $\mathcal{G}_\delta$, the set C_f is a $\mathcal{G}_\delta$.

To prove the converse, let H be any subset of $\mathbb{R}$ of type $\mathcal{G}_\delta$. Then H can be expressed in the form

$$H = \bigcap_{k=1}^{\infty} G_k$$

with each of the sets G_k being open. We may assume without loss of generality that $G_1 = \mathbb{R}$ and that $G_i \supset G_{i+1}$ for each $i \in \mathbb{N}$. (Verify this.)

Let $\{\alpha_k\}$ and $\{\beta_k\}$ be sequences of positive numbers, each converging to zero, with

$$\alpha_k > \beta_k > \alpha_{k+1},$$

for all $k \in \mathbb{N}$. Define a function $f:\mathbb{R} \to \mathbb{R}$ by

$$f(x) = \begin{cases} 0 & \text{if } x \in H \\ \alpha_k & \text{if } x \in (G_k \setminus G_{k+1}) \cap \mathbb{Q} \\ \beta_k & \text{if } x \in (G_k \setminus G_{k+1}) \cap (\mathbb{R} \setminus \mathbb{Q}). \end{cases}$$

We show that f is continuous at each point of H and discontinuous at each point of $\mathbb{R} \setminus H$.

Let $x_0 \in H$ and let $\varepsilon > 0$. Choose n such that $\alpha_n < \varepsilon$. Since

$$x_0 \in H = \bigcap_{k=1}^{\infty} G_k,$$

we see that $x_0 \in G_n$. The set G_n is open, so there exists $\delta > 0$ such that $(x_0 - \delta, x_0 + \delta) \subset G_n$. From the definition of G_n, we see that

$$0 \leq f(x) \leq \alpha_n < \varepsilon$$

for all $x \in (x_0 - \delta, x_0 + \delta)$. Thus

$$|f(x) - f(x_0)| = |f(x) - 0| = |f(x)| < \varepsilon$$

if $|x - x_0| < \delta$, so f is continuous at x_0.

Now let $x_0 \in \mathbb{R} \setminus H$. Then there exists $k \in \mathbb{N}$ such that x_0 belongs to the set $G_k \setminus G_{k+1}$. Thus $f(x_0) = \alpha_k$ or $f(x_0) = \beta_k$. Let us suppose that $f(x_0) = \alpha_k$. If x_0 is an interior point of $G_k \setminus G_{k+1}$, then x_0 is a limit point of

$$\{x : x \in (G_k \setminus G_{k+1}) \cap (\mathbb{R} \setminus \mathbb{Q})\} = \{x : f(x) = \beta_k\},$$

so f is discontinuous at x_0.

The argument is similar if x_0 is a boundary point of $G_k \setminus G_{k+1}$. Again, assume $f(x_0) = \alpha_k$. Arbitrarily close to x_0 there are points of the set

$$\mathbb{R} \setminus (G_k \setminus G_{k+1}).$$

At these points, f takes on values in the set

$$S = \{0\} \cup \bigcup_{i \neq k} \alpha_i \cup \bigcup_{j \neq k} \beta_j.$$

The only limit point of this set is zero and so S is closed. In particular, α_k is *not* a limit point of this set and does not belong to the set. Let ε be half the distance from the point α_k to the closed set S; that is, let

$$\varepsilon = \frac{1}{2} d(\alpha_k, S).$$

Arbitrarily close to x_0 there are points x such that $f(x) \in S$. For such a point,

$$|f(x) - f(x_0)| = |f(x) - \alpha_k| > \varepsilon,$$

so f is discontinuous at x_0. ■

Observe that Theorem 6.28 answers a question we asked earlier: Is there a function f continuous on $\mathbb{Q}$ and discontinuous at every point of $\mathbb{R} \setminus \mathbb{Q}$? The answer is negative, since $\mathbb{Q}$ is not of type $\mathcal{G}_\delta$.

Exercises

6.7.6 In the second part of the proof of Theorem 6.28 we provided a construction for a function f with $C_f = H$, where H is an arbitrary set of type $\mathcal{G}_\delta$. Exhibit explicitly sets G_k that will give rise to a function f such that $C_f = \mathbb{R} \setminus \mathbb{Q}$. Can you do this in such a way that the resulting function is the one we obtained at the beginning of this section?

6.7.7 In the proof of Theorem 6.28 we took $\varepsilon = \frac{1}{2} d(\alpha_k, S)$. Show that this number equals

$$\frac{1}{2} \min_{i \neq k} \{\min\{|\alpha_i - \alpha_k|, |\beta_i - \beta_k|\}\}.$$

6.8 Sets of Measure Zero

✂ *Advanced*

In analysis there are a number of ways in which a set might be considered as "small." For example, the Cantor set is not small in the sense of counting: It is uncountable. It is small in another different sense: It is nowhere dense, that is there is no interval at all in which it is dense. Now we turn to another way in which the Cantor set can be considered small: It has "zero length."

Example 6.29 Suppose we wish to measure the "length" of the Cantor set. Since the Cantor set is rather bizarre, we might look instead at the sequence of intervals that have been removed. There is no difficulty in assigning a meaning of length to an interval; the length of (a, b) is $b - a$. What is the total length of the intervals removed in the construction of the Cantor set? From the interval $[0, 1]$ we remove first a middle third interval of length $1/3$, then two middle third intervals of length $1/9$, and so on so that at the nth stage we remove 2^{n-1} intervals each of length 3^{-n}. The sum of the lengths of all intervals so removed is

$$1/3 + 2(1/9) + 4(1/27) + \cdots =$$

$$1/3\,(1 + 2/3 + (2/3)^2 + (2/3)^3 + \dots) = 1.$$

From the interval $[0, 1]$ we appear to have removed all of the length. What is left over, the Cantor set, must have length zero.

This method of computing lengths has some merit but it is not the one we wish to adopt here. Another approach to "measuring" the length of the Cantor set is to consider the length that *remains* at each stage. At the first stage the Cantor set is contained inside the union

$$[0, 1/3] \cup [2/3, 1],$$

which has length $2(1/3)$. At the next stage it is contained inside a union of four intervals, with total length $4(1/9)$. Similarly, at the nth stage the Cantor set is contained inside the union of 2^n intervals each of length 3^{-n}. The sum of the lengths of all these intervals is $(2/3)^n$, and this tends to zero as n gets large. Thus, as before, it seems we should assign zero length to the Cantor set. ◀

We convert the second method of the example into a definition of what it means for a set to be of measure zero. "Measure" is the technical term used to describe the "length" of sets that need not be intervals. In the example we used closed intervals while in our definition we have employed open intervals. There is no difference (see Exercise 6.8.13). In the example we covered the Cantor set with a finite sequence of intervals while in our definition we have employed an infinite sequence. For the Cantor set there is no difference but for other sets (sets that are not bounded or are not closed) there is a difference.

Definition 6.30 Let E be a set of real numbers. Then E is said to have *measure zero* if for every $\varepsilon > 0$ there is a finite or infinite sequence

$$(a_1, b_1), (a_2, b_2), (a_3, b_3), (a_4, b_4), \dots$$

of open intervals covering the set E so that

$$\sum_{k=1}^{\infty} (b_k - a_k) \le \varepsilon.$$

Note. In the definition of measure zero sets is there a change if we insist on an *infinite* sequence of intervals, disallowing finite sequences? Suppose that the sequence

$$(a_1, b_1), (a_2, b_2), (a_3, b_3), (a_4, b_4), \ldots (a_N, b_N)$$

of open intervals covers the set E so that

$$\sum_{k=1}^{N} (b_k - a_k) < \varepsilon/2.$$

Then to satisfy the definition we could add in some further intervals that do not amount in length to more than $\varepsilon/2$. For example, take

$$(a_{N+p}, b_{N+p}) = (0, \varepsilon/2^{p+1})$$

for $p = 1, 2, 3, \ldots$. Then

$$\sum_{k=1}^{\infty} (b_k - a_k) = \sum_{k=1}^{N} (b_k - a_k) + \sum_{p=1}^{\infty} \varepsilon/2^{p+1} < \varepsilon.$$

Thus the definition would not be changed if we had required infinite coverings.

Here are some examples of sets of measure zero.

Example 6.31 Every finite set has measure zero. The empty set is easily handled. If

$$E = \{x_1, x_2, \ldots x_N\}$$

and $\varepsilon > 0$, then the sequence of intervals

$$\left(x_i - \frac{\varepsilon}{2N}, x_i + \frac{\varepsilon}{2N}\right) \quad i = 1, 2, 3, \ldots, N$$

covers the set E and the sum of all the lengths is ε. ◀

Example 6.32 Every infinite, countable set has measure zero. If

$$E = \{x_1, x_2, \ldots\}$$

and $\varepsilon > 0$, then the sequence of intervals

$$\left(x_i - \frac{\varepsilon}{2^{i+1}}, x_i + \frac{\varepsilon}{2^{i+1}}\right) \quad i = 1, 2, 3, \ldots$$

covers the set E and, since

$$\sum_{k=1}^{\infty} 2\left(\frac{\varepsilon}{2^{k+1}}\right) = \sum_{k=1} \varepsilon 2^{-k} = \varepsilon.,$$

sum of all the lengths is ε. ◀

Example 6.33 The Cantor set has measure zero. Let $\varepsilon > 0$. Choose n so that $(2/3)^n < \varepsilon$. Then the nth stage intervals in the construction of the Cantor set give us 2^n closed intervals each of length $(1/3)^n$. This covers the Cantor set with 2^n closed intervals of total length $(2/3)^n$, which is less than ε. If the closed intervals trouble you (the definition requires open intervals), see Exercise 6.8.13 or argue as follows. Since $(2/3)^n < \varepsilon$ there is a positive number δ so that

$$(2/3)^n + \delta < \varepsilon.$$

Enlarge each of the closed intervals to form a slightly larger open interval, but change the length of each only enough so that the sum of the lengths of all the 2^n closed intervals does not increase by more than δ. The resulting collection of open intervals also covers the Cantor set, and the sum of the length of these intervals is less than ε. ◀

One of the most fundamental of the properties of sets having measure zero is how sequences of such sets combine. We recall that the union of any sequence of countable sets is also countable. We now prove that the union of any sequence of measure zero sets is also a measure zero set.

Theorem 6.34 *Let E_1, E_2, E_3, ... be a sequence of sets of measure zero. Then the set E formed by taking the union of all the sets in the sequence is also of measure zero.*

Proof Let $\varepsilon > 0$. We shall construct a cover of E consisting of a sequence of open intervals of total length less than ε. Since E_1 has measure zero, there is a sequence of open intervals

$$(a_{11}, b_{11}), (a_{12}, b_{12}), (a_{13}, b_{13}), (a_{14}, b_{14}), \ldots$$

covering the set E_1 and so that the sum of the lengths of these intervals is smaller than $\varepsilon/2$. Since E_2 has measure zero, there is a sequence of open intervals

$$(a_{21}, b_{21}), (a_{22}, b_{22}), (a_{23}, b_{23}), (a_{24}, b_{24}), \ldots$$

covering the set E_2 and so that the sum of the lengths of these intervals is smaller than $\varepsilon/4$. In general, for each $k = 1, 2, 3, \ldots$ there is a sequence of open intervals

$$(a_{k1}, b_{k1}), (a_{k2}, b_{k2}), (a_{k3}, b_{k3}), (a_{k4}, b_{k4}), \ldots$$

covering the set E_k and so that the sum of the lengths of these intervals is smaller than $\varepsilon/2^k$. The totality of all these intervals can be arranged into a single sequence of open intervals that covers every point in the union of the sequence $\{E_k\}$. The sum of the lengths of all the intervals in the large sequence is smaller than

$$\varepsilon/2 + \varepsilon/4 + \varepsilon/8 + \cdots = \varepsilon.$$

It follows that E has measure zero. ■

Let us return to the situation for the Cantor set once again. For each $\varepsilon > 0$ we were able to choose a finite cover of open intervals with total length less than ε. This is not the case for all sets of measure zero. For example, the set of all rational numbers on the real line is countable and hence also of measure zero. Any finite collection of intervals must fail to cover that set, in fact cannot come close to covering all rational numbers. For what sets is it possible to select finite coverings of small length? The answer is that this is possible for compact sets of measure zero.

Theorem 6.35 *Let E be a compact set of measure zero. Then for every $\varepsilon > 0$ there is a finite collection of open intervals*

$$(a_1, b_1), (a_2, b_2), (a_3, b_3), (a_4, b_4), \ldots (a_N, b_N)$$

that covers the set E and so that

$$\sum_{k=1}^{N}(b_k - a_k) < \varepsilon.$$

Proof Since E has measure zero, it is certainly possible to select a sequence of open intervals

$$(a_1, b_1), (a_2, b_2), (a_3, b_3), (a_4, b_4), \ldots$$

that covers the set E and so that

$$\sum_{k=1}^{\infty}(b_k - a_k) < \varepsilon.$$

But how can we reduce this collection to a finite one that also covers the set E? If you studied the Heine-Borel theorem (Theorem 4.33), then you know how.

We shall present here a proof that uses the Bolzano-Weierstrass theorem instead. We claim that we can find an integer N so that all points of E are in one of the intervals

$$(a_1, b_1), (a_2, b_2), (a_3, b_3), (a_4, b_4), \ldots (a_N, b_N).$$

This will prove the theorem.

We prove this by contradiction. If this is not so, then for each integer $k = 1, 2, 3, \ldots$ we must be able to find a point $x_k \in E$ but x_k is not in any of the intervals

$$(a_1, b_1), (a_2, b_2), (a_3, b_3), (a_4, b_4), \dots (a_k, b_k).$$

The sequence $\{x_k\}$ is bounded because E is bounded. By the Bolzano-Weierstrass theorem the sequence has a convergent subsequence $\{x_{n_j}\}$. Let z be the limit of the convergent subsequence. Since E is closed z is in E. The original sequence of intervals covers all of E and so there must be an interval (a_M, b_M) that contains z. For large values of j the points x_{n_j} also belong to (a_M, b_M). But this is impossible since x_{n_j} cannot belong to the interval (a_M, b_M) for $n_j \geq M$. Since this is a contradiction, the proof is done. ■

Exercises

6.8.1 Show that every subset of a set of measure zero also has measure zero.

6.8.2 If E has measure zero, show that the translated set

$$E + \alpha = \{x + \alpha : x \in E\}$$

also has measure zero.

6.8.3 If E has measure zero, show that the expanded set

$$cE = \{cx : x \in E\}$$

also has measure zero for any $c > 0$.

6.8.4 If E has measure zero, show that the reflected set

$$-E = \{-x : x \in E\}$$

also has measure zero.

6.8.5 Without referring to the proof of Theorem 6.34, show that the union of any two sets of measure zero also has measure zero.

6.8.6 If $E_1 \subset E_2$ and E_1 has measure zero but E_2 has not, what can you say about the set $E_2 \setminus E_1$?

6.8.7 Show that any interval (a, b) or $[a, b]$ is not of measure zero.

6.8.8 Give an example of a set that is not of measure zero and does not contain any interval $[a, b]$.

6.8.9 A careless student claims that if a set E has measure zero, then it is obviously true that the closure $\overline{E}$ must also have measure zero. Is this correct?

6.8.10 If a set E has measure zero what can you say about interior points of that set?

6.8.11 Suppose that a set E has the property that $E \cap [a, b]$ has measure zero for every compact interval $[a, b]$. Must E also have measure zero?

6.8.12 Show that the set of real numbers in the interval $[0,1]$ that do not have a 7 in their infinite decimal expansion is of measure zero.

6.8.13 In Definition 6.30 show that closed intervals may be used without changing the definition.

6.8.14 Describe completely the class of sets E with the following property: For every $\varepsilon > 0$ there is a *finite* collection of open intervals

$$(a_1,b_1),(a_2,b_2),(a_3,b_3),(a_4,b_4),\dots(a_N,b_N)$$

that covers the set E and so that

$$\sum_{k=1}^{N}(b_k - a_k) < \varepsilon.$$

(These sets are said to have *zero content.*)

6.8.15 Show that a set E has measure zero if and only if there is a sequence of intervals

$$(a_1,b_1),(a_2,b_2),(a_3,b_3),(a_4,b_4),\dots$$

so that every point in E belongs to infinitely many of the intervals and $\sum_{k=1}^{\infty}(b_k - a_k)$ converges.

6.8.16 By altering the construction of the Cantor set, construct a nowhere dense closed subset of $[0,1]$ so that the sum of the lengths of the intervals removed is not equal to 1. Will this set have measure zero?

6.9 Challenging Problems for Chapter 6

6.9.1 Show that a function is discontinuous except at the points of a first category set if and only if it is continuous at a dense set of points.

6.9.2 Let $f : \mathbb{R} \to \mathbb{R}$ be a continuous function. Assume that for every positive number ε the sequence $\{f(n\varepsilon)\}$ converges to zero as $n \to \infty$. Prove that

$$\lim_{x\to\infty} f(x) = 0.$$

6.9.3 Let f_n be a sequence of continuous functions defined on an interval $[a,b]$ such that $\lim_{n\to\infty} f_n(x) = 0$ for each $x \in [a,b]$. Show that for any $\varepsilon > 0$ there is an interval $[c,d] \subset [a,b]$ and an integer N so that

$$|f_n(x)| < \varepsilon$$

for every $n \geq N$ and every $x \in [c,d]$. Show that this need not be true for $[c,d] = [a,b]$.

6.9.4 Let f_n be a sequence of continuous functions defined on an interval $[a,b]$ such that $\lim_{n\to\infty} f_n(x) = \infty$ for each $x \in [a,b]$. Show that for any $M > 0$ there is an interval $[c,d] \subset [a,b]$ and an integer N so that

$$f_n(x) > M$$

for every $n \geq N$ and every $x \in [c,d]$. Show that this need not be true for $[c,d] = [a,b]$.

Chapter 7

DIFFERENTIATION

7.1 Introduction

Calculus courses succeed in conveying an idea of what a derivative is, and the students develop many technical skills in computations of derivatives or applications of them. We shall return to the subject of derivatives but with a different objective.

Now we wish to see a little deeper and to understand the basis on which that theory develops. Much of this chapter will appear to be a review of the subject of derivatives with more attention paid to the details now and less to the applications. Some of the more advanced material will be, however, completely new.

We start at the beginning, at the rudiments of the theory of derivatives.

7.2 The Derivative

Let f be a function defined on an interval I and let x_0 and x be points of I. Consider the *difference quotient* determined by the points x_0 and x:

$$\frac{f(x)-f(x_0)}{x-x_0}, \tag{1}$$

representing the average rate of change of f on the interval with endpoints at x and x_0.

In Figure 7.1 this difference quotient represents the slope of the chord (or secant line) determined by the points $(x, f(x))$ and $(x_0, f(x_0)$. This same picture allows a physical interpretation. If $f(x)$ represents the distance a point moving on a straight line has moved from some fixed point in time x, then $f(x)-f(x_0)$ represents the (net) distance it has moved in the time interval $[x_0, x]$, and the difference quotient (1) represents the average velocity in that time interval.

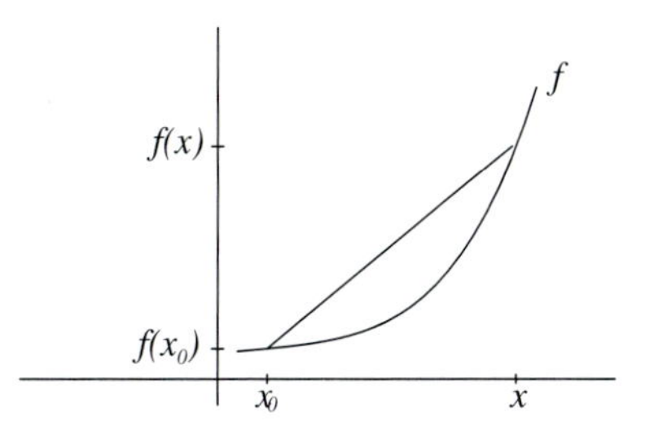

Figure 7.1. The chord determined by $(x, f(x))$ and $(x_0, f(x_0))$.

Suppose now that we fix x_0, and allow x to approach x_0. We learn in elementary calculus that if

$$\lim_{x \to x_0} \frac{f(x) - f(x_0)}{x - x_0}$$

exists, then the limit represents the slope of the tangent line to the graph of the function f at the point $(x_0, f(x_0))$. In the setting of motion, the limit represents instantaneous velocity at time x_0.

The derivative owes its origins to these two interpretations in geometry and in the physics of motion, but now completely transcends them; the derivative finds applications in nearly every part of mathematics and the sciences.

We shall study the structure of derivatives, but with less concern for computations and applications than we would have seen in our calculus courses. Now we wish to understand the notion and see why it has the properties used in the many computations and applications of the calculus.

7.2.1 Definition of the Derivative

We begin with a familiar definition.

Definition 7.1 Let f be defined on an interval I and let $x_0 \in I$. The *derivative* of f at x_0, denoted by $f'(x_0)$, is defined as

$$f'(x_0) = \lim_{x \to x_0} \frac{f(x) - f(x_0)}{x - x_0}, \tag{2}$$

provided either that this limit exists or is infinite. If $f'(x_0)$ is finite we say that f is *differentiable* at x_0. If f is differentiable at every point of a set $E \subset I$, we say that f is *differentiable* on E. When E is all of I, we simply say that f is a *differentiable* function.

Note. We have allowed infinite derivatives and they do play a role in many studies, but differentiable always refers to a finite derivative. Normally the phrase "a derivative exists" also means that that derivative is finite.

Example 7.2 Let $f(x) = x^2$ on $\mathbb{R}$ and let $x_0 \in \mathbb{R}$. If $x \in \mathbb{R}$, $x \neq x_0$, then

$$\frac{f(x) - f(x_0)}{x - x_0} = \frac{x^2 - x_0^2}{x - x_0} = \frac{(x - x_0)(x + x_0)}{(x - x_0)}.$$

Since $x \neq x_0$, the last expression equals $x + x_0$, so

$$\lim_{x \to x_0} \frac{f(x) - f(x_0)}{x - x_0} = \lim_{x \to x_0} (x + x_0) = 2x_0,$$

establishing the formula, $f'(x_0) = 2x_0$ for the function $f(x) = x^2$. ◀

Let us take a moment to clarify the definition when the interval I contains one or both of its endpoints. Suppose $I = [a, b]$. For $x_0 = a$ (or $x_0 = b$), the limit in (2) is just a one-sided, or unilateral, limit. The function f is defined only on $[a, b]$ so we cannot consider points outside of that interval.

This brings us to another point. It can happen that a function that is *not* differentiable at a point x_0 does satisfy the requirement of (2) from one side of x_0. This means that the limit in (2) exists as $x \to x_0$ from that side. We present a formal definition.

Definition 7.3 Let f be defined on an interval I and let $x_0 \in I$. The *right-hand derivative* of f at x_0, denoted by $f'_+(x_0)$ is the limit

$$f'_+(x_0) = \lim_{x \to x_0+} \frac{f(x) - f(x_0)}{x - x_0},$$

provided that one-sided limit exists or is infinite. Similarly, the *left-hand derivative* of f at x_0, $f'_-(x_0)$, is the limit

$$f'_-(x_0) = \lim_{x \to x_0-} \frac{f(x) - f(x_0)}{x - x_0}.$$

Observe that, if x_0 is an interior point of I, then $f'(x_0)$ exists if and only if $f'_+(x_0) = f'_-(x_0)$. (See Exercise 7.2.8)

Example 7.4 Let $f(x) = |x|$ on $\mathbb{R}$. Let us consider the differentiability of f at $x_0 = 0$. The difference quotient (1) becomes

$$\frac{f(x) - f(0)}{x - 0} = \frac{|x|}{x} = \begin{cases} 1, & \text{if } x > 0 \\ -1, & \text{if } x < 0. \end{cases}$$

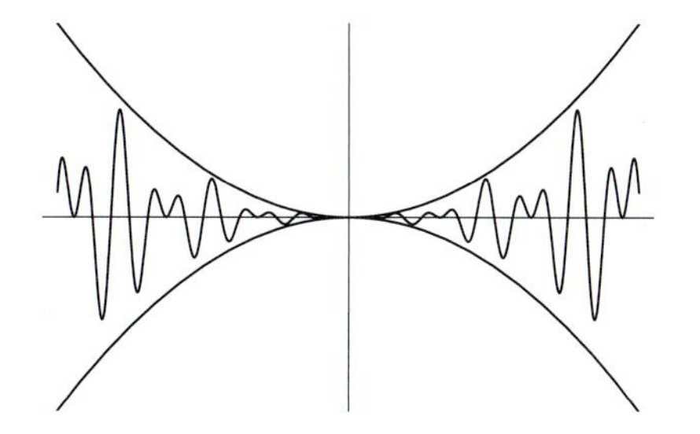

Figure 7.2. A function trapped between x^2 and $-x^2$.

Thus

$$f'_+(0) = \lim_{x \to x_0+} \frac{|x|}{x} = 1$$

while

$$f'_-(0) = \lim_{x \to x_0-} \frac{|x|}{x} = -1.$$

The function has different right-hand and left-hand derivatives at $x_0 = 0$ so is not differentiable at $x_0 = 0$. ◀

Example 7.5 (A "trapping principle")

Let f be any function defined in a neighborhood I of zero. Suppose f satisfies the inequality $|f(x)| \leq x^2$ for all $x \in I$. Thus, the graph of f is "trapped" between the parabolas $y = x^2$ and $y = -x^2$. In particular, $f(0) = 0$. The difference quotient computed for $x_0 = 0$ becomes

$$\frac{f(x) - f(0)}{x - 0} = \frac{f(x)}{x},$$

from which we calculate

$$\left| \frac{f(x)}{x} \right| \leq \left| \frac{x^2}{x} \right| = |x|$$

so

$$\lim_{x \to 0} \left| \frac{f(x)}{x} \right| \leq \lim_{x \to 0} |x| = 0.$$

Thus

$$\lim_{x \to 0} \frac{f(x)}{x} = 0.$$

As a result, $f'(0) = 0$. Figure 7.2 illustrates the principle. ◀

Higher-Order Derivatives When a function f is differentiable on I, it is possible that its derivative f' is also differentiable. When this is the case, the function $f'' = (f')'$ is called the *second derivative* of the function f. Inductively, we can define derivatives of all orders: $f^{(n+1)} = (f^{(n)})'$ (provided $f^{(n)}$ is differentiable). When n is small, it is customary to use the convenient notation f'' for $f^{(2)}$, f''' for $f^{(3)}$ etc.

Notation It is useful to have other notations for the derivative of a function f. Common notations are $\frac{df}{dx}$ and $\frac{dy}{dx}$ (when the function is expressed in the form $y = f(x)$). Another notation that is useful is Df. These alternate notations along with slight variations are useful for various calculations. You are no doubt familiar with such uses—the convenience of writing

$$\frac{dy}{dx} = \frac{dy}{du}\frac{du}{dx}$$

when using the chain rule, or viewing D as an operator in solving linear differential equations. Notation *can* be important at times. Consider, for example, how difficult it would be to perform a simple arithmetic calculation such as the multiplication (104)(90) using Roman numerals (CIV)(XC)!

Exercises

7.2.1 You might be familiar with a slightly different formulation of the definition of derivative. If x_0 is interior to I, then for h sufficiently small, the point $x_0 + h$ is also in I. Show that expression (2) then reduces to

$$f'(x_0) = \lim_{h \to 0} \frac{f(x_0 + h) - f(x_0)}{h}.$$

Repeat Examples 7.2 and 7.4 using this formulation of the derivative.

7.2.2 Let $c \in \mathbb{R}$. Calculate the derivatives of the functions $g(x) = c$ and $k(x) = x$ directly from the definition of derivative.

7.2.3 Check the differentiability of each of the functions below at $x_0 = 0$.

(a) $f(x) = x|x|$

(b) $f(x) = x \sin x^{-1}$ $(f(0) = 0)$

(c) $f(x) = x^2 \sin x^{-1}$ $(f(0) = 0)$

(d) $f(x) = \begin{cases} x^2, & \text{if } x \text{ rational} \\ 0, & \text{if } x \text{ irrational} \end{cases}$

7.2.4 Let $f(x) = \begin{cases} x^2, & \text{if } x \geq 0 \\ ax, & \text{if } x < 0 \end{cases}$

(a) For which values of a is f differentiable at $x = 0$?

(b) For which values of a is f continuous at $x = 0$?

(c) When f is differentiable at $x = 0$, does $f''(0)$ exist?

7.2.5 For what positive values of p is the function $f(x) = |x|^p$ differentiable at 0?

7.2.6 A function f has a *symmetric derivative* at a point if

$$f'_s(x) = \lim_{h \to 0} \frac{f(x+h) - f(x-h)}{2h}$$

exists. Show that $f'_s(x) = f'(x)$ at any point at which the latter exists but that $f'_s(x)$ may exist even when f is not differentiable at x.

7.2.7 Find all points where $f(x) = \sqrt{1 - \cos x}$ is not differentiable and at those points find the one-sided derivatives.

7.2.8 Prove that if x_0 is an interior point of an interval I, then $f'(x_0)$ exists or is infinite if and only if $f'_+(x_0) = f'_-(x_0)$.

7.2.9 Let a function $f : \mathbb{R} \to \mathbb{R}$ be defined by setting $f(1/n) = c_n$ for $n = 1$, 2, 3, ... where $\{c_n\}$ is a given sequence and elsewhere $f(x) = 0$. Find a condition on that sequence so that $f'(0)$ exists.

7.2.10 Let a function $f : \mathbb{R} \to \mathbb{R}$ be defined by setting $f(1/n^2) = c_n$ for $n = 1$, 2, 3, ... where $\{c_n\}$ is a given sequence and elsewhere $f(x) = 0$. Find a condition on that sequence so that $f'(0)$ exists.

7.2.11 Give an example of a function with an infinite derivative at some point. Give an example of a function f with $f'_+(x_0) = \infty$ and $f'_-(x_0) = -\infty$ at some point x_0.

7.2.12 If $f'(x_0) > 0$ for some point x_0 in the interior of the domain of f show that there is a $\delta > 0$ so that

$$f(x) < f(x_0) < f(y)$$

whenever $x_0 - \delta < x < x_0 < y < x_0 + \delta$. Does this assert that f is increasing in the interval $(x_0 - \delta, x_0 + \delta)$?

7.2.13 Let f be increasing and differentiable on an interval. Does this imply that $f'(x) \geq 0$ on that interval? Does this imply that $f'(x) > 0$ on that interval?

7.2.14 Suppose that two functions f and g have the following properties at a point x_0: $f(x_0) = g(x_0)$ and $f(x) \leq g(x)$ for all x in an open interval containing the point x_0. If both $f'(x_0)$ and $g'(x_0)$ exist show that they must be equal. How does this compare to the trapping principle used in Example 7.5, where it seems much more is assumed about the function f.

7.2.15 Suppose that f is a function defined on the real line with the property that $f(x + y) = f(x)f(y)$ for all x, y. Suppose that f is differentiable at 0 and that $f'(0) = 1$. Show that f must be differentiable everywhere and that $f'(x) = f(x)$.

7.2.2 Differentiability and Continuity

A continuous function need not be differentiable (Example 7.4) but the converse is true. Every differentiable function is continuous.

Theorem 7.6 *Let f be defined in a neighborhood I of x_0. If f is differentiable at x_0, then f is continuous at x_0.*

Proof It suffices to show that $\lim_{x \to x_0} (f(x) - f(x_0)) = 0$. For $x \neq x_0$,

$$f(x) - f(x_0) = \left(\frac{f(x) - f(x_0)}{x - x_0} \right) (x - x_0).$$

Now

$$\lim_{x \to x_0} \frac{f(x) - f(x_0)}{x - x_0} = f'(x_0)$$

and $\lim_{x \to x_0} (x - x_0) = 0$. We then obtain

$$\lim_{x \to x_0} (f(x) - f(x_0)) = (f'(x_0))(0) = 0$$

by the product rule for limits. ■

We can use this theorem in two ways. If we know that a function has a discontinuity at a point, then we know immediately that there is no derivative there. On the other hand, if we have been able to determine by some means that a function is differentiable at a point then we know automatically that the function must also be continuous at that point.

Exercises

7.2.16 Construct a function on the interval $[0, 1]$ that is continuous and is not differentiable at each point of some infinite set.

7.2.17 Suppose that a function has both a right-hand and a left-hand derivative at a point. What, if anything, can you conclude about the continuity of that function at that point?

7.2.18 Suppose that a function has an infinite derivative at a point. What, if anything, can you conclude about the continuity of that function at that point?

7.2.19 Show that if a function f has a symmetric derivative $f'_s(x_0)$ (see Exercise 7.2.6), then f must be symmetrically continuous at x_0 in the sense that $\lim_{h \to 0}[f(x_0 + h) - f(x_0 - h)] = 0$. Must f in fact be continuous?

7.2.20 If $f'(x_0) = \infty$, does it follow that f must be continuous at x_0 on one side at least?

7.2.21 Find an example of an everywhere differentiable function f so that f' is not everywhere continuous.

7.2.22 Show that a function f that satisfies an inequality of the form

$$|f(x) - f(y)| \leq M\sqrt{|x-y|}$$

for some constant M and all x, y must be everywhere continuous but need not be everywhere differentiable.

7.2.23 The Dirichlet function (see Section 5.2.6) is discontinuous at each rational number. By Theorem 7.6 it follows that this function has no derivative at any rational number. Does it have a derivative at any irrational number?

7.2.3 The Derivative as a Magnification

Enrichment

We offer now one more interpretation of the derivative, this time as a magnification factor. In elementary calculus one often makes use of the geometric content of the graph of a function f. In particular, we can view the derivative in terms of slopes of tangent lines to the graph. But the graph of f is a subset of two-dimensional space, while the range of f is a subset of one-dimensional space and, as such, has some additional geometric content.

Suppose f is differentiable on an interval I, and let J be a closed subinterval of I. The range of f on J will also be a closed interval, because f is differentiable and hence continuous on J, and continuous functions map closed intervals onto closed intervals (Exercise 5.8.9). The expression

$$\frac{|f(J)|}{|J|}$$

represents the amount that the interval J has been expanded (or contracted) under the mapping f.

For example, if $f(x) = x^2$ and $J = [2, 3]$, then

$$\frac{|f(J)|}{|J|} = \frac{|[4, 9]|}{|[2, 3]|} = \frac{5}{1} = 5.$$

Thus the interval $[2, 3]$ has been expanded by f to an interval of 5 times its size. If we look only at small intervals then the derivative offers a clue to the size of the magnification factor.

If J is a sufficiently small interval having x_0 as an endpoint, then the ratio $|f(J)|/|J|$ is approximately $|f'(x_0)|$, the approximation becoming "exact in the limit." Thus $|f'(x_0)|$ can be viewed as a "magnification factor" of small intervals containing the point x_0. In our illustration with the function $f(x) = x^2$, the magnification factor at $x_0 = 2$ is $f'(2) = 4$. Small intervals about x_0 are magnified by a factor of about 4. At the other endpoint $x_0 = 3$, small intervals about x_0 are magnified by a factor of about 6.

In Exercise 7.2.26 we ask you to prove a precise statement covering the preceding discussion.

Exercises

7.2.24 What is the ratio

$$\frac{|f(J)|}{|J|}$$

for the function $f(x) = x^2$ if $J = [2, 2.001]$, $J = [2, 2.0001]$, $J = [2, 2.00001]$?

7.2.25 In this section we have interpreted $f'(x_0)$ as a magnification factor. If $f'(x_0) = 0$, does this mean that small intervals containing the point x_0 are magnified by a factor of 0 when mapped by f?

7.2.26 Let f be differentiable on an interval I and let x_0 be an interior point of I. Make precise the following statement and prove it:

$$\lim_{J \to x_0} \frac{|f(J)|}{|J|} = |f'(x_0)|.$$

7.3 Computations of Derivatives

Example 7.2 provides a calculation of the derivative of the function $f(x) = x^2$. The calculation involved direct evaluation of the limit of an appropriate difference quotient. For the function $f(x) = x^2$, this evaluation was straightforward. But limits of difference quotients can be quite complicated. You are familiar with certain rules that are useful in calculating derivatives of functions that are "built up" from functions whose derivatives are known.

In this section we review some of the calculus rules that are commonly used to compute derivatives. We need first to prove the algebraic rules: The sum rule, the product rule, and the quotient rule. Then we turn to the chain rule. Finally, we look at the power rule. Our viewpoint here is not to practice the computation of derivatives but to build up the theory of derivatives, making sure to see how it depends on work on limits that we proved earlier on.

The various rules we shall obtain in this section should be viewed as aids for computations of derivatives. An understanding of these rules is, of course, necessary for various calculations. But they in no way can substitute for an understanding of the derivative. And they might not be useful in calculating certain derivatives. (For example, derivatives of the functions of Exercise 7.2.3 cannot be calculated at $x_0 = 0$ by using these rules.)

Nonetheless, it is true that one often has a function that can be expressed in terms of several functions via the operations we considered in this section, functions whose derivatives we know. In those cases, the techniques of this section might be useful.

7.3.1 Algebraic Rules

Functions can be combined algebraically by multiplying by constants, by addition and subtraction, by multiplication, and by division. To each of these there is a calculus rule for computing the derivative. We recall that the limit of a sum (a difference, a product, a quotient) is the sum (difference, product, quotient) of the limits. Perhaps we might have thought the same kind of rule would apply to derivatives. The derivative of the sum is indeed the sum of the derivatives, but the derivative of the product is not the product of the derivatives. Nor do quotients work in such a simple way. The reasons for the form of the various rules can be found by writing out the definition of the derivative and following through on the computations.

Theorem 7.7 *Let f and g be defined on an interval I and let $x_0 \in I$. If f and g are differentiable at x_0 then $f+g$ and fg are differentiable at x_0. If $g(x_0) \neq 0$, then f/g is differentiable at x_0. Furthermore, the following formulas are valid:*

(i) $(cf)'(x) = cf'(x)$ *for any real number c.*

(ii) $(f+g)'(x_0) = f'(x_0) + g'(x_0)$.

(iii) $(fg)'(x_0) = f(x_0)g'(x_0) + g(x_0)f'(x_0)$.

(iv) $\left(\dfrac{f}{g}\right)'(x_0) = \dfrac{g(x_0)f'(x_0) - f(x_0)g'(x_0)}{(g(x_0))^2}$ *(if $g(x_0) \neq 0$).*

Proof Parts (i) and (ii) follow easily from the definition of the derivative and appropriate limit theorems.

To verify part (iii), let $h = fg$. Then for each $x \in I$ we have

$$h(x) - h(x_0) = f(x)[g(x) - g(x_0)] + g(x_0)[f(x) - f(x_0)]$$

so

$$\frac{h(x) - h(x_0)}{x - x_0} = f(x)\frac{g(x) - g(x_0)}{x - x_0} + g(x_0)\frac{f(x) - f(x_0)}{x - x_0}. \tag{3}$$

As $x \to x_0$, $f(x) \to f(x_0)$ since f being differentiable is also continuous. By the definition of the derivative we also know that

$$\frac{g(x) - g(x_0)}{x - x_0} \to g'(x_0)$$

and

$$\frac{f(x) - f(x_0)}{x - x_0} \to f'(x_0)$$

as $x \to x_0$. We now see from equation (3) that

$$\lim_{x \to x_0} \frac{h(x) - h(x_0)}{x - x_0} = f(x_0)g'(x_0) + g(x_0)f'(x_0),$$

verifying part (iii).

Finally, to establish part (iv) of the theorem, let $h = f/g$. Straightforward algebraic manipulations show that

$$\frac{h(x) - h(x_0)}{x - x_0} =$$

$$\frac{1}{g(x)g(x_0)} \left[g(x_0) \left(\frac{f(x) - f(x_0)}{x - x_0} \right) - f(x) \left(\frac{g(x) - g(x_0)}{x - x_0} \right) \right]. \tag{4}$$

Now let $x \to x_0$. Since f and g are continuous at x_0, $f(x) \to f(x_0)$ and $g(x) \to g(x_0)$. Thus part (iv) of the theorem follows from equation (4), the definition of derivative, and basic limit theorems. ■

Example 7.8 To calculate the derivative of $h(x) = (x^3+1)^2$ we have several ways to proceed.

1. Apply the definition of derivative. You may wish to set up the difference quotient and see that a calculation of its limit is a formidable task.

2. Write $h(x) = x^6 + 2x^3 + 1$ and apply the formula $\frac{d}{dx}x^n = nx^{n-1}$ (Exercise 7.3.5) and the rule for sums. Thus we get

$$h'(x) = 6x^5 + 6x^2.$$

3. Use the product rule to obtain

$$h'(x) = (x^3 + 1)\frac{d}{dx}(x^3 + 1) + (x^3 + 1)\frac{d}{dx}(x^3 + 1).$$

   Then, again, use the formula $\frac{d}{dx}x^n = nx^{n-1}$ and the rule for sums to continue:

$$h'(x) = (x^3 + 1)3x^2 + (x^3 + 1)3x^2 = 6x^5 + 6x^2.$$

◀

Exercises

7.3.1 Give the details needed in the proof of Theorem 7.7 for the sum rule for derivatives; that is, $(f+g)'(x_0) = f'(x_0) + g'(x_0)$.

7.3.2 The table shown in Figure 7.3 gives the values of two functions f and g at certain points. Calculate $(f+g)'(1)$, $(fg)'(1)$ and $(f/g)'(1)$. What can you assert about $(f/g)'(3)$? Is there enough information to calculate $f''(3)$?

x	$f(x)$	$f'(x)$	$g(x)$	$g'(x)$
1	3	3	2	2
2	4	4	4	0
3	6	1	1	0
4	-1	0	1	1
5	2	5	3	3

Figure 7.3. Values of f and g at several points.

7.3.3 Obtain the rule

$$\frac{d}{dx}\frac{1}{f(x)} = -\frac{f'(x)}{f(x)^2}$$

from Theorem 7.7 and also directly from the definition of the derivative.

7.3.4 Obtain the rule for

$$\frac{d}{dx}\left(f(x)\right)^2 = 2f(x)f'(x)$$

from Theorem 7.7 and also directly from the definition of the derivative.

7.3.5 Obtain the formula

$$\frac{d}{dx}x^n = nx^{n-1}$$

for $n = 1, 2, 3, \ldots$ by induction.

7.3.6 State and prove a theorem that gives a formula for $f'(x_0)$ when

$$f = f_1 + f_2 + \cdots + f_n$$

and each of the functions $f_1, \ldots, f_n$ is differentiable at x_0.

7.3.7 State and prove a theorem that gives a formula for $f'(x_0)$ when

$$f = f_1 f_2 \ldots f_n$$

and each of the functions $f_1, \ldots, f_n$ is differentiable at x_0.

7.3.8 Show that

$$(fg)''(x_0) = f''(x_0)g(x_0) + 2f'(x_0)g'(x_0) + f(x_0)g''(x_0)$$

under appropriate hypotheses.

7.3.9 Extend Exercise 7.3.8 by obtaining a similar formula for $(fg)'''(x_0)$.

7.3.10 Obtain a formula for $(fg)^{(n)}(x_0)$ valid for $n = 1, 2, 3, \ldots$.

7.3.2 The Chain Rule

There is another, nonalgebraic, interpretation of Example 7.8 that you may recall from calculus courses.

Example 7.9 We can view the function $h(x) = (x^3 + 1)^2$ as a *composition* of the function $f(x) = x^3 + 1$ and $g(u) = u^2$. Thus

$$h(x) = g \circ f(x).$$

You are familiar with the *chain rule* that is useful in calculating derivatives of composite functions. In this case the calculation would lead to

$$\begin{aligned} h'(x) &= g'(f(x))f'(x) = g'(x^3 + 1)3x^2 \\ &= 2(x^3 + 1)3x^2 = 6x^5 + 6x^2. \end{aligned}$$

In elementary calculus you might have preferred to obtain

$$\frac{dy}{dx} = \frac{dy}{du}\frac{du}{dx} = 2(x^3 + 1)(3x^2) = 6x^5 + 6x^2$$

by making the substitution $u = x^3 + 1, y = u^2$. ◀

The chain rule is the familiar calculus formula

$$\frac{d}{dx}g(f(x)) = g'(f(x))f'(x)$$

for the differentiation of the composition of two functions $g \circ f$ under appropriate assumptions. Calculus students often memorize this in the form

$$\frac{dy}{dx} = \frac{dy}{du}\frac{du}{dx}$$

by using the new variables $y = g(u)$ and $u = f(x)$.

Let us first try to see why the chain rule should work. Then we'll provide a precise statement and proof of the chain rule. Perhaps the easiest way to "see" the chain rule is by interpreting the derivative as a magnification factor.

Let f be defined in a neighborhood of x_0 and let g be defined in a neighborhood of $f(x_0)$. If f is differentiable at x_0, then f maps each small interval J containing x_0 onto an interval $f(J)$ containing $f(x_0)$ with $|f(J)|/|J|$ approximately $|f'(x_0)|$. If, also, g is differentiable at $f(x_0)$, then g will map a small interval $f(J)$ containing $f(x_0)$ onto an interval $g(f(J))$ with $|g(f(J))|/|f(J)|$ approximately $|g'(f(x_0))|$. Thus $h = g \circ f$ maps J onto the interval $h(J) = g(f(J))$ and

$$\frac{|h(J)|}{|J|} = \frac{|g(f(J))|}{|f(J)|}\frac{|f(J)|}{|J|}$$

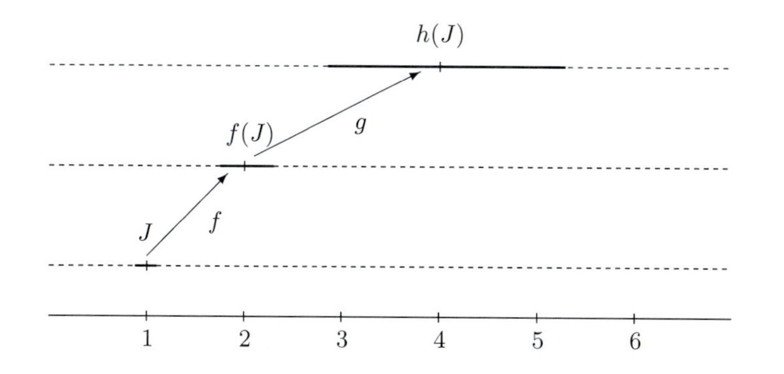

Figure 7.4. f maps J to $f(J)$ and g maps that to $h(J)$. Here $h = g \circ f$, $x_0 = 1$, and $J = [.9, 1.1]$.

and this is approximately equal to

$$|g'(f(x_0))||f'(x_0)|.$$

In short, the magnification factors $|f'(x_0)|$ and $|g'(f(x_0))|$ multiply to give the magnification factor $|h'(x_0)|$.

Example 7.10 Let us relate this discussion to our example $h(x) = (x^3+1)^2$. Here $f(x) = x^3 + 1$, $g(x) = x^2$. At $x_0 = 1$ we obtain $f(x_0) = 2$, $f'(x_0) = 3$, $g(f(x_0)) = 4$ and $g'(f(x_0)) = 4$. The function f maps small intervals about $x_0 = 1$ onto ones about three times as long, and in turn, the function g maps those intervals onto ones about four times as long, so the total magnification factor for the function $h = g \circ f$ is about 12 at $x_0 = 1$ (Fig. 7.4). ◀

Proof of the Chain Rule If we wished to formulate a proof of the chain rule based on the preceding discussion we could begin by writing

$$\frac{g(f(x)) - g(f(x_0))}{x - x_0} = \frac{g(f(x)) - g(f(x_0))}{f(x) - f(x_0)} \frac{f(x) - f(x_0)}{x - x_0} \tag{5}$$

which compares to our formula

$$\frac{|h(J)|}{|J|} = \frac{|g(f(J))|}{|f(J)|} \frac{|f(J)|}{|J|}.$$

If we let $x \to x_0$ in (5), we would expect to get the desired result

$$(g \circ f)'(x_0) = g'(f(x_0))f'(x_0).$$

And this argument would be valid if f were, for example, increasing. But in order for equation (5) to be valid, we must have $x \neq x_0$ and $f(x) \neq f(x_0)$.

When computing the limit of a difference quotient, we can assume $x \neq x_0$, but we can't assume, without additional hypotheses, that if $x \neq x_0$ then $f(x) \neq f(x_0)$. Yet the chain rule applies nonetheless.

The proof is clearer if we separate these two cases. In the simpler case the function does not repeat the value $f(x_0)$ in some neighborhood of x_0. In the harder case the function repeats the value $f(x_0)$ in *every* neighborhood of x_0. Exercise 7.3.11 shows that in that case we must have $f'(x_0) = 0$ and so the chain rule reduces to showing that the composite function $g \circ f$ also has a zero derivative.

Theorem 7.11 (Chain Rule) *Let f be defined on a neighborhood U of x_0 and let g be defined on a neighborhood V of $f(x_0)$ for which*

$$f(x_0) \in f(U) \subset V.$$

Suppose f is differentiable at x_0 and g is differentiable at $f(x_0)$. Then the composite function $h = g \circ f$ is differentiable at x_0 and

$$h'(x_0) = (g \circ f)'(x_0) = g'(f(x_0))f'(x_0).$$

Proof Consider any sequence of distinct points x_n converging to x_0. If we can show that the sequence

$$S_n = \frac{g(f(x_n)) - g(f(x_0))}{x_n - x_0}$$

converges to $g'(f(x_0))f'(x_0)$ for every such sequence then we have obtained our required formula.

Note that if $f(x_n) \neq f(x_0)$, then we can write $y_n = f(x_n)$, $y_0 = f(x_0)$ and display S_n as

$$S_n = \frac{g(y_n) - g(y_0)}{y_n - y_0} \times \frac{f(x_n) - f(x_0)}{x_n - x_0}. \tag{6}$$

Seen in this form it becomes obvious that

$$S_n \to g'(y_0)f'(x_0) = g'(f(x_0))f'(x_0)$$

except for the problem that we cannot (as we remarked before beginning our proof) assume that in all cases $f(x_n) \neq f(x_0)$.

Thus we consider two cases. In the first case we assume that for any sequence of distinct points x_n converging to x_0 there cannot be infinitely many terms with $f(x_n) = f(x_0)$. In that case the chain rule formula is evidently valid.

In the second case we assume that there does exist a sequence of distinct points x_n converging to x_0 with $f(x_n) = f(x_0)$ for infinitely many terms.

In that case (Exercise 7.3.11) we must have $f'(x_0) = 0$ and so, to establish the chain rule, we need to prove that $h'(x_0) = 0$. But in this case for any sequence x_n converging to x_0 either $S_n = 0$ [when $f(x_n) = f(x_0)$] or else S_n can be written in the form of equation (6) [when $f(x_n) \neq f(x_0)$]. It is then clear than $S_n \to 0$ and the proof is complete. ■

Exercises

7.3.11 Show that if for each neighborhood U of x_0 there exists $x \in U$, $x \neq x_0$ for which $f(x_n) = f(x_0)$, then either $f'(x_0)$ does not exist or else $f'(x_0) = 0$.

7.3.12 Give an explicit example of functions f and g such that the "proof" of the chain rule based on equation (5) fails.

7.3.13 The heuristic discussion preceding Theorem 7.11 dealt with $|h'(x_0)|$, not with $h'(x_0)$. Explain how the signs of $f'(x_0)$ and $g'(f(x_0))$ affect the discussion. In particular, how can we modify the discussion to get the correct sign for $h'(x_0)$?

7.3.14 Most calculus texts use a proof of Theorem 7.11 based on the following ideas. Define a function G in the neighborhood V of $f(x_0)$ by

$$G(v) = \begin{cases} [g(v) - g(f(x_0))]/[v - f(x_0)], & \text{if } v \neq f(x_0) \\ g'(f(x_0)), & \text{if } v = f(x_0). \end{cases} \tag{7}$$

(a) Show that G is continuous at $f(x_0)$.

(b) Show that $G(v)(v-f(x_0)) = g(v)-g(f(x_0))$ for every $v \in V$, regardless of whether or not $f(x_0) = v$.

(c) Prove that $\lim_{x\to x_0} \frac{h(x)-h(x_0)}{x-x_0} = g'(f(x_0))f'(x_0)$.

7.3.15 State and prove a theorem that gives a formula for $f'(x_0)$ when

$$f = f_n \circ f_{n-1} \circ \cdots \circ f_2 \circ f_1.$$

(Be sure to state all the hypotheses that you need.)

7.3.16 The table in Figure 7.5 gives the values of two functions f and g at certain points. Calculate $(f \circ g)'(1)$ and $(g \circ f)'(1)$. Is there enough information to calculate $(f \circ g)'(3)$ and/or $(g \circ f)'(3)$? How about $\frac{d}{dx}(f^2)(1)$ and $(f \circ f)'(1)$?

x	$f(x)$	$f'(x)$	$g(x)$	$g'(x)$
1	3	3	2	2
2	4	4	4	0
3	6	1	1	0
4	-1	0	1	1
5	2	5	3	3

Figure 7.5. Values of f and g at several points.

7.3.3 Inverse Functions

Suppose that a function $f : I \to J$ has an inverse. This simply means that there is a function g (called the *inverse* of f) that reverses the mapping: If $f(a) = b$ then $g(b) = a$. We can assume that I and J are intervals. Thus f maps the interval I onto the interval J and the inverse function g then maps J back to I. Not all functions have an inverse, but we are supposing that this one does.

Suppose too that f is differentiable at a point $x_0 \in I$. Then we would expect from geometric considerations that that the inverse function g should be differentiable at the image point $z_0 = f(x_0) \in J$.

This is entirely elementary. The connection between a function f and its inverse g is given by

$$f(g(x)) = x \text{ for all } x \in J$$

or

$$g(f(x)) = x \text{ for all } x \in I.$$

Using the chain rule on the second of these immediately gives

$$g'(f(x))f'(x) = 1$$

and hence we have the connection

$$g'(f(x)) = \frac{1}{f'(x)},$$

which a geometrical argument could also have found.

Example 7.12 Suppose that the exponential function e^x has been developed and that we have proved that it is differentiable for all values of x and we have the usual formula $\frac{d}{dx}e^x = e^x$. Then, provided we can be sure there is an inverse, a formula for the derivative of that inverse can be found. Let $L(x)$ be the inverse function of $f(x) = e^x$. Then, since we know that $f'(x) = f(x)$

$$L'(f(x)) = \frac{1}{f'(x)} = \frac{1}{f(x)}$$

or, replacing $f(x)$ by another letter, say z, we have

$$L'(z) = \frac{1}{z}.$$

This must be valid for every value z in the domain of L, that is, for every value in the range of f. You should recognize the derivative of the function $\ln z$ here. Even so, we would still need to justify the existence of the inverse function before we could properly claim to have proved this formula. ◀

We would like a better way to handle inverse functions than presented here. Our observations here allow us to compute the derivative of an inverse but do not assure us that an inverse will exist. For a theorem that allows us merely to look at the derivative and determine that an inverse exists and has a derivative, see Theorem 7.32.

Exercises

7.3.17 Find a formula for the derivative of the function $\sin^{-1} x$ assuming that the usual formula for
$$\frac{d}{dx} \sin x = \cos x$$
has been found.

7.3.18 Find a formula for the derivative of the function $\tan^{-1} x$ assuming that the usual formula for $\frac{d}{dx} \tan x = \sec^2 x$ has been found.

7.3.19 Give a geometric interpretation of the relationship between the slope of the tangent at a point (x_0, y_0) on the graph of $y = f(x)$ and the slope of the tangent at the point (y_0, x_0) on the graph of $y = g(x)$ where g is the inverse of f.

7.3.20 What facts about the function $f(x) = e^x$ would need to be established in order to claim that there is indeed an inverse function? What is the domain and range of that inverse function?

7.3.4 The Power Rule

The power rule is the formula
$$\frac{d}{dx} x^p = p x^{p-1}$$
which is the basis for many calculus problems. We have already shown (in Exercise 7.3.5) that
$$\frac{d}{dx} x^n = n x^{n-1}$$
for $n = 1, 2, 3, \ldots$ and for every value of x.

This is easy enough to extend to negative integers. Just interpret for $n = 1, 2, 3, \ldots$ and for every value of $x \neq 0$,
$$\frac{d}{dx} x^{-n} = \frac{d}{dx} \frac{1}{x^n}$$
and, using the quotient rule, we find that again the power rule formula is valid for $p = -1, -2, -3, \ldots$ and any value of x other than 0.

The formula also works for $p = 0$ since we interpret x^0 as the constant 1 (although for $x = 0$ we prefer not to make any claims). Is the formula indeed valid for every value of p, not just for integer values?

Example 7.13 We can verify the power rule formula for $p = 1/2$; that is, we prove that

$$\frac{d}{dx}\sqrt{x} = \frac{d}{dx}x^{1/2} = \frac{1}{2}x^{1/2-1} = \frac{1}{2\sqrt{x}}.$$

First we must insist that $x > 0$ otherwise $\sqrt{x}$ and the fraction in our formula would not be defined. Now interpret $\sqrt{x}$ as the inverse of the square function $f(x) = x^2$. Specifically let $f(x) = x^2$ for $x > 0$ and $g(x) = \sqrt{x}$ for $x > 0$ and note that $f(g(x)) = g(f(x)) = x$. Thus

$$\frac{d}{dx}f(g(x)) = \frac{d}{dx}x = 1$$

and so, since $f'(x) = 2x$ and $f'(g(x))g'(x) = 1$ we obtain $2\sqrt{x}g'(x) = 1$ and finally that

$$g'(x) = \frac{1}{2\sqrt{x}}$$

as required if the power rule formula is valid. ◄

Is the power rule

$$\frac{d}{dx}x^p = px^{p-1}$$

valid for all rational values of p? We can handle the case $p = m/n$ for integer m and n by essentially the same methods. We state this as a theorem whose proof is left as an exercise. For irrational p there is also a discussion in the exercises.

Theorem 7.14 *Let $f(x) = x^{\frac{m}{n}}$ for $x > 0$ and integers m, n. Then*

$$f'(x) = \frac{m}{n}x^{\frac{m}{n}-1}.$$

Example 7.15 Every polynomial is differentiable on $\mathbb{R}$ and its derivative can be calculated via term by term differentiation; that is,

$$\frac{d}{dx}(a_0 + a_1x + a_2x^2 + \cdots + a_nx^n) = a_1 + 2a_2x + \cdots + na_nx^{n-1}.$$

This follows from the power rule formula and the rule for sums. Note that the derivative of a polynomial is again a polynomial. ◄

Example 7.16 A rational function is a function $R(x)$ that can be expressed as the quotient of two polynomials,

$$R(x) = \frac{p(x)}{q(x)}.$$

This would be defined at every point at which the denominator $q(x)$ is not equal to zero. Every rational function is differentiable except at those points at which the denominator vanishes. This follows from the previous example, which showed how to differentiate a polynomial, and from the quotient rule. Thus

$$\frac{d}{dx}\left(\frac{p(x)}{q(x)}\right) = \frac{p'(x)q(x) - p(x)q'(x)}{q^2(x)}.$$

Notice that the derivative is another rational function with the same domain since both numerator and denominator are again polynomials. ◀

Exercises

7.3.21 Prove Theorem 7.14.

7.3.22 Show that the power formula is available for all values of p once the formula $\frac{d}{dx}e^x = e^x$ is known.

7.3.23 Let

$$p(x) = a_0 + a_1x + a_2x^2 + \cdots + a_nx^n.$$

Compute the sequence of values $p(0)$, $p'(0)$, $p''(0)$, $p'''(0)$, $\ldots$.

7.3.24 Determine the coefficients of the polynomial

$$p(x) = (1+x)^n = a_0 + a_1x + a_2x^2 + \cdots + a_nx^n$$

by using the formulas that you obtained in Exercise 7.3.23.

7.4 Continuity of the Derivative?

We have already observed (Theorem 7.6) that if a function f is differentiable on an interval I, then f is also continuous on I. This statement should not be confused with the (incorrect) statement that the derivative, f', is continuous.

Example 7.17 Consider the function f defined on $\mathbb{R}$ by

$$f(x) = \begin{cases} x^2 \sin x^{-1}, & \text{if } x \neq 0 \\ 0, & \text{if } x = 0. \end{cases}$$

Since $|\sin x^{-1}| \leq 1$ for all $x \neq 0$, $|f(x)| \leq x^2$ for all $x \in \mathbb{R}$. We can now conclude (e.g., from Example 7.5) that $f'(0) = 0$. For $x \neq 0$, we can calculate, as in elementary calculus, that

$$f'(x) = -\cos x^{-1} + 2x \sin x^{-1}.$$

This function f' is continuous at every point $x_0 \neq 0$. At $x_0 = 0$ it is discontinuous. To see this we need only consider an appropriate sequence

$x_n \to 0$ and see what happens to $f'(x_n)$. For example, try the sequence

$$x_n = \frac{1}{\pi n}.$$

Since

$$\cos\left(\frac{1}{x_n}\right) = \cos(\pi n)$$

and these numbers are alternately $+1$ and -1 it is clear that $f'(x_n)$ cannot converge. Consequently, f' is discontinuous at 0. ◀

Observe that the function f provides an example of a function that is differentiable on all of $\mathbb{R}$, yet f' is discontinuous at a point. It is possible to modify this function to obtain a differentiable function g whose derivative g' is discontinuous at infinitely many points, and even at all the points of the Cantor set (see Exercise 7.4.2).

You might wonder, then, if anything positive could be said about the properties of a derivative f'. It is possible for the derivative of a differentiable function to be discontinuous on a dense set[1]: An example is given later in Section 9.7. We will also show, in Section 7.9, that the function f', while perhaps discontinuous, nonetheless shares one significant property of continuous functions: It has the intermediate value property (Darboux property).

Exercises

7.4.1 Give a simple example of a function f differentiable in a deleted neighborhood of x_0 such that $\lim_{x\to x_0} f'(x)$ does not exist.

✂ **7.4.2** Let P be a Cantor subset of $[0,1]$ (i.e., P is a nonempty, nowhere dense perfect subset of $[0,1]$) and let $\{(a_n, b_n)\}$ be the sequence of intervals complementary to P in $(0,1)$. (See Section 6.5.1.)

(a) On each interval $[a_n, b_n]$ construct a differentiable function such that

$$f_n(a_n) = f_n(b_n) = (f_n')_+(a_n) = (f_n')_-(b_n) = 0,$$

$$\limsup_{x\to a_n^+} f'(x) = \limsup_{x\to b_n^-} f_n'(x) = 1,$$

$$\liminf_{x\to a_n^+} f'(x) = \liminf_{x\to b_n^-} f_n'(x) = -1,$$

and $|f_n(x)| \le (x-a_n)^2(x-b_n)^2$ and $|f_n'(x)|$ is bounded by 1 in each interval $[a_n, b_n]$.

[1]It is not possible for a derivative to be discontinuous at every point. See Corollary 9.40.

(b) Let g be defined on $[0,1]$ by

$$g(x) = \begin{cases} f_n(x), & \text{if } x \in (a_n, b_n), n = 1, 2, \ldots \\ 0, & \text{if } x \in P. \end{cases}$$

Sketch a picture of the graph of g.

(c) Prove that g is differentiable on $[0,1]$.

(d) Prove that $g'(x) = 0$ for each $x \in P$.

(e) Prove that g' is discontinuous at every point of P.

7.5 Local Extrema

We have seen in Section 5.7 that a continuous function defined on a closed interval $[a,b]$ achieves an absolute maximum value and an absolute minimum value on the interval. So there must be points where the maximum and minimum are attained. But how do we go about finding such points?

A familiar process studied in elementary calculus is sometimes useful for locating these extrema when the function is differentiable on (a,b): We look for critical points (i.e., points where the derivative is zero). We begin with the theorem that forms the basis for this process.

Theorem 7.18 *Let f be defined on an interval I. If f has a local extremum at a point x_0 in the interior of I and f is differentiable at x_0, then $f'(x_0) = 0$.*

Proof Suppose f has a local maximum at x_0 in the interior of I, the proof for a local minimum being similar. Then there exists $\delta > 0$ such that

$$[x_0 - \delta, x_0 + \delta] \subset I$$

and

$$f(x) \leq f(x_0)$$

for all $x \in [x_0 - \delta, x_0 + \delta]$. Thus

$$\frac{f(x) - f(x_0)}{x - x_0} \leq 0 \quad \text{for} \quad x \in (x_0, x_0 + \delta) \tag{8}$$

and

$$\frac{f(x) - f(x_0)}{x - x_0} \geq 0 \quad \text{for} \quad x \in (x_0 - \delta, x_0). \tag{9}$$

If $f'(x_0)$ exists, then

$$f'(x_0) = \lim_{x \to x_0+} \frac{f(x) - f(x_0)}{x - x_0} = \lim_{x \to x_0-} \frac{f(x) - f(x_0)}{x - x_0}. \tag{10}$$

By (8), the first of these limits is at most zero; by (9), the second is at least zero. By (10), these limits are equal and are therefore equal to zero. ■

It follows from Theorem 7.18 that a function f that is continuous on $[a, b]$ must achieve its maximum at one (or more) of these types of points:

1. Points $x_0 \in (a, b)$ at which $f'(x_0) = 0$
2. Points $x_0 \in (a, b)$ at which f is not differentiable
3. The points a or b

We leave it to you to provide simple examples of each of these possibilities.

The usual process for locating extrema in elementary calculus thus involves locating points at which f has a zero derivative and comparing the values of f at those points and the points of nondifferentiability (if any) and at the endpoints a and b. In the setting of elementary calculus the situation is usually relatively simple: The function is differentiable, the set on which $f'(x) = 0$ is finite (or contains an interval), and the equation $f'(x) = 0$ is easily solved. Much more complicated situations can occur, of course. The following exercises provide some examples and theorems that indicate just how complicated the set of extrema can be.

Exercises

7.5.1 Give an example of a differentiable function on $\mathbb{R}$ for which $f'(0) = 0$ but 0 is not a local maximum or minimum of f.

7.5.2 Let

$$f(x) = \begin{cases} x^4(2 + \sin x^{-1}), & \text{if } x \neq 0 \\ 0, & \text{if } x = 0. \end{cases}$$

(a) Prove that f is differentiable on $\mathbb{R}$.

(b) Prove that f has an absolute minimum at $x = 0$.

(c) Prove that f' takes on both positive and negative values in every neighborhood of 0.

✂ **7.5.3** Let K be the Cantor set and let $\{(a_k, b_k)\}$ be the sequence of intervals complementary to K in $[0, 1]$. For each k, let $c_k = (a_k + b_k)/2$. Define f on $[0, 1]$ to be zero on K, $1/k$ at c_k, linear and continuous on each of the intervals. (See Figure 7.6.)

(a) Write equations that represent f on the intervals $[a_k, c_k]$ and $[c_k, b_k]$.

(b) Show that f is continuous on $[0, 1]$.

(c) Verify that f has minimum zero, achieved at each $x \in K$.

(d) Verify that f has a local maximum at each of the points c_k.

(e) Modify f to a differentiable function with the same set of extrema.

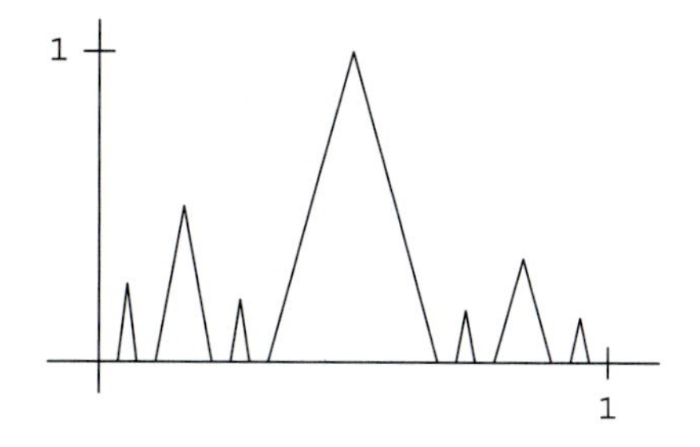

Figure 7.6. Part of the graph of the function in Exercise 7.5.3.

7.5.4 Find all local extrema of the Dirichlet function (see Section 5.2.6) defined on $[0,1]$ by

$$f(x) = \begin{cases} 0, & \text{if } x \text{ is irrational or } x = 0 \\ 1/q, & \text{if } x = p/q,\ \ p,q \in \mathbb{N},\ \ p/q \text{ in lowest terms.} \end{cases}$$

7.5.5 Show that the functions in Exercises 7.5.3 and 7.5.4 have infinitely many maxima, all of them strict. Show that the sets of points at which these functions have a strict maximum is countable.

7.5.6 Prove that if $f:\mathbb{R}\to\mathbb{R}$, then $\{x : f \text{ achieves a strict maximum at } x\}$ is countable.

7.5.7 Let $f:\mathbb{R}\to\mathbb{R}$ have the following property: For each $x \in \mathbb{R}$, f achieves a local maximum (not necessarily strict) at x.

(a) Give an example of such an f whose range is infinite.

(b) Prove that for every such f, the range is countable.

7.5.8 There are continuous functions $f:\mathbb{R}\to\mathbb{R}$, even differentiable functions, that are nowhere monotonic. This means that there is *no interval* on which the function is increasing, decreasing, or constant. For such functions, the set of maxima as well as the set of minima is dense in $\mathbb{R}$. Construction of such functions is given later in Section 13.14.2. Show that such a function f maps its set of extrema onto a dense subset of the range of f.

7.6 Mean Value Theorem

There is a close connection between the values of a function and the values of its derivative. In one direction this is trivial since the derivative is defined in terms of the values of the function. The other direction is more subtle. How does information about the derivative provide us with information about the function? One of the keys to providing that information is the mean value theorem.

Suppose f is continuous on an interval $[a, b]$ and is differentiable on (a, b). Consider a point x in (a, b). For $y \in (a, b)$, $y \neq x$, the difference quotient

$$\frac{f(y) - f(x)}{y - x}$$

represents the slope of the chord determined by the points $(x, f(x))$ and $(y, f(y))$. This slope may or may not be a good approximation to $f'(x)$. If y is sufficiently near x, the approximation will be good; otherwise it may not be. The mean value theorem asserts that somewhere in the interval determined by x and y there will be a point at which the derivative is exactly the slope of the given chord. It is the existence of such a point that provides a connection between the values of the function [in this case the value $(f(y)-f(x))/(y-x)$] and the value of the derivative (in this case the value at *some* point between x and y).

7.6.1 Rolle's Theorem

We begin with a preliminary theorem that provides a special case of the mean value theorem. This derives its name from Michel Rolle (1652–1719) who has little claim to fame other than this. Indeed Rolle's name was only attached to this theorem because he had published it in a book in 1691; the method itself he did not discover. Perhaps his greatest real contribution is the invention of the notation $\sqrt[n]{x}$ for the nth root of x.

Theorem 7.19 (Rolle's Theorem) *Let f be continuous on $[a, b]$ and differentiable on (a, b). If $f(a) = f(b)$ then there exists $c \in (a, b)$ such that $f'(c) = 0$.*

Proof If f is constant on $[a, b]$, then $f'(x) = 0$ for all $x \in (a, b)$, so c can be taken to be any point of (a, b).

Suppose then that f is not constant. Because f is continuous on the compact interval $[a, b]$, f achieves a maximum value M and a minimum value m on $[a, b]$ (Theorem 5.49). Because f is not constant, one of the values M or m is different from $f(a)$ and $f(b)$, say $M > f(a)$. Choose $c \in (a, b)$ such that $f(c) = M$. By Theorem 7.18, $f'(c) = 0$. Since $M > f(a) = f(b)$, $c \neq a$ and $c \neq b$, so $c \in (a, b)$. ■

Observe that Rolle's theorem asserts that under our hypotheses, there is a point at which the tangent to the graph of the function is horizontal, and therefore has the same slope as the chord determined by the points $(a, f(a))$ and $(b, f(b))$. (See Figure 7.7.)

There may, of course, be many such points; Rolle's theorem just guarantees the existence of at least one such point. Observe also that we did not

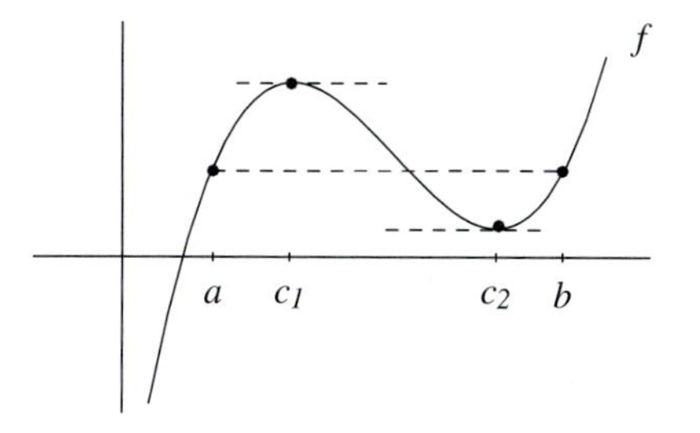

Figure 7.7. Rolle's theorem [note that $f(a) = f(b)$].

require that f be differentiable at the endpoints a and b. The theorem applies to such functions as $f(x) = x \sin x^{-1}$, $f(0) = 0$, on the interval $[0, 1/\pi]$. This function is not differentiable at zero, but it does have an infinite number of points between 0 and $1/\pi$ where the derivative is zero.

Exercises

7.6.1 Apply Rolle's theorem to the function $f(x) = \sqrt{1 - x^2}$ on $[-1, 1]$. Observe that f fails to be differentiable at the endpoints of the interval.

7.6.2 Use Rolle's theorem to explain why the cubic equation

$$x^3 + \alpha x^2 + \beta = 0$$

cannot have more than one solution whenever $\alpha > 0$.

7.6.3 If the nth-degree equation

$$p(x) = a_0 + a_1 x + a_2 x^2 + \cdots + a_n x^n = 0$$

has n distinct real roots, then how many distinct real roots does the $(n-1)$st degree equation $p'(x) = 0$ have?

7.6.4 Suppose that $f'(x) > c > 0$ for all $x \in [0, \infty)$. Show that $\lim_{x\to\infty} f(x) = \infty$.

7.6.5 Suppose that $f : \mathbb{R} \to \mathbb{R}$ and both f' and f'' exist everywhere. Show that if f has three zeros, then there must be some point ξ so that $f''(\xi) = 0$.

7.6.6 Let f be continuous on an interval $[a, b]$ and differentiable on (a, b) with a derivative that never is zero. Show that f maps $[a, b]$ one-to-one onto some other interval.

7.6.7 Let f be continuous on an interval $[a, b]$ and twice differentiable on (a, b) with a second derivative that never is zero. Show that f maps $[a, b]$ two-one onto some other interval; that is, there are at most two points in $[a, b]$ mapping into any one value in the range of f.

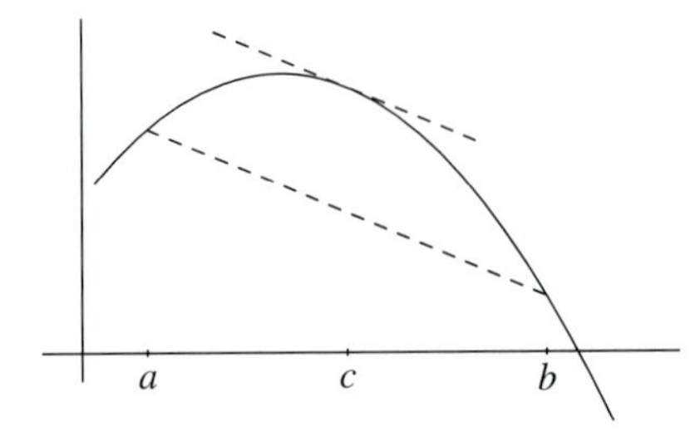

Figure 7.8. Mean value theorem [$f'(c)$ is slope of the chord].

7.6.2 Mean Value Theorem

If we drop the requirement in Rolle's theorem that $f(a) = f(b)$, we now obtain the result that there is a $c \in (a, b)$ such that

$$f'(c) = \frac{f(b) - f(a)}{b - a}.$$

Geometrically, this states that there exists a point $c \in (a, b)$ for which the tangent to the graph of the function at $(c, f(c))$ is parallel to the chord determined by the points $(a, f(a))$ and $(b, f(b))$. (See Figure 7.8.)

This is the mean value theorem, also known as the law of the mean or the first mean value theorem (because there are other mean value theorems).

Theorem 7.20 (Mean Value Theorem) *Suppose that f is a continuous function on the closed interval $[a, b]$ and differentiable on (a, b). Then there exists $c \in (a, b)$ such that*

$$f'(c) = \frac{f(b) - f(a)}{b - a}.$$

Proof We prove this theorem by subtracting from f a function whose graph is the straight line determined by the chord in question and then applying Rolle's theorem. Let

$$L(x) = f(a) + \frac{f(b) - f(a)}{b - a}(x - a).$$

We see that $L(a) = f(a)$ and $L(b) = f(b)$. Now let

$$g(x) = f(x) - L(x). \tag{11}$$

Then g is continuous on $[a, b]$, differentiable on (a, b), and satisfies the condition $g(a) = g(b) = 0$.

By Rolle's theorem, there exists $c \in (a,b)$ such that $g'(c) = 0$. Differentiating (11), we see that $f'(c) = L'(c)$. But

$$L'(c) = \frac{f(b) - f(a)}{b - a},$$

so

$$f'(c) = \frac{f(b) - f(a)}{b - a},$$

as was to be proved. ∎

Rolle's theorem and the mean value theorem were easy to prove. The proofs relied on the geometric content of the theorems. We suggest that you take the time to understand the geometric interpretation of these theorems.

Exercises

7.6.8 A function f is said to satisfy a *Lipschitz condition* on an interval $[a,b]$ if

$$|f(x) - f(y)| \leq M|x - y|$$

for all x, y in the interval. Show that if f is assumed to be continuous on $[a,b]$ and differentiable on (a,b) then this condition is equivalent to the derivative f' being bounded on (a,b).

7.6.9 Suppose f satisfies the hypotheses of the mean value theorem on $[a,b]$. Let S be the set of all slopes of chords determined by pairs of points on the graph of f and let

$$D = \{f'(x) : x \in (a,b)\}.$$

(a) Prove that $S \subset D$.

(b) Give an example to show that D can contain numbers not in S.

7.6.10 Interpreting the slope of a chord as an average rate of change and the derivative as an instantaneous rate of change, what does the mean value theorem say? If a car travels 100 miles in 2 hours, and the position $s(t)$ of the car at time t satisfies the hypotheses of the mean value theorem, can we be sure that there is at least one instant at which the velocity is 50 mph?

7.6.11 Give an example to show that the conclusion of the mean value theorem can fail if we drop the requirement that f be differentiable at every point in (a,b). Give an example to show that the conclusion can fail if we drop the requirement of continuity at the endpoints of the interval.

7.6.12 Suppose that f is differentiable on $[0,\infty)$ and that

$$\lim_{x\to\infty} f'(x) = C.$$

Determine

$$\lim_{x\to\infty} [f(x+a) - f(x)].$$

7.6.13 Suppose that f is continuous on $[a,b]$ and differentiable on (a,b). If

$$\lim_{x\to a+} f'(x) = C$$

what can you conclude about the right-hand derivative of f at a?

7.6.14 Suppose that f is continuous and that

$$\lim_{x\to x_0} f'(x)$$

exists. What can you conclude about the differentiability of f? What can you conclude about the continuity of f'?

7.6.15 Let $f : [0,\infty) \to \mathbb{R}$ so that f' is decreasing and positive. Show that the series

$$\sum_{i=1}^{\infty} f'(i)$$

is convergent if and only if f is bounded.

7.6.16 Prove a second-order version of the mean value theorem.

Let f be continuous on $[a,b]$ and twice differentiable on (a,b). Then there exists $c \in (a,b)$ such that

$$f(b) = f(a) + (b-a)f'(a) + (b-a)^2\frac{f''(c)}{2!}.$$

7.6.17 Determine all functions $f : \mathbb{R} \to \mathbb{R}$ that have the property that

$$f'\left(\frac{x+y}{2}\right) = \frac{f(x)-f(y)}{x-y}$$

for every $x \neq y$.

7.6.18 A function is said to be *smooth* at a point x if

$$\lim_{h\to 0} \frac{f(x+h)+f(x-h)-2f(x)}{h^2} = 0.$$

Show that a smooth function need not be continuous. Show that if f'' is continuous at x, then f is smooth at x.

7.6.3 Cauchy's Mean Value Theorem

✂ *Enrichment*

We can generalize the mean value theorem to curves given parametrically. Suppose f and g are continuous on $[a,b]$ and differentiable on (a,b). Consider the curve given parametrically by

$$x = g(t)\,, \qquad y = f(t) \qquad (t \in [a,b]).$$

As t varies over the interval $[a,b]$, the point (x,y) traces out a curve C joining the points $(g(a), f(a))$ and $(g(b), f(b))$. If $g(a) \neq g(b)$, the slope of the chord determined by these points is

$$\frac{f(b)-f(a)}{g(b)-g(a)}.$$

Cauchy's form of the mean value theorem asserts that there is a point (x, y) on C at which the tangent is parallel to the chord in question. We state and prove this theorem.

Theorem 7.21 (Cauchy Mean Value Theorem) *Let f and g be continuous on $[a, b]$ and differentiable on (a, b). Then there exists c such that*

$$[f(b)-f(a)]g'(c) = [g(b)-g(a)]f'(c). \tag{12}$$

Proof Let

$$\phi(x) = [f(b)-f(a)]g(x) - [g(b)-g(a)]f(x).$$

Then ϕ is continuous on $[a, b]$ and differentiable on (a, b). Furthermore,

$$\phi(a) = f(b)g(a) - f(a)g(b) = \phi(b).$$

By Rolle's theorem, there exists $c \in (a, b)$ for which $\phi'(c) = 0$. It is clear that this point c satisfies (12). ■

Exercises

7.6.19 Use Cauchy's mean value theorem to prove any simple version of L'Hôpital's rule that you can remember from calculus.

7.6.20 Show that the conclusion of Cauchy's mean value can be put into determinant form as

$$\begin{vmatrix} f(a) & g(a) & 1 \\ f(b) & g(b) & 1 \\ f'(c) & g'(c) & 0 \end{vmatrix} = 0.$$

7.6.21 Formulate and prove a generalized version of Cauchy's mean value whose conclusion is the existence of a point c such that

$$\begin{vmatrix} f(a) & g(a) & h(a) \\ f(b) & g(b) & h(b) \\ f'(c) & g'(c) & h'(c) \end{vmatrix} = 0.$$

7.7 Monotonicity

In elementary calculus one learns that if $f' \geq 0$ on an interval I, then f is nondecreasing on I. We use this and related results for a variety of purposes: sketching graphs of functions, locating extrema, etc. In this section we take a closer look at what's involved. We recall some definitions.

Definition 7.22 Let f be real valued on an interval I.

1. If $f(x_1) \leq f(x_2)$ whenever x_1 and x_2 are points in I with $x_1 < x_2$, we say f is *nondecreasing* on I.

2. If the strict inequality $f(x_1) < f(x_2)$ holds, we say f is *increasing*.

A similar definition was given for *nonincreasing* and *decreasing* functions.

Note. Some authors prefer the terms "increasing" and "strictly increasing" for what we would call nondecreasing and increasing. This has the unfortunate result that constant functions are then considered to be both increasing and decreasing. According to our definition we must say that they are both nondecreasing and non-increasing, which sounds more plausible—if something stays constant it is neither going up nor going down). The disadvantage of our usage is the discomfort you may at first feel in using the terms (which disappears with practice). It is always safe to say "strictly increasing" for increasing even though it is redundant according to the definition.

By a monotonic function we mean a function that is increasing, decreasing, nondecreasing, or nonincreasing.

The theorems involving monotonicity of functions that one encounters in elementary calculus usually are stated for differentiable functions. But a monotonic function need not be differentiable, or even continuous.

Example 7.23 For example, if

$$f(x) = \begin{cases} x, & \text{for } x < 0 \\ x+1, & \text{for } x \geq 0, \end{cases}$$

then f is increasing on $\mathbb{R}$, but is not continuous at $x = 0$. (For more on discontinuities of monotonic functions, see Section 5.9.2.) ◄

Let us now address the role of the derivative in the study of monotonicity. We prove a familiar theorem that is the basis for many calculus applications. Note that the proof is an easy consequence of the mean value theorem.

Theorem 7.24 *Let f be differentiable on an interval I.*

(i) *If $f'(x) \geq 0$ for all $x \in I$, then f is nondecreasing on I.*

(ii) *If $f'(x) > 0$ for all $x \in I$, then f is increasing on I.*

(iii) *If $f'(x) \leq 0$ for all $x \in I$, then f is nonincreasing on I.*

(iv) *If $f'(x) < 0$ for all $x \in I$, then f is decreasing on I.*

(v) *If $f'(x) = 0$ for all $x \in I$, then f is constant on I.*

Proof To prove (i), let $x_1, x_2 \in I$ with $x_1 < x_2$. By the mean value theorem (7.20) there exists $c \in (x_1, x_2)$ such that

$$f(x_2) - f(x_1) = f'(c)(x_2 - x_1).$$

If $f'(c) \geq 0$, then $f(x_2) \geq f(x_1)$. Thus, if $f'(x) \geq 0$ for all $x \in I$, f is nondecreasing on I.

Parts (ii), (iii) and (iv) have similar arguments, and (v) follows immediately from parts (i) and (iii). ■

Exercises

7.7.1 Establish the inequality $e^x < \frac{1}{1-x}$ for all $x < 1$.

7.7.2 Suppose that f and g are differentiable functions such that $f' = g$ and $g' = -f$. Show that there exists a number C with the property that

$$[f(x)]^2 + [g(x)]^2 = C$$

for all x.

7.7.3 Suppose f is continuous on (a, c) and $a < b < c$. Suppose also that f is differentiable on (a, b) and on (b, c). Prove that if $f' < 0$ on (a, b) and $f' > 0$ on (b, c), then f has a minimum at b.

7.7.4 The hypotheses of Theorem 7.24 require that f be differentiable on all of the interval I. You might think that a positive derivative at a single point also implies that the function is increasing, at least in a neighborhood of that point. This is not true. Consider the function

$$f(x) = \begin{cases} x/2 + x^2 \sin x^{-1}, & \text{if } x \neq 0 \\ 0, & \text{if } x = 0. \end{cases}$$

(a) Show that the function $g(x) = x^2 \sin x^{-1}$ ($g(0) = 0$) is everywhere differentiable and that $g'(0) = 0$.

(b) Show that g' is discontinuous at $x = 0$ and that g' takes on values close to ± 1 arbitrarily near 0.

(c) Show that f' takes on both positive and negative values in every neighborhood of zero.

(d) Show that $f'(0) = \frac{1}{2} > 0$ but that f is not increasing in any neighborhood of zero.

(e) Prove that if a function F is differentiable on a neighborhood of x_0 with $F'(x_0) > 0$ and F' is *continuous* at x_0, then F is increasing on some neighborhood of x_0.

(f) Why does the example $f(x)$ given here not contradict part (e)?

7.7.5 Let f be differentiable on $[0, \infty)$ and suppose that $f(0) = 0$ and that the derivative f' is an increasing function on $[0, \infty)$. Show that

$$\frac{f(x)}{x} < \frac{f(y)}{y}$$

for all $0 < x < y$.

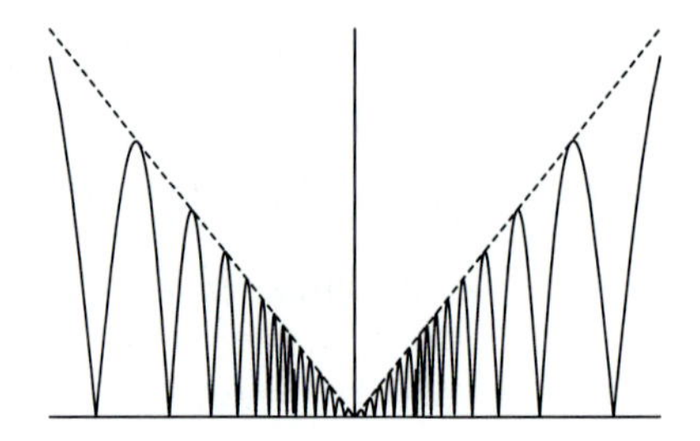

Figure 7.9. Graph of $f(x) = |x \cos x^{-1}|$.

7.7.6 Suppose that f, $g : \mathbb{R} \to \mathbb{R}$ and both have continuous derivatives and the determinant

$$\phi(x) = \left| \begin{array}{cc} f(x) & g(x)) \\ f'(x) & g'(x) \end{array} \right|$$

is never zero. Show that between any two zeros of f there must be a zero of g.

7.8 Dini Derivates

Advanced

We observed in Example 7.4 that the function $f(x) = |x|$ does not have a derivative at the point $x = 0$ but does have the one-sided derivatives $f'_+(0) = 1$ and $f'_-(0) = -1$. It is not difficult to construct continuous functions that don't have even one-sided derivatives at a point.

Example 7.25 Consider the function

$$f(x) = \begin{cases} |x| \left|\cos x^{-1}\right|, & \text{if } x \neq 0 \\ 0, & \text{if } x = 0. \end{cases}$$

(See Figure 7.9). Since $|\cos x^{-1}| \leq 1$ for all $x \neq 0$,

$$\lim_{x \to 0} f(x) = 0 = f(0)$$

so f is continuous at $x = 0$. It is clear that f is continuous at all other points in $\mathbb{R}$, so f is a continuous function.

The oscillatory behavior of f is such that the sets

$$\left\{x : \left|\cos x^{-1}\right| = 1\right\} \quad \text{and} \quad \left\{x : \left|\cos x^{-1}\right| = 0\right\}$$

both have zero as a two-sided limit point. Thus each of the sets

$$\{x : f(x) = |x|\} \quad \text{and} \quad \{x : f(x) = 0\}$$

has zero as two-sided limit point. Inspection of the difference quotient reveals that

$$\limsup_{x\to 0+} \frac{f(x)-f(0)}{x-0} = 1, \text{ while } \liminf_{x\to 0+} \frac{f(x)-f(0)}{x-0} = 0,$$

so f'_+ does not exist at $x = 0$. Similarly, $f'_-(0)$ does not exist. The limits that are required to exist for f to have a derivative, or a one-sided derivative, don't exist at $x = 0$. ◀

Example 7.26 A function defined on an interval I may fail to have a derivative, even a one-sided derivative, at every point. Let

$$g(x) = \begin{cases} 0, & \text{if } x \text{ is rational,} \\ 1, & \text{if } x \text{ is irrational.} \end{cases}$$

Since g is everywhere discontinuous on both sides, g has no derivative and no one-sided derivative at any point. ◀

There are, also, *continuous* functions that fail to have a one-sided derivative, finite or infinite, at even a single point. Such functions are difficult to construct, the first construction having been given by Besicovitch in 1925.

Now the derivative, when it exists, plays an important role in analysis, and it is useful to have a substitute when it doesn't exist. Many good substitutes have been developed for certain situations. Perhaps the simplest such substitutes are the Dini derivates. These exist at every point for every function defined on an open interval. They are named after the Italian mathematician Ulisse Dini (1845–1918).

Definition 7.27 Let f be real valued in a neighborhood of x_0. We define the four *Dini derivates* of f at x_0 by

1. [Upper right Dini derivate]

$$D^+f(x_0) = \limsup_{x\to x_0+} \frac{f(x)-f(x_0)}{x-x_0}$$

2. [Lower right Dini derivate]

$$D_+f(x_0) = \liminf_{x\to x_0+} \frac{f(x)-f(x_0)}{x-x_0}$$

3. [Upper left Dini derivate]

$$D^-f(x_0) = \limsup_{x\to x_0-} \frac{f(x)-f(x_0)}{x-x_0}$$

4. [Lower left Dini derivate]

$$D_-f(x_0) = \liminf_{x \to x_0-} \frac{f(x) - f(x_0)}{x - x_0}.$$

Example 7.28 For the function $f(x) = |x| \left|\cos x^{-1}\right|$, $f(0) = 0$, we calculate that

$$D^+f(0) = 1,\ D_+f(0) = 0,\ D^-f(0) = 0,\ D_-f(0) = -1.$$

Elsewhere $f'(x)$ exists and all four Dini derivatives have that value. ◀

Example 7.29 The function

$$g(x) = \begin{cases} 0, & \text{if } x \text{ is rational,} \\ 1, & \text{if } x \text{ is irrational.} \end{cases}$$

has at every rational x

$$D^+g(x) = 0, D_+g(x) = -\infty, D^-g(x) = \infty, D_-g(x) = 0.$$

For x irrational there are similar values for the Dini derivates (see Exercise 7.8.1a). ◀

It is easy to check that a function f has a derivative at a point x_0 if and only if all four Dini derivates are equal at that point, and a one-sided derivative at x_0 if the two Dini derivates from that side are equal (see Exercise 7.8.2).

We end this section with an illustration of the way in which knowledge about a Dini derivate can substitute for that of the ordinary derivative. We prove a theorem about monotonicity. You are familiar with the fact that if f is differentiable on an interval $[a, b]$ and $f'(x) > 0$ for all $x \in [a, b]$, then f is an increasing function on $[a, b]$. (We provided a formal proof in Section 7.7.)

Here is a generalization of that theorem.

Theorem 7.30 *Let f be continuous on $[a, b]$. If $D^+f(x) > 0$ at each point $x \in [a, b)$, then f is increasing on $[a, b]$.*

Proof Let us first show that f is nondecreasing on $[a, b]$. We prove this by contradiction. If f fails to be nondecreasing on $[a, b]$, there exist points c and d such that $a \leq c < d \leq b$ and $f(c) > f(d)$. Let y be any point in the interval $(f(d), f(c))$.

Since f is continuous on $[a, b]$, it possesses the intermediate value property. Thus from Theorem 5.52 [or more precisely from the version of that theorem given as Exercise 5.8.8(a)] there exists a point $t \in (c, d)$ such that $f(t) = y$. Thus the set $\{x : f(x) = y)\} \cap [c, d]$ is nonempty. Let

$x_0 = \sup\{x : c \le x \le d \text{ and } f(x) = y\}$. Now, $f(d) < y$ and f is continuous, from which it follows that $x_0 < d$. Thus $f(x) < y$ for $x \in (x_0, d]$. Furthermore, the set $\{x : f(x) = y\}$ is closed (because f is continuous), so $f(x_0) = y$.

But this implies that $D^+f(x_0) \le 0$.This contradicts our hypothesis that $D^+f(x) > 0$ for all $x \in [a, b)$. This contradiction completes the proof that f is nondecreasing.

Now we wish to show that it is in fact increasing. If not, then there must be some subinterval in which the function is constant. But at every point interior to that interval we would have $f'(x) = 0$ and so it would be impossible for $D^+f(x) > 0$ at such points. ■

Exercises

7.8.1 Calculate the four Dini derivates for each of the following functions at the given point.

(a)
$$g(x) = \begin{cases} 1, & \text{if } x \text{ is rational} \\ 0, & \text{if } x \text{ is irrational} \end{cases}$$
for $x = \pi$.

(b) $h(x) = x \sin x^{-1}$ $(h(0) = 0)$ at $x = 0$

(c) $f(x) = x \sin x^{-1}$ $(f(0) = 5)$ at $x = 0$

(d)
$$u(x) = \begin{cases} x^2, & \text{if } x \text{ is rational} \\ 0, & \text{if } x \text{ is irrational} \end{cases}$$
at $x = 0$ and at $x = 1$

7.8.2 Prove that f has a derivative at x_0 if and only if
$$D^+f(x_0) = D_+f(x_0) = D^-f(x_0) = D_-f(x_0).$$
In that case, $f'(x_0)$ is the common value of the Dini derivates at x_0. (We assume that f is defined in a neighborhood of x_0.)

7.8.3 **(Derived Numbers)** The Dini derivates are sometimes called "extreme unilateral derived numbers." Let $\lambda \in [-\infty, \infty]$. Then λ is a *derived number* for f at x_0 if there exists a sequence $\{x_k\}$ with $\lim_{k\to\infty} x_k = x_0$ such that
$$\lambda = \lim_{k\to\infty} \frac{f(x_k) - f(x_0)}{x_k - x_0}.$$

(a) For the function $f(x) = |x \cos x^{-1}|$, $f(0) = 0$, show that every number in the interval $[-1, 1]$ is a derived number for f at $x = 0$. Show that the two extreme derived numbers from the right are 0 and 1, and the two from the left are -1 and 0.

(b) Show that a function has a derivative at a point if and only if all derived numbers at that point coincide.

(c) Let $f : \mathbb{R} \to \mathbb{R}$ and let $x_0 \in \mathbb{R}$. Prove that if f is continuous on $\mathbb{R}$, then the set of derived numbers of f at x_0 consists of either one or two closed intervals (that might be degenerate or unbounded). Give examples to illustrate the various possibilities.

7.8.4 Let $f, g : \mathbb{R} \to \mathbb{R}$.

(a) Prove that $D^+(f+g)(x) \leq D^+f(x) + D^+g(x)$.

(b) Give an example to illustrate that the inequality in (a) can be strict.

(c) State and prove the analogue of part (a) for the lower right derivate D_+f.

7.8.5 Generalize Theorem 7.18 to the following:

If f achieves a local maximum at x_0, then $D^+f(x_0) \leq 0$ and $D_-f(x_0) \geq 0$.

Illustrate the result with a function that is not differentiable at x_0.

7.8.6 Prove a variant of Theorem 7.30 that assumes that, for all x in $[a, b)$ except for x in some countable set, the Dini derivate $D^+f(x) > 0$.

7.8.7 Prove a variant of Theorem 7.30: If f is continuous and $D^+f(x) \geq 0$ for all $x \in [a, b)$, then f is nondecreasing on $[a, b]$.

7.8.8 Prove yet another (more subtle) variant of Theorem 7.30: If f is continuous and $D^+f(x) > 0$ for all $x \in [a, b)$ except for x in some countable set, then f is increasing on $[a, b]$.

7.8.9 Prove that no continuous function can have $D^+f(x) = \infty$ for all $x \in \mathbb{R}$. Give an example of a function $f : \mathbb{R} \to \mathbb{R}$ such that $D^+f(x) = \infty$ for all $x \in \mathbb{R}$.

7.8.10 Show that the set

$$\{x : D^+f(x) < D_-f(x)\}$$

cannot be uncountable. Give an example of a function f such that $D^+f < D_-f$ on an infinite set.

7.9 The Darboux Property of the Derivative

Suppose f is differentiable on an interval $[a, b]$. We argued in the proof of Rolle's theorem (7.19) that if $f(a) = f(b)$, then there exists a point $c \in (a, b)$ at which f achieves an extremum. At this point c we have $f'(c) = 0$.

A different hypothesis can lead to the same conclusion. Suppose f is differentiable on $[a, b]$ and $f'(a) < 0 < f'(b)$ (or $f'(b) < 0 < f'(a)$). Once again, the extreme value f achieves must occur at a point c in the *interior* of $[a, b]$, (why?), and at this point we must have $f'(c) = 0$. This observation is a special case of the following theorem first proved by Darboux in 1875.

Theorem 7.31 *Let f be differentiable on an interval I. Suppose $a, b \in I$, $a < b$, and $f'(a) \neq f'(b)$. Let γ be any number between $f'(a)$ and $f'(b)$. Then there exists $c \in (a, b)$ such that $f'(c) = \gamma$.*

Proof Let $g(x) = f(x) - \gamma x$. If $f'(a) < \gamma < f'(b)$, then $g'(a) = f'(a) - \gamma < 0$ and $g'(b) = f'(b) - \gamma > 0$. The discussion preceding the statement of the theorem shows that there exists $c \in (a, b)$ such that $g'(c) = 0$. For this c we have

$$f'(c) = g'(c) + \gamma = \gamma,$$

completing the proof for the case $f'(a) < f'(b)$.

The proof when $f'(a) > f'(b)$ is similar. ■

You might have noted that Theorem 7.31 is exactly the statement that the derivative of a differentiable function has the Darboux property (i.e., the intermediate value property) that we established for continuous functions in Section 5.8. The derivative f' of a differentiable function f need not be continuous, of course. The result does imply, however, that f' cannot have jump discontinuities and cannot have removable discontinuities.

Both the mean value theorem and Theorem 7.31 give information about the range of the derivative f' of a differentiable function f. The mean value theorem implies that the range of f' includes all slopes of chords determined by the graph of f on the interval of definition of f. Theorem 7.31 tells us that this range is actually an interval. This interval may be unbounded and, if bounded, may or may not contain its endpoints. (See Exercise 7.9.1.)

Derivative of an Inverse Function Theorem 7.31 allows us to establish a familiar theorem about differentiating inverse functions.

Theorem 7.32 *Suppose f is differentiable on an interval I and for each $x \in I$, $f'(x) \neq 0$. Then*

(i) *f is one-to-one on I,*

(ii) *f^{-1} is differentiable on $J = f(I)$,*

(iii) $(f^{-1})'(f(x)) = \dfrac{1}{f'(x)}$ *for all $x \in I$.*

Proof By Theorem 7.31 either $f'(x) > 0$ for all $x \in I$ or $f'(x) < 0$ for all $x \in I$. In either case, f is either increasing or decreasing on I, and is thus one-to-one, establishing (i).

To verify (ii) and (iii), observe first that f^{-1} is continuous, since f is continuous and strictly monotonic (see Exercise 5.9.16). Let $y_0 \in J$ and let $x_0 = f^{-1}(y_0)$. We wish to show that $(f^{-1})'(y_0)$ exists and has value $1/(f'(x_0))$. For $x \in I$, write $y = f(x)$, so $x = f^{-1}(y)$.

Consider the difference quotient

$$\frac{f^{-1}(y) - f^{-1}(y_0)}{y - y_0} = \frac{x - x_0}{f(x) - f(x_0)}.$$

As $y \to y_0$, $x \to x_0$, because the function f^{-1} is continuous. Thus

$$\lim_{y \to y_0} \frac{f^{-1}(y) - f^{-1}(y_0)}{y - y_0} = \lim_{x \to x_0} \frac{1}{\left(\frac{f(x)-f(x_0)}{x-x_0}\right)} = \frac{1}{f'(x_0)}.$$

■

Exercises

7.9.1 Let f be differentiable on $[a, b]$ and let $\mathcal{R}(f')$ denote the range of f' on $[a, b]$. Give examples to illustrate that $\mathcal{R}(f')$ can be

(a) a closed interval

(b) an open interval

(c) a half-open interval

(d) an unbounded interval

7.9.2 Give an example of a differentiable function f such that

$$f'(x_0) \neq \lim_{x \to x_0} f'(x).$$

Show that if f is defined in a neighborhood of x_0 and $\lim_{x \to x_0} f'(x)$ exists and is finite, then f is differentiable at x_0 and f' is continuous at x_0.

7.9.3 Most classes of functions we have encountered are closed under the operations of addition and multiplication (e.g., polynomials, continuous functions, differentiable functions). The class of derivatives is closed under addition, but behaves badly with respect to multiplication. Consider, for example, the pair of functions F and G defined on $\mathbb{R}$ by

$$\begin{aligned} F(x) &= x^2 \sin \frac{1}{x^3}, \ (F(0) = 0), \text{ and} \\ G(x) &= x^2 \cos \frac{1}{x^3}, \ (G(0) = 0). \end{aligned}$$

Verify each of the following statements:

(a) F and G are differentiable on $\mathbb{R}$.

(b) The functions FG' and GF' are bounded functions.

(c) $F(x)G'(x) - F'(x)G(x) = \begin{cases} 3, & \text{if } x \neq 0 \\ 0, & \text{if } x = 0. \end{cases}$

(d) At least one of the functions FG' or GF' must fail to be a derivative.

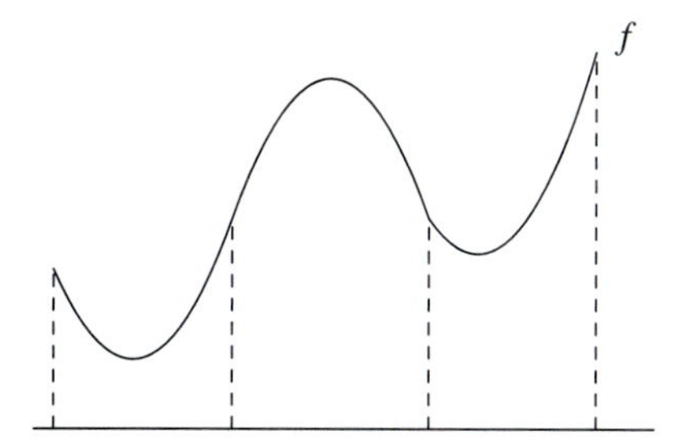

Figure 7.10. Concave up/down/up.

Thus, even the product of a differentiable function F with a derivative G' need not be a derivative.

7.9.4 Show, in contrast to Exercise 7.9.3, that if a function f has a continuous derivative on $\mathbb{R}$ and g is differentiable, then fg' is a derivative.

7.9.5 Let f be a differentiable function on an interval $[a, b]$. Show that f' is continuous if and only if the set

$$E_\alpha = \{x : f'(x) = \alpha\}$$

is closed for each real number α.

7.9.6 Let $f : [0, 1] \to \mathbb{R}$ be a continuous function that is differentiable on $(0, 1)$ and with $f(0) = 0$ and $f(1) = 1$. Show there must exist distinct numbers ξ_1 and ξ_2 in that interval such that

$$f'(\xi_1)f'(\xi_2) = 1.$$

7.9.7 Prove or disprove that if $f : \mathbb{R} \to \mathbb{R}$ is differentiable and monotonic, then f' must be continuous on $\mathbb{R}$.

7.10 Convexity

In elementary calculus one studies functions that are concave-up or concave-down on an interval. A knowledge of the intervals on which a function is concave-up or concave-down is useful for such purposes as sketching the graph of the equation $y = f(x)$ and studying extrema of the function (Fig. 7.10).

In the setting of elementary calculus the functions usually have second derivatives on the intervals involved. In that setting we define a function as being concave-up on an interval I if $f'' \geq 0$ on I, and concave-down if $f'' \leq 0$ on I. Definitions involving the first derivative, but not the second, can also be given: f is concave-up on I if f' is increasing on I, concave-down if f' is decreasing on I. Equivalently, f is concave-up if the graph of f lies "above" (more precisely "not below") each of its tangent lines, concave-down if the graph lies below (not above) each of its tangent lines.

The geometric properties we wish to capture when we say a function is concave-up or concave-down do not depend on differentiability properties. The condition is that the graph should lie below (or above) all its chords. The following definitions make this concept precise. We shall follow the common practice of using the terms "convex" and "concave" in place of the terms "concave-up" and "concave-down."

Definition 7.33 Let f be defined on an interval I. If for all $x_1, x_2 \in I$ and $\alpha \in [0,1]$ the inequality

$$f(\alpha x_1 + (1-\alpha)x_2) \leq \alpha f(x_1) + (1-\alpha) f(x_2) \tag{13}$$

is satisfied, we say that f is *convex* on I. If the reverse inequality in (13) applies, we say that f is *concave* on I. If the inequalities are strict for all $\alpha \in (0,1)$ we say f is *strictly convex* or *strictly concave* on I.

For example, the function $f(x) = |x|$ is convex, but not strictly convex on $\mathbb{R}$. Strict convexity implies that the graph of f has no line segments in it. Note that the function $f(x) = |x|$ is not differentiable at $x = 0$.

The geometric condition defining convexity does imply a great deal of regularity of a function. Our first objective is to address this issue. We begin with some simple geometric considerations.

Suppose f is convex on an open interval I. Let x_1 and x_2 be points in I with $x_1 < x_2$. The chord determined by the points $(x_1, f(x_1))$ and $(x_2, f(x_2))$ defines a linear function M on $[x_1, x_2]$: If $x = \alpha x_1 + (1-\alpha)x_2$, then

$$M(x) = \alpha f(x_1) + (1-\alpha) f(x_2).$$

The definition of "convex" states that

$$f(x) \leq M(x)$$

for all $x \in [x_1, x_2]$ and that

$$M(x_1) = f(x_1) \quad \text{and} \quad M(x_2) = f(x_2).$$

Now let $z \in (x_1, x_2)$ Then

$$\frac{f(z) - f(x_1)}{z - x_1} \leq \frac{M(z) - M(x_1)}{z - x_1} = \frac{M(x_2) - M(z)}{x_2 - z} \leq \frac{f(x_2) - f(z)}{x_2 - z} \tag{14}$$

(Fig. 7.11).

Thus, the chord determined by f and the points x_1 and x_2 has a slope between the slopes of the chord determined by x_1 and z and the chord determined by z and x_2.

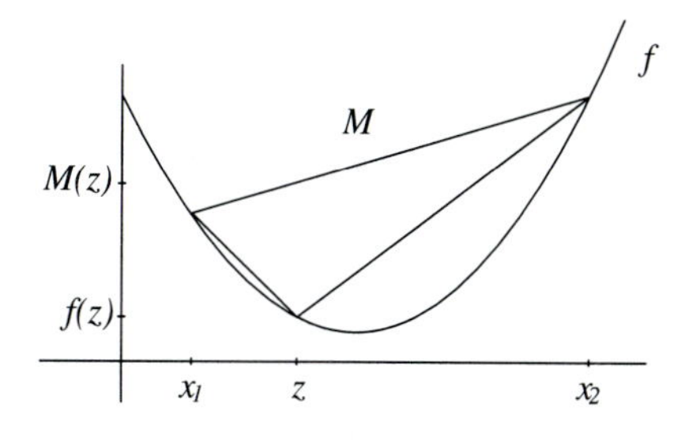

Figure 7.11. Comparison of the three slopes in the inequalities (14).

The inequalities (14) have a number of useful consequences:

1. For fixed $x \in I$,
$$(f(x+h) - f(x))/h$$
is a nondecreasing function of h on some interval $(0, \delta)$ Thus
$$\lim_{h \to 0+} \frac{f(x+h) - f(x)}{h} = \inf_{h>0} \frac{f(x+h) - f(x)}{h}$$
exists or possibly is $-\infty$. That it is in fact finite can be shown by using (14) again to get a finite lower bound, since
$$(f(x') - f(x))/(x' - x) \leq (f(x+h) - f(x))/h$$
for any $x' \in I$ with $x' < x$. Thus f has a right-hand derivative $f'_+(x)$ at x. Similarly, f has a finite left-hand derivative at x.

2. If $x, y \in I$ and $x < y$, then
$$f'_+(x) \leq f'_+(y).$$
From observation 1 we infer that
$$(f(x+h) - f(x))/h \leq (f(y+h) - f(y))/h$$
whenever $h > 0$ and $x + h, y + h$ are in I. Thus f'_+ is a nondecreasing function. Similarly, f'_- is a nondecreasing function.

3. It is also clear from (14) that
$$f'_-(x) \leq f'_+(x)$$
for all $x \in I$.

4. f is continuous on I. To see this, observe that since both one-sided derivatives exist at every point the function must be continuous on both sides, hence continuous.

We summarize the preceding discussion as a theorem.

Theorem 7.34 *Let f be convex on an open interval I. Then*

(i) *f has finite left and right derivatives at each point of I. Each of these one-sided derivatives is a nondecreasing function of x on I, and*

$$f'_-(x) \leq f'_+(x) \quad \text{for all } x \in I. \tag{15}$$

(ii) *f is continuous on I.*

Note. If f is convex on a *closed* interval $[a, b]$, some of the results do not apply at the endpoints a and b. (See Exercise 7.10.8.) Note, too, that the corresponding results are valid for concave functions on I, the one-sided derivatives now being nonincreasing functions of x and the inequality in (15) being reversed.

We can now obtain the characterizations of convex functions familiar from elementary calculus.

Corollary 7.35 *Let f be defined on an open interval I.*

(i) *If f is differentiable on I, then f is convex on I if and only if f' is nondecreasing on I.*

(ii) *If f is twice differentiable on I, then f is convex on I if and only if $f'' \geq 0$ on I.*

We leave the verification of Corollary 7.35 as Exercise 7.10.9.

Exercises

7.10.1 Show that a function f is convex on an interval I if and only if the determinant

$$\begin{vmatrix} 1 & x_1 & f(x_1) \\ 1 & x_2 & f(x_2) \\ 1 & x_3 & f(x_3) \end{vmatrix}$$

is nonnegative for any choices of $x_1 < x_2 < x_3$ in the interval I.

7.10.2 If f and g are convex on an interval I, show that any linear combination $\alpha f + \beta g$ is also convex provided α and β are nonnegative.

7.10.3 If f and g are convex functions, can you conclude that the composition $g \circ f$ is also convex?

7.10.4 Let f be convex on an open interval (a, b). Show that then there are only two possibilities. Either (i) f is nonincreasing or nondecreasing on the entire interval (a, b) or else (ii) there is a number c so that f is nonincreasing on $(a, c]$ and nondecreasing on $[c, b)$.

7.10.5 Suppose f is convex on an open interval I. Prove that f is differentiable except on a countable set.

7.10.6 Suppose f is convex on an open interval I. Prove that if f is differentiable on I, then f' is continuous on I.

7.10.7 Let f be convex on an open interval that contains the closed interval $[a, b]$. Let

$$M = \max\{f'_+(a), f'_-(b)\}.$$

Show that

$$|f(x) - f(y)| \leq M|x - y|$$

for all x, $y \in [a, b]$.

7.10.8 Theorem 7.34 pertains to functions that are convex on an open interval. Discuss the extent to which the results of the theorem hold when f is convex on a *closed* interval $[a, b]$. In particular, determine whether continuity of f at the endpoints of the interval follows from the definition. Must $f'_+(a)$ and $f'_-(b)$ be finite?

7.10.9 Prove Corollary 7.35.

7.10.10 Let f be convex on an open interval (a, b). Must f be bounded above? Must f be bounded below?

7.10.11 Let f be convex on an open interval (a, b). Show that f does not have a strict maximum value.

7.10.12 Let f be defined and continuous on an open interval (a, b). Show that f is convex there if and only if there do not exist real numbers α and β such that the function $f(x) + \alpha x + \beta$ has a strict maximum value in (a, b).

7.10.13 Let $A = \{a_1, a_2, a_3, \dots\}$ be any countable set of real numbers. Let ✂

$$f(x) = \sum_1^\infty \frac{|x - a_k|}{10^k}.$$

Prove that f is convex on $\mathbb{R}$, differentiable on the set $\mathbb{R} \setminus A$, and nondifferentiable on the set A.

7.10.14 **(Inflection Points)** In elementary calculus one studies inflection points. ✂ The definitions one finds try to capture the idea that at such a point the sense of concavity changes from strict "up to down" or vice versa. Here are three common definitions that apply to differentiable functions. In each case f is defined on an open interval (a, b) containing the point x_0. The point x_0 is an *inflection point* for f if there exists an open interval $I \subset (a, b)$ such that on I

(Definition A) f' increases on one side of x_0 and decreases on the other side.

(Definition B) f' attains a strict maximum or minimum at x_0.

(Definition C) The tangent line to the graph of f at $(x_0, f(x_0))$ lies below the graph of f on one side of x_0 and above on the other side.

(a) Prove that if f satisfies Definition A at x_0, then it satisfies Definition B at x_0.

(b) Prove that if f satisfies Definition B at x_0, then it satisfies Definition C at x_0.

(c) Give an example of a function satisfying Definition B at x_0, but not satisfying Definition A.

(d) Give an example of an infinitely differentiable function satisfying Definition C at x_0, but not satisfying Definition B.

(e) Which of the three definitions states that the sense of concavity of f is "up" on one side of x_0 and "down" on the other?

7.10.15 (Jensen's Inequality) Let f be a convex function on an interval I, let $x_1, x_2, \dots, x_n$ be points of I and let $\alpha_1, \alpha_2, \dots \alpha_n$ be positive numbers satisfying

$$\sum_{k=1}^{n} \alpha_k = 1.$$

Show that

$$f\left(\sum_{k=1}^{n} \alpha_k x_k\right) \leq \sum_{k=1}^{n} \alpha_k f(x_k).$$

7.10.16 Show that the inequality is strict in Jensen's inequality (Exercise 7.10.15) except in the case that f is linear on some interval that contains the points $x_1, x_2, \dots, x_n$.

7.11 L'Hôpital's Rule

✂ *Enrichment*

Suppose that f and g are defined in a deleted neighborhood of x_0 and that

$$\lim_{x\to x_0} f(x) = A \text{ and } \lim_{x\to x_0} g(x) = B.$$

According to our usual theory of limits, we then have

$$\lim_{x\to x_0} \frac{f(x)}{g(x)} = \frac{\lim_{x\to x_0} f(x)}{\lim_{x\to x_0} g(x)} = \frac{A}{B},$$

unless $B = 0$.

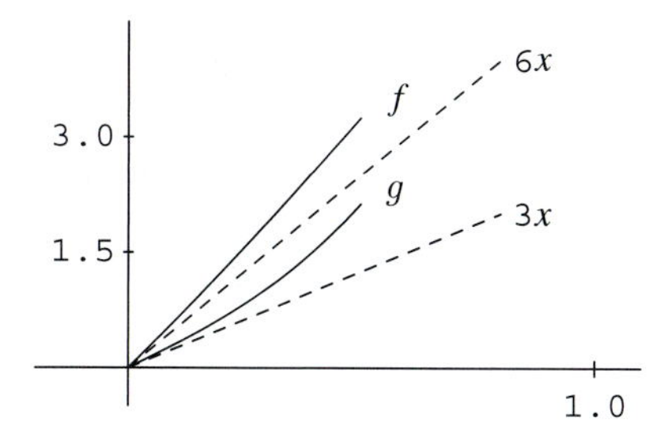

Figure 7.12. Comparison of the rates in Example 7.37.

But what happens if $B = 0$, which is often the case? A number of possibilities exist: If $B = 0$ and $A \neq 0$, then the limit does not exist. The most interesting case remains: If both A and B are zero, then the limiting behavior depends on the rates at which $f(x)$ and $g(x)$ approach zero.

Example 7.36 Consider

$$\lim_{x \to 0} \frac{6x}{3x} = \lim_{x \to 0} \frac{6}{3} = 2.$$

Look at this simple example geometrically. For $x \neq 0$, the height $6x$ is twice that of the height $3x$. The straight line $y = 6x$ approaches zero at twice the rate that the line $y = 3x$ does. ◀

Example 7.37 Now consider the slightly more complicated limit

$$\lim_{x \to 0} \frac{f(x)}{g(x)} = \lim_{x \to 0} \frac{6x + x^2}{3x + 5x^3}.$$

If we divide the numerator and denominator by $x \neq 0$, we see that the limit is the same as

$$\lim_{x \to 0} \frac{6 + x}{3 + 5x^2}.$$

This last limit can be calculated by our usual elementary methods as equaling $6/3 = 2$. Here, for $x \neq 0$ near zero, the height $f(x) = 6x + x^2$ is approximately $6x$, while the height of $g(x) = 3x + 5x^3$ is approximately $3x$, that is, the desired ratio is approximately 2. Again, the numerator approaches zero at about twice the rate that the denominator does.

We can be more precise by calculating these rates exactly. Let

$$f(x) = 6x + x^2 \text{ and } g(x) = 3x + 5x^3.$$

Then

$$f'(x) = 6 + 2x, \quad f'(0) = 6$$
$$g'(x) = 3 + 5x^2, \quad g'(0) = 3.$$

This makes precise our statement that the numerator approaches zero twice as fast as the denominator does. (See Figure 7.12 where there is an illustration showing the graphs of the functions f and g compared to the lines $y = 6x$ and $y = 3x$.) ◀

Let us try to generalize from these two examples. Suppose f and g are differentiable in a neighborhood of $x = a$ and that $f(a) = g(a) = 0$. Consider the following calculations and what conditions on f and g are required to make them valid.

$$\frac{f(x)}{g(x)} = \frac{f(x) - f(a)}{g(x) - g(a)} = \frac{\left(\frac{f(x)-f(a)}{x-a}\right)}{\left(\frac{g(x)-g(a)}{x-a}\right)} \overset{x \to a}{\longrightarrow} \frac{f'(a)}{g'(a)} = \lim_{x \to a} \frac{f'(x)}{g'(x)}. \tag{16}$$

If these calculations are valid, they show that under these assumptions ($f(a) = g(a) = 0$ and both $f'(a)$ and $g'(a)$ exist) we should be able to claim that

$$\lim_{x \to a} \frac{f(x)}{g(x)} = \lim_{x \to a} \frac{f'(x)}{g'(x)}.$$

You should check the various conditions that must be met to justify the calculations: $g(x)$ cannot equal zero at any point of the neighborhood in question (other than a); nor can $g(x) = g(a)$, (for $x \neq a$); $f(a)$ and $g(a)$ must equal zero (for the first equality), and f'/g' must be continuous at $x = a$ (for the last equality).

The calculations (16) provide a simple proof of a rudimentary form for a method of computing limits known as L'Hôpital's rule. We say "rudimentary" because some of the conditions we assumed are not needed for the conclusion

$$\lim_{x \to a} \frac{f(x)}{g(x)} = \lim_{x \to a} \frac{f'(x)}{g'(x)}.$$

7.11.1 L'Hôpital's Rule: $\frac{0}{0}$ Form

✂ *Enrichment*

Our first theorem provides a version of the rule identical with our introductory remarks but under weaker assumptions.

Theorem 7.38 (L'Hôpital's Rule: $\frac{0}{0}$ Form) *Suppose that the functions f and g are differentiable in a deleted neighborhood N of $x = a$. Suppose*

(i) $\lim_{x \to a} f(x) = 0$,

(ii) $\lim_{x\to a} g(x) = 0$,

(iii) *For every* $x \in N$, $g'(x) \neq 0$, *and*

(iv) $\lim_{x\to a} \frac{f'(x)}{g'(x)}$ *exists.*

Then $\displaystyle\lim_{x\to a} \frac{f(x)}{g(x)} = \lim_{x\to a} \frac{f'(x)}{g'(x)}$.

Proof Our hypotheses do not require f and g to be defined at $x = a$. But we can in any case define (or redefine) f and g at $x = a$ by $f(a) = g(a) = 0$. Because of assumptions (i) and (ii), this results in continuous functions defined on the full neighborhood $N \cup \{a\}$ of the point $x = a$. We can now apply Cauchy's form of the mean value theorem (7.21).

Suppose $x \in N$ and $a < x$. By Theorem 7.21 there exists $c = c_x$ in (a, x) such that

$$[f(x) - f(a)]g'(c_x) = [g(x) - g(a)]f'(c_x). \tag{17}$$

Since $f(a) = g(a) = 0$, (17) becomes

$$f(x)g'(c_x) = g(x)f'(c_x). \tag{18}$$

Equation (18) is valid for $x > a$ in N. We would like to express (18) in the form

$$\frac{f(x)}{g(x)} = \frac{f'(c_x)}{g'(c_x)}. \tag{19}$$

To justify (19) we show that $g(x)$ is never zero in $N \cap \{x : x > a\}$. (That $g'(c_x)$ is never zero in N is our hypothesis (iii).) If for some $x \in N$, $x > a$, we have $g(x) = 0$, then by Rolle's theorem there would exist a point $t \in (a, x)$ such that $g'(t) = 0$, contradicting hypothesis (ii). Thus equation (19) is valid for all $N \cap \{x : x > a\}$. A similar argument shows that if $x \in N$, $x < a$, then there exists $c_x \in (x, a)$ such that (19) holds.

Now as $x \to a$, c_x also approaches a, since c_x is between a and x. Thus

$$\lim_{x\to a} \frac{f(x)}{g(x)} = \lim_{x\to a} \frac{f(c_x)}{g(c_x)} = \lim_{x\to a} \frac{f'(x)}{g'(x)},$$

since the last limit exists by hypothesis (iv). ■

Note. Observe that we did not require f to be defined at $x = a$, nor did we require that f'/g' be continuous at $x = a$. It is also important to observe that L'Hôpital's rule does *not* imply that, under hypotheses (i), (ii), and (iii) of Theorem 7.38, if $\lim_{x\to 0} f(x)/g(x)$ exists, then $\lim_{x\to 0} f'(x)/g'(x)$ must also exist. Exercise 7.11.5 provides an example to illustrate this.

Example 7.39 Let us use L'Hôpital's rule to evaluate

$$\lim_{x\to 0} \ln(1+x)/x.$$

Let $f(x) = \ln(1+x)$, $g(x) = x$. Then

$$\lim_{x\to 0} f(x) = \lim_{x\to 0} g(x) = 0,\ f'(x) = \frac{1}{1+x}, \quad \text{and} \quad g'(x) = 1.$$

Thus

$$\lim_{x\to 0} \frac{\ln(1+x)}{x} = \lim_{x\to 0} \frac{1}{1+x} = 1.$$

◀

We refer to this theorem as the "$\frac{0}{0}$ form" for obvious reasons. There is also a version of the form $\frac{\infty}{\infty}$ (see Theorem 7.42). In addition, other modifications are possible. The point a can be replaced with $a = \infty$ or $a = -\infty$, (Theorem 7.41), and the results are valid for one-sided limits. (Our proof of Theorem 7.38 actually established that fact since we considered the case $x > a$ and $x < a$ separately.) Various other "indeterminate forms," ones for which the limit depends on the rates at which component parts approach their separate limits, can be manipulated to make use of L'Hôpital's rule possible.

Here is an example in which the forms "1^∞" and "$1^{-\infty}$" come into play. Observe that the function whose limit we wish to calculate is of the form $f(x)^{g(x)}$ where $f(x) \to 1$ as $x \to a$ but $g(x) \to \infty$ as $x \to a+$ and $g(x) \to -\infty$ as $x \to a-$.

Example 7.40 Evaluate $\lim_{x\to 0}(1+x)^{2/x}$. This expression is of the form 1^∞ (when $x > 0$). To calculate $\lim_{x\to 0}(1+x)^{2/x}$, write

$$y = (1+x)^{2/x},\ z = \ln y = \frac{2}{x}\ln(1+x).$$

Now the numerator and denominator of the function z satisfy the hypotheses of L'Hôpital's rule. Thus

$$\lim_{x\to 0} z = \lim_{x\to 0} \frac{2\ln(1+x)}{x} = \lim_{x\to 0} \frac{2}{1+x} = 2.$$

Since $\lim_{x\to 0} z = 2$, $\lim_{x\to 0} y = e^2$. ◀

7.11.2 L'Hôpital's Rule as $x \to \infty$

✂ *Enrichment*

We proved Theorem 7.38 under the assumption that $a \in \mathbb{R}$, but the theorem is valid when $a = -\infty$ or $a = +\infty$. In this case we are, of course, dealing with one-sided limits. As before, the relation

$$\lim_{x \to \infty} \frac{f'(x)}{g'(x)} = L$$

implies something about relative rates of growth of the functions $f(x)$ and $g(x)$ as $x \to \infty$. We can base a proof of the versions of L'Hôpital's rule that have $a = \infty$ (or $-\infty$) on Theorem 7.38 by a simple transformation.

Theorem 7.41 *Let f, g be differentiable on some interval $(-\infty, b)$. Suppose*

(i) $\lim_{x \to -\infty} f(x) = 0$,

(ii) $\lim_{x \to -\infty} g(x) = 0$,

(iii) *For every $x \in (-\infty, b)$, $g'(x) \neq 0$, and*

(iv) $\lim_{x \to -\infty} \frac{f'(x)}{g'(x)}$ *exists.*

Then

$$\lim_{x \to -\infty} \frac{f(x)}{g(x)} = \lim_{x \to -\infty} \frac{f'(x)}{g'(x)}.$$

A similar result holds when we replace ∞ by $-\infty$ in the hypotheses.

Proof Let $x = -1/t$. Then, as $t \to 0+$, $x \to -\infty$ and vice-versa. Define functions F and G by

$$F(t) = f\left(-\frac{1}{t}\right) \quad \text{and} \quad G(t) = g\left(-\frac{1}{t}\right).$$

Both functions F and G are defined on some interval $(0, \delta)$. We verify easily that

$$\lim_{t \to 0+} F(t) = \lim_{t \to 0+} G(t) = 0$$

and that

$$\lim_{t \to 0+} \frac{F'(t)}{G'(t)} = \lim_{x \to -\infty} \frac{f'(x)}{g'(x)}. \tag{20}$$

Using Theorem 7.38, we infer

$$\lim_{t \to 0+} \frac{F'(t)}{G'(t)} = \lim_{t \to 0+} \frac{F(t)}{G(t)} = \lim_{t \to 0+} \frac{f(-\frac{1}{t})}{g(-\frac{1}{t})} = \lim_{x \to -\infty} \frac{f(x)}{g(x)}. \tag{21}$$

The result follows from (20) and (21) ■

7.11.3 L'Hôpital's Rule: $\frac{\infty}{\infty}$ Form

Enrichment

When $f(x) \to \infty$ and $g(x) \to \infty$ as $x \to a$ we obtain the indeterminate form $\frac{\infty}{\infty}$. L'Hôpital's theorem then takes the form given in Theorem 7.42. Note, however, that we don't require $f(x) \to \infty$ in our hypotheses, or even that $f(x)$ approaches any limit.

Theorem 7.42 *Let f and g be differentiable on a deleted neighborhood N of $x = a$. Suppose that*

(i) $\lim_{x\to a} g(x) = \infty$.

(ii) *For every $x \in N$ $g'(x) \neq 0$.*

(iii) $\lim_{x\to a} f'(x)/g'(x)$ *exists.*

Then $\displaystyle\lim_{x\to a}\frac{f(x)}{g(x)} = \lim_{x\to a}\frac{f'(x)}{g'(x)}$. *The analogous statements are valid if $a = \pm\infty$ or* $\lim_{x\to a} g(x) = -\infty$.

Proof We prove the main part of Theorem 7.42 under the assumption that

$$\lim_{x\to a} f'(x)/g'(x)$$

is finite. The case that the limit is infinite as well as variants are left as Exercises 7.11.6 and 7.11.7. It suffices to consider the case of right-hand limits, the proof for left-hand limits being similar. Let

$$L = \lim_{x\to a+} f'(x)/g'(x).$$

We will show that if $p < L < q$, then there exists $\delta > 0$ such that

$$p < f(x)/g(x) < q$$

for $x \in (a, a+\delta)$. Since p and q are arbitrary (subject to the restriction $p < L < q$), we can then conclude

$$\lim_{x\to a+} f'(x)/g'(x) = L$$

as required.

Choose $r \in (L, q)$. By (iii) and the definition of L there exists δ_1 such that $f'(x)/g'(x) < r$ whenever $x \in (a, a+\delta_1)$. If $a < x < y < a+\delta_1$, then we infer from Theorem 7.21, Cauchy's form of the mean value theorem, and our assumption (ii) that there exists $c \in (x, y)$ such that

$$\frac{f(x)-f(y)}{g(x)-g(y)} = \frac{f'(c)}{g'(c)} < r. \tag{22}$$

Fix y in (22). Since $\lim_{x \to a+} g(x) = \infty$, there exists $\delta_2 > 0$ such that $a + \delta_2 < y$ and such that $g(x) > g(y)$ and $g(x) > 0$ if $a < x < a + \delta_2$. We then have

$$(g(x) - g(y))/g(x) > 0$$

for $x \in (a, a + \delta_2)$, so we can multiply both sides of the inequality (22) by $(g(x) - g(y))/g(x)$, obtaining

$$\frac{f(x)}{g(x)} < r - r\frac{g(y)}{g(x)} + \frac{f(y)}{g(x)} \text{ for } x \in (a, a + \delta_2). \tag{23}$$

Now let $x \to a+$. Then $g(x) \to \infty$ as $x \to a+$ by assumption (i). Since r, $g(y)$, and $f(y)$ are constants, the second and third terms on the right side of (23) approach zero. It now follows from the inequality $r < q$ that there exists $\delta_3 \in (0, \delta_2)$ such that

$$\frac{f(x)}{g(x)} < q \quad \text{whenever} \quad a < x < a + \delta_3. \tag{24}$$

In a similar fashion we find a $\delta_4 > 0$ such that

$$\frac{f(x)}{g(x)} > p \quad \text{whenever} \quad a < x < a + \delta_4.$$

If we let $\delta = \min(\delta_3, \delta_4)$, we have shown that

$$p < \frac{f(x)}{g(x)} < q \quad \text{whenever} \quad x \in (a, a + \delta).$$

Since p and q were arbitrary numbers satisfying $p < L < q$, our conclusion

$$\lim_{x \to a+} \frac{f(x)}{g(x)} = L = \lim_{x \to a+} \frac{f'(x)}{g'(x)}$$

follows. ■

Exercises

7.11.1 Consider the function $f(x) = (3^x - 2^x)/x$ defined everywhere except at $x = 0$.

(a) What value should be assigned to $f(0)$ in order that f be everywhere continuous?

(b) Does $f'(0)$ exist if this value is assigned to $f(0)$?

(c) Would it be correct to calculate $f'(0)$ by computing instead $f'(x)$ by the usual rules of the calculus and finding $\lim_{x \to 0} f'(x)$.

7.11.2 Suppose that f and g are defined in a deleted neighborhood of x_0 and that

$$\lim_{x\to x_0} f(x) = A \neq 0 \text{ and } \lim_{x\to x_0} g(x) = 0.$$

Show that

$$\lim_{x\to x_0} \left|\frac{f(x)}{g(x)}\right| = \infty.$$

7.11.3 Discuss the limiting behavior as $x \to 0$ for each of the following functions.

(a) $\dfrac{1}{x}$ (b) $\dfrac{1}{x^2}$

(c) $\dfrac{1}{\sin x}$ (d) $\dfrac{1}{x \sin x^{-1}}$

7.11.4 Evaluate each of the following limits.

(a) $\displaystyle\lim_{x\to 0} \frac{e^x - \cos x}{x}$

(b) $\displaystyle\lim_{t\to 0} \frac{\sin t - t}{t^3}$

(c) $\displaystyle\lim_{u\to 1} \frac{u^5 + 5u - 6}{2u^5 + 8u - 10}$

7.11.5 Let $f(x) = x^2 \sin x^{-1}$, $g(x) = x$. Show that

$$\lim_{x\to 0} \frac{f(x)}{g(x)} = 0$$

but that

$$\lim_{x\to 0} \frac{f'(x)}{g'(x)}$$

does not exist.

7.11.6 The proof we provided for Theorem 7.42 required that $\lim_{x\to a} f'(x)/g'(x)$ be finite. Prove that the result holds if this limit is infinite.

7.11.7 Prove the part of Theorem 7.42 dealing with $a = \pm\infty$ or $\lim_{x\to a} g(x) = -\infty$.

7.11.8 Evaluate the following limits.

(a) $\displaystyle\lim_{x\to\infty} \frac{x^3}{e^x}$

(b) $\displaystyle\lim_{x\to\infty} \frac{\ln x}{x}$

(c) $\displaystyle\lim_{x\to 0+} x \ln x$

(d) $\displaystyle\lim_{x\to 0+} x^x$

7.11.9 This exercise gives information about the relative rates of increase of certain types of functions. Prove that for each positive number p,

$$\lim_{x\to\infty} \frac{\ln x}{x^p} = \lim_{x\to\infty} \frac{x^p}{e^x} = 0.$$

7.11.10 Give an example of functions f and g defined on $\mathbb{R}$ such that

$$\lim_{x\to\infty} g(x) = \infty,\ \limsup_{x\to\infty} f(x) = \infty,\ \liminf_{x\to\infty} f(x) = -\infty$$

and Theorem 7.42 applies.

7.12 Taylor Polynomials

✂ *Enrichment*

Suppose f is continuous on an open interval I and $c \in I$. The constant function $g(x) = f(c)$ approximates f closely when x is sufficiently close to the point c, but may or may not provide a good approximation elsewhere. If f is differentiable on I, then we see from the mean value theorem (Theorem 7.20) that for each $x \in I$ $(x \neq c)$ there exists z between x and c such that

$$f(x) = f(c) + f'(z)(x - c).$$

The expression $R_0(x) = f'(z)(x - c) = f(x) - f(c)$ provides the size of the error obtained in approximating the function f by a constant function $P_0(x) = f(c)$. We can think of this as approximation by a zero-degree polynomial.

We do not expect a constant function to be a good approximation to a given continuous function in general. But our acquaintance with Taylor series (as presented in elementary calculus courses) suggests that if a function is sufficiently differentiable, it can be approximated well by polynomials of sufficiently high degree.

Suppose we wish to approximate f by a polynomial P_n of degree n. In order for the polynomial P_n to have a chance to approximate f well in a neighborhood of a point c, we should require

$$P_n(c) = f(c), P_n'(c) = f'(c), \ldots, P_n^{(n)}(c) = f^{(n)}(c).$$

In that case we at least guarantee that P_n "starts out" with the correct value, the correct rate of change, etc. to give it a chance to approximate f well in some neighborhood I of c. The test however is this. Write

$$f(x) = P_n(x) + R_n(x).$$

Is it true that the "error" or "remainder" $R_n(x)$ is small when $x \in I$?

In order to answer this sort of question, it would be useful to have workable forms for this error term $R_n(x)$. We present two forms for the remainder. The first is due to Joseph-Louis Lagrange (1736–1813), who obtained Theorem 7.43 in 1797. He used integration methods to prove the theorem. We provide a popular and more modern proof based on the mean value theorem.

Theorem 7.43 (Lagrange) *Let f possess at least $n+1$ derivatives on an open interval I and let $c \in I$. Let*

$$P_n(x) = f(c) + f'(c)(x-c) + \frac{f''(c)}{2!}(x-c)^2 + \cdots + \frac{f^{(n)}(c)}{n!}(x-c)^n$$

and let $R_n(x) = f(x) - P_n(x)$. Then for each $x \in I$ there exists z between x and c ($z = c$ if $x = c$) such that

$$R_n(x) = \frac{f^{(n+1)}(z)}{(n+1)!}(x-c)^{n+1}.$$

Proof Fix $x \in I$. Then there is a number M (depending on x, of course) such that

$$f(x) = P_n(x) + M(x-c)^{n+1}.$$

We wish to show that $M = (f^{(n+1)}(z))/n!$ for some z between x and c.

Consider the function g defined on I by

$$\begin{aligned} g(t) &= f(t) - P_n(t) - M(t-c)^{n+1} \\ &= R_n(t) - M(t-c)^{n+1}. \end{aligned}$$

Now P_n is a polynomial of degree at most n, so $P_n^{(n+1)}(t) = 0$ for all $t \in I$. Thus

$$g^{(n+1)}(t) = f^{(n+1)}(t) - (n+1)!M \quad \text{for all } t \in I. \tag{25}$$

Also, since $f^{(k)}(c) = P_n^{(k)}(c)$ for $k = 1, 2, \ldots, n$, we readily see that

$$g^{(k)}(c) = 0 \quad \text{for } k = 0, 1, 2, \ldots, n. \tag{26}$$

Suppose now that $x > c$, the case $x < c$ having a similar proof, and the case $x = c$ being obvious. We have chosen M in such a way that $g(x) = 0$ and, by (26), we see that $g(c) = 0$. Thus g satisfies the hypotheses of Rolle's theorem on the interval $[c, x]$. Therefore there exists a point $z_1 \in (c, x)$ such that $g'(z_1) = 0$.

Now apply Rolle's theorem to g' on the interval $[c, z_1]$, obtaining a point $z_2 \in (c, z_1)$ such that $g''(z_2) = 0$.

Continuing in this way we use (26) and Rolle's theorem repeatedly to obtain a point $z_n \in (c, z_{n-1})$ such that $g^{(n)}(z_n) = 0$. Finally, we apply the Rolle's theorem to the function $g^{(n)}$ on the interval $[c, z_n]$. We obtain a point $z \in (c, z_n)$ such that $g^{(n+1)}(z) = 0$. From (25) we deduce

$$f^{(n+1)}(z) = (n+1)!M,$$

completing the proof. ■

Note. The function P_n is called the *nth Taylor polynomial for f*. You will recognize P_n as the nth partial sum of the Taylor series studied in elementary calculus. (See also Chapter 10.) The function R_n is called the *remainder* or *error* function between f and P_n. If P_n is to be a good approximation to f, then R_n must be small in absolute value.

Observe that $P_n(c) = f(c)$ and that

$$P_n^{(k)}(c) = f^{(k)}(c) \quad \text{for} \quad k = 0, 1, 2, \ldots, n.$$

Observe also that the mean value theorem is the special case of Theorem 7.43 obtained by taking $n = 0$: on the interval $[c, x]$ there is a point z with

$$f(x) - f(c) = f'(z)(x - c).$$

Lagrange's result expresses the error term R_n in a particular way. It provides a sense of the error in approximating f by P_n. Note that we do not get an exact statement of the error term since it is given in terms of the value $f^{(n+1)}(z)$ at *some* point z. But if we know a little bit about the function $f^{(n+1)}$ on the interval in question, we might be able to say that this error is not very large.

Example 7.44 Suppose we wish to approximate the function $f(x) = \sin x$ on the interval $[-a, a]$ by a Taylor polynomial of degree 3, with $c = 0$. Here

$$f'(x) = \cos x \;,\; f''(x) = -\sin x \;,\; f'''(x) = -\cos x \;\text{ and }\; f^{(4)}(x) = \sin x.$$

Thus

$$\begin{aligned} P_3(x) &= \cos(0)x - \frac{\sin(0)}{2!}x^2 - \frac{\cos(0)}{3!}x^3 = x - \frac{x^3}{3!} \quad \text{and} \\ R_3(x) &= \frac{\sin z}{4!}x^4 \quad \text{for some } z \text{ in } [-a, a]. \end{aligned}$$

The exact error depends on which z makes this all true. But since $|\sin z| \leq 1$ for all z, we get immediately that

$$|R_3(x)| \leq a^4/4! = a^4/24,$$

so P_3 approximates f to within $a^4/24$ on the interval $[-a, a]$. For a small, the approximation should be sufficient for the purposes at hand. For large a, a higher-degree polynomial can produce the desired accuracy, since

$$|R_n(x)| \leq \frac{|x^{n+1}|}{(n+1)!}.$$

◀

Various other forms for the error term R_n are useful. The integral form is one of them. We state this form without proof. We assume that you are familiar with the integral as studied in calculus courses.

Theorem 7.45 (Integral Form of Remainder) *Suppose f possesses at least $n+1$ derivatives on an open interval I and $f^{(n+1)}$ is Riemann integrable on every closed interval contained in I. Let $c \in I$. Then*

$$R_n(x) = \frac{1}{n!}\int_c^x f^{(n+1)}(t)(x-t)^n\,dt \quad \text{for all } x \in I.$$

We shall see this form of the remainder again in Chapter 10 when we study Taylor series.

Exercises

7.12.1 Exhibit the Taylor polynomial about $x = 0$ of degree n for the function $f(x) = e^x$. Find n so that $|R_n(x)| \leq .0001$ for all $x \in [0, 2]$.

7.12.2 Show that if f is a polynomial of degree n, then it is its own Taylor polynomial of degree n with $c = 0$.

7.12.3 Calculate the Taylor polynomial of degree 5 with $c = 1$ for the functions $f(x) = x^5$ and $g(x) = \ln x$.

7.12.4 Let $f(x) = \frac{1}{x+2}$, $c = -1$, and $n = 2$. Show that

$$\frac{1}{x+2} = 1 - (x+1) + (x+1)^2 + R_3$$

where, for some z between x and -1,

$$R_3 = -\frac{(x+1)^3}{(2+z)^4}.$$

7.12.5 Let $f(x) = \ln(1+x)$, $c = 0$, and $(x > -1)$. Show that

$$f(x) = x - \frac{1}{2}x^2 + \frac{1}{3}x^3 + \cdots + (-1)^{n-1}\frac{x^n}{n} + R_n$$

where

$$R_n = \frac{(-1)^n}{n+1}\left(\frac{x}{1+z}\right)^{n+1}$$

for some z between 0 and x. Estimate R_n on the interval $[0, 1/10]$.

7.12.6 Just because a function possesses derivatives of all orders on an interval I does not guarantee that some Taylor polynomial approximates f in a neighborhood of some point of I. Let

$$f(x) = \begin{cases} e^{-\frac{1}{x^2}}, & \text{if } x \neq 0 \\ 0, & \text{if } x = 0. \end{cases}$$

(a) Show that f has derivatives of all orders and that $f^{(k)}(0) = 0$ for each $k = 0, 1, 2, \ldots$.

(b) Write down the polynomial P_n with $c = 0$.

(c) Write down Lagrange's form for the remainder of order n. Observe its magnitude and take the time to understand why P_n is not a good approximation for f on any interval I, no matter how large n is.

7.13 Challenging Problems for Chapter 7

7.13.1 **(Straddled derivatives)** Let $f:\mathbb{R}\to\mathbb{R}$ and let $x_0 \in \mathbb{R}$. Prove that f is differentiable at x_0 if and only if

$$\lim_{u\to x_0-,\, v\to x_0+} \frac{f(v)-f(u)}{v-u}$$

exists (finite), and, in this case, $f'(x_0)$ equals this limit.

7.13.2 **(Unstraddled Derivatives)** Let $f:\mathbb{R}\to\mathbb{R}$ and let $x_0 \in \mathbb{R}$. We say f is strongly differentiable at x_0 if

$$\lim_{u\to x_0,\, v\to x_0,\, u\neq v} \frac{f(v)-f(u)}{v-u}$$

exists.

(a) Show that a differentiable function need not be strongly differentiable everywhere.

(b) Show that a strongly differentiable function must be differentiable.

(c) If f is strongly differentiable at a point x_0 and differentiable in a neighborhood of x_0, show that f' must be continuous there.

7.13.3 Let p be a polynomial of the nth degree that is everywhere nonnegative. Show that

$$p(x) + p'(x) + p''(x) + \cdots + p^{(n)}(x) \geq 0$$

for all x.

7.13.4 Suppose that f is continuous on $[0,1]$, differentiable on $(0,1)$, and $f(0)=0$ and $f(1)=1$. For every integer n show that there must exist n distinct points $\xi_1, \xi_2, \dots, \xi_n$ in that interval so that

$$\sum_{k=1}^{n} \frac{1}{f'(\xi_k)} = n.$$

7.13.5 Show that there exists precisely one real number α with the property that for every function f differentiable on $[0,1]$ and satisfying $f(0) = 0$ and $f(1) = 1$ there exists a number ξ in $(0,1)$ (which depends, in general, on f) so that

$$f'(\xi) = \alpha\xi.$$

7.13.6 Let f be a continuous function. Show that the set of points where f is differentiable but not strongly differentiable (as defined in Exercise 7.13.2) is of the first category.

7.13.7 Let f be a continuous function on an open interval I. Show that f is convex on I if and only if

$$f\left(\frac{x+y}{2}\right) \leq \frac{f(x)+f(y)}{2}.$$

7.13.8 **(Wronskians)** The Wronskian of two differentiable functions f and g is the determinant

$$W(f,g) = \begin{vmatrix} f(x) & g(x) \\ f'(x) & g'(x) \end{vmatrix}.$$

Prove that if $W(f,g)$ does not vanish on an interval I and $f(x_1) = f(x_2) = 0$ for points $x_1 < x_2$ in I, then there exists $x_3 \in (x_1, x_2)$ such that $g(x_3) = 0$. [The functions $f(x) = \sin x, g(x) = \cos x$ furnish an example.]

✂ **7.13.9** Let f be a continuous function on an open interval I. Show that f is convex if and only if

$$\limsup_{h \to 0} \frac{f(x+h) + f(x-h) - 2f(x)}{h^2} \geq 0$$

for every $x \in I$.

✂ **7.13.10** Let f be continuous on an interval (a,b).

(a) Prove that the four Dini derivates of f and the difference quotient $\frac{f(y)-f(x)}{y-x}$ $(x \neq y \in (a,b))$ have the same bounds.

(b) Prove that if one of the Dini derivates is continuous at a point x_0, then f is differentiable at x_0.

(c) Show by example that the statements in the first two parts can fail for discontinuous functions.

✂ **7.13.11** **(Denjoy-Young-Saks Theorem)** The theorem with this name is a far-reaching theorem relating the four Dini derivates D^+f, D_+f, D^-f and D_-f. It was proved independently by an English mathematician, Grace Chisolm Young (1868–1944), and a French mathematician, Arnaud Denjoy (1884–1974), for continuous functions in 1916 and 1915 respectively. Young then extended the result to a larger class of functions called measurable functions. Finally, the Polish mathematician Stanislaw Saks (1897–1942) proved the theorem for all real-valued functions in 1924. Here is their theorem.

> **Theorem (Denjoy-Young-Saks)** Let f be an arbitrary finite function defined on $[a,b]$. Then except for a set of measure zero every point $x \in [a,b]$ is in one of four sets:
>
> (1) A_1 on which f has a finite derivative.
>
> (2) A_2 on which $D^+f = D_-f$ (finite), $D^-f = \infty$ and $D_+f = -\infty$.
>
> (3) A_3 on which $D^-f = D_+f$ (finite), $D^+f = \infty$ and $D_-f = -\infty$.
>
> (4) A_4 on which $D^-f = D^+f = \infty$ and $D_-f = D_+f = -\infty$.

(a) Sketch a picture illustrating points in the sets A_2, A_3 and A_4. To which set does $x = 0$ belong when $f(x) = \sqrt{|x|}\sin x^{-1}$, $f(0) = 0$?

(b) Use the Denjoy-Young-Saks theorem to prove that an increasing function f has a finite derivative except on a set of measure zero.

(c) Use the Denjoy-Young-Saks theorem to show that if all derived numbers of f are finite except on a set of measure zero, then f is differentiable except on a set of measure zero.

(d) Use the Denjoy-Young-Saks theorem to show that, for every finite function f, the set $\{x : f'(x) = \infty\}$ has measure zero.

7.13.12 Let f be a continuous function on an interval $[a, b]$ with a second derivative at all points in (a, b). Let $a < x < b$. Show that there exists a point $\xi \in (a, b)$ so that

$$\frac{\frac{f(x)-f(a)}{x-a} - \frac{f(b)-f(a)}{b-a}}{x-b} = \tfrac{1}{2}f''(\xi).$$

7.13.13 Let $f : \mathbb{R} \to \mathbb{R}$ be a differentiable function with $f(0) = 0$ and suppose that $|f'(x)| \leq |f(x)|$ for all $x \in \mathbb{R}$. Show that f is identically zero.

7.13.14 Let $f : \mathbb{R} \to \mathbb{R}$ have a third derivative that exists at all points. Suppose that

$$\lim_{x\to\infty} f(x)$$

exists and that

$$\lim_{x\to\infty} f'''(x) = 0.$$

Show that

$$\lim_{x\to\infty} f'(x) = \lim_{x\to\infty} f''(x) = 0.$$

7.13.15 Let f be defined on an interval I of length at least 2 and suppose that f'' exists there. If $|f(x)| \leq 1$ and $|f''(x)| \leq 1$ for all $x \in I$ show that $|f'(x)| \leq 2$ on the interval.

7.13.16 Let $f : \mathbb{R} \to \mathbb{R}$ be infinitely differentiable and suppose that

$$f\left(\frac{1}{n}\right) = \frac{n^2}{n^2+1}$$

for all $n = 1, 2, 3, \ldots$. Determine the values of

$$f'(0),\ f''(0),\ f'''(0),\ f^{(4)}(0), \ldots.$$

7.13.17 Let $f : \mathbb{R} \to \mathbb{R}$ have a third derivative that exists at all points. Show that there must exist at least one point ξ for which

$$f(\xi)f'(\xi)f''(\xi)f'''(\xi) \geq 0.$$

Chapter 8

THE INTEGRAL

✂ For a short course the integral as conceived by Cauchy can be introduced and the material on Riemann's integral omitted or abridged. The study of the Riemann integral introduces new techniques and ideas that may not be needed for some courses.

8.1 Introduction

Calculus students learn two processes, both of which are described as "integration." The following two examples should be familiar:

$$\int x^3\,dx = x^4/4 + C$$

and

$$\int_1^2 x^3\,dx = 2^4/4 - 1^4/4 = 16/4 - 1/4 = 15/4.$$

The first is called an *indefinite integral* or *antiderivative* and the second a *definite integral.* The use of nearly identical notation, terminology, and methods of computation does a lot to confuse the underlying meanings. Many calculus students would be hard pressed to make a distinction.

Indeed even for many eighteenth-century mathematicians these two different procedures were not much distinguished. It was a great discovery that the computation of an area could be achieved by finding an antiderivative. It is attributed to Newton, but vague ideas along this line can be found in the thinking of earlier authors. For these mathematicians a definite integral was defined directly in terms of the antiderivative.

This is most unfortunate for the development of a rigorous theory, as recognized by Cauchy. He saw clearly that it was vital that the meaning of the definite integral be separated from the indefinite integral and given a precise definition independent of it. For this he turned to the geometry of the Greeks, who had long ago described a method for computing areas

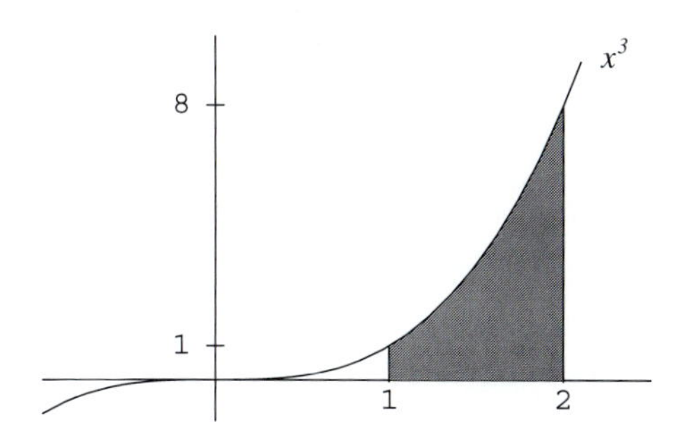

Figure 8.1. Region bounded by $x = 1$, $x = 2$, $y = x^3$, and $y = 0$.

of regions enclosed by curves. This method, the so-called method of exhaustion, involves computing the areas of simpler figures (squares, triangles, rectangles) that approximate the area of the region.

We return to the example

$$\int_1^2 x^3\, dx$$

interpreted as an area. The region is that bounded on the left and right by the lines $x = 1$ and $x = 2$, below by the line $y = 0$, and above by the curve $y = x^3$. (See Figure 8.1.)

Using the method of exhaustion, we may place this figure inside a collection of rectangles by dividing the interval $[1, 2]$ into n equal sized subintervals each of length $1/n$. This means selecting the points

$$1,\ 1 + 1/n,\ 1 + 2/n,\ \ldots\ 1 + (n-1)/n$$

and constructing rectangles with vertices at these points. The total area of these rectangles exceeds the true area and is precisely

$$\sum_{k=1}^{n} (1 + (k)/n)^3 (1/n).$$

The method of exhaustion requires a lower estimate as well and the true area of the region must be greater than

$$\sum_{k=1}^{n} (1 + (k-1)/n)^3 (1/n).$$

(See Figure 8.2 for an illustration with $n = 4$.)

The method of exhaustion requires us to show that as n increases *both* approximations, the upper one and the lower one, approach the same number. Cauchy saw that, because of the continuity of the function $f(x) = x^3$,

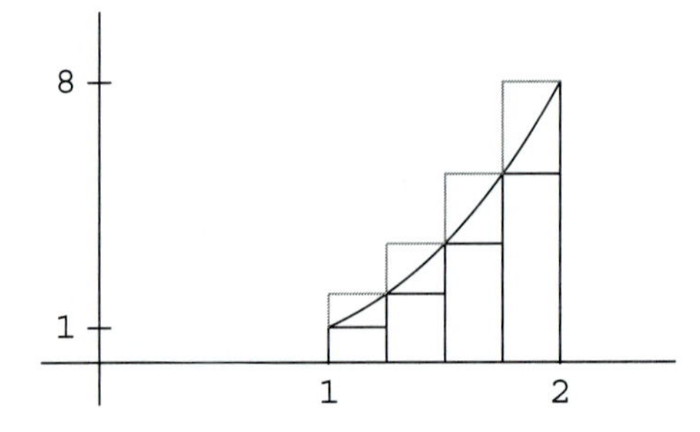

Figure 8.2. Method of exhaustion ($n = 4$).

these limits would be the same. More than that, any choice of points ξ_k from the interval $[1 + (k-1)/n, 1 + (k)/n]$ would have the property that

$$\lim_{n\to\infty} \sum_{k=1}^{n} {\xi_k}^3(1/n)$$

would exist.

This procedure, borrowed heavily from the Greeks, will work for any continuous function and thus it offered to Cauchy a way to *define* the integral

$$\int_a^b f(x)\,dx$$

for any function f, continuous on an interval $[a, b]$, without any reference whatsoever to notions of derivatives or antiderivatives. The key ingredients here are first dividing the interval $[a, b]$ by a finite sequence of points

$$a = x_0 < x_1 < x_2 < x_3 < \cdots < x_{n-1} < x_n = b,$$

thus forming a collection of nonoverlapping subintervals called *a partition* of $[a, b]$

$$[x_0, x_1],\ [x_1, x_2],\ \ldots,\ [x_{n-1}, x_n],$$

(it is not important that they have equal size, just that they get small). Then we form the sums

$$\sum_{k=1}^{n} f(\xi_k)(x_k - x_{k-1}) \tag{1}$$

with respect to this partition. The only constraint on the choice of the points ξ_k is that each is taken from the appropriate interval $[x_{k-1}, x_k]$ of the partition; these are often called the *associated points.* It is an unfortunate trick of fate that the sums (1) that originated with Cauchy are called *Riemann sums* because of Riemann's later (much later) use of them in defining his integral.

In this chapter we start with Cauchy's methods of integration and proceed to Riemann. The important thing for you to keep track of is how this theory develops in a manner that assigns meaning to the integral of various classes of functions in a way distinct from how we would compute an integral in a calculus course. It is much easier to compute that $\int_1^2 x^3\,dx = 15/4$ in the familiar way, rather than as a limit of Riemann sums; but the meaning of this statement is properly given in this more difficult way.

8.2 Cauchy's First Method

Cauchy's first goal in defining an integral was to give meaning to the integral for continuous functions. The integral is defined as the limit of Riemann sums. Before such a definition is valid we must show that the limit exists. Thus the first step is the proof of the following theorem.

Theorem 8.1 (Cauchy) *Let f be a continuous function on an interval $[a,b]$. Then there is a number I, called the* definite integral *of f on $[a,b]$, such that for each $\varepsilon > 0$ there is a $\delta > 0$ so that*

$$\left|\sum_{k=1}^{n} f(\xi_k)(x_k - x_{k-1}) - I\right| < \varepsilon$$

whenever $[x_0,x_1]$, $[x_1,x_2]$, ... , $[x_{n-1},x_n]$ is a partition of the interval $[a,b]$ into subintervals of length less than δ and each ξ_k is a point in the interval $[x_{k-1},x_k]$.

Once the theorem is proved, then we can safely define the definite integral of a continuous function as that number I guaranteed by the theorem. Loosely speaking, we say that the integral is defined as a limit of Riemann sums (1).

Definition 8.2 Let f be a continuous function on an interval $[a,b]$. Then we define

$$\int_a^b f(x)\,dx$$

to be that number I whose existence is proved in Theorem 8.1.

Now we must prove Theorem 8.1.

Proof For any particular partition (let us call it π)

$$[x_0,x_1],\ [x_1,x_2],\ \ldots,\ [x_{n-1},x_n]$$

of the interval $[a, b]$ write

$$M(\pi) = \sum_{k=1}^{n} \max\{f(x) : x \in [x_{k-1}, x_k]\}(x_k - x_{k-1})$$

and

$$m(\pi) = \sum_{k=1}^{n} \min\{f(x) : x \in [x_{k-1}, x_k]\}(x_k - x_{k-1}).$$

Here $M(\pi)$ and $m(\pi)$ depend on the partition π. These are called the *upper sums* and *lower sums* for the partition. Note that any Riemann sum over this partition must lie somewhere between the lower sum and the upper sum.

Since f is continuous on $[a, b]$ it is uniformly continuous there (Theorem 5.47). Thus for every $\varepsilon > 0$ there is a $\delta > 0$ that depends on ε so that

$$|f(x) - f(y)| < \frac{\varepsilon}{b - a}$$

if $|x - y| < \delta$. Since we shall need to find a different δ for many choices of ε, let us write it as $\delta(\varepsilon)$.

Thus if the partition we are using has the property that every interval is shorter than $\delta(\varepsilon)$, we must have

$$\max\{f(x) : x \in [x_{k-1}, x_k]\} - \min\{f(x) : x \in [x_{k-1}, x_k]\} < \frac{\varepsilon}{b - a}$$

It follows that for such partitions $0 < M(\pi) - m(\pi) < \varepsilon$.

Select a sequence of partitions $\{\pi_n\}$, each one containing all the points of the previous partition, and such that every interval in the nth partition π_n is shorter than $\delta(1/n)$. If $M(\pi_n)$ and $m(\pi_n)$ denote the corresponding sums for the nth partition of our sequence of partitions, then

$$0 < M(\pi_n) - m(\pi_n) < 1/n.$$

One more technical point needs to be raised. As we add points to a partition the upper sums cannot increase nor can the lower sums decrease. Thus $M(\pi_n) \geq M(\pi_{n+1})$ while $m(\pi_n) \leq m(\pi_{n+1})$. (The details just require some inequality work and are left as Exercise 8.2.17.)

Thus the intervals

$$[m(\pi_n), M(\pi_n)]$$

form a descending sequence with lengths shrinking to zero. By Cantor's intersection property (see Section 4.5.2) there is a number I so that $m(\pi_n) \to I$ and $M(\pi_n) \to I$ as $n \to \infty$. We shall show that I has the property of the theorem.

Now let $\varepsilon > 0$ and choose any partition π with the property that every interval is shorter than $\delta(\varepsilon/2)$. By what we have seen, the interval $[m(\pi), M(\pi)]$ has length smaller than $\varepsilon/2$.

Any Riemann sum over the partition π must evidently belong to the interval $[m(\pi), M(\pi)]$. Let $N > 2/\varepsilon$. Suppose for a moment that the intervals $[m(\pi), M(\pi)]$ and $[m(\pi_N), M(\pi_N)]$ intersect at some point. In that case the Riemann sum over the partition π and the value I, which is inside the interval $[m(\pi_N), M(\pi_N)]$, must be closer together than $\varepsilon/2 + 1/N$, which is smaller than ε. As this is precisely what we want to prove, we are done.

It remains to check that the two intervals

$$[m(\pi), M(\pi)] \text{ and } [m(\pi_N), M(\pi_N)]$$

intersect at some point. To find a point common to these two intervals combine the two partitions π and π_N to form a partition containing all points in either partition. The Riemann sum over such a partition belongs to the interval $[m(\pi), M(\pi)]$ and also to the interval $[m(\pi_N), M(\pi_N)]$. This completes the proof. (That I is unique is left as Exercise 8.2.2.) ■

A special case of this definition and this theorem allows us to compute an integral as a limit of a sequence. In practice this is seldom the best way to compute it, but it is interesting and useful in some parts of the theory.

Corollary 8.3 *Let f be a continuous function on an interval $[a, b]$ Then*

$$\int_a^b f(x)\,dx = \lim_{n\to\infty} \frac{b-a}{n} \sum_{k=0}^{n-1} f\left(a + \frac{k}{n}(b-a)\right).$$

8.2.1 Scope of Cauchy's First Method

✂ *Enrichment*

It is natural to ask whether this method of Cauchy for describing the integral of a continuous function would apply to a larger class of functions. But Cauchy did not ask this question. His goal was to assign a meaning for continuous functions, a class of functions that was large enough for most applications. The only limitation he might have seen was that this method would fail for functions having *infinite singularities* (i.e., discontinuity points where the function is unbounded). Thus he was led to the method we discuss in Section 8.4 as Cauchy's second method. Cauchy and other mathematicians of his time were sufficiently confused as to the meaning of the word "function" that they might never have asked such a question.

But we can. And many years later Riemann did too, as we shall see in Section 8.6. In the exercises you are asked to prove the following:

> *The first method of Cauchy will fail if applied to an unbounded function f on an interval $[a, b]$.*

> *The first method of Cauchy succeeds if applied to any function f that is bounded on an interval $[a, b]$ and has only finitely many discontinuities there.*

The first statement shows that the method used here to define an integral is severely limited. It can never be used for unbounded functions. Since we have restricted it here to continuous functions that is no problem; any function continuous on an interval $[a, b]$ is bounded there.

The second statement shows that the method is not, however, limited only to continuous functions even though that was Cauchy's intention. Later we will use the method to define Riemann's integral which applies to a large class of (bounded) functions that are permitted to have many, even infinitely many, points of discontinuity.

Exercises

8.2.1 To complete the computations in the introduction to this chapter, show that

$$\lim_{n\to\infty} \sum_{k=1}^{n} (1 + (k)/n)^3 (1/n) = 15/4.$$

This computation alone should be enough to convince you that the definition is intended theoretically and hardly ever used to compute integrals.

8.2.2 Show that the number I in the statement of Theorem 8.1 is unique; that is, that there cannot be two numbers that would be assigned to the symbol $\int_a^b f(x)\,dx$.

8.2.3 If f is constant and $f(x) = \alpha$ for all x in $[a, b]$ show that

$$\int_a^b f(x)\,dx = \alpha(b - a).$$

8.2.4 If f is continuous and $f(x) \geq 0$ for all x in $[a, b]$ show that

$$\int_a^b f(x)\,dx \geq 0.$$

8.2.5 If f is continuous and $m \leq f(x) \leq M$ for all x in $[a, b]$ show that

$$m(b - a) \leq \int_a^b f(x)\,dx \leq M(b - a).$$

8.2.6 Calculate $\int_0^1 x^p\,dx$ (for whatever values of p you can manage) by partitioning $[0, 1]$ into subintervals of equal length.

8.2.7 Calculate $\int_a^b x^p\,dx$ (for whatever values of p you can manage) by partitioning $[a, b]$ into subintervals $[a, aq]$, $[aq, aq^2]$, ... $[aq^{n-1}, b]$ where $aq^n = b$. (Note that the subintervals are not of equal length, but that the lengths form a geometric progression.)

8.2.8 Use the method of the preceding exercise to show that

$$\int_1^2 \frac{dx}{x^2} = \frac{1}{2}$$

and check it by the usual calculus method.

8.2.9 Compute the Riemann sums for the integral $\int_a^b x^{-2}\,dx$ $(a > 0)$ taken over a partition

$$[x_0, x_1],\ [x_1, x_2],\ \ldots,\ [x_{n-1}, x_n]$$

of the interval $[a, b]$ and with associated points $\xi_i = \sqrt{x_i x_{i-1}}$. What can you conclude from this?

8.2.10 Compute the Riemann sums for the integral $\int_a^b x^{-1/2}\,dx$ $(a > 0)$ taken over a partition

$$[x_0, x_1],\ [x_1, x_2],\ \ldots,\ [x_{n-1}, x_n]$$

of the interval $[a, b]$ and with associated points

$$\xi_i = \left(\frac{\sqrt{x_i} + \sqrt{x_{i-1}}}{2}\right)^2.$$

What can you conclude from this?

8.2.11 Show that

$$\lim_{n\to\infty} n\left\{\frac{1}{(n+1)^2} + \frac{1}{(n+2)^2} + \frac{1}{(n+3)^2} + \cdots \frac{1}{(2n)^2}\right\} = \frac{1}{2}.$$

8.2.12 Calculate

$$\lim_{n\to\infty} \frac{e^{1/n} + e^{2/n} + \cdots + e^{(n-1)/n} + e^{n/n}}{n}$$

by expressing this limit as a definite integral of some continuous function and then using calculus methods.

8.2.13 Express

$$\lim_{n\to\infty} \frac{1}{n}\sum_{k=1}^{n} f\left(\frac{k}{n}\right)$$

as a definite integral where f is continuous on $[0, 1]$.

8.2.14 Prove that the conclusion of Theorem 8.1 is false if f is not bounded.

8.2.15 Prove that the conclusion of Theorem 8.1 is true if f is continuous at all but a finite number of points in the interval $[a, b]$ and is bounded.

8.2.16 Prove that the conclusion of Theorem 8.1 is true for the function f defined on the interval $[0, 1]$ as follows: $f(0) = 0$ and $f(x) = 2^{-n}$ for each

$$2^{-n-1} < x \leq 2^{-n} \qquad (n = 0, 1, 2, 3, \ldots).$$

How many points of discontinuity does f have in the interval $[0, 1]$? What is the value of the number I in this case?

8.2.17 For a bounded function f and any partition π

$$[x_0, x_1],\ [x_1, x_2],\ \ldots,\ [x_{n-1}, x_n]$$

of the interval $[a, b]$ write

$$M(f, \pi) = \sum_{k=1}^{n} \max\{f(x) : x \in [x_{k-1}, x_k]\}(x_k - x_{k-1})$$

and

$$m(f,\pi) = \sum_{k=1}^{n} \min\{f(x) : x \in [x_{k-1}, x_k]\}(x_k - x_{k-1})$$

These are called the *upper sums* and *lower sums* for the partition for the function f and were used in the proof of Theorem 8.1.

(a) Show that if π_2 contains all of the points of the partition π_1, then

$$m(f,\pi_1) \le m(f,\pi_2) \le M(f,\pi_2) \le M(f,\pi_1).$$

(b) Show that if π_1 and π_2 are arbitrary partitions and f is any bounded function, then

$$m(f,\pi_1) \le M(f,\pi_2).$$

(c) Show that if π is any arbitrary partition and f is any bounded function on $[a,b]$ then

$$c(b-a) \le m(f,\pi) \le M(f,\pi) \le C(b-a)$$

where $C = \sup f$ and $c = \inf f$.

(d) Show that with any choice of associated points the Riemann sum over a partition π is in the interval $[m(f,\pi), M(f,\pi)]$.

(e) Show that, if f is continuous, every value in the interval between $m(f,\pi)$ and $M(f,\pi)$ is equal to some particular Riemann sum over the partition π with an appropriate choice of associated points ξ_k.

(f) Show that if f is not continuous the preceding assertion may be false.

8.3 Properties of the Integral

The integral has thus far been defined just for continuous functions. We ask what properties it must have. Later we shall have to extend the scope of the integral to much broader classes of functions. It will be important to us then that the collection of elementary properties here will still be valid.

These properties exhibit the structure of the integral. They are the most vital tools to use in handling integrals both for theoretical and practical matters. Since we are restricted to continuous functions in this section, the proofs are simple. As we enlarge the scope of the integral the proofs may become more difficult, and subtle differences in assertions may arise.

Note. All functions f and g appearing in the statements are assumed to be continuous on the intervals $[a,b]$, $[b,c]$, $[a,c]$ in the statements. Thus the integrals all have meaning. This means we do not have to prove that any of these integrals exist: They do. It is the stated identity that needs to be proved in each case. To prove the identity, we consider a sequence of partitions π_n chosen so that the points in the partition are closer together than $1/n$. Let us use the notation $S(\pi_n, f)$ to denote a Riemann sum taken over this partition for the function f with associated points chosen (say) at the left-hand endpoint of the corresponding intervals. Then

$$\lim_{n\to\infty} S(\pi_n, f) = \int_a^b f(x)\,dx.$$

We shall use this idea in the proofs.

8.4 (Additive Property) *Let f be continuous on $[a,c]$ and suppose that $a < b < c$. Then*

$$\int_a^b f(x)\,dx + \int_b^c f(x)\,dx = \int_a^c f(x)\,dx.$$

Proof For our sequence of partitions we choose π_n to be a partition of $[a,c]$ chosen so that the points in the partition are closer together than $1/n$ and so that the point b is one of the points. Each partition π_n splits into two parts; π_n' and π_n'' where the former is a partition of $[a,b]$ and where the latter is a partition of $[b,c]$. Note that

$$S(\pi_n, f) = S(\pi_n', f) + S(\pi_n'', f)$$

by elementary arithmetic. If we let $n \to \infty$ in this identity we obtain immediately the identity in the statement we wish to prove. ■

8.5 (Linear Property) *Let f and g be continuous on $[a,b]$. Then, for all α, $\beta \in \mathbb{R}$,*

$$\int_a^b [\alpha f(x) + \beta g(x)]\,dx = \alpha \int_a^b f(x)\,dx + \beta \int_a^b g(x)\,dx.$$

Proof Again consider a sequence of partitions of $[a,b]$, π_n chosen so that the points in the partition are closer together than $1/n$. If $S(\pi_n, f)$ denotes a Riemann sum taken over this partition for the function f, then

$$\lim_{n\to\infty} S(\pi_n, f) = \int_a^b f(x)\,dx.$$

In the same way for g we would have

$$\lim_{n\to\infty} S(\pi_n, g) = \int_a^b g(x)\,dx.$$

But it is easy to check that

$$S(\pi_n, \alpha f + \beta g) = \alpha S(\pi_n, f) + \beta S(\pi_n, g)$$

and taking $n \to \infty$ in this identity gives exactly the statement in the property. Note that we do not have to prove that

$$S(\pi_n, \alpha f + \beta g) \to \int_a^b [\alpha f(x) + \beta g(x)]\,dx.$$

This follows from Theorem 8.1 because the integrand is continuous. ■

8.6 (Monotone Property) *Let f and g be continuous on $[a,b]$. Then, if $f(x) \leq g(x)$ for all $a \leq x \leq b$,*

$$\int_a^b f(x)\,dx \leq \int_a^b g(x)\,dx.$$

Proof Consider a sequence of partitions π_n chosen so that the points in the partition are closer together than $1/n$. If $S(\pi_n, f)$ denotes a Riemann sum taken over this partition for the function f, then

$$\lim_{n\to\infty} S(\pi_n, f) = \int_a^b f(x)\,dx.$$

In the same way for g we would have

$$\lim_{n\to\infty} S(\pi_n, g) = \int_a^b g(x)\,dx.$$

But since $f(x) \leq g(x)$ for all x we must have

$$S(\pi_n, f) \leq S(\pi_n, g).$$

Taking limits as $n \to \infty$ in this inequality yields the property. ■

8.7 (Absolute Property) *Let f be continuous on $[a,b]$. Then*

$$-\int_a^b |f(x)|\,dx \leq \int_a^b f(x)\,dx \leq \int_a^b |f(x)|\,dx$$

or, equivalently,

$$\left|\int_a^b f(x)\,dx\right| \leq \int_a^b |f(x)|\,dx.$$

Proof This follows immediately from the monotone property because

$$-|f(x)| \leq f(x) \leq |f(x)|.$$

■

Fundamental Theorem of Calculus The next two properties are known together as the fundamental theorem of calculus. They establish the close relationship between differentiation and integration and offer, to the calculus student, a useful method for the computation of integrals. This method reduces the computational problem of integration (i.e., computing a limit of Riemann sums) to the problem of finding an antiderivative.

8.8 (Differentiation of the Indefinite Integral) *BST WARN Let f be continuous on $[a,b]$. Then the function*

$$F(x) = \int_a^x f(t)\,dt$$

has a derivative on $[a,b]$ and $F'(x) = f(x)$ at each point.

Proof Let $h > 0$ and $x \in [a, b)$. We compute

$$F(x+h) - F(x) - hf(x) = \int_x^{x+h} (f(t) - f(x))\, dt$$

provided only that $x + h \leq b$. Thus, using Exercise 8.3.1, we have

$$|F(x+h) - F(x) - hf(x)| \leq h \max\{|f(t) - f(x)| : t \in [x, x+h]\}$$

and hence that

$$\left| \frac{F(x+h) - F(x)}{h} - f(x) \right| \leq \max\{|f(t) - f(x)| : t \in [x, x+h]\}.$$

As f is continuous at x

$$\max\{|f(t) - f(x)| : t \in [x, x+h]\} \to 0$$

as $h \to 0+$ and this inequality shows that the right-hand derivative of F at $x \in [a, b)$ is exactly $f(x)$.

A similar argument would show that the left-hand derivative of F at $x \in (a, b]$ is exactly $f(x)$. This proves the property. ∎

8.9 (Integral of a Derivative) *If the function F has a continuous derivative on $[a, b]$, then*

$$\int_a^b F'(x)\, dx = F(b) - F(a).$$

Proof Given any $\varepsilon > 0$ there is a $\delta > 0$ so that any Riemann sum for the continuous function F' over a partition of $[a, b]$ into intervals of length less than δ is within ε of $\int_a^b F'(x)\, dx$. If

$$[x_0, x_1],\ [x_1, x_2],\ \ldots,\ [x_{n-1}, x_n]$$

is such a partition then observe that, if we choose the associated points $\xi_k \in [x_{k-1}, x_k]$ by the mean value theorem in such a way that

$$F(x_k) - F(x_{k-1}) = F'(\xi_k)(x_k - x_{k-1})$$

then we will have

$$F(b) - F(a) = \sum_{k=1}^{n} F(x_k) - F(x_{k-1}) = \sum_{k=1}^{n} F'(\xi_k)(x_k - x_{k-1}).$$

Since the right side of the identity is within ε of $\int_a^b F'(x)\, dx$ so too must be the value $F(b) - F(a)$. But this is true for any $\varepsilon > 0$ and hence it follows that these must be equal; that is, that

$$\int_a^b F'(x)\, dx = F(b) - F(a).$$

∎

Exercises

8.3.1 If f is continuous on an interval $[a,b]$ and

$$M = \max\{|f(x)| : x \in [a,b]|\}$$

show that

$$\left|\int_a^b f(x)\,dx\right| \leq M(b-a).$$

8.3.2 (Mean Value Theorem for Integrals) If f is continuous show that there is a point ξ in (a,b) so that

$$\int_a^b f(x)\,dx = f(\xi)(b-a).$$

8.3.3 If f is continuous and $m \leq f(x) \leq M$ for all x in $[a,b]$ show that

$$m\int_a^b g(x)\,dx \leq \int_a^b f(x)g(x)\,dx \leq M\int_a^b g(x)\,dx$$

for any continuous, nonnegative function g.

8.3.4 If f is continuous and nonnegative on an interval $[a,b]$ and

$$\int_a^b f(x)\,dx = 0$$

show that f is identically equal to zero there.

8.3.5 (Second Mean Value Theorem for Integrals) If f and g are continuous on an interval $[a,b]$ and g is nonnegative, show that there is a number $\xi \in (a,b)$ such that

$$\int_a^b f(x)g(x)\,dx = f(\xi)\int_a^b g(x)\,dx.$$

8.3.6 If f is continuous on an interval $[a,b]$ and

$$\int_a^b f(x)g(x)\,dx = 0$$

for every continuous function g on $[a,b]$ show that f is identically equal to zero there.

8.3.7 (Integration by parts) Suppose that f, g, f' and g' are continuous on $[a,b]$. Establish the *integration by parts* formula

$$\int_a^b f(x)g'(x)\,dx = [f(b)g(b) - f(a)g(a)] - \int_a^b f'(x)g(x)\,dx.$$

8.3.8 (Integration by substitution) State conditions on f and g so that the *integration by substitution* formula

$$\int_a^b f(g(x))g'(x)\,dx = \int_{g(a)}^{g(b)} f(s)\,ds$$

is valid.

8.3.9 State conditions on f, g and h so that the *integration by substitution* formula

$$\int_a^b f(g(h(x)))g'(h(x))h'(x)\,dx = \int_{g(h(a))}^{g(h(b))} f(s)\,ds$$

is valid.

8.3.10 If f and g are continuous on an interval $[a, b]$ show that

$$\left(\int_a^b f(x)g(x)\,dx\right)^2 \leq \left(\int_a^b [f(x)]^2\,dx\right)\left(\int_a^b [g(x)]^2\,dx\right).$$

8.4 Cauchy's Second Method

Defining an integral only for continuous functions, as we did in the preceding section, is far too limiting. Even in the early nineteenth century the need for considering more general functions was apparent. For Cauchy this meant handling functions that have discontinuities. But Cauchy would not have felt any need to handle badly discontinuous functions, indeed he may not even have considered such objects as functions. In our terminology we could say that Cauchy was interested in extending his integral from continuous functions to functions possessing isolated discontinuities (i.e., the set of discontinuity points contains only isolated points).

We have already noted in Section 8.2.1 that *bounded* functions with finitely many discontinuities present no difficulties. Cauchy's first method can be applied to them. It is the case of *unbounded* functions that offers real resistance. What should we mean by the integral

$$\int_0^1 \frac{dx}{\sqrt{x}}?$$

While the integrand has only one discontinuity (at $x = 0$) the function is unbounded and Cauchy's first method cannot be applied. If the integral did make sense, then we would expect that the function

$$F(\delta) = \int_\delta^1 \frac{dx}{\sqrt{x}}$$

would be defined and continuous everywhere on the interval $[0, 1]$ and the value $F(0)$ would equal our integral. But here $F(x)$ is not defined at $x = 0$ although it is defined for all x in $(0, 1]$ since the integrand is continuous on any interval $[x, 1]$ for $x > 0$. If we compute it we see that

$$F(\delta) = \int_\delta^1 \frac{dx}{\sqrt{x}} = 2 - 2\sqrt{\delta}.$$

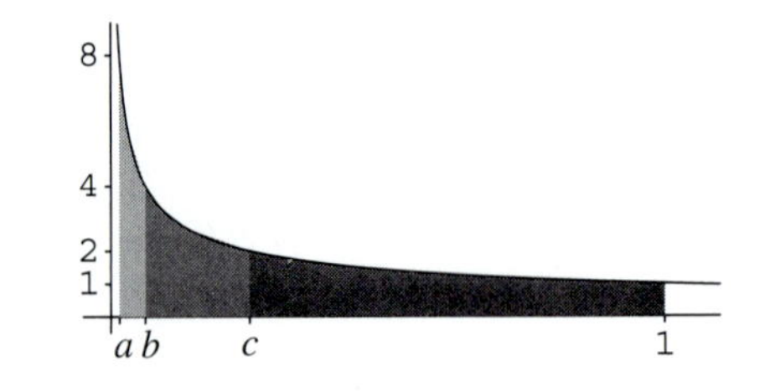

Figure 8.3. Computation of $\int_0^1 x^{-1/2}\,dx = 2$.

While we cannot take $F(0)$ itself (it is not defined), we can take the limit,

$$\lim_{\delta\to 0+} F(\delta) = \lim_{\delta\to 0+} \int_\delta^1 \frac{dx}{\sqrt{x}} = 2,$$

as a perfectly reasonable value for the integral.

Indeed if we consider this as a problem in determining the area of the unbounded region in Figure 8.3 we can see graphically why the answer should be 2. Note that in the figure $0 < a < b < c < 1$ and these numbers have the values $a = 1/64$, $b = 1/16$, and $c = 1/4$. The integrals have values

$$\int_c^1 x^{-1/2}\,dx = 1, \ \int_b^c x^{-1/2}\,dx = 1/2, \quad \text{and} \quad \int_a^b x^{-1/2}\,dx = 1/4$$

and so we would expect

$$\int_0^1 x^{-1/2}\,dx = 1 + 1/2 + 1/4 + \cdots = 2$$

as indeed this method does give.

This is precisely Cauchy's second method. If you understand this example, you understand the method. Any general write-up of the method might obscure this simple idea. We need some language, however. The procedure of taking a limit to obtain the final value of the integral may or may not work. If the limit does exist, we say that the integral *converges*, or is a *convergent integral*, and we say that the function f is *integrable by Cauchy's second method* or simply *integrable* if the context is clear. Otherwise the integral is said to be *divergent*.

We give a formal definition valid just in the case that the function has one point of unboundedness, and that point occurs at the left-hand endpoint of the interval. For more than one point, or for a point not at an endpoint, the definition is best generalized by splitting the integral into separate integrals, each of which can be handled one at a time in this fashion. (See Exercise 8.4.3.)

Definition 8.10 Let f be a continuous function on an interval $(a, b]$ that is unbounded in every interval $(a, a + \delta)$. Then we define

$$\int_a^b f(x)\,dx$$

to be

$$\lim_{\delta \to 0+} \int_{a+\delta}^b f(x)\,dx$$

if this limit exists, and in this case the integral is said to be *convergent.* If both integrals

$$\int_a^b f(x)\,dx \quad \text{and} \quad \int_a^b |f(x)|\,dx$$

converge the integral is said to be *absolutely convergent.*

The role of the extra condition of absolute convergence is much like its role in the study of infinite series. You will recall that absolutely convergent series are more "robust" in the sense that they can be rather freely manipulated, unlike the nonabsolutely convergent series, which are rather fragile. The same is true here of absolutely convergent integrals. Note that the integral

$$\int_0^1 x^{-1/2}\,dx$$

is both convergent and absolutely convergent merely because the integrand is nonnegative.

Exercises

8.4.1 Formulate a definition of the integral $\int_a^b f(x)\,dx$ for a function continuous on $[a, b)$ and unbounded at the right-hand endpoint. Supply an example.

8.4.2 Formulate a definition of the integral $\int_a^b f(x)\,dx$ for a function continuous on $[a, c)$ and on $(c, b]$ and unbounded in every interval containing c. Supply an example.

8.4.3 How would an integral of the form

$$\int_0^3 \frac{f(x)}{\sqrt{|x(x-1)(x-2)(x-3)|}}\,dx$$

be interpreted, where f is continuous?

8.4.4 Let f and g be continuous on $(a, b]$ and such that $|f(x)| \leq |g(x)|$ for all $a < x \leq b$. If the integral $\int_a^b g(x)\,dx$ is absolutely convergent, show that so also is the integral $\int_a^b f(x)\,dx$.

8.4.5 For what continuous functions f must the integral

$$\int_{-1}^{1} \frac{f(x)}{\sqrt{1-x^2}}\,dx$$

converge?

8.4.6 Let f be a bounded function, continuous on $(a,b]$ and that is discontinuous at the endpoint a. Show that if the second method of Cauchy is applied to f then the result is the same as applying the first method to the entire interval $[a,b]$ [regardless of the value assigned to $f(a)$].

8.4.7 Suppose that f is continuous on $[-1,1]$ except for an isolated discontinuity at $x=0$. If the limit

$$\lim_{\delta\to 0+}\left(\int_{-1}^{-\delta} f(x)\,dx + \int_{\delta}^{1} f(x)\,dx\right)$$

exists does it follow that f is integrable on $[-1,1]$?

8.4.8 As a project determine which of the properties of the integral in Section 8.3 (which apply only to continuous functions on an interval $[a,b]$) can be extended to functions that are integrable by Cauchy's second method on $[a,b]$. Give proofs.

8.5 Cauchy's Second Method (Continued)

The same idea that Cauchy used to assign meaning to the integral of unbounded functions he also used to handle functions on unbounded intervals. How should we interpret the integral

$$\int_{1}^{\infty} \frac{dx}{x^2}?$$

We might try first to form a partition of the unbounded interval $[1,\infty)$ and seek some kind of limit of Riemann sums. A much simpler idea is to adapt Cauchy's second method to this in the obvious way.

$$\int_{1}^{\infty} \frac{dx}{x^2} = \lim_{X\to\infty} \int_{1}^{X} \frac{dx}{x^2} = \lim_{X\to\infty}\left(1-\frac{1}{X}\right) = 1.$$

In Figure 8.4 we show graphically how to compute the area that is represented by $\int_1^\infty x^{-2}\,dx$. Note that

$$\int_{1}^{2} x^{-2}\,dx = 1/2,\ \int_{2}^{4} x^{-2}\,dx = 1/4,\ \int_{4}^{8} x^{-2}\,dx = 1/8$$

and so we would expect

$$\int_{1}^{\infty} x^{-2}\,dx = 1/2 + 1/4 + 1/8 + \cdots = 1$$

as indeed this method does give. (See Figure 8.4.)

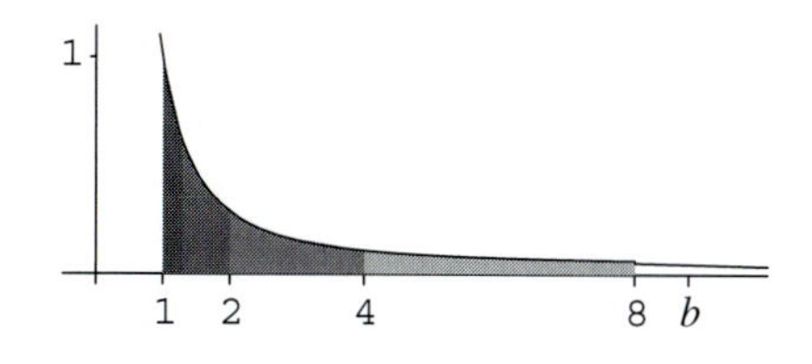

Figure 8.4. Computation of $\int_1^\infty x^{-2}\,dx = 1$.

This is precisely Cauchy's second method applied to unbounded intervals. Again, if you understand this example, you understand the method.

We give a formal definition valid just for an infinite interval of the form $[a,\infty)$. The case $(-\infty,b]$ is similar. The case $(-\infty,+\infty)$ is best split up into the sum of two integrals, from $(-\infty,a]$ and $[a,\infty)$, each of which can be handled in this fashion. (See Exercise 8.5.2.)

Definition 8.11 Let f be a continuous function on an interval $[a,\infty)$. Then we define

$$\int_a^\infty f(x)\,dx$$

to be

$$\lim_{X\to\infty}\int_a^X f(x)\,dx$$

if this limit exists, and in this case the integral is said to be *convergent.* If both integrals

$$\int_a^\infty f(x)\,dx \quad\text{and}\quad \int_a^\infty |f(x)|\,dx$$

converge the integral is said to be *absolutely convergent.*

Again, the role of the extra condition of absolute convergence is much like its role in the study of infinite series. Note that the integral $\int_1^\infty x^{-2}\,dx$ is convergent and also absolutely convergent merely because the integrand is nonnegative.

Exercises

8.5.1 Formulate a definition of the integral $\int_{-\infty}^b f(x)\,dx$ for a function continuous on $(-\infty,b]$. Supply examples of convergent and divergent integrals of this type.

8.5.2 Formulate a definition of the integral $\int_{-\infty}^\infty f(x)\,dx$ for a function continuous on $(-\infty,\infty)$. Supply examples of convergent and divergent integrals of this type.

8.5.3 For what values of p is the integral $\int_1^\infty x^{-p}\,dx$ convergent?

8.5.4 Show that

$$\int_0^\infty x^n e^{-x}\,dx = n!.$$

8.5.5 Let f be a continuous function on $[1,\infty)$ such that $\lim_{x\to\infty} f(x) = \alpha$. Show that if the integral $\int_1^\infty f(x)\,dx$ converges, then α must be 0.

8.5.6 Let f be a continuous function on $[1,\infty)$ such that the integral $\int_1^\infty f(x)\,dx$ converges. Can you conclude that $\lim_{x\to\infty} f(x) = 0$?

8.5.7 Let f be a continuous, decreasing function on $[1,\infty)$. Show that the integral $\int_1^\infty f(x)\,dx$ converges if and only if the series $\sum_{n=1}^\infty f(n)$ converges.

8.5.8 Give an example of a function f continuous on $[1,\infty)$ so that the integral $\int_1^\infty f(x)\,dx$ converges but the series $\sum_{n=1}^\infty f(n)$ diverges.

8.5.9 Give an example of a function f continuous on $[1,\infty)$ so that the integral $\int_1^\infty f(x)\,dx$ diverges but the series $\sum_{n=1}^\infty f(n)$ converges.

8.5.10 Show that

$$\int_0^\infty \frac{\sin x}{x}\,dx$$

is convergent but not absolutely convergent.

8.5.11 **(Cauchy Criterion for Convergence)** Let $f : [a,\infty) \to \mathbb{R}$ be a continuous function. Show that the integral $\int_a^\infty f(x)\,dx$ converges if and only if for every $\varepsilon > 0$ there is a number M so that, for all $M < c < d$,

$$\left| \int_c^d f(x)\,dx \right| < \varepsilon.$$

8.5.12 **(Cauchy Criterion for Absolute Convergence)** Let $f : [a,\infty) \to \mathbb{R}$ be a continuous function. Show that the integral $\int_a^\infty f(x)\,dx$ converges absolutely if and only if for every $\varepsilon > 0$ there is a number M so that, for all $M < c < d$,

$$\int_c^d |f(x)|\,dx < \varepsilon.$$

8.5.13 As a project determine which of the properties of the integral in Section 8.3 (which apply only to continuous functions on a finite interval) can be extended to integrals on an infinite interval $[a,\infty]$. Give proofs.

8.6 The Riemann Integral

Thus far in our discussion of the integral we have defined the meaning of the symbol

$$\int_a^b f(x)\,dx$$

first for all continuous functions, by Cauchy's first method, and then for functions that may have a finite number of discontinuities at which the function is unbounded, by Cauchy's second method.

Let us return to Cauchy's first method. We ask just how far this method can be applied. It can be applied to all continuous functions; that was the content of Theorem 8.1. It can be applied to all bounded functions with finitely many discontinuities (Exercise 8.2.15). It can be applied to *some* bounded functions with infinitely many discontinuities (Exercise 8.2.16).

Rather than search for broader classes of functions to which this method applies, we adopt the viewpoint that was taken by Riemann. We simply *define* the class of all functions to which Cauchy's first method can be applied and then seek to characterize that class. This represents a much more modern point of view than Cauchy would have taken with his much more limited idea of what a function is. Note that we need only turn Theorem 8.1 into a definition.

Definition 8.12 Let f be a function on an interval $[a,b]$. Suppose that there is a number I such that for all $\varepsilon > 0$ there is a $\delta > 0$ so that

$$\left| \sum_{k=1}^{n} f(\xi_k)(x_k - x_{k-1}) - I \right| < \varepsilon$$

whenever $[x_0, x_1], [x_1, x_2], \ldots, [x_{n-1}, x_n]$, is a partition of the interval $[a,b]$ into subintervals of length less than δ and each ξ_k is a point in the interval $[x_{k-1}, x_k]$. Then f is said to be *Riemann integrable* on $[a,b]$ and we write

$$\int_a^b f(x)\,dx$$

for that number I.

We can call the set of points

$$\pi = \{x_0, x_1, x_2, \ldots, x_{n-1}, x_n\}$$

the *partition* of the interval $[a,b]$ or, equivalently, if it is more convenient, the set of intervals

$$[x_0, x_1],\ [x_1, x_2],\ \ldots,\ [x_{n-1}, x_n]$$

can be called the partition. The points ξ_k that are chosen from each interval $[x_{k-1}, x_k]$ are called the *associated points* of the partition. Notice that in the definition the associated points can be freely chosen inside the intervals of the partition.

Loosely a function f is Riemann integrable if the limit of the Riemann sums for f exists over that interval. The program now is to determine what classes of functions are Riemann integrable and to obtain characterizations of Riemann integrability. We shall investigate this in the remainder of this section.

We need also to find out whether the properties of the integral that hold for continuous functions now continue to hold for all Riemann integrable functions. We shall consider that in the next section.

Two observations are immediate from our earlier work:

All continuous functions are Riemann integrable.

All Riemann integrable functions are bounded.

In light of this last statement we see that the Riemann integral is somewhat limited in that it will not do anything to handle unbounded functions. For that we must still return to Cauchy's second method. But, as we shall see, the Riemann integral will handle many bounded functions that are badly discontinuous (but not too badly). As research progressed in the nineteenth century the Riemann integral became the standard tool for discussing integrals of bounded functions. For unbounded functions Cauchy's second method continued to be employed, although other methods emerged.

By the early twentieth century the Riemann integral was abandoned by all serious analysts in favor of Lebesgue's integral. The Riemann integral survives in texts such as this mainly because of the technical difficulties of Lebesgue's better, but more difficult, methods.

8.6.1 Some Examples

All Riemann integrable functions are bounded. All continuous functions are Riemann integrable. In order to obtain some insight into the question as to what functions are Riemann integrable we present some examples, first of a bounded function that is not integrable and then of a badly discontinuous function that is integrable.

Example 8.13 Here is an example of a function that is bounded but "too discontinuous" to be Riemann integrable. On the interval $[0, 1]$ let f be the function equal to 1 for x rational and to 0 for x irrational. Let

$$[x_0, x_1],\ [x_1, x_2],\ \ldots,\ [x_{n-1}, x_n]$$

be any partition. If we choose associated points $\xi_k \in [x_{k-1}, x_k]$ so that ξ_k is rational, then the Riemann sum

$$\sum_{k=1}^{n} f(\xi_k)(x_k - x_{k-1}) = \sum_{k=1}^{n}(x_k - x_{k-1}) = 1 \tag{2}$$

while if we choose associated points $\eta_k \in [x_{k-1}, x_k]$ so that η_k is irrational

$$\sum_{k=1}^{n} f(\eta_k)(x_k - x_{k-1}) = 0. \tag{3}$$

Because of (2) and (3), the integral $\int_0^1 f(x)\,dx$ cannot exist. ◀

Example 8.14 Recall the Dirichlet function (Section 5.2.6) which provides an example of a function that is discontinuous at every rational number and continuous at every irrational. We show that this function is Riemann integrable. On the interval $[0,1]$ let f be the function equal to $1/q$ for $x = p/q$ rational (assuming that p/q has been expressed in its lowest terms) and to 0 for x irrational.

Let $\varepsilon > 0$. Let q_0 be any positive integer larger than $2/\varepsilon$. We count the number of points x in $[0,1]$ at which $f(x) > 1/q_0$. There are finitely many of these, say M of them. Choose δ_1 sufficiently small so that any two of these points are further apart than $2\delta_1$. Choose $\delta < \delta_1$ so that (for reasons that become clear only after all our computations are done) $M\delta < \varepsilon/2$. This will allow us to use the inequality

$$M\delta + 1/q_0 < \varepsilon. \tag{4}$$

Let

$$[x_0, x_1],\ [x_1, x_2],\ \ldots,\ [x_{n-1}, x_n]$$

be any partition chosen so that each of the intervals is shorter than δ. For any choice of associated points $\xi_k \in [x_{k-1}, x_k]$ we note that either

(i) $f(\xi_k) = 0$ if ξ_k is irrational, or

(ii) $f(\xi_k) > 1/q_0$ if ξ_k is one of the M points counted previously, or

(iii) $0 < f(\xi_k) \leq 1/q_0$ if ξ_k is any other rational point.

We can estimate the Riemann sum

$$\sum_{k=1}^{n} f(\xi_k)(x_k - x_{k-1})$$

by considering separately these three cases. Case (i) evidently contributes nothing to this sum. Case (ii) can contribute at most $M\delta$ to this sum since each interval in the partition can contain at most one of the points of type (ii) and there are only M such points. Finally, case (iii) can contribute in total no more than $1/q_0$. Thus, using the inequality (4), we have

$$0 \leq \sum_{k=1}^{n} f(\xi_k)(x_k - x_{k-1}) \leq M\delta + 1/q_0 < \varepsilon.$$

This proves that the integral $\int_0^1 f(x)\,dx = 0$. Considering just how discontinuous this function is (it has a dense set of discontinuities) it is startling that it is nonetheless integrable. ◀

8.6.2 Riemann's Criteria

✂ *Advanced*

What bounded functions then are Riemann integrable? The answer is that such functions must be "mostly" continuous. The example of the very discontinuous function in Example 8.13 suggests this. On the other hand, Example 8.14 shows that the discontinuities of a Riemann integrable function might even be dense. Riemann first analyzed this by using the oscillation of the function f on an interval. We recall (Definition 6.24) that this is defined as

$$\omega f([c,d]) = \sup_{x\in[c,d]} f(x) - \inf_{x\in[c,d]} f(x).$$

This measures how much the function f changes in the interval $[c,d]$. For a continuous function this is just the difference between the maximum and minimum values of f on $[c,d]$ and will be small if the interval $[c,d]$ is small enough.

Theorem 8.15 (Riemann) *A function f defined on an interval $[a,b]$ is Riemann integrable if and only if for every $\varepsilon > 0$ there is a $\delta > 0$ so that*

$$\sum_{k=1}^{n} \omega f([x_{k-1},x_k])(x_k - x_{k-1}) < \varepsilon$$

whenever $[x_0,x_1]$, $[x_1,x_2]$, ... , $[x_{n-1},x_n]$, is a partition of the interval $[a,b]$ into subintervals of length less than δ.

Proof If f is Riemann integrable on $[a,b]$ with integral I, then for any $\varepsilon > 0$ there must be a $\delta > 0$ so that any two Riemann sums taken over a partition with intervals smaller than δ are both within $\varepsilon/4$ of I. In particular, we have

$$\left|\sum_{k=1}^{n} f(\xi_k)(x_k - x_{k-1}) - \sum_{k=1}^{n} f(\eta_k)(x_k - x_{k-1})\right| < \varepsilon/2$$

whenever $[x_0,x_1]$, $[x_1,x_2]$, ... , $[x_{n-1},x_n]$, is a partition of the interval $[a,b]$ into subintervals of length less than δ. Here ξ_k and η_k are any choices from $[x_{k-1},x_k]$. We rewrite this as

$$\left|\sum_{k=1}^{n} [f(\xi_k) - f(\eta_k)](x_k - x_{k-1})\right| \le \varepsilon/2 < \varepsilon. \qquad (5)$$

Now notice that

$$\sup_{\eta,\xi\in[x_{k-1},x_k]} (f(\xi) - f(\eta)) = \omega f([x_{k-1}, x_k]).$$

Thus we see that the criterion follows immediately on taking sups over these choices of ξ_k and η_k in the inequality 5.

The other direction of the theorem can be interpreted as a "Cauchy criterion" and proved in a manner similar to all our other Cauchy criteria so far in the text (indeed similar to the proof of Theorem 8.1). We omit the details. ■

Theorem 8.15 offers an interesting necessary and sufficient condition for integrability. It is awkward to use the sufficiency criterion here since it demands that we check that *all* small partitions have a certain property. The following variant is a little easier to apply since we need find only *one* partition for each positive ε.

Theorem 8.16 *A function f on an interval $[a, b]$ is Riemann integrable if and only if for every $\varepsilon > 0$ there is at least one partition $[x_0, x_1]$, $[x_1, x_2]$, ... , $[x_{n-1}, x_n]$, of the interval $[a, b]$ so that*

$$\sum_{k=1}^{n} \omega f([x_{k-1}, x_k])(x_k - x_{k-1}) < \varepsilon.$$

Proof By Theorem 8.15 we see that if f is Riemann integrable there would have to exist such a partition.

In the opposite direction we must show that the condition here implies integrability. Certainly this condition implies that f is bounded (or else this sum would be infinite) and so we may assume that $|f(x)| \leq M$ for all x. This gives us a useful, if crude, estimate on the size of the oscillation on any interval $[c, d]$:

$$\omega f([c, d]) \leq 2M.$$

Let $\varepsilon > 0$. We shall find a number δ so that the criterion of Theorem 8.15 is satisfied. Let $[x_0, x_1]$, $[x_1, x_2]$, ... , $[x_{n-1}, x_n]$ be the partition whose existence is given. We use that to find our δ. Choose δ sufficiently small so that

$$2Mn\delta < \varepsilon.$$

Now let

$$[y_0, y_1],\ [y_1, y_2],\ \ldots,\ [y_{m-1}, y_m],$$

be any partition of the interval $[a, b]$ into subintervals of length less than δ. These intervals are of two types: Type (i) are those that are contained entirely inside intervals of our original partition, and type (ii) are those that include as interior points one of the points x_k for $k = 1, 2, \ldots, n-1$. In any case there are only $n-1$ of these intervals and each is of length less than δ. Thus, using just a crude estimate on each of these terms, the intervals of type (ii) contribute to the sum

$$\sum_{k=1}^{m} \omega f([y_{k-1}, y_k])(y_k - y_{k-1})$$

no more than $(2M)n\delta$. The sum taken over all the type (i) intervals must be smaller than

$$\sum_{k=1}^{m} \omega f([x_{k-1}, x_k])(x_k - x_{k-1}) < \varepsilon.$$

Thus the total sum

$$\sum_{k=1}^{m} \omega f([y_{k-1}, y_k])(y_k - y_{k-1}) < 2Mn\delta + \varepsilon < 2\varepsilon.$$

It follows by the criterion in Theorem 8.15 that f is Riemann integrable as required. ■

8.6.3 Lebesgue's Criterion

✂ *Advanced*

Theorem 8.16 is beautiful and seemingly characterizes the class of Riemann integrable functions in a meaningful way. But at the time of Riemann there was only an imperfect understanding of sets of real numbers and so it did not occur to Riemann that the property of Riemann integrability for a bounded function f depended exclusively on the nature of the set of points of discontinuity of f. Indeed the condition

$$\sum_{k=1}^{n} \omega f([x_{k-1}, x_k])(x_k - x_{k-1}) < \varepsilon$$

on the oscillation of the function suggests that something more subtle than just this is happening.

In 1901 Henri Lebesgue completed this theorem by using the notion of a set of measure zero. Recall (from Section 6.8) that a set E of real numbers is of measure zero if for every $\varepsilon > 0$ there is a sequence of intervals $\{(c_i, d_i)\}$ covering all points of E and with total length $\sum_{i=1}^{\infty}(d_i - c_i) < \varepsilon$. The exact characterization of Riemann integrable functions is precisely this: They are bounded (as we already well know) and they are continuous at all points except perhaps at the points of a set of measure zero. (In modern language they are said to be continuous *almost everywhere* or continuous a.e.)

Theorem 8.17 (Riemann-Lebesgue) *A function f on an interval $[a, b]$ is Riemann integrable if and only if f is bounded and the set of points in $[a, b]$ at which f is not continuous is a set of measure zero.*

Proof The necessity is not difficult to prove but is the least important part for us. The sufficiency is more important and harder to prove. Throughout the proof we require a familiarity with the notion of the oscillation $\omega_f(x)$ of a function f at a point x as discussed in Section 6.7. Recall that this value is positive if and only if f is discontinuous at x.

Let us suppose that f is Riemann integrable. Certainly f is bounded. Fix $e > 0$ and consider the set $N(e)$ of points x such that the oscillation of f at x is greater than e; that is, so that

$$\omega_f(x) > e.$$

Any interval (c, d) that contains a point $x \in N(e)$ will certainly have

$$\omega f([c, d]) \geq e.$$

Let $\varepsilon > 0$ and use Theorem 8.15 to find intervals

$$[x_0, x_1],\ [x_1, x_2],\ \ldots,\ [x_{n-1}, x_n],$$

forming a partition of the interval $[a, b]$ and such that

$$\sum_{k=1}^{n} \omega f([x_{k-1}, x_k](x_k - x_{k-1}) < \varepsilon e/2.$$

Select from this collection just those intervals that contain a point from $N(e)$ in their interior. The total length of these intervals cannot exceed $(e\varepsilon)/(2e)$ since for each such interval $[x_{k-1}, x_k]$ we must have $\omega f([x_{k-1}, x_k]) \geq e$.

Thus we have succeeded in covering the set $N(e)$ by a sequence of open intervals (x_{k-1}, x_k) of total length less than $\varepsilon/2$, except for an oversight. One or more of the points $\{x_i\}$ might be in the set $N(e)$, and we have neglected to cover it. Since there are only finitely many such points, we can add a few sufficiently short intervals to our collection.

Thus we have proved that for any $\varepsilon > 0$ the set $N(e)$ can be covered by a collection of open intervals of total length less than ε. It follows that $N(e)$ has measure zero. But the set of points of discontinuity of f is the union of the sets $N(1)$, $N(1/2)$, $N(1/4)$, $N(1/8)$, ... Since each of these is measure zero, it follows from Theorem 6.34 that the set of points of discontinuity of f has measure zero too as required.

This proves the theorem in one direction. In the other suppose that f is bounded, say that $|f(x)| \leq M$ for all x and that the set E of points in $[a, b]$ at which f is not continuous is a set of measure zero. Let $\varepsilon > 0$. By Theorem 8.16 we need to find at least one partition

$$[x_0, x_1],\ [x_1, x_2],\ \ldots,\ [x_{n-1}, x_n]$$

of the interval $[a,b]$ so that

$$\sum_{k=1}^{n} \omega f([x_{k-1},x_k])(x_k - x_{k-1}) < \varepsilon.$$

Let E_1 denote the set of points x in $[a,b]$ at which the oscillation is greater than or equal to $\varepsilon/(2(b-a))$; that is, for which

$$\omega_f(x) \geq \varepsilon/(2(b-a)).$$

This set is closed (see Theorem 6.27) and, being a subset of E, it must have measure zero. Now closed sets of measure zero can be covered by a *finite* number of small open intervals of total length smaller than

$$\varepsilon/(4M+1).$$

(See Theorem 6.35.) We can assume that these open intervals do not have endpoints in common. Note that, at points in the intervals that remain, the oscillation of f is smaller than $\varepsilon/(2(b-a))$. Consequently, these intervals may be subdivided into smaller intervals on which the oscillation is at least that small (Exercise 8.6.6).

Thus we may construct a partition

$$[x_0,x_1],\ [x_1,x_2],\ \ldots,\ [x_{n-1},x_n],$$

of the interval $[a,b]$ consisting of two kinds of closed intervals: (i) the first kind cover all the points of E_1 and have total length smaller than $\varepsilon/(4M+1)$ and (ii) the remaining kind contain no points of E_1 and the oscillation of f on each of these intervals is smaller than $\varepsilon/(2(b-a))$; that is,

$$\omega f([x_{k-1},x_k]) < \varepsilon/(2(b-a)).$$

The sum

$$\sum_{k=1}^{n} \omega f([x_{k-1},x_k])(x_k - x_{k-1})$$

splits into two sums depending on the intervals of type (i) or type (ii). The former sum contributes no more than

$$(2M) \times \varepsilon/(4M+1) < \varepsilon/2$$

while the latter sum contributes no more than

$$\varepsilon/(2(b-a)) \times (b-a) < \varepsilon/2.$$

Altogether, then,

$$\sum_{k=1}^{n} \omega f([x_{k-1},x_k])(x_k - x_{k-1}) < \varepsilon$$

and the proof is complete. ■

8.6.4 What Functions Are Riemann Integrable?

✂ *Advanced*

Theorem 8.17 exactly characterizes those functions that are Riemann integrable as the class of bounded functions that do not have too many points of discontinuity. We should recognize immediately that certain types of functions that we are used to working with are also integrable. We express these as corollaries to our theorem. (Recall that step functions were defined in Section 5.2.6.)

Corollary 8.18 *Every step function on an interval is Riemann integrable there.*

Proof A step function is bounded and has only finitely many discontinuities. Thus the set of discontinuities has measure zero. Consequently, the corollary follows from Theorem 8.17. ■

Corollary 8.19 *Every bounded function with only countably many points of discontinuity in an interval is Riemann integrable there.*

Proof The corollary follows directly from Theorem 8.17 since countable sets have measure zero. ■

Corollary 8.20 *Every function monotonic on an interval is Riemann integrable there.*

Proof A monotonic function is bounded and has only countably many discontinuities. Consequently, this corollary follows from the preceding corollary. ■

Corollary 8.21 *If a function f is Riemann integrable on an interval $[a, b]$ then so too is the function $|f|$ on that interval.*

Proof The corollary follows directly from Theorem 8.17 since if f is Riemann integrable on $[a, b]$ it must be bounded and continuous at every point except a set of measure zero. Exercise 8.6.7 shows that $|f|$ has precisely the same properties. ■

Exercises

8.6.1 Show directly from Theorem 8.16 that the characteristic function of the rationals is not Riemann integrable on any interval.

8.6.2 Show that the product of two Riemann integrable functions is itself Riemann integrable.

8.6.3 If f is Riemann integrable on an interval and f is never zero, does it follow that $1/f$ is Riemann integrable there? What extra hypothesis could we invoke to make this so?

8.6.4 If f is Riemann integrable on an interval $[a,b]$ show that for every $\varepsilon > 0$ there are a pair of step functions

$$L(x) \leq f(x) \leq U(x)$$

so that

$$\int_a^b (U(x) - L(x))\,dx < \varepsilon.$$

8.6.5 Let f be a function on an interval $[a,b]$ with the property that for every $\varepsilon > 0$ there are a pair of step functions $L(x) \leq f(x) \leq U(x)$ so that

$$\int_a^b (U(x) - L(x))\,dx < \varepsilon.$$

Show that f is Riemann integrable.

8.6.6 Suppose that the oscillation $\omega_f(x)$ of a function f is smaller than η at each point x of an interval $[c,d]$. Show that there must be a partition $[x_0, x_1]$, $[x_1, x_2]$, ... , $[x_{n-1}, x_n]$, of $[c,d]$ so that the oscillation

$$\omega f([x_{k-1}, x_k]) < \eta$$

on each member of the partition.

8.6.7 Show that the set of points at which a function F is discontinuous includes all points at which $|F|$ is discontinuous but not conversely. Deduce Corollary 8.21 as a result of this observation from Theorem 8.17.

8.6.8 Deduce Corollary 8.18 directly from Theorem 8.15 rather than from Theorem 8.17.

8.6.9 Deduce Corollary 8.19 directly from Theorem 8.15.

8.6.10 Deduce Corollary 8.20 directly from Theorem 8.15.

8.6.11 Show that the converse of Corollary 8.21 does not hold.

8.7 Properties of the Riemann Integral

✂ The proofs in this section make use of the Lebesgue criterion for integrability. You may skip the proofs and just see how the properties are essentially unchanged from Section 8.3 for Cauchy's original integral.

The Riemann integral is an extension of Cauchy's first integral from continuous functions to a larger class of bounded functions—those that are bounded and continuous except at the points of a small set (a set of measure zero). We have enlarged the class of functions to which the notion of an integral may be applied. Have we lost any of our crucial properties of Section 8.3?

These properties express how we expect integration to behave; it would be distressing to lose any of them. In some cases they remain completely unchanged. In some cases they need to be modified slightly. But our goal was never simply to integrate as many functions as possible; it is to preserve the theory of the integral and to apply that theory sufficiently broadly to handle all necessary applications. If we lose our basic properties we have lost too much. Fortunately the Riemann integral keeps all of the basic properties of the integral of continuous functions. The few differences should be carefully noted. Note especially how some of the properties must be rephrased.

8.22 (Additive Property) *If f is Riemann integrable on both intervals $[a, b]$ and $[b, c]$ then it is Riemann integrable on $[a, c]$ and*

$$\int_a^b f(x)\,dx + \int_b^c f(x)\,dx = \int_a^c f(x)\,dx.$$

Proof The proof of the identity need not change from the way we handled it for continuous functions (check this). It is the first assertion in the statement that must be verified. We prove that f is Riemann integrable on $[a, c]$.

By Theorem 8.17 if f is Riemann integrable on both of these intervals it is bounded on both and the set of points of discontinuity in each interval has measure zero. It follows that f is bounded on $[a, c]$. Also, its set of points of discontinuity in $[a, c]$ is the union of the set of points of discontinuity in $[a, b]$ and $[b, c]$ together with (possibly) the point b itself. Thus the set of points of discontinuity in $[a, c]$ is also of measure zero. Consequently, by Theorem 8.17, f is Riemann integrable. ■

8.23 (Linear Property) *If f and g are both Riemann integrable on $[a, b]$, then so too is any linear combination $\alpha f + \beta g$ and*

$$\int_a^b [\alpha f(x) + \beta g(x)]\,dx = \alpha \int_a^b f(x)\,dx + \beta \int_a^b g(x)\,dx.$$

Proof Again the proof of the identity does not change from the way we handled it for continuous functions (check this). It is the first assertion in the statement that needs to be verified. We must prove that $\alpha f + \beta g$ is Riemann integrable on on $[a, b]$.

The points of discontinuity of the function function $\alpha f + \beta g$ are either points of discontinuity of f or else they are points of discontinuity of g. If both functions f and g are Riemann integrable, then they are both bounded and continuous except at the points of a set of measure zero. It follows that $\alpha f + \beta g$ is bounded and continuous except at the points of a set of measure zero. Hence, by Theorem 8.17, $\alpha f + \beta g$ is Riemann integrable. ■

8.24 (Monotone Property) *If f and g are both Riemann integrable on $[a,b]$, then, if $f(x) \leq g(x)$ for all $a \leq x \leq b$,*

$$\int_a^b f(x)\,dx \leq \int_a^b g(x)\,dx.$$

Proof The proof for continuous functions works equally well here. ■

8.25 (Absolute Property) *If f is Riemann integrable on $[a,b]$, then so too is $|f|$ and*

$$-\int_a^b |f(x)|\,dx \leq \int_a^b f(x)\,dx \leq \int_a^b |f(x)|\,dx$$

or, equivalently,

$$\left|\int_a^b f(x)\,dx\right| \leq \int_a^b |f(x)|\,dx.$$

Proof The proof for continuous functions works equally well here because we have already shown, in Corollary 8.21 that if f is Riemann integrable on $[a,b]$, then so too is $|f|$. ■

Fundamental Theorem of Calculus The next two properties, 8.26 and 8.27, are important. They show how the processes of integration and differentiation are inverses of each other. Together they are known as the fundamental theorem of calculus for the Riemann integral. You should note, however, a weakness in this theory. If we compute F' we cannot immediately conclude from 8.27 that $\int_a^b F'(x)\,dx = F(b) - F(a)$. We need first to check that F' is Riemann integrable. This may not always be easy. Worse yet, it may be false, even for bounded derivatives (see the discussion in Section 9.7). It was this failure of the Riemann integral to integrate *all* derivatives that Lebesgue claimed was his motivation to look for a more general theory of integration.

8.26 (Differentiation of the Indefinite Integral) *If f is Riemann integrable on $[a,b]$ then the function*

$$F(x) = \int_a^x f(t)\,dt$$

is continuous on $[a,b]$ and $F'(x) = f(x)$ at each point x at which the function f is continuous.

Proof Once again, the proof for continuous functions works equally well here. Note, however, that we are no longer trying to prove that $F'(x) = f(x)$ at every point x, only at those points x where f is continuous.

It is left to you to check the proof and verify that it works here, unchanged. ■

8.27 (Integral of a Derivative) *Suppose that the function F is differentiable on $[a,b]$. Provided it is also true that F' is Riemann integrable on $[a,b]$, then*

$$\int_a^b F'(x)\,dx = F(b) - F(a).$$

Proof Again the proof for continuous functions works here. ∎

Exercises

8.7.1 Give a set of conditions under which the integration by substitution formula

$$\int_a^b f(\phi(t))\phi'(t)\,dt = \int_{\phi(a)}^{\phi(b)} f(x)\,dx$$

holds.

8.7.2 Give a set of conditions under which the integration by parts formula

$$\int_a^b f(t)g'(t)\,dt = f(b)g(b) - f(a)(g(a) - \int_a^b f'(t)g(t)\,dt$$

holds.

8.7.3 Suppose that f is Riemann integrable on $[a,b]$ and define the function

$$F(x) = \int_a^x f(t)\,dt.$$

(a) Show that F satisfies a Lipschitz condition on $[a,b]$; that is, that there exists $M > 0$ such that for every $x, y \in [a,b]$,

$$|F(y) - F(x)| \leq M|y-x|.$$

(b) If x is a point at which f is not continuous is it still possible that $F'(x) = f(x)$?

(c) Is it possible that $F'(x)$ exists but is not equal to $f(x)$?

(d) Is it possible that $F'(x)$ fails to exist?

8.7.4 The function

$$F(x) = \int_0^x \sin(1/t)\,dt$$

has a derivative at every point where the integrand is continuous. Does it also have a derivative at $x = 0$?

8.7.5 Improve Property 8.27 by assuming that F is continuous on $[a,b]$ and allowing that F' exists at all points of $[a,b]$ with finitely many exceptions.

8.7.6 Do much better than the preceding exercise and improve Property 8.27 by assuming that F is continuous on $[a,b]$ and allowing that F' exists at all points of $[a,b]$ with countably many exceptions. ✂

8.7.7 If f and g are Riemann integrable on an interval $[a, b]$ show that

$$\left(\int_a^b f(x)g(x)\,dx\right)^2 \leq \left(\int_a^b [f(x)]^2\,dx\right)\left(\int_a^b [g(x)]^2\,dx\right).$$

This extends the Cauchy-Schwarz inequality of Exercise 8.3.10.

8.7.8 Show that the integration by parts formula of Exercise 8.3.7 extends to the case where f and g are continuous and f' and g' are Riemann integrable.

✂ **8.7.9 (More on the Fundamental Theorem of Calculus)** Let f be bounded on $[a, b]$ and continuous a.e. on $[a, b]$. Suppose that F is defined on $[a, b]$ and that $F'(x) = f(x)$ for all x in $[a, b]$ except at the points of some set of measure zero.

(a) Is it necessarily true that $F(x) - F(a) = \int_a^x f(t)\,dt$ for every $x \in [a, b]$?

(b) Same question as in (a) but assume also that F is continuous.

(c) Same question, but this time assume that F is a Lipschitz function. You may assume the nonelementary fact that a Lipschitz function H with $H' = 0$ a.e. must be constant.

(d) Give an example of a Lipschitz function F such that F is differentiable, F' is bounded, but F' is not integrable.

8.8 The Improper Riemann Integral

✂ *Enrichment*

The Riemann integral applies only to bounded functions. What should we mean by the integral

$$\int_0^1 \frac{dx}{\sqrt{x}}?$$

Since the integrand is unbounded on $[0, 1]$, it is not Riemann integrable even though the integrand is continuous at all but one point. There is not much else for us to do but to back track by several decades and return to Cauchy's second method; namely, we compute

$$\lim_{\delta\to 0+} F(\delta) = \lim_{\delta\to 0+} \int_\delta^1 \frac{dx}{\sqrt{x}} = 2.$$

What we should probably do now is to create a new hybrid integral by combining the Riemann integral with Cauchy's second method. This is often called the *improper Riemann integral.* As before we give a definition that considers only one point of unboundedness (at the left endpoint of the interval) with the understanding that the ideas can be applied to any finite number of such points.

Definition 8.28 Let f be a function on an interval $(a,b]$ that is Riemann integrable on $[a+\delta,b]$ and that is unbounded in the interval $(a,a+\delta)$ for every $0<\delta<b-a$. Then we define

$$\int_a^b f(x)\,dx$$

to be

$$\lim_{\delta\to 0+}\int_{a+\delta}^b f(x)\,dx$$

if this limit exists, and in this case the integral is said to be *convergent.* If both integrals

$$\int_a^b f(x)\,dx \quad\text{and}\quad \int_a^b |f(x)|\,dx$$

converge the integral is said to be *absolutely convergent.*

In the same way we also extend the Riemann integral from bounded intervals to unbounded ones. How should we interpret the integral

$$\int_1^\infty \frac{dx}{x^2}?$$

This cannot exist as a Riemann integral since the definition is clearly restricted to finite intervals and would not allow any easy interpretation for infinite intervals. As before we use Cauchy's second method to obtain

$$\int_1^\infty \frac{dx}{x^2} = \lim_{X\to\infty}\int_1^X \frac{dx}{x^2} = \lim_{X\to\infty} 1-\frac{1}{X} = 1.$$

We give a formal definition valid just for an infinite interval of the form $[a,\infty)$. The case $(-\infty,b]$ is similar. The case $(-\infty,+\infty)$ is best split up into the sum of two integrals, from $(-\infty,a]$ and $[a,\infty)$, each of which can be handled in this fashion.

Definition 8.29 Let f be a function on an interval $[a,\infty)$ that is Riemann integrable on every interval $[a,b]$ for $a<b<\infty$. Then we define

$$\int_a^\infty f(x)\,dx$$

to be

$$\lim_{X\to\infty}\int_a^X f(x)\,dx$$

if this limit exists, and in this case the integral is said to be *convergent.* If both integrals

$$\int_a^\infty f(x)\,dx \quad\text{and}\quad \int_a^\infty |f(x)|\,dx$$

converge the integral is said to be *absolutely convergent.*

Both of these definitions extend the Riemann integral to a more general concept. Note that in any applications using an improper Riemann integral of either type, we are obliged to announce whether the integral is convergent or divergent, and frequently whether it is absolutely or nonabsolutely convergent.

It might seem that this theory would be important to master and represents the final word on the subject of integration. By the end of the nineteenth century it had become increasingly clear that this theory of the Riemann integral itself was completely inadequate to handle the bounded functions that were arising in many applications. The extra step here, using Cauchy's second method, designed to handle unbounded functions, also proved far too restrictive. The modern theory of integration was developed in the first decades of the twentieth century. The methods are different and even the language needs many changes.

Thus, the material in these last few sections has largely a historical interest. Some mathematicians claim it has only that, others that learning this material is a good preparation for learning the more advanced material.

Exercises

8.8.1 For what values of p, q are the integrals

$$\int_0^1 \frac{\sin x}{x^p}\,dx \text{ and } \int_0^1 \frac{(\sin x)^q}{x}\,dx$$

ordinary Riemann integrals, convergent improper Riemann integrals, or divergent improper Riemann integrals?

8.9 More on the Fundamental Theorem of Calculus

✂ *Enrichment*

The Riemann integral does not integrate all bounded derivatives and so the fundamental theorem of calculus for this integral assumes the awkward form

$$\int_a^b F'(x)\,dx = F(b) - F(a)$$

provided F is differentiable on $[a,b]$ and *the derivative F' is Riemann integrable there.*

The emphasized phrase is unfortunate. It means we have a limited theory and it also means that, in practice, we must always check to be sure that a derivative F' is integrable before proceeding to integrate it. In Section 9.7 we shall show how to construct a function F that is everywhere differentiable on an interval and whose derivative F' is bounded but not itself Riemann integrable on any subinterval.

Let us take another look at the integrability of derivatives to see if we can discover what goes wrong. We take a completely naive approach and start with the definition of the derivative itself. If $F' = f$ everywhere, then, at each point ξ and for every $\varepsilon > 0$, there is a $\delta > 0$ so that

$$|F(x'') - F(x') - f(\xi)(x'' - x')| < \varepsilon(x'' - x') \tag{6}$$

for $x' \leq \xi \leq x''$ and $0 < x'' - x' < \delta$.

A careless student might argue that one can recover $F(b) - F(a)$ as a limit of Riemann sums for f as follows. Let

$$a = x_0 < x_1 < x_2 \ldots x_n = b$$

be a partition of $[a, b]$, and let $\xi_i \in [x_{i-1}, x_i]$. Then

$$F(b) - F(a) = \sum_{i=1}^{n} (F(x_{i-1}) - F(x_i)) = \sum_{i=1}^{n} f(\xi_i)(x_i - x_{i-1}) + R$$

where

$$R = \sum_{i=1}^{n} (F(x_i) - F(x_{i-1}) - f(\xi_i)(x_i - x_{i-1})).$$

Thus $F(b) - F(a)$ has been given as a Riemann sum for f plus some error term R. But it appears now that, if the partition is finer than the number δ so that (6) may be used, we have

$$|R| \leq \sum_{i=1}^{n} \left| F(x_i) - F(x_{i-1}) - f(\xi_i)(x_i - x_{i-1}) \right|$$

$$< \sum_{i=1}^{n} \varepsilon(x_i - x_{i-1}) = \varepsilon(b - a).$$

Evidently, then, *if there are no mistakes here* it follows that f is Riemann integrable and that $\int_a^b f(t)\, dt = F(b) - F(a)$.

This is false, as we have mentioned, and we leave if for you in Exercise 8.9.1 to spot the error. But, instead of abandoning the argument, we can change the definition of the Riemann integral to allow this argument to work. The definition changes to look like this.

Definition 8.30 A function f is *generalized Riemann integrable* on $[a, b]$ with value I if for every ε there is a positive function δ on $[a, b]$ so that

$$\left| \sum_{i=1}^{n} f(\xi_i)(x_i - x_{i-1}) - I \right| < \varepsilon$$

whenever

$$a = x_0 < x_1 < x_2 < \cdots < x_n = b$$

is a partition of $[a, b]$ with $\xi_i \in [x_{i-1}, x_i]$ and $0 < x_i - x_{i-1} < \delta(\xi_i)$.

The integral $\int_a^b f(x)\,dx$ is taken as this number I that exists. It is easy to check that if f is Riemann integrable it is also generalized Riemann integrable and the integrals have the same value. Thus this new integral is an extension of the old one. To justify the definition requires knowing that such partitions actually exist for any such positive function δ; this is supplied by the Cousin theorem (Lemma 4.26).

This defines a Riemann-type integral that includes the usual Riemann integral and integrates *all* derivatives. The generalized Riemann integral was discovered in the 1950s, independently, by R. Henstock and J. Kurzweil, and these ideas have led to a number of other integration theories that exploit the geometry of the underlying space in the same way that this integral exploits the geometry of derivatives on the real line.

We shall not carry these ideas any further but refer you to the monographs of Pfeffer[1] or Gordon.[2]

Exercises

8.9.1 Spot the error in the careless student argument given in the text.

8.9.2 Develop the elementary properties of the generalized Riemann integral directly from its definition.

8.9.3 Show directly from the definition that the characteristic function of the rationals is not Riemann integrable but is generalized Riemann integrable on any interval and that $\int_0^1 f(x)\,dx = 0$.

8.9.4 Show that the generalized Riemann integral is closed under the extension procedure of Cauchy from Section 8.4.

8.10 Challenging Problems for Chapter 8

8.10.1 Let $m(f,\pi)$ and $M(f,\pi)$ denote the upper and lower sums over a partition π for a bounded function f. Define the upper and lower integrals as

$$\overline{\int_a^b} f(x)\,dx = \inf M(f,\pi) \quad \text{and} \quad \underline{\int_a^b} f(x)\,dx = \inf m(f,\pi)$$

where the inf and sup are taken over all possible partitions π of the interval $[a,b]$. We say f is *Darboux integrable* if the upper and lower integrals are equal.

(a) Show that $\underline{\int_a^b} f(x)\,dx \le \overline{\int_a^b} f(x)\,dx$.

[1]W. F. Pfeffer, *The Riemann Approach to Integration: Local Geometric Theory.* Cambridge University Press (1993).

[2]R. A. Gordon, *The Integrals of Lebesgue, Denjoy, Perron and Henstock.* Grad. Studies in Math., 4, Amer. Math. Soc. (1994).

(b) Show that every Riemann integrable function is Darboux integrable.

(c) Show that every Darboux integrable function is Riemann integrable.

(d) Show that if f is Riemann integrable, then

$$\overline{\int_a^b} f(x)\,dx = \underline{\int_a^b} f(x)\,dx = \int_a^b f(x)\,dx.$$

(e) Show that

$$\overline{\int_a^b} (f(x) + g(x))\,dx \le \overline{\int_a^b} f(x)\,dx + \overline{\int_a^b} g(x)\,dx$$

with strict inequality possible.

8.10.2 Let $f : [0,1] \to \mathbb{R}$ be a differentiable function such that $|f'(x)| \le M$ for all $x \in (0,1)$. Show that

$$\left| \int_0^1 f(x)\,dx - \frac{1}{n}\sum_{k=1}^{n} f\left(\frac{k}{n}\right) \right| \le \frac{M}{n}.$$

Chapter 9

SEQUENCES AND SERIES OF FUNCTIONS

✂ If the material on series in Chapter 3 was omitted in a first reading, then Sections 3.4, 3.5, and parts of 3.6 should be studied before attempting this chapter.

9.1 Introduction

We have seen that a function f that is the sum of two or more functions will share certain desirable properties with those functions. For example, our study of continuity, differentiation, and integration allows us to state that if

$$f = f_1 + f_2 + \cdots + f_n$$

on an interval $I = [a, b]$, then

(1) If $f_1, f_2, \ldots, f_n$ are continuous on I, so is f.

(2) If $f_1, f_2, \ldots, f_n$ are differentiable on I, so is f, and

$$f' = f_1' + f_2' + \cdots + f_n'.$$

(3) If $f_1, f_2, \ldots, f_n$ are integrable on I, so is f, and

$$\int_a^b f(x)\, dx = \int_a^b f_1(x)\, dx + \int_a^b f_2(x)\, dx + \cdots + \int_a^b f_n(x)\, dx.$$

It is natural to ask whether the corresponding results hold when f is the sum of an *infinite* series of functions,

$$f = \sum_{k=0}^{\infty} f_k.$$

If each term of the series is continuous, is the sum function also continuous? Can the derivative be obtained by summing the derivatives? Can the integral be obtained by summing the integrals? We study such questions in this chapter.

These problems are of considerable practical importance. For example, if we are allowed to take limits, integrate, and differentiate freely, then the computations in the following example would all be valid.

Example 9.1 From the formula for the sum of a geometric series we know that

$$\frac{1}{1+x} = 1 - x + x^2 - x^3 + x^4 - x^5 + \ldots \tag{1}$$

on the interval $(-1, 1)$. Differentiation of both sides of (1) leads immediately to

$$\frac{-1}{(1+x)^2} = -1 + 2x - 3x^2 + 4x^3 - 5x^4 + \ldots.$$

Repeated differentiation would give formulas for $(1+x)^{-n}$ for all positive integers n.

On the other hand, integration of both sides of (1) from 0 to t leads immediately to

$$\ln(1+t) = t - \frac{1}{2}t^2 + \frac{1}{3}t^3 - \frac{1}{4}t^4 + \frac{1}{5}t^5 - \ldots.$$

Taking limits as $t \to 1$ in the latter yields the intriguing formula for the sum of the alternating harmonic series:

$$\ln 2 = 1 - \frac{1}{2} + \frac{1}{3} - \frac{1}{4} + \frac{1}{5} - \ldots$$

◀

The conclusions in the example are all true and useful. But have we used illegitimate means to find them? If we use such methods freely might we find situations where our conclusions are wrong?

We first formulate our questions in the language of sequences of functions (rather than series). We do this in Section 9.2, where we see that the answer to our questions is "not necessarily." Then in Sections 9.3–9.6 we see that if we require a bit more of convergence, the answer to each of our questions is "yes."

9.2 Pointwise Limits

Suppose $f_1, f_2, f_3, \ldots$ is a sequence of functions, each of which is defined on a common domain D. What should we mean by the sum $f = \sum_{k=0}^{\infty} f_k$? Perhaps the simplest notion for the sum is to extend the definition of finite sum using our familiar interpretation of convergence of an infinite series of numbers as a limit of the sequence of partial sums. We consider this idea first.

Definition 9.2 For each x in D and $n \in \mathbb{N}$ let

$$S_n(x) = f_1(x) + \cdots + f_n(x).$$

If $\lim_{n\to\infty} S_n(x)$ exists (as a real number), we say the series $\sum_{k=1}^{\infty} f_k$ *converges at* x and we write

$$\sum_{k=1}^{\infty} f_k(x)$$

for $\lim_{n\to\infty} S_n(x)$. If the series converges for all $x \in D$, we say the series *converges pointwise* on D to the function f defined by

$$f(x) = \sum_{k=1}^{\infty} f_k(x) \quad (= \lim_{n\to\infty} \sum_{k=1}^{n} f_k(x)).$$

We would like such infinite sums of functions to behave like finite sums of functions (as our three questions in Section 9.1 suggest): If $f = \sum_1^{\infty} f_k$ on an interval $I = [a, b]$, is it true that

(1) If f_k is continuous on I for all $k \in \mathbb{N}$, then so is f?

(2) If f_k is differentiable on I for all $k \in \mathbb{N}$, then so is f and

$$f'(x) = \sum_{k=1}^{\infty} f_k'(x)?$$

(3) If f_k is integrable on I for all $k \in \mathbb{N}$, then so is f, and

$$\int_a^b f(x)\, dx = \sum_{k=1}^{\infty} \int_a^b f_k(x)\, dx?$$

Let us reformulate our questions in the language of sequences.

Definition 9.3 Let $\{f_n\}$ be a sequence of functions defined on a common domain D. If $\lim_{n\to\infty} f_n(x)$ exists (as a real number) for all $x \in D$, we say that the sequence $\{f_n\}$ *converges pointwise* on D. This limit defines a function f on D by the equation

$$f(x) = \lim_n f_n(x).$$

We write $\lim_n f_n = f$ or $f_n \to f$.

In the special case that D is an interval $I = [a, b]$ our questions then become the following: Is it true that

1. If f_n is continuous on I for all n, then is f continuous on I?

2. If f_n is differentiable on I for all n, then is f differentiable on I and, if so, does $f' = \lim_n f_n'$?

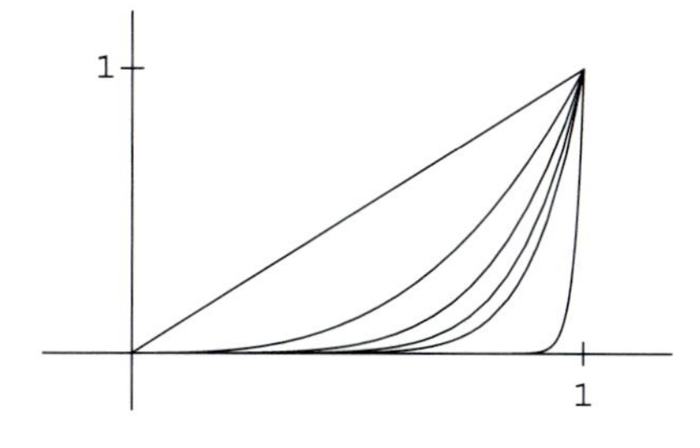

Figure 9.1. Graphs of x^n on $[0, 1]$ for $n = 1, 3, 5, 7, 9$, and 50.

3. If f_n is integrable on I for all n, then is f integrable on I and, if so, does $\int_a^b f(x)\, dx = \lim_n \int_a^b f_n(x)\, dx$?

These questions have *negative* answers in general, as the three examples that follow show.

Example 9.4 (A discontinuous limit of continuous functions) For each $n \in \mathbb{N}$ and $x \in [0, 1]$, let $f_n(x) = x^n$. Each of the functions is continuous on $[0, 1]$. Notice, however, that for each $x \in (0, 1)$, $\lim_n f_n(x) = 0$ and yet $\lim_n f_n(1) = 1$. This is easy to see, but it is instructive to check the details since we can use them later to see what is going wrong in this example. At the right-hand endpoint it is clear that, for $x = 1$, $\lim_n f_n(x) = 1$. For $0 < x_0 < 1$ and $\varepsilon > 0$, let $N \geq \ln \varepsilon / \ln x_0$. Then $(x_0)^N \leq \varepsilon$, so for $n \geq N$

$$|f_n(x_0) - 0| = (x_0)^n < (x_0)^N \leq \varepsilon.$$

Thus

$$f(x) = \lim_n f_n(x) = \begin{cases} 0 & \text{if } 0 \leq x < 1 \\ 1 & \text{if } x = 1, \end{cases}$$

so the pointwise limit f of the sequence of continuous functions $\{f_n\}$ is discontinuous at $x = 1$. (Figure 9.1 shows the graphs of several of the functions in the sequence.) ◀

Example 9.5 (The derivative of the limit is not the limit of the derivative.) Let $f_n(x) = x^n/n$. Then $f_n \to 0$ on $[0, 1]$. Now $f_n'(x) = x^{n-1}$, so by the previous example, Example 9.4,

$$\lim_n f_n'(x) = x^{n-1} = \begin{cases} 0 & \text{if } 0 \leq x < 1 \\ 1 & \text{if } x = 1, \end{cases}$$

while the derivative of the limit function, $f \equiv 0$, equals zero on $[0, 1]$. Thus

$$\lim_{n\to\infty} \frac{d}{dx}(f_n(x)) \neq \frac{d}{dx}\left(\lim_{n\to\infty} f_n(x)\right)$$

at $x = 1$. ◀

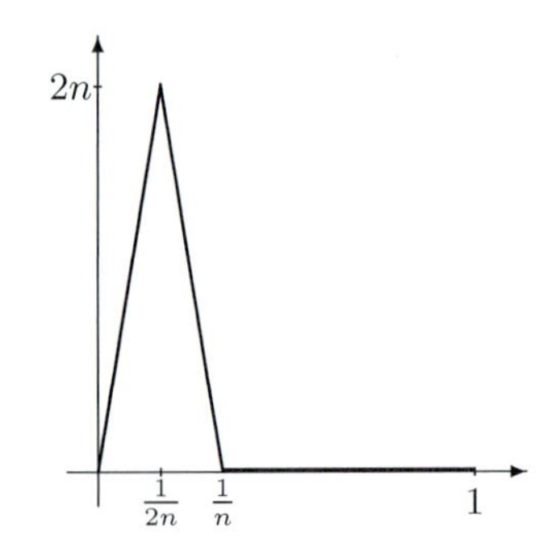

Figure 9.2. Graph of $f_n(x)$ on $[0,1]$ in Example 9.6.

Example 9.6 (The integral of the limit is not the limit of the integrals.) In this example we consider a sequence of continuous functions, each of which has the same integral over the domain. For each $n \in \mathbb{N}$ let f_n be defined on $[0,1]$ as follows: $f_n(0) = 0$, $f_n(1/(2n)) = 2n$, $f_n(1/n) = 0$, f_n is linear on $[0, 1/(2n)]$ and on $[1/(2n), 1/n]$, and $f_n = 0$ on $[1/n, 1]$. (See Figure 9.2.)

It is easy to verify that $f_n \to 0$ on $[0,1]$. Now, for each $n \in \mathbb{N}$,

$$\int_0^1 f_n(x)\, dx = 1.$$

But

$$\int_0^1 (\lim_n f_n(x))\, dx = \int_0^1 0\, dx = 0.$$

Thus

$$\lim_n \int_0^1 f_n(x)\, dx \neq \int_0^1 \lim_n f_n(x)\, dx$$

so that the limit of the integrals is not the integral of the limit. ◀

These examples show that the answer to each of our three questions is negative, in general. We present some additional examples that illustrate similar phenomena in the exercises.

We shall see in the next few sections that by replacing pointwise convergence in appropriate places with a stronger form of convergence, the answers to our questions become affirmative. The form of convergence in question is called *uniform convergence.*

Interchange of Limit Operations Before turning to uniform convergence, let us first try to get an insight into a difficulty we must overcome if we wish affirmative answers to our questions. If f_n is a sequence of continuous functions converging to a function f, must f be continuous? Continuity of f at a point x_0 would mean that

$$\lim_{x\to x_0} f(x) = f(x_0)$$

and this would require that

$$\lim_{x\to x_0} \left(\lim_{n\to\infty} f_n(x) \right) = \lim_{n\to\infty} f_n(x_0) = \lim_{n\to\infty} \left(\lim_{x\to x_0} f_n(x) \right).$$

Apparently, to verify the continuity of f at x_0 we need to use two limit operations and be assured that the order of passing to the limits is immaterial.

You will remember situations in which two limit operations are involved and the order of taking the limit does not affect the result. For example, in elementary calculus one finds conditions under which the value of a double integral can be obtained by iterating "single integrals" in either order. By way of contrast, we present an example in the setting of double sequences in which the order of taking limits *is* important.

Example 9.7 In this example we illustrate that an interchange of limit operations may not give a correct result. Let

$$S_{mn} = \begin{cases} 0, & \text{if } m \le n \\ 1, & \text{if } m > n. \end{cases}$$

Viewed as a matrix,

$$[S_{mn}] = \begin{bmatrix} 0 & 0 & 0 & \cdots \\ 1 & 0 & 0 & \cdots \\ 1 & 1 & 0 & \cdots \\ \vdots & \vdots & \vdots & \ddots \end{bmatrix}$$

where we are placing the entry S_{mn} in the mth row and nth column. For each row m, we have $\lim_{n\to\infty} S_{mn} = 0$, so

$$\lim_{m\to\infty} \left(\lim_{n\to\infty} S_{mn} \right) = 0.$$

On the other hand, for each column n, $\lim_{m\to\infty} S_{mn} = 1$, so

$$\lim_{n\to\infty} \left(\lim_{m\to\infty} S_{mn} \right) = 1.$$

◀

Exercises

9.2.1 Examine the pointwise limiting behavior of the sequence of functions

$$f_n(x) = \frac{x^n}{1+x^n}.$$

9.2.2 Show that the logarithm function can be expressed as the pointwise limit of a sequence of "simpler" functions,

$$\ln x = \lim_{n\to\infty} n \left(\sqrt[n]{x} - 1 \right)$$

for every point in its domain. If the answer to our three questions for this particular limit is affirmative, what can you say about the continuity of the logarithm function? What would be its derivative? What would be $\int_1^2 \ln x\, dx$?

9.2.3 Let $x_1, x_2, \ldots$ be an enumeration of $\mathbb{Q}$, let

$$f_n(x) = \begin{cases} 1, & \text{if } x \in \{x_1, \ldots, x_n\} \\ 0, & \text{otherwise,} \end{cases}$$

and let

$$f(x) = \begin{cases} 1, & \text{if } x \in \mathbb{Q} \\ 0, & \text{otherwise.} \end{cases}$$

Show that $f_n \to f$ pointwise on $[0,1]$, but $\int_0^1 f_n(x)\, dx = 0$ for all $n \in \mathbb{N}$, while f is not integrable on $[0,1]$.

9.2.4 Let $f_n(x) = \sin nx/\sqrt{n}$. Show that $\lim_n f_n = 0$ but $\lim_n f_n'(0) = \infty$.

9.2.5 Each of Examples 9.4, 9.5 and 9.6 can be interpreted as a statement that the order of taking the limit operation does matter. Verify this.

9.2.6 Refer to Example 9.7. What should we mean by the statement that a "double sequence" $\{t_{mn}\}$ converges; that is, that

$$\lim_{m\to\infty, n\to\infty} t_{mn}$$

exists? Does the double sequence $\{S_{mn}\}$ of Example 9.7 converge?

9.2.7 Let $f_n \to f$ pointwise at every point in the interval $[a,b]$. We have seen that even if each f_n is continuous it does not follow that f is continuous. Which of the following statements are true?

(a) If each f_n is increasing on $[a,b]$, then so is f.

(b) If each f_n is nondecreasing on $[a,b]$, then so is f.

(c) If each f_n is bounded on $[a,b]$, then so is f.

(d) If each f_n is everywhere discontinuous on $[a,b]$, then so is f.

(e) If each f_n is constant on $[a,b]$, then so is f.

(f) If each f_n is positive on $[a,b]$, then so is f.

(g) If each f_n is linear on $[a,b]$, then so is f.

(h) If each f_n is convex on $[a,b]$, then so is f.

9.2.8 A careless student[1] once argued as follows: "It seems to me that one can construct a curve without a tangent in a very elementary way. We divide the diagonal of a square into n equal parts and construct on each subdivision

[1]In this case the "careless student" was the great Russian analyst N. N. Luzin (1883–1950), who recounted in a letter [reproduced in *Amer. Math. Monthly*, 107, (2000), pp. 64–82] how he offered this argument to his professor after a lecture on the Weierstrass continuous nowhere differentiable function.

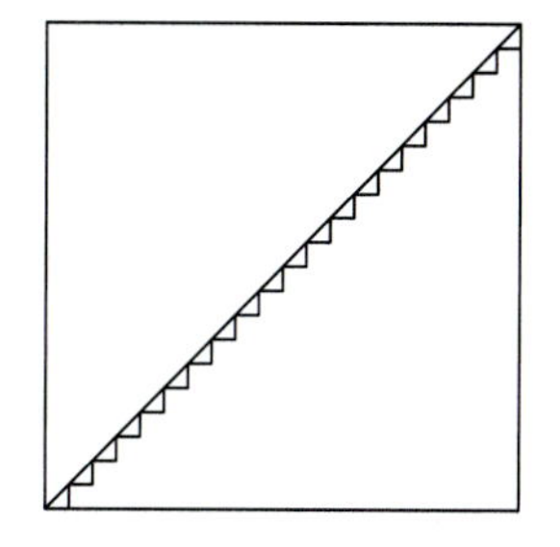

Figure 9.3. Construction in Exercise 9.2.8.

as base a right isoceles triangle. In this way we get a kind of delicate little saw. Now I put $n = \infty$. The saw becomes a continuous curve that is infinitesimally different from the diagonal. But it is perfectly clear that its tangent is alternately parallel now to the x-axis, now to the y-axis." What is the error? (Figure 9.3 illustrates the construction.)

9.2.9 As yet another illustration that some properties are not preserved in the limit, compute the length of the curves in Exercise 9.2.8 (Fig. 9.3) and compare with the length of the limiting curve.

9.2.10 If $f_n \to f$ pointwise at every real number, then prove that

$$\{x : f(x) > \alpha\} = \bigcup_{m=1}^{\infty} \bigcup_{r=1}^{\infty} \bigcap_{n=r}^{\infty} \{x : f_n(x) \geq \alpha + 1/m\}.$$

9.2.11 Let $\{f_n\}$ be a sequence of real functions. Show that the set E of points of convergence of the sequence can be written in the form

$$E = \bigcap_{k=1}^{\infty} \bigcup_{N=1}^{\infty} \bigcap_{n=N}^{\infty} \bigcap_{m=N}^{\infty} \left\{x : |f_n(x) - f_m(x)| \leq \tfrac{1}{k}\right\}.$$

9.3 Uniform Limits

Pointwise limits do not allow the interchange of limit operations. In many situations, uniform limits will. To see how the definition of a uniform limit needs to be formulated, let us return to the sequence of Example 9.4. That sequence illustrated the fact that a pointwise limit of continuous functions need not be continuous. The difficulty there was that

$$\lim_{x\to 1-} \left(\lim_{n\to\infty} f_n(x)\right) \neq \lim_{n\to\infty} \left(\lim_{x\to 1-} f_n(x)\right).$$

A closer look at the limits involved here shows what went wrong and suggests what we need to look for in order to allow an interchange of limits.

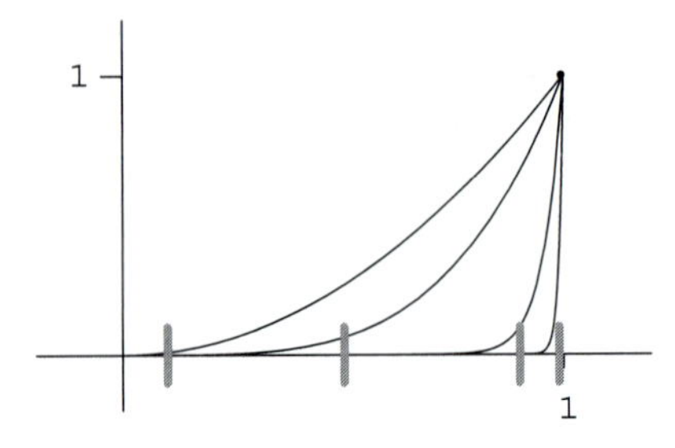

Figure 9.4. The sequence $\{x^n\}$ converges infinitely slowly on $[0,1]$. The functions $y = x^n$ are shown with $n = 2$, 4, 22, and 100, with $x_0 = .1$, .5, .9, and .99, and with $\varepsilon = .1$.

Example 9.8 Consider again the sequence $\{f_n\}$ of functions $f_n(x) = x^n$. We saw that $f_n \to 0$ pointwise on $[0,1)$, and that for every fixed $x_0 \in (0,1)$ and $\varepsilon > 0$,

$$|x_0|^n < \varepsilon \quad \text{if and only if} \quad n > \ln\varepsilon / \ln x_0.$$

Now fix ε but let the point x_0 vary. Observe that, when x_0 is relatively small in comparison with ε, the number $\ln x_0$ is large in absolute value compared with $\ln\varepsilon$, so relatively small values of n suffice for the inequality $|x_0|^n < \varepsilon$. On the other hand, when x_0 is near 1, $\ln x_0$ is small in absolute value, so $\ln\varepsilon/\ln x_0$ will be large. In fact,

$$\lim_{x_0 \to 1-} \frac{\ln\varepsilon}{\ln x_0} = \infty. \tag{2}$$

The following table illustrates how large n must be before $|x_0^n| < \varepsilon$ for $\varepsilon = .1$.

x_0	n
.1	2
.5	4
.9	22
.99	230
.999	2,302
.9999	23,025

Note that for $\varepsilon = .1$, there is no single value of N such that $|x_0|^n < \varepsilon$ for every value of $x_0 \in (0,1)$ and $n > N$. (Figure 9.4 illustrates this.) ◀

Some nineteenth-century mathematicians would have described the varying rates of convergence in the example by saying that "the sequence $\{x^n\}$ converges *infinitely slowly* on $(0,1)$." Today we would say that this sequence, which does converge pointwise, does *not* converge uniformly. Our definition is formulated precisely to avoid this possibility of infinitely slow convergence.

Definition 9.9 Let $\{f_n\}$ be a sequence of functions defined on a common domain D. We say that $\{f_n\}$ *converges uniformly* to a function f on D if, for every $\varepsilon > 0$, there exists $N \in \mathbb{N}$ such that

$$|f_n(x) - f(x)| < \varepsilon \text{ for all } n \geq N \text{ and } x \in D.$$

We write

$$f_n \to f \text{ [unif] on } D \text{ or } \lim_n f_n = f \text{ [unif] on } D$$

to indicate that the sequence $\{f_n\}$ converges uniformly to f on D. If the domain D is understood from the context, we may delete explicit reference to D and write

$$f_n \to f \text{ [unif] or } \lim_n f_n = f \text{ [unif]}.$$

Uniform convergence plays an important role in many parts of analysis. In particular, it figures in questions involving the interchanging of limit processes such as those we discussed in Section 9.2. This was not apparent to mathematicians in the early part of the nineteenth century. As late as 1823, Cauchy believed a convergent series of continuous functions could be integrated term by term. Similarly, Cauchy believed that a convergent series of continuous functions has a continuous sum. Abel provided a counterexample in 1826. It may have been Weierstrass who first recognized the importance of uniform convergence in the middle of the nineteenth century.[2]

Example 9.10 Let $f_n(x) = x^n$, $D = [0, \eta]$, $0 < \eta < 1$. We observed that the sequence $\{f_n\}$ converges pointwise, but not uniformly, on $(0, 1)$ (or on $[0, 1]$). We realized that the difficulty arises from the fact that the convergence near 1 is very "slow." But for any fixed η with $0 < \eta < 1$, the convergence *is* uniform on $[0, \eta]$.

To see this, observe that for $0 \leq x_0 < \eta$, $0 \leq (x_0)^n < \eta^n$. Let $\varepsilon > 0$. Since $\lim_n \eta^n = 0$, there exists N such that if $n \geq N$, then $0 < \eta^n < \varepsilon$. Thus, if $n \geq N$, we have

$$0 \leq x_0^n < \eta^n < \varepsilon,$$

so the same N that works for $x = \eta$, also works for all $x \in [0, \eta)$. ◀

Suppose that $f_n \to f$ on $[0, 1]$. It follows easily from the definition that the convergence is uniform on any *finite* subset D of $[0, 1]$ (Exercise 9.3.3). Thus given any $\varepsilon > 0$ and any finite set $x_1, x_2, \ldots, x_m$ in [0,1], we can find $n \in \mathbb{N}$ such that

$$|f_n(x_i) - f(x_i)| < \varepsilon$$

for all $n \geq N$ and all $i = 1, 2, \ldots, m$. (Figure 9.5 illustrates this.)

[2]More on the history of uniform convergence can be found in Thomas Hawkins' interesting historical book *Lebesgue's Theory of Integration: Its Origins and Development,* Univ. of Wisconsin Press (Madison, 1970)

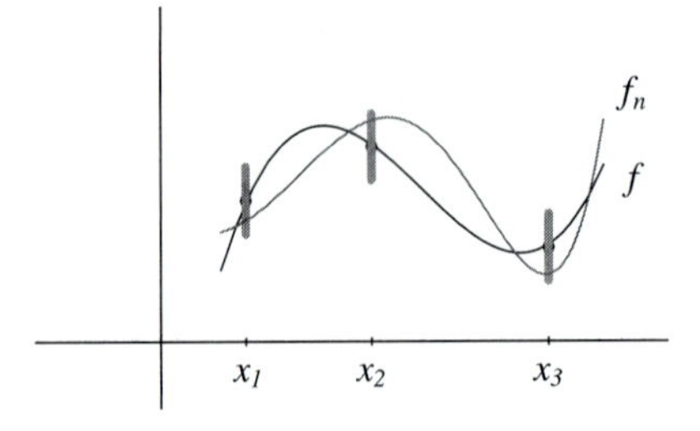

Figure 9.5. Uniform convergence on the finite set $\{x_1, x_2, x_3\}$.

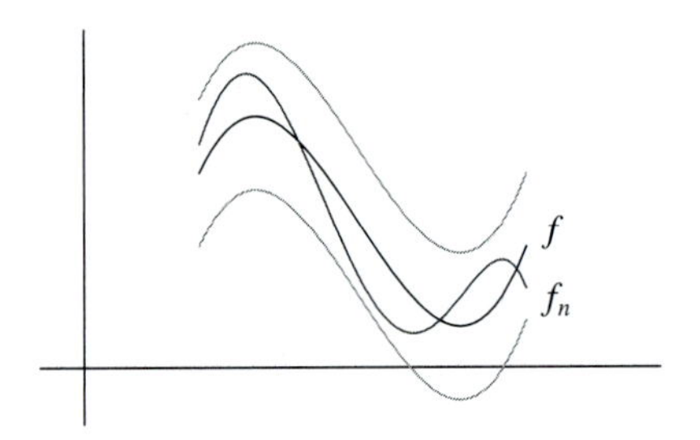

Figure 9.6. Uniform convergence on the whole interval.

The vertical line segments over the points $x_1, \ldots, x_m$ are centered on the graph of f and are of length 2ε. In simple geometric language, we can go sufficiently far out in the sequence to guarantee that the graphs of all the functions f_n intersect all of these finitely many vertical segments.

In contrast, uniform convergence on $[0, 1]$ requires that we can go sufficiently far out in the sequence to guarantee that the graphs of the functions go through such vertical segments at *all* points of $[0, 1]$; that is, that the graph of f_n for n sufficiently large lies in the "ε-band" centered on the graph of f. (See Figure 9.6.)

9.3.1 The Cauchy Criterion

Suppose now that we are given a sequence of functions $\{f_n\}$ on an interval I, and we wish to know whether it converges uniformly to some function on I. We are not told what that limit function might be. The problem is similar to one we faced for a sequence of *numbers* $\{a_n\}$ in our study of sequences. There we saw that $\{a_n\}$ converges if and only if it is a Cauchy sequence. We can formulate a similar criterion for uniform convergence of a sequence of functions.

Definition 9.11 Let $\{f_n\}$ be a sequence of functions defined on a set D. The sequence is said to be *uniformly Cauchy* on D if for every $\varepsilon > 0$ there exists $N \in \mathbb{N}$ such that if $n \geq N$ and $m \geq N$, then $|f_m(x) - f_n(x)| < \varepsilon$ for all $x \in D$.

Theorem 9.12 (Cauchy Criterion) *Let $\{f_n\}$ be a sequence of functions defined on a set D. Then there exists a function f defined on D such that $f_n \to f$ uniformly on D if and only if $\{f_n\}$ is uniformly Cauchy.*

Proof We leave the proof of Theorem 9.12 as Exercise 9.3.15. ■

Example 9.13 In Example 9.10 we showed that the sequence $f_n(x) = x^n$ converges uniformly on any interval $[0, \eta]$, for $0 < \eta < 1$. Let us prove this again, but using the Cauchy criterion.

Fix $n \geq m$ and compute

$$\sup_{x\in[0,\eta]} |x^n - x^m| \leq \eta^m. \tag{3}$$

Let $\varepsilon > 0$ and choose an integer N so that $\eta^N < \varepsilon$. Equivalently we require that $N > \ln \varepsilon / \ln \eta$. Then it follows from (3) for all $n \geq m \geq N$ and all $x \in [0, \eta]$ that

$$|x^n - x^m| \leq \eta^m < \varepsilon.$$

We conclude, by the Cauchy criterion, that the sequence $f_n(x) = x^n$ converges uniformly on any interval $[0, \eta]$, for $0 < \eta < 1$. Here there was no computational advantage over the argument in Example 9.10. Frequently, though, we do not know the limit function and *must* use the Cauchy criterion rather than the definition. ◀

Cauchy Criterion for Series The Cauchy criterion can be expressed for uniformly convergent series too. We say that a series $\sum_1^\infty f_k$ converges *uniformly* to the function f on D if the sequence $\{S_n\} = \{\sum_{k=1}^n f_k\}$ of partial sums converges uniformly to f on D.

Theorem 9.14 (Cauchy Criterion) *Let $\{f_n\}$ be a sequence of functions defined on a set D. Then the series $\sum_1^\infty f_k$ converges uniformly to some function f on D if and only if for every $\varepsilon > 0$ there is an integer N so that*

$$\left|\sum_{j=m}^{n} f_j(x)\right| < \varepsilon$$

for all $n \geq m \geq N$ and all $x \in D$.

Proof This follows immediately from Theorem 9.12. ■

Example 9.15 Let us show that the series

$$1 + x + x^2 + x^3 + x^4 + \ldots$$

converges uniformly on any interval $[0, \eta]$, for $0 < \eta < 1$. Our computations could be based on the fact that the sum of this series is known to us; it is $(1-x)^{-1}$. We could prove the uniform convergence directly from the definition. Instead let us use the Cauchy criterion.

Fix $n \geq m$ and compute

$$\sup_{x\in[0,\eta]} \left|\sum_{j=m}^{n} x^j\right| \leq \sup_{x\in[0,\eta]} \left|\frac{x^m}{1-x}\right| \leq \frac{\eta^m}{1-\eta}. \tag{4}$$

Let $\varepsilon > 0$. Since

$$\eta^m(1-\eta)^{-1} \to 0$$

as $m \to \infty$ we may choose an integer N so that

$$\eta^N(1-\eta)^{-1} < \varepsilon.$$

Then it follows from (4) for all $n \geq m \geq N$ and all $x \in [0, \eta]$ that

$$\left|x^m + x^{m+1} + \cdots + x^n\right| \leq \frac{\eta^m}{1-\eta} < \varepsilon.$$

It follows now, by the Cauchy criterion, that the series converges uniformly on any interval $[0, \eta]$, for $0 < \eta < 1$. Observe, however, that the series does not converge uniformly on $(-1, 1)$, though it does converge pointwise there. (See Exercise 9.3.16.) ◀

9.3.2 Weierstrass M-Test

It is not always easy to determine whether a sequence of functions is uniformly convergent. In the settings of *series* of functions, a certain simple test is often useful. This will certainly become one of the most frequently used tools in your study of uniform convergence.

Theorem 9.16 (M-Test) *Let $\{f_k\}$ be a sequence of functions defined on a set D and let $\{M_k\}$ be a sequence of positive constants. If*

$$\sum_{0}^{\infty} M_k < \infty$$

and if

$$|f_k(x)| \leq M_k$$

for each $x \in D$ and $k = 0, 1, 2, \ldots$, then the series $\sum_0^\infty f_k$ converges uniformly on D.

Proof Let $S_n(x) = \sum_{k=0}^{n} f_k(x)$. We show that $\{S_n\}$ is uniformly Cauchy on D. Let $\varepsilon > 0$. For $m < n$ we have

$$S_n(x) - S_m(x) = f_{m+1}(x) + \cdots + f_n(x),$$

so

$$|S_n(x) - S_m(x)| \leq M_{m+1} + \cdots + M_n.$$

Since the series of constants $\sum_{k=0}^{\infty} M_k$ converges by hypothesis, there exists an integer N such that if $n > m \geq N$,

$$M_{m+1} + \cdots + M_n < \varepsilon.$$

This implies that for $n > m \geq N$,

$$|S_n(x) - S_m(x)| < \varepsilon$$

for all $x \in D$. Thus the sequence $\{S_n\}$ is uniformly convergent on D; that is, the series $\sum_0^{\infty} f_k$ is uniformly convergent on D. ■

Example 9.17 Consider again the geometric series $1 + x + x^2 + \ldots$ on the interval $[-a, a]$, for any $0 < a < 1$ (as we did in Example 9.15). Then $|x^k| \leq a^k$ for every $k = 0, 1, 2 \ldots$ and $x \in [-a, a]$. Since $\sum_{k=0}^{\infty} a^k$ converges, by the M-test the series $\sum_{k=0}^{\infty} x^k$ converges uniformly on $[-a, a]$. ◀

Example 9.18 Let us investigate the uniform convergence of the series

$$\sum_{k=1}^{\infty} \frac{\sin k\theta}{k^p}$$

for values of $p > 0$. The crudest estimate on the size of the terms in this series is obtained just by using the fact that the sine function never exceeds 1 in absolute value. Thus

$$\left| \frac{\sin k\theta}{k^p} \right| \leq \frac{1}{k^p} \quad \text{for all } \theta \in \mathbb{R}.$$

Since the series $\sum_{k=1}^{\infty} 1/k^p$ converges for $p > 1$, we obtain immediately by the M-test that our series converges uniformly (and absolutely) for all real θ provided $p > 1$. In particular, as we shall see in subsequent sections, this series represents a continuous function, one that could be integrated term by term in any bounded interval.

We seem to have been particularly successful here, but a closer look also reveals a limitation in the method. The series is also pointwise convergent for $0 < p \leq 1$ (use the Dirichlet test) for all values of θ, but it converges nonabsolutely. The M-test cannot be of any help in this situation since it can address only absolutely convergent series. ◀

Because of the remark at the end of this example, it is perhaps best to conclude, when using the M-test, that the series tested "converges absolutely and uniformly" on the set given. This serves, too, to remind us to use a different method for checking uniform convergence of nonabsolutely convergent series (see the next section).

9.3.3 Abel's Test for Uniform Convergence

Advanced

The M-test is a highly useful tool for checking the uniform convergence of a series. By its nature, though, it clearly applies only to absolutely convergent series. For a more delicate test that will apply to some nonabsolutely convergent series we should search through our methods in Chapter 3 for tests that handled nonabsolute convergence. Two of these, the Dirichlet test and Abel's test, can be modified so as to give uniform convergence.

A number of nineteenth century authors (including Abel, Dirichlet, Dedekind, and du Bois-Reymond) arrived at similar tests for uniform convergence. We recall that Abel's test for convergence of a series $\sum_{k=1}^{\infty} a_k b_k$ required the sequence $\{b_k\}$ to be convergent and monotone and for the series $\sum_{k=1}^{\infty} a_k$ to converge. Dirichlet's variant weakened the latter requirement so that $\sum_{k=1}^{\infty} a_k$ had bounded partial sums but required of the sequence $\{b_k\}$ that it converge monotonically to zero. Here we seek similar conditions on a series

$$\sum_{k=1}^{\infty} a_k(x) b_k(x)$$

of functions in order to obtain uniform convergence. The next theorem is one variant; others can be found in the Exercises.

Theorem 9.19 (Abel) *Let $\{a_k\}$ and $\{b_k\}$ be sequences of functions on a set $E \subset \mathbb{R}$. Suppose that there is a number M so that*

$$-M \leq s_N(x) = \sum_{k=1}^{N} a_k(x) \leq M$$

for all $x \in E$ and every $N \in \mathbb{N}$. Suppose that the sequence of functions $\{b_k\} \to 0$ converges monotonically to zero at each point and that this convergence is uniform on E. Then the series

$$\sum_{k=1}^{\infty} a_k b_k$$

converges uniformly on E.

Proof We will use the Cauchy criterion applied to the series to obtain uniform convergence. We may assume that the $b_k(x)$ are nonnegative and decrease to zero. Let $\varepsilon > 0$. We need to estimate the sum

$$\left| \sum_{k=m}^{n} a_k(x) b_k(x) \right| \tag{5}$$

for large n and m and all $x \in E$. Since the sequence of functions $\{b_k\}$ converges uniformly to zero on E, we can find an integer N so that for all $k \geq N$ and all $x \in E$

$$0 \leq b_k(x) \leq \frac{\varepsilon}{2M}.$$

The key to estimating the sum (5), now, is the summation by parts formula that we have used earlier (see Section 3.2). This is just the elementary identity

$$\sum_{k=m}^{n} a_k b_k = \sum_{k=m}^{n} (s_k - s_{k-1}) b_k$$

$$= s_m(b_m - b_{m+1}) + s_{m+1}(b_{m+1} - b_{m+2}) \cdots + s_{n-1}(b_{n-1} - b_n) + s_n b_n.$$

This provides us with

$$\left| \sum_{k=m}^{n} a_k(x) b_k(x) \right| \leq 2M \left(\sup_{x \in E} |b_m(x)| \right) < \varepsilon$$

for all $n \geq m \geq N$ and all $x \in E$ which is exactly the Cauchy criterion for the series and proves the theorem. ■

It is worth pointing out that in many applications of this theorem the sequence $\{b_k\}$ can be taken as a sequence of numbers, in which case the statement and the conditions that need to be checked are simpler.

Corollary 9.20 *Let $\{a_k\}$ be a sequence of functions on a set $E \subset \mathbb{R}$. Suppose that there is a number M so that*

$$\left| \sum_{k=1}^{N} a_k(x) \right| \leq M$$

for all $x \in E$ and every integer N. Suppose that the sequence of real numbers $\{b_k\}$ converges monotonically to zero. Then the series

$$\sum_{k=1}^{\infty} b_k a_k$$

converges uniformly on E.

Proof Consider that $\{b_k\}$ is a sequence of constant functions on E and then apply the theorem. ■

In the exercises there are several other variants of Theorem 9.19, all with similar proofs and all of which have similar applications.

Example 9.21 As an interesting application of Theorem 9.19, consider a series that arises in Fourier analysis:

$$\sum_{k=1}^{\infty} \frac{\sin k\theta}{k}.$$

It is possible by using Dirichlet's test (see Section 3.6.13) to prove that this series converges for all θ.

Questions about the uniform convergence of this series are intriguing. In Figure 9.7 we have given a graph of some of the partial sums of the series.

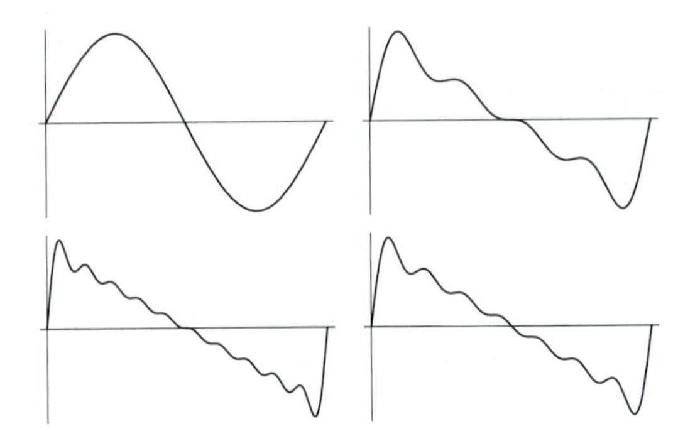

Figure 9.7. Graph of $\sum_{k=1}^{n} (\sin k\theta)/k$ on $[0, 2\pi]$ for, clockwise from upper left, $n = 1$, 4, 7, and 10.

The behavior near $\theta = 0$ is most curious. Apparently, if we can avoid that point (more precisely if we can stay a small distance away from that point) we should be able to obtain uniform convergence. Theorem 9.19 will provide a proof. We apply that theorem with $b_k(\theta) = 1/k$ and $a_k(\theta) = \sin k\theta$. All that is required is to obtain an estimate for the sums

$$\left| \sum_{k=1}^{n} \sin k\theta \right|$$

for all n and all θ in an appropriate set. Let $0 < \eta < \pi/2$ and consider making this estimate on the interval $[\eta, 2\pi - \eta]$. From Exercise 3.2.11 we obtain the formula

$$\sin\theta + \sin 2\theta + \sin 3\theta + \sin 4\theta + \cdots + \sin n\theta = \frac{\cos\theta/2 - \cos(2n+1)\theta/2}{2\sin\theta/2}$$

and using this we can see that

$$\left| \sum_{k=1}^{n} \sin k\theta \right| \leq \frac{1}{\sin(\eta/2)}.$$

Now Theorem 9.19 immediately shows that

$$\sum_{k=1}^{\infty} \frac{\sin k\theta}{k}$$

converges uniformly on $[\eta, 2\pi - \eta]$.

Figure 9.7 illustrates graphically why the convergence cannot be expected to be uniform near to 0. A computation here is instructive. To check the

Cauchy criterion on $[0,\pi]$ we need to show that the sums

$$\sup_{\theta\in[0,\pi]}\left|\sum_{k=m}^{n}\frac{\sin k\theta}{k}\right|$$

are small for large m, n. But in fact

$$\sup_{\theta\in[0,\pi]}\left|\sum_{k=m}^{2m}\frac{\sin k\theta}{k}\right| \geq \sum_{k=m}^{2m}\frac{\sin(k/2m)}{k} \geq \sum_{k=m}^{2m}\frac{\sin 1/2}{2m} > \frac{\sin 1/2}{2},$$

obtained by checking the value at points $\theta = 1/2m$. Since this is not arbitrarily small, the series cannot converge uniformly on $[0,\pi]$. ◀

Exercises

9.3.1 Examine the uniform limiting behavior of the sequence of functions

$$f_n(x) = \frac{x^n}{1+x^n}.$$

On what sets can you determine uniform convergence?

9.3.2 Examine the uniform limiting behavior of the sequence of functions

$$f_n(x) = x^2 e^{-nx}.$$

On what sets can you determine uniform convergence? On what sets can you determine uniform convergence for the sequence of functions $n^2 f_n(x)$?

9.3.3 Prove that if $f_n \to f$ pointwise on a finite set D, then the convergence is uniform.

9.3.4 Prove that if $f_n \to f$ uniformly on a set E_1 and also on a set E_2, then $f_n \to f$ uniformly on $E_1 \cup E_2$.

9.3.5 Prove or disprove that if $f_n \to f$ uniformly on each set E_1, E_2, E_3, $\dots$, then $f_n \to f$ uniformly on the union of all these sets $\bigcup_{k=1}^{\infty} E_k$.

9.3.6 Prove that if $f_n \to f$ uniformly on a set E, then $f_n \to f$ uniformly on every subset of E.

9.3.7 Prove or disprove that if $f_n \to f$ uniformly on each set $E \cap [a,b]$ for every interval $[a,b]$, then $f_n \to f$ uniformly on E.

9.3.8 Prove or disprove that if $f_n \to f$ uniformly on each closed interval $[a,b]$ contained in an open interval (c,d), then $f_n \to f$ uniformly on (c,d).

9.3.9 Prove that if $\{f_n\}$ and $\{g_n\}$ both converge uniformly on a set D, then so too does the sequence $\{f_n + g_n\}$.

9.3.10 Prove or disprove that if $\{f_n\}$ and $\{g_n\}$ both converge uniformly on a set D, then so too does the sequence $\{f_n g_n\}$.

9.3.11 Prove or disprove that if f is a continuous function on $(-\infty, \infty)$, then
$$f(x + 1/n) \to f(x)$$
uniformly on $(-\infty, \infty)$. (What extra condition, stronger than continuity, would work if not?)

9.3.12 Prove that $f_n \to f$ converges uniformly on D if and only if
$$\lim_n \sup_{x \in D} |f_n(x) - f(x)| = 0.$$

9.3.13 Show that a sequence of functions $\{f_n\}$ fails to converge to a function f uniformly on a set E if and only if there is some positive ε_0 so that a sequence $\{x_k\}$ of points in E and a subsequence $\{f_{n_k}\}$ can be found such that
$$|f_{n_k}(x_k) - f(x_k)| \geq \varepsilon_0.$$

9.3.14 Apply the criterion in the preceding exercise to show that the sequence $f_n(x) = x^n$ does not converge uniformly to zero on $(0, 1)$.

9.3.15 Prove Theorem 9.12.

9.3.16 Verify that the geometric series $\sum_{k=0}^{\infty} x^k$, which converges pointwise on $(-1, 1)$, does not converge uniformly there.

9.3.17 Do the same for the series obtained by differentiating the series in Exercise 9.3.16; that is, show that $\sum_{k=1}^{\infty} kx^{k-1}$ converges pointwise but not uniformly on $(-1, 1)$. Show that this series does converges uniformly on every closed interval $[a, b]$ contained in $(-1, 1)$.

9.3.18 Verify that the series
$$\sum_{k=1}^{\infty} \frac{\cos kx}{k^2}$$
converges uniformly on all of $\mathbb{R}$.

9.3.19 If $\{f_n\}$ is a sequence of functions converging uniformly on a set E to a function f, what conditions on the function g would allow you to conclude that $g \circ f_n$ converges uniformly on E to $g \circ f$?

9.3.20 Prove that the series $\displaystyle\sum_{k=0}^{\infty} \frac{x^k}{k}$ converges uniformly on $[0, b]$ for every $b \in [0, 1)$ but does not converge uniformly on $[0, 1)$.

9.3.21 Prove that if $\sum_0^{\infty} f_k$ converges uniformly on a set D, then the sequence of terms $\{f_k\}$ converges uniformly on D.

9.3.22 A sequence of functions $\{f_n\}$ is said to be *uniformly bounded* on an interval $[a, b]$ if there is a number M so that
$$|f_n(x)| \leq M$$
for every n and also for every $x \in [a, b]$. Show that a uniformly convergent sequence $\{f_n\}$ of continuous functions on $[a, b]$ must be uniformly bounded. Show that the same statement would not be true for pointwise convergence.

9.3.23 Suppose that $f_n \to f$ on $(-\infty, +\infty)$. What conditions would allow you to compute that

$$\lim_{n\to\infty} f_n(x+1/n) = f(x)?$$

9.3.24 Suppose that $\{f_n\}$ is a sequence of continuous functions on the interval $[0,1]$ and that you know that $f_n \to f$ uniformly on the set of rational numbers inside $[0,1]$. Can you conclude that $f_n \to f$ uniformly on $[0,1]$? (Would this be true without the continuity assertion?)

9.3.25 Prove the following variant of the Weierstrass M-test: Let $\{f_k\}$ and $\{g_k\}$ be sequences of functions on a set $E \subset \mathbb{R}$. Suppose that $|f_k(x)| \le g_k(x)$ for all k and $x \in E$ and that $\sum_{k=1}^{\infty} g_k$ converges uniformly on E. Then the series $\sum_{k=1}^{\infty} f_k$ converges uniformly on E.

9.3.26 Prove the following variant on Theorem 9.19: Let $\{a_k\}$ and $\{b_k\}$ be sequences of functions on a set $E \subset \mathbb{R}$. Suppose that $\sum_{k=1}^{\infty} a_k(x)$ converges uniformly on E. Suppose that $\{b_k\}$ is monotone for each $x \in E$ and uniformly bounded on E. Then the series $\sum_{k=1}^{\infty} a_k b_k$ converges uniformly on E.

9.3.27 Prove the following variant on Theorem 9.19: Let $\{a_k\}$ and $\{b_k\}$ be sequences of functions on a set $E \subset \mathbb{R}$. Suppose that there is a number M so that

$$\left|\sum_{k=1}^{N} a_k(x)\right| \le M$$

for all $x \in E$ and every integer N. Suppose that

$$\sum_{k=1}^{\infty} |b_k - b_{k+1}|$$

converges uniformly on E and that $b_k \to 0$ uniformly on E. Then the series $\sum_{k=1}^{\infty} a_k b_k$ converges uniformly on E.

9.3.28 Prove the following variant on Theorem 9.19: Let $\{a_k\}$ and $\{b_k\}$ be sequences of functions on a set $E \subset \mathbb{R}$. Suppose that $\sum_{k=1}^{\infty} a_k$ converges uniformly on E. Suppose that the series

$$\sum_{k=1}^{\infty} |b_k - b_{k+1}|$$

has uniformly bounded partial sums on E. Suppose that the sequence of functions $\{b_k\}$ is uniformly bounded on E. Then the series $\sum_{k=1}^{\infty} a_k b_k$ converges uniformly on E.

9.3.29 Suppose that $\{f_n\}$ is a sequence of continuous functions on an interval $[a,b]$ converging uniformly to a function f on the open interval (a,b). If f is also continuous on $[a,b]$, show that the convergence is uniform on $[a,b]$.

9.3.30 Suppose that $\{f_n\}$ is a sequence of functions converging uniformly to zero on an interval $[a, b]$. Show that $\lim_{n\to\infty} f_n(x_n) = 0$ for every convergent sequence $\{x_n\}$ of points in $[a, b]$. Give an example to show that this statement may be false if $f_n \to 0$ merely pointwise.

9.3.31 Suppose that $\{f_n\}$ is a sequence of functions on an interval $[a, b]$ with the property that $\lim_{n\to\infty} f_n(x_n) = 0$ for every convergent sequence $\{x_n\}$ of points in $[a, b]$. Show that $\{f_n\}$ converges uniformly to zero on $[a, b]$.

9.4 Uniform Convergence and Continuity

We can now address the questions we asked at the beginning of this chapter. We begin with continuity. We know that the pointwise limit of a sequence of continuous functions need not be continuous. We now show that the *uniform* limit of a sequence of continuous functions must be continuous.

Theorem 9.22 *Let $\{f_n\}$ be a sequence of functions defined on an interval I, and let $x_0 \in I$. If the sequence $\{f_n\}$ converges uniformly to some function f on I and if each of the functions f_n is continuous at x_0, then the function f is also continuous at x_0. In particular, if each of the functions f_n is continuous on I, then so too is f.*

Proof Let $\varepsilon > 0$. We must show there exists $\delta > 0$ such that

$$|f(x) - f(x_0)| < \varepsilon$$

if $|x - x_0| < \delta$, $x \in I$. For each $x \in I$ we have

$$f(x) - f(x_0) = (f(x) - f_n(x)) + (f_n(x) - f_n(x_0)) + (f_n(x_0) - f(x_0)),$$

so

$$|f(x) - f(x_0)| \leq |f(x) - f_n(x)| + |f_n(x) - f_n(x_0)| + |f_n(x_0) - f(x_0)|. \quad (6)$$

Since $f_n \to f$ *uniformly*, there exists $N \in \mathbb{N}$ such that

$$|f_n(x) - f(x)| < \frac{\varepsilon}{3} \quad (7)$$

for all $x \in I$ and all $n \geq N$. We infer from inequalities (6) and (7) that

$$|f(x) - f(x_0)| < |f_N(x) - f_N(x_0)| + \frac{2}{3}\varepsilon. \quad (8)$$

We now use the continuity of the function f_N. We choose $\delta > 0$ such that if $x \in I$ and $|x - x_0| < \delta$, then

$$|f_N(x) - f_N(x_0)| < \frac{\varepsilon}{3}. \quad (9)$$

Combining (8) and (9), we have

$$|f(x) - f(x_0)| < \frac{\varepsilon}{3} + \frac{2}{3}\varepsilon = \varepsilon,$$

for each $x \in I$ for which $|x - x_0| < \delta$, as was to be shown. ■

Note. Let us look a bit more closely at the proof of Theorem 9.22. We first obtained $N \in \mathbb{N}$ such that the function f_N approximated f closely (within $\varepsilon/3$) on all of I. This function f_N served as a "stepping stone" toward verifying the continuity of f at x_0. There are three small "steps" involved:

1. $|f_N(x) - f(x)|$ is small (*for all* $x \in I$) because of uniform convergence.
2. $|f_N(x) - f_N(x_0)|$ is small (for all x near x_0) because of the continuity of f_N.
3. $|f_N(x_0) - f(x_0)|$ is small because $\{f_n(x_0)\} \to f(x_0)$.

If we tried to imitate the proof under the assumption of pointwise convergence, the first of these steps would fail. You may wish to observe the failure by working Exercise 9.4.2.

Theorem 9.22 can be stated in terms of series. Recall that a series $\sum_1^\infty f_k$ converges uniformly to the function f on D if the sequence

$$\{S_n\} = \{\sum_{k=1}^{n} f_k\}$$

of partial sums converges uniformly to f on D.

Corollary 9.23 *If $\sum_1^\infty f_k$ converges uniformly to f on an interval I and if each of the functions f_k is continuous on I, then f is continuous on I.*

Proof This follows immediately from Theorem 9.22. ■

9.4.1 Dini's Theorem

✂ *Advanced*

Observe that Theorem 9.22 provides a *sufficient* condition for continuity of the limit function f. The condition is not necessary. (The sequence in Example 9.6 converges to the zero function, which is continuous, even though the convergence is not uniform.)

Under certain circumstances, however, uniform convergence *is* necessary, as Theorem 9.24 shows. (See also Exercise 9.4.6.) This theorem is due to Ulisse Dini (1845–1918) and gives a condition under which pointwise convergence of a sequence of continuous functions to a continuous function must be uniform.

Theorem 9.24 (Dini) *Let $\{f_n\}$ be a sequence of continuous functions on an interval $[a, b]$. Suppose for each $x \in [a, b]$ and for all $n \in \mathbb{N}$,*

$$f_n(x) \geq f_{n+1}(x).$$

Suppose in addition that for all $x \in [a, b]$

$$f(x) = \lim_n f_n(x).$$

If f is continuous, then the convergence is uniform.

Proof Suppose the convergence were *not* uniform. Then

$$\max_{x\in[a,b]}(f_n(x)-f(x))$$

does not approach zero as $n\to\infty$ (see Exercise 9.3.12). Hence there exists $c>0$ such that for infinitely many $n\in\mathbb{N}$,

$$\max_{x\in[a,b]}(f_n(x)-f(x))>c>0.$$

Now, for each $n\in\mathbb{N}$, f_n-f is continuous, so it achieves a maximum value at a point $x_n\in[a,b]$. By the Bolzano-Weierstrass theorem we can thus choose a subsequence $\{x_{n_k}\}$ of the sequence $\{x_n\}$ such that $\{x_{n_k}\}$ converges to a point $x_0\in[a,b]$. Note that we must have

$$f_{n_k}(x_{n_k})-f(x_{n_k})>c$$

for all $k\in\mathbb{N}$.

Because of our assumption that $f_n(x)\geq f_{n+1}(x)$ for all $n\in\mathbb{N}$ and $x\in[a,b]$, we infer

$$f_n(x_{n_k})-f(x_{n_k})>c \text{ for each } n\leq n_k.$$

Now fix n and let $k\to\infty$. Using the continuity of the functions f_n-f we obtain $f_n(x_0)-f(x_0)\geq c$ for all $n\in\mathbb{N}$. But this is impossible since $f_n(x_0)\to f(x_0)$ by hypothesis. Thus our assumption that the convergence is not uniform has led to a contradiction. ■

Example 9.25 The sequence of continuous functions $f_n(x)=x^n$ is converging monotonically to a function f on the interval $[0,1]$. But that function f is (as we have seen before) discontinuous at $x=1$, so immediately we know that the convergence cannot be uniform. Dini's theorem implies that the convergence is uniform on $[0,b]$ for any $0<b<1$ since the function f is continuous there. ◀

Exercises

9.4.1 Can a sequence of discontinuous functions converge uniformly on an interval to a continuous function?

9.4.2 Let $f_n(x)=x^n$, $0\leq x\leq 1$. Try to imitate the proof of Theorem 9.22 for $x_0=1$ and observe where the proof breaks down.

9.4.3 Let $\{f_n\}$ be a sequence of functions each of which is uniformly continuous on an open interval (a,b). If $f_n\to f$ uniformly on (a,b) can you conclude that f is also uniformly continuous on (a,b)?

9.4.4 Give an example of a sequence of continuous functions $\{f_n\}$ on the interval $(0,1)$ that is monotonic decreasing and converges pointwise to a continuous function f on $(0,1)$ but for which the convergence is not uniform. Why does this not contradict Theorem 9.24?

9.4.5 Give an example of a sequence of continuous functions $\{f_n\}$ on the interval $[0, \infty)$ that is monotonic decreasing and converges pointwise to a continuous function f on $[0, \infty)$ but for which the convergence is not uniform. Why does this not contradict Theorem 9.24?

9.4.6 Let $\{f_n\}$ be a sequence of continuous nondecreasing functions defined on an interval $[a, b]$. Suppose $f_n \to f$ pointwise on $[a, b]$. Prove that if f is continuous on $[a, b]$, then the convergence is uniform. Observe that, in this exercise, the *functions* are assumed monotonic, whereas in Theorem 9.24 it is the *sequence* that we assume monotonic.

9.4.7 The proof of Theorem 9.24 depends on the compactness of the interval $[a, b]$. The compactness argument used here relied on the Bolzano-Weierstrass theorem. Attempt another proof using one of our other strategies from Section 4.5.

9.4.8 Prove this variant on Dini's theorem (Theorem 9.24). Let $\{f_n\}$ be a sequence of continuous functions on an interval $[a, b]$. Suppose for each $x \in [a, b]$ and for all $n \in \mathbb{N}$, $f_n(x) \leq f_{n+1}(x)$. Suppose in addition that for all $x \in [a, b]$ $\lim_n f_n(x) = \infty$. Show that for all $M > 0$ there is an integer N so that $f_n(x) > M$ for all $x \in [a, b]$ and all $n \geq N$. Show that this conclusion would not be valid without the monotonicity assumption.

9.4.9 Show that if, in Exercise 9.4.8, the interval $[a, b]$ is replaced by the unbounded interval $[0, \infty)$ or the nonclosed interval $(0, 1)$, then the conclusion need not be valid.

9.4.10 Let $\{f_n\}$ be a sequence of Lipschitz functions on $[a, b]$ with common Lipschitz constant M. (This means that $|f_n(x) - f_n(y)| \leq M|x - y|$ for all $n \in \mathbb{N}$, $x, y \in [a, b]$.)

(a) If $f = \lim_n f_n$ pointwise, then f is continuous and, in fact, satisfies a Lipschitz condition with constant M.

(b) If $f = \lim_n f_n$ pointwise the convergence is uniform.

(c) Show by example that the results in (a) and (b) fail if we weaken our hypotheses by requiring only that each function is a Lipschitz function. (Here, the constant M may depend on n.)

9.4.11 Give an example to show that the analogue of Theorem 9.24 fails if $[a, b]$ is replaced with an interval that is not closed or is not bounded.

9.4.12 **(Continuous convergence and uniform convergence)** A sequence of functions $\{f_n\}$ defined on an interval I is said to *converge continuously* to the function f if $f_n(x_n) \to f(x_0)$ whenever $\{x_n\}$ is a sequence of points in the interval I that converges to a point x_0 in I. Prove the following theorem:

> *Let $\{f_n\}$ be a sequence of continuous functions on an interval $[a, b]$. Then $\{f_n\}$ converges continuously on $[a, b]$ if and only if $\{f_n\}$ converges to f uniformly on $[a, b]$.*

Does the theorem remain true if the interval $[a,b]$ is replaced with (a,b) or $[a,\infty)$?

9.4.13 Show that the sequence $f_n(x) = x^n/n$ converges uniformly on [0,1]:

(a) By direct computation using the definition of uniform convergence

(b) By using Theorem 9.24

(c) By using Exercise 9.4.6

(d) By using Exercise 9.4.12

9.5 Uniform Convergence and the Integral

Calculus students often learn the following simple computation. The geometric series

$$\frac{1}{1-t} = \sum_{k=0}^{\infty} t^k \tag{10}$$

is valid on the interval $(-1,1)$. An integration of both sides for t in the interval $[0,x]$, and any choice of $x < 1$ will yield

$$-\log(1-x) = \int_0^x \frac{1}{1-t}\,dt = \sum_{k=0}^{\infty} \frac{x^{k+1}}{k+1}.$$

Indeed this identity is valid and provides a series expansion for the logarithm function. But can this really be justified?

In general, do we know that if $f(x) = \sum_0^\infty f_n(x)$ on an interval $[a,b]$, then

$$\int_a^b f(x)\,dx = \sum_0^\infty \int_a^b f_n(x)\,dx?$$

In fact, we already observed in Section 9.3 that during the early part of the nineteenth century, some prominent mathematicians took for granted the permissibility of term by term integration of convergent infinite series of functions. This was true of Fourier, Cauchy, and Gauss. Example 9.6, cast in the setting of sequences of integrable functions, shows that these mathematicians were mistaken.

9.5.1 Sequences of Continuous Functions

Around the middle of the nineteenth century, Weierstrass showed that term by term integration is permissible when the series of integrable functions converges *uniformly*. Let us first verify this result for sequences of continuous functions.

Theorem 9.26 *Suppose that $f(x) = \lim_{n\to\infty} f_n(x)$ for all $x \in [a,b]$, that each function f_n is continuous on $[a,b]$, and that the convergence is uniform. Then*

$$\int_a^b f(x)\,dx = \lim_{n\to\infty} \int_a^b f_n(x)\,dx.$$

Proof By Theorem 9.22, f is continuous, so $\int_a^b f(x)\,dx$ exists. We must show that $\int_a^b f_n(x)\,dx \to \int_a^b f(x)\,dx$.

Let $\varepsilon > 0$. We wish to obtain $N \in \mathbb{N}$ such that

$$\left| \int_a^b f_n(x)\,dx - \int_a^b f(x)\,dx \right| < \varepsilon \text{ for all } n \geq N.$$

We calculate that for any $n \in \mathbb{N}$

$$\left| \int_a^b f_n(x)\,dx - \int_a^b f(x)\,dx \right| = \left| \int_a^b [f_n(x) - f(x)]\,dx \right|$$

$$\leq \int_a^b |f_n(x) - f(x)|\,dx \leq \int_a^b \max_{x\in[a,b]} |f_n(x) - f(x)|\,dx$$

$$\leq (b-a)\left(\max_{x\in[a,b]} |f_n(x) - f(x)| \right).$$

Since $f_n \to f$ uniformly on $[a,b]$, there exists $N \in \mathbb{N}$ such that

$$\max_{x\in[a,b]} |f_n(x) - f(x)| < \frac{\varepsilon}{b-a} \text{ for all } n \geq N.$$

Thus, for $n \geq N$, we have

$$\left| \int_a^b f_n(x)\,dx - \int_a^b f(x)\,dx \right| \leq (b-a)\frac{\varepsilon}{b-a} = \varepsilon$$

as was to be shown. ■

Applying the theorem to the partial sums S_n of a series allows us to express this result for series.

Corollary 9.27 *If an infinite series of continuous functions $\sum_0^\infty f_k$ converges uniformly to a function f on an interval $[a,b]$, then f is also continuous and*

$$\int_a^b f(x)\,dx = \sum_0^\infty \int_a^b f_k(x)\,dx.$$

Example 9.28 Let us justify the computations that we made in our introduction to this topic. The geometric series

$$\frac{1}{1-t} = \sum_{k=0}^\infty t^k \tag{11}$$

converges pointwise on the interval $(-1, 1)$. Let $0 < x < 1$. By the M-test we see that this series converges uniformly on $[0, x]$. Each of the terms in the sum is continuous. As a result we may apply our theorem to integrate term by term just as we might have seen in a calculus course. Thus

$$\int_0^x \frac{1}{1-t}\, dt = \sum_{k=0}^{\infty} \frac{x^{k+1}}{k+1}.$$

◀

9.5.2 Sequences of Riemann Integrable Functions

Enrichment

In Theorem 9.26 we required that the functions f_n be continuous. Suppose we now weaken our hypotheses for these functions by requiring only that they be integrable, but still requiring the sequence $\{f_n\}$ to converge uniformly to f. We note that in all respects the proof is the same. Thus, if a uniformly convergent sequence of integrable functions converges to an integrable function, we can integrate the sequence term by term. Our next theorem shows that a uniform limit of integrable functions must be integrable and so we have the following extension of Theorem 9.26.

Theorem 9.29 *Let $\{f_n\}$ be a sequence of functions Riemann integrable on an interval $[a, b]$. If $f_n \to f$ uniformly on $[a, b]$, then f is Riemann integrable on $[a, b]$ and*

$$\int_a^b f(x)\, dx = \lim_n \int_a^b f_n(x)\, dx.$$

Proof Because of the preceding development, we need only show that the limit function f is integrable on $[a, b]$.

One proof (see Exercise 9.5.7) would be to show that f is bounded and continuous everywhere except at a set of measure zero. It follows by Theorem 8.17 that f is Riemann integrable.

We can also give a proof by constructing, for any $\varepsilon > 0$, step functions having the property of Exercise 8.6.5. Since this proof is one that was available to nineteenth-century mathematicians, who would not have known about sets of measure zero, this is worth presenting, if only for historical reasons.

Let $\varepsilon > 0$. We wish to find step functions L and U such that

$$L(x) \leq f(x) \leq U(x)$$

for all $x \in [a, b]$, with

$$\int_a^b [U(x) - L(x)]\, dx < \varepsilon.$$

We shall obtain the functions L and U in three natural steps:

1. We approximate f by one of the functions f_N.

2. We obtain corresponding step functions L_N and U_N approximating f_N.

3. We modify L_N and U_N to obtain L and U.

We proceed according to the aforementioned plan.
(i) Since $f_n \to f$ *uniformly*, there exists $N \in \mathbb{N}$ such that

$$|f_N(x) - f(x)| \leq \frac{\varepsilon}{4(b-a)} \text{ for all } x \in [a,b].$$

(ii) Since f_N is integrable by hypothesis, there exist step functions L_N and U_N such that

$$L_N(x) \leq f_N(x) \leq U_N(x)$$

for all $x \in [a,b]$ and

$$\int_a^b [U_N(x) - L_N(x)]\, dx < \frac{\varepsilon}{2}.$$

(iii) Let us define the step functions U and L by

$$L(x) = L_N(x) - \frac{\varepsilon}{4(b-a)}, \quad U(x) = U_N(x) + \frac{\varepsilon}{4(b-a)}$$

for all $x \in [a,b]$.

We then have

$$L(x) < L_N(x) + |f(x) - f_N(x)| \leq f(x) \leq U_N(x) + |f(x) - f_N(x)| < U(x)$$

and

$$\begin{aligned}
\int_a^b [U(x) - L(x)]\, dx &= \int_a^b \left\{ [U_N(x) - L_N(x)] + \frac{\varepsilon}{2(b-a)} \right\} dx \\
&= \int_a^b [U_N(x) - L_N(x)]\, dx + \int_a^b \frac{\varepsilon}{2(b-a)}\, dx \\
&< \frac{\varepsilon}{2} + \frac{\varepsilon}{2} = \varepsilon,
\end{aligned}$$

as was to be shown. ■

Corollary 9.30 *If an infinite series of integrable functions $\sum_0^\infty f_k$ converges uniformly to a function f on an interval $[a,b]$, then f is also integrable and*

$$\int_a^b f(x)\, dx = \sum_0^\infty \int_a^b f_k(x)\, dx.$$

Example 9.31 Let $f_n(x) = e^{-nx^2}$. Then for each $x \in [1,2]$ and for every $n \in \mathbb{N}$, $0 < e^{-nx^2} \le e^{-n}$ and $e^{-n} \to 0$, so $f_n \to 0$ uniformly on [1,2]. It follows that

$$\lim_n \int_1^2 e^{-nx^2}\,dx = \int_1^2 0\,dx = 0.$$

◀

Note. We end this section with a short note that considers whether our main theorem would be true under weaker hypotheses than uniform convergence.

It is possible for a sequence $\{f_n\}$ of functions to converge pointwise (but *not* uniformly) to a function f on $[a,b]$ and still have

$$\lim_n \int_a^b f_n(x)\,dx = \int_a^b f(x)\,dx. \tag{12}$$

For example, suppose we modify the functions of Example 9.6 so that $f_n(1/(2n)) = 1$ instead of $f_n(1/(2n)) = 2n$. We still have $f_n \to 0$ pointwise (but not uniformly), but now

$$\int_0^1 f_n(x)\,dx \to 0.$$

These functions form a *uniformly bounded* sequence of functions : that is, there exists a constant M ($M = 1$ in this case) such that $|f_n(x)| \le M$ for all $n \in \mathbb{N}$ and all $x \in [0,1]$. A theorem (whose proof is beyond the scope of this chapter) asserts that if a uniformly bounded sequence of integrable functions $\{f_n\}$ converges pointwise to an integrable function f on $[a,b]$, then the identity (12) holds. We cannot drop the hypothesis of integrability of f in this theorem. If, for example, $\{r_n\}$ is an enumeration of the rationals in $[0,1]$ and

$$f_n(x) = \begin{cases} 1, & \text{if } x = r_1, r_2, \dots, r_n \\ 0, & \text{otherwise,} \end{cases}$$

then

$$\lim_{n\to\infty} f_n(x) = f(x) = \begin{cases} 1, & \text{if } x \in \mathbb{Q} \cap [0,1] \\ 0, & \text{if } x \in [0,1] \setminus \mathbb{Q} \end{cases}$$

and f is not integrable on $[0,1]$ by Exercise 9.2.3.

9.5.3 Sequences of Improper Integrals

Thus far we have studied limits of ordinary integrals, either of continuous functions on a finite interval $[a,b]$ or Riemann integrable functions on such an interval. What if the integrals are of unbounded functions so that they must be taken in the Cauchy (improper) sense? What if the integrals are to be taken on an infinite interval?

More narrowly, let us just ask for the validity of the formulas:

$$\lim_{n\to\infty} \int_a^\infty f_n(t)\,dt = \int_a^\infty f(t)\,dt$$

in case $f_n \to f$ or

$$\sum_{k=1}^{\infty} \int_a^{\infty} g_k(t)\,dt = \int_a^{\infty} f(t)\,dt$$

in case $f = \sum_{k=1}^{\infty} g_k$. A fast and glib answer would be that we hardly expect these to be true for pointwise convergence but certainly uniform convergence will suffice.

But these integrals involve an extra limit operation and we therefore need extra caution. Indeed the following example shows that uniform convergence is far from enough. It is not just the "smoothness" of the convergence that is an issue here.

Example 9.32 Let $f_n(x)$ be defined as $1/n$ for all values of $x \in [0, n]$ but as zero for $x > n$. Then the sequence $\{f_n\}$ converges to zero uniformly on the interval $[0, \infty)$. But the integrals do not converge to zero (as we would have hoped) since

$$\int_0^{\infty} f_n(t)\,dt = 1$$

for all n. ◀

What further condition can we impose so that, together with uniform convergence, we will be able to take the limit operation inside the integral

$$\lim_{n\to\infty} \int_0^{\infty} f_n(t)\,dt?$$

The condition we impose in the theorem just requires that all the functions are controlled or dominated by some function that is itself integrable. In Example 9.32 note that there is no possibility of an integrable function g on $[0, \infty)$ such that $f_n(x) \leq g(x)$ for all n and x. Theorems of this kind are called *dominated convergence theorems.*

Theorem 9.33 *Suppose that $\{f_n\}$ is a sequence of continuous functions on the interval $[a, \infty)$ such that $f_n \to f$ uniformly on any interval $[a, b]$. If there is a continuous function g on $[a, \infty)$ such that*

$$|f_n(x)| \leq g(x)$$

for all $a \leq x$ and such that the integral

$$\int_a^{\infty} g(x)\,dx$$

exists, then

$$\lim_{n\to\infty} \int_a^{\infty} f_n(t)\,dt = \int_a^{\infty} f(t)\,dt.$$

Proof As a first step let us show that f is integrable on $[a, \infty)$. Certainly f is continuous since it is a uniform limit of a sequence of continuous functions. Since each $|f_n(x)| \leq g(x)$ it follows that $|f(x)| \leq g(x)$. We check then

$$\left| \int_c^d f(t)\,dt \right| \leq \int_c^d |f(t)|\,dt \leq \int_c^d g(t)\,dt.$$

Since g is integrable, it follows by the Cauchy criterion for improper integrals (see Exercise 8.5.11) that the integral $\int_c^d g(t)\,dt$ can be made arbitrarily small for large c and d. But then so also is the integral $\int_c^d f(t)\,dt$, and a further application of the Cauchy criterion for improper integrals shows that f is integrable. (Indeed this argument shows that f is absolutely integrable in fact.)

Now let $\varepsilon > 0$. Choose L_0 so large that

$$\int_{L_0}^{\infty} g(t)\,dt < \varepsilon/4.$$

Choose N so large that

$$|f_n(t) - f(t)| < \frac{\varepsilon}{2(L_0 - a)}$$

if $n \geq N$ and $t \in [a, L_0]$. This is possible because $f_n \to f$ uniformly on $[a, L_0]$. Then we have

$$\left| \int_a^{\infty} f_n(t)\,dt - \int_a^{\infty} f(t)\,dt \right| \leq \int_a^{L_0} |f_n(t) - f(t)|\,dt + \int_{L_0}^{\infty} 2g(t)\,dt$$

$$\leq \frac{\varepsilon}{2(L_0 - a)}(L_0 - a) + \frac{2\varepsilon}{4} = \varepsilon$$

for all $n \geq N$. This proves the assertion of the theorem. ■

Exercises

9.5.1 Prove that

$$\lim_{n\to\infty} \int_{\frac{\pi}{2}}^{\pi} \frac{\sin nx}{nx}\,dx = 0.$$

9.5.2 Prove that

$$\int_0^{\pi} \sum_{n=1}^{\infty} \frac{\sin nx}{n^2}\,dx = \sum_{n=1}^{\infty} \frac{2}{(2n-1)^3}.$$

9.5.3 Show that if $f_n \to f$ uniformly on $[a, b]$ and each f_n is continuous then the sequence of functions

$$F_n(x) = \int_a^x f_n(t)\,dt$$

also converges uniformly on $[a, b]$.

9.5.4 Show that if $f_n \to f$ uniformly on $[a,b]$ and each f_n is continuous then

$$\lim_{n\to\infty} \int_a^b \left(\int_a^x f_n(t)\,dt \right) dx = \int_a^b \left(\int_a^x f(t)\,dt \right) dx.$$

9.5.5 Show that the series

$$\sum_{k=0}^{\infty} \frac{x^k}{k!}$$

converges uniformly on $[-a,a]$ for every $a \in \mathbb{R}$ but does not converge uniformly on all of the real line. (Does it converge pointwise on the real line?) Obtain a series representation for

$$\int_{-a}^{a} \sum_{k=0}^{\infty} \frac{x^k}{k!}\,dx.$$

9.5.6 Let $\{f_n\}$ be a sequence of continuous functions on an interval $[a,b]$ that converges uniformly to a function f. What conditions on g would allow you to conclude that

$$\lim_{n\to\infty} \int_a^b f_n(t)g(t)\,dt = \int_a^b f(t)g(t)\,dt?$$

9.5.7 Let $\{f_n\}$ be a sequence of bounded functions each continuous on an interval $[a,b]$ except at a set of measure zero. Show that if $f_n \to f$ uniformly on $[a,b]$, then the function f is also bounded and continuous on $[a,b]$ except at a set of measure zero. Conclude that a uniformly convergent sequence of Riemann integrable functions must converge to a function that is also Riemann integrable.

9.5.8 Let $p > -1$. Show that

$$\lim_{n\to\infty} \int_1^n \left(1 - \frac{t}{n}\right)^n t^p\,dt = \int_1^\infty e^{-t} t^p\,dt.$$

9.5.9 Formulate and prove a version of the dominated convergence theorem (Theorem 9.33) that would apply to improper integrals on an interval $[a,b]$ where the point of unboundedness is at the endpoint a.

9.5.10 Compute the limit of the improper integral

$$\lim_{n\to\infty} \int_0^1 \frac{e^{-nt}}{\sqrt{t}}\,dt.$$

9.6 Uniform Convergence and Derivatives

We saw in Section 9.5 that a uniformly convergent sequence (or series) of continuous functions can be integrated term by term. This allows an easy proof of a theorem on term by term differentiation.

Theorem 9.34 *Let $\{f_n\}$ be a sequence of functions each with a continuous derivative on an interval $[a,b]$. If the sequence $\{f'_n\}$ of derivatives converges uniformly to a function on $[a,b]$ and the sequence $\{f_n\}$ converges pointwise to a function f, then f is differentiable on $[a,b]$ and*

$$f'(x) = \lim_n f'_n(x) \quad \text{for all } x \in [a,b].$$

Proof Let $g = \lim_n f'_n$. Since each of the functions f'_n is assumed continuous and the convergence is uniform, the function g is also continuous (Theorem 9.22). From Theorem 9.29 we infer

$$\int_a^x g\,(t)\;dt = \lim_n \int_a^x f'_n\,(t)\;dt \text{ for all } x \in [a,b]. \tag{13}$$

Applying the fundamental theorem of calculus (Theorem 8.9), we see that

$$\int_a^x f'_n\,(t)\;dt = f_n(x) - f_n(a) \text{ for all } x \in [a,b] \tag{14}$$

for all $n \in \mathbb{N}$.

But $f_n(x) \to f(x)$ for all $x \in [a,b]$ by hypothesis, so letting $n \to \infty$ in equation (14) and noting (13), we obtain

$$\int_a^x g\,(t)\;dt = f(x) - f(a)$$

or

$$f(x) = \int_a^x g\,(t)\;dt + f(a).$$

It follows from the continuity of g and the fundamental theorem of calculus (Theorem 8.8) that f is differentiable and that

$$f'(x) = g(x)$$

for all $x \in [a,b]$. ■

For series, the theorem takes the following form:

Corollary 9.35 *Let $\{f_k\}$ be a sequence of functions each with a continuous derivative on $[a,b]$ and suppose $f = \sum_0^\infty f_k$ on $[a,b]$. If the series $\sum_{k=0}^\infty f'_k$ converges uniformly on $[a,b]$, then $f' = \sum_{k=0}^\infty f'_k$ on $[a,b]$.*

Example 9.36 Starting with the geometric series

$$\frac{1}{1-x} = \sum_{k=0}^{\infty} x^k \quad \text{on } (-1,1), \tag{15}$$

we obtain from Corollary 9.35 that

$$\frac{1}{(1-x)^2} = \sum_{k=1}^{\infty} kx^{k-1} \quad \text{on } (-1,1) \tag{16}$$

To justify (16) we observe first that the series (15) converges pointwise on $(-1,1)$. Next we note (Exercise 9.3.17) that the series (16) converges pointwise on $(-1,1)$ and uniformly on any closed interval $[a,b] \subset (-1,1)$. Thus, if $x \in (-1,1)$ and $-1 < a < x < b < 1$, then (16) converges uniformly on $[a,b]$, so (16) holds at x. ◀

9.6.1 Limits of Discontinuous Derivatives

✂ *Enrichment*

The hypotheses of Theorem 9.34 are somewhat more restrictive than necessary for the conclusion to hold. We need not assume that $\{f_n\}$ converges on all of $[a,b]$; convergence at a single point suffices. Nor need we assume that each of the derivatives f_n' is continuous. (We cannot, however, replace uniform convergence of the sequence $\{f'_n\}$ with pointwise convergence, as Example 9.5 shows.) The theorem that follows applies in a number of cases in which Theorem 9.34 does not apply.

Theorem 9.37 *Let $\{f_n\}$ be a sequence of continuous functions defined on an interval $[a,b]$. Suppose that $f_n'(x)$ exists for each n and each $x \in [a,b]$. Suppose that the sequence $\{f'_n\}$ of derivatives converges uniformly on $[a,b]$ and that there exists a point $x_0 \in [a,b]$ such that the sequence of numbers $\{f_n(x_0)\}$ converges. Then the sequence $\{f_n\}$ converges uniformly to a function f on the interval $[a,b]$, f is differentiable, and*

$$f'(x) = \lim_{n\to\infty} f_n'(x)$$

at each point $x \in [a,b]$.

Proof Let $\varepsilon > 0$. Since the sequence of derivatives converges uniformly on $[a,b]$, there is an integer N_1 so that

$$|f_n'(x) - f_m'(x)| < \varepsilon$$

for all n, $m \geq N_1$ and all $x \in [a,b]$. Also, since the sequence of numbers $\{f_n(x_0)\}$ converges, there is an integer $N > N_1$ so that

$$|f_n(x_0) - f_m(x_0)| < \varepsilon$$

for all n, $m \geq N$. Let us, for any $x \in [a,b]$, $x \neq x_0$, apply the mean value theorem to the function $f_n - f_m$ on the interval $[x_0, x]$ (or on the interval $[x, x_0]$ if $x < x_0$). This gives us the existence of some point ξ between x and x_0 so that

$$f_n(x) - f_m(x) - [f_n(x_0) - f_m(x_0)] = (x - x_0)[f_n'(\xi) - f_m'(\xi)]. \qquad (17)$$

From this we deduce that

$$\begin{aligned} |f_n(x) - f_m(x)| &\leq |f_n(x_0) - f_m(x_0)| + |(x - x_0)(f_n'(\xi) - f_m'(\xi)| \\ &< \varepsilon(1 + (b - a)) \end{aligned}$$

for any $n, m \geq N$. Since this N depends only on ε this assertion is true for all $x \in [a, b]$ and we have verified that the sequence of continuous functions $\{f_n\}$ is uniformly Cauchy on $[a, b]$ and hence converges uniformly to a continuous function f on $[a, b]$.

Let us now show that $f'(x_0)$ is the limit of the derivatives $f_n'(x_0)$. Again, for any $\varepsilon > 0$, equation (17) implies that

$$|f_n(x) - f_m(x) - [f_n(x_0) - f_m(x_0)]| \leq |x - x_0|\varepsilon \tag{18}$$

for all $n, m \geq N$ and any $x \neq x_0$ in the interval $[a, b]$. In this inequality let $m \to \infty$ and, remembering that $f_m(x) \to f(x)$ and $f_m(x_0) \to f(x_0)$, we obtain

$$|f_n(x) - f_n(x_0) - [f(x) - f(x_0)]| \leq |x - x_0|\varepsilon \tag{19}$$

if $n \geq N$. Let C be the limit of the sequence of numbers $\{f_n'(x_0)\}$. Thus there exists $M > N$ such that

$$|f_M'(x_0) - C| < \varepsilon. \tag{20}$$

Since the function f_M is differentiable at x_0, there exists $\delta > 0$ such that if $0 < |x - x_0| < \delta$, then

$$\left|\frac{f_M(x) - f_M(x_0)}{x - x_0} - f_M'(x_0)\right| < \varepsilon. \tag{21}$$

From Equation (19) and the fact that $M > N$, we have

$$\left|\frac{f_M(x) - f_M(x_0)}{x - x_0} - \frac{f(x) - f(x_0)}{x - x_0}\right| < \varepsilon.$$

This, together with the inequalities (20) and (21), shows that

$$\left|\frac{f(x) - f(x_0)}{x - x_0} - C\right| < 3\varepsilon$$

for $0 < |x - x_0| < \delta$. This proves that $f'(x_0)$ exists and is the number C, which we recall is $\lim_{n\to\infty} f_n'(x_0)$.

In this argument x_0 may be taken as any point inside the interval $[a, b]$ and so the theorem is proved. ■

For infinite series Theorem 9.37 takes the following form:

Corollary 9.38 *Let $\{f_k\}$ be a sequence of differentiable functions on an interval $[a, b]$. Suppose that the series $\sum_{k=0}^{\infty} f_k'$ converges uniformly on $[a, b]$. Suppose also that there exists $x_0 \in [a, b]$ such that the series $\sum_{k=0}^{\infty} f_k(x_0)$ converges. Then the series $\sum_{k=0}^{\infty} f_k(x)$ converges uniformly on $[a, b]$ to a function F, F is differentiable, and*

$$F'(x) = \sum_{k=0}^{\infty} f_k'(x)$$

for all $a \leq x \leq b$.

Note. In the statement of Theorem 9.37 we hypothesized the existence of a single point x_0 at which the sequence $\{f_n(x_0)\}$ converges. It then followed that the sequence $\{f_n\}$ converges on all of the interval I. If we drop that requirement but retain the requirement that the sequence $\{f_n'\}$ converges uniformly to a function g on I, we cannot conclude that $\{f_n\}$ converges on I [e.g. let $f_n(x) \equiv n$], but we can still conclude that there exists f such that $f' = g = \lim_n f_n'$ on I. (To see this, fix $x_0 \in I$, let $F_n = f_n - f_n(x_0)$ and apply Theorem 9.37 to the sequence $\{F_n\}$.) Thus, the uniform limit of a sequence of derivatives $\{f_n'\}$ is a derivative even if the sequence of primitives $\{f_n\}$ does not converge.

Exercises

9.6.1 Can the sequence of functions $f_n(x) = \dfrac{\sin nx}{n^3}$ be differentiated term by term?

How about the series $\displaystyle\sum_{k=1}^{\infty} \frac{\sin kx}{k^3}$?

9.6.2 Verify that the function

$$y(x) = 1 + \frac{x^2}{1!} + \frac{x^4}{2!} + \frac{x^6}{3!} + \frac{x^8}{4!} + \dots$$

is a solution of the differential equation $y' = 2xy$ on $(-\infty, \infty)$ without first finding an explicit formula for $y(x)$.

9.7 Pompeiu's Function

✂ *Enrichment*

By the end of the nineteenth century analysts had developed enough tools to begin constructing examples of functions that challenged the then prevailing views. One famous mathematician, Henri Poincaré, complained that

> Before when one would invent a new function it was to some practical end; today they are invented to demonstrate the errors in the reasoning of our fathers ....

Many mathematicians were both shocked and appalled that functions could be constructed which possessed, to them, bizarre and unnatural properties. The beautiful and elegant theories of the nineteenth century were being torn to pieces by pathological examples.

Perhaps the earliest shock was the construction by Weierstrass and others of continuous functions that had derivatives at no points. This did indeed demonstrate some earlier errors because not a few mathematicians thought they had succeeded in proving that continuous functions could not be like this. Another famous example is due to Vito Volterra (1860–1940), who produced a differentiable function F with a bounded derivative F' that was not Riemann integrable.

In this section we present an example due to D. Pompeiu in 1906. This function h is differentiable and has the remarkable property that h' is discontinuous on a dense set and h' is zero on another dense set. We shall see that this implies that h is a differentiable function that, like Volterra's example, has a derivative that is not Riemann integrable. In fact, it is integrable on *no* interval while Volterra's example is integrable on many subintervals.

The example makes use of many theorems that we have established to this point and so offers an excellent review of our techniques. We present the example in a series of steps, each of which is left as a relatively easy exercise. (Exercise 9.7.4 is plausible but messy to verify, and you may prefer not to check the details.)

To begin the example we observe that the function

$$f(x) = \sqrt[3]{x-a}$$

has an infinite derivative at $x = a$ and a finite derivative elsewhere. Let $q_1, q_2, q_3, \ldots$ be an enumeration of $\mathbb{Q} \cap [0,1]$. Let

$$f(x) = \sum_{k=1}^{\infty} \frac{\sqrt[3]{x-q_k}}{10^k}.$$

The Pompeiu function is the inverse of this function, $h = f^{-1}$.

The details appear in the exercises. Note especially that our main goal is to prove that h is differentiable, h' is bounded, $h' = 0$ on a dense set and h' is positive and discontinuous on another dense set, and h' is not Riemann integrable.

For comparisons let us recall that in Exercise 7.4.2, we provided an example of a differentiable function g with g' bounded but discontinuous on a nowhere-dense perfect set P. Because of Section 8.6.3, we know that if P does not have measure zero, such a function g' will not be integrable, so we cannot write

$$g(x) - g(a) = \int_a^x g'(t)\, dt,$$

that is, the fundamental theorem of calculus does not hold for the function g and its derivative g'. This is essentially how Volterra constructed his example, by ensuring that the set P does not have measure zero.

We also mentioned in Section 7.4 that it is possible for a differentiable function f to have f' discontinuous on a dense set, and so Pompeiu's function justifies this comment.

Exercises

9.7.1 Show that the function $f(x) = (x-a)^{\frac{1}{3}}$ has an infinite derivative at $x = a$ and a finite derivative elsewhere.

9.7.2 Let $q_1, q_2, q_3, \ldots$ be an enumeration of $\mathbb{Q} \cap [0,1]$. For each $k \in \mathbb{N}$ let $f_k(x) = (x - q_k)^{\frac{1}{3}}$. Let

$$f(x) = \sum_{k=1}^{\infty} \frac{f_k(x)}{10^k}.$$

Show that the series defining f converges uniformly.

9.7.3 Show that f is continuous on $[0,1]$.

9.7.4 Check that, for all $x \in \mathbb{R}$,

$$f'(x) = \sum_{k=1}^{\infty} \frac{f_k'(x)}{10^k} = \sum_{k=1}^{\infty} \frac{(x - q_k)^{-\frac{2}{3}}}{3 \times 10^n}.$$

(This part is messy to prove. Indicate why it is that we cannot simply apply Corollary 9.38 and differentiate term by term.)

9.7.5 Show that $f'(x) = \infty$ for all $x \in \mathbb{Q} \cap [0,1]$. (There are also other points at which f' is infinite; see Exercise 9.7.17.)

9.7.6 Show that $f([0,1])$ is an interval. Call it $[a,b]$.

9.7.7 Let $S = f(\mathbb{Q} \cap [0,1])$. Show that S is dense in $[a,b]$.

9.7.8 Show that f has an inverse.

9.7.9 Let $h = f^{-1}$. Show that h is continuous and strictly increasing on $[a,b]$.

9.7.10 Show that $h' = 0$ on the dense set S.

9.7.11 Show that there exists $\gamma > 0$ such that $f'(x) \geq \gamma$ for all $x \in [0,1]$.

9.7.12 Show that h is differentiable and that h' is bounded.

9.7.13 Show that $h' > 0$ on a dense subset of $[a,b]$.

9.7.14 Show that h' is discontinuous on a dense subset of $[a,b]$.

9.7.15 Thus far we know that h is differentiable, has a bounded derivative, $h' = 0$ on a dense set and h' is positive and discontinuous on another dense set. Show that h' is not Riemann integrable.

9.7.16 Show that $\{x : h'(x) \neq 0\}$ does not have measure zero.

9.7.17 Show that there exists $x \notin S$ such that $h'(x) = 0$ and that there exists $t \notin \mathbb{Q}$ such that $f'(t) = \infty$.

9.7.18 Show that the function h is not convex or concave in any interval. Which of the definitions of inflection point given as Exercise 7.10.14 apply to the points x at which $h'(x) = 0$? Do you think that such a point should be called an inflection point?

9.8 Continuity and Pointwise Limits

✂ *Advanced*

Much of this chapter focused on the concept of uniform convergence because of its role in providing affirmative answers to the questions we raised in Section 9.1. In particular, we saw in Section 9.2 that a pointwise limit of a sequence of continuous functions need not be continuous. On the other hand, these problems will not occur if the convergence is uniform.

There are, however, many situations in which pointwise convergence arises naturally, but uniform convergence doesn't. Consider, for example, a function F that is differentiable on $\mathbb{R}$. Then for $x \in \mathbb{R}$,

$$F'(x) = \lim_{n\to\infty} \frac{F(x+\frac{1}{n}) - F(x)}{\frac{1}{n}}.$$

If we define functions f_n by

$$f_n(x) = \frac{F(x+\frac{1}{n}) - F(x)}{\frac{1}{n}},$$

then each of the functions f_n is continuous on $\mathbb{R}$ and $f_n \to F'$ pointwise.

Now, we have seen examples of derivatives that are discontinuous at many points. For example, the function h' in Section 9.7 is discontinuous on a set that is dense in $[0,1]$ and does not have measure zero. Similarly, Exercise 7.4.2 provides an example of a differentiable function g whose derivative g' is discontinuous at every point of a Cantor set that does not have measure zero. We might ask the question, "Can the derivative of a differentiable function be discontinuous everywhere?" We shall see that the answer is "no." In fact, the set of points of continuity must be large in the sense of category—this set must be dense and of type $\mathcal{G}_\delta$, therefore residual (Theorem 6.17).

We actually prove a more general theorem.

Theorem 9.39 *Let $\{g_n\}$ be a sequence of continuous functions defined on an interval I and converging pointwise to a function g on I. Then the set of points of continuity of g forms a dense set of type $\mathcal{G}_\delta$ in I.*

Proof Let us first outline the idea of the proof, leaving the formal proof for a moment. In Section 6.7 we defined the oscillation $\omega_f(x_0)$ of a function f at a point x_0 and showed (Theorem 6.25) that f is continuous at x_0 if and only if $\omega_f(x_0) = 0$. We now show that under the hypotheses of Theorem 9.39, $\omega_g(x)$ will be zero on a dense set. That will imply that g is continuous on a dense set. This set must be of type $\mathcal{G}_\delta$ (by Theorem 6.28).

We will argue by contradiction. We suppose that g is discontinuous at every point of some subinterval J. We will then use the Baire category theorem (Theorem 6.11) to show that there exists $n \in \mathbb{N}$ and an interval $H \subset J$ such that $\omega_g(x) \geq 1/n$ at every point of H. (This argument is valid for any function discontinuous at every point of an interval J.) We then use our hypotheses on g to show this is impossible. We do this by applying the Baire category theorem once again to obtain a subinterval K of H that g maps onto a set of diameter less than $1/n$. This implies that $\omega_f(x) < 1/n$ for every $x \in K$, a contradiction.

Now we can begin a formal proof of Theorem 9.39.

In order to obtain a contradiction, we suppose that g is discontinuous everywhere on some interval $J \subset I$. For each $n \in \mathbb{N}$, let

$$E_n = \{x \in J : \omega_g(x) \geq 1/n\}.$$

Each of the sets E_n is closed (by Theorem 6.27)and $J = \bigcup_{n=1}^{\infty} E_n$.

By the Baire category theorem there exists $n \in \mathbb{N}$ and an interval $H \subset J$ such that E_n is dense in H. The interval H has the property that g maps every subinterval of H onto a set of diameter at least $1/n$. We now show this not possible for g, a pointwise limit of continuous functions.

Let $\{I_k = (a_k, b_k)\}$ be a sequence of intervals, each of length less than $1/n$, such that

$$g(H) \subset \bigcup_{k=1}^{\infty} I_k.$$

For each k, let $H_k = g^{-1}(I_k) \cap H$. Then $H = \bigcup_{k=1}^{\infty} H_k$, but none of the sets H_k can contain an interval [since each H_k has length less than $1/n$, but $\omega_g(x) \geq 1/n$ for all $x \in H$].

Now

$$H_k = \{x : g(x) < b_k\} \cap \{x : g(x) > a_k\}.$$

By Exercise 9.8.4, each of these sets is of type $\mathcal{F}_\sigma$, thus $H_k = \bigcup_{j=1}^{\infty} H_{kj}$, with each of the sets H_{kj} closed. It follows that

$$H = \bigcup_{k=1}^{\infty} H_k = \bigcup_{k=1}^{\infty} \bigcup_{j=1}^{\infty} H_{kj}.$$

The interval H is expressed as a countable union of closed sets. It follows from the Baire category theorem that at least one of these sets, say H_{ij}, is dense in some interval $K \subset H$. Since H_{ij} is closed, $H_{ij} \supset K$. But this implies that $H_i \supset K$, which we have seen is not possible (since each of the sets H_k contains no intervals). This contradiction completes the proof. ∎

Corollary 9.40 *Let f be differentiable on an interval (a, b). Then f' is continuous on a residual subset of (a, b). Thus the set of points of continuity of f' must be dense in (a, b).*

Note. Theorem 9.39 and Exercise 9.8.4 describe two important properties of functions that are pointwise limits of sequences of continuous functions. Each such function f is continuous on a residual set, and every set of the form $\{x : f(x) > a\}$ or $\{x : f(x) < a\}$ is of type $\mathcal{F}_\sigma$.

Theorem 9.39 can be generalized. If P is a nonempty closed subset of the domain of f, then the function $f|P$ is continuous on a dense $\mathcal{G}_\delta$ subset of P.

The converses are also true[3]: A function f is a pointwise limit of a sequence of continuous functions on an interval I if and only if for every closed set $P \subset I$, f considered as a function defined on the set P is continuous on a dense $\mathcal{G}_\delta$ in P, and this happens if and only if every set of the form $\{x : f(x) > a\}$ or $\{x : f(x) < a\}$ is of type $\mathcal{F}_\sigma$.

These theorems have many applications. Functions that are pointwise limits of sequences of continuous functions are called *Baire* 1 functions. We have seen that this class of functions contains the class of derivatives. It also contains all monotonic functions and many other important classes of functions that arise in analysis.

The following exercises may be instructive. You may need to use one of the unproved statements in this section to work some of these exercises.

Exercises

9.8.1 Give an example of a function F that is differentiable on $\mathbb{R}$ such that the sequence

$$f_n(x) = n(F(x + 1/n) - F(x)),$$

converges pointwise but not uniformly to F'.

9.8.2 Give an example of a function f that is Baire 1 and a real number a so that the sets $\{x : f(x) > a\}$ and $\{x : f(x) < a\}$ are not open. Show that, for your example, these sets are of type $\mathcal{F}_\sigma$.

9.8.3 Give an example of a function f that is Baire 1 and a real number a so that the sets $\{x : f(x) \geq a\}$ and $\{x : f(x) \leq a\}$ are not closed. Show that, for your example, these sets are of type $\mathcal{G}_\delta$.

9.8.4 Show that for any f that is Baire 1 and any real number a the sets

$$\{x : f(x) > a\} \quad \text{and} \quad \{x : f(x) < a\}$$

are of type $\mathcal{F}_\sigma$.

9.8.5 If f has only countably many discontinuities on an interval I, then f is a Baire 1 function. In particular, this is true for every monotonic function.

9.8.6 Let K be the Cantor set in $[0, 1]$. Define

$$f(x) = \begin{cases} 1, & \text{if } x \in K \\ 0, & \text{elsewhere;} \end{cases}$$

and

[3]Proofs of these statements can be found in I. P. Natanson, *Theory of Functions of a Real Variable*, Vol. II, Chapter XV, Fredrick Ungar Pub. Co., New York (1955) [English translation].

$$g(x) = \begin{cases} 1, & \text{if } x \text{ is a one-sided limit point of } K \\ 0, & \text{elsewhere.} \end{cases}$$

(a) Show that f and g have exactly the same set of continuity points.

(b) Show that f is a Baire 1 function but g is not.

9.8.7 Let f be the characteristic function of the rationals. Show that f is not a Baire 1 function. Show that f is a pointwise limit of a sequence of Baire 1 functions. (Such functions are called functions of Baire class 2.)

9.9 Challenging Problems for Chapter 9

9.9.1 Let f_n be a sequence of functions converging pointwise to a function f on the interval $[0,1]$. Suppose that each function f_n is convex on $[0,1]$. Show that the convergence is uniform on any interval $[a,b] \subset (0,1)$. Need it be uniform on $[0,1]$?

9.9.2 Let $f_n : [0,1] \to \mathbb{R}$ be a sequence of continuous functions converging pointwise to a function f. If the convergence is uniform, prove that there is a finite number M so that $|f_n(x)| < M$ for all n and all $x \in [0,1]$. Even if the convergence is not uniform, show that there must be a subinterval $[a,b] \subset [0,1]$ and a finite number M so that $|f_n(x)| < M$ for all n and all $x \in [a,b]$.

9.9.3 Let E be a set of real numbers, fixed throughout this exercise. For any function f defined on E write

$$\|f\|_\infty = \sup_{x \in E} |f(x)|.$$

Show that

(a) $\|f\|_\infty = 0$ if and only if f is identically zero on E.

(b) $\|cf\|_\infty = |c|\|f\|_\infty$ for any real number c.

(c) $\|f+g\|_\infty \le \|f\|_\infty + \|g\|_\infty$ for any functions f and g.

(d) $f_n \to f$ uniformly on E if and only if $\|f - f_n\|_\infty \to 0$ as $n \to \infty$.

(e) f_n converges uniformly on E if and only if $\|f_m - f_n\|_\infty \to 0$ as $n, m \to \infty$.

(f) Using $E = (0,1)$ and $f_n(x) = x^n$, compute $\|f_n\|_\infty$ and, hence, show that $\{f_n\}$ is not converging uniformly to zero on $(0,1)$.

Chapter 10

POWER SERIES

✂ If the material on lim sups and lim infs in Section 2.13 of Chapter 2 was omitted, that should be studied before attempting this chapter. The notion of a radius of convergence depends naturally on these concepts.

10.1 Introduction

One of the simplest and, arguably, the most important type of series of functions is the power series. This is a series of the form

$$\sum_{0}^{\infty} a_k x^k$$

or, more generally,

$$\sum_{0}^{\infty} a_k (x-c)^k.$$

It represents the notion of an "infinitely long" polynomial

$$a_0 + a_1 x + a_2 x^2 + \cdots + a_k x^k + \ldots.$$

The material we developed in Chapter 9 will allow us to show in this chapter that power series can be treated very much as if they were indeed polynomials in the sense that they can be integrated and differentiated term by term.

The main reason for developing this theory is that it allows a representation for functions as series. This enlarges considerably the class of functions that we can work with. Not all functions that arise in applications can be expressed as finite combinations of the elementary functions (i.e., as combinations of e^x, x^p, $\sin x$, $\cos x$, etc.). Thus, if we remain at the level of an elementary calculus class, we would be unable to solve many problems since we cannot express the functions needed for the solution in any way. For a large and important class of problems, the functions that can be represented as power series (the so-called analytic functions) are precisely the functions needed.

10.2 Power Series: Convergence

We begin with the formal definition of power series.

Definition 10.1 Let $\{a_k\}$ be a sequence of real numbers and let $c \in \mathbb{R}$. A series of the form

$$\sum_0^\infty a_k(x-c)^k = a_0 + a_1(x-c) + a_2(x-c)^2 + \dots$$

is called a *power series* centered at c. The numbers a_k are called the *coefficients* of the power series.

What can we say about the set of points on which the power series $\sum_0^\infty a_k(x-c)^k$ converges? It is immediately clear that the series converges at its center c. What possibilities are there? A collection of examples illustrates the methods and also essentially all of the possibilities.

Example 10.2 The simple example

$$\sum_1^\infty k^k x^k = x + 4x^2 + 27x^3 + \dots$$

shows that a power series can diverge at every point other than its center. Observe that in this example $k^k x^k = (kx)^k$ does not approach 0 unless $x = 0$, so the series diverges for every $x \neq 0$ by the trivial test. Thus the set of convergence of this series is the set $\{0\}$. ◀

Example 10.3 The familiar geometric series

$$\sum_{k=0}^\infty x^k$$

should be considered the most elementary of all power series. We know that this series converges precisely on the interval $(-1, 1)$ and diverges everywhere else. ◀

Example 10.4 The series

$$\sum_{k=1}^\infty \frac{x^k}{k}$$

has as coefficients $a_k = 1/k$ and the root test[1] supplies

$$s = \limsup_{k\to\infty} \sqrt[k]{|x|^k/k} = |x|.$$

[1]The form of the root test needed to discuss power series uses the limit superior. For that the study of Section 2.13 may be required.

(Verify this!) Thus the series converges on $(-1, 1)$ and diverges for $|x| > 1$. At the two endpoints of the interval $(-1, 1)$ a different test is required. We see that for $x = 1$ this is the familiar harmonic series and so diverges, while for $x = -1$ this is the familiar alternating harmonic series and so converges nonabsolutely. The interval of convergence is $[-1, 1)$. Observe that the series converges at only one of the two endpoints of the interval. ◀

Example 10.5 The series

$$\sum_{k=1}^{\infty} \frac{x^k}{k^2}$$

converges on $[-1, 1]$ and diverges otherwise. Again the root test (or the ratio test) is helpful here. Simpler, though, is to notice that

$$\left| \frac{x^k}{k^2} \right| \leq \frac{1}{k^2}$$

for all $|x| \leq 1$ and so obtain convergence on $[-1, 1]$ by a comparison test with the convergent series $\sum_{k=0}^{\infty} 1/k^2$. If $|x| > 1$ the terms $|x^k/k^2| \to \infty$ and so, trivially, the series diverges. Note here that the series converges on the interval $[-1, 1]$ and is absolutely convergent there. ◀

Example 10.6 The root test applied to the series

$$\sum_{k=1}^{\infty} \frac{x^k}{k^k}$$

gives

$$\lim_{k\to\infty} \sqrt[k]{\frac{|x|^k}{k^k}} = |x| \lim_{k\to\infty} \frac{1}{k} = 0.$$

(The ratio test can also be used here.) It follows that the series converges for all $x \in \mathbb{R}$. Perhaps an easier method in this particular example is to use the comparison test and the fact that

$$\left| \frac{x}{k} \right|^k < \frac{1}{2^k} \text{ when } k \geq 2|x|.$$

Thus the series converges at any x by comparison with a geometric series. Thus the set of convergence of this series is $(-\infty, \infty)$, again as in the previous examples an interval. ◀

In general, as these examples seem to suggest, the set of points of convergence of a power series forms an interval and an application of the root test is an essential tool in determining that interval. Let us apply this test to the series

$$\sum_{0}^{\infty} a_k (x - c)^k.$$

Let

$$s = \limsup_{k\to\infty} \sqrt[k]{|a_k|}.$$

Then

$$\limsup_{k\to\infty} \sqrt[k]{|a_k||x-c|^k} = \limsup_{k\to\infty} \sqrt[k]{|a_k|}|x-c| = s|x-c|.$$

By the root test the series converges absolutely if $s|x-c| < 1$ and diverges if $s|x-c| > 1$.

If $0 < s < \infty$, then the series converges on the interval

$$(c - 1/s, c + 1/s)$$

and diverges for x outside the interval

$$[c - 1/s, c + 1/s].$$

The root test is inconclusive about convergence at the endpoints

$$x = c \pm 1/s$$

of these intervals. The interval of convergence is thus one of the four possibilities

$$(c - 1/s, c + 1/s) \text{ or } [c - 1/s, c + 1/s) \text{ or}$$
$$(c - 1/s, c + 1/s] \text{ or } [c - 1/s, c + 1/s].$$

If $s = 0$, then the series converges for all values of x. We could say that the interval of convergence is $(-\infty, \infty)$ in this case. If $s = \infty$, then the series converges for no values of x other than the trivial value $x = c$. We could say that the interval of convergence is the degenerate "interval" $\{c\}$.

Thus the set of convergence is an interval centered at c. This interval might be degenerate (consisting of only the center), might be all of the real line, and might contain none, one, or both of its endpoints.

Our next theorem summarizes the discussion of convergence to this point. We first give a formal definition.

Definition 10.7 Let $\sum_0^\infty a_k(x-c)^k$ be a power series. Then the number

$$R = \frac{1}{\limsup_{k\to\infty} \sqrt[k]{|a_k|}}$$

is called the *radius of convergence* of the series. Here we interpret $R = \infty$ if

$$\limsup_{k\to\infty} \sqrt[k]{|a_k|} = 0$$

and $R = 0$ if

$$\limsup_{k\to\infty} \sqrt[k]{|a_k|} = \infty.$$

Note. This book deals with *real* analysis, but a full theory of power series fits more naturally into the setting of *complex* analysis. In that setting, a power series converges in a "circle of convergence" centered at a complex number c in the complex plane and with radius

$$R = \frac{1}{\limsup_k \sqrt[k]{|a_k|}}.$$

This explains the origin of the term "radius of convergence."

Theorem 10.8 *Let $\sum_0^\infty a_k(x-c)^k$ be a power series with radius of convergence R.*

1. *If $R = 0$, then the series converges only at $x = c$.*
2. *If $R = \infty$, then the series converges absolutely for all x.*
3. *If $0 < R < \infty$, then the series converges absolutely for all x satisfying $|x-c| < R$ and diverges for all x satisfying $|x-c| > R$.*

Proof We first consider the case $R = 0$. Here $\limsup_k \sqrt[k]{|a_k|} = \infty$ so, for $x \neq c$,

$$\limsup_k \sqrt[k]{|a_k||x-c|^k} = |x-c| \limsup_k \sqrt[k]{|a_k|} = \infty.$$

By the root test the series cannot converge for $x \neq c$. The other cases are similarly established by the root test as in the discussion following our examples. ■

In general, a power series

$$\sum_{k=0}^\infty a_k x^k$$

with a finite radius of convergence R must have as its set of convergence one of the four intervals

$$(-R, R), \quad [-R, R], \quad (-R, R] \text{ or } [-R, R).$$

As we have seen from the examples, each of these four cases can occur. The other possibilities are for series with radius of convergence $R = 0$, in which case the set of convergence is trivially $\{0\}$, or with radius of convergence $R = \infty$, in which case the set of convergence is the entire real line. Note too that if the series converges absolutely at $x = R$ or at $x = -R$, then it must converge absolutely on all of $[-R, R]$. It is possible, though, for the series to converge nonabsolutely at one endpoint but not at the other.

Exercises

10.2.1 Find the radius of convergence for each of the following series.

(a) $\sum_{k=0}^{\infty}(-1)^k x^{2k}$

(b) $\sum_{k=0}^{\infty}\frac{x^k}{k!}$

(c) $\sum_{k=0}^{\infty} kx^k$

(d) $\sum_{k=0}^{\infty} k!x^k$

10.2.2 If the limit

$$\lim_{k\to\infty}\left|\frac{a_k}{a_{k+1}}\right|$$

exists or equals ∞, then show that the following expression also gives the radius of convergence of a power series:

$$R = \lim_{k\to\infty}\left|\frac{a_k}{a_{k+1}}\right|.$$

10.2.3 For the examples

$$\sum_{k=0}^{\infty} x^k, \quad \sum_{k=1}^{\infty}\frac{x^k}{k}, \quad \text{and} \sum_{k=1}^{\infty}\frac{x^k}{k^2}$$

verify in each case that

$$R = \lim_{k}\left|\frac{a_k}{a_{k+1}}\right| = 1.$$

10.2.4 For the series

$$\sum_{k=1}^{\infty} k^k x^k \quad \text{and} \quad \sum_{k=1}^{\infty}\frac{x^k}{k^k}$$

check that the radius of convergence is $R = 0$ and ∞, respectively.

10.2.5 Give an example of a power series $\sum_0^{\infty} a_k x^k$ for which the radius of convergence R satisfies

$$R = \frac{1}{\lim_{k\to\infty}\sqrt[k]{|a_k|}}$$

but

$$\lim_{k\to\infty}\left|\frac{a_k}{a_{k+1}}\right|$$

does not exist.

10.2.6 Give an example of a power series $\sum_0^\infty a_k x^k$ for which the radius of convergence R satisfies

$$\liminf_k \left|\frac{a_{k+1}}{a_k}\right| < R < \limsup_k \left|\frac{a_{k+1}}{a_k}\right|.$$

10.2.7 Give an example of a power series $\sum_0^\infty a_k x^k$ with radius of convergence 1 that is nonabsolutely convergent at both endpoints 1 and -1 of the interval of convergence.

10.2.8 Give an example of a power series $\sum_0^\infty a_k x^k$ with interval of convergence exactly $[-\sqrt{2}, \sqrt{2})$.

10.2.9 If the power series $\sum_0^\infty a_k x^k$ has a radius of convergence R, what must be the radius of convergence of the series

$$\sum_{k=0}^\infty k a_k x^k \quad \text{and} \quad \sum_{k=1}^\infty k^{-1} a_k x^k?$$

10.2.10 If the coefficients $\{a_k\}$ of a power series $\sum_0^\infty a_k x^k$ form a bounded sequence show that the radius of convergence is at least 1.

10.2.11 If the power series $\sum_0^\infty a_k x^k$ has a radius of convergence R_a and the power series $\sum_0^\infty b_k x^k$ has a radius of convergence R_b and $|a_k| \le |b_k|$ for all k sufficiently large, what relation must hold between R_a and R_b?

10.2.12 If the power series $\sum_0^\infty a_k x^k$ has a radius of convergence R, what must be the radius of convergence of the series $\sum_{k=0}^\infty a_k x^{2k}$?

10.2.13 If the power series $\sum_0^\infty a_k x^k$ has a finite positive radius of convergence show that the radius of convergence of the series $\sum_{k=0}^\infty a_k x^{k^2}$ is 1.

10.2.14 Find the radius of convergence of the series

$$\sum_{k=0}^\infty \frac{(\alpha k)!}{(k!)^\beta} x^k,$$

where α and β are positive and α is an integer.

10.2.15 Let $\{a_k\}$ be a sequence of real numbers and let $x_0 \in \mathbb{R}$. Suppose there exists $M > 0$ such that $|a_k x_0^k| \le M$ for all $k \in \mathbb{N}$. Prove that $\sum_0^\infty a_k x^k$ converges absolutely for all x satisfying the inequality $|x| < |x_0|$. What can you say about the radius of convergence of this series?

10.3 Uniform Convergence

So far we have reached a complete understanding of the nature of the set of convergence of any power series. In order to apply many of our theorems of Chapter 9 to questions concerning term by term integration or differentiation of power series, we need to check questions related to the *uniform* convergence of power series. Our next theorem does this and also summarizes the discussion of convergence to this point.

We repeat the convergence results of Theorem 10.8 but now add a discussion of uniform convergence.

Theorem 10.9 *Let $\sum_0^\infty a_k(x-c)^k$ be a power series with radius of convergence R.*

1. *If $R = 0$, then the series converges only at $x = c$.*
2. *If $R = \infty$, then the series converges absolutely and uniformly on any compact interval $[a, b]$.*
3. *If $0 < R < \infty$, then the series converges absolutely and uniformly on any interval $[a, b]$ contained entirely inside the interval $(c - R, c + R)$.*

Proof To verify (2) and (3), let us choose $0 < \rho < R$ so that the interval $[a, b]$ is contained inside the interval $(c - \rho, c + \rho)$. Fix $\rho_0 \in (\rho, R)$. Then

$$\limsup_k \sqrt[k]{|a_k|} = \frac{1}{R} < \frac{1}{\rho_0}.$$

Thus there exists $N \in \mathbb{N}$ such that

$$\sqrt[k]{|a_k|} < \frac{1}{\rho_0} \quad \text{for all } k \geq N. \tag{1}$$

For $k \geq N$ and $|x - c| \leq \rho$ we calculate

$$|a_k(x-c)^k| \leq |a_k|\rho^k < \left(\frac{\rho}{\rho_0}\right)^k,$$

the last inequality following from (1).

Now since $\rho/\rho_0 < 1$, it follows that

$$\sum_{k=0}^{\infty} \left(\frac{\rho}{\rho_0}\right)^k < \infty.$$

It now follows from the Weierstrass M-test (Theorem 9.16) that the series converges absolutely and uniformly on the set $\{x : |x - c| < \rho\}$ and hence also on the subset $[a, b]$. ■

If the interval of convergence of a power series is $(-R, R)$, then certainly the assertion (3) of Theorem 10.9 is the best that can be made. (See Exercise 10.3.3.) The geometric series $\sum_{n=0}^{\infty} x^n$ furnishes the clearest example of this. This series converges on $(-1, 1)$ but does not converges uniformly on the entire interval of convergence $(-1, 1)$. It does, however, converge uniformly on any $[a, b] \subset (-1, 1)$.

To improve on this we can ask the following: If R is the radius of convergence of a power series and the interval of convergence is $[-R, R]$ or $(-R, R]$ or $[-R, R)$, can uniform convergence be extended to the endpoints? If the convergence at an endpoint R (or $-R$) is absolute, then an application of the Weierstrass M-test shows immediately that the convergence is absolute and uniform on $[-R, R]$. For nonabsolute convergence a more delicate test is needed and we need to appeal to material developed in Section 9.3.3. The following theorem contains, for easy reference, a repeat of the third assertion in Theorem 10.9.

Theorem 10.10 *Suppose that the power series $\sum_0^\infty a_k(x-c)^k$ has a finite and positive radius of convergence R and an interval of convergence I.*

1. *If $I = [c-R, c+R]$, then the series converges uniformly (but not necessarily absolutely) on $[c-R, c+R]$.*

2. *If $I = (c-R, c+R]$, then the series converges uniformly (but not necessarily absolutely) on any interval $[a, c+R]$ for all*
$$c - R < a < c + R.$$

3. *If $I = [c-R, c+R)$, then the series converges uniformly (but not necessarily absolutely) on any interval $[c-R, b]$ for all*
$$c - R < b < c + R.$$

4. *If $I = (c-R, c+R)$, then the series converges uniformly and absolutely on any interval $[a, b]$ for $c - R < a < b < c + R$.*

Proof For the purposes of the proof we can take $c = 0$. Let us examine the case
$$I = (c-R, c+R] = (-R, R]$$
which is typical. Consider the intervals $[a, 0]$ for $-R < a < 0$ and $[0, R]$. The uniform convergence of the series on $[a, 0]$ is clear since this is contained entirely inside the interval of convergence.

Now we examine uniform convergence on $[0, R]$. We consider the series
$$\sum_{k=0}^{\infty} a_k x^k = \sum_{k=0}^{\infty} A_k t^k,$$
where $A_k = a_k R^k$ and $t = (x/R)$. The series $\sum_{k=0}^\infty A_k t^k$ converges for $0 \le t \le 1$ by our assumptions. Note that $\sum_{k=0}^\infty A_k$ is convergent while the sequence $\{t^k\}$ converges monotonically on the interval $[0, 1]$. By a variant of Theorem 9.19 (Exercise 9.3.26) this series converges uniformly for t in the interval $[0, 1]$. This translates easily to the assertion that our original series

converges uniformly for $x \in [0, R]$. Thus since the series converges uniformly on $[a, 0]$ and on $[0, R]$ we have obtained the uniform convergence on $[a, R]$ as required. The other cases are similarly handled. ■

Exercises

10.3.1 Characterize those power series $\sum_0^\infty a_k(x-c)^k$ that converge uniformly on $(-\infty, \infty)$.

10.3.2 Show that if $\sum_{k=0}^\infty a_k x^k$ converges absolutely at a point $x_0 > 0$, then the convergence of the series is uniform on $[-x_0, x_0]$.

10.3.3 Show that if $\sum_{k=0}^\infty a_k x^k$ converges uniformly on an interval $(-r, r)$, then it must in fact converge uniformly on $[-r, r]$. Deduce that if the interval of convergence is exactly of the form $(-R, R)$, or $[-R, R)$ or $[-R, R)$, then the series cannot converge uniformly on the entire interval of convergence.

10.4 Functions Represented by Power Series

Suppose now that a power series $\sum_0^\infty a_k(x-c)^k$ has positive or infinite radius of convergence R. Then this series represents a function f on (at least) the interval $(c-R, c+R)$:

$$f(x) = \sum_0^\infty a_k(x-c)^k \quad \text{for } |x-c| < R. \tag{2}$$

If the series converges at one or both endpoints, then this represents a function on $[c-R, c+R)$ or $(c-R, c+R]$ or $[c-R, c+R]$.

What can we say about the function f? In terms of the questions that have motivated us throughout Chapter 9 we can ask

1. Is the function f continuous on its domain of definition?

2. Can f be differentiated by termwise differentiation of its series?

3. Can f be integrated by termwise integration of its series?

We address each of these questions and find that generally the answer to each is yes.

10.4.1 Continuity of Power Series

A power series may represent a function on an interval. Is that function necessarily continuous?

Theorem 10.11 *A function f represented by a power series*

$$f(x) = \sum_{0}^{\infty} a_k (x-c)^k \quad \text{for } |x-c| < R. \tag{3}$$

is continuous on its interval of convergence.

Proof This follows from Theorem 10.10. For example, if the interval of convergence is $(c-R, c+R]$, then we can show that f is continuous at each point of this interval. Since convergence is uniform on $[c, c+R]$ and since each of the functions $a_k(x-c)^k$ is continuous on $[c, c+R]$, the same is true of the function f (Corollary 9.23). For any point $x_0 \in (c-R, c)$ we can similarly prove that f is continuous at x_0 in the same way by noting that the series converges uniformly on an interval $[a, c]$, where a is chosen so that $c - R < a < x_0 < c$. ■

Example 10.12 The series

$$f(x) = \sum_{k=1}^{\infty} \frac{x^k}{k}$$

converges at every point of the interval $[-1, 1)$. Consequently, this function is continuous at every point of that interval. We shall show in the next section that the identity

$$\log(1-x) = \sum_{k=1}^{\infty} \frac{x^k}{k}$$

holds for all $x \in (-1, 1)$ (by integrating the geometric series term by term). Since we are also assured of continuity at the endpoint $x = -1$ we can conclude that

$$\log 2 = \sum_{k=1}^{\infty} \frac{(-1)^k}{k}.$$

◀

10.4.2 Integration of Power Series

If a function is represented by a power series, is it possible to integrate that function by integrating the power series term by term?

Theorem 10.13 *Let a function f be represented by a power series*

$$f(x) = \sum_{0}^{\infty} a_k (x-c)^k$$

with an interval of convergence I. Then for every point x in that interval f is integrable on $[c, x]$ (or $[x, c]$ if $x < c$) and

$$\int_c^x f(t)\, dt = \sum_{k=0}^{\infty} \frac{a_k}{k+1}(x-c)^{k+1}.$$

Proof Let x be a point in the interval of convergence. The convergence of the series $\sum_0^{\infty} a_k(x-c)^k$ is uniform on $[c, x]$ (or on $[x, c]$ if $x < c$), so the series can be integrated term by term (Theorem 9.29). ■

Example 10.14 The geometric series

$$\frac{1}{1-x} = \sum_{k=0}^{\infty} x^k$$

has radius of convergence 1 and so can be integrated term by term provided we stay inside the interval $(-1, 1)$. Thus

$$-\log(1-x) = \int_0^x \frac{1}{1-t}\, dt = \sum_{k=0}^{\infty} \frac{1}{k+1} x^{k+1}$$

for all $-1 < x < 1$. We would not be able to conclude from this theorem that the integral can be extended to the endpoints of $(-1, 1)$. The new series, however, also converges at $x = -1$ and so we can apply Theorem 10.11 to show that the identity just proved is actually valid on $[-1, 1)$. ◀

10.4.3 Differentiation of Power Series

If a function is represented by a power series, is it possible to differentiate that function by differentiating the power series term by term?

Note that for continuity and integration we were able to prove Theorems 10.11 and 10.13 immediately from general theorems on uniform convergence. To prove a theorem on term by term differentiation, we need to check uniform convergence of the series of *derivatives*. The following lemma gives us what we need.

Lemma 10.15 *Let $\sum_0^{\infty} a_k(x-c)^k$ have radius of convergence R. Then the series*

$$\sum_{k=1}^{\infty} k a_k (x-c)^{k-1}$$

obtained via term by term differentiation also has the same radius of convergence R.

Proof The radius of convergence of the series is given by

$$R = \frac{1}{\limsup_k \sqrt[k]{|a_k|}}.$$

The radius of convergence of the differentiated series is given by

$$R' = \frac{1}{\limsup_k \sqrt[k]{|ka_k|}}.$$

But since $\sqrt[k]{k} \to 1$ as $k \to \infty$ we see immediately that these two expressions are equal. (They may be both zero or both infinite.) ■

Theorem 10.16 *Let $\sum_0^\infty a_k(x-c)^k$ have radius of convergence $R > 0$, and let*

$$f(x) = \sum_0^\infty a_k(x-c)^k$$

whenever $|x-c| < R$. Then f is differentiable on $(c-R, c+R)$ and

$$f'(x) = \sum_{k=1}^\infty ka_k(x-c)^{k-1}$$

for each $x \in (c-R, c+R)$.

Proof It follows from the preceding lemma that the series

$$\sum_{k=1}^\infty ka_k(x-c)^{k-1}$$

has radius of convergence R. Thus this series converges uniformly on any compact interval $[a,b]$ contained in $(c-R, c+R)$. Since each value of x in $(c-R, c+R)$ can be placed inside some such interval $[a,b]$ it now follows immediately from Corollary 9.35 that $f'(x) = \sum_{k=1}^\infty ka_k(x-c)^{k-1}$ whenever $|x-c| < R$. ■

We can apply the same argument to the differentiated series and differentiate once more. From the expansion

$$f'(x) = \sum_{k=1}^\infty ka_k(x-c)^{k-1}$$

we obtain a formula for $f''(x)$:

$$f''(x) = \sum_2^\infty k(k-1)a_k(x-c)^{k-2}.$$

Let us express explicitly the formulas of $f(x)$, $f'(x)$, and $f''(x)$.

$$\begin{aligned} f(x) &= a_0 + a_1(x-c) + a_2(x-c)^2 + a_3(x-c)^3 + \dots \\ f'(x) &= a_1 + 2a_2(x-c) + 3a_3(x-c)^2 + \dots \\ f''(x) &= 2a_2 + 3 \cdot 2a_3(x-c) + \dots \end{aligned}$$

These expressions are valid in the interval $(c-R, c+R)$. For $x = c$ we obtain

$$\begin{aligned} f(c) &= a_0 \\ f'(c) &= a_1 \\ f''(c) &= 2a_2. \end{aligned}$$

If we continue in this way, we can obtain power series expansions for all the derivatives of f. This results in the following theorem. The proof (which requires mathematical induction) is left as Exercise 10.4.1.

Theorem 10.17 *Let $\sum_0^\infty a_k(x-c)^k$ have radius of convergence $R > 0$. Then the function*

$$f(x) = \sum_0^\infty a_k(x-c)^k$$

has derivatives of all orders and these derivatives can be calculated by repeated term by term differentiation. The coefficients a_k are related to the derivatives of f at $x = c$ by the formula

$$a_k = \frac{f^{(k)}(c)}{k!}.$$

Uniqueness of Power Series From Theorem 10.17 we deduce that any two power series representations of a function must be identical. Note that the centers have to be the same for this to be true.

Corollary 10.18 *Suppose two power series*

$$f(x) = \sum_0^\infty a_k(x-c)^k$$

and

$$g(x) = \sum_0^\infty b_k(x-c)^k$$

agree on some interval centered at c, that is $f(x) = g(x)$ for $|x-c| < \rho$ and some positive ρ. Then $a_k = b_k$ for all $k = 0, 1, 2, \ldots$.

Proof It follows immediately from Theorem 10.17 that

$$a_k = \frac{f^{(k)}(c)}{k!} = \frac{g^{(k)}(c)}{k!} = b_k$$

for all $k = 0, 1, 2, \ldots$. ■

Example 10.19 The series for the exponential function

$$e^x = \sum_{k=0}^\infty \frac{x^k}{k!}$$

reveals one of the key facts about the exponential function, namely that it is its own derivative. Note simply that

$$\frac{d}{dx}e^x = \frac{d}{dx}\sum_{k=0}^{\infty}\frac{x^k}{k!} = \sum_{k=0}^{\infty}\frac{d}{dx}\frac{x^k}{k!} = \sum_{k=1}^{\infty}\frac{x^{k-1}}{k-1!} = e^x.$$

◀

Example 10.20 The material in this section can also be used to obtain the power series expansion of the exponential function. Suppose that we know that the exponential function $f(x) = e^x$ does in fact have a power series expansion

$$f(x) = \sum_{k=0}^{\infty} a_k x^k.$$

Then the coefficients must be given by the formulas we have obtained, namely

$$a_k = \frac{f^{(k)}(0)}{k!}.$$

But for $f(x) = e^x$ it is clear that $f^{(k)}(0) = 1$ for all k and so the series must be indeed be given by $a_k = 1/k!$ as we well know. But how can we be assured that the exponential function does have a power series expansion? This argument shows only that if there is a series, then that series is precisely $\sum_{k=0}^{\infty}\frac{x^k}{k!}$. There remains the possibility that there may not be a series after all. This is the subject of the next section. ◀

10.4.4 Power Series Representations

Corollary 10.18 shows that if we can obtain a power series representation for a function f by any means whatsoever, then that series must have its coefficients given by the equations $a_k = f^{(k)}(c)/k!$. In particular, a power series representation for f about a given point must be unique.

Example 10.21 For example, the formula for the sum of a geometric series can be used to show that

$$\frac{1}{1+x^2} = 1 - x^2 + x^4 - \cdots + (-1)^j x^{2j} + \ldots.$$

Thus this series represents the function $f(x) = \frac{1}{1+x^2}$ on the interval $(-1, 1)$. Note that the coefficients a_k are zero if k is odd and that $a_{2j} = (-1)^j$ for $k = 2j$ even. It now follows *automatically* that for even integers $k = 2j$

$$\frac{f^{(k)}(0)}{k!} = a_k = (-1)^j$$

while all the odd derivatives are zero. Thus

$$\frac{d^k}{dx^k}\left(\frac{1}{1+x^2}\right) = 0 \quad \text{at } x = 0$$

if k is odd and, if $k = 2j$ is even,

$$\frac{d^k}{dx^k}\left(\frac{1}{1+x^2}\right) = (-1)^j(2j)! \quad \text{at } x = 0.$$

◄

Note. There is a curious fact here that should be puzzled upon. The formula

$$\frac{1}{1+x^2} = 1 - x^2 + x^4 - \cdots + (-1)^j x^{2j} + \dots$$

is valid precisely for $-1 < x < 1$. But the function on the left-hand side of this identity is defined for all values of x. We might have hoped for a representation valid for all x, but we do not obtain one!

Sometimes the easiest way to obtain a power series expansion formula for a function is by using the formula

$$a_k = \frac{f^{(k)}(c)}{k!}.$$

For example, this is how we obtained the power series for $f(x) = e^x$. We compute $f^{(k)}(x) = e^x$ for $k = 0, 1, 2 \dots$, so $f^{(k)}(0) = 1$ for all k. Thus the series expansion for this function (if it has a series expansion) would have to be

$$e^x = 1 + x + \frac{x^2}{2!} + \frac{x^3}{3!} + \cdots = \sum_0^\infty \frac{x^k}{k!}. \tag{4}$$

Note that the series converges for all $x \in \mathbb{R}$. In the next section we will show how to verify that the equality holds for all x.

If we had wanted a formula for $g(x) = e^{x^2}$ we might have used the same idea and determined all the derivatives $g^{(k)}(0)$. It would be simplest, however, to just substitute x^2 for x in the expansion (4), obtaining

$$e^{x^2} = 1 + x^2 + \frac{x^4}{2!} + \frac{x^6}{3!} + \cdots = \sum_0^\infty \frac{x^{2k}}{k!}. \tag{5}$$

Also, from this expansion we can readily obtain an expansion for $2xe^{x^2}$ in either of two ways: We can multiply the expansion in (5) by $2x$ giving

$$2xe^{x^2} = 2x + 2x^3 + \frac{2x^5}{2!} + \frac{2x^7}{3!} + \cdots = \sum_0^\infty \frac{2x^{2k+1}}{k!}.$$

Alternatively, we can use Theorem 10.16 and differentiate (5) term by term, giving

$$2xe^{x^2} = \frac{d}{dx}e^{x^2} = 2x + \frac{4x^3}{2!} + \frac{6x^5}{3!} + \frac{8x^7}{4!} \cdots = \sum_0^\infty \frac{2x^{2k+1}}{k!}.$$

You may wish instead to try to obtain these expansions directly by using the formula $a_k = f^{(k)}(c)/k!$.

Exercises

10.4.1 Provide the details in the proof of Theorem 10.17.

10.4.2 Obtain expansions for

$$\frac{x}{1+x^2} \quad \text{and} \quad \frac{x}{(1+x^2)^2}.$$

10.4.3 Obtain expansions for

$$\frac{1}{1+x^3} \quad \text{and} \quad \frac{x^2}{1+x^3}.$$

10.4.4 Find a power series expansion about $x = 0$ for the function

$$f(x) = \int_0^1 \frac{1 - e^{-sx}}{s}\, ds.$$

10.4.5 The function

$$J_0(x) = \sum_{k=0}^\infty (-1)^k \frac{x^{2k}}{(k!)^2 2^{2k}}$$

is called a Bessel function of order zero of the first kind. Show that this is defined for all values of x. The function $J_1(x) = -J_0'(x)$ is called a Bessel function of order one of the first kind. Find a series expansion for $J_1(x)$.

10.4.6 Let

$$f(x) = \sum_{k=0}^\infty a_k x^k$$

have a positive radius of convergence. If the function f is *even* [i.e., if it satisfies $f(-x) = f(x)$ for all x], what can you deduce about the coefficients a_k? What can you deduce if the function is *odd* (i.e., if $f(-x) = -f(x)$ for all x)?

10.4.7 Let

$$f(x) = \sum_{k=0}^\infty a_k x^k$$

have a positive radius of convergence. If zero is a critical point (i.e., if $a_1 = 0$) and if $a_2 > 0$, then the point $x = 0$ is a strict local minimum. Prove this and also formulate and prove a generalization of this that would allow

$$a_2 = a_3 = a_4 = \cdots = a_{N-1} = 0 \text{ and } a_N \neq 0.$$

10.5 The Taylor Series

We have seen that if a power series $\sum_0^\infty a_k(x-c)^k$ converges on an interval I, then the series represents a function f that has derivatives of all orders. In particular, the coefficients a_k are related to the derivatives of f at c:

$$a_k = \frac{f^{(k)}(c)}{k!}.$$

We then call the series the *Taylor series* for f about the point $x = c$.

Let us turn the question around:

> What functions f have a Taylor series representation in their domain?

We see immediately that such a function must be infinitely differentiable in a neighborhood of c since for such a series to be valid we know that all of the derivatives $f^{(k)}(c)$ must exist. But is that enough?

If we start with a function that has derivatives of all orders on an interval I containing the point c and write the series

$$\sum_{k=0}^{\infty} \frac{f^{(k)}(c)}{k!}(x-c)^k,$$

we might expect that this is exactly the representation we want. Indeed *if there is a valid representation*, then this must be the one, since such representations are unique. But can we be sure the series converges to f on I? Or even that the series converges at all on I. The answer to both questions is "no."

Example 10.22 Consider, for example, the function

$$f(x) = 1/(1+x^2).$$

This function is infinitely differentiable on all of the real line. Its Taylor series about $x = 0$ is, as we have seen in Example 10.21,

$$1 - x^2 + x^4 - x^6 + \cdots = \sum_{k=0}^{\infty}(-1)^k x^{2k}.$$

This series converges for $|x| < 1$ but diverges for $|x| \geq 1$. It does represent f on the interval $(-1, 1)$ but not on the full domain of f. Indeed there can be no series of the form $\sum_{k=0}^{\infty} a_k x^k$ that represents f on $(-\infty, \infty)$ since that series would agree with this present series on $(-1, 1)$ and so could not be any different. ◀

Worse situations are possible. For example, there are infinitely differentiable functions whose Taylor series have zero radius of convergence for every c; for these functions

$$\sum_{k=0}^{\infty} \frac{f^{(k)}(c)}{k!}(x-c)^k$$

diverges except at $x = c$ and this is true for all $c \in \mathbb{R}$.[2] For these functions the Taylor series cannot represent the function.

Another unpleasant situation occurs when a Taylor series converges to the wrong function. This possibility seems even more startling!

Example 10.23 Consider the function

$$f(x) = \begin{cases} 0, & \text{if } x = 0 \\ e^{-1/x^2}, & \text{if } x \neq 0. \end{cases}$$

Exercise 10.5.4 provides an outline for showing that f is infinitely differentiable on the real line, and that $f^{(k)}(0) = 0$ for $k = 1, 2, 3, \ldots$. Thus the Taylor series for f about $x = 0$ takes the form $\sum_{k=0}^{\infty} 0x^k$ with all coefficients equal to zero. This series converges to the zero function on the real line, so it does not represent f except at the origin, even though the series converges for all x. ◀

10.5.1 Representing a Function by a Taylor Series

The preceding discussion shows that we should not automatically assume that a Taylor series for a function f represents f. It is true, however, that the developments in the earlier sections of this chapter help us see that many of the familiar Taylor series encountered in elementary calculus are valid.

Example 10.24 For example, starting with the geometric series

$$\frac{1}{1+x} = \sum_{k=0}^{\infty} (-1)^k x^k,$$

we can apply Theorem 10.13 on integrating a power series term by term to obtain, for $|x| < 1$,

$$\begin{aligned} \ln(1+x) &= \int_1^x \frac{1}{1+t}\,dt = \sum_{k=0}^{\infty} \int_0^x (-1)^k t^k\,dt \\ &= \sum_{k=0}^{\infty} \frac{(-1)^k}{k+1} x^{k+1} = x - \frac{x^2}{2} + \frac{x^3}{3} - \ldots. \end{aligned}$$

We can notice that the integrated series converges at $x = 1$ and so the convergence is uniform on $[0, 1]$. It follows that the representation is valid

[2] See D. Morgenstern, *Math. Nach.*, **12** (1954), p. 74. We find here that in a certain sense "most" infinitely differentiable functions have this property!

for $x \in (-1, 1]$ but for no other points. In this case we obtained a valid Taylor series expansion by integrating a series expansion that we already knew to be valid. ◀

To study the convergence of a Taylor series in general, let us return to fundamentals. Let f be infinitely differentiable in a neighborhood of c. For $n = 0, 1, 2, \ldots$ let

$$P_n(x) = \sum_{k=0}^{n} \frac{f^{(k)}(c)}{k!}(x-c)^k.$$

The polynomial P_n is called the *nth Taylor polynomial* of f at c. The difference $R_n(x) = f(x) - P_n(x)$ is called the *nth remainder function.* In order for the Taylor series about c to converge to f on an interval I, it is necessary and sufficient that $R_n \to 0$ pointwise on I.

Example 10.25 We know that the geometric series represents the function $f(x) = (1-x)^{-1}$ on the interval $(-1, 1)$. We could also prove this result by relying on the remainder term. For $x \neq 1$ and $n = 0, 1, 2, \ldots$ we have

$$\frac{1}{1-x} = 1 + x + x^2 + \cdots + x^n + \frac{x^{n+1}}{1-x}.$$

Here

$$P_n(x) = 1 + x + x + x^2 + \cdots + x^n$$

and

$$R_n(x) = \frac{x^{n+1}}{1-x}.$$

For $|x| < 1$, $R_n(x) \to 0$ as $n \to \infty$. But we have

$$f(x) = P_n(x) + R_n(x)$$

and so the Taylor series for $f(x) = 1/(1-x)$ represents f on the interval $(-1, 1)$. For $|x| \geq 1$, the remainder term does not tend to zero. As before, we see that the representation is confined to the interval $(-1, 1)$. ◀

In a more general situation than this example we would not have an explicit formula for the remainder term. How should we be able to show that the remainder term tends to zero? For functions that are infinitely differentiable in a neighborhood I of c, the various expressions we obtained in Section 7.12 for the remainder functions R_n can be used. In particular, Lagrange's form of the remainder allows us to write for $n = 0, 1, 2, 3, \ldots$

$$f(x) = P_n(x) + \frac{f^{(n+1)}(z)}{(n+1)!}(x-c)^{n+1},$$

where z is between x and c. With some information on the size of the derivatives $f^{(n+1)}(z)$ we might be able to show that this remainder term tends to zero. The integral form of the remainder term, gives us

$$f(x) = P_n(x) + \frac{1}{n!}\int_c^x (x-t)^n f^{(n+1)}(t)\,dt.$$

Again information on the size of the derivatives $f^{(n+1)}(t)$ might show that this remainder term tends to zero.

Example 10.26 Let us justify the familiar Taylor series for $\sin x$:

$$\sin x = \sum_0^\infty \frac{(-1)^k}{(2k+1)!}x^{2k+1}. \tag{6}$$

The remainder term is not expressible in any simple way but can be estimated by using the Lagrange's form of the remainder. The coefficients

$$\frac{(-1)^k}{(2k+1)!}$$

are easily verified by calculating successive derivatives of $f(x) = \sin x$ and using the formulas

$$a_k = \frac{f^{(k)}(0)}{k!}.$$

To check convergence of the series, apply Lagrange's form for $R_n(x)$: For each $x \in \mathbb{R}$, there exists z such that

$$R_n(x) = \frac{f^{(n+1)}(z)}{(n+1)!}x^{n+1}.$$

Now $|f^{(n+1)}(z)|$ equals either $|\cos z|$ or $|\sin z|$ (depending on n) so, in either case, $|f^{(n+1)}(z)| \le 1$, and

$$|R_n(x)| \le |x|^{n+1}/(n+1)!.$$

Since $|x|^{n+1}/(n+1)! \to 0$ as $n \to \infty$ for all $x \in \mathbb{R}$, we can see that the remainder term $|R_n(x)| \to 0$ as $n \to \infty$ for all $x \in \mathbb{R}$. Thus the series representation is completely justified for all real x.

Observe that our estimate for $|R_n(x)|$,

$$|R_n(x)| \le |x|^{n+1}/(n+1)!,$$

gives also a sense of the *rate* of convergence of the series for fixed x. For example, for $|x| \le 1$ we find

$$|R_n(x)| \le 1/(n+1)!.$$

Thus, if we want to calculate $\sin x$ on $(-1, 1)$ to within .01, we need take only the first five terms of the series ($n = 4$) to achieve that degree of accuracy.

Had we used the integral form for $R_n(x)$ we would have obtained a similar estimate. We leave that calculation as Exercise 10.5.1. ◀

10.5.2 Analytic Functions

The class of functions that can be represented as power series is not large. As we have remarked, the class of infinitely differentiable functions is much larger. The terminology that is commonly used for this very special class of functions is given by the following definition.

Definition 10.27 A function f whose Taylor series converges to f in a neighborhood of c is said to be *analytic at c*.

The functions commonly encountered in elementary calculus are generally analytic except at certain obviously nonanalytic points. For example, $|x|$ is not analytic at $x = 0$, and $1/(1-x)$ is not analytic at $x = 1$. These functions fail to have even a first derivative at the point in question. Similarly a function such as $f(x) = |x|^3$ cannot be analytic at $x = 0$ because, while $f'(0)$ and $f''(0)$ exist, $f^{(3)}(0)$ does not. It is not possible to write the complete Taylor series for such a function since some of the derivatives fail to exist.

Even if a function has infinitely many derivatives at a point it need not be analytic there. We would be able to write the complete Taylor expansion but, as we have already noted, the resulting series might not converge to f on any interval. In this connection, it is instructive to work Exercise 10.5.4.

In Example 10.26 we justified the Taylor expansion for $\sin x$. Part of the justification involved the fact that $\sin x$ and all of its derivatives are bounded on the real line. This suggests a general result.

Theorem 10.28 *Let f be infinitely differentiable in a neighborhood I of c. Suppose $x \in I$ and there exists $M > 0$ such that*

$$|f^{(m)}(t)| \leq M$$

for all $m \in I\!N$ and $t \in [c, x]$ (or $[x, c]$ if $x < c$). Then

$$\lim_{n\to\infty} R_n(x) = 0.$$

Thus, f is analytic at c.

Proof We prove the theorem for $x > c$. We leave the case $x < c$ as Exercise 10.5.5.

We use the integral form of the remainder (Theorem 7.45), obtaining

$$|R_n(x)| = \left| \frac{1}{n!} \int_c^x (x-t)^n f^{(n+1)}(t)\, dt \right|. \tag{7}$$

Using our hypothesis that $|f^{(m)}(t)| \le M$ for all $t \in [c, x]$, we infer from (7) that

$$\begin{aligned} |R_n(x)| &\le \frac{M}{n!} \int_c^x (x-t)^n \, dt \\ &= \frac{M}{n!} \frac{(x-t)^{n+1}}{n+1} \Big|_c^x \\ &= \frac{M}{(n+1)!}(x-c)^{n+1} \end{aligned}$$

For fixed x and c, $(x - c)$ is just a constant, so

$$\frac{M(x-c)^{n+1}}{(n+1)!} \to 0.$$

Thus $|R_n(x)| \to 0$ and f is analytic at c. ■

Example 10.29 Let us show that the function $f(x) = e^x$ is analytic at $x = 0$. It is infinitely differentiable, but we need to prove more. The fact that f is analytic at $x = 0$ follows from the previous theorem: We choose, say, the interval $[-1, 1]$ and note that $|f^{(n)}(x)| = |e^x| \le e$ for all $x \in (-1, 1)$ and $n \in \mathbb{N}$. A similar observation applies to the analyticity of f at any point $c \in \mathbb{R}$. ◀

Exercise 10.5.6 provides another theorem similar to Theorem 10.28.

Exercises

10.5.1 Justify formula (6) for $\sin x$ using the integral form of the remainder $R_n(x)$.

10.5.2 Show that

$$f(x) = \sum_{k=0}^{n} \frac{f^k(0)}{k!} x^k + \frac{x^{n+1}}{n!} \int_0^1 f^{(n+1)}(sx)(1-s)^n \, ds$$

under appropriate assumptions on f.

10.5.3 Show that

$$\int_0^1 f^{(n+1)}(sb)(1-s)^n \, ds \le \frac{n! f(b)}{b^{n+1}}$$

if f and all of its derivatives exist and are nonnegative on the interval $[0, b]$.

10.5.4 Let

$$f(x) = \begin{cases} 0, & \text{if } x = 0 \\ e^{-1/x^2}, & \text{if } x \ne 0. \end{cases}$$

Prove that f is infinitely differentiable on the real line. Show that $f^{(k)}(0) = 0$ for all $k \in \mathbb{N}$. Explain why the Taylor series for f about $x = 0$ does not represent f in any neighborhood of zero. Is f analytic at $x = c$ for $c \ne 0$?

10.5.5 Prove Theorem 10.28 for $x < c$.

10.5.6 Prove Bernstein's Theorem: If f is infinitely differentiable on an interval I, and $f^{(n)}(x) \geq 0$ for all $n \in \mathbb{N}$ and $x \in I$, then f is analytic on I. Apply this result to $f(x) = e^x$.

10.5.7 Use the results of this section to verify that each of the following functions is analytic at $x = 0$, and write the Taylor series about $x = 0$.

(a) $\cos x^2$

(b) e^{-x^2}

10.5.8 Show that if f and g are analytic functions at each point of an interval (a, b), then so too is any linear combination $\alpha f + \beta g$.

10.6 Products of Power Series

Suppose that we have two power series representations

$$f(x) = \sum_{k=0}^{\infty} a_k(x - x_0)^k$$

and

$$g(x) = \sum_{k=0}^{\infty} b_k(x - x_0)^k$$

valid in the intervals $(-R_f, R_f)$ and $(-R_g, R_g)$, respectively. How should we obtain a power series representation for the product $f(x)g(x)$? We might merely compute all the derivatives of this function and so construct its Taylor series. But is this the easiest or most convenient method? How do we know that such a representation would be valid?

The most direct approach to this problem is to apply here our study of products of series from Section 3.8. We know when such a product would be valid. Indeed, from that theory, we know immediately that

$$f(x)g(x) = \sum_{k=0}^{\infty} c_k(x - x_0)^k$$

would hold in the interval $(-R, R)$, where $R = \min\{R_f, R_g\}$ and the coefficients are given by the formulas

$$c_k = \sum_{j=0}^{k} a_j b_{k-j}.$$

Example 10.30 The product of the series

$$\frac{1}{1-x} = 1 + x + x^2 + x^3 + \ldots$$

and the series

$$f(x) = \sum_{k=0}^{\infty} a_k x^k$$

gives the representation

$$\frac{f(x)}{1-x} = \sum_{k=0}^{\infty}(a_0 + a_1 + a_2 + \cdots + a_k)x^k.$$

Where would this be valid? ◀

Example 10.31 A representation for the function $e^x \sin x$ might be most easily obtained by forming the product

$$\left(1 + x + \frac{1}{2}x^2 + \frac{1}{6}x^3 + \dots\right)\left(x - \frac{1}{3}x^3 + \frac{1}{5}x^5 - \dots\right) = x + x^2 + \frac{1}{6}x^3 + \dots$$

and the series continued as far as is needed for the application at hand. ◀

10.6.1 Quotients of Power Series

Suppose that we have power series representations of two functions

$$f(x) = \sum_{k=0}^{\infty} a_k x^k \quad \text{and} \quad g(x) = \sum_{k=0}^{\infty} b_k x^k$$

both valid in some interval $(-r, r)$ at least. Can we find a representation of the quotient function $f(x)/g(x)$? Certainly we must demand that $g(0) \neq 0$, which amounts to asking for the leading coefficient in the series for g (the term b_0) not to be zero.

If there is a representation, say a series $\sum_{k=0}^{\infty} c_k x^k$, then, evidently, we require that

$$\frac{\sum_{k=0}^{\infty} a_k x^k}{\sum_{k=0}^{\infty} b_k x^k} = \sum_{k=0}^{\infty} c_k x^k.$$

This merely means that we want

$$\left(\sum_{k=0}^{\infty} b_k x^k\right)\left(\sum_{k=0}^{\infty} c_k x^k\right) = \sum_{k=0}^{\infty} a_k x^k.$$

The conditions for this are known to us since we have already studied how to form the product of two power series. For this to hold the coefficients $\{c_k\}$ (which, at the moment, we do not know how to determine) should satisfy the equations

$$b_0 c_0 = a_0$$

$$b_0 c_1 + b_1 c_0 = a_1$$

$$b_0 c_2 + b_1 c_1 + b_2 c_0 = a_2$$

and, in general,

$$b_0c_k + b_1c_{k-1} + b_2c_{k-2} + \cdots + b_kc_0 = a_k.$$

Since we know all the a_k's and b_k's, we can readily solve these equations, one at a time starting from the first to obtain the coefficients for the quotient series. This algorithm (for that is what it is) for determining the c_k's is precisely "long division." Simply divide formally the expression (the denominator)

$$b_0 + b_1x + b_2x^2 + b_3x^3 + \ldots$$

into the expression (the numerator)

$$a_0 + a_1x + a_2x^2 + a_3x^3 + \ldots$$

and you will find yourself solving exactly these equations in our algorithm.

But what have we determined? We have shown that if there is a series representation for $f(x)/g(x)$, then this method will determine it. We do not, however, have any assurances in advance that there is such a series. We offer the next theorem, without proof, for those assurances. Alternatively, in any computation we could construct the quotient series (all terms!) and determine that it has a positive radius of convergence. That, too, would justify the method although it is not likely the most practical approach.

Theorem 10.32 *Suppose that there are power series representations for two functions*

$$f(x) = \sum_{k=0}^{\infty} a_kx^k \quad \textit{and} \quad g(x) = \sum_{k=0}^{\infty} b_kx^k$$

both valid in some interval $(-r, r)$ at least and that $b_0 \neq 0$. Then there is some positive $\delta > 0$ so that the function $f(x)/g(x)$ is analytic at zero and a quotient series can be found.

The proper setting for a proof of Theorem 10.32 is complex analysis, where it is proved that a quotient of complex analytic functions is analytic if the denominator is not zero.

Exercises

10.6.1 Show that if f and g are analytic functions at each point of an interval (a, b), then so too is the product fg.

10.6.2 Under what conditions on the functions f and g on an interval (a, b) can you conclude that the quotient f/g is analytic?

10.6.3 Using long division, find severalt terms of the power series expansion of

$$\frac{x + 2}{x^2 + x + 1}$$

centered at $x = 0$. What other method would work?

10.6.4 Using long division and the power series expansions for $\sin x$ and $\cos x$, find the first few terms of the power series expansion of $\tan x$ centered at $x = 0$. What other method would have given you these same numbers?

10.6.5 Find a power series expansion centered at $x = 0$ for the function

$$\frac{\sin 2x}{\sin x}.$$

Did the fact that $\sin x = 0$ at $x = 0$ make you modify the method here?

10.6.6 Show that if

$$\frac{1}{\sum_{k=0}^{\infty} b_k x^k} = \sum_{k=0}^{\infty} c_k x^k$$

is valid, then

$$c_k = \frac{(-1)^k}{b_0^{k+1}} \begin{vmatrix} b_1 & b_0 & 0 & 0 & \dots & 0 \\ b_2 & b_1 & b_0 & 0 & \dots & 0 \\ b_3 & b_2 & b_1 & b_0 & \dots & 0 \\ \dots & & & & \dots & \dots \\ b_k & b_{k-1} & b_{k-2} & b_{k-3} & \dots & b_1 \end{vmatrix}.$$

10.7 Composition of Power Series

Suppose that we wished to obtain a power series expansion for the function $e^{\sin x}$ using the two series expansions

$$e^x = 1 + x + \frac{1}{2}x^2 + \frac{1}{6}x^3 + \dots$$

and

$$\sin x = x - \frac{1}{3}x^3 + \frac{1}{5}x^5 - \dots.$$

Without pausing to decide if this makes any sense let us simply insert the series for $\sin x$ in the appropriate positions in the series for e^x. Then we might hope to justify that

$$e^{\sin x} = 1 + \left(x - \frac{1}{3}x^3 + \frac{1}{5}x^5 - \dots \right)$$

$$+ \frac{1}{2}\left(x - \frac{1}{3}x^3 + \frac{1}{5}x^5 - \dots \right)^2 + \frac{1}{6}\left(x - \frac{1}{3}x^3 + \frac{1}{5}x^5 - \dots \right)^3 + \dots$$

and expand, grouping terms in the obvious way, getting (at least for a start)

$$e^{\sin x} = 1 + x + \frac{1}{2}x^2 - \frac{1}{6}x^3 + \dots.$$

Is this method valid?

To justify this method we state (without proof) a theorem giving some conditions when this could be verified. Note that the conditions are as we should expect for a composition of functions $f(g(x))$. The series for $g(x)$ is expanded about a point x_0. That is inserted into a series expanded about the value $g(x_0)$, thus obtaining a series for $f(g(x))$ expanded about the point x_0. The proof is not difficult if approached within a course in complex variables but would be mysterious if attempted as a real variable theorem.

Theorem 10.33 *Suppose that there are power series representations for two functions*

$$g(x) = C + \sum_{k=1}^{\infty} a_k(x - x_0)^k \quad \text{and} \quad f(x) = \sum_{k=0}^{\infty} b_k(x - C)^k$$

both valid in some nondegenerate intervals about their centers. Then there is a power series expansion for

$$f(g(x)) = \sum_{k=0}^{\infty} c_k(x - x_0)^k$$

with a positive radius of convergence whose coefficients can be obtained by inserting the series for $g(x) - C$ into the series for f, that is, by expanding

$$f(g(x)) = \sum_{k=0}^{\infty} b_k \left(\sum_{j=1}^{\infty} a_j(x - x_0)^j \right)^k$$

formally.

Exercises

10.7.1 Under what conditions on the functions f and g on an interval (a, b) can you conclude that the composition $f \circ g$ is analytic?

10.7.2 Find several terms in the power series expansion of $e^{\sin x}$ by a method different from that in this section.

10.7.3 Find several terms in the power series expansion of $e^{\tan x}$ using the method discussed in this section.

10.8 Trigonometric Series

✂ *Enrichment*

In this section we present a short introduction to another way of representing functions, namely as trigonometric series or Fourier series. There are deep connections between power series and Fourier series so this theory does belong in this chapter (see Exercise 10.8.1).

The origins of the subject go back to the middle of the eighteenth century. Certain problems in mathematical physics seemed to require that an arbitrary function f with a fixed period (taken here as 2π) be represented in the form of a trigonometric series

$$f(t) = \tfrac{1}{2}a_0 + \sum_{j=1}^{\infty} (a_j \cos jt + b_j \sin jt), \tag{8}$$

and mathematicians such as Daniel Bernoulli, d'Alembert, Lagrange, and Euler had debated whether such a thing should be possible. Bernoulli maintained that this would always be possible, while Euler and d'Alembert argued against it.

Joseph Fourier (1768–1830) saw the utility of these representations and, although he did nothing to verify his position other than to perform some specific calculations, claimed that the representation in (8) would be available for every function f and gave the formulas

$$a_j = \frac{1}{\pi}\int_{-\pi}^{\pi} f(t)\cos jt\,dt \quad \text{and} \quad b_j = \frac{1}{\pi}\int_{-\pi}^{\pi} f(t)\sin jt\,dt$$

for the coefficients.

While his mathematical reasons were not very strong and much criticized at the time, his instincts were correct, and series of this form with coefficients computed in this way are now known as *Fourier series.* The a_j and b_j are called the *Fourier coefficients* of f.

10.8.1 Uniform Convergence of Trigonometric Series

✂ *Enrichment*

For a first taste of this theory we prove an interesting theorem that justifies some of Fourier's original intuitions. We show that if a trigonometric series converges *uniformly* to a function f, then necessarily those coefficients given by Fourier are the correct ones.

Theorem 10.34 *Suppose that*

$$f(t) = \tfrac{1}{2}a_0 + \sum_{j=1}^{\infty} (a_j \cos jt + b_j \sin jt), \tag{9}$$

with uniform convergence on the interval $[-\pi, \pi]$. Then it follows that the function f is continuous and the coefficients are given by Fourier's formulas:

$$a_j = \frac{1}{\pi}\int_{-\pi}^{\pi} f(t)\cos jt\,dt \quad \textit{and} \quad b_j = \frac{1}{\pi}\int_{-\pi}^{\pi} f(t)\sin jt\,dt$$

Proof Fix $j \geq 1$, choose $n > j$ and write

$$s_n(t) = \tfrac{1}{2}a_0 + \sum_{k=1}^{n} (a_k \cos kt + b_k \sin kt)$$

that is, the partial sums of the series. A straightforward, if tiresome, calculation shows that for $j \geq 1$, and for $n > j$,

$$\int_{-\pi}^{\pi} s_n(t) \cos jt \, dt = \int_{-\pi}^{\pi} a_j (\cos jt)^2 \, dt = a_j \pi. \tag{10}$$

This is, remember, just a finite sum. The orthogonality formulas in Exercise 10.8.3 assist in this computation.

We are assuming that $s_n \to f$ uniformly and so it follows too, since $\cos jt$ is bounded that $s_n(t) \cos jt \to f(t) \cos jt$ uniformly for $t \in [-\pi, \pi]$. It follows, since all functions here are continuous, that

$$\lim_{n\to\infty} \int_{-\pi}^{\pi} s_n(t) \cos jt \, dt = \int_{-\pi}^{\pi} f(t) \cos jt \, dt.$$

In view of (10) this proves the formula for a_j and $j \geq 1$. The formulas for a_0 and b_j for $j \geq 1$ can be obtained by an identical method. ■

10.8.2 Fourier Series

✂ *Enrichment*

Emboldened by the theorem we have just proved we make a dramatic move, the same move that Fourier made. We start with the function f (not the series) and construct a trigonometric series by using these coefficient formulas.

Note the twist in the logic. *If* there is a trigonometric series converging uniformly to a continuous function f, then it would have to be given by the formulas of Theorem 10.34. Why not start with the series even if we have no knowledge that the series will converge uniformly, even if we do not know whether it will converge uniformly to the function we started with, indeed even if the series diverges?

Definition 10.35 Let f be a Riemann integrable function on the interval $[-\pi, \pi]$ and let

$$a_j = \frac{1}{\pi} \int_{-\pi}^{\pi} f(t) \cos jt \, dt \quad \text{and} \quad b_j = \frac{1}{\pi} \int_{-\pi}^{\pi} f(t) \sin jt \, dt.$$

Then the series

$$\tfrac{1}{2} a_0 + \sum_{j=1}^{\infty} (a_j \cos jt + b_j \sin jt) \tag{11}$$

is called the *Fourier series* of f.

There is a mild understanding here that the series should be somehow related to f and there is a hope that the series can be used as a "representation" of f. But, in general, uniform convergence is out of the question.

Indeed, even pointwise convergence is too much to hope for. To emphasize that this relation is not one of equality, we usually write

$$f(t) \sim \tfrac{1}{2}a_0 + \sum_{j=1}^{\infty} (a_j \cos jt + b_j \sin jt).$$

Exercises

10.8.1 Let $f(z) = \sum_{k=0}^{\infty} \alpha_k z^k$ be a complex power series with a radius of convergence larger than 1. By setting $z = e^{it}$ find a connection between complex power series and trigonometric series.

10.8.2 Explain why it is that for any Riemann integrable function f we can claim that the integrals defining the Fourier coefficients of f exist.

10.8.3 Check the so-called *orthogonality relations* by computing that for integers $k \neq j$ and all i

$$\int_{-\pi}^{\pi} \sin(kt)\sin(jt)\,dt = 0, \quad \int_{-\pi}^{\pi} \cos(kt)\sin(it)\,dt = 0,$$

and

$$\int_{-\pi}^{\pi} \cos(kt)\cos(jt)\,dt = 0.$$

10.8.4 Check that for integers i, $k \neq 0$,

$$\int_{-\pi}^{\pi} (\sin kt)^2\,dt = \pi \quad \text{and} \quad \int_{-\pi}^{\pi} (\cos it)^2\,dt = \pi.$$

10.8.3 Convergence of Fourier Series

✂ *Enrichment*

The theory of Fourier Series would have a much simpler, if less fascinating, development if the Fourier series of every continuous function converged uniformly to the original function. Not only is this false, but the Fourier series of a continuous function can diverge at a large set of points. This leaves us with a serious difficulty. The Fourier series of a function is expected to represent the function, but how? If it does not converge to the function, how can it be used as a representation?

There is a mistake in our reasoning. We know that if a series converges to a function in suitable ways, then the function may be integrated and differentiated by termwise integration and differentiation of the series. But it is possible that a series may be manipulated in these ways *even if the series diverges* at some points. A representation need not be a pointwise or uniform representation to be useful.

In our next theorem we show that the Cesàro sums of the Fourier series of a suitable function do converge uniformly to the function even if the series itself is divergent. You should review the topic of Cesàro summability in

Section 3.9.1. A young Hungarian mathematician, Leopold Fejér (1880–1959), obtained this theorem in 1900.

Theorem 10.36 (Fejér) *Let f be a continuous function on $[-\pi, \pi]$ for which $f(-\pi) = f(\pi)$. Then the sequence of Cesàro means of the partial sums of the Fourier series for f converges uniformly to f on $[-\pi, \pi]$.*

Proof Throughout the proof we may consider that f is defined on all of $\mathbb{R}$ and is 2π-periodic. We write

$$s_n(x) = \tfrac{1}{2}a_0 + \sum_{k=1}^{n} (a_k \cos kx + b_k \sin kx)$$

for the partial sums of the Fourier series of f (this means the coefficients a_j, b_j are determined by using Fourier's formulas). Then we write

$$\sigma_n(x) = \frac{s_0(x) + s_1(x) + s_2(x) + \cdots + s_n(x)}{n+1}$$

for the sequence of averages (Cesàro means).

Our task is to prove that $\sigma_n \to f$ uniformly. Looking back we see that each $\sigma_n(x)$ is a finite sum of terms $s_k(x)$ and each $s_k(x)$ is a finite sum of terms involving a_j, b_j. In turn, each of these terms is expressible as an integral involving f and sin's and cos's. Thus after some considerable but routine computations, we arrive at a formula

$$\sigma_n(x) = \frac{1}{\pi}\int_{-\pi}^{\pi} \frac{1}{2}\left(f(x+t) + f(x-t)\right) K_n(t)\, dt$$

or the equivalent formula

$$\sigma_n(x) = \frac{1}{\pi}\int_{0}^{\pi} \left(f(x+t) + f(x-t)\right) K_n(t)\, dt. \qquad (12)$$

Here K_n is called the *Fejér kernel* and for each n,

$$K_n(t) = \frac{1}{2(n+1)}\left(\frac{\sin\left(\frac{1}{2}(n+1)t\right)}{\sin \frac{1}{2}t}\right)^2.$$

You can just accept the computations for the purposes of our short introduction to the subject.

The Fejér kernel of order n enjoys the following properties, each of which is evident from its definition:

1. Each $K_n(t)$ is a nonnegative, continuous function.

2. For each n,

$$\frac{1}{\pi}\int_{-\pi}^{\pi} K_n(t)\, dt = \frac{2}{\pi}\int_{0}^{\pi} K_n(t)\, dt = 1.$$

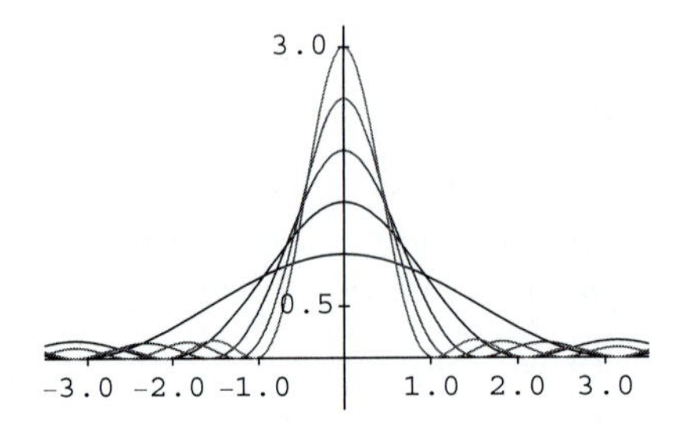

Figure 10.1. Fejér kernel $K_n(t)$ for $n = 1, 2, 3, 4,$ and 5 on $[-\pi, \pi]$.

3. For each n and $0 < |t| < \pi$,

$$0 \leq K_n(t) \leq \frac{\pi}{(n+1)t^2}.$$

Figure 10.1 illustrates the graph of this function for $n = 1, 2, 3, 4,$ and 5.

Let $\varepsilon > 0$, and choose $\delta > 0$ so that

$$|f(x+t) + f(x-t) - 2f(x)| < \varepsilon$$

for every $0 \leq t \leq \delta$. This uses the uniform continuity of f. We note that

$$\frac{2}{\pi} \int_0^\pi f(x) K_n(t)\, dt = f(x)$$

by using property 2. Thus we have

$$\begin{aligned} |\sigma_n(x) - f(x)| &\leq \frac{1}{\pi} \int_0^\pi |f(x+t) + f(x-t) - 2f(x)|\, K_n(t)\, dt \\ &\leq I_1 + I_2, \end{aligned}$$

where I_1 is the integral taken over $[0, \delta]$ and I_2 is the integral taken over $[\delta, \pi]$. Since K_n is nonnegative, we did not need to keep it inside the absolute value in the integral. The part I_1 will be small (for all n) because the expression in the absolute values is small for t in the interval $[0, \delta]$. The part I_2 will be small (for large n) because of the bound on the size of K_n for t away from zero in property 3. Here are the details: For I_1 we have

$$I_1 \leq \frac{\varepsilon}{\pi} \int_0^\delta K_n(t)\, dt \leq \varepsilon.$$

For I_2, let

$$\kappa_n = \sup\{K_n(t) : \delta \leq t \leq \pi\},$$

and note that property 3 supplies us with the fact that $\kappa_n \to 0$ as $n \to \infty$. Now we have

$$I_2 \leq \frac{\kappa_n \varepsilon}{\pi} \int_\delta^\pi \left(|f(x+t)| + |f(x-t)| + 2|f(x)|\right)\, dt$$

so that we can make I_2 as small as we please by choosing n large enough.

It follows, since ε and x are arbitrary, that $\lim_{n\to\infty} \sigma_n(x) = f(x)$, uniformly for $x \in [-\pi, \pi]$ as required. ■

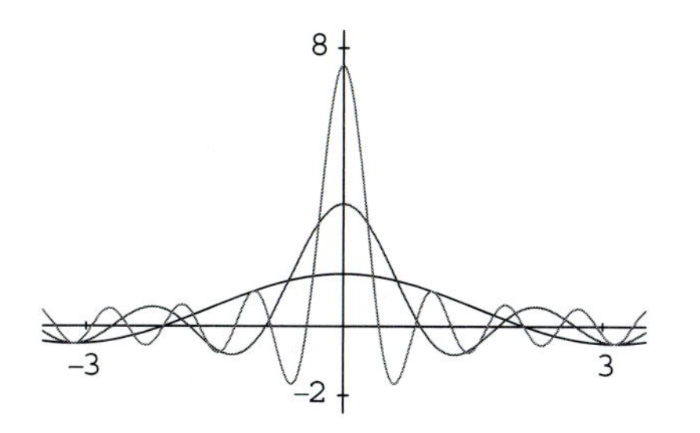

Figure 10.2. Dirichlet kernel $D_n(t)$ for $n = 1$, 3, and 7 on $[-\pi, \pi]$.

Exercises

10.8.5 Let $s_n(x)$ be the sequence of partial sums of the Fourier series for a 2π-periodic integrable function f. Show that

$$s_n(x) = \frac{1}{\pi}\int_{-\pi}^{\pi} \frac{1}{2}\left(f(x+t) + f(x-t)\right) D_n(t)\, dt$$

and

$$s_n(x) = \frac{1}{\pi}\int_{0}^{\pi} \left(f(x+t) + f(x-t)\right) D_n(t)\, dt,$$

where

$$D_n(t) = \frac{1}{2} + \sum_{k=1}^{n} \cos kt$$

is called the Dirichlet kernel.

Figure 10.2 illustrates the graph of this function for $n = 1$, 3, and 7. It should be contrasted with Figure 10.1.

10.8.6 Establish the following properties of the Dirichlet kernel $D_n(t)$ for each n:

(a) $D_n(t)$ is a continuous, 2π-periodic function.

(b) $D_n(t)$ is an even function.

(c) $\displaystyle \frac{1}{\pi}\int_{-\pi}^{\pi} D_n(t)\, dt = \frac{2}{\pi}\int_{0}^{\pi} D_n(t)\, dt = 1.$

(d) $\displaystyle D_n(t) = \frac{\sin\left(n + \frac{1}{2}\right)t}{2\sin\frac{1}{2}t}.$

(e) $D_n(0) = n + \frac{1}{2}$.

(f) For all t, $|D_n(t)| \le n + \frac{1}{2}$.

(g) For all $0 < |t| < \pi$, $|D_n(t)| \le \frac{\pi}{2|t|}$.

10.8.7 Let

$$K_n(t) = \frac{1}{n+1}\sum_{j=0}^{n} D_j(t),$$

where D_j are the Dirichlet kernels. Show that the formula for the averages σ_n given in the proof of Theorem 10.36 is correct.

10.8.4 Weierstrass Approximation Theorem

Enrichment

Fejér's theorem allows us to prove the famous Weierstrass approximation theorem. Note that a consequence of Fejér's theorem is that continuous, 2π-periodic functions can be uniformly approximated by trigonometric polynomials. The Weierstrass theorem asserts that continuous functions on a compact interval can be uniformly approximated by ordinary polynomials.

Theorem 10.37 (Weierstrass) *Let f be a continuous function on an interval $[a,b]$, and let $\varepsilon > 0$. Then there is a polynomial*

$$g(x) = \alpha_n x^n + \alpha_{n-1} x^{n-1} + \cdots + \alpha_1 x + \alpha_0$$

so that

$$|f(x) - g(x)| < \varepsilon$$

for all $x \in [a,b]$.

Proof It is more convenient for this proof to assume that $[a,b] = [0,1]$. The general case can be obtained from this.

Let f be a continuous function on $[0,1]$, let $\varepsilon > 0$, and write

$$F(t) = f(|\cos t|).$$

Then F is a continuous, 2π-periodic function and can be approximated by a trigonometric polynomial within ε. This is because, in view of Theorem 10.36, for large enough n the Cesàro means $\sigma_n(f)$ are uniformly close to f.

Since F is even [i.e., $F(t) = F(-t)$] we can figure out what form that trigonometric polynomial may take. All the coefficients b_k involving $\sin kt$ in the Fourier series for F must be zero. Thus when we form the averages of the partial sums we obtain only sums of cosines. Consequently, we can find $c_0, c_1, c_2, \ldots c_n$ so that

$$\left| F(t) - \sum_0^n c_j \cos jt \right| < \varepsilon \tag{13}$$

for all t. Each $\cos jt$ can be written using elementary trigonometric identities as $T_j(\cos t)$ for some jth-order (ordinary) polynomial T_j, and so, by setting $x = \cos t$ for any $x \in [0,1]$, we have

$$\left| f(x) - \sum_0^n c_j T_j(x) \right| < \varepsilon,$$

which is exactly the polynomial approximation that we need. ■

The polynomials T_j that appear in the proof are well known as the Chebychev polynomials and are easily generated (see Exercise 10.8.9). They are named after the Russian mathematician Pafnuty Lvovich Chebychev (1821–1894).

Exercises

10.8.8 Show that once Theorem 10.37 is proved for the interval $[0,1]$ it can be deduced for any interval $[a,b]$.

10.8.9 Define the Chebychev polynomials by requiring T_j to be a polynomial so that

$$\cos jt = T_j(\cos t)$$

identically. Show that $T_0(x) = 1$, $T_1(x) = x$, and

$$T_n(x) = 2xT_{n-1}(x) - T_{n-2}(x).$$

Generate the first few of these polynomials.

10.8.10 Show that Theorem 10.37 can be interpreted as asserting that for any continuous function on an interval $[a,b]$ there is a sequence of polynomials p_n converging to f uniformly on $[a,b]$.

10.8.11 Does Exercise 10.8.10 also imply that there must be a power series expansion converging to f uniformly on $[a,b]$?

10.8.12 Let f be a continuous function on an interval $[a,b]$, and let $\varepsilon > 0$. Show that there must exist a polynomial p with rational coefficients so that

$$|f(x) - p(x)| < \varepsilon$$

for all $x \in [a,b]$.

10.8.13 Let $f : \mathbb{R} \to \mathbb{R}$ be a continuous function and let $\varepsilon > 0$. Must there exist a polynomial p so that $|f(x) - p(x)| < \varepsilon$ for all $x \in \mathbb{R}$?

10.8.14 Let $f : (0,1) \to \mathbb{R}$ be a continuous function and let $\varepsilon > 0$. Must there exist a polynomial p so that $|f(x) - p(x)| < \varepsilon$ for all $x \in (0,1)$?

10.8.15 Let $f : [0,1] \to \mathbb{R}$ be a continuous function with the property that

$$\int_0^1 f(x)x^n\,dx = 0$$

for all $n = 0, 1, 2, \ldots$. What can you conclude about the function f?

10.8.16 Let $f : [0,1] \to \mathbb{R}$ be a continuous function with the property that $f(0) = 0$ and

$$\int_0^1 f(x)\sin \pi nx\,dx = 0$$

for all $n = 1, 2, 3, \ldots$. What can you conclude about the function f?

Chapter 11

THE EUCLIDEAN SPACES $\mathbb{R}^n$

In our study of analysis we have observed that the real number system $\mathbb{R}$ is more than just a collection of points. It has various algebraic, metric, and order structures that play a role in the development of the subject. The same is true of many other important sets of mathematical objects. Among the most important are the euclidean spaces $\mathbb{R}^n$, which we study in this chapter. The collection of elements of the set $\mathbb{R}^n$ is easy to describe: It is just the set of all n-tuples of real numbers

$$\mathbf{x} = (x_1, \ldots, x_n).$$

We'll spend some time developing a natural algebraic structure (the space $\mathbb{R}^n$ forms a vector space) and a natural metric structure (there are natural notions of distance) that we'll exploit in this and the next chapter. (While we can also impose order structures on these spaces, such structures are not natural for our purposes and will not enter our discussions.)

11.1 The Algebraic Structure of $\mathbb{R}^n$

The space $\mathbb{R}^n$ consists of all n-tuples $\mathbf{x} = (x_1, \ldots, x_n)$ of real numbers. That is, $\mathbb{R}^n$ is the cartesian product of n copies of $\mathbb{R}$:

$$\mathbb{R}^n = \mathbb{R} \times \mathbb{R} \times \cdots \times \mathbb{R} \quad (n \text{ factors}).$$

The members $\mathbf{x}$ of $\mathbb{R}^n$ are called the points of $\mathbb{R}^n$. They can also be viewed as *vectors*, and $\mathbb{R}$ as a field of *scalars*. We define addition of vectors in $\mathbb{R}^n$ as coordinate-wise addition: If $\mathbf{x} = (x_1, \ldots, x_n)$ and $\mathbf{y} = (y_1, \ldots, y_n)$, then

$$\mathbf{x} + \mathbf{y} = (x_1 + y_1, \ldots, x_n + y_n). \tag{1}$$

For $\alpha \in \mathbb{R}$ and $\mathbf{x} = (x_1, \ldots, x_n) \in \mathbb{R}^n$ we define

$$\alpha\mathbf{x} = (\alpha x_1, \ldots, \alpha x_n). \tag{2}$$

We refer to such multiplication of a vector by a scalar as *scalar multiplication.*

With these notions of addition and scalar multiplication, $\mathbb{R}^n$ possesses the algebraic structure of a *vector space.* This means that the properties indicated in Theorem 11.1 are valid.

Theorem 11.1 *If addition and scalar multiplication on $\mathbb{R}^n$ are defined by (1) and (2), respectively, then the following are true for all $\mathbf{x}$, $\mathbf{y}$, $\mathbf{z} \in \mathbb{R}^n$ and all α, $\beta \in \mathbb{R}$:*

$\mathbf{x}+\mathbf{y}$ *and* $\alpha\mathbf{x}$ *are in* $\mathbb{R}^n$	*(closure of* $\mathbb{R}^n$ *under addition and scalar multiplication)*
$\mathbf{x}+\mathbf{y}=\mathbf{y}+\mathbf{x}$	*(commutative law)*
$(\mathbf{x}+\mathbf{y})+\mathbf{z}=\mathbf{x}+(\mathbf{y}+\mathbf{z})$	*(associative law)*
$\mathbf{x}+\mathbf{0}=\mathbf{x}$	*where* $\mathbf{0}=(0,\ldots,0)$
$\alpha(\mathbf{x}+\mathbf{y})=\alpha\mathbf{x}+\alpha\mathbf{y}$	*(distributive*
$(\alpha+\beta)\mathbf{x}=\alpha\mathbf{x}+\beta\mathbf{x}$	*laws)*
$\alpha(\beta x)=(\alpha\beta)\mathbf{x}$	
$0\mathbf{x}=\mathbf{0}$	
$1\mathbf{x}=\mathbf{x}.$	

The proof of Theorem 11.1 is just a simple exercise of checking each statement. We leave it as Exercise 11.1.1. We also observe that by defining subtraction by

$$\mathbf{x}-\mathbf{y}=\mathbf{x}+(-1)\mathbf{y},$$

then $\mathbf{x}-\mathbf{x}=\mathbf{0}$. We can also write $-\mathbf{x}$ for $(-1)\mathbf{x}$ and $\mathbf{x}/\alpha$ for $(1/\alpha)\mathbf{x}$.

The Dot Product We shall have need for a notion of *dot product* of two vectors $\mathbf{x}$ and $\mathbf{y}$. This is defined by

$$\mathbf{x}\cdot\mathbf{y}=x_1y_1+\cdots+x_ny_n, \tag{3}$$

where $\mathbf{x}=(x_1,\ldots,x_n)$ and $\mathbf{y}=(y_1,\ldots,y_n)$. Observe that the dot product (3) is a scalar. For that reason $\mathbf{x}\cdot\mathbf{y}$ is sometimes called the *scalar product* of $\mathbf{x}$ and $\mathbf{y}$. It is also sometimes called the *inner product* of $\mathbf{x}$ and $\mathbf{y}$. The dot product satisfies certain conditions that we summarize in Theorem 11.2.

Theorem 11.2 *If* $\mathbf{x}$, $\mathbf{y}$, $\mathbf{z} \in \mathbb{R}^n$ *and* $\alpha \in \mathbb{R}$, *then*

$\mathbf{x}\cdot\mathbf{y}=\mathbf{y}\cdot\mathbf{x}$	*(commutative law)*
$\mathbf{x}\cdot(\mathbf{y}+\mathbf{z})=\mathbf{x}\cdot\mathbf{y}+\mathbf{x}\cdot\mathbf{z}$	*(distributive law)*
$(\alpha\mathbf{x})\cdot\mathbf{y}=\alpha(\mathbf{x}\cdot\mathbf{y}).$	

We leave the proof as Exercise 11.1.2.

Observe that $\mathbf{0}\cdot\mathbf{x}=0$, but that $\mathbf{x}\cdot\mathbf{y}$ can equal 0 without either $\mathbf{x}$ or $\mathbf{y}$ being 0 [E.g., in $\mathbb{R}^2$, $(1,1)\cdot(1,-1)=1-1=0$]. Two nonzero vectors $\mathbf{x}$ and $\mathbf{y}$ for which $\mathbf{x}\cdot\mathbf{y}=0$ are said to be *orthogonal* vectors. Geometrically, two orthogonal vectors in $\mathbb{R}^n$ are perpendicular.

Exercises

11.1.1 Prove Theorem 11.1.

11.1.2 Prove Theorem 11.2.

11.1.3 Verify that $\mathbf{x} \cdot \mathbf{x} = 0$ if and only if $\mathbf{x} = \mathbf{0}$.

11.1.4 Let $\mathcal{C}[a,b]$ be the set of continuous real functions on $[a,b]$. Verify that $\mathcal{C}[a,b]$ is a vector space under the usual notions of addition and scalar multiplication.

11.1.5 Let $\mathcal{D}[a,b]$ be the set of functions with the intermediate value property (IVP) on $[a,b]$. Is $\mathcal{D}$ a vector space under the usual notions of addition and scalar multiplication?

11.1.6 For f, g in $\mathcal{C}[0,2\pi]$, let $f \cdot g = \int_0^{2\pi} f(t)g(t)\,dt$.

(a) Show that "·" satisfies the conditions of Theorem 11.2.

(b) Show that the set of functions

$$T = \{\sin nx, \cos mx : (n = 1, 2, \ldots \; m = 1, 2, \ldots\}$$

form an orthogonal set of functions in $\mathcal{C}$, that is, $f \cdot g = 0$ if $f, g \in T$, with $f \neq g$.

11.2 The Metric Structure of $\mathbb{R}^n$

The spaces $\mathbb{R}^n$ possess a natural notion of distance between two points. The notion of distance is fundamental to such notions as convergence of sequences, limits, continuity, and differentiation.

Our starting point is that of the norm $\|\mathbf{x}\|$ of a vector in $\mathbb{R}^n$.

Definition 11.3 Let $\mathbf{x} = (x_1, \ldots, x_n) \in \mathbb{R}^n$. We define the *euclidean norm* by

$$\|\mathbf{x}\| = \sqrt{\sum_{i=1}^{n} x_i^2}.$$

Note that $\|\mathbf{x}\| \geq 0$ for all $x \in \mathbb{R}^n$.

In the familiar cases of $n = 1$, 2 or 3, $\|\mathbf{x}\|$ is just the distance between $\mathbf{x}$ and the vector $\mathbf{0}$. [For example, in $\mathbb{R}^2$ with $\mathbf{z} = (x,y)$ and $\|\mathbf{z}\| = \sqrt{x^2+y^2}$.]

This suggests the following definition of distance between the vectors $\mathbf{x}$ and $\mathbf{y}$ in $\mathbb{R}^n$ (Fig. 11.1).

Definition 11.4 For $\mathbf{x}, \mathbf{y} \in \mathbb{R}^n$, the *euclidean distance* between $\mathbf{x}$ and $\mathbf{y}$ is

$$d(x,y) = \|\mathbf{x} - \mathbf{y}\|.$$

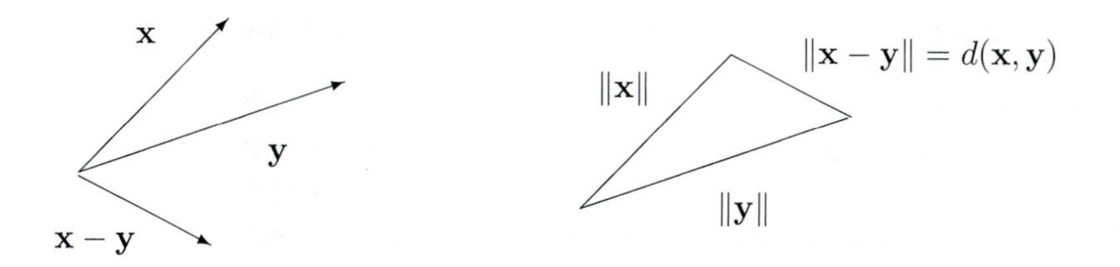

Figure 11.1. Vectors, distance, and norm in $\mathbb{R}^2$.

In terms of coordinates, with $\mathbf{x} = (x_1, \ldots, x_n)$ and $\mathbf{y} = (y_1, \ldots, y_n)$

$$\|\mathbf{x} - \mathbf{y}\| = \sqrt{(x_1 - y_1)^2 + \cdots + (x_n - y_n)^2}.$$

Observe that we can express the norm via the dot product:

$$\|\mathbf{x}\|^2 = \mathbf{x} \cdot \mathbf{x} \quad \text{or} \quad \|\mathbf{x}\| = \sqrt{\mathbf{x} \cdot \mathbf{x}}.$$

Because of this, many norm computations are simplified by employing the dot product. An important relation between the dot product and norms is given by the Cauchy-Schwarz inequality.

Theorem 11.5 (Cauchy-Schwarz) *For* $\mathbf{x}, \mathbf{y} \in \mathbb{R}^n$,

$$|\mathbf{x} \cdot \mathbf{y}| \leq \|\mathbf{x}\| \, \|\mathbf{y}\|.$$

Proof First observe that if $t \in \mathbb{R}$, then

$$p(t) = (\mathbf{x} - t\mathbf{y}) \cdot (\mathbf{x} - t\mathbf{y}) \geq 0. \tag{4}$$

Using Theorem 11.2, we can rewrite (4) as

$$p(t) = \mathbf{x} \cdot \mathbf{x} - 2(\mathbf{x} \cdot \mathbf{y})t + (\mathbf{y} \cdot \mathbf{y})t^2 \geq 0. \tag{5}$$

The function $p(t)$ is a quadratic in t. Since it is nonnegative for all t, it cannot have two distinct roots. Hence its discriminant cannot be positive, that is,

$$4(\mathbf{x} \cdot \mathbf{y})^2 - 4(\mathbf{x} \cdot \mathbf{x})(\mathbf{y} \cdot \mathbf{y}) \leq 0. \tag{6}$$

It now follows from (6) that

$$(\mathbf{x} \cdot \mathbf{y})^2 \leq (\mathbf{x} \cdot \mathbf{x})^2 (\mathbf{y} \cdot \mathbf{y})^2$$

and so

$$|\mathbf{x} \cdot \mathbf{y}| \leq \|\mathbf{x}\| \, \|\mathbf{y}\|$$

as was to be proved. ■

The Cauchy-Schwarz inequality allows us to obtain some important properties of the norm and of distance (or *metric* as it is often known).

Theorem 11.6 (Norm Properties) *Let* $\mathbf{x}, \mathbf{y} \in \mathbb{R}^n$, *and* $\alpha \in \mathbb{R}$. *Then*

(i) $\|\mathbf{x}\| \geq 0$, *and* $\|\mathbf{x}\| = 0$ *if and only if* $\mathbf{x} = 0$.

(ii) $\|\alpha\mathbf{x}\| = |\alpha|\,\|\mathbf{x}\|$.

(iii) $\|\mathbf{x} + \mathbf{y}\| \leq \|\mathbf{x}\| + \|\mathbf{y}\|$ *(triangle inequality).*

Proof The proofs of (i) and (ii) are immediate. To prove (iii), we calculate

$$\begin{aligned}\|\mathbf{x} + \mathbf{y}\|^2 &= (\mathbf{x} + \mathbf{y}) \cdot (\mathbf{x} + \mathbf{y}) = \|\mathbf{x}\|^2 + 2\mathbf{x} \cdot \mathbf{y} + \|\mathbf{y}\|^2 \\ &\leq \|\mathbf{x}\|^2 + 2\|\mathbf{x}\|\|\mathbf{y}\| + \|\mathbf{y}\|^2 = (\|\mathbf{x}\|^2 + \|\mathbf{y}\|^2)^2,\end{aligned}$$

the inequality following from the Cauchy-Schwarz inequality. ■

Theorem 11.7 (Metric Properties) *The distance function*

$$d : \mathbb{R}^n \times \mathbb{R}^n \to \mathbb{R}$$

satisfies the following conditions for all $\mathbf{x}, \mathbf{y}, \mathbf{z} \in \mathbb{R}^n$.

(i) $d(\mathbf{x}, \mathbf{y}) \geq 0$, *and* $d(\mathbf{x}, \mathbf{y}) = 0$ *if and only if* $\mathbf{x} = \mathbf{y}$.

(ii) $d(\mathbf{x}, \mathbf{y}) = d(\mathbf{y}, \mathbf{x})$ *(symmetry).*

(iii) $d(\mathbf{x}, \mathbf{z}) \leq d(\mathbf{x}, \mathbf{y}) + d(\mathbf{y}, \mathbf{z})$ *(triangle inequality).*

Proof Again, the proofs of (i) and (ii) are immediate. To verify (iii), let $\mathbf{u} = \mathbf{x} - \mathbf{y}$, $\mathbf{v} = \mathbf{y} - \mathbf{z}$. Then $\mathbf{u} + \mathbf{v} = \mathbf{x} - \mathbf{z}$. From (iii) of Theorem 11.6 we see that

$$\begin{aligned}\|\mathbf{u} + \mathbf{v}\| &\leq \|\mathbf{u}\| + \|\mathbf{v}\|, \text{ or} \\ \|\mathbf{x} - \mathbf{z}\| &\leq \|\mathbf{x} - \mathbf{y}\| + \|\mathbf{y} - \mathbf{z}\|,\end{aligned}$$

which is just the statement $d(\mathbf{x}, \mathbf{z}) \leq d(\mathbf{x}, \mathbf{y}) + d(\mathbf{y}, \mathbf{z})$. ■

In Exercises 11.2.6 and 11.2.7 we introduce two further norms on $\mathbb{R}^n$ and find some comparisons among them. Later, Exercises 11.3.8 and 11.4.1 show some ways in which these norms can be used interchangeably when dealing with concepts directly related to open and closed sets and convergence. Thus even though the euclidean norm encountered in this section is our main focus, other norms can be used at the same time to simplify and clarify the arguments.

Exercises

11.2.1 Establish the identity

$$\left(\sum_{i=1}^{n} x_i y_i\right)^2 = \left(\sum_{i=1}^{n} x_i^2\right)\left(\sum_{i=1}^{n} y_i^2\right) - \sum_{i<j} (x_i y_j - x_j y_i)^2$$

and use it to obtain another proof of the Cauchy-Schwarz inequality.

11.2.2 Verify the validity of parts (i) and (ii) in Theorems 11.6 and 11.7.

11.2.3 Prove that for $\mathbf{x}, \mathbf{y} \in \mathbb{R}^n$, $\|\mathbf{x} - \mathbf{y}\| \geq |\,\|\mathbf{x}\| - \|\mathbf{y}\|\,|$.

11.2.4 Let $\mathbf{x}, \mathbf{y}$ be vectors in $\mathbb{R}^2$, and let θ be the angle between them. Use the law of cosines to verify that

$$\mathbf{x} \cdot \mathbf{y} = \|\mathbf{x}\|\,\|\mathbf{y}\| \cos\theta.$$

Thus, the dot product $\mathbf{x} \cdot \mathbf{y}$ equals the length of $\mathbf{x}$ times the (signed) length of the projection of $\mathbf{y}$ onto the vector $\mathbf{x}$.

11.2.5 The result of Exercise 11.2.4 suggests a definition for the cosine of an angle between two nonzero vectors in $\mathbb{R}^n$:

$$\cos\theta = \frac{\mathbf{x} \cdot \mathbf{y}}{\|\mathbf{x}\|\,\|\mathbf{y}\|} \qquad (0 \leq \theta \leq \pi).$$

The preceding definition of $\cos\theta$ is given via dot products and norms. Why does that imply $|\cos\theta| \leq 1$?

11.2.6 The euclidean norm $\|\mathbf{x}\| = \sqrt{x_1^2 + \cdots + x_n^2}$ is not the only norm that shall concern us. In the next chapter we shall find it computationally convenient to use the norm $\|\mathbf{x}\|_1 = |x_1| + \cdots + |x_n|$.

(a) Prove that $\|\mathbf{x}\|_1$ satisfies the conditions (i), (ii), (iii) of Theorem 11.6.

(b) If $d_1(\mathbf{x}, \mathbf{y}) = \|\mathbf{x} - \mathbf{y}\|_1$, show that $d_1(\mathbf{x}, \mathbf{y})$ satisfies conditions (i), (ii), (iii) of Theorem 11.7.

(c) Show that for each $x \in \mathbb{R}^n$, $\|\mathbf{x}\| \leq \|\mathbf{x}\|_1 \leq \sqrt{n}\|\mathbf{x}\|$.

11.2.7 There are many norms on $\mathbb{R}^n$ other than the euclidean norm and the norm $\|\cdot\|_1$ of Exercise 11.2.6. For "$\|\cdot\|$" to be a norm means that the three conditions of Theorem 11.6 are satisfied.

(a) Show that $\|\mathbf{x}\|_\infty = \max(|x_1|, \ldots, |x_n|)$ is a norm on $\mathbb{R}^n$.

(b) Of the three norms $\|\cdot\|$, $\|\cdot\|_1$, and $\|\cdot\|_\infty$, for which is the Cauchy-Schwarz inequality valid?

(c) For which of these norms is it true that $|\mathbf{x} \cdot \mathbf{x}| = \|\mathbf{x}\|^2$?

11.2.8 The identity

$$\|\mathbf{x} + \mathbf{y}\|^2 + \|\mathbf{x} - \mathbf{y}\|^2 = 2(\|\mathbf{x}\|^2 + \|\mathbf{y}\|^2)$$

is known as the *parallelogram law.*

(a) Prove the identity is valid for all $x, y \in \mathbb{R}^n$.

(b) Interpret this identity as a statement about the sides and the diagonals of a parallelogram.

(c) Is the identity valid if we replace $\|\cdot\|$ by $\|\cdot\|_1$ or $\|\cdot\|_\infty$?

11.3 Elementary Topology of $\mathbb{R}^n$

By now you understand the fundamental role that the notion of distance plays in the study of $\mathbb{R}$. Indeed, the concept of open set is defined directly in terms of the distance notion. Various other important concepts, such as closed set, dense set, and accumulation point, depend directly or indirectly on the notion of distance. The same is true in $\mathbb{R}^n$. In this section we define some of these important concepts. We shall see that the definitions for $\mathbb{R}^n$ are the same as those for $\mathbb{R}$ if we replace the usual distance function on $\mathbb{R}$ by the euclidean distance function on $\mathbb{R}^n$.

Let $\mathbf{x_0} \in \mathbb{R}$, and let $r > 0$. The set

$$B(\mathbf{x_0}, r) = \{\mathbf{x} : \|\mathbf{x} - \mathbf{x_0}\| < r\}$$

is called the *open ball of radius* r *centered at* $\mathbf{x_0}$. It is also called the *r-neighborhood* of $\mathbf{x_0}$. It obviously consists of those points in $\mathbb{R}^n$ whose distance from $\mathbf{x_0}$ is less than r. We define a number of terms, familiar in our study of $\mathbb{R}$, in terms of this simple concept.

1. A set E in $\mathbb{R}^n$ is *open* if to each $\mathbf{x_0} \in E$ there corresponds an $\varepsilon > 0$ such that $B(\mathbf{x_0}, \varepsilon) \subset E$. An open set containing a point $\mathbf{x_0}$ is also called a *neighborhood* of $\mathbf{x_0}$.
2. A set E in $\mathbb{R}^n$ is *closed* if $\mathbb{R}^n \setminus E$ is open.
3. A point $\mathbf{x_0} \in E \subset \mathbb{R}^n$ is an *accumulation point*, or *limit point*, of E if every open ball centered at $\mathbf{x_0}$ contains points of E other than $\mathbf{x_0}$. We shall use the terms "accumulation point" and "limit point" interchangeably.
4. A point $\mathbf{x_0} \in \mathbb{R}^n$ is a *boundary point* of a set $E \subset \mathbb{R}^n$ provided that for each $\varepsilon > 0$, $B(\mathbf{x_0}, \varepsilon) \cap E \neq \emptyset$ and $B(\mathbf{x_0}, \varepsilon) \setminus E \neq \emptyset$.
5. A set $E \subset \mathbb{R}^n$ is *bounded* if there exists $M > 0$ such that
$$E \subset B(\mathbf{0}, M).$$

We list in the next four examples some subsets of $\mathbb{R}^2$ and indicate which of the preceding properties they possess. We leave verification to you.

Example 11.8 The x-axis, $E_1 = \{(x, y) \in \mathbb{R}^2 : y = 0\}$. E_1 is not open; E_1 is closed; every point of E_1 is an accumulation point and a boundary point of E_1; E_1 is not bounded. ◄

Example 11.9 Let $E_2 = \{(x, y) : x^2 + y^2 \leq 1\}$. This set is not open; it is closed; every point of E_2 is an accumulation point of E_2; the boundary points of E_2 are those for which $x^2 + y^2 = 1$. E_2 is bounded. ◄

Example 11.10 The set

$$E_3 = \mathbb{Q} \times \mathbb{Q} = \{(x,y) : \ x \text{ and } y \text{ are rational}\}$$

is neither open nor closed; every point in $\mathbb{R}^2$ is an accumulation point of E_3 and a boundary point of E_3; E_3 is not bounded in $\mathbb{R}^2$. ◀

Example 11.11 The annular region

$$E_4 = \{(x,y) : 1 < x^2 + y^2 \leq 2\}$$

is neither open nor closed; every point in $\{(x,y) : 1 \leq x^2 + y^2 \leq 2\}$ is an accumulation point of E_4; the boundary points of E_4 are the points (x,y) such that $x^2 + y^2 = 1$ or $x^2 + y^2 = 2$; E_4 is bounded. ◀

Other terms we defined for $\mathbb{R}$ can be defined in the same way for $\mathbb{R}^n$. For example, the concepts of isolated point, dense set, and nowhere dense set can be defined in $\mathbb{R}^n$ in exactly the same way as they are in $\mathbb{R}$. Our development in the rest of this chapter and the next will deal primarily with open or closed sets in $\mathbb{R}^n$. We leave discussions of these and other concepts to the exercises. The notion of compactness will also be important to our development. We shall discuss compactness in Section 11.8.

A number of facts that relate the concepts in this section appear in the exercises. These are analogous to ones we already noted for $\mathbb{R}$.

Exercises

11.3.1 Show that a set $E \subset \mathbb{R}^n$ is closed if and only if it contains all of its limit points.

11.3.2 The *closure* $\overline{E}$ of a set E in $\mathbb{R}^n$ is the union of E and its accumulation points.

(a) Prove that E is closed if and only if $E = \overline{E}$.

(b) Prove that $\overline{B(\mathbf{x}, \varepsilon)} = \{\mathbf{y} : \|\mathbf{y} - \mathbf{x}\| \leq \varepsilon\}$.

11.3.3 List all subsets of $\mathbb{R}^n$ that are both open and closed.

11.3.4 Prove the following analogues of Theorems 4.17 and 4.18 in the setting of $\mathbb{R}^n$.

(a) The sets $\emptyset$ and $\mathbb{R}^n$ are both open and closed.

(b) Any intersection of a finite number of open sets is open.

(c) Any union of an arbitrary collection of open sets is open.

(d) The complement of an open set is closed.

(e) Any union of a finite number of closed sets is closed.

(f) Any intersection of an arbitrary collection of closed sets is closed.

(g) The complement of a closed set is open.

11.3.5 Let I be an open interval in $\mathbb{R}$, and let J be a closed interval in $\mathbb{R}$. We can view I and J as subsets of $\mathbb{R}^2$ by defining

$$I_1 = \{(x,0) : x \in I\} \quad \text{and} \quad J_1 = \{(x,0) : x \in J\}.$$

Is I_1 open in $\mathbb{R}^2$? Is J_1 closed in $\mathbb{R}^2$?

✂ **11.3.6** Provide definitions for *isolated point*, *dense set*, and *nowhere dense set* in the setting of $\mathbb{R}^n$. The definitions should be analogous to those that apply in $\mathbb{R}$ so that the definitions in $\mathbb{R}$ are just special cases of those for $\mathbb{R}^n$.

(a) For each of the sets in Examples 11.8–11.11 indicate whether they have isolated points and whether they are dense or nowhere dense.

(b) Prove that E is dense if and only if $\overline{E} = \mathbb{R}^n$. (See Exercise 11.3.2.)

(c) Prove that E is nowhere dense if and only if every point of $\overline{E}$ is a boundary point of $\overline{E}$.

11.3.7 An open set D in $\mathbb{R}^n$ is called *connected* if to each pair of points $\mathbf{x}$ and $\mathbf{y}$ in D corresponds a "polygonal arc" in D, that is, a path consisting of a finite number of line segments joined end to end consecutively. Which of the following subsets of $\mathbb{R}^2$ are connected?

(a) The open unit ball centered at $\mathbf{0}$

(b) The set $D = \{(x,y) : xy > 0\}$

(c) The set $E = \{(x,y) : |y| > x^2\}$

11.3.8 Suppose we had defined $B_1(\mathbf{x_0}, \varepsilon) = \{\mathbf{x} : \|\mathbf{x} - \mathbf{x_0}\|_1 < \varepsilon\}$, replacing the norm $\|\cdot\|$ by $\|\cdot\|_1$.

(a) Sketch a picture of $B_1(\mathbf{0}, 1)$ in $\mathbb{R}^2$.

(b) Prove that a set $E \subset \mathbb{R}^n$ is open if and only if to each $\mathbf{x_0} \in E$ there corresponds $\varepsilon > 0$ such that $B_1(\mathbf{x_0}, \varepsilon) \subset E$. (Thus $\mathbb{R}^n$ has exactly the same open sets when we use $\|\cdot\|_1$ as when we use $\|\cdot\|$ in the definition of open sets.)

(c) Are the closed sets exactly the same?

(d) Does the status of a point (as an accumulation point or boundary point or isolated point of a set) depend on whether we use $\|\cdot\|$ or $\|\cdot\|_1$ in our definition of open set?

(e) What about the status of a set as dense, nowhere dense, or bounded?

11.4 Sequences in $\mathbb{R}^n$

Much of what we have studied about sequences of real numbers carries over to the case of sequences in $\mathbb{R}^n$. Indeed, many of the proofs for sequences in $\mathbb{R}^n$ are virtually identical to the proofs of corresponding statements for sequences in $\mathbb{R}$. We shall leave most of these proofs to the exercises. When a proof requires a fresh idea, we shall provide the proof in detail.

Definition 11.12 A *sequence of points in* $\mathbb{R}^n$ is a function

$$f : \mathbb{N} \to \mathbb{R}^n.$$

As was the case for sequences in $\mathbb{R}$, we will frequently use notation such as $\{\mathbf{x_k}\}$ to denote a sequence in $\mathbb{R}^n$. Here each vector $\mathbf{x_k}$ can be written as an n-tuple using double subscript notation:

$$\mathbf{x_k} = (x_{k1}, \ldots, x_{kn}).$$

Definition 11.13 A sequence $\{\mathbf{x_k}\}$ in $\mathbb{R}^n$ is *bounded* if there exists $M \in \mathbb{N}$ such that $\|\mathbf{x_k}\| \leq M$ for all $k \in \mathbb{N}$.

We can define convergence of a sequence $\{\mathbf{x_k}\}$ in $\mathbb{R}^n$ in the same way as we did for sequences of real numbers (Definition 2.6). Note that here the norm plays the same role that the absolute value did earlier.

Definition 11.14 Let $\{\mathbf{x_k}\}$ be a sequence in $\mathbb{R}^n$. We say $\{\mathbf{x_k}\}$ *converges* to a point $\mathbf{x}$ and write

$$\lim_{k\to\infty} \mathbf{x_k} = \mathbf{x} \quad \text{or}$$

$$\mathbf{x_k} \to \mathbf{x} \text{ as } k \to \infty$$

provided that for each $\varepsilon > 0$ there exists $N \in \mathbb{N}$ such that $\|\mathbf{x_k} - \mathbf{x}\| < \varepsilon$ whenever $k \geq N$.

This definition is equivalent to the requirement that every open ball centered at $\mathbf{x}$ contains all but a finite number of terms of the sequence: For each $\varepsilon > 0$ there exists N such that $\mathbf{x_k} \in B(\mathbf{x}, \varepsilon)$ for all $k \geq N$. It is also equivalent to the requirements

$$\|\mathbf{x_k} - \mathbf{x}\| \to 0$$

or

$$d(\mathbf{x_k}, \mathbf{x}) \to 0$$

as $k \to \infty$, where d is the euclidean distance.

Coordinate-Wise Convergence We can also describe convergence in $\mathbb{R}^n$ in terms of coordinate-wise convergence. To see this, observe first that for $\mathbf{x} = (x_1, \ldots, x_n)$ and $j = 1, \ldots, n$,

$$x_j^2 \leq \sum_{i=1}^{n} x_i^2 = \|\mathbf{x}\|^2 \leq \left(\sum_{i=1}^{n} |x_i|\right)^2 .$$

Thus

$$|x_j| \leq \|\mathbf{x}\| \leq \sum_{i=1}^{n} |x_i|. \tag{7}$$

Now let

$$\{\mathbf{x_k}\} = \{(x_{k1}, \ldots, x_{kn})\}$$

be a sequence in $\mathbb{R}^n$, and let $\mathbf{x} = (x_1, \ldots, x_n)$ be a point in $\mathbb{R}^n$. By (7)

$$|x_{kj} - x_j| \le \|\mathbf{x_k} - \mathbf{x}\| \le \sum_{i=1}^{n} |x_{ki} - x_i|. \tag{8}$$

If $\mathbf{x_k} \to \mathbf{x}$ as $k \to \infty$, that is,

$$\|\mathbf{x_k} - \mathbf{x}\| \to 0 \quad \text{as } k \to \infty,$$

then we see from (8) that for each $j = 1, \ldots, n$, $|x_{kj} - x_j| \to 0$ as $k \to \infty$. Conversely, if for each $j = 1, \ldots, n$ we have $|x_{kj} - x_j| \to 0$ as $k \to 0$, then

$$\sum_{i=1}^{n} |x_{ki} - x_i| \to 0$$

as $k \to \infty$, so we see once again from (8) that $\|\mathbf{x_k} - \mathbf{x}\| \to 0$ as $k \to \infty$. We summarize this discussion as a theorem.

Theorem 11.15 *Let*

$$\{\mathbf{x_k}\} = \{(x_{k1}, \ldots, x_{kn})\}$$

be a sequence in $\mathbb{R}^n$ *and let* $\mathbf{x} = (x_1, \ldots, x_n) \in \mathbb{R}^n$. *Then*

$$\lim_{k\to\infty} \mathbf{x_k} = \mathbf{x}$$

if and only if for each $j = 1, \ldots, n$, $\lim_{k\to\infty} x_{kj} = x_j$.

Algebraic Properties Our next result shows that limits combine as expected with respect to addition, multiplication by a scalar, and the dot product operation. We leave the proof as Exercise 11.4.2.

Theorem 11.16 *Let* $\{\mathbf{x}_k\}$ *and* $\{\mathbf{y}_k\}$ *be sequences in* $\mathbb{R}^n$, *and let* $\alpha \in \mathbb{R}$. *If* $\lim_{k\to\infty} \mathbf{x_k} = \mathbf{x}$ *and* $\lim_{k\to\infty} \mathbf{y_k} = \mathbf{y}$, *then*

(i) $\displaystyle\lim_{k\to\infty} (\alpha \mathbf{x_k}) = \alpha \mathbf{x}$.

(ii) $\displaystyle\lim_{k\to\infty} (\mathbf{x_k} + \mathbf{y_k}) = \mathbf{x} + \mathbf{y}$.

(iii) $\displaystyle\lim_{k\to\infty} (\mathbf{x_k} \cdot \mathbf{y_k}) = \mathbf{x} \cdot \mathbf{y}$.

Limit Points As was true in $\mathbb{R}$, we can characterize limit points in $\mathbb{R}^n$ in the language of sequences.

Theorem 11.17 *Let* $E \subset \mathbb{R}^n$ *and let* $\mathbf{x} \in \mathbb{R}^n$. *Then* $\mathbf{x}$ *is a limit point of* E *if and only if there exists a sequence* $\{\mathbf{x}_k\}$ *of distinct points of* E *such that* $\lim_{k\to\infty} \mathbf{x_k} = \mathbf{x}$.

Proof Suppose $\mathbf{x}$ is a limit point of E. Then for each $\varepsilon > 0$ there exists $\mathbf{x_k} \neq \mathbf{x}$ in E such that $\|\mathbf{x_k} - \mathbf{x}\| < \varepsilon$. Choose $\mathbf{x_1} \in E$ such that $0 < \|\mathbf{x_1} - \mathbf{x}\| < 1$. Inductively, having chosen $\mathbf{x_k}$, choose $\mathbf{x_{k+1}}$ in E such that

$$0 < \|\mathbf{x_{k+1}} - \mathbf{x}\| < \tfrac{1}{2}\|\mathbf{x_k} - \mathbf{x}\|.$$

Then $\lim_{k\to\infty} \mathbf{x_k} = \mathbf{x}$, $\mathbf{x_k} \neq \mathbf{x}$, and if $k \neq j$, $\mathbf{x_k} \neq \mathbf{x_j}$. The converse is obvious. ∎

Theorem 11.18 *A set $E \subset \mathbb{R}^n$ is closed if and only if it contains all its limit points.*

The proof is left as Exercise 11.4.3.

Bolzano-Weierstrass Theorem The Bolzano-Weierstrass Theorem in one dimension asserts that bounded sequences of real numbers must have convergent subsequences. This theorem carries over to $\mathbb{R}^n$ and will play an important role in our study of continuous functions on $\mathbb{R}^n$. The proof involves using the one-dimensional version n times. While the idea of the proof is straightforward, the notation is messy, involving subsequences of subsequences of subsequences, n times. We shall adopt more condensed notation in order to avoid multiple levels of subscripts. It is worth studying the proof to see how messy details can be simplified by an appropriate notation.

Theorem 11.19 (Bolzano-Weierstrass) *Every bounded sequence $\{\mathbf{x}_k\}$ in $\mathbb{R}^n$ contains a convergent subsequence.*

Proof Let $\{\mathbf{x}_k\}$ be a bounded sequence in $\mathbb{R}^n$, say $\|\mathbf{x_k}\| \leq M$ for all $k \in \mathbb{N}$. For each k, let $\mathbf{x_k} = (x_{k1}, \ldots, x_{kn})$. Then for each $k \in \mathbb{N}$ and $j = 1, \ldots, n$, $|x_{kj}| \leq M$. Thus, for all j, the sequence of real numbers $\{x_{kj}\}$ is bounded.

Let $j = 1$. By the Bolzano-Weierstrass theorem applied to the bounded sequence $\{x_{kj}\}$ $(k = 1, 2, 3, \ldots)$ there exists a sequence of integers

$$1 \leq k(1,1) < k(1,2) < \ldots$$

and a number x_1 such that

$$x_{k(1,i),1} \to x_1 \text{ as } i \to \infty.$$

Observe that the sequence $\{x_{k(1,i),1}\}$ is just a subsequence of the sequence $\{x_{k1}\}$.

Next let $j = 2$. The sequence $\{x_{k(1,i),2}\}$ is also bounded by M, so again, by the Bolzano-Weierstrass theorem there is a subsequence $\{k(2,i)\}$ of $\{k(1,i)\}$ such that $x_{k(2,i),2}$ converges. Let

$$x_2 = \lim_{i\to\infty} x_{k(2,i),2}.$$

Now, the sequence $\{x_{k(1,i),1}\}$ converges to x_1, so the same is true of any subsequence of this sequence. In particular $\{x_{k(2,i),1}\} \to x_1$ as $i \to \infty$.

We continue this process. Having obtained a sequence $\{x_{k(j,i),j}\}$ with

$$\{x_{k(m,i),m}\} \to x_m \quad \text{for all } m \leq j,$$

we use the Bolzano-Weierstrass theorem yet again to obtain a further subsequence $x_{k(j+1,i),j+1}\}$ that converges to a point $x_{j+1} \in \mathbb{R}$. The process stops when $j+1=n$. Letting $k_i = k(n,i)$ and $\mathbf{x} = (x_1,\ldots,x_n)$, we have $\lim_{i\to\infty} \mathbf{x}_{k_i} = \mathbf{x}$ by Theorem 11.15. ■

Exercises

11.4.1 For $\mathbf{x} = (x_1,\ldots,x_n) \in \mathbb{R}^n$ define the norms

$$\|\mathbf{x}\| = \sqrt{x_1^2 + \cdots + x_n^2},$$
$$\|\mathbf{x}\|_1 = |x_1| + \cdots + |x_m|,$$
$$\|\mathbf{x}\|_\infty = \max\{|x_i| : i = 1,\ldots,n\}.$$

Prove that the following are equivalent for sequences in $\mathbb{R}^n$.

(a) $\lim_{k\to\infty} \mathbf{x_k} = \mathbf{x}$

(b) $\|\mathbf{x_k} - \mathbf{x}\|_1 \to 0$ as $k \to \infty$

(c) $\|\mathbf{x_k} - \mathbf{x}\|_\infty \to 0$ as $k \to \infty$

Thus convergence does not depend on which of the norms $\|\cdot\|$, $\|\cdot\|_1$, $\|\cdot\|_\infty$ we use.

11.4.2 Prove Theorem 11.16.

11.4.3 Prove Theorem 11.18.

11.4.4 Prove that every convergent sequence in $\mathbb{R}^n$ is bounded.

11.4.5 Prove that a set $E \subset \mathbb{R}^n$ is closed and bounded if and only if every sequence $\{\mathbf{x}_k\}$ of points of E has a subsequence converging to a point in E.

11.4.6 State and prove the analogue in $\mathbb{R}^n$ to Cauchy's criterion for the convergence of real sequences (Theorem 2.41).

11.4.7 Which of the following sequences $\{\mathbf{x_k}\}$ in $\mathbb{R}^3$ or $\mathbb{R}^2$ are convergent?

(a) $\mathbf{x_k} = (\cos\frac{\pi}{k}, \sin\frac{\pi}{k}, \frac{k}{1+k})$

(b) $\mathbf{x_k} = (\cos\pi k, \sin\pi k, \frac{k}{k+1})$

(c) $\mathbf{x_k} = (e^{-k}, \frac{1}{k!}, k)$

(d) $\mathbf{x_k} = (\frac{\ln k}{k}, \frac{k^2}{k^2+1}, (-1)^k)$

(e) $\mathbf{x_k} = (\frac{\sin k}{k}, k\sin\frac{1}{k})$

(f) $\mathbf{x_k} = (\sqrt{k+1} - \sqrt{k}, 7)$

11.4.8 Which of the sequences in Exercise 11.4.7 have convergent subsequences?

11.4.9 Prove that if $\{\mathbf{x}_k\} \to \mathbf{x}$ as $k \to \infty$, then $\|\mathbf{x_k}\| \to \|\mathbf{x}\|$ as $k \to \infty$. Does this conclusion remain valid if the euclidean norm is replaced by the norms $\|\cdot\|_1$ or $\|\cdot\|_\infty$ of Exercise 11.4.1?

11.4.10 Prove the version of the Bolzano-Weierstrass theorem that applies to sets: Every infinite bounded subset of $\mathbb{R}^n$ has a point of accumulation in $\mathbb{R}^n$.

11.5 Functions and Mappings

When we dealt with functions in the preceding chapters, we were concerned primarily with functions whose domains were subsets of $\mathbb{R}$ and whose ranges were in $\mathbb{R}$; these functions have traditionally been known as "real-valued functions of a real variable." The domains were usually intervals. In this chapter and the next we are concerned with functions whose domains and ranges are in the spaces $\mathbb{R}^n$. Usually, the domain will be an open subset of $\mathbb{R}^n$. As was the case with real functions of a real variable, our study focuses on questions concerned with limits, continuity, and differentiability. In the present section we deal only with some definitions, some examples, and a bit about notation.

11.5.1 Functions from $\mathbb{R}^n \to \mathbb{R}$

Let us begin with functions $f : E \to \mathbb{R}$, where E is a subset of $\mathbb{R}^n$. Such functions are sometimes called "real functions of several variables." We present some examples.

There is a special and traditional notational feature for the case $n = 2$ (i.e., for functions $f : \mathbb{R}^2 \to \mathbb{R}$). The two variables needed to represent a point $(x_1, x_2) \in \mathbb{R}^2$ are written as (x, y). In all discussions the x refers to the first variable and the y to the second. Similarly for functions $f : \mathbb{R}^3 \to \mathbb{R}$ we often use (x, y, z) rather than (x_1, x_2, x_3) to represent points in $\mathbb{R}^3$. We will usually make use of this convention, especially in discussions of derivatives in the next chapter. For $n \geq 4$ the subscript notation is used.

Example 11.20 Let $f(x, y) = x^2 + xy + y^2 + 5$. A natural domain for f is all of $\mathbb{R}^2$, so $f : \mathbb{R}^2 \to \mathbb{R}$ is a function of two variables. The function f is an example of a polynomial in two variables. The general polynomial in two variables can be written as

$$f(x, y) = \sum_{i,j} a_{ij} x^i y^j \tag{9}$$

where the coefficients a_{ij} are real numbers, the numbers i and j are non-negative integers, and the sum has finitely many terms. Figure 11.2 shows the graph of this polynomial f in $\mathbb{R}^3$; you should be familiar with such representations from calculus classes. For much of our study we shall not be

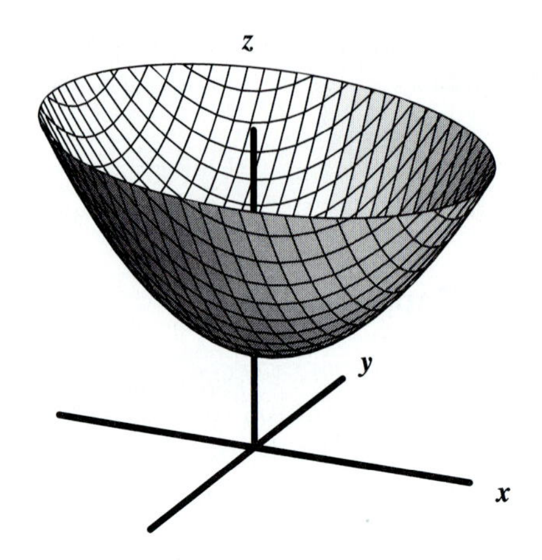

Figure 11.2. Graph of the polynomial $f(x, y) = x^2 + xy + y^2 + 5$.

able to rely on any meaningful pictures since we study functions of many variables, and pictures outside of $\mathbb{R}^2$ and $\mathbb{R}^3$ are not possible. Even so, we rely on these special cases to help develop our intuition as to what is going on in more general situations. ◀

Example 11.21 Let $g(x_1, x_2, x_3) = \sqrt{1 - x_1^2 - x_2^2 - x_3^2}$. Here a natural domain for g is the set

$$E = \{(x_1, x_2, x_3) : x_1^2 + x_2^2 + x_3^2 \leq 1\}.$$

You will recognize E as the closed unit ball $B(\mathbf{0}, 1)$ in the space $\mathbb{R}^3$. ◀

Example 11.22 Let $h(x, y) = \ln xy$. A natural domain for h is the set $E = \{(x, y) : xy > 0\}$. Thus E consists of the union of the first and third open quadrants of the xy plane. ◀

Example 11.23 To each triangle T contained in $\mathbb{R}^2$ let $A(T)$ denote the area of the triangle. We can view A as a function whose domain is the set of triangles in $\mathbb{R}^2$. But we can also express a formula for A as a function of six variables. If (a, b), (c, d), (e, f) are the vertices of T, then we can express the area as a function $g : \mathbb{R}^6 \to \mathbb{R}$ where $g(a, b, c, d, e, f) = A(T)$. (Exercise 11.5.2 asks for an analytic formula for the area.) ◀

Sometimes it is more convenient to represent a function without referring directly to the coordinates.

Example 11.24 Let $\mathbf{a} \in \mathbb{R}^n$ and let $f : \mathbb{R}^n \to \mathbb{R}$ be defined by $f(\mathbf{x}) = \mathbf{a} \cdot \mathbf{x}$. If $\mathbf{a} = (a_1, \dots, a_n)$ and $\mathbf{x} = (x_1, \dots, x_n)$, then

$$f(\mathbf{x}) = f(x_1, \dots, x_n) = a_1 x_1 + \cdots + a_n x_n.$$

Functions of this type are *linear* functions. This means they satisfy the condition

$$f(\alpha \mathbf{x} + \beta \mathbf{y}) = \alpha f(\mathbf{x}) + \beta f(\mathbf{y}) \quad \text{for all } \alpha, \beta \in \mathbb{R} \text{ and all } \mathbf{x}, \mathbf{y} \in \mathbb{R}^n.$$

In fact *every* linear function $f : \mathbb{R}^n \to \mathbb{R}$ can be represented in the form $f(\mathbf{x}) = \mathbf{a} \cdot \mathbf{x}$ for some $\mathbf{a} \in \mathbb{R}^n$ and all $\mathbf{x} \in \mathbb{R}^n$. We leave verification as Exercise 11.5.3. ◀

Example 11.25 For each $\mathbf{x} \in \mathbb{R}^n$, let $f(\mathbf{x}) = \|\mathbf{x}\|$. Then $f : \mathbb{R}^n \to \mathbb{R}$. In terms of coordinates

$$f(\mathbf{x}) = f(x_1, \dots, x_n) = \sqrt{x_1^2 + \cdots + x_n^2}.$$

◀

Exercises

11.5.1 Give a definition for a polynomial of n-variables. Write the definition in a form analogous to equation (9).

11.5.2 Refer to Example 11.23. Find a formula for $A(T)$ as a function of six variables.

11.5.3 Refer to Example 11.24. Prove that the function $f : \mathbb{R}^n \to \mathbb{R}$ is linear if and only if there exists a unique $\mathbf{a} \in \mathbb{R}^n$ such that $f(\mathbf{x}) = \mathbf{a} \cdot \mathbf{x}$ for all $\mathbf{x} \in \mathbb{R}^n$.

11.5.2 Functions from $\mathbb{R}^n \to \mathbb{R}^m$

When dealing with functions with domain in $\mathbb{R}^n$ and range in $\mathbb{R}^m$, we often use the term *mapping* or *transformation* or *operator* in place of "function." The term *vector-valued function* is also in common use.

We have been denoting vectors or points in $\mathbb{R}^m$ $(m \geq 2)$ with **bold** letters. We shall also denote functions with range in $\mathbb{R}^m$ $(m \geq 2)$ with **bold** letters. Suppose $E \subset \mathbb{R}^n$ and $\mathbf{f} : E \to \mathbb{R}^m$. Then for all $\mathbf{x} \in E$, $\mathbf{f}(\mathbf{x}) \in \mathbb{R}^m$. If we express $\mathbf{x}$ and $\mathbf{f}(\mathbf{x})$ in terms of coordinates, say $\mathbf{x} = (x_1, \dots, x_n)$ and $\mathbf{f}(\mathbf{x}) = (y_1, \dots, y_m)$, then

$$\mathbf{f}(x_1, \dots, x_n) = (y_1, \dots, y_m).$$

The numbers $y_1, \dots, y_m$ depend on $\mathbf{x} = (x_1, \dots, x_n)$, of course. Thus there exist functions $f^1, \ldots, f^m : E \to \mathbb{R}$ called the *coordinate functions* such that

$$\begin{aligned} y_1 &= f^1(\mathbf{x}) = f^1(x_1, \ldots, x_n) \\ y_2 &= f^2(\mathbf{x}) = f^2(x_1, \ldots, x_n) \\ &\vdots \\ y_m &= f^m(\mathbf{x}) = f^m(x_1, \ldots, x_n) \end{aligned}$$

We use superscripts for the functions f^j instead of subscripts in order to avoid confusion with the subscript notation we shall use in Chapter 12 for partial derivatives.

Example 11.26 Define $\mathbf{f} : \mathbb{R}^2 \to \mathbb{R}^2$ by

$$\mathbf{f}(r, \theta) = (r\cos\theta, r\sin\theta), \quad r, \theta \in \mathbb{R}.$$

Then $f^1(r, \theta) = r\cos\theta$, $f^2(r, \theta) = r\sin\theta$. For $r \geq 0$, this mapping $\mathbf{f}$ is familiar in connection with transforming polar coordinates to rectangular coordinates. Here we usually write x for f^1 and y for f^2, so we get

$$x = r\cos\theta \ , \ y = r\sin\theta.$$

◀

Example 11.27 (Linear functions) Suppose $\mathbf{A} : \mathbb{R}^n \to \mathbb{R}^m$ is given by the equations

$$\begin{aligned} y_1 &= a_{11}x_1 + \cdots + a_{1n}x_n \\ y_2 &= a_{21}x_1 + \cdots + a_{2n}x_n \\ &\vdots \\ y_m &= a_{m1}x_1 + \cdots + a_{mn}x_n \end{aligned}$$

where $a_{ij} \in \mathbb{R}$, $i = 1, \ldots, m$, $j = 1, \ldots, n$ and $\mathbf{x} = (x_1, \ldots, x_n)$. The transformation $\mathbf{A}$ can be represented by a matrix $A = (a_{ij})$, the m by n matrix of coefficients. We then have

$$\mathbf{A}(\mathbf{x}) = A\mathbf{x},$$

the product being matrix multiplication:

$$\begin{pmatrix} a_{11} & \ldots & a_{1n} \\ \vdots & & \\ a_{m1} & \ldots & a_{mn} \end{pmatrix} \begin{pmatrix} x_1 \\ \vdots \\ \vdots \\ x_n \end{pmatrix}$$

It is easy to check that $\mathbf{A}$ is linear, that is, that

$$\mathbf{A}(\alpha\mathbf{x_1} + \beta\mathbf{x_2}) = \alpha\mathbf{A}(\mathbf{x_1}) + \beta\mathbf{A}(\mathbf{x_2})$$

for all $\alpha, \beta \in \mathbb{R}$ and all $\mathbf{x_1}, \mathbf{x_2} \in \mathbb{R}^n$. (See Exercise 11.5.4.) ◀

Observe that in Example 11.27 we used capital letters to denote linear transformations from $\mathbb{R}^n \to \mathbb{R}^m$. This is a common practice that is also frequently used when dealing with linear transformations on other vector spaces (see Chapter 13). The use of capital letters for transformations from $\mathbb{R}^n$ to $\mathbb{R}^m$ is also sometimes preferred even when the functions are not linear. We shall use whatever notation is convenient for our purposes. One more word about notation. It is sometimes convenient, particularly when dealing with linear transformations, to use notation such as $\mathbf{Ax}$ in place of $\mathbf{A}(\mathbf{x})$. We shall use whichever of these notations is convenient at the time.

Example 11.28 Let $\mathbf{T}(x,y) = (u,v)$ where $u = e^x \cos y$, $v = e^x \sin y$. Thus $u = T^1(x,y) = e^x \cos y$ and $v = T^2(x,y) = e^x \sin y$. We can study some geometric properties of the mapping $\mathbf{T}$.

Observe that $u^2 + v^2 = e^{2x}$, and $v \cos y = u \sin y$. For fixed x_0, (u,v) is a point on the circle having center $(0,0)$ and radius e^{x_0}. As y increases over an interval of length 2π, (u,v) traverses that circle one time. Thus $\mathbf{T}$ maps a line $x = x_0$ infinitely often around that circle. Similarly, $\mathbf{T}$ maps a line $y = y_0$ onto a ray having slope $\tan y_0$ and emanating from the origin.

We leave verification of this and some other mapping properties of $\mathbf{T}$ to Exercise 11.5.5. Note that this function offers us some problems should we wish a graphical representation. The graph would be a subset of $\mathbb{R}^4$ and so beyond our abilities to represent. Figure 11.3 can be used to study this function; you should be familiar with such representations from calculus classes. This picture carries the information disussed previously about lines mapping into circles. ◀

In Chapter 12 we shall be concerned with differentiability of functions of several variables and of mappings. Central to our work will be the use of linear transformations to approximate such functions or mappings.

Exercises

11.5.4 Refer to Example 11.27. Prove that $\mathbf{A}$ is linear.

11.5.5 Refer to Example 11.28. Verify that $\mathbf{T}$ has the mapping properties stated in that example. Show also that if S is the strip in the xy plane between the lines $y = 0$ and $y = 2\pi$, then $\mathbf{T}(S) = \mathbb{R}^2 \setminus (0,0)$.

11.5.6 Let $\mathbf{H} : \mathbb{R}^2 \to \mathbb{R}^2$ be defined by $\mathbf{H}(x,y) = (x^2 - y^2, 2xy)$. Write $u = x^2 - y^2$ and $v = 2xy$, so that $\mathbf{H}(x,y) = (u,v)$.

(a) Describe the sets in the xy-plane that $\mathbf{H}$ maps onto horizontal lines in the uv-plane. Do the same for vertical lines.

(b) Determine a set $S \subset \mathbb{R}^2$ such that $\mathbf{H}(S) = [1,2] \times [4,5]$.

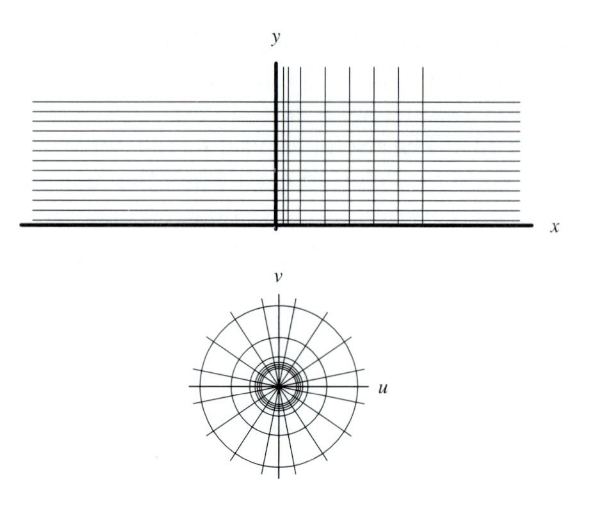

Figure 11.3. A representation of the function $\mathbf{T}(x, y) = (u, v)$, where $u = e^x \cos y$, $v = e^x \sin y$.

11.6 Limits of Functions from $\mathbb{R}^n \to \mathbb{R}^m$

The concept of limit for a function $\mathbf{f} : E \to \mathbb{R}^m$, where $E \subset \mathbb{R}^n$ is entirely analogous to the concept of limit for a real function $f : E \to \mathbb{R}$, where $E \subset \mathbb{R}$. We wish to capture the idea that $\mathbf{f}(\mathbf{x})$ is near $\mathbf{y_0} \in \mathbb{R}^m$ when $\mathbf{x}$ is near $\mathbf{x_0}$ in E. Our definition and some basic results and their proofs are essentially identical to their counterparts for real functions.

11.6.1 Definition

We begin with the ε-δ definition of the limit for functions from $\mathbb{R}^n$ to $\mathbb{R}^m$. The equivalence with a sequence version is proved in Lemma 11.30 which follows.

Definition 11.29 Let $\mathbf{f} : E \to \mathbb{R}^m$, where E is a subset of $\mathbb{R}^n$ and let $\mathbf{x_0}$ be a limit point of E. We say

$$\lim_{\mathbf{x} \to \mathbf{x_0}} \mathbf{f}(\mathbf{x}) = \mathbf{y_0}$$

if for every $\varepsilon > 0$ there exists $\delta > 0$ such that $\|\mathbf{f}(\mathbf{x}) - \mathbf{y_0}\| < \varepsilon$ whenever $0 < \|\mathbf{x} - \mathbf{x_0}\| < \delta$ and $\mathbf{x} \in E$.

Observe that the existence of the limit, and its value, does not depend on whether $\mathbf{f}$ is or is not defined at $\mathbf{x_0}$ or on its value at $\mathbf{x_0}$. In loose geometric language, $\lim_{\mathbf{x} \to \mathbf{x_0}} \mathbf{f}(\mathbf{x}) = \mathbf{y_0}$ means that to every neighborhood V of $\mathbf{y_0}$, no matter how small, corresponds a sufficiently small neighborhood U of $\mathbf{x_0}$ such that $\mathbf{f}$ maps $(U \cap E) \setminus \{\mathbf{x_0}\}$ into V. (This definition and this discussion should be compared with the treatment of limits for functions of a single real variable given in Chapter 5.)

Our lemma relates this notion of limit with the sequential one (which is the analogue of the material in Section 5.1.2 for functions of one variable).

Lemma 11.30 *Let $E \subset \mathbb{R}^n$ and let $\mathbf{x_0}$ be an accumulation point of E. Suppose $\mathbf{f} : E \to \mathbb{R}^m$. Then $\lim_{\mathbf{x}\to\mathbf{x_0}} \mathbf{f}(\mathbf{x}) = \mathbf{y_0}$ if and only if for every sequence $\{\mathbf{x_k}\} \subset E$ such that $\mathbf{x_k} \to \mathbf{x_0}$, $\mathbf{x_k} \neq \mathbf{x_0}$, we have $\mathbf{f}(\mathbf{x_k}) \to y_0$.*

Proof Suppose $\lim_{\mathbf{x}\to\mathbf{x_0}} \mathbf{f}(\mathbf{x}) = \mathbf{y_0}$. Let $\mathbf{x_k} \to \mathbf{x_0}$ $(\mathbf{x_k} \neq \mathbf{x_0})$, $\mathbf{x_k} \in E$ and let $\varepsilon > 0$. Then there exists $\delta > 0$ such that if $\mathbf{x} \in E$ and $0 < \|\mathbf{x} - \mathbf{x_k}\| < \delta$,

$$\|\mathbf{f}(\mathbf{x}) - \mathbf{y_0}\| < \varepsilon.$$

Since $\mathbf{x} \to \mathbf{x_0}$, there exists $N \in \mathbb{N}$ such that $\|\mathbf{x_0} - \mathbf{x_k}\| < \delta$ if $k \geq N$. Thus, for $k \geq N$,

$$\|\mathbf{f}(\mathbf{x_k}) - \mathbf{y_0}\| < \varepsilon,$$

and $\mathbf{f}(\mathbf{x_k}) \to \mathbf{y_0}$.

To prove the converse, suppose $\mathbf{y_0}$ is *not* the limit of $\mathbf{f}(\mathbf{x})$ as $\mathbf{x} \to \mathbf{x_0}$. Then there exists $\varepsilon > 0$ such that for each $k \in \mathbb{N}$ there exists $\mathbf{x_k} \in E$ such that

$$0 < \|\mathbf{x_k} - \mathbf{x_0}\| < \frac{1}{k}$$

and

$$\|\mathbf{f}(\mathbf{x}) - \mathbf{y_0}\| \geq \varepsilon.$$

The sequence $\{\mathbf{x_k}\}$ converges to $\mathbf{x_0}$ but the sequence $\mathbf{f}(\mathbf{x_k})$ does not converge to $\mathbf{y_0}$. ■

Example 11.31 Let

$$f(x,y) = xy\left(\frac{x^2 - y^2}{x^2 + y^2}\right), \quad f(0,0) = 7.$$

Then $f : \mathbb{R}^2 \to \mathbb{R}$. We show that $\displaystyle\lim_{(x,y)\to(0,0)} f(x,y) = 0$. Let $\varepsilon > 0$. We have

$$|f(x,y) - 0| \leq |xy|\frac{|x^2 - y^2|}{|x^2 + y^2|} \leq |xy|$$

when $(x,y) \neq (0,0)$. Now $|xy| \leq \sqrt{x^2 + y^2}$ for all $(x,y) \in \mathbb{R}^2$, so $|f(x,y)| < \varepsilon$ whenever $x^2 + y^2 < \delta = \sqrt{\varepsilon}$, $(x,y) \neq (0,0)$. Thus

$$\lim_{(x,y)\to(0,0)} f(x,y) = 0.$$

Note that the fact that $f(0,0) = 7$ does not affect the limit. ◀

When a limit exists, it must be unique (Exercise 11.6.2). We can sometime use this fact to show that a limit does not exist.

Example 11.32 Let

$$f(x,y) = \frac{2xy}{x^2+y^2}, \quad f(0,0) = 0.$$

Here $\lim_{y\to 0} f(0,y) = \lim_{x\to 0} f(x,0) = 0$, but $\lim_{(x,y)\to(0,0)} f(x,y)$ does not exist. To see this, observe that

$$\lim_{(t,t)\to(0,0)} f(t,t) = \lim_{t\to 0} \frac{2t^2}{2t^2} = 1.$$

Thus, approaching $(0,0)$ along the coordinate axes requires the limit to be 0, but approaching $(0,0)$ along the line $y = x$ requires the limit to be 1. As a result, for $0 < \varepsilon < 1$, there is no $y_0 \in \mathbb{R}$ and $\delta > 0$ such that $|f(x,y) - y_0| < \varepsilon$ whenever $0 < \sqrt{x^2+y^2} < \varepsilon$. ◀

Example 11.33 Let $\mathbf{f}(r,\theta) = (r\cos\theta, r\sin\theta)$. Thus $f : \mathbb{R}^2 \to \mathbb{R}^2$. We show that

$$\lim_{(r,\theta)\to(0,0)} \mathbf{f}(r,\theta) = (0,0).$$

Here

$$\|\mathbf{f}(r,\theta) - (0,0)\| = \sqrt{r^2\cos^2\theta + r^2\sin^2\theta} = |r|.$$

Thus $\|\mathbf{f}(r,\theta) - (0,0)\| < \varepsilon$ whenever $|r| < \varepsilon$, independent of θ. In particular if

$$0 < \|(r,\theta)\| = \sqrt{r^2+\theta^2} < \delta = \varepsilon,$$

then $|r| < \varepsilon$, so

$$\lim_{(r,\theta)\to(0,0)} \mathbf{f}(r,\theta) = (0,0).$$

◀

Exercises

11.6.1 Establish Examples 11.31 and 11.32 by using Lemma 11.30 rather than an ε-δ argument.

11.6.2 (Uniqueness of Limits) Let $E \subset \mathbb{R}^n$ and let $\mathbf{x_0}$ be an accumulation point of E. Suppose $\mathbf{f} : E \to \mathbb{R}^m$. If

$$\lim_{\mathbf{x}\to\mathbf{x_0}} \mathbf{f}(\mathbf{x}) = \mathbf{y_0} \quad\text{and}\quad \lim_{\mathbf{x}\to\mathbf{x_0}} \mathbf{f}(\mathbf{x}) = \mathbf{z_0},$$

then $\mathbf{y_0} = \mathbf{z_0}$.

11.6.3 Let $E \subset \mathbb{R}^n$ and let $\mathbf{x_0}$ be an accumulation point of E. Prove that if $\lim_{\mathbf{x}\to\mathbf{x_0}} \mathbf{f}(\mathbf{x})$ exists, then there exist $\varepsilon > 0$ and $M > 0$ such that $\|\mathbf{f}(\mathbf{x})\| < M$ for all $\mathbf{x} \in B(\mathbf{x_0}, \varepsilon)$ at which $\mathbf{f}$ is defined.

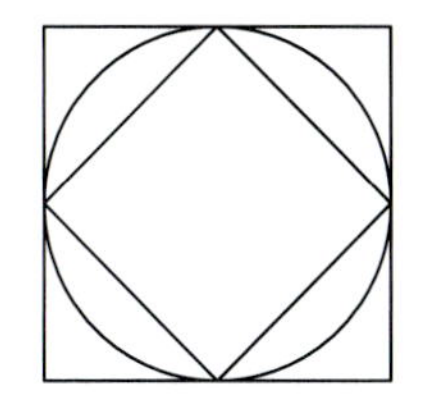

Figure 11.4. Three unit balls centered at the origin for the three norms $\|\cdot\|_1$, $\|\cdot\|$, and $\|\cdot\|_\infty$ (from innermost to outermost).

11.6.2 Coordinate-Wise Convergence

We have seen earlier (Exercises 11.2.6, 11.2.7, 11.3.8, and 11.4.1) that there are further norms on $\mathbb{R}^n$ that can be used to study convergence. Let us briefly discuss the use of the norm $\|\cdot\|_\infty$ in connection with the limit concept. This will give us some insight into the nature of the limit concept and, in particular, allows us to show that convergence can be interpreted coordinate-wise.

We start by recalling the definition of the infinity norm and comparing it to the euclidean norm. For $\mathbf{x} = (x_1, \ldots, x_n) \in \mathbb{R}^n$ we defined

$$\|\mathbf{x}\|_\infty = \max\{|x_1|, |x_2, \ldots, |x_n|\}.$$

For simplicity, consider first the case of $\mathbb{R}^2$, using the notation (x, y) for members of $\mathbb{R}^2$. Then

$$\|(x,y)\| = \sqrt{x^2+y^2} \quad \text{while} \quad \|(\mathbf{x},\mathbf{y})\|_\infty = \max\{|x|, |y|)\}.$$

The open euclidean ball $B((0,0),r)$ is an open disk centered at $(0,0)$ with radius r. The open ball $B_\infty((0,0),r)$ is an open square with center at $(0,0)$ having sides parallel to the coordinate axes and side length $2r$. Note that

$$B((0,0),r) \subset B_\infty((0,0),r),$$

and there exists $s > 0$ such that

$$B_\infty((0,0),s) \subset B((0,0),r).$$

Figure 11.4 shows the three unit balls centered at the origin for the three norms $\|\cdot\|_1$, $\|\cdot\|$, and $\|\cdot\|_\infty$. A similar comment applies to balls centered at other points of $\mathbb{R}^2$.

For an arbitrary $n \in \mathbb{N}$, something similar is true. For every $r > 0$ and $\mathbf{x_0} \in \mathbb{R}^n$, we have

$$\{\mathbf{x} \in \mathbb{R}^n : \|\mathbf{x} - \mathbf{x_0}\| < r\} \subset \{\mathbf{x} \in \mathbb{R}^n : \|\mathbf{x} - \mathbf{x_0}\|_\infty < r\} \tag{10}$$

and there exists $s > 0$ such that

$$\{\mathbf{x} \in \mathbb{R}^n : \|\mathbf{x} - \mathbf{x_0}\|_\infty < s\} \subset \{\mathbf{x} \in \mathbb{R}^n : \|\mathbf{x} - \mathbf{x_0}\| < r\}. \tag{11}$$

From these inclusions we can see that we could substitute $\|\cdot\|_\infty$ for $\|\cdot\|$ in Definition 11.14 and Definition 11.29 without changing any conclusions. That is, the definition of the sequence limit is equivalent to the following version with a different norm.

Lemma 11.34 *The limit* $\mathbf{x_k} \to \mathbf{x}$ *is valid in the sense of the norm* $\|\cdot\|$ *if and only if for all* $\varepsilon > 0$*, there exists an integer* N *such that*

$$\|\mathbf{x_k} - \mathbf{x}\|_\infty < \varepsilon$$

whenever $k \geq N$.

Proof Suppose $\mathbf{x_k} \to \mathbf{x}$ and $\varepsilon > 0$. Then there exists $N \in \mathbb{N}$ such that $\|\mathbf{x_k} - \mathbf{x}\| < \varepsilon$ for all $k \geq N$. From (10) we see $\|\mathbf{x_k} - \mathbf{x}\|_\infty < \varepsilon$ for all $k \geq N$. Thus $\mathbf{x_k} \to \mathbf{x}$ relative to $\|\cdot\|_\infty$.

Conversely, suppose $\mathbf{x_k} \to \mathbf{x}$ relative to $\|\cdot\|_\infty$. Let $\varepsilon > 0$. Using (11), choose $\varepsilon' < \varepsilon$ such that if $\|\mathbf{z} - \mathbf{x}\|_\infty < \varepsilon'$, then $\|\mathbf{z} - \mathbf{x}\| < \varepsilon$. Now, there exists $N \in \mathbb{N}$ such that if $k \geq N$, then $\|\mathbf{x_k} - \mathbf{x}\|_\infty < \varepsilon'$. Thus for $k \geq N$, $\|\mathbf{x_k} - \mathbf{x}\| < \varepsilon$. This shows that $\mathbf{x_k} \to \mathbf{x}$ relative to $\|\cdot\|$. ■

Similarly, the function limit can also be written using the other norm.

Lemma 11.35 $\lim_{\mathbf{x} \to \mathbf{x_0}} \mathbf{f}(\mathbf{x}) = \mathbf{y_0}$ *if and only if all* $\varepsilon > 0$*, there exists a* $\delta > 0$ *such that*

$$\|\mathbf{f}(\mathbf{x}) - \mathbf{y_0}\|_\infty < \varepsilon$$

whenever $0 < \|\mathbf{x} - \mathbf{x_0}\|_\infty < \delta$ *and* $\mathbf{x} \in E$.

We leave the proof as Exercise 11.6.5.

Coordinate-Wise Convergence As an application of the preceding remarks we show that convergence of maps from $\mathbb{R}^n$ to $\mathbb{R}^m$ reduces to coordinate-wise convergence.

Theorem 11.36 *Let* $E \subset \mathbb{R}^n$, $\mathbf{f} : E \to \mathbb{R}^m$, $\mathbf{f} = (f^1, \ldots, f^m)$. *Let* $\mathbf{x_0} = (x_1, \ldots, x_n)$ *be a limit point of* E. *Let* $\mathbf{y_0} = (y_1, \ldots, y_n) \in \mathbb{R}^m$. *Then* $\lim\limits_{\mathbf{x} \to \mathbf{x_0}} \mathbf{f}(\mathbf{x}) = \mathbf{y_0}$ *if and only if*

$$\lim_{\mathbf{x} \to \mathbf{x_0}} f^j(\mathbf{x}) = y_j \text{ for all } j = 1, \ldots, m.$$

Proof The proof follows immediately from Lemma 11.35 and the observation that for any $p \in \mathbb{N}$ and any vector $\mathbf{z} = (z_1, \ldots, z_p) \in \mathbb{R}^p$, $|z_i| \leq \|\mathbf{z}\|$ for all $i = 1, \ldots, p$, so $\|\mathbf{z}\|_\infty \to 0$ is therefore equivalent to

$$\lim_{\mathbf{z} \to \mathbf{z_0}} z_i = 0 \text{ for all } i = 1, \ldots, p.$$

Thus, for example,

$$\lim_{\mathbf{x} \to \mathbf{x_0}} \|\mathbf{f}(\mathbf{x}) - \mathbf{y}\|_\infty = 0$$

is equivalent to

$$\lim_{\mathbf{x}\to\mathbf{x_0}} |f^j(\mathbf{x}) - y_i| = 0 \text{ for all } j = 1, \ldots, m.$$

■

Exercises 11.2.6, 11.2.7, 11.3.8, and 11.4.1 are all concerned with comparing several norms for $\mathbb{R}^n$. In essence, these exercises suggested that the norms $\|\cdot\|$, $\|\cdot\|_1$, and $\|\cdot\|_\infty$ could be used interchangeably when dealing with concepts directly related to open sets and to convergence.

Exercises

11.6.4 Verify the correctness of (10) and (11).

11.6.5 Prove Lemma 11.35.

11.6.3 Algebraic Properties

An immediate consequence of Theorem 11.36 is that some of the algebraic properties of limits we obtained in Section 5.2.3 for real functions of one real variable are also valid in our present setting.

Theorem 11.37 *Suppose that $E \subset \mathbb{R}^n$,*

$$\mathbf{f} : E \to \mathbb{R}^m, \ \mathbf{g} : E \to \mathbb{R}^m,$$

$\alpha \in \mathbb{R}$, and $\mathbf{x_0}$ is an accumulation point of E. If $\lim_{\mathbf{x}\to\mathbf{x_0}} \mathbf{f}(\mathbf{x}) = \mathbf{y_0}$ and $\lim_{\mathbf{x}\to\mathbf{x_0}} \mathbf{g}(\mathbf{x}) = \mathbf{z_0}$, then

1. *$\lim_{\mathbf{x}\to\mathbf{x_0}}(\mathbf{f}(\mathbf{x}) + \mathbf{g}(\mathbf{x})) = \mathbf{y_0} + \mathbf{z_0}$ and*

2. *$\lim_{\mathbf{x}\to\mathbf{x_0}} \mathbf{f}(\alpha\mathbf{x}) = \alpha\mathbf{y_0}$.*

We leave the proof to the exercises. Regarding products, the expected product rule for dot products is valid.

Theorem 11.38 *Let $E \subset \mathbb{R}^n$, let $\mathbf{x_0}$ be an accumulation point of E and let $\mathbf{f}, \mathbf{g} : E \to \mathbb{R}^m$. If*

$$\lim_{\mathbf{x}\to\mathbf{x_0}} \mathbf{f}(\mathbf{x}) = \mathbf{y_0} \text{ and } \lim_{\mathbf{x}\to\mathbf{x_0}} \mathbf{g}(\mathbf{x}) = \mathbf{z_0},$$

then

$$\lim_{\mathbf{x}\to\mathbf{x_0}} (\mathbf{f}(\mathbf{x}) \cdot \mathbf{g}(\mathbf{x})) = \mathbf{y_0} \cdot \mathbf{z_0}.$$

Proof We apply Lemma 11.30. Let $\mathbf{x_k} \to \mathbf{x_0}$, $\mathbf{x_k} \in E$, $\mathbf{x_k} \neq \mathbf{x_0}$. For $k \in \mathbb{N}$, let $\mathbf{u_k} = \mathbf{f}(\mathbf{x_k})$, $\mathbf{v_k} = \mathbf{g}(\mathbf{x_k})$. Then $\mathbf{u_k} \to \mathbf{y_0}$ and $\mathbf{v_k} \to \mathbf{z_0}$, so $\mathbf{u_k} \cdot \mathbf{v_k} \to \mathbf{y_0} \cdot \mathbf{z_0}$ by Theorem 11.16(iii). ■

Exercises

11.6.6 Prove Theorem 11.37.

11.6.7 For those of you familiar with the vector cross product in $\mathbb{R}^3$ (see Section 11.10 for an introduction), does the analogue of Theorem 11.38 hold if $m = 3$ and the dot product is replaced by the cross product?

11.7 Continuity of Functions from $\mathbb{R}^n$ to $\mathbb{R}^m$

Now that we have the concept of limit for vector-valued functions, we can introduce the notion of continuity in just the same way we did for real functions in Section 5.4.2. The proofs in this section are virtually identical to the corresponding proofs in the one variable case and so we omit them.

Definition 11.39 Let $E \subset \mathbb{R}^n$, $\mathbf{f} : E \to \mathbb{R}^m$, $\mathbf{x_0} \in E$. We say $\mathbf{f}$ is *continuous* at $\mathbf{x_0}$ provided that for every $\varepsilon > 0$ there exists $\delta > 0$ such that $\|\mathbf{f}(\mathbf{x}) - \mathbf{f}(\mathbf{x_0})\| < \varepsilon$ for all $\mathbf{x} \in E$ for which $\|\mathbf{x} - \mathbf{x_0}\| < \delta$.

Observe that with this definition every function $\mathbf{f}$ is continuous at each isolated point of E, that is, any point of E that is not a limit point of E. For a limit point $\mathbf{x_0}$ of E, it is easy to see that $\mathbf{f}$ is continuous at $\mathbf{x_0}$ if and only if

$$\lim_{\mathbf{x} \to \mathbf{x_0}} \mathbf{f}(\mathbf{x}) = \mathbf{f}(\mathbf{x_0}).$$

Because of Lemma 11.30, a definition of continuity in terms of sequences is also immediately available: $\mathbf{f}$ is continuous at $\mathbf{x_0}$ if and only if for every sequence $\{\mathbf{x_k}\} \to \mathbf{x_0}$ $(\mathbf{x_k} \in E)$,

$$\lim_{k \to \infty} \mathbf{f}(\mathbf{x_k}) = \mathbf{f}(\mathbf{x_0}).$$

In terms of neighborhoods, we find that $\mathbf{f}$ is continuous at $\mathbf{x_0}$ provided that for every neighborhood V of $\mathbf{f}(\mathbf{x_0})$ there exists a neighborhood U of $\mathbf{x_0}$ such that $\mathbf{f}(U \cap E) \subset V$.

In short, all these definitions for continuity are equivalent and state in precise language that all points in E that are near $\mathbf{x_0}$ map into points in $\mathbb{R}^m$ that are near $\mathbf{f}(\mathbf{x_0})$.

Definition 11.40 Let $E \subset \mathbb{R}^n$ and let $\mathbf{f} : E \to \mathbb{R}^m$. If $\mathbf{f}$ is continuous at all points of E, we say that $\mathbf{f}$ is *continuous on* E. When E is understood from the context we say simply that f is *continuous*.

A global characterization of continuity analogous to Theorem 5.36 for real functions on subsets of $\mathbb{R}$ is also available.

Theorem 11.41 *Let $E \subset \mathbb{R}^n$ and let $\mathbf{f} : E \to \mathbb{R}^m$. Then $\mathbf{f}$ is continuous (on E) if and only if for every open set $V \subset \mathbb{R}^m$, the set*

$$\mathbf{f}^{-1}(V) = \{\mathbf{x} \in E : \mathbf{f}(\mathbf{x}) \in V\}$$

is open (relative to E). [This means that $\mathbf{f}^{-1}(V)$ is the intersection of E with an open subset of $\mathbb{R}^n$.] In particular, if E is open, $\mathbf{f}$ is continuous if and only if for every open set $V \subset \mathbb{R}^m$, the set $\mathbf{f}^{-1}(V)$ is open in $\mathbb{R}^n$.

The expected rules of continuity for sums, multiplication by a scalar, and dot products of functions are valid for vector-valued functions; their proofs involve no more than invoking the corresponding limit laws from Theorems 11.37 and 11.38.

From Theorem 11.36 we infer that a vector-valued function $\mathbf{f}$ is continuous at a point $\mathbf{x_0}$ if and only if all of its coordinate functions f^i are continuous real-valued functions at $\mathbf{x_0}$. This statement should not be confused with the *incorrect* statement that a real-valued function of several variables that is "continuous in each variable separately" is continuous.

Example 11.42 Consider once again Example 11.32,

$$f(x,y) = \frac{2xy}{x^2+y^2}\ ,\ f(0,0) = 0.$$

For each fixed $x = x_0$, $f(x_0, \cdot)$ is continuous on $\mathbb{R}$. The same is true of the functions $f(\cdot, y_0) : \mathbb{R} \to \mathbb{R}$. But f is not continuous at $(0,0)$ since

$$\lim_{(x,y)\to(0,0)} f(x,y)$$

does not exist. See also Exercises 11.7.12 and 11.7.13. ◀

Exercises

11.7.1 Prove that all the definitions for continuity of a function at a point are equivalent.

11.7.2 Show that continuity of $\mathbf{f}$ at $\mathbf{x_0}$ does not depend on which of the norms $\|\cdot\|$, $\|\cdot\|_1$, $\|\cdot\|_\infty$ is in use.

11.7.3 Prove Theorem 11.41.

11.7.4 Let $E \subset \mathbb{R}^n$, $\mathbf{f}, \mathbf{g} : E \to \mathbb{R}^m$ and $\alpha \in \mathbb{R}$. Prove that if $\mathbf{f}$ and $\mathbf{g}$ are continuous at $\mathbf{x_0}$, then so are $\mathbf{f} + \mathbf{g}$ and $\alpha\mathbf{f}$.

11.7.5 State carefully a theorem whose conclusion is that the composition of two continuous functions is continuous.

11.7.6 Let

$$f(x,y) = xy\left(\frac{x^2-y^2}{x^2+y^2}\right)\ ,\ f(0,0) = 7.$$

(This function was discussed in Example 11.31 of Section 11.6.) Is it possible to change the value of f at $(0,0)$ in such a way that the resulting function is continuous at $(0,0)$?

11.7.7 Let $\mathbf{f}(r,\theta) = (r\cos\theta, r\sin\theta)$. (This function was discussed in Example 11.33 of Section 11.6.) Is $\mathbf{f}$ continuous at $(0,0)$? Is $\mathbf{f}$ continuous on all of $\mathbb{R}^2$?

11.7.8 Use Exercise 11.7.4 to prove that any polynomial of n-variables is continuous on $\mathbb{R}^n$. (See Exercise 11.5.1.)

11.7.9 Let $g(x_1,x_2,x_3) = \sqrt{1-x_1^2-x_2^2-x_3^2}$, where g is defined on the closed unit ball $B(\mathbf{0},1) \subset \mathbb{R}^3$. (This function was discussed in Example 11.21 of Section 11.5.) Is g continuous at $(1,0,0)$?

11.7.10 Prove that $\|\cdot\|$ is a continuous function on $\mathbb{R}^m$.

11.7.11 Prove that if $\mathbf{f} : \mathbb{R}^n \to \mathbb{R}^m$ is continuous, then $\|\mathbf{f}(\mathbf{x})\|$ is continuous on $\mathbb{R}^n$.

11.7.12 Let

$$f(x,y) = \frac{2x^2y}{x^4+y^2}$$

with $f(0,0) = 0$. Show

(a) The limit of $f(x,y)$ as $(x,y) \to (0,0)$ along any straight line is $f(0,0)$.

(b) f is discontinuous at $(0,0)$.

11.7.13 A careless student claims to have a proof of the *incorrect* statement,

> *If $f : \mathbb{R}^2 \to \mathbb{R}$ is continuous in each variable separately, then f is continuous.*

Proof: Let $(x_0,y_0) \in \mathbb{R}^2$. For $(x,y) \in \mathbb{R}^2$,

$$|f(x,y) - f(x_0,y_0)| \le |f(x,y) - f(x,y_0)| + |f(x,y_0) - f(x_0,y_0)|.$$

There exists δ_1 such that if $|y-y_0| < \delta_1$, then

$$|f(x,y) - f(x,y_0)| < \varepsilon/2$$

and δ_2 such that if $|x-x_0| < \delta_2$, then

$$|f(x,y_0) - f(x_0,y_0)| < \varepsilon/2.$$

Thus if $\delta = \min(\delta_1,\delta_2)$ and $|x-x_0| < \delta$, $|y-y_0| < \delta$, then

$$|f(x,y) - f(x_0,y_0)| < \varepsilon.$$

(a) What is the flaw in the "proof?"

(b) Find an added hypothesis that would make the quoted statement correct and the preceding outline of a proof valid.

(c) Show that if f is continuous in each variable separately on $\mathbb{R}^2$, then for every $(x_0, y_0) \in \mathbb{R}^2$,

$$\lim_{x\to x_0} (\lim_{y\to y_0} f(x,y)) \quad \text{and} \quad \lim_{y\to y_0} (\lim_{x\to x_0} f(x,y))$$

exist. Does this imply the existence of

$$\lim_{(x,y)\to(x_0,y_0)} f(x,y)?$$

11.8 Compact Sets in $\mathbb{R}^n$

In Section 4.5 we introduced the important notion of compactness for subsets of $\mathbb{R}$. In this section we extend this notion to subsets of $\mathbb{R}^n$. We shall see in the next section that properties of continuous functions defined on compact subsets of $\mathbb{R}$ extend to continuous functions defined on compact subsets of $\mathbb{R}^n$.

There are many equivalent definitions we can give for "compact set" in the setting of $\mathbb{R}^n$. (In Section 4.5 we saw that a set in $\mathbb{R}$ was compact if it was closed and bounded, or if it had the Bolzano-Weierstrass property, or if it had the Heine-Borel property.) Since we have already obtained the Bolzano-Weierstrass theorem (Theorem 11.19), it is natural to base our definition on the notion of sequences.

Definition 11.43 A set $E \subset \mathbb{R}^n$ is *compact* if every sequence in E has a convergent subsequence whose limit is in E.

Theorem 11.44 *A set $E \subset \mathbb{R}^n$ is compact if and only if E is closed and bounded.*

Proof Suppose E is closed and bounded, and $\{\mathbf{x}_k\}$ is a sequence of points in E. Since E is bounded, $\{\mathbf{x}_k\}$ has a convergent subsequence $\{\mathbf{x}_{k_j}\}$ by Theorem 11.19, Section 11.4. Let $\mathbf{x_0} = \lim_{j\to\infty} \mathbf{x}_{k_j}$. Then $\mathbf{x_0}$ is a limit point of E. Since E is closed, $\mathbf{x_0} \in E$ (by Exercise 11.3.1).

Conversely, suppose E is compact. We show that E is closed and bounded by contradiction. If E is not closed, then E has a limit point $\mathbf{x_0} \notin E$. This means there exists a sequence $\{\mathbf{x}_k\}$ in E converging to $\mathbf{x_0}$. Every subsequence of this sequence also converges to $\mathbf{x_0}$, so there is no subsequence of $\{\mathbf{x}_k\}$ converging *to a point of* E. Thus E must be closed.

If E is not bounded, there exists a sequence $\{\mathbf{x}_k\}$ in E such that $\|\mathbf{x_k}\| \geq k$ for all $k \in \mathbb{N}$. This sequence has no convergent subsequence. Thus E must be bounded. ■

Corollary 11.45 *A set $E \subset \mathbb{R}^n$ is closed and bounded if and only if every infinite subset of E has a point of accumulation that belongs to E.*

The proof follows immediately from Theorem 11.44. The Heine-Borel Theorem also carries over to $\mathbb{R}^n$. We leave this as Exercise 11.8.2.

Exercises

11.8.1 Prove Corollary 11.45.

✂ **11.8.2** Provide definitions for "open cover" and "Heine-Borel property" for subsets of $\mathbb{R}^n$. Your definitions should be consistent with the one-dimensional version in Section 4.5.4. State and prove the analogue of Theorem 4.33 for subsets of $\mathbb{R}^n$.

11.9 Continuous Functions on Compact Sets

We turn now to the behavior of continuous functions defined on compact sets of $\mathbb{R}^n$. There were three fundamental properties possessed by every continuous real-valued function defined on a compact subset of $\mathbb{R}$: The range of the function was bounded, the function attained its maximum and minimum values, and the continuity was uniform. Each of these extends to vector-valued functions with few complications.

Theorem 11.46 *If $E \subset \mathbb{R}^n$ is compact and $\mathbf{f} : E \to \mathbb{R}^m$ is continuous on E, then the set $\mathbf{f}(E)$ is compact.*

Proof Let $\{\mathbf{y}_k\}$ be a sequence in the set $\mathbf{f}(E)$. We show that $\{\mathbf{y}_k\}$ has a convergent subsequence with limit in $\mathbf{f}(E)$. For each $k \in \mathbb{N}$ there exists $\mathbf{x_k} \in E$ such that $\mathbf{f}(\mathbf{x_k}) = \mathbf{y_k}$. Since E is compact, the sequence $\{\mathbf{x}_k\}$ has a convergent subsequence $\{\mathbf{x_{k_j}}\}$ converging to a point $\mathbf{x_0} \in E$. Since $\mathbf{f}$ is continuous at $\mathbf{x_0}$, $\mathbf{f}(\mathbf{x_0}) = \lim_{j\to\infty} \mathbf{f}(\mathbf{x_{k_j}})$, and since $\mathbf{x_0} \in E$, $\mathbf{f}(\mathbf{x_0}) \in \mathbf{f}(E)$.

The sequence $\{\mathbf{y_{k_j}}\} = \{\mathbf{f}(\mathbf{x_{k_j}})\}$ is thus a convergent subsequence of $\{\mathbf{y}_k\}$ that converges to the point $\mathbf{f}(\mathbf{x_0}) \in \mathbf{f}(E)$. ■

Corollary 11.47 *Let $E \subset \mathbb{R}^n$ be compact and let $f : E \to \mathbb{R}$ be continuous. Then f achieves an absolute maximum and absolute minimum on E.*

Proof The set $f(E)$ is a compact subset of $\mathbb{R}$ by Theorem 11.46. Thus $f(E)$ is closed and bounded by Theorem 11.44. Let $[c, d]$ be the smallest closed interval containing $f(E)$. Such an interval exists since $f(E)$ is bounded. Since $f(E)$ is closed, $c \in f(E)$ and $d \in f(E)$. The number c is the absolute minimum of f on E, and d is the absolute maximum of f on E. ■

Uniform Continuity Our definition of uniform continuity extends the concept from Section 5.6 to vector-valued functions.

Definition 11.48 Let $E \subset \mathbb{R}^n$ and let $\mathbf{f} : E \to \mathbb{R}^m$. We say $\mathbf{f}$ is *uniformly continuous* (on E) if for every $\varepsilon > 0$ there exists $\delta > 0$ such that if $\mathbf{x}, \mathbf{y} \in E$ and $\|\mathbf{x} - \mathbf{y}\| < \delta$, then $\|\mathbf{f}(\mathbf{x}) - \mathbf{f}(\mathbf{y})\| < \varepsilon$.

Theorem 11.49 *If $E \subset \mathbb{R}^n$ is compact and $\mathbf{f} : E \to \mathbb{R}^m$ is continuous on E, then $\mathbf{f}$ is uniformly continuous on E.*

The proof is identical to the proof of Theorem 5.47 when norms replace absolute value signs.

Exercises

11.9.1 Prove that if $\mathbf{f} : E \to \mathbb{R}^m$ is uniformly continuous on E, a subset of $\mathbb{R}^n$, and E is bounded, then $\mathbf{f}$ is bounded on E.

11.9.2 Prove Theorem 11.49 using the result of Exercise 11.8.2.

11.9.3 Show that the function $f(\mathbf{x}) = \|\mathbf{x}\|$ is uniformly continuous on $\mathbb{R}^n$.

11.9.4 Let $E \subset \mathbb{R}^n$ be compact and let $\mathbf{f} : E \to \mathbb{R}^m$. Prove that if $\mathbf{f}$ is continuous on E and one-to-one, then $\mathbf{f}^{-1}$ is continuous on the set $\mathbf{f}(E)$. Show that this conclusion might fail if E is not compact.

11.9.5 Let $S = [0,1] \times [0,1]$, $f : S \to \mathbb{R}$. Prove that if f is continuous on S and $g : [0,1] \to \mathbb{R}$ is defined by

$$g(x) = \max_{0 \leq y \leq 1} f(x,y),$$

then g is continuous on $[0,1]$.

11.9.6 Let E be a compact subset of $\mathbb{R}^n$ and let $\mathbf{x} \in \mathbb{R}^n$. Let

$$\operatorname{dist} \mathbf{x} = \inf\{\|\mathbf{x} - \mathbf{y}\| : y \in E\}.$$

(a) Prove that "inf" can be replaced by "min" in the definition of "dist."

(b) Prove that "dist" is a continuous function on $\mathbb{R}^n$.

(c) Are parts (a) and (b) valid if E is assumed closed but not bounded? What if E is assumed bounded but not closed?

11.10 Additional Remarks

✂ *Enrichment*

We mention in this section some items that we won't need in the next chapter but that may be of interest.

On Open and Closed Sets in $\mathbb{R}^n$ We saw in Chapter 4 that open sets in $\mathbb{R}$ have a particularly simple structure. Every nonempty open subset of $\mathbb{R}$ can be expressed as a finite or countably infinite union of pairwise disjoint open intervals (Theorem 4.15). Closed sets in $\mathbb{R}$ are relatively easy to visualize. They are made up of intervals, isolated points, limits of isolated points, etc., and Cantor sets.

In contrast, open or closed sets in $\mathbb{R}^n$, $n \geq 2$, can be more complicated. It is easy, for example, to think of a closed set that is the common boundary of two disjoint open sets—a circle will do the job. Think of the inside and the outside as the two disjoint open sets. Can you construct a closed set that is the common boundary of *three* pairwise disjoint open sets? Of five? Of infinitely many? Constructions of these types of sets are not easy, but they exist, and can arise in natural ways.[1]

The open sets must be very "wiggly" sets and their common boundary B is connected (in the sense introduced in Definition 11.50), yet B contains no arcs! Can you visualize such a set B?

The Four-Color Problem This famous problem[2] was finally settled affirmatively in 1976, after many mathematicians had tried in vain to solve it over a period of more than 100 years.

In loose language, the four-color problem, originally posed in 1852, asks whether "the regions of any map in the plane can be colored using no more than four colors, in such a way that those regions that have common boundaries consisting of more than one point have different colors." The solution that, in 1976, was finally announced for this problem remains controversial, however, since the proof required hundreds of pages and thousands of hours of computer verification.

Our preceding remarks about the boundaries of sets in higher-dimensional spaces shows the importance of being precise in stating problems. Suppose that $E_1, E_2, \ldots, E_5$ are bounded open sets with a common boundary B. Then five colors would be needed to color the resulting map. This would violate the four-color theorem as we loosely stated it. The actual four-color theorem is carefully stated, of course, so that the types of regions allowed and their common boundaries are limited appropriately.

Cross Products We have made considerable use of the dot product $\mathbf{x} \cdot \mathbf{y}$ of two vectors in $\mathbb{R}^n$. For the special case of $n = 3$, there is another important product, the *cross product* or *vector product*, denoted by $\mathbf{x} \times \mathbf{y}$. For any two vectors $\mathbf{x} = (x_1, x_2, x_3)$ and $\mathbf{y} = (y_1, y_2, y_3)$ we define

$$\mathbf{x} \times \mathbf{y} = (x_2 y_3 - x_3 y_2, x_3 y_1 - x_1 y_3, x_1 y_2 - x_2 y_1).$$

The cross product is important in various parts of mathematics and physics. We shall have no need of it in the chapters that follow, so we shall not provide a development here. We mention only that $\mathbf{x} \times \mathbf{y}$ is a vector that is

[1] An interesting short discussion and original references can be found in J. H. Hubbard, The Forced Damped, Pendulum: Chaos, Complication and Control, *Amer. Math. Monthly*, **106**, No. 8 (Oct. 1999), pp. 745–747.

[2] An interesting book on the subject, written for a lay readership, is *The Four Color Theorem,* R. and G. Fritsch (Springer-Verlag, 1998).

perpendicular to the plane determined by the vectors $\mathbf{x}$ and $\mathbf{y}$, and

$$\|\mathbf{x} \times \mathbf{y}\| = \|\mathbf{x}\| \, \|\mathbf{y}\| \sin \theta,$$

where $0 \le \theta \le \pi$ is the angle determined by $\mathbf{x}$ and $\mathbf{y}$. Geometrically, $\|\mathbf{x} \times \mathbf{y}\|$ is the area of the parallelogram of which $\mathbf{x}$ and $\mathbf{y}$ are adjacent sides.

Connectedness in $\mathbb{R}^n$ In Exercise 11.3.7 we defined connectedness for open sets in $\mathbb{R}^n$. An open set D in $\mathbb{R}^n$ is connected provided that each pair of points in D can be joined by a polygonal path lying in D. The notion of connectedness captures the idea of a set being in "one piece." What about sets that are not open? A closed disk, the graph of a continuous function $f : \mathbb{R} \to \mathbb{R}$, and the set

$$A = \{(x, y) : x^2 \le y \le 2x^2\}$$

are subsets of $\mathbb{R}^2$ that are not open but do seem to be in one piece. The definition involving polygonal paths applies to a closed disk and to the graphs of some, but not all, continuous functions. It does not apply to the set A since no polygonal path can join the origin to some other point of A. Here is a more general definition of connectedness in $\mathbb{R}^n$.

Definition 11.50 Let $S \subset \mathbb{R}^n$. If there exist disjoint open sets U and V such that $S \subset U \cup V$, $S \cap U \ne \emptyset$ and $S \cap V \ne \emptyset$, then U and V are said to *separate* S. The set S is *connected* if there are no open sets U, V that separate S.

With this definition, the previously mentioned sets are all connected. See the exercises for more examples and for consistency of this definition with the definition given earlier for open sets.

Exercises

11.10.1 Let D be an open set in $\mathbb{R}^n$. Prove that D is connected according to Definition 11.50 if and only if D is connected according to the definition in Exercise 11.3.7.

11.10.2 Verify that the sets (i) any closed disk, (ii) the graph of a continuous function $f : \mathbb{R} \to \mathbb{R}$ and (iii) the set $A = \{(x, y) : x^2 \le y \le 2x^2\}$ mentioned in our discussion of connectedness are all connected according to Definition 11.50.

11.10.3 Let $E \subset \mathbb{R}^n$, $\mathbf{f} : E \to \mathbb{R}^m$, $\mathbf{f}$ continuous on E. Show that if E is connected, then $\mathbf{f}(E)$ is connected.

11.10.4 State and prove an intermediate value property for continuous real-valued functions defined on connected sets.

11.10.5 Let $f : [a, b] \to \mathbb{R}$. Prove that f is continuous if and only if the graph of f is closed and connected (in $\mathbb{R}^2$).

11.10.6 Prove that the intersection of a descending sequence of compact connected sets in $\mathbb{R}^n$ is connected. Give an example to show that the statement is false if we drop the hypothesis of compactness.

11.10.7 Look at Hubbard's article cited in footnote 1. Observe that the common boundary of the open sets can be expressed as the intersection of a descending sequence of compact connected sets and is therefore connected by Exercise 11.10.6.

11.10.8 Verify the properties of the cross product $\mathbf{x} \times \mathbf{y}$ as we have defined it in this section. Show that $\mathbf{x} \times \mathbf{y}$ is orthogonal to $\mathbf{x}$ and to $\mathbf{y}$.

Chapter 12

DIFFERENTIATION ON $\mathbb{R}^n$

12.1 Introduction

We shall assume that you are familiar with notions related to differentiation of functions of several variables. This familiarity should include some understanding of partial and directional derivatives, their roles in obtaining tangent lines and tangent planes when dealing with functions of two variables, and some comfort in performing calculations with partials.

Students in elementary calculus can usually master some of the simpler concepts related to differentiation of functions of several variables, but may have more difficulty understanding the meaning of differentiability (beyond that it implies existence of a tangent plane for $n = 2$). Concepts related to the differential or to various chain rules are often difficult to grasp.

In this chapter we study concepts related to differentiation of functions of several variables. We begin with fixing notation for partial and directional derivatives, and then proceed in a leisurely fashion to discuss a number of topics. When our experiences tell us that students find a topic difficult, we devote a section to "setting up" the material, leaving the proofs to the following section, or we begin with special cases, deferring the general situation until we believe you feel comfortable with the topic.

A case in point is the differential. If you are well-prepared, you can proceed rapidly to Section 12.8, browsing quickly through much of the earlier material. A much gentler approach is to use the earlier sections as a means for familiarizing yourself with the concept of differentiability, differential, and chain rule before attacking the more abstract setting of mappings from $\mathbb{R}^n$ to $\mathbb{R}^m$. We have labeled that section with ✂, indicating that it can be omitted, because it involves a bit of linear algebra—previous sections involve only a minimal use of linear algebra.

12.2 Partial and Directional Derivatives

For a function of two variables $f(x, y)$ there is an obvious first attempt at the study of derivatives. The founders of our subject (Newton, Leibniz, and others) carried calculus techniques over to such functions merely by differentiating according to the usual rules of calculus, treating each variable separately. The example and its notation should be familiar to the student from earlier courses.

Example 12.1 Let $f(x, y) = (\sqrt{36 - 4x^2 - y^2})/3$. You may recognize the surface corresponding to this function as half of an ellipsoid. The partial derivatives of f are

$$\frac{\partial}{\partial x} f(x, y) = \frac{-4x}{3\sqrt{36 - 4x^2 - y^2}} \quad \text{and} \quad \frac{\partial}{\partial y} f(x, y) = \frac{-y}{3\sqrt{36 - 4x^2 - y^2}}.$$

◀

A more formal way of saying what is happening here is to compare with the usual definition of the derivative of a function $f(x)$,

$$\frac{d}{dx} f(x) = \lim_{h \to 0} \frac{f(x + h) - f(x)}{h}.$$

The process here for $f(x, y)$ is to perform the same computation but holding y as a fixed constant throughout:

$$\frac{\partial}{\partial x} f(x, y) = \lim_{h \to 0} \frac{f(x + h, y) - f(x, y)}{h}$$

and then holding x as a fixed constant,

$$\frac{\partial}{\partial y} f(x, y) = \lim_{h \to 0} \frac{f(x, y + h) - f(x, y)}{h}.$$

It is this that we will take as our definition for functions of two variables; for more variables the extension is similar. We shall prefer a subscript notation in place of the $\frac{\partial}{\partial x}$ notation when there are more than two or three variables.

The partials $f_1 = \frac{\partial f}{\partial x}$ and $f_2 = \frac{\partial f}{\partial y}$ at a point (x_0, y_0) provide slopes of two specific tangent lines to the surface $z = f(x, y)$. These are the lines in the planes $x = x_0$ and $y = y_0$ that go through the point (x_0, y_0, z_0) and are tangent to the surface. Figure 12.1 shows these tangent lines for the function of Example 12.1 at the point $(x_0, y_0, z_0) = (1, 3, \sqrt{23}/3)$.

These first simple steps at a theory of differentiation for functions of two or more variables served the pioneers of the calculus well. As we shall see, however, a deeper look at differentiation is needed to advance much further.

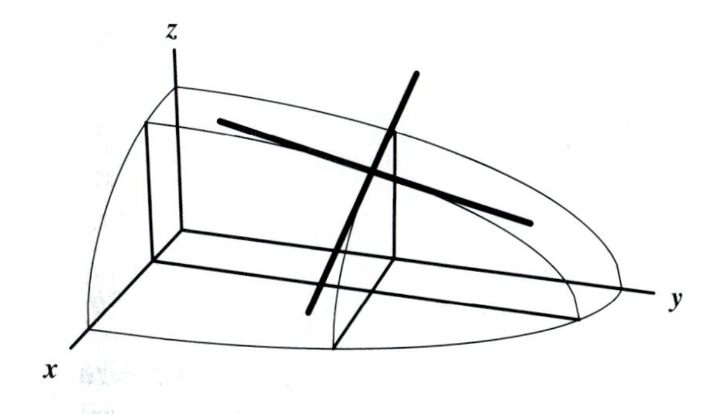

Figure 12.1. Tangent lines for the function $f(x, y) = \frac{1}{3}\sqrt{36 - 4x^2 - y^2}$ at the point $(1, 3, \sqrt{23}/3)$.

12.2.1 Partial Derivatives

Suppose $f : D \to R$, where D is an open subset of $\mathbb{R}^n$. Let

$$\mathbf{x} = (x_1, \ldots, x_n) \in D.$$

If we allow one of the coordinates of $\mathbf{x}$, say x_i, to vary while the others are fixed, we obtain a function of the one variable x_i. We may denote this function as

$$f(x_1, x_2, \ldots, x_{i-1}, \cdot, x_{i+1}, \ldots, x_n). \tag{1}$$

The dot in the ith position indicates that x_i may vary subject to the constraint that the point $(x_1, x_2, \ldots, x_{i-1}, x_i, x_{i+1}, \ldots, x_n)$ is in D.

If the derivative of the function (1) exists at $\mathbf{x}$ we obtain the *partial derivative* of f with respect to x_i at $\mathbf{x}$. We denote this partial derivative at $\mathbf{x}$ by $f_i(\mathbf{x})$ or $f_i(x_1, \ldots, x_n)$. Thus

$$f_i(\mathbf{x}) = \lim_{h \to 0} \frac{f(x_1, \ldots, x_i + h, \ldots, x_n) - f(x_1, \ldots, x_i, \ldots, x_n)}{h}. \tag{2}$$

Note. As was true about ordinary derivatives, it is true here too that different notations are sometimes useful, and many notations are in use. For example, $f_i(\mathbf{x})$ is often denoted by $\frac{\partial f}{\partial x_i}(\mathbf{x})$ or $f_{x_i}(\mathbf{x})$. If we write $u = f(x_1, \ldots, x_n)$ we often use the notations

$$u_i, \quad \frac{\partial u}{\partial x_i}, \quad \text{or} \quad u_{x_i}$$

in place of

$$f_i, \quad \frac{\partial f}{\partial x_i}, \quad \text{or} \quad f_{x_i}.$$

Of course, it is often convenient when we deal with only a few variables to denote these variables by x, y, z, instead of x_1, x_2, x_3.

Second-Order Partial Derivatives The partial derivatives f_i of a function $f(\mathbf{x})$ are themselves functions of $\mathbf{x}$. If they are defined at all points of a neighborhood of $\mathbf{x}$, they might themselves have partial derivatives at $\mathbf{x}$. These are called the *second partial derivatives* of f at $\mathbf{x}$ or the partial derivatives of the *second order*.

There can be as many as n^2 second partials. For example, for $n = 2$ four second partials can be defined:

$$\frac{\partial}{\partial x}\left(\frac{\partial f}{\partial x}\right) = \frac{\partial^2 f}{\partial x^2}, \qquad \frac{\partial}{\partial y}\left(\frac{\partial f}{\partial y}\right) = \frac{\partial^2 f}{\partial y^2},$$

$$\frac{\partial}{\partial x}\left(\frac{\partial f}{\partial y}\right) = \frac{\partial^2 f}{\partial x \partial y}, \qquad \frac{\partial}{\partial y}\left(\frac{\partial f}{\partial x}\right) = \frac{\partial^2 f}{\partial y \partial x}.$$

The two partials that involve both x and y are often called *cross partials* or *mixed partials* . We shall see presently that it is often, but not always, true that the mixed partials

$$\frac{\partial^2 f}{\partial y \partial x} \quad \text{and} \quad \frac{\partial^2 f}{\partial x \partial y}$$

are equal. In any case be sure to grasp that they mean two different things and to sort out from the notation which derivative is performed first.

A typical second partial for a function f of n variables can be denoted as

$$\frac{\partial}{\partial x_i}\left(\frac{\partial f}{\partial x_j}\right) = \frac{\partial^2 f}{\partial x_i \partial x_j}.$$

Once again, note which derivation is being performed first.

What notation should we use for second partials if we write f_1 for $\frac{\partial f}{\partial x}$? It seems natural to write

$$f_{12} \quad \text{for} \quad \frac{\partial}{\partial y}\left(\frac{\partial f}{\partial x}\right)$$

because we would like f_{12} to be short for $(f_1)_2$. If we do so, then we have preserved order in one sense. But we have reversed it in another sense, since for f_{12} we compose from left to right $((f_1)_2)$ whereas in $\frac{\partial^2 f}{\partial y \partial x}$ we compose from right to left. Nonetheless, we shall use f_{12} to mean $(f_1)_2$ because it's a bit simpler to manipulate. In any case, there should be no confusion since there are few places in which both the "f_{ij}" notation and the "$\frac{\partial^2 f}{\partial x_i \partial x_j}$" notation are used simultaneously.

Note. We mention that some authors reverse the notation so that, for example, $\frac{\partial^2 f}{\partial x \partial y}$ means $\frac{\partial}{\partial y}\left(\frac{\partial f}{\partial x}\right)$. We prefer our notation because it preserves the order of composition, but you should be aware when reading material involving mixed partials (in other books) that the other convention might be in force.

Higher-Order Partial Derivatives We can also consider partial derivatives of higher orders. Thus, for example

$$\frac{\partial^3 f}{\partial x \partial y \partial y} \text{ would mean } \frac{\partial}{\partial x}\left(\frac{\partial^2 f}{\partial y^2}\right),$$

which we could also write as f_{221} meaning $(f_{22})_1$. As before the order in which the derivatives are taken is given by the notation and the convention must be remembered.

Exercises

12.2.1 Calculate $f_1(1, 2\pi)$, $f_2(1, 2\pi)$, $f_{12}(1, 2\pi)$, and $f_{21}(1, 2\pi)$ for the function

$$f(x, y) = x \sin y.$$

12.2.2 Let $u(x, y) = y^3 - 3x^2y$ and $v(x, y) = x^3 - 3xy^2$.

(a) Show that $\dfrac{\partial^2 u}{\partial x^2} + \dfrac{\partial^2 u}{\partial y^2} = 0$ on $\mathbb{R}^2$.

(b) Show that $\dfrac{\partial^2 v}{\partial x^2} + \dfrac{\partial^2 v}{\partial y^2} = 0$ on $\mathbb{R}^2$.

(c) Show that $\dfrac{\partial u}{\partial x} = \dfrac{\partial v}{\partial y}$ and $\dfrac{\partial u}{\partial y} = -\dfrac{\partial v}{\partial x}$ on $\mathbb{R}^2$.

12.2.3 (Harmonic Functions) A function h defined on a region D of $\mathbb{R}^2$ is called *harmonic* if h has continuous partials of the first and second order and $h_{11} + h_{22} = 0$ at all points of D. This equation is called *Laplace's equation.* Verify that each of the following is harmonic on all of $\mathbb{R}^2$ and, for each, verify that $f_{12} = f_{21}$ and $f_{112} = f_{121} = f_{211}$.

(a) $e^x \cos y$

(b) $y^2 - 3x^2y$

(c) $x^2 - y^2 + 2y$

12.2.4 (Cauchy-Riemann Equations) The two equations introduced in Exercise 12.2.2(c)

$$\frac{\partial u}{\partial x} = \frac{\partial v}{\partial y} \text{ and } \frac{\partial u}{\partial y} = -\frac{\partial v}{\partial x} \tag{3}$$

are called the Cauchy-Riemann equations and are fundamental in complex analysis. This is so because a necessary and sufficient condition that a continuous function of a complex variable defined in a neighborhood N of a point in the complex plane be analytic is that its real and imaginary parts satisfy the Cauchy-Riemann equations. This means that if

$$f(x, y) = (u(x, y), v(x, y)),$$

then both equations in (3) are valid for all $(x, y) \in N$. [In complex notation we write $z = x + iy$ and $f(z) = u + iv$.] Show that each of the following functions $f = u + iv$ is analytic in a neighborhood of the origin.

(a) $u = x^2 - y^2$ and $v = 2xy$

(b) $u = e^x \cos y$ and $v = e^x \sin y$

(c) $u = 3x + y$ and $v = 3y - x$

12.2.5 Suppose u and v have all second-order partials on an open set $D \subset \mathbb{R}^2$ and that $v_{xy} = v_{yx}$ on D. Prove that if u and v satisfy the Cauchy-Riemann equations [equation (3) of Exercise 12.2.4] on D, then u is a *harmonic* function on D (see Exercise 12.2.3).

12.2.2 Directional Derivatives

Consider a function f of two variables. The partials f_1 and f_2 at a point (x_0, y_0) provide slopes of the two tangent lines to the surface in two directions. (Again see Figure 12.1.) There may be tangent lines corresponding to other vertical planes through (x_0, y_0, z_0), planes that are not parallel to the coordinate planes $x = 0$ or $y = 0$. The slopes of these tangent lines at (x_0, y_0, z_0) can be obtained by calculating *directional derivatives.*

Definition 12.2 Let $f : \mathbb{R}^2 \to \mathbb{R}$ and let $\mathbf{u} = (u_1, u_2)$ be a unit vector in $\mathbb{R}^2$, that is,

$$\|\mathbf{u}\| = \sqrt{u_1^2 + u_2^2} = 1.$$

The *directional* derivative of f in the direction $\mathbf{u}$ at the point (x_0, y_0) is

$$D_{\mathbf{u}}f(x_0, y_0) = \lim_{t \to 0} \frac{f(x_0 + tu_1, y_0 + tu_2) - f(x_0, y_0)}{t} \tag{4}$$

if the limit exists.

Just as the partial derivatives measure the rate of change of f in the x and y directions, $D_{\mathbf{u}}f(x_0, y_0)$ measures the rate of change of f in the direction of the vector $\mathbf{u}$. Indeed, we should be aware that the partials *are themselves* directional derivatives: f_1 and f_2 are directional derivatives in the directions $(1, 0)$ and $(0, 1)$ and

$$f_1 = D_{(1,0)} \text{ and } f_2 = D_{(0,1)}.$$

There is even a kind of converse: Later we will see that, under appropriate additional hypotheses, the derivatives in all directions can be written as soon as we know just the partials.

Observe that if $D_{\mathbf{u}}f(x_0, y_0)$ exists, then so too does the derivative in the opposite direction $D_{-\mathbf{u}}f(x_0, y_0)$, and

$$D_{-\mathbf{u}}f(x_0, y_0) = -D_{\mathbf{u}}f(x_0, y_0).$$

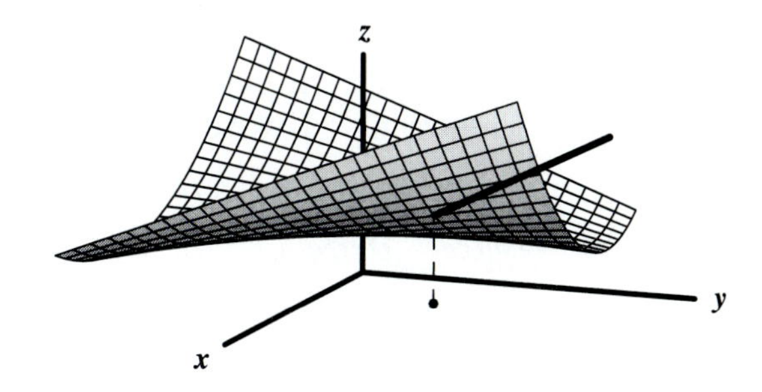

Figure 12.2. A directional derivative for $f(x, y) = x^2 + xy + 2$ at the point $(1, 1)$ in the direction $(1/2, \sqrt{3}/2)$.

Example 12.3 Let $f(x, y) = x^2 + xy + 2$ and $\mathbf{u} = (1/2, \sqrt{3}/2)$. To compute $D_{\mathbf{u}}f(1, 1)$ we calculate

$$\frac{f(1 + \frac{1}{2}t, 1 + \frac{\sqrt{3}}{2}t) - f(1, 1)}{t} =$$

$$\frac{(\frac{3}{2} + \frac{\sqrt{3}}{2})t + (\frac{1}{4} + \frac{\sqrt{3}}{4})t^2}{t} \to \frac{3}{2} + \frac{\sqrt{3}}{2} \text{ as } t \to 0.$$

We shall see later that under suitable hypotheses, we can calculate

$$D_{\mathbf{u}}f = (f_1, f_2) \cdot (u_1, u_2) = f_1u_1 + f_2u_2.$$

In this case $f_1(1, 1) = 3,\ \ f_2(1, 1) = 1$, so

$$(3, 1) \cdot (\tfrac{1}{2}, \tfrac{\sqrt{3}}{2}) = \tfrac{3}{2} + \tfrac{\sqrt{3}}{2}$$

as we calculated before. Figure 12.2 illustrates the graph of f and shows the tangent line in the direction $(1/2, \sqrt{3}/2)$ at the point $(1, 1)$. ◀

Exercises

12.2.6 Define directional derivatives for real valued functions of n variables.

12.2.7 Verify that f_1 and f_2 are directional derivatives in the directions $(1, 0)$ and $(0, 1)$ respectively (i.e., that $f_1 = D_{(1,0)}$ and $f_2 = D_{(0,1)}$). Check that $D_{-\mathbf{u}}f(x_0, y_0) = -D_{\mathbf{u}}f(x_0, y_0)$.

12.2.3 Cross Partials

Let f be defined on a region $D \subset \mathbb{R}^2$, and suppose both partial derivatives

$$\frac{\partial f}{\partial x}(x, y) \quad \text{and} \quad \frac{\partial f}{\partial y}(x, y)$$

exist at each point (x, y) in D. These partial derivatives are themselves functions in D and, as such, might in turn possess partial derivatives. In this case, there are four possible partial derivatives:

$$\frac{\partial^2 f}{\partial x^2} = \frac{\partial^2 f}{\partial x \partial x}, \frac{\partial^2 f}{\partial y \partial x}, \frac{\partial^2 f}{\partial x \partial y} \text{ and } \frac{\partial^2 f}{\partial y \partial y} = \frac{\partial^2 f}{\partial y^2}.$$

We recall, for example, that our notation requires

$$\frac{\partial^2 f}{\partial y \partial x} = \frac{\partial}{\partial y}\left(\frac{\partial f}{\partial x}\right).$$

The two partials $\frac{\partial^2 f}{\partial y \partial x}$ and $\frac{\partial^2 f}{\partial x \partial y}$ are often called *cross partials* or *mixed partials.*

Students in elementary calculus frequently believe there are only three second-order partials because it is usually true that

$$\frac{\partial^2 f}{\partial y \partial x} = \frac{\partial^2 f}{\partial x \partial y}. \tag{5}$$

In this section we determine conditions under which (5) is valid. We begin with an example that illustrates that (5) is not always valid.

Example 12.4 Let

$$f(x, y) = \begin{cases} \frac{xy(x^2-y^2)}{x^2+y^2}, & \text{if } (x, y) \neq (0, 0) \\ 0, & \text{if } (x, y) = (0, 0) \end{cases}.$$

Using the subscript notation, we compute the partial derivatives

$$f_1 = \frac{\partial f}{\partial x}, \ f_2 = \frac{\partial f}{\partial y}, \ f_{12} = \frac{\partial^2 f}{\partial y \partial x}, \ f_{21} = \frac{\partial^2 f}{\partial x \partial y}$$

at the point $(0, 0)$. Now

$$f_1(0, y) = \lim_{h \to 0} \frac{f(h, y) - f(0, y)}{h} = \lim_{h \to 0} \frac{\left(\frac{hy(h^2-y^2)}{h^2+y^2}\right)}{h} = -y.$$

Similarly, $f_2(x, 0) = x$. It follows that

$$\begin{aligned} f_{12}(0, 0) &= \lim_{k \to 0} \frac{f_1(0, k) - f_1(0, 0)}{k} = \frac{-k - 0}{k} = -1 \text{ and} \\ f_{21}(0, 0) &= \lim_{h \to 0} \frac{f_2(h, 0) - f_2(0, 0)}{h} = \lim_{h \to 0} \frac{h - 0}{h} = 1. \end{aligned}$$

Thus the two cross partials are not equal. ◄

What is it that "went wrong" here? Consider the two identical expressions involving the function f of Example (12.4):

$$\frac{1}{k}\left[\frac{f(h,k)-f(0,k)}{h}-\frac{f(h,0)-f(0,0)}{h}\right], \tag{6}$$

$$\frac{1}{h}\left[\frac{f(h,k)-f(h,0)}{k}-\frac{f(0,k)-f(0,0)}{k}\right]. \tag{7}$$

If we first let $h \to 0$ in (6), and then let $k \to 0$, we obtain $f_{12}(0,0)$. On the other hand, letting $k \to 0$ in (7) and then letting $h \to 0$, we obtain $f_{21}(0,0)$. As we saw, these two iterated limits were not equal.

Note. You might recall that we have seen this phenomenon before, that two limits might be different if performed in different orders. In Example 9.7 we saw that the two iterated limits of a double sequence $\{s_{nm}\}$ might not be equal. See also Exercise 9.2.8 and its hint for a discussion of this phenomenon. Here we once again have a situation that involves iterated limits.

We can explain now what went wrong in our example: The failure of the two mixed partial derivatives f_{12} and f_{21} to be equal can be traced to the fact that the limit of these expressions (6) or (7) as $(h,k) \to 0$ (i.e., the *double limit*) fails to exist.

A simple sufficient condition for the existence of this double limit appears in the following theorem.

Theorem 12.5 *Let f be defined in a neighborhood of $(x_0, y_0) \in \mathbb{R}^2$. Suppose f has partial derivatives f_1, f_2, f_{12} and f_{21} in this neighborhood and that the mixed partials f_{12} and f_{21} are continuous at (x_0, y_0). Then $f_{12}(x_0, y_0) = f_{21}(x_0, y_0)$.*

Proof Consider the two equal expressions

$$\frac{1}{k}\left[\frac{f(x_0+h,y_0+k)-f(x_0,y_0+k)}{h}-\frac{f(x_0+h,y_0)-f(x_0,y_0)}{h}\right] \tag{8}$$

$$\frac{1}{h}\left[\frac{f(x_0+h,y_0+k)-f(x_0+h,y_0)}{k}-\frac{f(x_0,y_0+k)-f(x_0,y_0)}{k}\right] \tag{9}$$

If we first let $h \to 0$ in (8) and then let $k \to 0$, we obtain $f_{12}(x_0, y_0)$. On the other hand, letting $k \to 0$ and then $h \to 0$ in (9), we obtain $f_{21}(x_0, y_0)$. We shall show that these two iterated limits are equal.

Observe first that the numerator D_{hk} in brackets in expression (8) can be written in the form

$$(f(x_0+h,y_0+k)-f(x_0+h,y_0))-(f(x_0,y_0+k)-f(x_0,y_0)). \tag{10}$$

Applying the mean value theorem (Theorem 7.20) to (10), we obtain

$$D_{hk} = (f_1(\xi, x_0+k) - f_1(\xi, y_0))h, \tag{11}$$

where ξ is between x_0 and $x_0 + h$. Applying the mean value theorem again, we obtain

$$D_{hk} = f_{12}(\xi, \zeta)hk,$$

where ζ is between y_0 and $y_0 + k$.

We can rewrite (10) in the form

$$(f(x_0 + h, y_0 + k) - f(x_0, y_0 + k)) - (f(x_0 + h, y_0) - f(x_0, y_0)).$$

Applying the same argument as before, we obtain

$$D_{hk} = f_{21}(\sigma, \tau)kh,$$

where σ is between x_0 and $x_0 + h$ and τ is between y_0 and $y_0 + k$.

Now let $(h, k) \to (0, 0)$. Then $(\xi, \zeta) \to (0, 0)$ and $(\sigma, \tau) \to (0, 0)$. Since the functions f_{12} and f_{21} are continuous at (x_0, y_0) (by hypothesis), we obtain

$$\lim_{(h,k)\to(0,0)} \frac{D_{hk}}{hk} = f_{21}(x_0, y_0) = f_{12}(x_0, y_0)$$

as required. ■

Note. Observe that our proof actually shows that the *double limit*

$$\lim_{(h,k)\to(0,0)} \frac{D_{hk}}{hk} = f_{21}(x_0, y_0) = f_{12}(x_0, y_0).$$

This is a stronger statement than the conclusion of the theorem. We will return to this idea later when we discuss differentiability.

Theorem 12.5 shows that continuity of the partials is a sufficient condition for f_{12} to equal f_{21}. This condition is often met but is not a necessary condition.

Example 12.6 We can construct an example by using the function

$$g(x) = x^2 \sin \frac{1}{x}, \qquad g(0) = 0,$$

which is differentiable on $\mathbb{R}$ but whose derivative g' is discontinuous at $x = 0$. It follows that the function

$$f(x, y) = yx^2 \sin \frac{1}{x}, \qquad f(0, y) = 0$$

has both partials f_{12} and f_{21} discontinuous at $(0, 0)$. Thus Theorem 12.5 does not apply here, yet the partials

$$f_{21}(0, 0) = f_{12}(0, 0) = 0$$

are equal. ◄

Other sufficient conditions for the equality of the mixed partials are known. For example, it suffices to assume existence of both mixed partials and continuity of only one of them at (x_0, y_0) or to assume that f_1 and f_2 are differentiable[1] at (x_0, y_0). We omit the proofs.

Theorem 12.5 extends readily to higher-order derivatives. For example, if f has partial derivatives of order one, two, and three, and all of these are continuous at (x_0, y_0), then

$$f_{221} = (f_2)_{21} = (f_2)_{12} = f_{212} = (f_{21})_2 = (f_{12})_2 = f_{122}.$$

Partial Derivatives of $f : \mathbb{R}^n \to \mathbb{R}$ Theorem 12.5 also extends to functions of more than two variables. The following theorem reduces to Theorem 12.5 by holding all of the variables except the ith and jth fixed in the proof. Only the notation becomes more complicated.

Theorem 12.7 *Let f be defined in some neighborhood of the point $\mathbf{x_0}$ in $\mathbb{R}^n$. Suppose f has partial derivatives f_i, f_j, f_{ij}, and f_{ji} in this neighborhood and that the mixed partials f_{ij} and f_{ji} are continuous at $\mathbf{x_0}$; then $f_{ij}(\mathbf{x_0}) = f_{ji}(\mathbf{x_0})$.*

Exercises

12.2.8 In general how many mixed partials of the third order does a function $f : \mathbb{R}^2 \to \mathbb{R}$ have? How many for a function $f : \mathbb{R}^n \to \mathbb{R}$?

12.2.9 For each of the harmonic functions of Exercise 12.2.3

(a) $f(x, y) = e^x \cos y$

(b) $f(x, y) = y^2 - 3x^2 y$

(c) $f(x, y) = x^2 - y^2 + 2y$

verify that $f_{12} = f_{21}$ and $f_{112} = f_{121} = f_{211}$.

12.2.10 Let $f(x, y) = x^2 \tan^{-1}(y/x) - y^2 \tan^{-1}(x/y)$ when $x \neq 0$ and $y \neq 0$, and $f(x, y) = 0$ if either x or y is zero. Compute $f_{12}(0, 0)$ and $f_{21}(0, 0)$. Does your result contradict Theorem 12.5?

12.2.11 **(Double limits and iterated limits, revisited)** A bit of care is needed with the statement "If a double limit exists, so do the two iterated limits, and the two iterated limits equal the double limit." Let

$$f(x, y) = \begin{cases} y + x \sin \frac{1}{y} & \text{if } y \neq 0 \\ 0 & \text{if } y = 0 \end{cases}.$$

(a) Show that $\lim_{(x,y)\to(0,0)} f(x, y) = 0$ but $\lim_{x\to 0} \lim_{y\to 0} f(x, y)$ does not exist.

---

[1]Differentiable in the sense defined later in Section 12.4.

(b) Prove that if $\lim_{(x,y)\to(x_0,y_0)} g(x,y)$ and $\lim_{y\to y_o} g(x,y)$ exist for all x, then

$$\lim_{x\to x_0}\lim_{y\to y_o} g(x,y) = \lim_{(x,y)\to(x_0,y_0)} g(x,y).$$

12.3 Integrals Depending on a Parameter

Let g be continuous on $[a,b]$ and let

$$G(y) = \int_a^y g(x)\,dx. \tag{12}$$

We recall from the fundamental theorem of calculus that $G'(y) = g(y)$ for all $y \in [a,b]$. This represents half of the inverse nature of differentiation and integration.

Let us now complicate things a bit. Let f be continuous on a rectangle

$$R = \{(x,y) \in \mathbb{R}^2 : a \le x \le b, c \le y \le d\}.$$

Define a function $F : [c,d] \to \mathbb{R}$ by

$$F(y) = \int_a^b f(x,y)\,dx. \tag{13}$$

To see that F is well defined, we need only note that for every fixed $y \in [c,d]$, $f(\cdot,y)$ is simply a continuous function on $[a,b]$ and therefore the integral does exist.

We might consider (13) as a "partial integral," in the same way in which we consider our derivatives as partial derivatives: In both cases an operation is performed on one variable, while the other variable is held fixed.

What can we say about the function F? Is F continuous on $[c,d]$? Is F differentiable? If so, how can we compute F'? We address such questions in this section.

Observe two differences between (12) and (13). In (12) we are dealing with a function g of *one* variable; in (13) f is a function of two variables. Furthermore, in (12) the upper limit is the variable y, whereas in (13) the upper limit is the constant b.

Integrals of the type $\int_a^b f(x,y)\,dx$ appear frequently in practice. We think of y as a parameter and we are often concerned with the question "How does a small change in the parameter affect the resulting integral?" We address this question first with respect to continuity (of F) and then differentiability.

Theorem 12.8 *If f is continuous on $R = [a,b] \times [c,d]$ and*

$$F(y) = \int_a^b f(x,y)\,dx,$$

then F is continuous on $[c,d]$.

Proof Let $y_0 \in [c, d]$. We prove F is continuous at y_0. Since R is closed and bounded, and therefore compact, f is uniformly continuous on R by Theorem 11.49. Thus, for $\varepsilon > 0$ there exists $\delta > 0$ such that for each $x \in [a, b]$

$$|f(x, y) - f(x, y_0)| < \frac{\varepsilon}{b - a}$$

whenever $|y - y_0| < \delta$. For $|y - y_0| < \delta$ we calculate

$$\begin{aligned} |F(y) - F(y_0)| &= \left| \int_a^b (f(x, y) - f(x, y_0))\, dx \right| \\ &\leq \int_a^b |f(x, y) - f(x, y_0)|\, dx \leq \int_a^b \frac{\varepsilon}{b - a}\, dx = \varepsilon. \end{aligned}$$

Thus $\lim_{y \to y_0}(F(y) - F(y_0)) = 0$ and F is continuous at y_0. ■

We turn now to the question of differentiability of F. We would like to find a formula that obtains F' via a "differentiation under the integral sign." But here f is a function of two variables, and a partial derivative is called for. Theorem 12.9 is often called Leibniz's rule.

Theorem 12.9 (Leibniz's Rule) *Let f be continuous on the rectangle*

$$R = [a, b] \times [c, d]$$

and let

$$F(y) = \int_a^b f(x, y)\, dx$$

for each $y \in [c, d]$. If the partial derivative $\frac{\partial f}{\partial y}$ exists and is continuous on R, then F is differentiable on $[c, d]$, and

$$F'(y) = \int_a^b \frac{\partial}{\partial y} f(x, y)\, dx.$$

Proof It suffices to show that for each $y \in [c, d]$

$$\lim_{h \to 0} \left(\frac{F(y + h) - F(y)}{h} - \int_a^b \frac{\partial}{\partial y} f(x, y)\, dx \right) = 0. \tag{14}$$

Now for $y + h \in [c, d]$,

$$F(y + h) - F(y) = \int_a^b [f(x, y + h) - f(x, y)]\, dx. \tag{15}$$

It follows from the mean value theorem (Theorem 7.20) that for fixed x, y, and h, there exists a number $\xi \in (0, 1)$ such that

$$f(x, y + h) - f(x, y) = h \frac{\partial}{\partial y} f(x, y + \xi h). \tag{16}$$

Substituting (16) in (15), we obtain

$$\frac{F(y+h)-F(y)}{h} - \int_a^b \frac{\partial}{\partial y} f(x,y)\,dx = \tag{17}$$
$$\int_a^b \left(\frac{\partial}{\partial y} f(x, y+\xi h) - \frac{\partial}{\partial y} f(x,y)\right) dx.$$

Since $\frac{\partial f}{\partial y}$ is continuous (by assumption), it is uniformly continuous on the compact set R. Thus, for $\varepsilon > 0$ there exists $\delta > 0$ such that if $(x,y) \in R$ and $(x',y') \in R$ with $|x - x'| < \delta$ and $|y - y'| < \delta$, then

$$\left|\frac{\partial}{\partial y} f(x,y) - \frac{\partial}{\partial y} f(x',y')\right| < \frac{\varepsilon}{b-a}.$$

In particular, if $|h| < \delta$, then

$$\left|\frac{\partial}{\partial y} f(x,y+\xi h) - \frac{\partial}{\partial y} f(x,y)\right| < \frac{\varepsilon}{b-a}. \tag{18}$$

From (17) and (18) we infer that for $|h| < \delta$,

$$\left|\frac{F(y+h)-F(y)}{h} - \int_a^b \frac{\partial}{\partial y} f(x,y)\,dx\right| \leq \int_a^b \frac{\varepsilon}{b-a}\,dx = \varepsilon$$

and (14) follows. ■

Example 12.10 If

$$F(y) = \int_a^b \sin(xy)\,dx,$$

we can conclude from Theorem 12.9 that

$$F'(y) = \int_a^b x\cos(xy)\,dx.$$

We simply differentiate "through" the integral sign. ◄

In this case, both integrals appearing can easily be integrated by elementary methods (and Leibniz's rule thus verified), but this need not be true in general. It is often not possible to evaluate one or both of the integrals

$$\int_a^b \frac{\partial}{\partial y} f(x,y)\,dx \quad \text{or} \quad \int_a^b f(x,y)\,dx$$

in terms of elementary functions. We continue with some examples of Leibniz's rule used in conjunction with other techniques.

Example 12.11 To compute the derivative of the function

$$F(y) = \int_0^{\pi/2} \frac{dx}{\sqrt{1 - y^2 \sin^2 x}}$$

for $0 < y < 1$ we simply apply Leibniz's rule to obtain

$$F'(y) = \int_0^{\pi/2} \frac{y \sin^2 x}{\sqrt{1 - y^2 \sin^2 x}}\, dx.$$

This is an example of an elliptic integral with parameter y. ◀

Example 12.12 The integral

$$F(y) = \int_0^{y^2} yx^3\, dx \tag{19}$$

and its derivative can be calculated easily by elementary means. As an exercise, we calculate F' using Leibniz's rule. We observe first that there is now a variable upper limit of integration, so more than a direct application of Leibniz's rule is required.

In this case, the parameter occurs in a limit of integration as well as in the integrand, so we need to use the fundamental theorem of calculus and the chain rule as well as Leibniz's rule. (The chain rule is proved in Theorem 12.34 of Section 12.5.4.)

Let

$$u = y^2, \quad v = y, \text{ and } G(u, v) = \int_0^u vx^3\, dx.$$

Then the fundamental theorem of calculus gives us

$$\frac{\partial G}{\partial u} = vu^3$$

and Leibniz's rule provides

$$\frac{\partial G}{\partial v} = \int_0^u x^3\, dx = \frac{u^4}{4}$$

and the chain rule that we remember from the calculus of functions of two variables gives us

$$\frac{d}{dy} F(y) = \frac{\partial G}{\partial u}\frac{du}{dy} + \frac{\partial G}{\partial v}\frac{dv}{dy}$$

$$= vu^3 2y + \frac{u^4}{4} \cdot 1 = 2y^8 + \frac{y^8}{4} = \frac{9}{4}y^8.,$$

which is the same result we would obtain by evaluating the integral (19) and differentiating. ◀

Some of the exercises involve the use of chain rules from calculus. These and other chain rules are discussed in greater detail and proved in Sections 12.5, 12.5.4, 12.5.6, and 12.8.

Exercises

12.3.1 Use methods of elementary calculus to verify directly the result in Example 12.12.

12.3.2 Calculate $F'(y)$ for each of the following functions F.

(a) $F(y) = \int_0^1 e^{-x^2y^2}\,dx$

(b) $F(y) = \int_0^y e^{-x^2y^2}\,dx$

(c) $F(y) = \int_0^{y^3} e^{-x^2y^2}\,dx$

12.3.3 Let u and v have continuous derivatives on $\mathbb{R}$ and let f be continuous with continuous partial derivatives on $\mathbb{R}^2$. Obtain a formula for $F'(y)$, where

$$F(y) = \int_{u(y)}^{v(y)} f(x,y)\,dx.$$

12.3.4 The integral

$$F(y) = \int_0^y \sqrt{1 - k^2\sin^2 t}\,dt$$

is called an *elliptic integral of the second kind.* (It arises in computing the length of an arc of an ellipse given parametrically by $x = a\cos t$, $y = b\sin t$, $0 < a < b$.) Calculate $F'(y)$.

12.3.5 Using elementary calculus techniques, show that the arc length of an ellipse given parametrically by

$$x = a\cos t,\ y = b\sin t, \qquad (0 < a < b)$$

is given by the integral

$$b\int_0^{2\pi} \sqrt{1 - k^2\sin^2 t}\,dt \ \text{ where } k^2 = \frac{b^2 - a^2}{b^2}.$$

12.4 Differentiable Functions

When a function f of one real variable has a finite derivative at a point x_0, we say that f is differentiable at x_0. This is simply the definition of differentiability. Differentiable functions of one variable have many useful properties. The most fundamental of these is that the tangent line at x_0 is a close approximation to the function near x_0.

We might be tempted to carry this language over to higher dimensions and say that a function of several variables that has finite partial derivatives with respect to each variable is "differentiable." But such a definition would not be useful. It would not, moreover, generalize this notion that the tangent line is a close approximation to a differentiable function.

12.4.1 Approximation by Linear Functions

Let's see what is involved by looking at the one variable situation more carefully. Suppose $f:\mathbb{R}\to\mathbb{R}$, $x_0 \in \mathbb{R}$, and

$$L(x) = a_0 x + a_1$$

is *any* affine function[2] such that $L(x_0) = f(x_0)$. If f is continuous at x_0, then for x near x_0 we will have $f(x)$ near $L(x)$. More precisely,

$$\lim_{x\to x_0} |f(x) - L(x)| = 0. \tag{20}$$

We could say L approximates f near x_0 in the sense of (20). Thus any affine function through the point $(x_0, f(x_0))$ approximates f near x_0 if f is continuous there.

If f is differentiable at x_0, we can obtain a better approximation using the tangent line T. This line has the equation

$$T(x) = f(x_0) + f'(x_0)(x - x_0).$$

Now, differentiability of f at x_0 means that

$$\left|\frac{f(x) - f(x_0)}{x - x_0} - f'(x_0)\right| \to 0 \text{ as } x \to x_0. \tag{21}$$

We can write (21) in the form

$$\left|\frac{f(x) - T(x)}{x - x_0}\right| \to 0 \text{ as } x \to x_0. \tag{22}$$

Comparing (22) with (20), we see the improvement in the approximation when we use the tangent line (when it exists): In (20) we divide $|f(x)-L(x)|$ by the constant 1 and obtain the limit 0; in (22) we divide $|f(x) - T(x)|$ by $|x - x_0|$, which approaches 0 as $x \to x_0$, and still obtain the limit 0. Thus not only is $T(x)$ near $f(x)$ when x is near x_0, but the distance between $T(x)$ and $f(x)$ is small in comparison with the distance between x and x_0.

Finally, we rewrite (22) in a form convenient for generalization to functions of several variables. If we write $h = x - x_0$ and manipulate (22) a bit, we arrive at the the equivalent formulation

$$f(x_0 + h) - f(x_0) = f'(x_0)h + \varepsilon h, \tag{23}$$

where $\varepsilon \to 0$ as $h \to 0$. Here ε depends on h [to satisfy the equality in (23)], and the requirement for differentiability (21) is that $\varepsilon \to 0$ as $h \to 0$.

---

[2] A real function L of a real variable is said to be *affine* if its graph is a straight line. Thus $L(x) = a_0 x + a_1$. In calculus courses such functions are usually said to be "linear," but our language must be rather more precise since we are using some linear algebra in our presentation. Note that it is the function $f(x) - f(x_0)$ that can be approximated by a *linear* function $a_0 x$.

12.4.2 Definition of Differentiability

We shall imitate the preceding for functions of several variables. To keep notation simple for the moment, let us begin by considering functions of two variables defined in a neighborhood of a point $(x_0, y_0) \in \mathbb{R}^2$. We wish to express the condition that the surface corresponding to the equation

$$z = f(x, y)$$

has a tangent plane at (x_0, y_0). Thus the tangent plane T replaces the tangent line for functions of one variable.

Recall from elementary calculus that the tangent plane, if it exists, can be expressed by the linear equation

$$T(x, y) = f(x_0, y_0) + f_1(x_0, y_0)(x - x_0) + f_2(x_0, y_0)(y - y_0).$$

In order for this plane to be the tangent plane, we require that the analogue of (22) be valid, namely that

$$\frac{|f(x, y) - T(x, y)|}{\sqrt{(x - x_0)^2 + (y - y_0)^2}} \to 0 \text{ as } (x, y) \to (x_0, y_0). \tag{24}$$

Manipulating (24) as we did (22) and writing $h = x - x_0$, $k = y - y_0$, we arrive at the correct formulation for the two-variable version.

Definition 12.13 A function $f : \mathbb{R}^2 \to \mathbb{R}$ is *differentiable* at (x_0, y_0) if

1. The partial derivatives f_1 and f_2 exist at (x_0, y_0).

2. It is possible to write

$$\begin{aligned} f(x_0 + h, y_0 + k) - f(x_0, y_0) = \\ f_1(x_0, y_0)h + f_2(x_0, y_0)k + \varepsilon(h, k)\sqrt{h^2 + k^2}, \end{aligned} \tag{25}$$

where $\varepsilon(h, k) \to 0$ as $\sqrt{h^2 + k^2} \to 0$ (with $\varepsilon(0, 0) = 0$).

Note. By observing that for $(h, k) \in \mathbb{R}^2$,

$$\sqrt{h^2 + k^2} \leq |h| + |k| \leq \sqrt{2}\sqrt{h^2 + k^2},$$

we can rewrite statement (2) in the definition in the simpler form

2′. It is possible to write

$$\begin{aligned} f(x_0 + h, y_0 + k) - f(x_0, y_0) = \\ f_1(x_0, y_0)h + f_2(x_0, y_0)k + \varepsilon(h, k)(|h| + |k|), \end{aligned} \tag{26}$$

where $\varepsilon(h, k) \to 0$ as $|h| + |k| \to 0$.

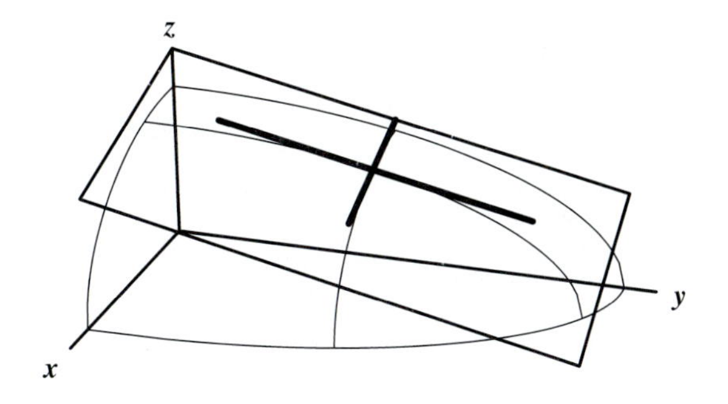

Figure 12.3. Tangent plane for the function $f(x,y) = \frac{1}{3}\sqrt{36-4x^2-y^2}$ at the point $(1,3,\sqrt{23}/3)$.

The condition $|h|+|k| \to 0$ is equivalent to $h \to 0$ and $k \to 0$, which is, in turn, equivalent to $\sqrt{h^2+k^2} \to 0$.

Let us return for a moment to the elementary geometry underlying the role of the tangent plane. This plane should approximate f near (x_0, y_0) in the sense of (24). Roughly, this requires that when (x,y) is near (x_0,y_0), then $|f(x,y) - T(x,y)|$ is small in comparison with the distance between (x,y) and (x_0,y_0). Mere existence of the partials at (x_0,y_0) does *not* guarantee this; that would guarantee only that the desired comparison is small when (x,y) is near (x_0,y_0) and $x = x_0$ (or $y = y_0$). Appropriate tangent lines exist in the intersection of the surface $z = f(x,y)$ with the plane $x = x_0$ (or $y = y_0$), but (24) need not hold when $(x,y) \to (x_0,y_0)$ in other ways.

Example 12.14 Consider again the function

$$f(x,y) = (\sqrt{36-4x^2-y^2})/3$$

from Example 12.1. In Figure 12.1 we illustrated the tangent lines at the point $(x_0,y_0,z_0) = (1,3,\sqrt{23}/3)$. With some more effort we could now show that this function is differentiable at that point by going through the computations in the definition to show that the tangent plane approximates the function in the correct sense. Figure 12.3 shows this tangent plane and illustrates that this approximation is plausible. In practice we would normally apply some theorem (such as Theorem 12.21) that would allow us to conclude differentiability without resorting to such computations. ◀

Example 12.15 Let $f(x,y) = \sqrt{|xy|}$. (This function is reminiscent of the function $|x| = \sqrt{|x^2|}$, which is continuous but not differentiable at $x = 0$.) It is clear that f is continuous at $(0,0)$ (Exercise 12.4.1). Since $f = 0$ on the

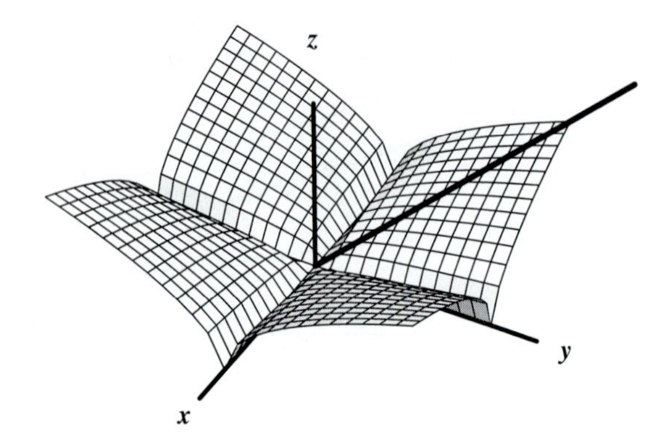

Figure 12.4. Graph of $f(x,y) = \sqrt{|xy|}$ which is continuous and nondifferentiable at $(0,0)$.

coordinate axes,

$$f_1(0,0) = f_2(0,0) = 0. \tag{27}$$

For f to be differentiable at $(0,0)$ it must be true that $\varepsilon(h,k) \to 0$ as $|h|+|k| \to 0$, where $\varepsilon(h,k)$ as defined by

$$\sqrt{|hk|} = f_1(0,0)h + f_2(0,0)k + \varepsilon(h,k)(|h|+|k|).$$

By (27), $\sqrt{|hk|} = \varepsilon(|h|+|k|)$.

Letting $h = -k$, for example, we obtain $|h| = 2\varepsilon|h|$, so $\varepsilon(h,-h) = \frac{1}{2}$. Thus it is not the case that $\varepsilon(h,k) \to 0$ as $|h|+|k| \to 0$. (Letting $h = -k$ is equivalent to having (x,y) approach (x_0,y_0) along the line $x+y=0$, $z=0$.) Figure 12.4 illustrates the graph of the function f, evidently shaped like a folded napkin with the folds along the two axes. Note especially that the information along the "folds" that $f_1(0,0) = f_2(0,0) = 0$ does not help to compute the directional derivative in the direction $(-1/\sqrt{2}, 1/\sqrt{2})$ (illustrated in the figure) and that corresponds to the case $h = -k$ discussed previously. ◀

Differentiability for $f : \mathbb{R}^n \to \mathbb{R}$. With these preliminaries we can now give a formal definition for *differentiability* of a function $f : \mathbb{R}^n \to \mathbb{R}$. This is the obvious generalization of Definition 12.13 that we have just given for the differentiability of a function of two variables.

Definition 12.16 Let f be defined on an open set $D \subset \mathbb{R}^n$ and let

$$\mathbf{x} = (x_1, \ldots, x_n) \in D.$$

Then f is *differentiable* at $\mathbf{x}$ if

1. The partial derivatives $f_1, \ldots, f_n$ all exist (finite) at $\mathbf{x}$, and

2. When $\varepsilon = \varepsilon(h_1, \ldots, h_n)$ is defined by

$$f(x_1 + h_1, \ldots, x_n + h_n) - f(x_1, \ldots, x_n) =$$
$$\sum_{i=1}^{n} f_i(x_1, \ldots, x_n)h_i + \varepsilon\sqrt{(|h_1|^2 + \cdots + |h_n|^2)}$$

we have $\varepsilon(h_1, \ldots, h_n) \to 0$ when $\sqrt{(|h_1|^2 + \cdots + |h_n|^2)} \to 0$ (with $\varepsilon(0, \ldots, 0) = 0$).

If f is differentiable at all points in D, we say f is *differentiable* on D.

Note. Statement (2) in the definition is given using the usual euclidean norm. As we have seen before in the note following Definition 12.13, that gives the $\mathbb{R}^2$ case, it often simplifies proofs involving differentiability to use the $\|\cdot\|_1$ norm. Exercise 12.4.2 shows that this is possible. Using the $\|\cdot\|_1$ norm, then statement (2) becomes

2′. When $\varepsilon = \varepsilon(h_1, \ldots, h_n)$ is defined by

$$f(x_1 + h_1, \ldots, x_n + h_n) - f(x_1, \ldots, x_n) =$$
$$\sum_{i=1}^{n} f_i(x_1, \ldots, x_n)h_i + \varepsilon(|h_1| + \cdots + |h_n|)$$

with $\varepsilon(0, \ldots, 0) = 0$, we have $\varepsilon \to 0$ when $|h_1| + \cdots + |h_n| \to 0$.

See also Exercise 12.4.3.

Exercises

12.4.1 Verify that the function $f(x, y) = \sqrt{|xy|}$ is continuous at $(0, 0)$.

12.4.2 Verify that for $h_1, \ldots, h_n$ real numbers

$$\sqrt{h_1^2 + \cdots + h_n^2} \le |h_1| + \cdots + |h_n| \le \sqrt{n}\sqrt{h_1^2 + \cdots + h_n^2}.$$

Use these inequalities to obtain another (equivalent) definition for *differentiability* of a function of n variables at a point $\mathbf{x_0}$ that indicates that $f(\mathbf{x}) - f(\mathbf{x_0})$ can be approximated near $\mathbf{x_0}$ by a linear function T in such a way that $|f(\mathbf{x}) - T(\mathbf{x})| \to 0$ as $\|\mathbf{x} - \mathbf{x_0}\|_1 \to 0$.

12.4.3 For $\mathbf{h} = (h_1, \ldots, h_n) \in \mathbb{R}^n$ write

$$\begin{aligned} \|\mathbf{h}\|_1 &= |h_1| + \cdots + |h_n| \\ \|\mathbf{h}\|_2 &= \|\mathbf{h}\| = \sqrt{h_1^2 + \cdots + h_n^2} \\ \|\mathbf{h}\|_\infty &= \max(|h_1|, \ldots, |h_n|). \end{aligned}$$

Show that the following four conditions are equivalent:

$$h_i \to 0 \text{ for all } i = 1, \dots, n,$$
$$\|\mathbf{h}\|_1 \to 0, \quad \|\mathbf{h}\|_2 \to 0, \quad \|\mathbf{h}\|_\infty \to 0$$

Provide yet another definition of *differentiability* using $\|\mathbf{h}\|_\infty$ in place of $\|\mathbf{h}\|_1$ or $\|\mathbf{h}\|_2$.

12.4.4 Show that

$$f(x,y) = \begin{cases} \frac{x^3 - xy^2}{x^2+y^2} & \text{if } (x,y) \neq (0,0) \\ 0, & \text{if } (x,y) = (0,0) \end{cases}$$

is continuous and has first-order partial derivatives on $\mathbb{R}^2$, but is not differentiable at $(0,0)$.

12.4.5 Some authors avoid the use of partial derivatives in defining differentiability. Definition 12.16 then takes the following form.

Definition. Let f be defined on an open set $D \subset \mathbb{R}^n$ and let

$$\mathbf{x} = (x_1, \dots, x_n) \in D.$$

Then f is differentiable at $\mathbf{x}$ if there exists a linear function $L : \mathbb{R}^n \to \mathbb{R}$ such that when ε is defined by

$$f(x_1 + h_1, \dots, x_n + h_n) - f(x_1, \dots, x_n) =$$
$$\sum_{i=1}^{n} L(\mathbf{x})h_i + \varepsilon\sqrt{(|h_1|^2 + \cdots + |h_n|^2)}$$

with $\varepsilon(\mathbf{0}) = 0$, we have $\varepsilon \to 0$ when $\sqrt{(|h_1|^2 + \cdots + |h_n|^2)} \to 0$. Prove that this definition is equivalent to Definition 12.16.

12.4.3 Differentiability and Continuity

We now show that differentiable functions are continuous. We first consider the case $n = 2$. We do this because the essentials of a proof are already contained in the special case $n = 2$, while the notation is simpler in $\mathbb{R}^2$. The relevant pictures that go with the proof are easier to visualize.

Theorem 12.17 *If f is differentiable at (x_0, y_0), then f is continuous at (x_0, y_0).*

Proof We must prove that

$$\lim_{(h,k)\to(0,0)} f(x_0 + h, y_0 + k) = f(x_0, y_0).$$

Since f is differentiable at (x_0, y_0), we can write

$$f(x_0 + h, y_0 + k) - f(x_0, y_0) = f_1(x_0, y_0)h + f_2(x_0, y_0)k + \varepsilon(|h| + |k|),$$

where $\varepsilon \to 0$ as $(h,k) \to (0,0)$. We obtain from the triangle inequality that

$$|f(x_0+h, y_0+k) - f(x_0,y_0)| \leq \qquad (28)$$
$$|f_1(x_0,y_0)||h| + |f_2(x_0,y_0)||k| + \varepsilon(|h|+|k|)$$

Now $f_1(x_0,y_0)$ and $f_2(x_0,y_0)$ are fixed numbers, so the right side of the inequality (28) approaches 0 as $(h,k) \to (0,0)$. Thus the left side also approaches 0. But that is an equivalent formulation of the requirement

$$\lim_{(h,k)\to(0,0)} f(x_0+h, y_0+k) = f(x_0,y_0).$$

■

Note. Observe that we needed only finiteness of f_1 and f_2 at (x_0,y_0) and that $\varepsilon(h,k)(|h|+|k|) \to 0$ as $(h,k) \to (0,0)$ to infer that the right side of (28) approaches 0 as $(h,k) \to (0,0)$. Differentiability gave us more than we needed, namely that $\varepsilon(h,k) \to 0$ as $(h,k) \to (0,0)$. (See Exercise 12.4.7.)

The proof of the generalization to $\mathbb{R}^n$ is essentially the same. We leave this as Exercise 12.4.6.

Exercises

12.4.6 Prove that if f is defined in a neighborhood of a point $\mathbf{x} \in \mathbb{R}^n$ and differentiable at $\mathbf{x}$, then f is continuous at $\mathbf{x}$.

12.4.7 Give an example of a function $f : \mathbb{R}^2 \to \mathbb{R}$ such that

$$|f(h,k) - f(0,0)| = f_1(0,0)h + f_2(0,0)k + \varepsilon(|h|+|k|)$$

where $\varepsilon(|h|+|k|) \to 0$ as $(h,k) \to (0,0)$ but $\varepsilon \not\to 0$ as $(h,k) \to (0,0)$.

12.4.8 Give an example of a differentiable function of two variables whose partials are discontinuous at $(0,0)$.

12.4.4 Directional Derivatives

In Example 12.15 we saw that the function $f(x,y) = \sqrt{|xy|}$ had partial derivatives equal to 0 at $(0,0)$, but was not differentiable. We identified a problem in one direction. The function does not have a directional derivative in that direction (Exercise 12.4.11). We next see that a differentiable function has directional derivatives in all directions and that, moreover, all directional derivatives may be computed from the partials by a simple formula.

Theorem 12.18 *Let f be differentiable at $(x_0,y_0) \in \mathbb{R}^2$ and let $\mathbf{u} = (u_1,u_2)$ be any unit vector. Then $D_{\mathbf{u}}f(x_0,y_0)$ exists and*

$$D_{\mathbf{u}}f(x_0,y_0) = f_1(x_0,y_0)u_1 + f_2(x_0,y_0)u_2. \qquad (29)$$

Proof Since f is differentiable at (x_0, y_0) we can write

$$\begin{aligned}f(x_0+h, y_0+k) - f(x_0, y_0) &= \\ &f_1(x_0, y_0)h + f_2(x_0, y_0)k + \varepsilon(|h|+|k|),\end{aligned}$$

where $\varepsilon \to 0$ as $(h, k) \to (0, 0)$. Thus

$$\begin{aligned}f(x_0+tu_1, y_0+tu_2) - f(x_0, y_0) &= \\ &f_1(x_0, y_0)tu_1 + f_2(x_0, y_0)tu_2 + \varepsilon(|tu_1|+|tu_2|)\end{aligned} \tag{30}$$

and $\varepsilon \to 0$ as $t \to 0$.

Dividing both sides of (30) by t and letting $t \to 0$, we obtain

$$\begin{aligned}\lim_{t\to 0} \frac{f(x_0+tu_1, y_0+tu_2) - f(x_0, y_0)}{t} &= \\ \lim_{t\to 0}(f_1(x_0, y_0)u_1 + f_2(x_0, y_0)u_2 + \varepsilon\frac{|t|}{t}(|u_1|+|u_2|)).\end{aligned}$$

Now, $|u_1|+|u_2|$ is a constant, $\left|\frac{|t|}{t}\right| = 1$, and $\varepsilon \to 0$ as $t \to 0$, so the limit equals $f_1(x_0, y_0)u_1 + f(x_0, y_0)u_2$ as required. ■

Theorem 12.18 is valid for functions on $\mathbb{R}^n$ for every n, the proof being similar. We leave this proof as Exercise 12.4.12.

The Gradient Observe that, for a differentiable function f, the identity

$$D_{\mathbf{u}}f(x_0, y_0) = f_1(x_0, y_0)u_1 + f_2(x_0, y_0)u_2$$

can be written in the form

$$D_{\mathbf{u}}f(x_0, y_0) = (f_1(x_0, y_0), f_2(x_0, y_0)) \cdot (u_1, u_2).$$

The vector (f_1, f_2) or $(\frac{\partial f}{\partial x}, \frac{\partial f}{\partial y})$, where the partials are evaluated at (x_0, y_0), is of sufficient importance to have a name. It is called the *gradient* of f at (x_0, y_0) and denoted by

$$\textbf{grad}\ f(x_0, y_0) \quad \text{or} \quad \nabla f(x_0, y_0).$$

By the law of cosines (or Exercise 11.2.4) we can write

$$\begin{aligned}D_{\mathbf{u}}f(x_0, y_0) &= (\nabla f(x_0, y_0)) \cdot \mathbf{u} \\ &= |\nabla f(x_0, y_0)|\, \|\mathbf{u}\| \cos\theta \\ &= |\nabla f(x_0, y_0)| \cos\theta\end{aligned}$$

where θ is the angle between the vectors $\nabla f(x_0, y_0)$ and $\mathbf{u}$. [Here we must have $\nabla f(x_0, y_0) \neq 0$, otherwise θ is not defined.] Thus the maximum rate of change of f is in the direction corresponding to $\theta = 0$ (where $\cos\theta$ achieves its maximum). This occurs when $\mathbf{u}$ and $\nabla f(x_0, y_0)$ have the same direction. The magnitude of the rate of increase of f in this direction is $|\nabla f(x_0, y_0)|$.

For reference we state this as a theorem, expressed for functions of several variables.

Theorem 12.19 *If $f : \mathbb{R}^n \to \mathbb{R}$ is a differentiable function on an open set D, then at each point $\mathbf{x}$ in D at which the gradient of f does not vanish the maximum rate of change of f is in the direction $\nabla f(\mathbf{x})$ and the magnitude of the rate of increase of f in this direction is $|\nabla f(\mathbf{x})|$.*

Exercises 12.4.17 and 12.4.18 show that the preceding discussion might fail if we weaken the assumption of differentiability to the mere existence of the partials at (x_0, y_0).

Exercises

12.4.9 What are the directional derivatives for a function $f : \mathbb{R}^n \to \mathbb{R}$ if $n = 1$?

12.4.10 Is it still true that the gradient is in the direction of greatest change for the function if $f : \mathbb{R}^n \to \mathbb{R}$ where $n = 1$?

12.4.11 Verify that the function $f(x, y) = \sqrt{|xy|}$ does not have a directional derivative in the direction $(\sqrt{2}/2, \sqrt{2}/2)$ at the point $(0, 0)$.

12.4.12 State and prove the extension of Theorem 12.18 to functions of n variables.

12.4.13 Verify that the function

$$f(x, y) = \frac{x^2 y}{x^4 + y^2}, \quad f(0, 0) = 0$$

is not differentiable at $(0, 0)$ and yet has directional derivatives in every direction.

12.4.5 An Example

We have already seen that a continuous function can possess partial derivatives at a point without being differentiable at that point. More remarkably, a continuous function can possess directional derivatives *in every direction* and still be nondifferentiable. Exercise 12.4.13 has exhibited a *discontinuous* function that possesses all directional derivatives; since it is not continuous it cannot be differentiable. In this section we shall describe geometrically a *continuous* function with these properties. The picture we'll describe should be instructive, while a computational analysis would be a bit tedious. (See Exercise 12.4.14.)

Example 12.20 We build the function f in stages. First we let $f(x, 0) = x$ for all $x \in \mathbb{R}$. This will guarantee that $f_1(0, 0) = 1$. Next, for $|y| \geq x^2$, we define $f(x, y) = 0$. This will guarantee that all other directional derivatives are 0 at $(0, 0)$.

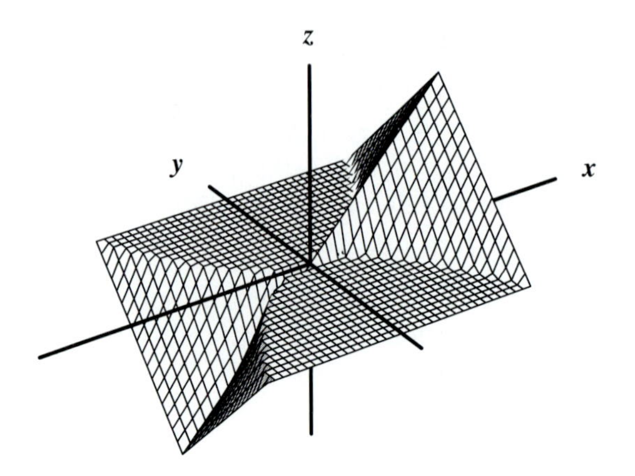

Figure 12.5. A continuous, nondifferentiable function with directional derivatives in every direction.

To see this, we need only observe that any line through $(0,0)$, other than the x-axis, intersects the set

$$\{(x,y) : |y| \geq x^2\}$$

on a line segment containing $(0,0)$.

Finally, on the set

$$\{(x,y) : |y| < x^2\}$$

we define f in any manner that makes f continuous on $\mathbb{R}^2$. For example, f can be linear on each vertical segment joining a point of the parabola $y = x^2$ or $y = -x^2$ to the x-axis, with the value 0 on the parabola and x on the x-axis. Figure 12.5 illustrates.

Thus f has directional derivatives 0 in all directions except for

$$(u_1, u_2) = (1, 0)$$

or

$$(u_1, u_2) = (-1, 0).$$

If f were differentiable at $(0,0)$, we would have, by Theorem 12.18, that for any unit vector $\mathbf{u} = (u_1, u_2)$,

$$D_{\mathbf{u}}f(0,0) = f_1(0,0)u_1 + f_2(0,0)u_2 = 1 \cdot u_1 + 0 \cdot u_2 = u_1.$$

But we have seen that $D_{\mathbf{u}}f(0,0) = 0$ unless $u_2 = 0$, so the equality

$$D_{\mathbf{u}}f(0,0) = u_1$$

is valid only for $u_1 = 0$ or $u_2 = 0$. Since Theorem 12.18 does not give the correct directional derivatives, f is not differentiable at $(0,0)$. ◀

Observe that for the function we described, all directional derivatives except the ones in the direction of the positive or negative x-axis require the tangent plane to be the plane $z = 0$. But the tangent line that lies in the plane $y = 0$ does not lie in this plane. In short, there can be no tangent plane. See also Exercise 12.4.15.

Example 12.20 also shows that Theorem 12.18 is not valid if we drop the requirement that f be differentiable. Directional derivatives cannot always be computed from formula (29) if that assumption is dropped. (But see Exercise 12.4.15.)

Exercises

12.4.14 An analytic representation of the example we gave in the text (Example 12.20) can take the form

$$f(x) = \begin{cases} 0, & \text{if } |y| \geq x^2 \\ x, & \text{if } y = 0 \\ -\frac{1}{x}(y - x^2), & \text{if } 0 < y < x^2 \\ \frac{1}{x}(y + x^2), & \text{if } -x^2 < y < 0 \end{cases}$$

(a) Verify analytically that f is continuous and has directional derivative $D_{\mathbf{u}}f(0,0) = 0$, for all directions with the exception of $(1,0)$ and $(-1,0)$.

(b) Calculate the partial derivatives f_1 and f_2, and show they are discontinuous at $(0,0)$. [That f_1 and f_2 are discontinuous at $(0,0)$ follows also from Theorem 12.21, proved later.]

12.4.15 A careless student states "The functions in Example 12.20 and Exercise 12.4.14 fail to be differentiable at $(0,0)$ even though all derivatives exist at $(0,0)$ because the tangent lines at $(0,0)$ don't all lie in the same plane, so there is no tangent plane. If all tangent lines at a point exist and *do* lie in the same plane, then that plane must be the tangent plane." Is the second statement correct? That is, if $f : \mathbb{R}^2 \to \mathbb{R}$ is continuous on $\mathbb{R}^2$ with $f(0,0) = 0$ and *every* directional derivative at $(0,0)$ is zero, then f is differentiable at $(0,0)$ and the xy-plane is the tangent plane at $(0,0,0)$.

12.4.6 Sufficient Conditions for Differentiability

It is not always easy to prove that a function is differentiable directly from Definition 12.16. The same was true in dealing with functions of one variable. There we obtained general theorems that often simplified our task. We next present a theorem that can often be applied to show differentiability for functions of several variables.

Theorem 12.21 *Let f be defined in a neighborhood D of (x_0, y_0). Suppose one of the partial derivatives is defined on D and is continuous at (x_0, y_0),* *while the other partial is defined at least at* (x_0, y_0) *(and finite there). Then* f *is differentiable at* (x_0, y_0).

Proof Suppose f_1 is continuous at (x_0, y_0), the proof being similar if it is f_2 that is continuous there. We wish to show that we can write

$$f(x_0 + h, y_0 + k) - f(x_0, y_0) = \\ f_1(x_0, y_0)h + f_2(x_0, y_0)k + \varepsilon(|h| + |k|), \tag{31}$$

where $\varepsilon \to 0$ as $(h, k) \to 0$.

We shall express the left side of (31) as a sum of two terms. We apply the mean value theorem to one of the terms and use the continuity of f_1 at (x_0, y_0) to obtain an estimate of the first term on the right side of (31). We then estimate the second term using only the existence of f_2 at (x_0, y_0). Manipulating these estimates will give the desired result.

To begin, write

$$f(x_0 + h, y_0 + k) - f(x_0, y_0) = \\ [f(x_0 + h, y_0 + k) - f(x_0, y_0 + k)] + [f(x_0, y_0 + k) - f(x_0, y_0)]. \tag{32}$$

Applying the mean value theorem (Theorem 7.20) to the first bracketed term, we obtain a number x' between x_0 and $x_0 + h$ such that

$$f(x_0 + h, y_0 + k) - f(x_0, y_0 + k) = f_1(x', y_0 + k)h. \tag{33}$$

Now f_1 is continuous at (x_0, y_0) so

$$\lim_{x' \to x_0, k \to 0} f_1(x', y_0 + k) = f_1(x_0, y_0).$$

Thus we can write

$$f_1(x', y_0 + k) = f_1(x_0, y_0) + \varepsilon_1 \tag{34}$$

where $\varepsilon_1 \to 0$ as $x' \to x_0$ and $k \to 0$. Note that $x' \to x_0$ as $h \to 0$, since x' is between x_0 and $x_0 + h$.

We now consider the second bracketed term. Since f_2 is finite at (x_0, y_0), we can write for $k \neq 0$

$$\frac{f(x_0, y_0 + k) - f(x_0, y_0)}{k} = f_2(x_0, y_0) + \varepsilon_2$$

where $\varepsilon_2 \to 0$ as $k \to 0$. This we can write in the form

$$f(x_0, y_0 + k) - f(x_0, y_0) = f_2(x_0, y_0)k + \varepsilon_2 k. \tag{35}$$

Substituting (33) and (35) into (32) and then using (34), we arrive at the equality

$$f(x_0 + h, y_0 + k) - f(x_0, y_0) = \\ f_1(x_0, y_0)h + f_2(x_0, y_0)k + (\varepsilon_1 h + \varepsilon_2 k) \tag{36}$$

where both ε_1 and ε_2 approach 0 as $(h, k) \to (0, 0)$.

Comparing (36) with the desired form (31), we see that we must replace the term $(\varepsilon_1 h + \varepsilon_2 k)$ by a term of the form $\varepsilon(|h| + |k|)$. Letting

$$\varepsilon = \frac{\varepsilon_1 h + \varepsilon_2 k}{|h| + k|}$$

does the job. Indeed, with this value for ε,

$$\varepsilon_1 h + \varepsilon_2 k = \varepsilon(|h| + |k|),$$

so (36) reduces to (30). Furthermore, $\varepsilon \to 0$ as $(h, k) \to (0, 0)$. To see this observe that

$$|\varepsilon| = \left| \frac{\varepsilon_1 h}{|h| + |k|} + \frac{\varepsilon_2 k}{|h| + |k|} \right| \leq \left| \frac{\varepsilon_1 h}{|h| + |k|} \right| + \left| \frac{\varepsilon_2 k}{|h| + |k|} \right| \leq |\varepsilon_1| + |\varepsilon_2|.$$

Since $\varepsilon_1 \to 0$ and $\varepsilon_2 \to 0$ as $(h, k) \to (0, 0)$, it follows that $\varepsilon \to 0$. ∎

The conditions of Theorem 12.21 are often met and may be easier to verify than verifying differentiability directly from the definition. As before, the general case for functions of n variables can be obtained with a similar proof (Exercise 12.4.16).

Exercises

12.4.16 State and prove the extension of Theorem 12.21 to functions of n variables.

12.4.17 Consider a function f constructed so as to be continuous and such that (i) $f(x, y) = 0$ unless $x > 0$ and $x^2 < y < 3x^2$, (ii) for each $x > 0$, $f(x, 2x^2) = x$, and (iii) $0 \leq f(x, y) \leq x$ for all (x, y) with $x > 0$. Then all directional derivatives vanish at $(0, 0)$ (this was given in the hint to Exercise 12.4.15). Show that this function has no direction of maximum change at $(0, 0)$. Modify this example to obtain a function g with

$$g_1(0, 0) = 1 = g_2(0, 0)$$

yet there is no direction of maximum change.

12.4.18 Give an example of a continuous function $f : \mathbb{R}^2 \to \mathbb{R}$ having partial derivatives at $(0, 0)$ with

$$f_1(0, 0) \neq 0\ ,\ f_2(0, 0) \neq 0$$

but the vector $(f_1(0, 0), f_2(0, 0))$ does not point in the direction of maximal change, even though there is such a direction.

12.4.7 The Differential

Suppose f is differentiable at a point $(x_0, y_0) \in \mathbb{R}^2$. Then we can write

$$f(x_0 + h, y_0 + k) - f(x_0, y_0) =$$
$$f_1(x_0, y_0)h + f_2(x_0, y_0)k + \varepsilon(|h| + |k|) \tag{37}$$

where $\varepsilon \to 0$ as $(h, k) \to (0, 0)$.

Let us rewrite this in a form that may be more familiar from elementary calculus. Let $z = f(x, y)$. Write $h = \triangle x$, $k = \triangle y$, $\triangle f = \triangle z$. Then

$$\triangle f = \triangle z = f(x_0 + \triangle x, y_0 + \triangle y) - f(x_0, y_0)$$

represents the change in f (or z) that corresponds to a change in x and y given by $\triangle x$ and $\triangle y$. With this notation, (37) becomes

$$\triangle f = f_1(x_0, y_0)\triangle x + f_2(x_0, y_0)\triangle y + \varepsilon(|\triangle x| + |\triangle y|). \tag{38}$$

The term $\varepsilon(|\triangle x|+|\triangle y|)$ represents the error in estimating $\triangle z$, the change in z, by the change in the tangent plane at (x_0, y_0) corresponding to changes of $\triangle x$ and $\triangle y$, respectively. It has been customary historically, when obvious limits are involved, to use notation such as

$$df = f_1 dx + f_2 dy \quad \text{or} \quad dz = \frac{\partial z}{\partial x}dx + \frac{\partial z}{\partial y}y.$$

What do we mean by such notation? To be precise, *df* is a function depending on x, y, dx, and dy:

$$df = df(x, y, dx, dy).$$

(So x, y, dx, and dy are independent variables and *df* is a function of these four variables.) This function *df* is of sufficient importance to deserve a name.

Definition 12.22 Let f be differentiable at a point (x, y). The function *df* defined by

$$df(x, y, dx, dy) = f_1 dx + f_2 dy$$

is called the *differential of f at (x, y)*.

Let's look at a simple example to illustrate the concepts.

Example 12.23 Let $f(x, y) = x^2y^3$, $x = 3$, $y = 1$, $\triangle x = \triangle y = .01$. We'll compute $\triangle f$ and *df*, the approximation to $\triangle f$. Here $x + \triangle x = 3.01$ and $y + \triangle y = 1.01$, so

$$\triangle f = f(3.01, 1.01) - f(3, 1) = 9.3346 - 9 = .3346$$

to four decimal places. Our approximation, $df(3, 1, .01, .01)$, becomes

$$df = f_1(3, 1)\, dx + f_2(3, 1)\, dy = (6 \times .01) + (27 \times .01) = .33.$$

Thus the error in using the differential *df* for estimating the actual change $\triangle f$ is only about .0046. Note that this error is small in comparison with $|\triangle x| + |\triangle y| = .02$. ◀

Functions of Several Variables When dealing with more than two variables, the situation is similar.

Definition 12.24 Let f be differentiable at $\mathbf{x} = (x_1, \ldots, x_n) \in \mathbb{R}^n$. The *differential of f at* $\mathbf{x}$ is the function of $2n$ variables given by

$$df(x_1, \ldots x_n, h_1, \ldots, h_n) = f_1 h_1 + \cdots + f_n h_n \tag{39}$$

As before, it is sometimes convenient to write (39) in the form

$$df = \frac{\partial f}{\partial x_1} dx_1 + \cdots + \frac{\partial f}{\partial x_n} dx_n. \tag{40}$$

As is often true with alternative notations when dealing with derivatives, the notation (40) is suggestive of various formulas. For example, if u and v are differentiable at a point $\mathbf{x} \in \mathbb{R}^n$, then

$$\begin{aligned} d(u+v) &= du + dv \ , \ d(uv) = u\,dv + v\,du \,, \\ d\left(\frac{u}{v}\right) &= \frac{v\,du - u\,dv}{v^2} \quad (v \neq 0). \end{aligned} \tag{41}$$

To check the product formula, for example, note that by definition each of the differentials can be written as

$$\begin{aligned} du &= \frac{\partial u}{\partial x_1} dx_1 + \cdots + \frac{\partial u}{\partial x_n} dx_n, \\ dv &= \frac{\partial v}{\partial x_1} dx_1 + \cdots + \frac{\partial v}{\partial x_n} dx_n, \\ d(uv) &= \frac{\partial (uv)}{\partial x_1} dx_1 + \cdots + \frac{\partial (uv)}{\partial x_n} dx_n. \end{aligned} \tag{42}$$

But the product rule clearly is valid for partial derivatives. Thus (42) becomes

$$d(uv) = \left[u\frac{\partial v}{\partial x_1} + v\frac{\partial u}{\partial x_1}\right] dx_1 + \cdots + \left[u\frac{\partial v}{\partial x_n} + v\frac{\partial u}{\partial x_n}\right] dx_n =$$

$$u\left[\frac{\partial v}{\partial x_1} dx_1 + \cdots + \frac{\partial v}{\partial x_n} dx_n\right] + \cdots + v\left[\frac{\partial u}{\partial x_1} dx_1 + \cdots + \frac{\partial u}{\partial x_n} dx_n\right]$$

$$= u\,dv + v\,du.$$

Similarly, differentials of elementary functions are as expected as the example now shows.

Example 12.25 If $u : \mathbb{R}^n \to \mathbb{R}$ is differentiable and we are allowed, for the moment, to assume that e^u is also differentiable, then

$$\begin{aligned} d(e^u) &= \frac{\partial e^u}{\partial x_1}dx_1 + \cdots + \frac{\partial e^u}{\partial x_n}dx_n \\ &= e^u\frac{\partial u}{\partial x_1}dx_1 + \cdots + e^u\frac{\partial u}{\partial x_n}dx_n \\ &= e^u\left[\frac{\partial u}{\partial x_1}dx_1 + \cdots + \frac{\partial u}{\partial x_n}dx_n\right] \\ &= e^u du \end{aligned}$$

thus assuming a familiar form. (See Exercise 12.8.7 for a discussion of whether e^u is differentiable.) ◀

Exercises

12.4.19 Calculate $\triangle f$ and *df* in Example 12.23 when $\triangle x = .001$ and $\triangle y = .002$. Compare the resulting error with $|\triangle x| + |\triangle y| = .003$.

12.4.20 Verify the formulas in (41) for $d(u+v)$ and $d\left(\frac{u}{v}\right)$.

12.4.21 Since the definition of differential involves differentiability of the functions, the formulas (41) require the differentiability of the functions u and v, uv and u/v. Prove that when u and v are differentiable at a point $(x_0, y_0) \in \mathbb{R}^2$, then so too are their sum, product, and quotient. [For the quotient assume also that $v(x_0, y_0) \neq 0$].

12.5 Chain Rules

We saw in Example 12.25 that if we have a formula such as $z = e^u$, where u is a real-valued function of several variables, we can compute its differential $dz = e^u du$, as we would for functions of one real variable. We can view this as an instance of a *chain rule*. Actually there are many chain rules. They involve computing differentials or partial derivatives of functions defined from other functions via composition. We discuss such situations in this section and show how a chain is created and what the resulting chain rule should be. We also give an indication of why the chain rule should work, and then proceed to a formal proof.

12.5.1 Preliminary Discussion

We begin with three examples, discussing each of them informally. To keep this informal discussion simple, we avoid technicalities such as the domains of definition of the functions and the precise hypotheses needed for the resulting chain rules.

Example 12.26 Let $u = u(x, y)$ and let $z = F(u)$. We can view z as a function G of x and y via the intermediate variable u. Thus

$$z = F(u) = F(u(x, y)) = G(x, y).$$

The dependencies of the variables can be described as "z depend on u, and u depends on x and y" and expressed schematically:

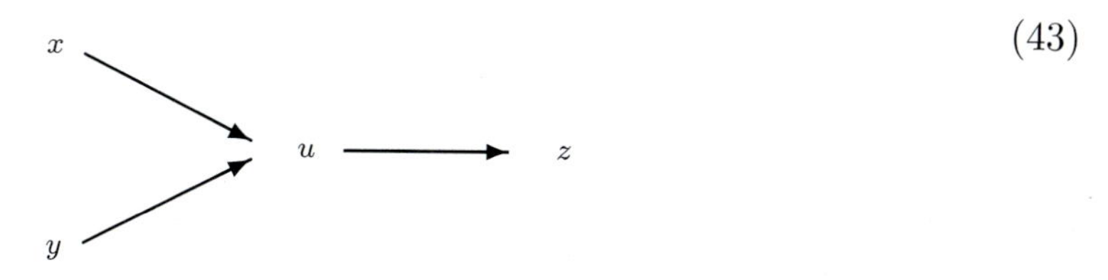

This has associated chain rules

$$\frac{\partial z}{\partial x} = \frac{\partial z}{\partial u}\frac{\partial u}{\partial x} \quad \text{and} \quad \frac{\partial z}{\partial y} = \frac{\partial z}{\partial u}\frac{\partial u}{\partial y} \tag{44}$$

that you may remember from calculus.

Ultimately z depends on the two variables x and y via the single variable u. Thus there are two partials to compute, $\frac{\partial z}{\partial x}$ and $\frac{\partial z}{\partial y}$, each involving a one-term chain as seen in (44). ◀

Example 12.27 For a concrete example, let $z = \sin(x^2 + y^3) = G(x, y)$, and write $u(x, y) = x^2 + y^3$ and $F(u) = \sin u$. This corresponds to the schema in (43). We would then conclude, using the chain rule (44), that

$$\frac{\partial z}{\partial x} = 2x\cos(x^2 + y^3) \quad \text{and} \quad \frac{\partial z}{\partial y} = 3y^2\cos(x^2 + y^3).$$

◀

Let us complicate things a bit.

Example 12.28 Let $x = f(t)$, $y = g(t)$, $z = F(x, y)$. Here we can view z as a function of t:

$$z = F(x, y) = F(f(t), g(t)) = G(t).$$

The dependencies of the variables can be described as "z depends on x and y, each of which depends on t" and expressed schematically:

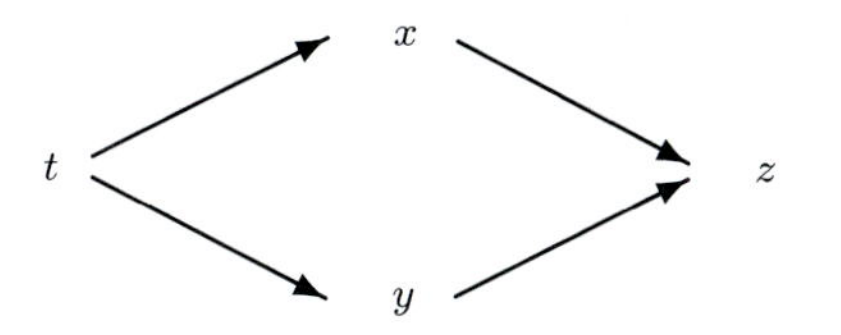

with associated chain rule

$$G'(t) = \frac{dz}{dt} = \frac{\partial z}{\partial x}\frac{dx}{dt} + \frac{\partial z}{\partial y}\frac{dy}{dt}. \tag{45}$$

Note that x and y are functions of t alone, as is G, so ordinary derivatives are involved. On the other hand, $F(x, y)$ is a function of two variables, so

$$\frac{\partial z}{\partial x} = F_1 \text{ and } \frac{\partial z}{\partial y} = F_2$$

are *partial* derivatives. In contrast with Example 12.26, there is now only one derivative we wish to calculate, $G'(t) = \frac{dz}{dt}$, but the chain rule involves *two* terms to be added, one term arising from each "path from t to z." ◀

Example 12.29 Again, for a concrete example, let

$$x = t^2, \ y = t^3, \text{ and } \ z = xy.$$

Then from (45),

$$\frac{dz}{dt} = y2t + x3t^2 = 2t^4 + 3t^4 = 5t^4,$$

as expected when we observe that $G(t) = t^5$. ◀

Let's complicate matters a bit more.

Example 12.30 Let $z = F(x, y)$, $x = f(s, t)$, $y = g(s, t)$. We can view z as a function of s and t via x and y:

$$z = F(x(s, t), y(s, t)) = G(s, t).$$

The dependencies of the variables can be described as "z depends on both x and y, and each in turn depends on both s and t" and expressed schematically:

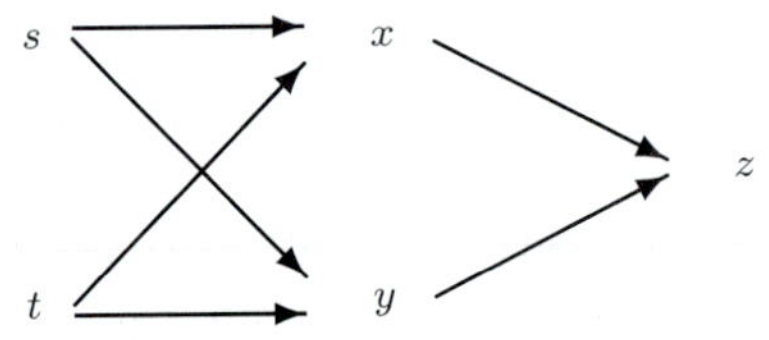

To compute $\frac{\partial z}{\partial s}$ we follow each path from s to z. Thus

$$\frac{\partial z}{\partial s} = \frac{\partial z}{\partial x}\frac{\partial x}{\partial s} + \frac{\partial z}{\partial y}\frac{\partial y}{\partial s} \tag{46}$$

and, similarly, following each path from t to z,

$$\frac{\partial z}{\partial t} = \frac{\partial z}{\partial x}\frac{\partial x}{\partial t} + \frac{\partial z}{\partial y}\frac{\partial y}{\partial t}.$$

◀

Example 12.31 Let

$$z = F(x,y) = x^2 + y^3,$$

$x = f(s,t) = st$, and $y = g(s,t) = e^{st}$. Then the chain rule in (46) gives

$$\frac{\partial z}{\partial s} = 2xt + 3y^2te^{st} = 2st^2 + 3e^{2st}te^{st} + 2st^2 + 3te^{3st}.$$

Again, we can check by observing that $G(s,t) = (st)^2 + e^{3st}$ and performing the straightforward calculations. ◀

Many other chains are possible, of course.

Example 12.32 The schema

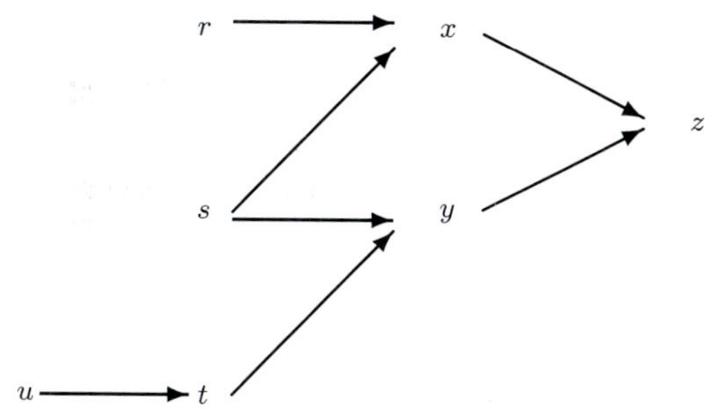

would lead to the chain rules

$$\frac{\partial z}{\partial r} = \frac{\partial z}{\partial x}\frac{\partial x}{\partial r}$$
$$\frac{\partial z}{\partial s} = \frac{\partial z}{\partial x}\frac{\partial x}{\partial s} + \frac{\partial z}{\partial y}\frac{\partial y}{\partial s}$$
$$\frac{\partial z}{\partial u} = \frac{\partial z}{\partial y}\frac{\partial y}{\partial t}\frac{\partial t}{\partial u}.$$

You should invent a concrete example of this schema and test out the chain rule for it. ◀

Exercises

12.5.1 Invent a concrete example to illustrate the chain rule for Example 12.32 and verify by direct computation.

12.5.2 Suppose that the dependencies of the variables can be described informally as "w depends on all three of x, y, and z and each in turn depends on both s and t." Express this schematically and write a chain rule for it.

12.5.3 Write chain rules for $\frac{\partial z}{\partial t}$ and $\frac{\partial z}{\partial s}$ that relate to the schema

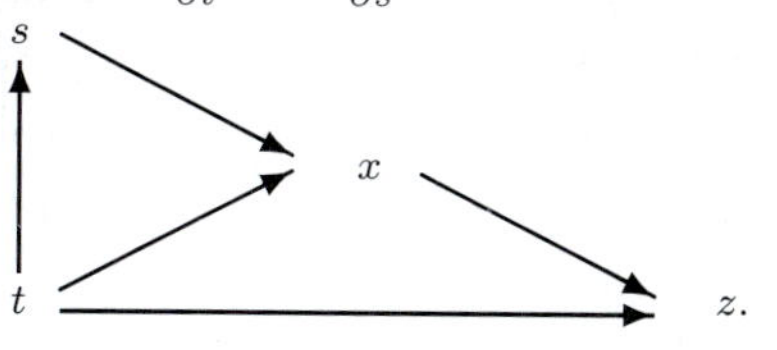

12.5.4 Let $z = F(x, y)$, $x = g(u, v, w)$, and $y = h(u, v)$.

(a) Make a path schema to show the dependencies among the variables z, x, y, u, v, and w.

(b) Write chain rules for $\frac{\partial z}{\partial u}$, $\frac{\partial z}{\partial v}$, and $\frac{\partial z}{\partial w}$.

12.5.2 Informal Proof of a Chain Rule

Let's try to see, informally, why a chain rule works. Consider, for example, the schema

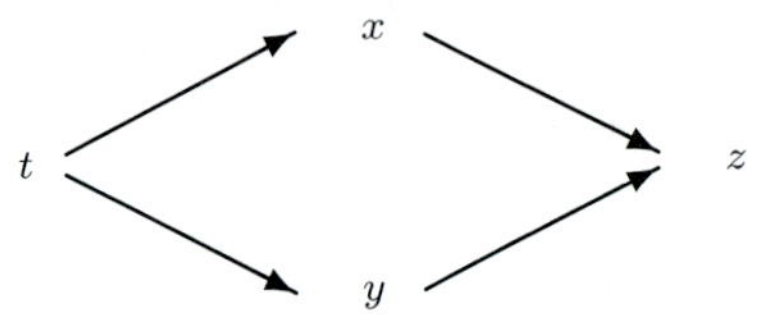

and its chain rule

$$\frac{dz}{dt} = \frac{\partial z}{\partial x}\frac{dx}{dt} + \frac{\partial z}{\partial y}\frac{dy}{dt}.$$

Assuming the needed differentiability requirements and using obvious notation, we can write

$$\triangle z = f_1 \triangle x + f_2 \triangle y + \varepsilon(|\triangle x| + |\triangle y|) \tag{47}$$

where $\varepsilon \to 0$ as $(\triangle x, \triangle y) \to (0, 0)$. Dividing by $\triangle t$, we get

$$\frac{\triangle z}{\triangle t} = f_1 \frac{\triangle x}{\triangle t} + f_2 \frac{\triangle y}{\triangle t} + \varepsilon \left(\frac{|\triangle x|}{\triangle t} + \frac{|\triangle y|}{\triangle t} \right). \tag{48}$$

Letting $\triangle t \to 0$ in (48), we note that since $\triangle x \to 0$ and $\triangle y \to 0$ as $\triangle t \to 0$, $\varepsilon \to 0$ as $\triangle t \to 0$.

Now consider the term

$$\varepsilon \left(\frac{|\triangle x|}{\triangle t} + \frac{|\triangle y|}{\triangle t} \right)$$

as $\triangle t \to 0$. The term in parentheses approaches

$$\varepsilon \left(\left| \frac{dx}{dt} \right| + \left| \frac{dy}{dt} \right| \right) \quad \text{when } \triangle t \to 0+ \text{ and}$$

$$-\varepsilon \left(\left| \frac{dx}{dt} \right| + \left| \frac{dy}{dt} \right| \right) \quad \text{when } \triangle t \to 0- .$$

Since $\varepsilon \to 0$, the product approaches 0 as $\triangle t \to 0$ from either side. Thus

$$\frac{dz}{dt} = \lim_{\triangle t \to 0} \frac{\triangle z}{\triangle t} = f_1 \frac{dx}{dt} + f_2 \frac{dy}{dt} + 0 = \frac{\partial z}{\partial x}\frac{dx}{dt} + \frac{\partial z}{\partial y}\frac{dy}{dt}.$$

You will note that this development presents merely the idea of a proof. We haven't spelled out various hypotheses. For example, (47) requires differentiability of z (as a function of x and y). At this point, we are trying only to set things up so we can proceed with rigorous proofs of some chain rules in the next subsection. There we shall prove a chain rule that covers Example 12.30. Proofs of other chain rules would follow similar patterns.

Exercises

12.5.5 Write up an informal proof of a different chain rule than the one here.

12.5.3 Notation of Chain Rules

A word of caution about notation is needed. Chains such as

$$\frac{\partial z}{\partial s} = \frac{\partial z}{\partial x}\frac{\partial x}{\partial s} + \frac{\partial z}{\partial y}\frac{\partial y}{\partial s}$$

are convenient to write using symbols such as $\frac{\partial z}{\partial x}$ and $\frac{\partial z}{\partial y}$ because the notation is familiar and suggests various "cancellations:" Thus we are tempted to say

$$\frac{\partial z}{\partial x}\frac{\partial x}{\partial s} \text{ “looks like” } \frac{\partial z}{\partial s}.$$

But care must be taken when variables appear at different levels.

Consider the schema

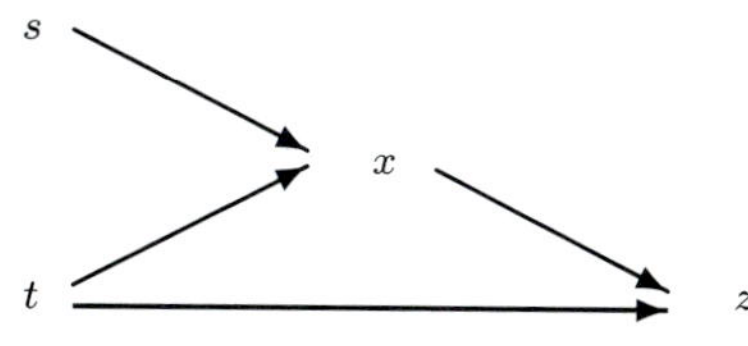

and its associated chain rule

$$\frac{\partial z}{\partial t} = \frac{\partial z}{\partial t} + \frac{\partial z}{\partial x}\frac{\partial x}{\partial t}. \tag{49}$$

The symbol $\frac{\partial z}{\partial t}$ has two different meanings here, one on the left side of (49), the other on the right side. Let's sort this out with a concrete example.

Example 12.33 Let $z = te^{st}$. If we compute $\frac{\partial z}{\partial t}$ directly,we obtain from the product rule that $\frac{\partial z}{\partial t} = e^{st} + ste^{st}$. Now let's view z as a composite function. Let $x = e^{st}$, so $z = tx$. From the chain rule (49) we calculate

$$\begin{aligned}\frac{\partial z}{\partial t} &= \frac{\partial z}{\partial t} + \frac{\partial z}{\partial x}\frac{\partial x}{\partial t} \\ &= x + t\frac{\partial x}{\partial t} = e^{st} + tse^{st}\end{aligned}$$

as before. The first occurrence of $\frac{\partial z}{\partial t}$ is what we are after. It represents the partial of z with respect to t when we view z as a function of s and t. The second occurrence represents the partial of z with respect to t when we view z as a function of x and t. ◀

If a complicated schema causes confusion, it would be preferable to use other notation that avoids ambiguity. We can avoid the ambiguous notation in (49) by introducing an additional variable $y = t$. The schema then becomes

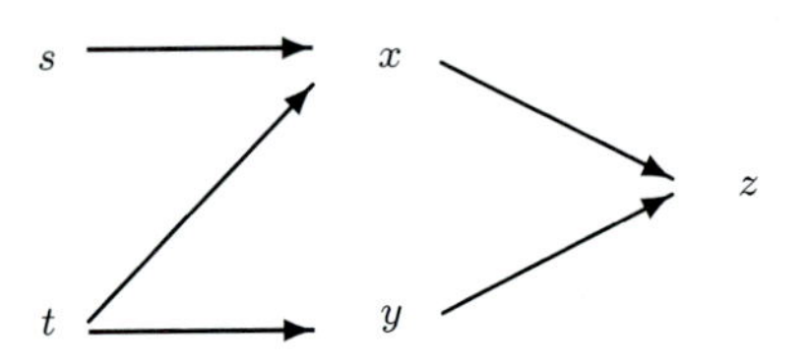

and the chain becomes

$$\frac{\partial z}{\partial t} = \frac{\partial z}{\partial x}\frac{\partial x}{\partial t} + \frac{\partial z}{\partial y}\frac{\partial y}{\partial t}$$

where

$$\frac{\partial y}{\partial t} = \frac{dy}{dt} = 1$$

since $y = t$. In this expression $\frac{\partial z}{\partial y}$ eliminates use of $\frac{\partial z}{\partial t}$ in an ambiguous manner. Or we could use unambiguous notation by writing

$$z = F(x,t) \ , \ x = f(s,t),$$

so $z = F(f(s,t),t)$. Then

$$\frac{\partial z}{\partial t} = F_1(f(s,t),t)f_2(s,t) + F_2(f(s,t),t).$$

Exercises

12.5.6 Suppose that $z = f(x,y)$ and $y = e^x$. Then by the chain rule

$$\frac{\partial z}{\partial x} = \frac{\partial z}{\partial x}\frac{\partial x}{\partial x} + \frac{\partial z}{\partial y}\frac{\partial y}{\partial x} = \frac{\partial z}{\partial x} + e^x\frac{\partial z}{\partial y}.$$

A careless student suggests that we cancel $\frac{\partial z}{\partial x}$ on both sides and get $e^x\frac{\partial z}{\partial y} = 0$ so $\frac{\partial z}{\partial y} = 0$. Do you agree with this? If not, what is the correct computation of $\frac{\partial z}{\partial y}$ and what are the appropriate hypotheses you need?

12.5.7 Let F be differentiable on $\mathbb{R}^2$. Let $x = r\cos\theta$, $y = r\sin\theta$ and let

$$G(r,\theta) = F(x,y) = F(r\cos\theta, r\sin\theta).$$

Thus F is transformed into G when we transform rectangular coordinates into polar coordinates. Show that

$$\frac{\partial G}{\partial r} = \frac{\partial F}{\partial x}\cos\theta + \frac{\partial F}{\partial y}\sin\theta$$

and

$$\frac{\partial G}{\partial \theta} = -\frac{\partial F}{\partial x} r\sin\theta + \frac{\partial F}{\partial y} r\cos\theta.$$

12.5.8 Consider a rectangle with horizontal side x and vertical side y. Its area is given by the formula $A(x,y) = xy$. Its perimeter is $P(x,y) = 2x + 2y$. We readily compute

$$\frac{\partial A}{\partial x} = y. \tag{50}$$

Now let us view A as a function of x and P.

$$A(x,P) = x\left(\frac{P}{2} - x\right) = \frac{Px}{2} - x^2.$$

Here $\frac{\partial A}{\partial x} = \frac{P}{2} - 2x$, which in terms of x and y gives

$$\frac{\partial A}{\partial x} = y - x. \tag{51}$$

We see that (50) and (51) don't agree.

(a) Explain the apparent discrepancy.

(b) Obtain (50) by viewing A as a function of x and y via the intermediate variable P.

12.5.4 Proofs of Chain Rules (I)

In Section 12.5.2 we provided a nonrigorous indication why chain rules work. We considered in Example 12.28 the case corresponding to the chain

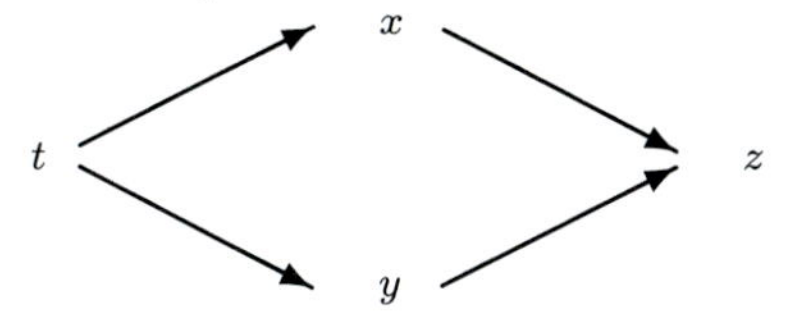

We now give a precise statement and proof of this chain rule that can serve as a model for how to proceed in general.

Theorem 12.34 *Let f and g be real-valued functions defined on a neighborhood of $t_0 \in \mathbb{R}$. Suppose both f and g are differentiable at t_0. Let F be a real-valued function defined on a neighborhood of $(x_0, y_0) \in \mathbb{R}^2$, where $x_0 = f(t_0)$ and $y_0 = g(t_0)$. Suppose F is differentiable at (x_0, y_0). Let $G(t) = F(f(t), g(t))$. Then G is differentiable at t_0 and*

$$G'(t_0) = F_1(x_0, y_0)f'(t_0) + F_2(x_0, y_0)g'(t_0).$$

Proof Write $x = f(t)$, $y = g(t)$. Then

$$\begin{aligned} G(t) - G(t_0) &= F(f(t), g(t)) - F(f(t_0), g(t_0)) \\ &= F(x, y) - F(x_0, y_0). \end{aligned}$$

Since F is differentiable at (x_0, y_0), we can write

$$\begin{aligned} \frac{G(t) - G(t_0)}{t - t_0} &= \frac{F(x, y) - F(x_0, y_0)}{t - t_0} \\ &= F_1(x_0, y_0)\frac{(x - x_0)}{t - t_0} + F_2(x_0, y_0)\frac{(y - y_0)}{t - t_0} \\ &+\varepsilon(x, y)\left(\frac{|x - x_0|}{t - t_0} + \frac{|y - y_0|}{t - t_0}\right) \end{aligned} \tag{52}$$

where

$$\varepsilon(x, y) \to 0 \text{ as } |x - x_0| \to 0 \text{ and } |y - y_0| \to 0. \tag{53}$$

We now let $t \to t_0$ and consider the three terms in lines 2 and 3 of (52). Since f and g are differentiable at t_0, they are continuous there, so

$$f(t) \to f(t_0) \text{ and } g(t) \to g(t_0) \text{ as } t \to t_0.$$

Thus

$$x \to x_0 \text{ and } y \to y_0 \text{ as } t \to t_0. \tag{54}$$

We also have

$$\frac{x - x_0}{t - t_0} = \frac{f(t) - f(t_0)}{t - t_0} \to f'(t_0) \text{ as } t \to t_0 \text{ and} \tag{55}$$

$$\frac{y - y_0}{t - t_0} = \frac{g(t) - g(t_0)}{t - t_0} \to g'(t_0) \text{ as } t \to t_0. \tag{56}$$

The last term of (52) is a product of two terms. The first term is $\varepsilon(x, y)$, which approaches zero as $t \to t_0$ [by (53) and (54)]. the second term is

$$\frac{|x - x_0|}{t - t_0} + \frac{|y - y_0|}{t - t_0},$$

which in absolute value is less than or equal to

$$\left|\frac{x - x_0}{t - t_0}\right| + \left|\frac{y - y_0}{t - t_0}\right|.$$

This last expression approaches $|f'(t_0)| + |g'(t_0)|$, so

$$\varepsilon(x, y)\left(\frac{|x - x_0|}{t - t_0} + \frac{|y - y_0|}{t - t_0}\right) \to 0 \text{ as } t \to t_0. \tag{57}$$

Thus, we see from (52), (55), (56), and (57) that

$$\lim_{t\to t_0} \frac{G(t)-G(t_0)}{t-t_0} = F_1(x_0,y_0)f'(t_0) + F_2(x_0,y_0)g'(t_0)$$

as was to be proved. ■

Exercises

12.5.9 Formulate and prove a version of Theorem 12.34 that would apply to a function $G(t) = F(f(t), g(t), h(t))$ and conclude that G is differentiable at t_0.

12.5.5 Mean Value Theorem

Theorem 12.34 allows us to obtain a two-dimensional analogue of the classical mean value theorem. Recall that the mean value theorem asserts, under appropriate differentiability assumptions on a function f, that

$$f(x_0+h) - f(x_0) = f'(\xi_0)h$$

for some ξ_0 between x_0+h and x_0.

Theorem 12.35 *Let F be defined on an open set $D \subset \mathbb{R}^2$. Let (x_0, y_0) and (x_0+h, y_0+k) be points in D, and suppose the line segment L determined by these points lies in D. Suppose F is differentiable on L. Then there exist ξ_0 between x_0 and x_0+h and η_0 between y_0 and y_0+k such that*

$$F(x_0+h, y_0+k) - F(x_0,y_0) = F_1(\xi_0,\eta_0)h + F_2(\xi_0,\eta_0)k. \tag{58}$$

Proof We begin by expressing the line segment L parametrically:

$$x = x_0 + th\,,\ \ y = y_0 + tk \quad (0 \le t \le 1).$$

We thus can view the Function F on L as a function of t on $[0,1]$. Let

$$G(t) = F(x_0+th, y_0+tk)\,, \quad (0 \le t \le 1). \tag{59}$$

By Theorem 12.34

$$G'(t) = hF_1(x_0+th, y_0+tk) + kF_2(x_0+th, y_0+tk). \tag{60}$$

By the mean value theorem (Theorem 7.20) for functions of one variable we obtain

$$G(1) - G(0) = G'(t_0) \ \text{ for some } \ t_0 \in (0,1).$$

Now, we see from (59) that $G(0) = F(x_0,y_0)$,

$$G(1) = F(x_0+h, y_0+k),$$

so

$$G'(t_0) = F(x_0+h, y_0+k) - F(x_0,y_0). \tag{61}$$

But from (60) we have

$$G'(t_0) = hF_1(x_0 + t_0h, y_0 + t_0k) + kF_2(x_0 + t_0h, y_0 + t_0k). \qquad (62)$$

Combining (61) and (62) we get

$$F(x_0 + h, y_0 + k) - F(x_0, y_0) = \\ hF_1(x_0 + t_0h, y_0 + t_0k) + kF_2(x_0 + t_0h, y_0 + t_0k).$$

To obtain (58), we simply let $\xi_0 = x_0 + t_0h$ and $\eta_0 = y_0 + t_0k$. ■

Exercises

12.5.10 State the mean value theorem (Theorem 12.35) in a form that gives the conclusion

$$F(P_2) - F(P_1) = \nabla F(P_0) \cdot (P_1 - P_2)$$

where P_1, P_2, and P_0 are points in $\mathbb{R}^2$ and ∇ is the gradient.

12.5.6 Proofs of Chain Rules (II)

We now turn to a statement and proof of a chain rule that corresponds to the schema in Example 12.30.

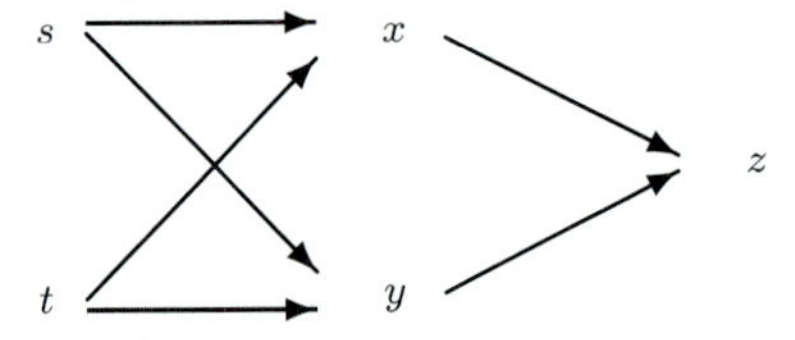

Theorem 12.36 *Let F be defined on an open set $D \subset \mathbb{R}^2$ and let (x_0, y_0) be a point in D. Suppose F is differentiable at (x_0, y_0). Let f and g be defined in a neighborhood of the point $(s_0, t_0) \in \mathbb{R}^2$. Suppose f and g have first partial derivatives at (s_0, t_0) and that $x_0 = f(s_0, t_0)$, $y_0 = g(s_0, t_0)$. Define G by $G(s, t) = F(f(s, t), g(s, t))$. Then G has first partial derivatives at (s_0, t_0) and*

$$\left.\begin{array}{lcl} G_1(s_0, t_0) & = & F_1(x_0, y_0)f_1(s_0, t_0) + F_2(x_0, y_0)g_1(s_0, t_0) \\ G_2(s_0, t_0) & = & F_1(x_0, y_0)f_2(s_0, t_0) + F_2(x_0, y_0)g_2(s_0, t_0) \end{array}\right\}. \qquad (63)$$

Before proving Theorem 12.36 we make several observations.

1. You should check that formulas (63) are just precise versions of the schema in Example 12.30. In particular, formulas (63) make it clear where the partials are evaluated.

2. In Theorem 12.34 we concluded differentiability of the function G. Here, we don't. Nor do we assume differentiability of f and g here. Theorem 12.36 is simply a theorem about first partial derivatives of G, their existence, and a formula for them. [It is true that if we assumed

differentiability of f and g at (s_0, t_0), then we could have concluded that G is differentiable, but we shall not prove this until Section 12.8.]

3. Note the similarity between the approaches to the proofs of Theorems 12.34 and 12.36.

Proof We shall establish the first of the rules (63), the proof of the second being similar (Exercise 12.5.11).

To compute $G_1(s_0, t_0)$, write $x = f(s, t_0)$, $y = g(s, t_0)$. Then by definition of G,

$$G(s, t_0) - G(s_0, t_0) = F(x, y) - F(x_0, y_0). \tag{64}$$

Since F is assumed differentiable at (x_0, y_0) we can write

$$\begin{aligned} F(x, y) - F(x_0, y_0) &= \\ &F_1(x_0, y_0)(x - x_0) + F_2(x_0, y_0)(y - y_0) \\ &+\varepsilon(x, y)(|x - x_0| + |y - y_0|) \end{aligned} \tag{65}$$

where $\varepsilon(x, y) \to 0$ as $(x, y) \to (x_0, y_0)$.

From (64) and (65) we see that, for $s \neq s_0$,

$$\begin{aligned} \frac{G(s, t_0) - G(s_0, t_0)}{s - s_0} &= \frac{F(x, y) - F(x_0, y_0)}{s - s_0} = \\ &F_1(x_0, y_0)\frac{x - x_0}{s - s_0} + F_2(x_0, y_0)\frac{y - y_0}{s - s_0} + \\ &\varepsilon(x, y)\left(\frac{|x - x_0|}{s - s_0} + \frac{|y - y_0|}{s - s_0}\right). \end{aligned} \tag{66}$$

We can now complete the proof (as we did in the proof of Theorem 12.34) by letting $s \to s_0$. The right side of (66) approaches

$$F_1(x_0, y_0) f_1(s_0, t_0) + F_2(x_0, y_0) g_1(s_0, t_0), \tag{67}$$

the fact that the remaining term

$$\varepsilon(x, y)\left(\frac{|x - x_0|}{s - s_0} + \frac{|y - y_0|}{s - s_0}\right)$$

approaches zero being similar to the corresponding part of the proof of Theorem 12.34. As a result, the left side of (66) approaches a limit as $s \to s_0$. That limit is, of course, $G_1(s_0, t_0)$. Thus

$$G_1 = F_1(x_0, y_0) f_1(s_0, t_0) + F_2(x_0, y_0) g_1(s_0, t_0).$$

■

Exercises

12.5.11 Prove the second of the chain rules (63).

12.5.12 Provide the details of the argument that the term

$$\varepsilon(x,y)\left(\frac{|x-x_0|}{s-s_0}+\frac{|y-y_0|}{s-s_0}\right)$$

in (66) approaches zero as $s \to s_0$.

12.5.13 State precisely and prove a chain rule for the schema

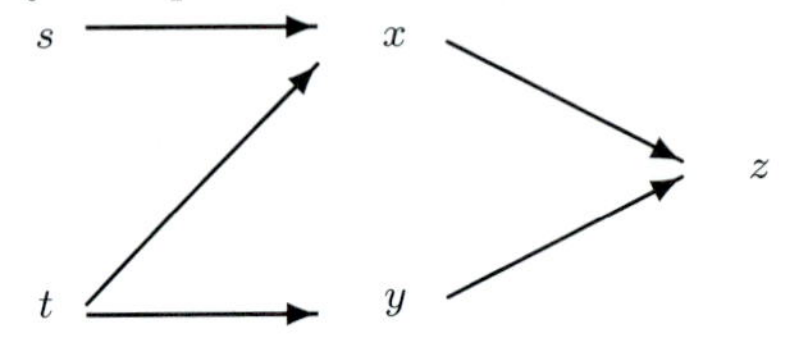

in two ways: First, by imitating the proofs of Theorems 12.34 and 12.36; then, by viewing this schema as a special case of the schema governing Theorem 12.36.

12.5.7 Higher Derivatives

It is sometimes necessary to use a chain rule to calculate higher partial derivatives. Consider, for example, Laplace's equation

$$\frac{\partial^2 z}{\partial x^2}+\frac{\partial^2 z}{\partial y^2}=0. \tag{68}$$

A twice differentiable function z satisfying this equation is called *harmonic*. Such functions are important in many parts of mathematics (such as complex analysis, applied mathematics) and physics. The expression

$$\frac{\partial^2 z}{\partial x^2}+\frac{\partial^2 z}{\partial y^2}$$

is called the *Laplacian* of the function z.

Suppose we wish to express this equation in polar coordinates. We let $x = r\cos\theta$, $y = r\sin\theta$, and consider the following schema.

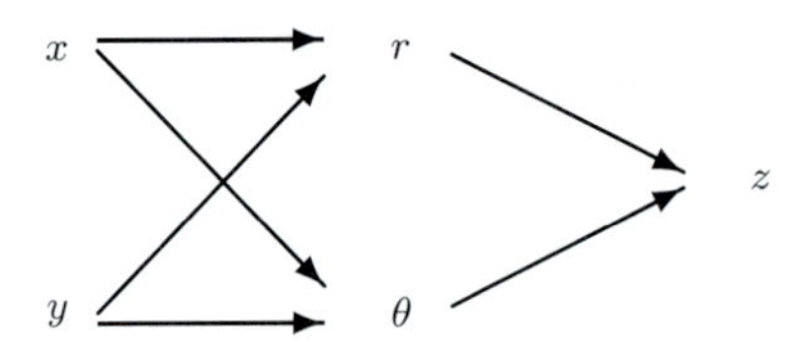

We want to express (68) in terms of r and θ by obtaining $\frac{\partial^2 z}{\partial x^2}$ and $\frac{\partial^2 z}{\partial y^2}$ as functions of r and θ. The calculations are messy and have been left as Exercise 12.5.14.

We shall instead consider a less messy schema first shown in Example 12.28.

$$(69)$$

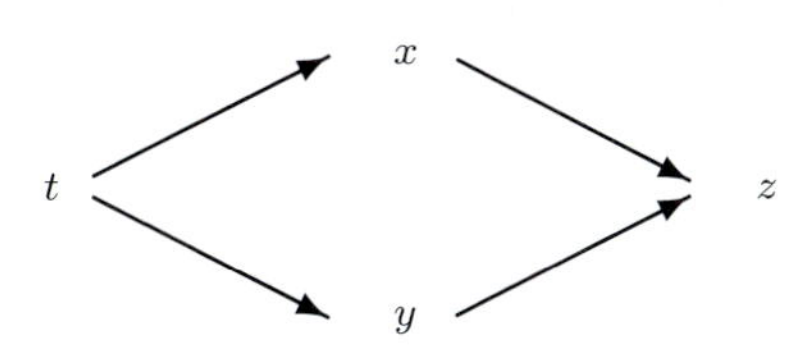

Here $z = F(x, y)$, while x and y are functions of t. The corresponding chain rule is

$$\frac{\partial z}{\partial t} = \frac{\partial z}{\partial x}\frac{dx}{dt} + \frac{\partial z}{\partial y}\frac{dy}{dt}. \tag{70}$$

We wish to compute $\frac{d^2z}{dt^2} = \frac{d}{dt}\left(\frac{dz}{dt}\right)$. From (70) we see that

$$\begin{aligned} \frac{\partial^2 z}{\partial t^2} &= \frac{d}{dt}\left(\frac{\partial z}{\partial x}\frac{dx}{dt}\right) + \frac{d}{dt}\left(\frac{\partial z}{\partial y}\frac{dy}{dt}\right) \\ &= \frac{d}{dt}\left(\frac{\partial z}{\partial x}\right)\frac{dx}{dt} + \frac{\partial z}{\partial x}\frac{d^2x}{dt^2} \\ &+ \frac{d}{dt}\left(\frac{\partial z}{\partial y}\right)\frac{dy}{dt} + \frac{\partial z}{\partial y}\frac{d^2y}{dt^2}. \end{aligned} \tag{71}$$

This last expression involves two terms that should be further developed, $\frac{d}{dt}\left(\frac{\partial z}{\partial x}\right)$, and $\frac{d}{dt}\left(\frac{\partial z}{\partial y}\right)$. Now $\frac{\partial z}{\partial x} = F_1$ and $\frac{\partial z}{\partial y} = F_2$. Both of these are functions of x and y. To obtain their derivatives with respect to t, we note that the schema (69) applies again. We obtain

$$\begin{aligned} \frac{d}{dt}\left(\frac{\partial z}{\partial x}\right) &= \frac{\partial}{\partial x}\left(\frac{\partial z}{\partial x}\right)\frac{dx}{dt} + \frac{\partial}{\partial y}\left(\frac{\partial z}{\partial y}\right)\frac{dy}{dt} \\ &= \frac{\partial^2 z}{\partial x^2}\frac{dx}{dt} + \frac{\partial^2 z}{\partial y \partial x}\frac{dy}{dt}. \end{aligned} \tag{72}$$

Similarly,

$$\frac{d}{dt}\left(\frac{\partial z}{\partial y}\right) = \frac{\partial^2 z}{\partial y^2}\frac{dy}{dt} + \frac{\partial^2 z}{\partial x \partial y}\frac{dx}{dt}. \tag{73}$$

If we assume continuity of all partials involved, the mixed partials appearing in (72) and (73) are equal (Theorem 12.5). Substituting (72) and (73) into (71) and rearranging the terms, leads to the formula

$$\frac{\partial^2 z}{\partial t^2} = \frac{\partial^2 z}{\partial x^2}\left(\frac{dx}{dt}\right)^2 + 2\frac{\partial^2 z}{\partial x \partial y}\frac{dx}{dt}\frac{dy}{dt} \tag{74}$$
$$+\frac{\partial^2 z}{\partial y^2}\left(\frac{dy}{dt}\right)^2 + \frac{\partial z}{\partial x}\frac{d^2 x}{dt^2} + \frac{\partial z}{\partial y}\frac{d^2 y}{dt^2}.$$

If we do not assume sufficient regularity of the partials to assure the mixed partials are equal, the second term of (74) must be expressed as the sum

$$\left(\frac{\partial^2 z}{\partial x \partial y} + \frac{\partial^2 z}{\partial y \partial x}\right)\frac{dx}{dt}\frac{dy}{dt}.$$

Example 12.37 Formula (74) can be illustrated with the simple example

$$z = x^2 y^3\ ,\ x = t^4\ ,\ y = t^5.$$

Thus directly we obtain $z = t^{23}$, so $\frac{d^2 z}{dt^2} = 506t^{21}$. Application of formula (74) gives the same result. ◀

Example 12.38 Consider the wave equation

$$c^2 \frac{\partial^2 f}{\partial x^2} = \frac{\partial^2 f}{\partial t^2},$$

where c is a constant. Here $f(x,t)$ describes the vertical displacement of a particle in a wave corresponding to the horizontal coordinate x at time t. We show that any function of the form

$$f(x,t) = g(x + ct),$$

where g is twice differentiable, satisfies this equation. By making the substitutions $u = x + ct$, $z = g(u)$ we arrive at the following schema, which is equivalent to schema (43) of Example 12.26.

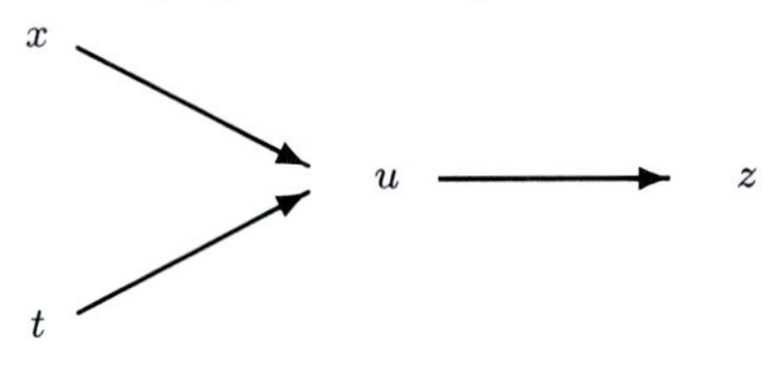

Thus the appropriate chain rule becomes

$$f_1(x,t) = g'(u)\frac{\partial u}{\partial x} = g'(x + ct)$$
$$f_2(x,t) = g'(u)\frac{\partial u}{\partial t} = cg'(x + ct)$$

Applying the chain rule again, we obtain

$$\begin{aligned} f_{11}(x,t) &= \frac{\partial}{\partial x}(g'(x+ct)) = g''(x+ct), \\ f_{22}(x,t) &= \frac{\partial}{\partial t}(cg'(x+ct)) = c^2 g''(x+ct). \end{aligned}$$

Thus $c^2 f_{11}(x,t) = f_{22}(x,t)$, or

$$c^2 \frac{\partial^2 f}{\partial x^2} = \frac{\partial^2 f}{\partial t^2},$$

verifying that the function f does satisfy the wave equation. ◀

Exercises

12.5.14 Verify that the Laplacian (68) transformed into polar coordinates via the substitution $x = r\cos\theta$, $y = r\sin\theta$ becomes

$$\frac{\partial^2 z}{\partial r^2} + \frac{1}{r^2}\frac{\partial^2 z}{\partial \theta^2} + \frac{1}{r}\frac{\partial z}{\partial r}.$$

12.5.15 The Laplacian for functions of three variables takes the form

$$\frac{\partial^2 u}{\partial x^2} + \frac{\partial^2 u}{\partial y^2} + \frac{\partial^2 u}{\partial^2 z}.$$

Transform this expression to one in cylindrical coordinates via the substitutions $x = r\cos\theta$, $y = r\sin\theta$, and $z = z$.

12.5.16 Use the result of Exercise 12.5.15 to obtain the Laplacian in spherical coordinates.

12.6 Implicit Function Theorems

Suppose that we are required to solve an equation of the form

$$F(x,y) = 0 \tag{75}$$

for y as a function of x. What we must mean by stating that some function $y = \phi(x)$ "solves" this equation for y in terms of x in a neighborhood I of a point x_0 is that $\phi : I \to \mathbb{R}$ and $F(x, \phi(x)) = 0$ for all $x \in I$. Since it is not often possible to solve such an equation explicitly [i.e., to find a formula for $\phi(x)$], we need to know conditions under which a solution does exist, conditions under which the solution is unique, and we need to know methods for obtaining the derivative $y' = \phi'(x)$ of the solution.

This should be a familiar problem. It is taught in most calculus courses as "implicit differentiation," where techniques for finding the derivative are obtained. Usually little attention is paid to the existence and uniqueness problems. An example should be enough to recall the ideas.

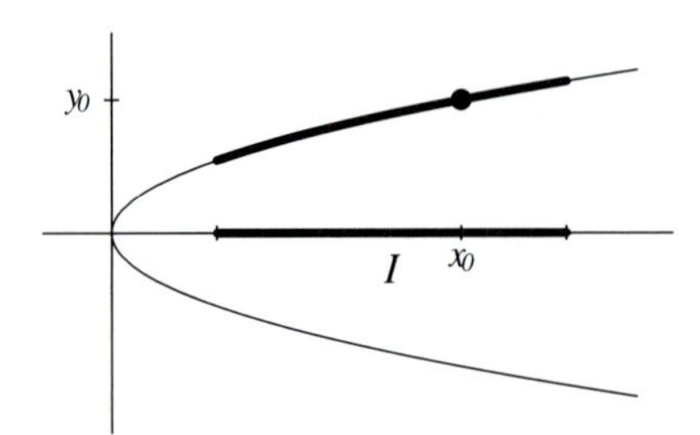

Figure 12.6. A solution of $F(x,y) = y^2 - x = 0$ in a neighborhood I of x_0.

Example 12.39 Consider the simple equation

$$F(x,y) = y^2 - x = 0. \tag{76}$$

If we solve this equation explicitly for x in terms of y, we obtain

$$x = y^2. \tag{77}$$

In elementary calculus it is said that equation (76) presents x as a function of y *implicitly*, and that (77) gives an *explicit* representation of x as a function of y.

If, instead, we attempt to solve (76) for y in terms of x, we obtain

$$y = \pm\sqrt{x}. \tag{78}$$

This does *not* present y as a function of x because for every value of $x > 0$ there exist *two* possible values of y. Nonetheless, if $x_0 > 0$ and y_0 is one of the values that allows (76) to be satisfied, there is a neighborhood I of x_0 in $\mathbb{R}$ such that one of the choices in equation (78) represents y as a continuous function of x in that neighborhood (Fig. 12.6). Observe that we cannot make the same statement for $x_0 = 0$. Observe also that at $(0,0)$ there is a vertical tangent and, for $F(x,y) = y^2 - x$, $F_2(0,0) = 0$. ◀

The implicit function theorem that we now proceed to state and prove exactly describes this situation and gives a condition under which a solution does exist. It also justifies the calculus technique of implicit differentiation that is used to obtain the derivative of the function defined implicitly.

12.6.1 One-Variable Case

Theorem 12.40 provides a condition under which an equation of the form $F(x,y) = 0$ can locally be solved uniquely. In addition, it shows that the regularity conditions we impose on F guarantee that the solution function will also be well behaved. We view this theorem as a "warm-'up" for the more general implicit function theorems we obtain in the next two subsections. Note the *local* character of the conclusion. We do not claim a global solution.

Theorem 12.40 *Let D be an open set in $\mathbb{R}^2$ and let $F : D \to \mathbb{R}$. Suppose F has continuous partial derivatives F_1 and F_2 on D. Let $(x_0, y_0) \in D$ be such that*

$$F(x_0, y_0) = 0 \text{ and } F_2(x_0, y_0) \neq 0.$$

Then there is an open interval $I_0 \in \mathbb{R}$ and a continuously differentiable function $\phi : I_0 \to \mathbb{R}$ such that $x_0 \in I_0$, $(x, \phi(x)) \in D$ for all $x \in I_0$, $\phi(x_0) = y_0$, and such that $F(x, \phi(x)) = 0$ for all $x \in I_0$. Furthermore, the formula

$$\phi'(x) = -\frac{F_1(x, \phi(x))}{F_2(x, \phi(x))}$$

is valid for all $x \in I_0$.

Proof Suppose $F_2(x_0, y_0) > 0$. [The proof is similar if $F_2(x_0, y_0) < 0$.] Since F_2 is continuous by assumption, there exists a neighborhood $N \subset D$ of (x_0, y_0) such that $F_2 > 0$ on N. We may take N to be rectangular of the form $N = I \times J$, with (x_0, y_0) the center of the rectangle. Suppose $J = [c, d]$. Since $F_2 > 0$ on N, the function $F(x_0, \cdot)$ is increasing on J. Since $F(x_0, y_0) = 0$, it follows that

$$F(x_0, c) < 0 < F(x_0, d).$$

The function F is continuous, so there exists an open interval $I_0 \subset I$ such that x_0 is the center of I_0 and

$$F(x, c) < 0 < F(x, d) \text{ for each } x \in I_0. \tag{79}$$

The continuity of F (as a function of two variables) implies that for each $x \in I_0$ the function $F(x, \cdot)$ is continuous in the second variable. It follows from (79) and the intermediate value property of continuous functions (which we discussed in Section 5.8) that for each $x \in I_0$ there exists at least one value $y \in (c, d)$ such that $F(x, y) = 0$. Moreover, since $F(x, \cdot)$ is also increasing, this value of y is unique. We can express this dependency of y on x by writing $y = \phi(x)$. Thus we have found a function ϕ such that for each $x \in I_0$,

$$F(x, \phi(x)) = 0 \tag{80}$$

and

$$c < \phi(x) < d. \tag{81}$$

We now show that ϕ has all the remaining properties claimed in the conclusion of the theorem. It is clear that $\phi(x_0) = y_0$ and that $(x, \phi(x)) \in D$ for each $x \in I_0$.

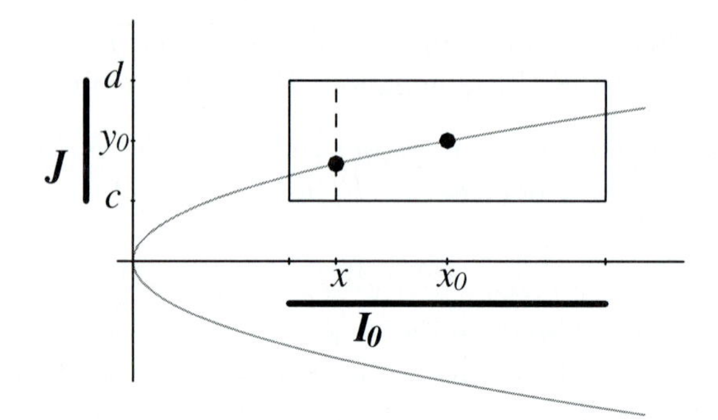

Figure 12.7. Construction of the rectangle $I_0 \times J$ in the proof.

We next check that ϕ is continuous on I_0. Observe first from (81) that if $x_1, x_2 \in I_0$, then $|\phi(x_1) - \phi(x_2)| < d - c$. The rectangle N could have been chosen as small as we like, a smaller rectangle leading perhaps to a shorter interval I_0. Thus, if x_1 is any point in I_0 and $\varepsilon > 0$, we can apply the same argument we gave previously on a neighborhood $I_1 \subset I_0$ of x_1 with $d_1 - c_1 < \varepsilon$. In Figure 12.7 we have added the rectangle $I_0 \times J$ centered at (x_0, y_0) to Figure 12.6. The result is that $|\phi(x) - \phi(x_1)| < \varepsilon$ on I_1. But this implies ϕ is continuous at x_1. Since x_1 is an arbitrary point of I_1, ϕ is continuous on I_0.

It remains to show that ϕ is continuously differentiable on I_0 and that the formula given for ϕ' is valid. Let $x, x+h \in I_0$. Since the graph of ϕ is contained in the rectangle N, the line segment L with endpoints $(x, \phi(x))$ and $(x+h, \phi(x+h))$ lies in N. By the mean value theorem (Theorem 12.35) there exists $(\xi, \eta) \in L$ such that

$$F(x+h, \phi(x+h)) - F(x, \phi(x)) = \\ F_1(\xi,\eta)h + F_2(\xi,\eta)[\phi(x+h) - \phi(x)].$$

But $F(x+h, \phi(x+h)) = F(x, \phi(x)) = 0$, so

$$F_1(\xi,\eta)h + F_2(\xi,\eta)[\phi(x+h) - \phi(x)] = 0.$$

Thus, if $F_2(\xi,\eta) \neq 0$, we can write

$$\frac{\phi(x+h) - \phi(x)}{h} = -\frac{F_1(\xi,\eta)}{F_2(\xi,\eta)}. \tag{82}$$

Since F_2 is never 0 in some neighborhood of $(x, \phi(x))$, equation (82) is valid in that neighborhood. As $h \to 0$, the left side of (82) approaches $\phi'(x)$, (ξ, η) approaches $(x, \phi(x))$, and, because of the continuity of F_1 and F_2, the right side of (82) approaches

$$-\frac{F_1(x, \phi(x))}{F_2(x, \phi(x))},$$

as was to be proved. ∎

Example 12.41 The example $F(x, y) = x^2 - y^4$ and $(x_0, y_0) = (1, 1)$ illustrates that while there is a solution valid in a neighborhood of $x_0 = 1$ there cannot exist a solution in too large an interval. Indeed there is no solution in any interval I_0 that contains $x = 0$. ◀

Example 12.42 We return to Example 12.39:

$$F(x, y) = y^2 - x = 0.$$

Here $F_1(x, y) = -1$, $F_2(x, y) = 2y$. The hypotheses of Theorem 12.40 are met provided that $y \neq 0$. If $(x_0, y_0) = (4, 2)$, then the resulting function is $\phi(x) = \sqrt{x}$ and $\phi'(x) = 1/2y$. In this case, the interval I_0 can be any interval containing $x_0 = 4$ but not containing 0. ◀

Exercises

12.6.1 Show that the equation $x^2y^2 + 2e^{xy} - 4 - 2e^2 = 0$ can be solved for y in terms of x in a neighborhood of the point $x = 1$ with $y(1) = 2$. Calculate $\frac{dy}{dx}$ when $x = 1$.

12.6.2 Do Exercise 12.6.1 via implicit differentiation as one would in an elementary calculus class. (There one usually simply *assumes* the existence of a solution.)

12.6.2 Several-Variable Case

Theorem 12.40 extends to situations involving more variables. Consider one equation in three variables. We wish to solve an equation

$$F(x, y, z) = 0$$

for z as a function of x and y in some neighborhood of a point (x_0, y_0). As before, the solution should be a function $z = \phi(x, y)$ so that

$$F(x, y, \phi(x, y)) = 0$$

for all (x, y) in that neighborhood.

Example 12.43 Consider the equation

$$x^2 + y^2 + z^2 - 9 = 0. \tag{83}$$

Equation (83) can easily be untangled to represent one of the variables in terms of the others. Thus

$$z = \pm\sqrt{9 - x^2 - y^2} \tag{84}$$

represents z in terms of x and y when $x^2+y^2 \leq 3$. Just as we saw in the one-variable case, z is not a function of x and y on the disk $x^2+y^2 \leq 9$ because the values of z are not uniquely determined by equation (84). Nonetheless, if (x_0, y_0) is a point inside this disk and and z_0 is one of the values that allows (83) to be satisfied, there is a neighborhood I of (x_0, y_0) in $\mathbb{R}^2$ such that one of the choices in equation (84) represents z as a continuous function of (x, y) in that neighborhood. ◀

Theorem 12.44 that follows is the analogue of Theorem 12.40 when we deal with one equation in more than two variables. We state the theorem for $n+1$ variables.

Theorem 12.44 *Let F have continuous first-order partial derivatives on an open set $D \subset \mathbb{R}^{n+1}$. Let $\mathbf{x_0} = (x_1^0, x_2^0, \ldots, x_n^0, z_0) \in D$. Suppose*

$$F(x_1^0, x_2^0, \ldots, x_n^0, z_0) = 0$$

and

$$F_{n+1}(x_1^0, x_2^0, \ldots, x_n^0, z_0) \neq 0.$$

Then there is a neighborhood $J_0 \subset \mathbb{R}$ of z_0 and a neighborhood $I_0 \subset \mathbb{R}^n$ of $(x_1^0, x_2^0, \ldots, x_n^0)$, and there is a unique function ϕ defined on I_0 such that

$$z_0 = \phi(x_1^0, x_2^0, \ldots, x_n^0)$$

and

$$F(x_1, \ldots, x_n, \phi(x_1, \ldots, x_n)) = 0$$

for all $(x_1, \ldots, x_n) \in I_0$. Furthermore, the function ϕ has continuous partial derivatives with respect to each of the variables $(x_1, \ldots, x_n)$ and

$$\phi_i(x_1, \ldots, x_n) = -\frac{F_i(x_1, \ldots, x_n, z)}{F_{n+1}(x_1, \ldots, x_n, z)} \tag{85}$$

for each $i = 1, \ldots, n$ and for all $(x_1, \ldots, x_n) \subset I_0$.

Proof Observe that by letting $z = \phi(x_1, \ldots, x_n)$, we have explicitly represented the variable z as a continuously differentiable function of $x_1, \ldots, x_n$ on a neighborhood of $(x_1^0, \ldots, x_n^0)$ once we specify the value of z_0.

We prove Theorem 12.44 for the case $n = 2$. The general proof needs no fresh ideas, but the notation is more messy and pictures are less easy to visualize. Observe the similarity of the ideas underlying our proof and the proof of Theorem 12.40. To simplify notation, let us denote the point in question by (x_0, y_0, z_0). We first obtain the neighborhoods J_0 and I_0 mentioned in the theorem.

Assume that $F_3(x_0, y_0, z_0) > 0$. [The case $F_3(x_0, y_0, z_0) < 0$ is similar.] Since F_3 is continuous in D, there is an open rectangular box, A, with center (x_0, y_0, z_0) and edges parallel to the coordinate axes such that $F_3 > 0$ on A. Let $2c$ be the height of this box. Then $F(x_0, y_0, \cdot)$ is an increasing function of z on the interval $[z_0 - c, z_0 + c]$. Since $F(x_0, y_0, z_0) = 0$ by assumption, we have

$$F(x_0, y_0, z_0 - c) < 0 < F(x_0, y_0, z_0 + c).$$

Now F is continuous on D since, by assumption, F has continuous first partial derivatives on D. We can thus obtain positive numbers a and b such that

$$F_1(x, y, z_0 - c) < 0 \text{ and } F(x, y, z_0 + c) > 0$$

if $|x - x_0| < a$ and $|y - y_0| < b$. We can choose a and b so small that the box

$$B = \{(x, y, z) : |x - x_0| < a, |y - y_0| < b, |z - z_0| < c\}$$

is contained in A. The required neighborhoods are thus

$$I_0 = \{(x, y) : |x - x_0| < a, |y - y_0| < b\}$$

and

$$J_0 = \{z : |z - z_0| < c\}.$$

We next obtain the function ϕ. Let $(x, y) \in I_0$. Since $F_3 > 0$ on the set $B = I_0 \times J_0$, it follows that $F(x, y, z)$ is a continuous increasing function of z on the interval $[z_0 - c, z_0 + c]$. It is positive at $(x, y, z_0 + c)$ and negative at $(x, y, z_0 - c)$, so there is a unique value of $z \in [z_0 - c, z_0 + c]$ such that $F(x, y, z) = 0$. Denote this value of z by $\phi(x, y)$. We have thus determined the function ϕ on I_0.

We now show that ϕ has the desired properties. We prove continuity as we did in Theorem 12.40 . Let $(x_1, y_1) \in I_0$ and let $\varepsilon > 0$. We may assume $\varepsilon \leq c$. Let $z_1 = \phi(x_1, y_1)$. Then $F(x_1, y_1, z_1) = 0$. Applying the same argument that we applied earlier to obtain the box B, we obtain a (possibly smaller) box B_1, centered at (x_1, y_1, z_1), whose height, $2c_1$, is as small as we like. Take $2c_1 < \varepsilon$ and let $B_1 = I_1 \times J_1$ were

$$J_1 = [z_1 - c_1, z_1 + c_1]$$

and

$$I_1 = \{(x, y) : |x - x_1| < a_1, |y - y_1| < b_1\}.$$

Our previous argument applied to B_1 gives rise to a function g defined on I_1 such that $F(x, y, g(x, y)) = 0$ on I_1 and

$$z_1 - c_1 < g(x, y) < z_1 + c_1$$

on I_1. Thus, if $(\xi, \eta) \in I_1$, then

$$|g(\xi, \eta) - g(x_1, y_1)| < 2c_1 < \varepsilon.$$

It follows that g is continuous at (x_1, y_1). But g was obtained from F in exactly the same way ϕ was, and the values are unique, so $g = \phi$ on I_1. Thus ϕ is continuous at (x_1, y_1). Since (x_1, y_1) is an arbitrary point of I_0, ϕ is continuous on I_0.

Finally, we verify that the formulas (85) for the partial derivatives are valid. We provide a proof for ϕ_1. The proof for ϕ_2 is the same.

Consider the function $F(\cdot, y, \cdot)$ as a function of x and z with y fixed. Write $z + k = \phi(x + h, y)$. [Here we assume, of course, that $(x, y) \in I_0$, and $(x + h, y) \in I_0$.] We may also assume $z + k \in J_0$, since ϕ is continuous on I_0. Now

$$F(x, y, z) = 0 = F(x + h, y, z + k).$$

By the mean value theorem (Theorem 12.35)

$$0 = F(x + h, y, z + k) - F(x, y, z) = F_1(\xi, y, \zeta)h + F_3(\xi, y, \zeta)k, \tag{86}$$

where ξ is between x and $x + h$ and ζ is between z and $z + k$. Thus

$$\begin{aligned}\phi_1(x, y) &= \lim_{h \to 0} \frac{\phi(x + h, y) - \phi(x, y)}{h} \\ &= \lim_{h \to 0} \frac{(z + k) - z}{h} = \lim_{h \to 0} \frac{k}{h}.\end{aligned}$$

By (86) this last limit is just

$$\lim_{h \to 0} \left(-\frac{F_1(\xi, y, \zeta)}{F_3(\xi, y, \zeta)} \right) = \frac{F_1(x, y, z)}{F_3(x, y, z)}$$

since $(\xi, \zeta) \to (x, z)$ as $h \to 0$ and the functions F_1 and F_3 are continuous by assumption. ■

Example 12.45 Returning to Example 12.43, where

$$x^2 + y^2 + z^2 - 9 = 0,$$

we find that

$$\begin{aligned}F_1(x, y, z) &= 2x, \\ F_2(x, y, z) &= 2y, \\ F_3(x, y, z) &= 2z.\end{aligned}$$

When $z \neq 0$ and $x^2 + y^2 + z^2 \leq 9$ we have $F_3(x, y, z) \neq 0$ and the hypotheses of theorem 12.44 are met. Given (x_0, y_0, z_0) such that $x_0^2 + y_0^2 + z_0^2 \leq 9$ and

$z_0 \neq 0$, there exists a neighborhood N of (x_0, y_0) and a function ϕ defined on that neighborhood such that

$$F(x, y, \phi(x, y)) = 0$$

or, equivalently,

$$x^2 + y^2 + [\phi(x, y)]^2 - 9 = 0.$$

The partial derivatives of the function ϕ are given, for all $(x, y) \in N$, by

$$\phi_1(x, y) = -x/z \quad \text{and} \quad \phi_2(x, y) = -y/z.$$

In fact we know that either

$$\phi(x, y) = \sqrt{9 - x^2 - y^2} \quad \text{or} \quad \phi(x, y) = -\sqrt{9 - x^2 - y^2}$$

depending on whether $z_0 > 0$ or $z_0 < 0$. ◀

Exercises

12.6.3 Compare the proof of Theorem 12.44 of this section with that of Theorem 12.40 of the previous section. Were any nonobvious new ideas needed to prove Theorem 12.44?

12.6.4 Prove Theorem 12.44 as stated, that is, for general $n \geq 2$.

12.6.5 Suppose that $F(x, y, z) = 0$ defines x as a function of y, z and also y as a function of x, z and also z as a function of x, y. Show that (under appropriate hypotheses on F)

$$\frac{\partial z}{\partial x}\frac{\partial z}{\partial y}\frac{\partial x}{\partial z} = -1.$$

12.6.6 Look ahead to Exercise 13.11.3 for a proof of an implicit function theorem using metric space methods.

12.6.3 Simultaneous Equations

In the preceding section we dealt with one equation in several variables, say $F(x, y, z) = 0$. We found conditions under which we could solve for z in terms of x and y in a neighborhood of a point (x_0, y_0), obtaining a function $z = \phi(x, y)$ continuously differentiable on that neighborhood. We turn now to a situation that occurs frequently involving *several* simultaneous equations, say m equations in $m + n$ variables.

Example 12.46 Here are two equations

$$x = r\cos\theta$$
$$y = r\sin\theta$$

involving the four variables x, y, r, and θ. These equations can be viewed as ones giving a change of coordinate systems from polar coordinates to rectangular coordinates. Alternatively, we can view them as presenting x and y explicitly in terms of r and θ, or r and θ implicitly in terms of x and y. A third perspective is to view them as defining a mapping of $\mathbb{R}^2$ onto $\mathbb{R}^2$: $F(r, \theta) = (r\cos\theta, r\sin\theta)$.

In each of these perspectives there is reason to wish to express r and θ as functions of x and y. Our first interpretation would provide the equations to transform from rectangular to polar coordinates. The second interpretation merely would provide an *explicit* representation of r and θ in terms of x and y. The third interpretation would provide an inverse function to F: A function G such that $G(x, y) = (r, \theta)$.

You may recall that the equations

$$\begin{aligned} r &= \sqrt{x^2 + y^2} \\ \theta &= \arctan\frac{y}{x} \end{aligned}$$

do the job when x and y are not both zero. ◀

Example 12.47 Consider the system

$$\left.\begin{aligned} A_1u + B_1v + C_1 &= 0 \\ A_2u + B_2v + C_2 &= 0 \end{aligned}\right\}, \tag{87}$$

where the A's, B's, and C's are functions of x and y. Here we have two equations in the four variables x, y, u, v.

Using ideas from elementary algebra, we can "solve" these equations for u and v in terms of x and y provided that the determinant

$$|J| = \begin{vmatrix} A_1 & B_1 \\ A_2 & B_2 \end{vmatrix} = A_1B_2 - A_2B_1 \neq 0. \tag{88}$$

The equations (87) take the form

$$\begin{aligned} F(x, y, u, v) &= 0 \\ G(x, y, u, v) &= 0. \end{aligned} \tag{89}$$

Note that

$$\begin{aligned} \frac{\partial F}{\partial u} &= A_1, & \frac{\partial F}{\partial v} &= B_1, \\ \frac{\partial G}{\partial u} &= A_2, & \frac{\partial G}{\partial v} &= B_2. \end{aligned}$$

The condition (88) takes the form

$$|J| = \begin{vmatrix} \frac{\partial F}{\partial u} & \frac{\partial F}{\partial v} \\ \frac{\partial G}{\partial u} & \frac{\partial G}{\partial v} \end{vmatrix} \neq 0. \tag{90}$$

◀

The determinant in (90) is called the *Jacobian determinant* or the determinant of the *Jacobian*

$$J = \begin{pmatrix} \frac{\partial F}{\partial u} & \frac{\partial F}{\partial v} \\ \frac{\partial G}{\partial u} & \frac{\partial G}{\partial v} \end{pmatrix}.$$

It is only one of many Jacobians that arise and play important roles in this section and in other parts of mathematics. We often write the Jacobian determinant in the form

$$\frac{\partial(F,G)}{\partial(u,v)}.$$

We are concerned with determining when solutions of the type Examples 12.46 and 12.47 suggest can be found, and in determining the partial derivatives of the functions obtained (for example, $\frac{\partial u}{\partial x}, \frac{\partial u}{\partial y}, \frac{\partial v}{\partial x}, \frac{\partial v}{\partial y}$ in Example 12.47).

We treat the case of a pair of simultaneous equations in four variables, as it is representative of the general situation.

Theorem 12.48 *Let F and G have continuous first partial derivatives on an open set $D \subset \mathbb{R}^4$ and let $\mathbf{p_0} = (x_0, y_0, u_0, v_0) \in D$ with*

$$\left.\begin{array}{rcl} F(x_0, y_0, u_0, v_0) & = & 0 \\ G(x_0, y_0, u_0, v_0) & = & 0 \end{array}\right\}. \tag{91}$$

Let

$$|J| = \frac{\partial(F,G)}{\partial(u,v)} = \begin{vmatrix} \frac{\partial F}{\partial u} & \frac{\partial F}{\partial v} \\ \frac{\partial G}{\partial u} & \frac{\partial G}{\partial v} \end{vmatrix}.$$

Suppose $|J|$ is not zero at $\mathbf{p_0}$. Then there are neighborhoods I_0 and J_0 of (x_0, y_0) and (u_0, v_0), respectively, such that

(i) *To each $(x, y) \in I_0$ there corresponds a unique $(u, v) \in J_0$ such that equations (91) are satisfied at (x, y, u, v). This correspondence defines u and v as functions on I_0 by*

$$u = \phi(x, y)\ ,\ v = \psi(x, y).$$

(ii) *The functions ϕ and ψ have continuous partial derivatives on I_0 given by the formulas*

$$\frac{\partial \phi}{\partial x} = -\frac{1}{|J|}\frac{\partial(F,G)}{\partial(x,v)}, \quad \frac{\partial \psi}{\partial x} = -\frac{1}{|J|}\frac{\partial(F,G)}{\partial(u,x)},$$

$$\frac{\partial \phi}{\partial y} = -\frac{1}{|J|}\frac{\partial(F,G)}{\partial(y,v)}, \quad \frac{\partial \psi}{\partial y} = -\frac{1}{|J|}\frac{\partial(F,G)}{\partial(u,y)}.$$

Proof We shall apply Theorem 12.44 twice, first to obtain v as a function of x, y, and u, and then to obtain u as a function of x and y. Since $|J| \neq 0$ at $\mathbf{p_0}$, at least one of the partials $\frac{\partial F}{\partial v}$, $\frac{\partial G}{\partial v}$ must be different from 0 at $\mathbf{p_0}$. We may assume $F_4 = \frac{\partial F}{\partial v} \neq 0$ at $\mathbf{p_0}$, the argument being similar if $\frac{\partial G}{\partial v} \neq 0$ at $\mathbf{p_0}$.

Now we apply Theorem 12.44 to the equation $F(x, y, u, v) = 0$. We obtain a function $v = g(x, y, u)$, defined in a neighborhood of (x_0, y_0, u_0) such that $F(x, y, u, g(x, y, u)) = 0$ on that neighborhood.

We next consider the function G. Let

$$H(x, y, u) = G(x, y, u, g(x, y, u)).$$

We have thus replaced the function G of four variables with the function H of three variables. Our task is to solve this equation for u in terms of x and y.

To do this, we use Theorem 12.44 once more. In order to check that Theorem 12.44 applies, we must show $H_3 \neq 0$ at (x_0, y_0, u_0). Applying the chain rule, we obtain

$$\begin{aligned} H_3 &= G_3 + G_4 g_3 = G_3 - G_4 \frac{F_3}{F_4} \\ &= \frac{G_3 F_4 - G_4 F_3}{F_4} = -\frac{|J|}{F_4}. \end{aligned} \tag{92}$$

Now $|J|$ and F_4 are different from 0 in some neighborhood of $\mathbf{p_0}$. Thus H_3 is finite and nonzero in that neighborhood. Applying Theorem 12.44 to the equation $H(x, y, u) = 0$, we obtain a solution

$$u = \phi(x, y) = g(x, y, \psi(x, y))$$

in some neighborhood of (x_0, y_0).

The functions $u = \phi(x, y)$ and $v = \psi(x, y)$ are solutions to the system of equations

$$\begin{aligned} F(x, y, u, v) &= 0 \\ G(x, y, u, v) &= 0 \end{aligned}$$

for (x, y) in some neighborhood I_0 of (x_0, y_0) and (u, v) in some neighborhood J_0 of (u_0, v_0). From Theorem 12.44 we are assured that the functions ϕ and ψ have continuous first partials. It remains to verify the formulas stated for these partials.

To check the formula for ϕ_1, recall that $u = \phi(x, y)$ is a solution of the equation $H(x, y, u) = 0$. Using (92), we obtain

$$\begin{aligned}
\phi_1 &= -\frac{H_1}{H_3} = -\frac{G_1 + G_4 g_1}{-\frac{|J|}{F_4}} \\
&= -\frac{G_1 - G_4\left(-\frac{F_1}{F_4}\right)}{-\frac{|J|}{F_4}} \\
&= -\frac{F_4 G_1 - G_4 F_1}{|J|} = -\frac{1}{|J|}\frac{\partial(F, G)}{\partial(x, v)}.
\end{aligned}$$

Formulas for the other partials are obtained similarly. ■

Note. We stated Theorem 12.48 for two equations in four variables. Theorem 12.44 of the previous section deals with the case of one equation in several variables. A general theorem involving m equations in $n + m$ variables follows similar lines. A typical one of the m equations is of the form

$$F^i(x_1, \ldots, x_n, u_1, \ldots, u_m) = 0.$$

(Here, the superscript identifies the function so as not to confuse the equation number with a partial derivative, which might occur if subscripts were used.) We assume that F^i has continuous partial derivatives in some region in $\mathbb{R}^{m+n}$ and that the appropriate Jacobian determinant,

$$\frac{\partial(F^1, \ldots, F^m)}{\partial(u_1, \ldots, u_m)},$$

is not zero. The conclusions are then the obvious analogues of the conclusion to Theorem 12.48.

Exercises

12.6.7 Verify those formulas not verified in the text for the partial derivatives in Theorem 12.48.

12.6.4 Inverse Function Theorem

We now apply Theorem 12.48 to obtain a theorem concerning inverses of mappings. Suppose we have equations of the form

$$\begin{aligned}
x &= f(u, v) \\
y &= g(u, v),
\end{aligned}$$

where f and g are defined on an open set $D \subset \mathbb{R}^2$. These equations determine a function $T : D \to \mathbb{R}^2$, $T(u, v) = (x, y)$. We often call such functions *mappings* or *transformations*. If we can solve these equations for u and v for values of (x, y) in some set $D' \subset \mathbb{R}^2$,

$$\begin{aligned} u &= \phi(x,y) \\ v &= \psi(x,y), \end{aligned} \tag{93}$$

then this determines a mapping $S : D' \to \mathbb{R}^2$ such that $T \circ S$ is the identity on D'. Thus S and T are, in some sense, inverses of each other. Some care must be taken in making this inverse relationship between S and T precise. The problem has to do with the domains of S and T on which $S \circ T$ and $T \circ S$ are the identity.

Example 12.49 Consider the function of one variable

$$u = f(x) = x^2$$

defined on $\mathbb{R}$. If we let $g(u) = \sqrt{u} \quad (u \geq 0)$, then

$$(f \circ g)(u) = (\sqrt{u})^2 = u,$$

so $f \circ g$ is the identity on the set $\{u : u \geq 0\}$. But

$$(g \circ f)(x) = \sqrt{x^2} = |x|,$$

so $g \circ f$ is the identity on the set $\{x : x \geq 0\}$ but not on all of $\mathbb{R}$. Since f is not one-to-one on $\mathbb{R}$, its domain must be reduced to a smaller one on which f is one-to-one. When this is done properly, f does have an inverse. If we had chosen a specific value of $x_0 \neq 0$, we would have been able to find a neighborhood I of x_0 and a neighborhood J of $u_0 = f(x_0)$ such that $f(I) = J$ and f is one-to-one on I. ◀

Theorem 12.50 provides conditions under which such a local inverse exists for mappings from $\mathbb{R}^2 \to \mathbb{R}^2$.

Theorem 12.50 (Inverse Function Theorem) *Let f and g have continuous first partial derivatives on an open set $D \subset \mathbb{R}^2$. Let $(u_0, v_0) \in D$ and suppose*

$$x_0 = f(u_0, v_0) \ , \ y_0 = g(u_0, v_0).$$

Suppose further that

$$|J| = \frac{\partial(f,g)}{\partial(u,v)} \neq 0 \text{ at } (u_0, v_0).$$

Then there are neighborhoods I_0 of (x_0, y_0) and J_0 of (u_0, v_0) such that for each $(x, y) \in I_0$ there corresponds a unique $(u, v) \in J_0$ with

$$x = f(u,v) \ , \ y = g(u,v).$$

This correspondence determines functions from I_0 into J_0

$$u = \phi(x,y) \ , \ v = \psi(x,y).$$

The functions ϕ and ψ have continuous partial derivatives given by

$$\frac{\partial \phi}{\partial x} = -\frac{1}{|J|}\frac{\partial g}{\partial v}, \quad \frac{\partial \phi}{\partial y} = -\frac{1}{|J|}\frac{\partial f}{\partial v},$$
$$\frac{\partial \psi}{\partial x} = -\frac{1}{|J|}\frac{\partial g}{\partial u}, \quad \frac{\partial \psi}{\partial y} = -\frac{1}{|J|}\frac{\partial f}{\partial u}.$$

Proof Theorem 12.50 is just a special case of Theorem 12.48. Let

$$\begin{aligned} F(x,y,u,v) &= f(u,v) - x \\ G(x,y,u,v) &= g(u,v) - y. \end{aligned}$$

Applying Theorem 12.48, we obtain the functions ϕ and ψ and the formulas for the partial derivatives. For example, to obtain $\frac{\partial \psi}{\partial y}$ we need only calculate, using Theorem 12.48,

$$\begin{aligned} \frac{\partial \psi}{\partial y} &= -\frac{1}{|J|}\frac{\partial(F,G)}{\partial(u,x)} = -\frac{1}{|J|}\begin{vmatrix} \frac{\partial F}{\partial u} & \frac{\partial F}{\partial x} \\ \frac{\partial G}{\partial u} & \frac{\partial G}{\partial x} \end{vmatrix} \\ &= -\frac{1}{|J|}\begin{vmatrix} \frac{\partial f}{\partial u} & -1 \\ \frac{\partial f}{\partial u} & 0 \end{vmatrix} = -\frac{1}{|J|}\frac{\partial f}{\partial u}. \end{aligned}$$

The other calculations are similar. ■

Example 12.51 Consider the equations in Example 12.46.

$$\begin{aligned} x &= r\cos\theta \\ y &= r\sin\theta \end{aligned}$$

The Jacobian determinant is

$$\begin{vmatrix} \cos\theta & -r\sin\theta \\ \sin\theta & r\cos\theta \end{vmatrix} = r(\cos^2\theta + \sin^2\theta) = r.$$

For $(r_0, \theta_0) \in \mathbb{R}^2$, $(r_0 \neq 0)$ there will be a neighborhood of $(r_0, \theta_0) \in \mathbb{R}^2$ on which an inverse to the transformation exists. There is no inverse on all of $\mathbb{R}^2$ or even on any D containing both of the points $(4,0)$ and $(4, 2\pi)$, because

$$(4\cos 0, 4\sin 0) = (4\cos 2\pi, 4\sin 2\pi) = (4,0).$$

At these points the Jacobian determinant does not equal zero, but the transformation is not one-to-one on D. ◀

Exercises

12.6.8 The standard formulas relating spherical to rectangular coordinates are

$$z = \rho\sin\phi\cos\theta\,, \; y = \rho\sin\theta\,, \; z = \rho\cos\phi.$$

Can these equations be solved for ρ, ϕ, and θ in terms of x, y, and z?

12.6.9 State in full detail the analogue of Theorem 12.48 arising from four equations in the six variables w, x, y, z, s, t that involve solving for s and t in terms of the other variables.

12.6.10 Verify those formulas not verified in the text for the partial derivatives in Theorem 12.50.

12.6.11 Does the pair of equations $x = u^2 - v^2$, $y = 2uv$, have an inverse on a neighborhood of $(0,0)$?

12.6.12 Does the pair of equations $x = r\cos\theta$, $y = r\sin\theta$, have an inverse on a neighborhood of $(0,0)$?

12.6.13 Show that the Jacobian determinant of the mapping

$$x = e^u \cos v\,, \;\; y = e^u \sin v$$

is never zero, yet the mapping is not one-to-one on all of $\mathbb{R}^2$.

12.6.14 Under the hypotheses of Theorem 12.50 show

$$\frac{\partial(u,v)}{\partial(x,y)} = \left[\frac{\partial(x,y)}{\partial(u,v)}\right]^{-1}.$$

12.6.15 Let f and g have continuous first partial derivatives on an open set $D \subset \mathbb{R}^2$ and let $T : D \to \mathbb{R}^2$ be defined by

$$T(u,v) = (f(u,v), g(u,v)).$$

Prove that if T is one-to-one on D, then the set $T(D)$ is open.

12.7 Functions From $\mathbb{R} \to \mathbb{R}^m$

We shall be concerned with differentiability of a function $\mathbf{f}$ defined on a subset of $\mathbb{R}$ with values in $\mathbb{R}^m$, $m > 1$. Suppose we are given two equations

$$\begin{array}{rcl} x &=& x(t) \\ y &=& y(t) \end{array} \qquad (a \le t \le b). \tag{94}$$

These equations determine a function $\mathbf{f} : [a,b] \to \mathbb{R}^2$ defined by

$$\mathbf{f}(t) = (x(t), y(t)).$$

In elementary calculus we often view the equations (94) as a parametric representation of a curve. As t moves from a to b, $\mathbf{f}(t)$ traces out a curve. Students learn that, under appropriate hypotheses, we can determine the slope of the tangent line at a point $(x(t), y(t))$ by calculating

$$\frac{dy}{dx} = \frac{y'(t)}{x'(t)}.$$

In this section we view the function $\mathbf{f}$ simply as a function from $[a,b]$ to $\mathbb{R}^2$. We shall define differentiability of $\mathbf{f}$ in terms of approximations of $\mathbf{f}$ by a linear function. Our requirement for differentiability at t is *not* the same as the requirement that the curve given parametrically by (94) has a tangent at $(x(t), y(t))$. (Indeed, Example 12.52 will show that a curve need not have such a tangent even when $\mathbf{f}$ is differentiable at t.)

How then should we define differentiability of $\mathbf{f}$? We take our cue from material in preceding sections. We wish to approximate a function $\mathbf{f}$ near a point t by a linear function $\mathbf{T}$. The approximation should be good in the sense that $\|\mathbf{f}(t) - \mathbf{T}(t)\|$ is small in comparison with the distance between t and t_0 when t is near t_0. Since

$$\mathbf{T}(t_0 + h) - \mathbf{T}(t_0) = \mathbf{T}(h)$$

for linear functions, we can write this in the form

$$\lim_{h \to 0} \frac{\|\mathbf{f}(t_0 + h) - \mathbf{f}(t_0) - \mathbf{T}(h)\|}{h} = 0. \tag{95}$$

Let's look at the familiar setting, $m = 2$, that we used as an introduction to this section:

$$\begin{aligned} x &= x(t) \\ y &= y(t) \end{aligned} \qquad (a \leq t \leq b),$$

where x and y are continuous functions on $[a, b]$. As we mentioned, these equations define a function $\mathbf{f} : [a, b] \to \mathbb{R}^2$ by

$$\mathbf{f}(t) = (x(t), y(t)).$$

We would like to define differentiability of $\mathbf{f}$ at a point $t \in [a, b]$ according to a criterion similar to (95).

Suppose for the moment that x and y are differentiable at t. Let

$$\mathbf{T}(h) = (x'(t), y'(t))h.$$

Then $\mathbf{f}(t + h) - \mathbf{f}(t) - \mathbf{T}(h)$ becomes

$$\begin{aligned} &(x(t + h) - x(t), y(t + h) - y(t)) - (x'(t)h, y'(t)h) = \\ &\quad (x(t + h) - x(t) - x'(t)h, y(t + h) - y(t) - y'(t)h). \end{aligned}$$

In looking at these expressions, several things are immediately apparent:

1. $\mathbf{f}(t + h) - \mathbf{f}(t) - \mathbf{T}(h)$ is a vector in $\mathbb{R}^2$, not a real number, so the numerator in (95) required a norm in place of the absolute value.

2. $\mathbf{T}(h) = (x'(t), y'(t))h$ is linear in h, as required. This means that, for all α, h_1, $h_2 \in \mathbb{R}$,

   $$\mathbf{T}(\alpha h) = \alpha \mathbf{T}(h)$$

   and

   $$\mathbf{T}(h_1 + h_2) = \mathbf{T}(h_1) + \mathbf{T}(h_2).$$

   (The point t is fixed in this discussion.)

3. Since x and y are differentiable (by assumption), then, as $h \to 0$,

$$\frac{x(t+h) - x(t) - x'(t)h}{h} \to 0,$$
$$\frac{y(t+h) - y(t) - y'(t)h}{h} \to 0,$$

and hence

$$\frac{\|\mathbf{f}(t+h) - \mathbf{f}(t) - \mathbf{T}(h)\|}{|h|} \to 0.$$

More generally, we obtain the following. Let $\mathbf{f}$ be defined on a neighborhood of a point $t \in \mathbb{R}$ with range in $\mathbb{R}^m$. We can write

$$\mathbf{f}(t) = (x_1(t), \ldots, x_m(t)),$$

where $x_1, \ldots, x_m$ are all defined on a neighborhood of t. If each of the functions $x_1, \ldots, x_m$ is differentiable at t, then

$$\lim_{h \to 0} \frac{\|\mathbf{f}(t+h) - \mathbf{f}(t) - \mathbf{T}(h)\|}{h} = 0, \tag{96}$$

where

$$\mathbf{T}(h) = (x_1'(t), \ldots, x_m'(t))h. \tag{97}$$

The function $\mathbf{T}$ is linear in h. (See Exercise 12.7.1.) We say $\mathbf{f}$ is *differentiable* at t and that the *differential* of $\mathbf{f}$ at t is the function $\mathbf{T}$ whose value at h is $\mathbf{T}(h)$. Note that $\mathbf{T}$ depends on the point t.

Note. Our definition of differentiability at t assumes the differentiability of each of the functions $x_1, \ldots, x_m$ at t. This assumption is similar to our definition for real-valued functions of several variables in which we assumed existence of the appropriate partial derivatives. This assumption may seem natural, but technically we are seeking *any* linear function $\mathbf{T}$ that satisfies (96). A priori, it need not satisfy (97).

It is a fact, however, that if $\mathbf{f}$ is differentiable at t, then the functions $x_1, \ldots, x_m$ are differentiable at t. Furthermore, if a linear function $\mathbf{T}$ satisfies (96), it must be given by (97).

To see this, suppose $\mathbf{f} : \mathbb{R} \to \mathbb{R}^m$ is differentiable at t and $\mathbf{T} = \mathbf{T}_t$ is a linear function of h. Then there exists $(a_1, \ldots, a_m) \in \mathbb{R}^m$ such that $\mathbf{T}(h) = (a_1, \ldots, a_m)h$ for all $h \in \mathbb{R}$. The ith component of the vector

$$\frac{\mathbf{f}(t+h) - \mathbf{f}(t) - \mathbf{T}(h)}{h}$$

is

$$\frac{x_i(t+h) - x_i(t)}{h} - a_i.$$

In order for (96) to hold, it is necessary and sufficient that

$$\lim_{h \to 0} \frac{x_i(t+h) - x_i(t)}{h} = a_i$$

for $i = 1, \ldots, m$, that is, that $a_i = x_i'(t)$, as was to be shown.

At this point we should contrast two statements that students sometimes confuse. Both statements are true.

A function $\mathbf{f} : \mathbb{R} \to \mathbb{R}^m$ is differentiable at t if and only if all the coordinate functions $x_1, \ldots, x_m$ are differentiable at t.

If $f : \mathbb{R}^n \to \mathbb{R}$ is differentiable at $\mathbf{x} \in \mathbb{R}^n$, then all the partials $\dfrac{\partial f}{\partial x_1}, \ldots, \dfrac{\partial f}{\partial x_n}$ exist at $\mathbf{x}$. The converse is false: All the partials can exist without f being differentiable.

The situation is similar to a corresponding one for continuity. $\mathbf{f} : \mathbb{R} \to \mathbb{R}^m$ is continuous at a point t if and only if all the coordinate functions are continuous at the point. But a functions $f : \mathbb{R}^n \to \mathbb{R}$ can be continuous in each variable separately at a point without being continuous there.

We end this section by noting that the reason for taking our definition of differentiability was to obtain the kind of linear approximation to $\mathbf{f}$ that we wanted—approximations consistent with the spirit of those in previous sections. It does *not* imply that the curve given parametrically by

$$\begin{aligned} x_1 &= x_1(t) \\ x_2 &= x_2(t) \\ &\vdots \\ x_m &= x_m(t) \end{aligned} \qquad a \le t \le b$$

has a tangent line at a point $(x_1(t), \ldots, x_m(t))$ just because all the $x_1, \ldots, x_m$ are differentiable at t.

Example 12.52 Define the curve

$$\begin{aligned} x(t) &= \begin{cases} t^2, & \text{if } 0 \le t \le 1 \\ -t^2, & \text{if } -1 \le t < 0 \end{cases} \\ y(t) &= t^2, \quad \text{if } -1 \le t \le 1. \end{aligned} \tag{98}$$

Then $x'(0) = y'(0) = 0$ so the function $\mathbf{f}(t) = (x(t), y(t))$ is differentiable at $t = 0$. But the curve given by (98) is just the curve $y = |x|$ which does not have a tangent at $(0, 0) = \mathbf{f}(0)$. ◀

Exercises

12.7.1 Show that $\mathbf{T}$, as given in (97) is linear in h, that is, $\mathbf{T}(\alpha h) = \alpha \mathbf{T}(h)$ and $\mathbf{T}(h_1 + h_2) = \mathbf{T}(h_1) + \mathbf{T}(h_2)$ whenever α, h_1, $h_2 \in \mathbb{R}$.

12.7.2 Let

$$\begin{array}{rcl} x(t) & = & t \\ y(t) & = & |t| \end{array} \qquad (-1 \leq t \leq 1).$$

Show that these equations define the same curve as the curve given by (98), yet the function $g(t) = (x(t), y(t))$ given here is not differentiable at $t = 0$. Show that there is *no* linear function $\mathbf{T}$ that satisfies equation (96).

12.8 Functions From $\mathbb{R}^n \to \mathbb{R}^m$

✂ *Advanced*

✂ This section requires some familiarity with linear algebra and may be omitted if necessary.

We have discussed thus far differentiability of functions from $\mathbb{R}^n$ to $\mathbb{R}$ and from $\mathbb{R}$ to $\mathbb{R}^m$. These are, of course, just special cases of the more general setting, functions from $\mathbb{R}^n$ to $\mathbb{R}^m$. We now turn to this general setting. Indeed much of what we covered so far was in an attempt to get to this level of abstraction in a gentle manner.

We first recall that our objective in Sections 12.4 and 12.7 was to approximate a function by a linear function in a certain sense. We have a similar objective here. The natural exposition of the material involves a linear algebraic viewpoint.

12.8.1 Review of Differentials and Derivatives

✂ *Enrichment*

Let's review our notion of the differential from the earlier settings but now using some language from linear algebra. For a differentiable function

$$f : \mathbb{R}^n \to \mathbb{R}$$

the differential at a point $\mathbf{x} = (x_1, \ldots, x_n)$ took the form

$$df(\mathbf{x}, \mathbf{h}) = \frac{\partial f}{\partial x_1} h_1 + \cdots + \frac{\partial f}{\partial x_n} h_n,$$

where $\mathbf{h} = (h_1, \ldots, h_n)$ and the partials are evaluated at $\mathbf{x}$. We can view this as a product of the $1 \times n$ matrix

$$\left(\frac{\partial f}{\partial x_1}, \ldots, \frac{\partial f}{\partial x_n} \right)$$

with the $n \times 1$ matrix

$$\begin{pmatrix} h_1 \\ \vdots \\ h_n \end{pmatrix},$$

so that the differential is the product of a row vector with a column vector. If we replace h_i by dx_i, $i = 1, \ldots, n$, in our notation and we define f' by

$$f'(\mathbf{x}) = \left(\frac{\partial f}{\partial x_1}, \ldots, \frac{\partial f}{\partial x_n} \right),$$

then we can write

$$d(f, d\mathbf{x}) = f'(\mathbf{x}) d\mathbf{x},$$

where

$$d\mathbf{x} = (dx_1, \ldots, dx_n).$$

(This notation is reminiscent of the notation in elementary calculus for the differential of a function from $\mathbb{R}$ to $\mathbb{R}$.)

Let us now look at the situation we discussed in Section 12.7, where we dealt with functions $\mathbf{f} : \mathbb{R} \to \mathbb{R}^m$ so that, in our usual notation,

$$\mathbf{f} = (f^1, f^2, f^3, \ldots, f^m).$$

There the differential took the form

$$d\mathbf{f}(x, h) = \left(\frac{df^1}{dx}, \frac{df^2}{dx}, \ldots, \frac{df^m}{dx} \right) h,$$

with the derivatives evaluated at x. We can write this in matrix notation as

$$d\mathbf{f}(x, h) = \begin{pmatrix} \frac{df^1}{dx} \\ \frac{df^2}{dx} \\ \vdots \\ \frac{df^m}{dx} \end{pmatrix} (h).$$

If we write

$$\mathbf{f}'(x) = \begin{pmatrix} \frac{df^1}{dx} \\ \frac{df^2}{dx} \\ \vdots \\ \frac{df^m}{dx} \end{pmatrix},$$

then the differential takes the form

$$d\mathbf{f}(x, h) = \mathbf{f}'(x) h \quad \text{or} \quad d\mathbf{f}(x, dx) = \mathbf{f}'(x)\, dx.$$

In both of these cases, we have a notion of a derivative $\mathbf{f}'(\mathbf{x})$ and of a differential

$$d\mathbf{f}(\mathbf{x}, \mathbf{h}) = \mathbf{f}'(\mathbf{x})\mathbf{h}$$

Domain	Range	$A = \mathbf{f}'(\mathbf{x}) =$	$d\mathbf{f}(\mathbf{x},\mathbf{h}) = \mathbf{f}'(\mathbf{x})\mathbf{h}$
$\mathbb{R}$	$\mathbb{R}$	$f'(x)$	$f'(x)h$
$\mathbb{R}^n$	$\mathbb{R}$	$\left(\frac{\partial f}{\partial x_1}, \ldots, \frac{\partial f}{\partial x_1}\right)$	$\left(\frac{\partial f}{\partial x_1}, \ldots, \frac{\partial f}{\partial x_n}\right)\begin{pmatrix} h_1 \\ \vdots \\ h_n \end{pmatrix}$
$\mathbb{R}$	$\mathbb{R}^m$	$\begin{pmatrix} \frac{df^1}{dx} \\ \frac{df^2}{dx} \\ \vdots \\ \frac{df^m}{dx} \end{pmatrix}$	$\begin{pmatrix} \frac{df^1}{dx} \\ \frac{df^2}{dx} \\ \vdots \\ \frac{df^m}{dx} \end{pmatrix}(h)$

Figure 12.8. The form of the derivative and differential.

or

$$d\mathbf{f}(\mathbf{x}, d\mathbf{x}) = \mathbf{f}'(\mathbf{x})d\mathbf{x}.$$

In each case we can view the derivative $\mathbf{f}'(\mathbf{x})$ as a matrix representing a certain linear transformation, $\mathbf{A}_{\mathbf{x}}$, and the differential as the result of applying this transformation to the vector $\mathbf{h}$:

$$d\mathbf{f}(\mathbf{x}, \mathbf{h}) = \mathbf{A}_{\mathbf{x}}(\mathbf{h}).$$

Summary Let us summarize the preceding with a chart. We have a function defined on an open set D in $\mathbb{R}$ or in $\mathbb{R}^n$, with range in $\mathbb{R}$ or $\mathbb{R}^m$. The table in Figure 12.8 indicates the form of the derivative and of the differential at point $\mathbf{x}$.

Note. We view $\mathbf{f}'(\mathbf{x})$ and $\mathbf{h}$ as matrices representing linear transformations, and $d\mathbf{f}(\mathbf{x}, \mathbf{h})$ as a product of these matrices. (When we are dealing with the case of a real function on $\mathbb{R}$, the product of two numbers can be viewed as the product of 1×1 matrices.)

12.8.2 Definition of the Derivative

✂ *Enrichment*

Let us now turn to the general case. Suppose we have m linear equations in n variables

$$\begin{aligned} y_1 &= a_{11}x_1 + a_{12}x_2 + \cdots + a_{1n}x_n \\ y_2 &= a_{21}x_1 + a_{22}x_2 + \cdots + a_{2n}x_n \\ &\vdots \\ y_m &= a_{m1}x_1 + a_{m2}x_2 + \cdots + a_{mn}x_n \end{aligned} \tag{99}$$

These equations define a linear transformation from $\mathbb{R}^n$ to $\mathbb{R}^m$. We can write these equations as a matrix equation

$$\mathbf{y} = \mathbf{A}\mathbf{x},$$

where $\mathbf{x} = (x_1, \ldots, x_n)$, $\mathbf{y} = (y_1, \ldots, y_m)$, and $\mathbf{A} = (a_{ij})$ is the $m \times n$ matrix of coefficients.

A general function from $\mathbb{R}^n$ to $\mathbb{R}^m$ can be expressed in a form similar to (99). For $\mathbf{f} = (f^1, \ldots, f^m)$ write

$$\begin{aligned} y_1 &= f^1(x_1, \ldots, x_n) \\ y_2 &= f^2(x_1, \ldots, x_n) \\ &\vdots \\ y_m &= f^m(x_1, \ldots, x_n) \end{aligned} \tag{100}$$

(Here, as before, superscripts are used to identify the functions so as not to confuse them with partial derivatives.)

Our task is to approximate a function given by (100) using a linear transformation given by (99). Specifically, we seek a linear transformation $\mathbf{T}$ from $\mathbb{R}^n$ to $\mathbb{R}^m$, with $\mathbf{T} = \mathbf{T_x}$ depending on $\mathbf{x}$, such that

$$\lim_{\|\mathbf{h}\| \to 0} \frac{\|\mathbf{f}(\mathbf{x}+\mathbf{h}) - \mathbf{f}(\mathbf{x}) - \mathbf{T}(\mathbf{h})\|}{\|\mathbf{h}\|} = 0.$$

The norms involved can be any of the norms for $\mathbb{R}^m$ for the numerator and any of the norms for $\mathbb{R}^n$ for the denominator $\|\mathbf{h}\|$. This is so because, as we saw in Section 11.6, convergence with respect to one norm is equivalent to convergence with respect to another norm. To be specific, however, we shall use $\|t\| = \|t\|_1 =$ the sum of the absolute values of the components of t.

Definition 12.53 Let D be an open set in $\mathbb{R}^n$, $\mathbf{x}$ a point in D and suppose

$$\mathbf{f} : D \to \mathbb{R}^m.$$

We say $\mathbf{f}$ is *differentiable* at $\mathbf{x}$ if there exists a linear transformation $\mathbf{T_x}$ from $\mathbb{R}^n$ to $\mathbb{R}^m$ such that

$$\lim_{\|\mathbf{h}\| \to 0} \frac{\|\mathbf{f}(\mathbf{x}+\mathbf{h}) - \mathbf{f}(\mathbf{x}) - \mathbf{T_x}(\mathbf{h})\|}{\|\mathbf{h}\|} = 0. \tag{101}$$

Definition 12.54 The *differential* of $\mathbf{f}$, denoted by $d\mathbf{f}$ is defined at $\mathbf{x}$ by

$$d\mathbf{f}(\mathbf{x},\mathbf{h}) = \mathbf{T}_{\mathbf{x}}(\mathbf{h})$$

for all $\mathbf{h} \in \mathbb{R}^n$.

Definition 12.55 The *derivative* of $\mathbf{f}$ at $\mathbf{x}$ is the function

$$\mathbf{f}'(\mathbf{x}) = \mathbf{T}_{\mathbf{x}}$$

defined on the set of points of differentiability of $\mathbf{f}$.

Observe that $\mathbf{f}'(\mathbf{x})$ is the linear transformation $\mathbf{T}_{\mathbf{x}}$. Its domain is $\mathbb{R}^n$ and its range is $\mathbb{R}^m$. When $\mathbf{f}$ is differentiable at $\mathbf{x}$ we can write (101) in the form

$$\lim_{||\mathbf{h}||\to 0} \frac{||\mathbf{f}(\mathbf{x}+\mathbf{h}) - \mathbf{f}(\mathbf{x}) - \mathbf{f}'(\mathbf{x})\mathbf{h}||}{||\mathbf{h}||} = 0. \qquad (102)$$

This form resembles the familiar form of differentiability for functions from $\mathbb{R}$ to $\mathbb{R}$, when absolute values are replaced by norms.

Exercises

12.8.1 Verify that the validity of (101) does not depend on the norms for $\mathbb{R}^n$ and $\mathbb{R}^m$ that are used.

12.8.2 Let $\mathbf{A}$ be a linear transformation from $\mathbb{R}^n$ to $\mathbb{R}^m$. Show that for each $\mathbf{x} \in \mathbb{R}^n$, $\mathbf{A}'(\mathbf{x}) = \mathbf{A}$.

12.8.3 Suppose $\mathbf{f},\mathbf{g} : \mathbb{R}^n \to \mathbb{R}^m$ and $\alpha \in \mathbb{R}$. Prove that if $\mathbf{f}$ and $\mathbf{g}$ are differentiable at $\mathbf{x}$, then so too are $\mathbf{f}+\mathbf{g}$ and $\alpha\mathbf{f}$. In addition,

$$(\mathbf{f}+\mathbf{g})'(\mathbf{x}) = \mathbf{f}'(\mathbf{x}) + \mathbf{g}'(\mathbf{x})$$

and

$$(\alpha\mathbf{f})'(\mathbf{x}) = \alpha\mathbf{f}'(\mathbf{x}).$$

12.8.4 Suppose $\mathbf{f} : \mathbb{R}^n \to \mathbb{R}$ and $\mathbf{g} : \mathbb{R}^n \to \mathbb{R}^m$. Prove that if $\mathbf{f}$ and $\mathbf{g}$ are differentiable at $\mathbf{x}$, then so too is the product function $\mathbf{fg}$ and that

$$(\mathbf{fg})'(\mathbf{x}) = \mathbf{f}'(\mathbf{x})\mathbf{g}(\mathbf{x}) + \mathbf{f}(\mathbf{x})\mathbf{g}'(\mathbf{x}).$$

12.8.3 Jacobians

✂ *Enrichment*

With these preliminaries completed, we ask our first question. If $\mathbf{f}$ is differentiable at $\mathbf{x}$, can we represent the linear transformation $\mathbf{f}'(\mathbf{x})$ by a matrix involving partial derivatives? For the cases of mappings from $\mathbb{R}^n$ to $\mathbb{R}$ or from $\mathbb{R}$ to $\mathbb{R}^m$, we saw how to do this (Fig. 12.8). These results suggest that a matrix representation of $\mathbf{f}'(\mathbf{x})$ would be an $m \times n$ Jacobian matrix. This is indeed the case.

Theorem 12.56 *Let D be an open subset of $\mathbb{R}^n$ and let $\mathbf{f} : D \to \mathbb{R}^m$. Let $(f^1, \ldots, f^m)$ be the component functions of $\mathbf{f}$ as in (100). Thus*

$$\mathbf{f} = (f^1, \ldots, f^m)$$

with $f^i : D \to \mathbb{R}$ for each $i = 1, \ldots, m$. Then $\mathbf{f}$ is differentiable at $\mathbf{x} \in D$ if and only if each of the functions f^i is differentiable at $\mathbf{x}$. When $\mathbf{f}$ is differentiable at $\mathbf{x}$, the derivative $\mathbf{f}'(\mathbf{x})$ can be represented by the Jacobian matrix

$$\mathbf{T_x} = \mathbf{f}'(\mathbf{x}) = \begin{pmatrix} \frac{\partial f^1}{\partial x_1} & \frac{\partial f^1}{\partial x_2} & \cdots & \frac{\partial f^1}{\partial x_n} \\ \frac{\partial f^2}{\partial x_1} & \frac{\partial f^2}{\partial x_2} & \cdots & \frac{\partial f^2}{\partial x_n} \\ \vdots & & & \\ \frac{\partial f^m}{\partial x_1} & \frac{\partial f^m}{\partial x_2} & \cdots & \frac{\partial f^m}{\partial x_n} \end{pmatrix}, \tag{103}$$

where the partial derivatives are evaluated at $\mathbf{x}$.

Proof The proof of this theorem involves no more than looking at the component functions f^i. Suppose first that $\mathbf{f}$ is differentiable at $\mathbf{x}$. This means that

$$\lim_{||\mathbf{h}|| \to 0} \frac{||\mathbf{f}(\mathbf{x}+\mathbf{h}) - \mathbf{f}(\mathbf{x}) - \mathbf{f}'(\mathbf{x})\mathbf{h}||}{||\mathbf{h}||} = 0. \tag{104}$$

Now $\mathbf{f}'(\mathbf{x})$, being a linear transformation from $\mathbb{R}^n$ to $\mathbb{R}^m$, has a matrix representation $\mathbf{A} = (a_{ij})$. Thus if $\mathbf{h} = (h_1, \ldots, h_n)$, then the ith component of $\mathbf{f}'(\mathbf{x})\mathbf{h}$ is $\sum_{j=1}^{n} a_{ij} h_j$, and the ith component of

$$\mathbf{f}(\mathbf{x}+\mathbf{h}) - \mathbf{f}(\mathbf{x})$$

is $f^i(\mathbf{x}+\mathbf{h}) - f^i(\mathbf{x})$. As a result, the ith component of

$$\mathbf{f}(\mathbf{x}+\mathbf{h}) - \mathbf{f}(\mathbf{x}) - \mathbf{f}'(\mathbf{x})\mathbf{h}$$

is

$$f^i(\mathbf{x}+\mathbf{h}) - f^i(\mathbf{x}) - \sum_{j=1}^{n} a_{ij} h_j.$$

Now

$$|f^i(\mathbf{x}+\mathbf{h}) - f^i(\mathbf{x}) - \sum_{j=1}^{n} a_{ij} h_j| \le ||\mathbf{f}(\mathbf{x}+\mathbf{h}) - \mathbf{f}(\mathbf{x}) - \mathbf{f}'(\mathbf{x})\mathbf{h}||$$

so, because of (104),

$$\lim_{||\mathbf{h} \to 0||} \frac{|f^i(\mathbf{x}+\mathbf{h}) - f^i(\mathbf{x}) - \sum_{j=1}^{n} a_{ij} h_j|}{||\mathbf{h}||} = 0.$$

Thus f^i is differentiable at $\mathbf{x}$. Furthermore,

$$a_{ij} = \frac{\partial f^i}{\partial x_j}. \tag{105}$$

The verification of (105) is similar to the proof of the corresponding fact in the previous section. (We leave it as Exercise 12.8.5.)

To prove the converse, assume each of the functions f^i is differentiable at $\mathbf{x}$. This means

$$\lim_{\|\mathbf{h}\|\to 0} \frac{|f^i(\mathbf{x}+\mathbf{h}) - f^i(\mathbf{x}) - \sum_{j=1}^n \frac{\partial f^i}{\partial x_j} h_j|}{\|\mathbf{h}\|} = 0. \tag{106}$$

Now (106) is valid for each $i = 1, \dots, m$. But the norm in the numerator of (104) is just the sum

$$\sum_{i=1}^m \left| f^i(\mathbf{x}+\mathbf{h}) - f^i(\mathbf{x}) - \sum_{j=1}^n \frac{\partial f^i}{\partial x_j} h_j \right|,$$

each term of which approaches zero when divided by $\|\mathbf{h}\|$. Hence so does the sum, and (104) is satisfied. Thus $\mathbf{f}$ is differentiable at $\mathbf{x}$. ■

Observe that the requirement for differentiability of $\mathbf{f}$ at $\mathbf{x}$ can be written in the form

$$\mathbf{f}(\mathbf{x}+\mathbf{h}) - \mathbf{f}(\mathbf{x}) = \mathbf{f}'(\mathbf{x})\mathbf{h} + \mathbf{E}(\mathbf{h})\|\mathbf{h}\|, \tag{107}$$

where the error term $\mathbf{E}$ satisfies

$$\lim_{\|\mathbf{h}\|\to 0} \|\mathbf{E}(\mathbf{h})\| = 0 \text{ and } \mathbf{E}(\mathbf{0}) = \mathbf{0}.$$

This observation shows that when $\mathbf{f}$ is differentiable at $\mathbf{x}$, $\mathbf{f}$ is also continuous at $\mathbf{x}$. It also highlights the sense in which the linear transformation $\mathbf{f}'(\mathbf{x})$ approximates $\mathbf{f}$ near $\mathbf{x}$. We see that (107) is just a compact way of expressing condition 2 of Definition 12.16 when we are dealing with functions from $\mathbb{R}^n$ to $\mathbb{R}$.

Example 12.57 An example will clarify (107). Let

$$\begin{aligned} u &= xy \\ v &= x + y \\ \mathbf{f}(x, y) &= (u, v). \end{aligned}$$

By (103)

$$\mathbf{f}'(\mathbf{x}) = \begin{pmatrix} \frac{\partial u}{\partial x} & \frac{\partial u}{\partial y} \\ \frac{\partial v}{\partial x} & \frac{\partial v}{\partial y} \end{pmatrix} = \begin{pmatrix} y & x \\ 1 & 1 \end{pmatrix}.$$

Then (107) becomes

$$((x+h)(y+k) - xy, (x+h) + (y+k) - (x+y)) =$$

$$\begin{pmatrix} y & x \\ 1 & 1 \end{pmatrix} \begin{pmatrix} h \\ k \end{pmatrix} + \mathbf{E}(h,k)(|h| + |k|).$$

Simplifying and solving for $\mathbf{E}(h,k)$ gives

$$\mathbf{E}(h,k) = \frac{(hk, 0)}{|h| + |k|}.$$

Thus

$$\|\mathbf{E}(h,k)\| = \frac{|hk|}{|h| + |k|}.$$

If $h = 0$ or $k = 0$, $\|\mathbf{E}(h,k)\| = 0$. Otherwise,

$$\|\mathbf{E}(h,k)\| = \frac{1}{|1/k| + |1/h|}.$$

Thus $\lim_{(h,k)\to(0,0)} \|\mathbf{E}(h,k)\| = 0$. ◀

Exercises

12.8.5 Verify equation (105) in the proof of Theorem 12.56.

12.8.4 Chain Rules

✂ *Enrichment*

In Section 12.5 we discussed chain rules and we indicated how employing a "schema" might be useful in writing down relevant chains. All such chain rules, as well as the familiar calculus chain rule (Theorem 7.11) involving functions from $\mathbb{R}$ to $\mathbb{R}$, are special cases of a general chain rule. We consider this chain rule next. Our framework of linear algebra together with the convenient notation we adopted in this section make this chain rule easy to prove and easy to remember.

Before we state and prove this chain rule, let us recall one of the chain rules (Theorem 12.36) from Section 12.5.6. This is the rule associated with the schema we have already seen in Example 12.30:

(108)

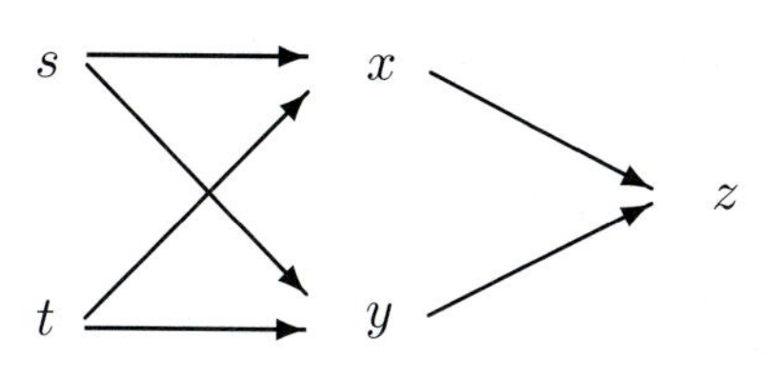

In the terminology of this section, we have here two functions, $\mathbf{f} : \mathbb{R}^2 \to \mathbb{R}^2$, with coordinate functions $\mathbf{x}$ and $\mathbf{y}$; and $g : \mathbb{R}^2 \to \mathbb{R}$ indicated in (108) as z, $z = g(\mathbf{x}, \mathbf{y})$. Assuming differentiability as needed, ignoring for the moment the points at which partials are evaluated, and using the notation suggested by (108), we find g' represented by the matrix $\mathbf{B} = \left(\frac{\partial z}{\partial x}, \frac{\partial z}{\partial y}\right)$ and $\mathbf{f}'$ by the matrix

$$\mathbf{A} = \begin{pmatrix} \frac{\partial x}{\partial s} & \frac{\partial x}{\partial t} \\ \frac{\partial y}{\partial s} & \frac{\partial y}{\partial t} \end{pmatrix}.$$

Thus

$$\begin{aligned} \mathbf{BA} &= \left(\frac{\partial z}{\partial x}, \frac{\partial z}{\partial y}\right) \begin{pmatrix} \frac{\partial x}{\partial s} & \frac{\partial x}{\partial t} \\ \frac{\partial y}{\partial s} & \frac{\partial y}{\partial t} \end{pmatrix} \\ &= \left(\frac{\partial z}{\partial x}\frac{\partial x}{\partial s} + \frac{\partial z}{\partial y}\frac{\partial y}{\partial s}, \frac{\partial z}{\partial x}\frac{\partial x}{\partial t} + \frac{\partial z}{\partial y}\frac{\partial y}{\partial t}\right). \end{aligned}$$

This last expression is just the derivative of the composite function $F = g \circ \mathbf{f}$, since $\mathbf{F}' = \left(\frac{\partial z}{\partial s}, \frac{\partial z}{\partial t}\right)$ and the components of $\mathbf{F}'$ are by Example 12.30 (or Theorem 12.36) the components of the vector $\mathbf{BA}$.

The preceding discussion is not precise, but it does suggest a chain rule that concludes with the appealing formula

$$(g \circ \mathbf{f})'(\mathbf{x}) = g'(\mathbf{f}(\mathbf{x}))\mathbf{f}'(\mathbf{x}).$$

Before we attempt a proof of this chain rule formula (which we do in Theorem 12.58), let us try to get a sense of why it works. In Section 7.3.2 we motivated the chain rule for functions from $\mathbb{R}$ to $\mathbb{R}$ by viewing the absolute value of the derivative at a point as a local magnification factor. When dealing with $\mathbf{f} : \mathbb{R}^n \to \mathbb{R}^n$ we can do something similar by replacing the absolute value of f' by the determinant of the Jacobian.

Suppose first that $\mathbf{A}$ is a linear transformation from $\mathbb{R}^2$ to $\mathbb{R}^2$, say $\mathbf{A}(u, v) = (x, y)$, where $x = au$ and $y = bv$. Then

$$\mathbf{A}'(u, v) = \begin{pmatrix} \frac{\partial x}{\partial u} & \frac{\partial x}{\partial v} \\ \frac{\partial y}{\partial u} & \frac{\partial y}{\partial v} \end{pmatrix} = \begin{pmatrix} a & 0 \\ 0 & b \end{pmatrix}.$$

If we view $\mathbf{A}$ as a mapping from $\mathbb{R}^2$ to $\mathbb{R}^2$, we see it stretches a set E horizontally by a factor of a and vertically by a factor of b. (For example, if $a = 2$ and $b = 3$, $\mathbf{A}$ maps the unit square $[0, 1] \times [0, 1]$ onto the rectangle $[0, 2] \times [0, 3]$.) Thus if E and $\mathbf{A}(E)$ have area,

$$\frac{\text{Area}\, \mathbf{A}(E)}{\text{Area}\, E} = ab.$$

The mapping $\mathbf{A}$ magnifies the area of a set by a factor of ab. This is reflected by calculating the determinant of $\mathbf{A}'$,

$$\begin{vmatrix} a & 0 \\ 0 & b \end{vmatrix} = ab.$$

Now take any $\mathbf{f} : \mathbb{R}^2 \to \mathbb{R}^2$ that is differentiable at a point $(u, v) \in \mathbb{R}^2$. If

$$\mathbf{f}'(u, v) = \begin{pmatrix} a & 0 \\ 0 & b \end{pmatrix},$$

then the mapping defined by $\mathbf{f}$ has a magnification factor ab in a limit sense near (u, v). In suggestive notation similar to that in Section 7.3.2 we could write

$$\lim_{J \to (u,v)} \frac{\text{Area}\,\mathbf{f}(J)}{\text{Area}\,J} = ab.$$

This is so because of the sense in which the linear transformation approximates $\mathbf{f}$ near a point of differentiability of $\mathbf{f}$.

Now consider another mapping $\mathbf{g} : \mathbb{R}^2 \to \mathbb{R}^2$ with derivative $\mathbf{g}'$ at $\mathbf{f}(\mathbf{x})$. Its derivative at $\mathbf{f}(\mathbf{x})$ is a linear transformation $\mathbf{B} = \mathbf{g}'(\mathbf{f}(\mathbf{x}))$. The magnification factor of $\mathbf{g}$ at $\mathbf{f}(\mathbf{x})$ is the same as the magnification factor of $\mathbf{B}$, namely the determinant of $\mathbf{B}$.

Thus if α is the magnification factor of $\mathbf{A}$ (and of $\mathbf{f}$ at $\mathbf{x}$) and β is the magnification factor of $\mathbf{B}$ (and of $\mathbf{g}$ at $\mathbf{f}(\mathbf{x})$), then $\beta\alpha$ is the magnification factor of $\mathbf{BA}$ (and of $\mathbf{g} \circ \mathbf{f}$ at $\mathbf{x}$). This suggests that the linear transformation $\mathbf{BA}$ is the derivative of $\mathbf{g} \circ \mathbf{f}$ at $\mathbf{x}$, at least for the case $n = m = p$. Theorem 12.58 says this is the case for all m, n, p.

We need to recall a bit of linear algebra that will appear in the proof of Theorem 12.58. Let $\mathbf{A}$ be a linear transformation from $\mathbb{R}^n$ to $\mathbb{R}^m$. Since $\mathbf{A}$ and the norm are continuous (see Exercise 11.7.10), $\|\mathbf{A}\mathbf{x}\|$ achieves a maximum on the compact set $\{x : \|\mathbf{x}\| = 1\}$. We denote this maximum by $\|\mathbf{A}\|$. Thus

$$\|\mathbf{A}\| = \max_{\|\mathbf{x}\|=1} \|\mathbf{A}\mathbf{x}\|.$$

We use this "norm" notation because $\|\cdot\|$ is actually a norm on the set of linear transformations from $\mathbb{R}^n$ to $\mathbb{R}^m$, but we will not need this fact. Observe that for all $\mathbf{x} \in \mathbb{R}^n$, $(\mathbf{x} \neq \mathbf{0})$,

$$\|\mathbf{A}\mathbf{x}\| = \left\| \mathbf{A}\left(\frac{\mathbf{x}}{\|\mathbf{x}\|}\right) \right\| \|\mathbf{x}\| \leq \|\mathbf{A}\| \, \|\mathbf{x}\|.$$

In the proof that follows there will be several different norms, all denoted by $\|\cdot\|$. Each of the spaces $\mathbb{R}^n$, $\mathbb{R}^m$,' and $\mathbb{R}^p$ have norms, as do the sets of linear transformations from one of the spaces to another. We could have used different notations for different norms, (e.g., $\|\|\cdot\|\|$) but this might cause eye strain. Instead, we follow the common practice of using only one symbol for all the norms, trusting that the context will make it clear which norm is meant.

Exercises

12.8.6 Show that the set of all linear transformations from $\mathbb{R}^n$ to $\mathbb{R}^m$ is a vector space (under appropriate interpretations of sum and scalar product) and show that

$$\|\mathbf{A}\| = \max_{\|\mathbf{x}\|=1} \|\mathbf{A}\mathbf{x}\|$$

defines a norm for elements of this space.

12.8.5 Proof of Chain Rule

Enrichment

We are now in a position to give a formal statement and proof of the general chain rule formula.

Theorem 12.58 *Let D be an open set in $\mathbb{R}^n$. Let $\mathbf{f} : D \to \mathbb{R}^m$ be differentiable at $\mathbf{x_0} \in D$. Let H be an open set containing $\mathbf{f}(D)$ and let $\mathbf{g} : H \to \mathbb{R}^p$ be differentiable at $\mathbf{f}(\mathbf{x_0})$. Then the function $\mathbf{F} : D \to \mathbb{R}^p$ defined by*

$$\mathbf{F}(\mathbf{x}) = \mathbf{g}(\mathbf{f}(\mathbf{x}))$$

is differentiable at $\mathbf{x_0}$ and

$$\mathbf{F}'(\mathbf{x_0}) = \mathbf{g}'(\mathbf{f}(\mathbf{x_0}))\mathbf{f}'(\mathbf{x_0}). \tag{109}$$

The product in (109) is a product of two linear transformations.

Proof Let

$$\mathbf{y_0} = \mathbf{f}(\mathbf{x_0}), \quad \mathbf{A} = \mathbf{f}'(\mathbf{x_0}), \quad \text{and} \quad \mathbf{B} = \mathbf{g}'(\mathbf{y_0}).$$

We wish to show that

$$\frac{\|\mathbf{F}(\mathbf{x_0}+\mathbf{h}) - \mathbf{F}(\mathbf{x_0}) - \mathbf{B}\mathbf{A}\,\mathbf{h}\|}{\|\mathbf{h}\|} \to 0 \text{ as } \mathbf{h} \to \mathbf{0}. \tag{110}$$

Since $\mathbf{f}$ is differentiable at $\mathbf{x_0}$ and $\mathbf{g}$ is differentiable at $\mathbf{y_0}$ we can, using (107), write for $\mathbf{x_0}+\mathbf{h} \in D$ and $\mathbf{y_0}+\mathbf{k} \in H$,

$$\begin{array}{rcl} \mathbf{f}(\mathbf{x_0}+\mathbf{h}) - \mathbf{f}(\mathbf{x_0}) - \mathbf{A}\,\mathbf{h} & = & \mathbf{E}_1(\mathbf{h})\|\mathbf{h}\|, \\ \mathbf{g}(\mathbf{y_0}+\mathbf{k}) - \mathbf{g}(\mathbf{y_0}) - \mathbf{B}\,\mathbf{k} & = & \mathbf{E}_2(\mathbf{k})\|\mathbf{k}\| \end{array}, \tag{111}$$

where

$$\|\mathbf{E}_1(\mathbf{h})\| \to 0 \text{ as } \mathbf{h} \to \mathbf{0} \text{ and } \|\mathbf{E}_2(\mathbf{k})\| \to 0 \text{ as } \mathbf{k} \to \mathbf{0}. \tag{112}$$

We now estimate the numerator in (110). For a given $\mathbf{h}$, let

$$\mathbf{k} = \mathbf{f}(\mathbf{x_0}+\mathbf{h}) - \mathbf{f}(\mathbf{x_0}). \tag{113}$$

From (111) we obtain

$$\begin{aligned} \|\mathbf{k}\| &= \|\mathbf{A}\,\mathbf{h} + \mathbf{E}_1(\mathbf{h})\,\|\mathbf{h}\|\,\| \\ &\le \|\mathbf{A}\|\,\|\mathbf{h}\| + \|\mathbf{E}_1(\mathbf{h})\|\,\|\mathbf{h}\| \\ &= (\|\mathbf{A}\| + \|\mathbf{E}_1(\mathbf{h})\|)\,\|\mathbf{h}\|. \end{aligned} \tag{114}$$

From (113) we see that

$$\mathbf{f}(\mathbf{x}_0 + \mathbf{h}) = \mathbf{f}(\mathbf{x}_0) + \mathbf{k} = \mathbf{y}_0 + \mathbf{k},$$

so

$$\begin{aligned} &\mathbf{F}(\mathbf{x}_0 + \mathbf{h}) - \mathbf{F}(\mathbf{x}_0) = \\ &\quad \mathbf{g}(\mathbf{f}(\mathbf{x}_0 + \mathbf{h})) - \mathbf{g}(\mathbf{y}_0) = \mathbf{g}(\mathbf{y}_0 + \mathbf{k}) - \mathbf{g}(\mathbf{y}_0). \end{aligned}$$

Thus

$$\mathbf{F}(\mathbf{x}_0 + \mathbf{h}) - \mathbf{F}(\mathbf{x}_0) - \mathbf{BAh} = \mathbf{g}(\mathbf{y}_0 + \mathbf{k}) - \mathbf{g}(\mathbf{y}_0) - \mathbf{BAh}. \tag{115}$$

Using (111), we express the right side of (115) as

$$\begin{aligned} &\mathbf{Bk} + \mathbf{E}_2(\mathbf{k})\,\|\mathbf{k}\| - \mathbf{BAh} = \\ &\quad \mathbf{B}(\mathbf{k} - \mathbf{Ah}) + \mathbf{E}_2(\mathbf{k})\,\|\mathbf{k}\| = \mathbf{BE}_1(\mathbf{h})\,\|\mathbf{h}\| + \mathbf{E}_2(\mathbf{k})\,\|\mathbf{k}\|. \end{aligned}$$

This estimate for

$$\mathbf{F}(\mathbf{x}_0 + \mathbf{h}) - \mathbf{F}(\mathbf{x}_0) - \mathbf{BAh}$$

together with (112) and (114), implies that for $\mathbf{h} \ne \mathbf{0}$,

$$\begin{aligned} &\frac{\|\mathbf{F}(\mathbf{x}_0 + \mathbf{h}) - \mathbf{F}(\mathbf{x}_0) - \mathbf{BAh}\|}{\|\mathbf{h}\|} \\ &\le \frac{\|\mathbf{BE}_1(\mathbf{h})\,\|\mathbf{h}\|\,\| + \|\mathbf{E}_2(\mathbf{k})\,\|\mathbf{k}\|\,\|}{\|\mathbf{h}\|} \\ &\le \|\mathbf{B}\|\,\mathbf{E}_1(\mathbf{h}) + \|\mathbf{E}_2(\mathbf{k})(\|\mathbf{A}\| + \|\mathbf{E}_1(\mathbf{h})\|)\|. \end{aligned} \tag{116}$$

Now $\|\mathbf{B}\|$ and $\|\mathbf{A}\|$ are fixed numbers, and

$$\mathbf{E}_1(\mathbf{h}) \to 0 \quad \text{as} \quad \mathbf{h} \to \mathbf{0}.$$

Furthermore, $\mathbf{k} \to \mathbf{0}$ as $\mathbf{h} \to \mathbf{0}$ by (113) and the continuity of $\mathbf{f}$ at $\mathbf{x}_0$, so

$$\mathbf{E}_2(\mathbf{k}) \to 0 \quad \text{as} \quad \mathbf{h} \to \mathbf{0}.$$

Thus the expression in (116) approaches zero as $\mathbf{h} \to \mathbf{0}$. But we have seen that this expression dominates the quotient in (110), so (110) is satisfied, and $\mathbf{F}$ is differentiable at $\mathbf{x}_0$ with derivative $\mathbf{BA}$. Since

$$\mathbf{BA} = \mathbf{g}'(\mathbf{f}(\mathbf{x}_0))\mathbf{f}'(\mathbf{x}_0),$$

our proof is complete. ■

Exercises

12.8.7 If $u : \mathbb{R}^n \to \mathbb{R}$ is differentiable, then we obtained earlier (Example 12.25) the formula

$$d(e^u) = e^u \, du$$

but we were required to assume that e^u is also differentiable. Use Theorem 12.58 to justify that assumption and to check the formula.

12.8.8 Use Theorem 12.58 to obtain an explicit formula for the schema

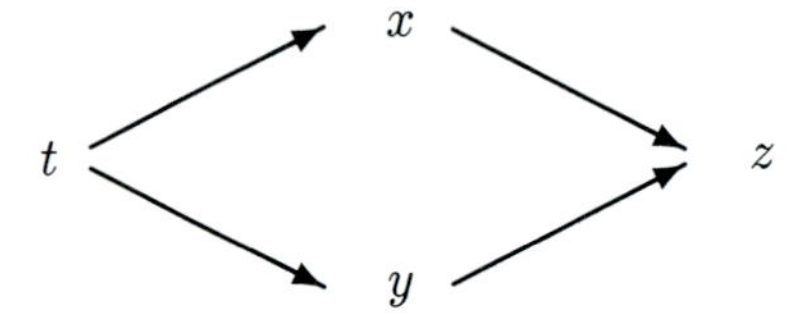

12.8.9 Use Theorem 12.58 to obtain an explicit formula for the schema

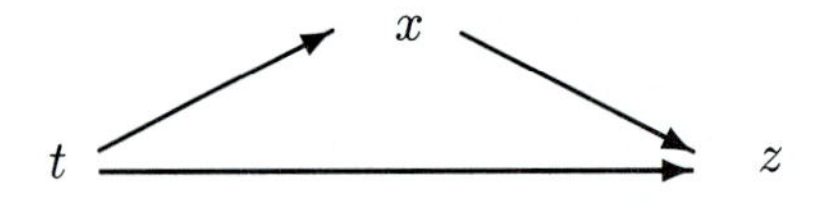

12.8.10 State a theorem that covers a situation of the form

$$\mathbf{F} = \mathbf{h} \circ \mathbf{g} \circ \mathbf{f}.$$

12.8.11 Let

$$\begin{array}{lll} x = r\cos\theta & u = x - y & f(r,\theta) = (x,y) \\ y = r\sin\theta & v = x + y & g(x,y) = (u,v) \ . \end{array}$$

(a) Calculate f', g' and $(g \circ f)'$ directly from the definition of derivative.

(b) Calculate $(g \circ f)'$ from the chain rule.

(c) Calculate the magnification factors for f at $(2,0)$, g at $f((2,0))$, and g at $(2,0)$.

12.8.12 Let

$$\begin{array}{lll} x = re^s & u = y^2 & f(r,s) = (x,y,z) \\ y = se^r & v = x + z & g(x,y,z) = (u,v) \\ z = r & & \end{array}$$

(a) Calculate f', g' and $(g \circ f)'$ directly from the definition.

(b) Calculate $(g \circ f)'$ from the chain rule and compare with the result in (a).

(c) Check that $(g \circ f)'(3,4) = \begin{pmatrix} 32e^6 & 8e^6 \\ e^4 + 1 & 3e^4 \end{pmatrix}$.

12.8.13 Prove the inverse function theorem of Section 12.6.4 using the chain rule of this section.

Chapter 13

METRIC SPACES

13.1 Introduction

The real number system $\mathbb{R}$ is more than just a set of objects. It has various structures we were able to exploit in the preceding chapters. There was an algebraic structure that allowed us to add, subtract, multiply and divide numbers. There were order structures on $\mathbb{R}$ given by "$<$" (or "$\leq$", "$>$", "$\geq$"). And there was a notion of distance between pairs of members of $\mathbb{R}$ given by $|x-y|$. These structures interacted in various ways. In fact, we were able to characterize the real number system $\mathbb{R}$ via these structures: Speaking loosely, we can say that $\mathbb{R}$ is the only complete ordered field (Exercise 1.11.3). Some simple ways in which these structures interact can be found in the exercises at the end of this section.

Now, there are many sets of objects besides $\mathbb{R}$ that are important in analysis. We have encountered several in earlier chapters (e.g., sets of sequences or sets of functions). Many of these have natural algebraic structures (not necessarily that they form a field), natural order structures (not necessarily satisfying the axioms found in Section 1.3), and one or more natural notions of distance.

Consider, for example, the set $\mathcal{C}[0,1]$ of continuous functions on $[0,1]$. With the usual definitions for addition and multiplication of functions, $\mathcal{C}[0,1]$ is not a field. It does satisfy the axioms for a different algebraic structure, an *algebra* (see Exercise 13.1.2). If we write $f \leq g$ whenever $f(x) \leq g(x)$ for every $x \in (0,1)$ then "$\leq$" provides an order structure, but this structure does not satisfy the axioms indicated in Section 1.4 for $\mathbb{R}$. It does, however, satisfy another set of axioms, that for a *partially ordered set.* (Exercises 13.1.2 and 13.1.3 make these last two statements precise.)

There are also several notions of "distance" that can be considered between pairs of continuous functions. We shall see some of them in this chapter. We shall develop an abstract structure on sets of objects based on the distance concept. The resulting structure is called a *metric space.*

This abstract structure can be imposed on any nonempty set. It is most efficient to develop the material in an abstract fashion, but we shall spend a good deal of time applying the theory to specific "concrete" metric spaces. Our concrete examples will be quite varied. In particular, our applications will include metric spaces whose elements consist of such diverse objects as sequences, functions and sets.

You may wish to review how the notions of distance between pairs of points in $\mathbb{R}$ (and in $\mathbb{R}^n$) figured in the preceding chapters. From the distance notion, we obtained other central concepts of elementary analysis (open set, sequence limits, neighborhood, limits, continuous functions, etc.). The entire subject of elementary real analysis in $\mathbb{R}$ relies fundamentally on the distance notion.

Exercises

13.1.1 Verify each of the following facts (or find them explicitly stated in earlier sections of the book). Each of these facts provides an example of ways in which the algebraic, order, and/or distance structures interact. In each case, indicate which structures appear in the statement.

(a) $|x - z| \leq |x - y| + |y - z|$ for all $x, y, z \in \mathbb{R}$.

(b) If $\{a_k\} \to a$ and $\{b_k\} \to b$, then $\{a_k + b_k\} \to a + b$.

(c) If $x > 0$ and $y > 0$, then there exists $n \in \mathbb{N}$ such that $nx > y$.

(d) For every $n \in \mathbb{N}$, $n^2 \geq n$.

(e) If $\{a_k\} \to a$ and $a_k \leq b_k \leq a$ for all $k \in \mathbb{N}$, then $b_k \to a$.

(f) Every nonempty bounded set $E \subset \mathbb{R}$ has a least upper bound.

13.1.2 Let $\mathcal{C}[0,1]$ denote the continuous functions on $[0,1]$ furnished with the usual notions of addition and multiplication of functions. Prove that $\mathcal{C}[0,1]$ is an *algebra* of functions, that is, that it satisfies the following axioms:

(a) $\mathcal{C}[0,1]$ is a vector space.

(b) $f(gh) = (fg)h$ for all $f, g, h \in \mathcal{C}[0,1]$.

(c) $f(g+h) = fg + fh$ for all $f, g, h \in \mathcal{C}[0,1]$.

(d) $(cf)g = c(fg)$ for all $c \in \mathbb{R}$ and $f, g \in \mathcal{C}[0,1]$.

13.1.3 Let S be a set. A relation $a \preceq b$ defined for certain pairs in S is called a *partial order* on S if it satisfies the following axioms.

(a) $a \preceq a$ for all $a \in S$.

(b) If $a \preceq b$ and $b \preceq a$, then $a = b$.

(c) If $a \preceq b$ and $b \preceq c$, then $a \preceq c$.

We then say that S *is partially ordered by* $\preceq$. Show that the set $\mathcal{C}[0,1]$ of continuous functions on $[0,1]$ is partially ordered by the relation $f \preceq g$ if $f(x) \leq g(x)$ for all $x \in [0,1]$.

13.1.4 Find a partial order on the set $\mathcal{C}[0,1]$ of continuous functions on $[0,1]$ that is different from the partial order in Exercise 13.1.3.

13.1.5 Define a relation $f \preceq g$ on the set $\mathcal{C}[0,1]$ of continuous functions on $[0,1]$ if

$$\int_0^1 f(t)\,dt \leq \int_0^1 g(t)\,dt.$$

Is this a partial order as defined in Exercise 13.1.3?

13.1.6 Let $\mathcal{S}$ denote the set of all subsets of $\mathbb{R}$. Show that $\mathcal{S}$ is partially ordered by "$\subset$".

13.1.7 Let $\mathcal{P}$ denote the polynomials defined on $\mathbb{R}$. For $p_1, p_2 \in \mathcal{P}$, write $p_2 \succ p_1$ if there exists $n \in \mathbb{N}$ such that $p_2(x) > p_1(x)$ for all $x \geq n$. Does "$\succ$" satisfy the order conditions that are given for "$<$" (relative to $\mathbb{R}$) in Section 1.4? What would your answer be if we replaced $\mathcal{P}$ by the set of continuous functions on $\mathbb{R}$?

13.2 Metric Spaces—Specific Examples

In Section 1.10 we introduced the metric structure of the real line $\mathbb{R}$ via the concept of absolute value. For $x, y \in \mathbb{R}$, we called

$$d(x,y) = |x - y|$$

the distance between x and y. We identified four properties that the distance function d possesses:

1. $d(x,y) \geq 0$

   [All distances are positive or zero.]

2. $d(x,y) = 0$ if and only if $x = y$

   [Different points are at a positive distance apart.]

3. $d(x,y) = d(y,x)$

   [Distance is symmetric, that is, the distance from x to y is the same as that from y to x.]

4. $d(x,y) \leq d(x,z) + d(z,y)$

   [The triangle inequality; it is no farther to go directly from x to y than to go from x to z and then to y.]

All of the work we did connected with the limit concept and convergence rested ultimately on these four properties of d. Readers who worked through Chapter 11 and 12 on the euclidean spaces $\mathbb{R}^n$ have seen that the euclidean distance between points $\mathbf{x}$ and $\mathbf{y}$,

$$d(x, y) = \|\mathbf{x} - \mathbf{y}\|,$$

obeyed the same properties.

With these facts in mind, we shall base our definition of metric space on a *metric* that possesses these properties. The following concepts are due to Maurice René Fréchet (1878–1973), who introduced them in his 1906 dissertation. The name "metric space" was not originally used but was introduced later in 1914, by Felix Hausdorff (1869–1942), who took Fréchet's ideas, built on them, and extended them to create a branch of mathematics now known as topology. Metric spaces play a fundamental role in the study of topology.

Definition 13.1 Let X be any nonempty set. A function

$$d : X \times X \to \mathbb{R}$$

is called a *metric* if it satisfies the following four conditions:

1. $d(x, y) \geq 0$ for all $x, y \in X$.

2. $d(x, y) = 0$ if and only if $x = y$.

3. $d(x, y) = d(y, x)$ for all $x, y \in X$.

4. $d(x, y) \leq d(x, z) + d(z, y)$ for all $x, y, z \in X$.

Example 13.2 We know already that on the real numbers $\mathbb{R}$ the function

$$d(x, y) = |x - y|$$

is a metric since it is our model on which we based the definition. This is frequently called the *usual metric* since in most (but not all) studies of real numbers this is the metric that is used. For a collection of interesting examples we can consider any subset $A \subset \mathbb{R}$. The metric is the same: More precisely the metric we take on A is the function d_A defined on $A \times A$ by $d_A(x, y) = |x - y|$. Usually this level of precision is not needed and we just refer to the "usual metric." Each of the following examples will prove useful in the sequel:

1. $A = \mathbb{Q}$, the set of rational numbers

2. $A = \mathbb{N}$, the set of natural numbers

3. $A = K$, the Cantor set

4. $A = \{1/k : k = 1, 2, 3, \dots\}$

◀

Example 13.3 (Discrete Space) Let X be any nonempty set. For points x and y in X let

$$d(x, y) = 1 \text{ if } x \neq y \quad (d(x, x) = 0).$$

Then d is a metric on X. (Check this.) This space may not appear to be of any interest, but it is useful to keep in mind because it can be helpful in avoiding certain misconceptions. (See Example 13.18.) ◀

Example 13.4 (The Euclidean Plane) Let

$$\mathbb{R}^2 = \mathbb{R} \times \mathbb{R} = \{(x_1, x_2) : x_1, x_2 \in \mathbb{R}\}.$$

For $x = (x_1, x_2)$ and $y = (y_1, y_2)$ define

$$d_2(x, y) = \sqrt{(x_1 - y_1)^2 + (x_2 - y_2)^2}.$$

You will recognize that this is the usual way in elementary geometry that distances in the plane are computed and can readily verify that d_2 is indeed a metric on $\mathbb{R}^2$. Condition (4) is merely the familiar triangle inequality. It states intuitively that the "shortest distance between two points" is along the line joining those points. A rigorous proof depends upon the Cauchy-Schwarz inequality, with which you may be familiar from the exercises (e.g., Exercises 3.5.13, 3.5.14, 8.3.10). This material was also covered in Chapter 11. ◀

Example 13.5 (Euclidean n-dimensional space) Let

$$\mathbb{R}^n = \mathbb{R} \times \mathbb{R} \times \cdots \times \mathbb{R}$$

$$= \{(x_1, x_2, \dots x_n) : x_i \in \mathbb{R}, i = 1, 2 \dots n\}.$$

For $x = (x_1, x_2, \dots, x_n)$ and $y = (y_1, y_2, \dots, y_n)$ define

$$d_2(x, y) = \sqrt{(x_1 - y_1)^2 + (x_2 - y_2)^2 + \cdots + (x_n - y_n)^2}.$$

Once again, d_2 is a metric on $\mathbb{R}^n$. The triangle inequality (4) follows from the Cauchy-Schwarz inequality. (See Section 11.2.) ◀

Metric Spaces When d is a metric on a set X we refer to X as a metric space furnished with the metric d. The precise language is in the definition, but you will likely find that lecturers have less formal ways of talking about metric spaces.

Definition 13.6 A *metric space* is an ordered pair (X, d), where X is a nonempty set and d is a metric on X.

Sometimes the metric on X is understood from the context. In this case we often shorten our notation by calling the metric space "X" instead of "(X, d)." In particular, when no confusion can arise, we assume that $\mathbb{R}$ is furnished with the usual euclidean metric

$$d(x, y) = |x - y|.$$

There can be many different metrics on a set X. If more than one metric is under consideration, we must make it clear which metric is considered at a particular point of the discussion. (See Exercise 13.2.6 for two further interesting metrics on the space $\mathbb{R}^2$. Exercise 13.6.31 illustrates how a sloppy treatment when two metrics are being used can lead to errors.)

We shall discuss a few additional metric spaces in the next section.

Exercises

13.2.1 Which of the following functions defined for pairs of numbers x and y are metrics on $\mathbb{R}$?

(a) $d(x, y) = |x| + |y|$

(b) $d(x, y) = (x - y)^2$

(c) $d(x, y) = \sqrt{|x - y|}$

(d) $d(x, y) = \min\{1, |x - y|\}$

(e) $d(x, y) = \frac{|x-y|}{1+|x-y|}$

(f) $d(x, y) = 1$ if $x \neq y$ and $d(x, y) = 0$ if $x = y$

13.2.2 Let $\mathbb{R}^+$ denote the set of positive real numbers. Show that

$$d(x, y) = \left| \frac{1}{x} - \frac{1}{y} \right|$$

is a metric on $\mathbb{R}^+$.

13.2.3 Let $\{x_1, x_2, \dots, x_m\}$ be a finite set of points in a metric space (X, d). Show that

$$d(x_1, x_m) \leq \sum_{i=1}^{m-1} d(x_i, x_{i+1}).$$

13.2.4 Show that a function

$$d : X \times X \to \mathbb{R}$$

is a metric if and only if it satisfies the following three conditions:

(a) $d(x, y) \geq 0$ for all $x, y \in X$

(b) $d(x, y) = 0$ if and only if $x = y$

(c) $d(x, y) \leq d(z, x) + d(z, y)$ for all $x, y, z \in X$

As a question of mathematical taste, would you prefer to use these conditions rather than the four in Definition 13.1 for the definition of this term?

13.2.5 Let $X = \{x_1, x_2, x_3, \ldots, x_n\}$ be a finite set and let d be a metric on X. Consider the $n \times n$ matrix whose i,j entry is $d(x_i, x_j)$. What properties must such a matrix have?

13.2.6 Let $X = \mathbb{R}^2$. For $x = (x_1, x_2)$, $y = (y_1, y_2)$ let

$$\begin{aligned} d_1(x, y) &= |x_1 - y_1| + |x_2 - y_2|, \\ d_\infty(x, y) &= \max\{|x_1 - y_1|, |x_2 - y_2|\}. \end{aligned}$$

Show that d_1 and d_∞ are metrics on X. Show that these are distinct from each other and from the metric d_2 used in Example 13.4 (which is the most commonly used metric in $\mathbb{R}^2$).

13.2.7 Let (X, d) be a metric space. Define a function $e : X \times X \to \mathbb{R}$ by

$$e(x, y) = \frac{d(x, y)}{1 + d(x, y)}.$$

Prove that e is a metric, that $e(x, y) \leq d(x, y)$, and that $e(x, y) \leq 1$ for all $x, y \in X$.

13.2.8 Let (X, d) be a metric space. Define a function $e : X \times X \to \mathbb{R}$ by

$$e(x, y) = \min\{1, d(x, y)\}.$$

Prove that e is a metric and that $e(x, y) \leq 1$ for all $x, y \in X$.

13.2.9 **(Product Spaces)** Given two metric spaces we can form a product metric space. Let (X_1, d_1) and (X_2, d_2) be metric spaces. The set

$$X_1 \times X_2 = \{(x_1, x_2) : x_1 \in X_1, x_2 \in X_2\}$$

is called the product or Cartesian product of X_1 and X_2. For

$$u = (x_1, x_2) \text{ and } v = (y_1, y_2) \text{ in } nX_1 \times X_2$$

define

$$d(u, v) = d_1(x_1, y_1) + d_2(x_2, y_2).$$

(a) Prove that d is a metric on $X_1 \times X_2$. (It is called the product metric.)

(b) Let $X_1, X_2 = \mathbb{R}$, $d_1 = d_2$ be the usual euclidean metric on $\mathbb{R}$. Calculate $d(u, v)$, where $u = (0, 1)$ and $v = (-3, 4)$ are points in $\mathbb{R}^2 = \mathbb{R} \times \mathbb{R}$.

13.2.10 **(Normed Vector Spaces)** Let X be a vector space. A *norm* on X is a real-valued function on X so that if x, $y \in X$ and $\alpha \in \mathbb{R}$, then

(i) $\|x\| \geq 0$, and $\|x\| = 0$ if and only if x is the zero of X,

(ii) $\|\alpha x\| = |\alpha| \, \|x\|$, and

(iii) $\|x + y\| \leq \|x\| + \|y\|$.

Show that a vector space equipped with a norm can be considered a metric space by defining $d(x, y) = \|x - y\|$.

13.3 Additional Examples

We have already mentioned that the elements of a metric space can be objects of various sets. In this section we introduce several important metric spaces whose elements are sequences or functions. Exercise 13.3.9 provides an example of an important metric space whose elements are sets—the closed nonempty subsets of $[0, 1]$.

13.3.1 Sequence Spaces

Example 13.7 (Hilbert Space—ℓ_2) An immediate generalization of euclidean n-dimensional space $\mathbb{R}^n$ arises when we replace n-tuples

$$x = (x_1, x_2, \ldots, x_n)$$

with sequences

$$x = \{x_1, x_2, \ldots\}.$$

Let ℓ_2 denote the set of all sequences of real numbers such that

$$\sum_1^{\infty} (x_k)^2 < \infty.$$

For

$$x = \{x_1, x_2, \ldots\} \text{ and } y = \{y_1, y_2, \ldots\}$$

define

$$d_2(x, y) = \sqrt{\sum_{k=1}^{\infty} (x_k - y_k)^2}.$$

The resulting metric space ℓ_2 is usually called classical (sequential) Hilbert space, named after one of the most important and influential mathematicians of his time, David Hilbert (1862–1943). Let us verify that d_2 is a metric on ℓ_2. It is clear that for $x, y \in \ell_2$, $d_2(x, y) \geq 0$, but we must check that $d_2(x, y)$ is finite. From the elementary inequality

$$(x_k \pm y_k)^2 \leq 2(x_k^2 + y_k^2)$$

we see that the convergence of the series $\sum_1^{\infty} x_k^2$ and $\sum_1^{\infty} y_k^2$ implies the convergence of the series $\sum_1^{\infty} (x_k - y_k)^2$. Thus $d_2(x, y) < \infty$ for all $x, y \in \ell_2$. It is clear that $d_2(x, y) = 0$ if and only if $x = y$. The symmetry condition (3) is also apparent.

Let us verify the triangle inequality. In this space, it takes the form

$$\sqrt{\sum_{k=1}^{\infty}(x_k - y_k)^2} \le \sqrt{\sum_{k=1}^{\infty}(x_k - z_k)^2} + \sqrt{\sum_{k=1}^{\infty}(z_k - y_k)^2}. \tag{1}$$

From the Cauchy-Schwarz inequality we can deduce that for all $n \in \mathbb{N}$,

$$\sqrt{\sum_{k=1}^{n}(x_k - y_k)^2} \le \sqrt{\sum_{k=1}^{n}(x_k - z_k)^2} + \sqrt{\sum_{k=1}^{n}(z_k - y_k)^2}. \tag{2}$$

We have left the deduction of (2) from the Cauchy-Schwarz inequality as an exercise. Letting $n \to \infty$ in (2), we obtain (1), verifying the triangle inequality for d_2. ◀

Note. We mention that Example 13.7 is known as *real* ℓ_2. The complex version, which is often studied, is applied to sequences of complex numbers $\{z_1, z_2, \dots\}$ satisfying $\sum_1^\infty |z_k|^2 < \infty$ and is the same except that here we must define

$$d_2(z, w) = \sqrt{\sum_1^\infty |z_k - w_k|^2}$$

for a pair of sequences of complex numbers, $z = \{z_1, z_2, \dots\}$ and $w = \{w_1, w_2, \dots\}$. The proof that this is again a metric is similar, using now properties of the complex modulus.

Example 13.8 Now let ℓ_1 be the set of sequences $\{x_k\}$ of real numbers for which $\sum_1^\infty |x_k| < \infty$, and let

$$d_1(x, y) = \sum_1^\infty |x_k - y_k|.$$

We verify easily that d_1 is a metric on ℓ_1. The metric space (ℓ_1, d_1) can be described loosely as the space of all absolutely convergent series. ◀

Example 13.9 Let ℓ_∞ denote the set of all bounded sequences of real numbers, let $x = \{x_1, x_2, \dots\}$ and $y = \{y_1, y_2 \dots\}$, and let

$$d_\infty(x, y) = \sup_k |x_k - y_k|.$$

Once again, it is easy to verify that (ℓ_∞, d_∞) is a metric space. (We leave this verification as Exercise 13.3.3.) This space ℓ_∞ (sometimes denoted by m) can be described loosely as the space of all bounded sequences. ◀

These three spaces ℓ_1, ℓ_2, and ℓ_∞ are related in a number of ways. For example, you should be able to use your study of series convergence to prove

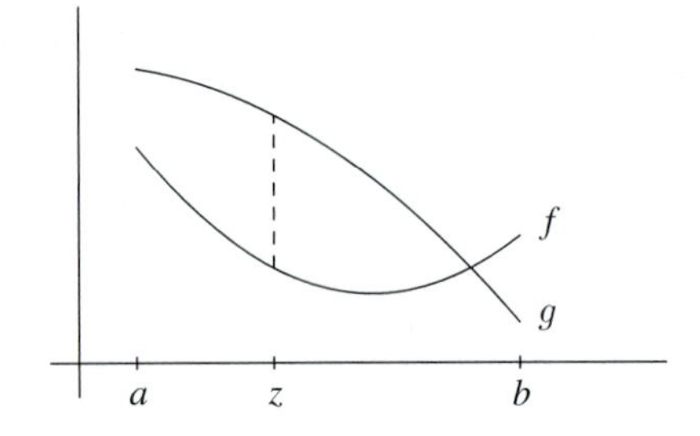

Figure 13.1. Distance between functions f and g in $\mathcal{C}[a,b]$.

that

$$\ell_1 \subset \ell_2 \subset \ell_\infty.$$

Thus, on the smallest of these spaces ℓ_1, we can define three metrics; all are different and all are important. For example, if

$$x = \{3,4,0,0,\dots\} \quad \text{and} \quad y = \{0,0,\dots\},$$

then $d_1(x,y) = 7$, $d_2(x,y) = 5$ and $d_\infty(x,y) = 4$. Note the inequalities

$$d_1(x,y) \geq d_2(x,y) \geq d_\infty(x,y)$$

for this example. Do you think these inequalities are valid in general whenever x and y are *any* absolutely convergent sequences?

13.3.2 Function Spaces

We often use the term "function space" when the elements of a metric space are functions. Function spaces have played a major role in the development of twentieth-century analysis.

Example 13.10 We use $\mathcal{C}[a,b]$ to denote the set of continuous functions on $[a,b]$. For a "distance" between two continuous functions f and g we measure the largest vertical distance between points on their graphs. Thus we shall employ the metric

$$d(f,g) = \max_{a \leq t \leq b} |f(t) - g(t)|.$$

Figure 13.1 illustrates a point z in the interval at which the maximum is attained. (We know there must be a maximum since $|f-g|$ is continuous.) It is easy to verify that d is a metric on $\mathcal{C}[a,b]$. ◀

Example 13.11 The space of continuous functions on $[a,b]$ can be enlarged by passing to the collection of all *bounded* functions on $[a,b]$. We use $M[a,b]$ to denote the set of bounded functions defined on $[a,b]$. For $f,g \in M[a,b]$

we define a metric in much the same way by measuring the largest vertical distance between the graphs of f and g, but this time we cannot be assured that a maximum is attained. Thus we write

$$d(f,g) = \sup_{a \leq t \leq b} |f(t) - g(t)|. \tag{3}$$

This is often called the *sup metric.* It is easy to verify that $(M[a,b],d)$ is a metric space. Note especially that if f and g are continuous functions, then the distance $d(f,g)$ assigned by this metric is the same as the distance assigned by the metric in the space $\mathcal{C}[a,b]$ of the preceding example. ◀

Subspaces These last two examples introduce the important concept of "subspace" of a metric space. The space of bounded functions $M[a,b]$ contains inside it the space of continuous functions $\mathcal{C}[a,b]$ and the metrics agree.

(Since our subject matter is metric spaces, we intend a subspace to be understood relative to that structure. You may also have studied "subspaces" in the context of vector spaces; this has an entirely different meaning and you should avoid confusing the two meanings.)

Definition 13.12 If (X,d) is a metric space and $A \subset X$, then (A,d) is called a *subspace* of (X,d).

Note that the metric must be the same for A and for X.

Example 13.13 For example, with $X = \mathbb{R}$ and the usual metric

$$d(x,y) = |x - y|,$$

any nonempty subset A of $\mathbb{R}$ becomes a subspace of (X,d) as long as we use the same metric $d(x,y) = |x-y|$ for $x,y \in A$ on the subspace. ◀

Example 13.14 Let X denote the continuous functions on $[a,b]$. Define

$$e(f,g) = \int_a^b |f(t) - g(t)|\,dt.$$

Again we verify easily that (X,e) is a metric space. Note that if $f(t) = t$ and $g(t) = t^2$ on $[0,1]$, then

$$e(f,g) = \int_0^1 |t - t^2|\,dt = \frac{1}{6}.$$

The metric from Example 13.10 is quite different since it would give

$$d(f,g) = \max_{0 \leq t \leq 1} |t - t^2| = 1/4.$$

Thus while X is contained inside the set $M[a,b]$ it cannot be considered a subspace of the space $M[a,b]$ since it has been equipped with a different metric. ◀

Exercises

13.3.1 Deduce for all $n \in \mathbb{N}$, that

$$\sqrt{\sum_{k=1}^{n}(x_k - y_k)^2} \leq \sqrt{\sum_{k=1}^{n}(x_k - z_k)^2} + \sqrt{\sum_{k=1}^{n}(z_k - y_k)^2} \tag{4}$$

from the Cauchy-Schwarz inequality.

13.3.2 Let ℓ_1 be the set of sequences $\{x_k\}$ of real numbers for which $\sum_1^\infty |x_k| < \infty$, and let

$$d_1(x,y) = \sum_1^\infty |x_k - y_k|.$$

Show that d_1 is a metric on ℓ_1.

13.3.3 Let ℓ_∞ denote the set of all bounded sequences of real numbers, let

$$x = \{x_1, x_2, \dots\} \text{ and } y = \{y_1, y_2 \dots\}$$

belong to ℓ_∞, and let

$$d_\infty(x,y) = \sup_k |x_k - y_k|.$$

Show that d_∞ is a metric on ℓ_1.

13.3.4 Let $\mathcal{C}[0,1]$ consist of the continuous functions on $[0,1]$ furnished with the metric

$$d(f,g) = \int_0^1 |f(t) - g(t)|\,dt.$$

Show that d is a metric on $\mathcal{C}[0,1]$ different from the one in Example 13.10.

13.3.5 Let $\mathcal{R}[0,1]$ consist of all Riemann integrable functions on $[0,1]$ (not necessarily continuous). Let

$$d(f,g) = \int_0^1 |f(t) - g(t)|\,dt.$$

Show that d is *not* a metric on $\mathcal{R}[0,1]$.

13.3.6 Verify the inclusions

$$\ell_1 \subset \ell_2 \subset \ell_\infty$$

for the sequence spaces in this section. Is any one of these a subspace of any other?

13.3.7 Let $M(\mathbb{R})$ denote the collection of all bounded real-valued functions on $\mathbb{R}$ and let

$$d(f,g) = \sup\{|f(t) - g(t)| : t \in \mathbb{R}\}.$$

Show that d is a metric on $M(\mathbb{R})$. Which of the following are subspaces of $M(\mathbb{R})$?

$\mathcal{A}$ = the constant functions on $\mathbb{R}$

$\mathcal{P}$ = the polynomials

$\mathcal{C}$ = the continuous functions

$\mathcal{S}$ = the set of functions f of the form $f(t) = a\sin(nt) + b\cos(nt)$ for $a, b \in \mathbb{R}$, $n \in \mathbb{N}$

13.3.8 Let $\mathcal{K}$ consist of the nonempty closed subsets of $[0,1]$. For $A, B \in \mathcal{K}$ let

$$\operatorname{dist}(A,B) = \inf\{|a-b| : a \in A, b \in B\}.$$

This is often called the "distance between A and B." Show that dist is *not* a metric on $\mathcal{K}$.

13.3.9 **(The Hausdorff Metric)** Since the "distance function" in Exercise 13.3.8 is not a true metric, let us define a metric on $\mathcal{K}$, the family of nonempty closed subsets of $[0,1]$, that captures the idea that the distance between two sets A and B in $\mathcal{K}$ is less than δ if every point of A is within δ of some point of B, and vice versa. For $A \in \mathcal{K}$ and $\delta > 0$, let A_δ be the union of all closed intervals of length 2δ centered at points of A. Define d by

$$d(A,B) = \inf\{\delta > 0 : A \subset B_\delta \text{ and } B \subset A_\delta\} \qquad (*).$$

(a) Verify that $d(A,B)$ measures the greatest distance that a point in A can be from the set B or a point in B can be from the set A.

(b) Show that d is a metric on $\mathcal{K}$. It is called the *Hausdorff metric* on the space of (nonempty) closed subsets of $[0,1]$.

(c) Let $A = \{1/n : n \in \mathbb{N}\} \cup \{0\}$, $B = [0,1]$ and $C = \{1/2\}$. Calculate $d(A,B)$, $d(B,C)$ and $d(A,C)$.

(d) If we replace the family $\mathcal{K}$ with the family of *all* nonempty subsets of $[0,1]$ and define d by $(*)$, we would not get a metric. Which of the conditions for a metric would fail? What would be the value of $d(A,B)$ if $A = \mathbb{Q} \cap [0,1]$ and $B = [0,1]$?

13.3.10 Let X be the set of continuous functions on $(0,1)$. For $x, y \in X$ let $U(x,y) = \{t \in (0,1) : x(t) \neq y(t)\}$. Then $U(x,y)$ is a disjoint union of intervals. Let $d(x,y)$ be the sum of the lengths of these intervals. Verify that (X,d) is a metric space.

13.4 Convergence

We recall that a sequence $\{x_k\}$ of real numbers converges to x_0 if and only if $|x_n - x_0|$ converges to zero, that is, if and only if

$$d(x_n, x_0) \to 0,$$

where d is the usual metric. This meaning of convergence carries over to any metric space.

Definition 13.15 Let (X,d) be a metric space. Let $\{x_k\}$ be a sequence of members of X and let $x_0 \in X$. If $\lim_{k\to\infty} d(x_k, x_0) = 0$ we say $\{x_k\}$ *converges to* x_0 and we write $\lim_{k\to\infty} x_k = x_0$ or $x_k \to x_0$.

It is clear from the way that sequence convergence in a metric space has been defined that convergence in $\mathbb{R}$ (with the usual metric) is precisely the notion of convergence as we have studied it so far. What sequences will converge in other metric spaces? The following examples and some of the exercises will illustrate.

Example 13.16 Let $\{x_n\}$ be a sequence of points in $\mathbb{R}^2$ equipped with the usual metric of Example 13.4. How can we recognize when the sequence $\{x_n\}$ converges to a point $z \in \mathbb{R}^2$? We recall that this means that

$$d_2(x_n, z) \to 0,$$

where d_2 is the usual metric in $\mathbb{R}^2$. It is instructive to see that this convergence is equivalent to coordinate-wise convergence.

If we write out the coordinates

$$x_n = (a_n, b_n), \quad z = (c, d),$$

then we can easily verify that this occurs precisely when the ordinary sequences $\{a_n\}$ and $\{b_n\}$ converge to c and d. Indeed just observe that

$$|a_n - c| \leq d_2(x_n, z) \leq |a_n - c| + |b_n - d|$$

and

$$|b_n - c| \leq d_2(x_n, z) \leq |a_n - c| + |b_n - d|$$

and this becomes obvious. (For a general version of this observation in the space $\mathbb{R}^n$, see Theorem 11.15.) ◀

Example 13.17 In the space $M[a, b]$ (Example 13.11), convergence reduces to what we called uniform convergence in Chapter 9. To see this, observe that for f, $g \in M[a, b]$

$$d(f, g) = \sup_{a \leq t \leq b} |f(t) - g(t)|.$$

Thus $d(f_k, g) \to 0$ if and only if

$$\sup_{a \leq t \leq b} |f_k(t) - g(t)| \to 0.$$

We saw (in Exercise 9.3.12) that this condition characterizes uniform convergence. ◀

Note. The definition of convergence implies that limit of the sequence must be a point in the space. Thus, for example, the sequence $\left\{\frac{1}{n}\right\}$ converges to 0 in $\mathbb{R}$, or in the subspace $[0, 1]$, but does *not* converge in the subspace $(0, 1)$ since 0 is not a point of (0,1). Similarly, if $x_n(t) = 1 + t + \cdots + t^n$, then $x_n \to x_0$, where $x_0(t) = \frac{1}{1-t}$, in the space $M\left[0, \frac{1}{2}\right]$ or in the subspace $\mathcal{C}\left[0, \frac{1}{2}\right]$, but does *not* converge in the subspace $\mathcal{P}\left[0, \frac{1}{2}\right]$ of polynomials on $\left[0, \frac{1}{2}\right]$ because the function x_0 is not a polynomial.

Exercises

13.4.1 Describe the convergent sequences in a metric space (X, d), where d is the discrete metric.

13.4.2 Describe the convergent sequences in the euclidean n-dimensional space of Example 13.5.

13.4.3 Example 13.5 and Exercise 13.4.2 suggest that the following should be true for sequences in Hilbert space ℓ_2 (Example 13.7). In order for a sequence of points

$$x^{(n)} = (x_1^{(n)}, x_2^{(n)}, x_3^{(n)}, \dots)$$

$n = 1, 2, 3, \dots$ in ℓ_2 to converge to a point

$$y = \{y_1, y_2, y_3, \dots\}$$

in ℓ_2 it is necessary and sufficient that each $x_k^{(n)}$ converges to y_k as $n \to \infty$ for $k = 1, 2, 3, \dots$. Is this true?

13.4.4 **(The Hilbert Cube)** Consider the following subspace of Hilbert space ℓ_2:

$$H = \{(x_1, x_2, \dots) \in \ell_2 : |x_i| \leq 1/i\}.$$

H is called the Hilbert cube. Show that, in contrast to Exercise 13.4.3, a sequence of points

$$x^{(n)} = (x_1^{(n)}, x_2^{(n)}, x_3^{(n)}, \dots)$$

in H converges if and only if each $x_k^{(n)}$ converges for $k = 1, 2, 3, \dots$.

13.4.5 Example 13.5 and Exercises 13.4.2 and 13.4.3 suggest that the following should be true for sequences in the function space $\mathcal{C}[0, 1]$ consisting of the continuous functions on $[0, 1]$ furnished with the metric

$$d(f, g) = \int_0^1 |f(t) - g(t)|\, dt.$$

It is a necessary (but perhaps not sufficient) condition for a sequence of functions $\{f_n\}$ to converge to a function g in $\mathcal{C}[0, 1]$ that $f_n(x) \to g(x)$ for each $x \in [0, 1]$. Is this true?

13.4.6 Establish some elementary properties of sequence convergence in a metric space:

(a) If $x_k \to x$ and $x_k \to y$, then $x = y$.

(b) If $x_k \to x$, then the sequence is bounded in the sense that

$$\sup\{d(x, x_k) : k = 1, 2, 3, 4, \dots\} < \infty.$$

(c) Are there any further elementary results of the theory of real sequences that can be formulated and proved in a general metric space?

13.4.7 If $\{x_n\}$ and $\{y_n\}$ are convergent sequences in a metric space (X, d), show that $\lim_{n\to\infty} d(x_n, y_n)$ exists.

13.4.8 If $\{x_n\}$ and $\{y_n\}$ are sequences in a metric space (X, d) and

$$\lim_{n\to\infty} d(x_n, y_n) = 0,$$

show that they are either both convergent or both divergent.

13.4.9 Show that a sequence of real numbers $\{x_n\}$ is convergent in $\mathbb{R}$ if and only if it is convergent when $\mathbb{R}$ is furnished with any of the following metrics:

(a) $d(x, y) = |x - y|$ (i.e., the usual metric)

(b) $e_1(x, y) = \min\{1, |x - y|\}$

(c) $e_2(x, y) = \frac{|x-y|}{1+|x-y|}$

(Thus, while the usual metric and the metrics e_1 and e_2 differ, they describe the same class of convergent sequences.)

13.4.10 Generalize Exercise 13.4.9 to an arbitrary metric space.

13.4.11 Show that a sequence of points in the euclidean plane (Example 13.4) is convergent if and only if it is convergent under either of the metrics d_1 and d_∞ of Exercise 13.2.6. (Thus while the metrics d_1 and d_2 and d_∞ differ, they describe the same class of convergent sequences.)

13.4.12 Let a set X be equipped with two metrics d_1 and d_2. Determine the relation between sequence convergence in the two spaces (X, d_1) and (X, d_2) if one of the following conditions holds for some positive numbers m, M:

(a) $d_1(x, y) \leq M d_2(x, y)$ for all x, $y \in X$.

(b) $m d_1(x, y) \leq d_2(x, y)$ for all x, $y \in X$.

(c) $m d_1(x, y) \leq d_2(x, y) \leq M d_1(x, y)$ for all x, $y \in X$.

13.4.13 Consider the set $\mathcal{C}[0, 1]$ of continuous functions on $[0, 1]$ with the two metrics (both of which are important in analysis):

$$d_1(x, y) = \max_{0\leq t\leq 1} |x(t) - y(t)|$$

$$d_2(x, y) = \int_0^1 |x(t) - y(t)|\, dt$$

(a) If a sequence $\{x_k\}$ from $(\mathcal{C}[0, 1], d_1)$ converges, must it also converge in $(\mathcal{C}[0, 1], d_2)$? If it does converge also in $(\mathcal{C}[0, 1], d_2)$, must it converge to the same limit?

(b) What are the answers in (a) if we interchange the roles of d_1 and d_2?

13.4.14 Let $\mathcal{C}^1[a, b]$ denote the continuously differentiable functions on $[a, b]$. Define d by

$$d(x, y) = \max_{a\leq t\leq b} |x(t) - y(t)| + \max_{a\leq t\leq b} |x'(t) - y'(t)|$$

Verify that d is a metric on $\mathcal{C}^1[a, b]$. Which of the following sequences converge in the space $\mathcal{C}^1[0, 1]$?

(a) $x_k = t^k$

(b) $x_k(t) = \dfrac{t}{k}$

(c) $x_k(t) = \dfrac{\sin(kt)}{k}$

(d) $x_k(t) = \displaystyle\int_0^{1/k} \sin(kst)\,ds$

13.4.15 In $\mathbb{R}$ we say that "addition is continuous," meaning that if $x_n \to x$ and $y_n \to y$, then $x_n + y_n \to x + y$. In each of the following spaces there is a natural way of defining addition. Define that addition and determine if addition is continuous in these spaces.

(a) ℓ_1

(b) ℓ_2

(c) ℓ_∞

(d) $\mathcal{C}[a, b]$

13.4.16 Let $\mathcal{K}$ denote the family of nonempty closed subsets of $[0, 1]$ furnished with the Hausdorff metric of Exercise 13.3.9. Determine whether the following sequences converge and if so to what they converge.

(a) $A_n = [0, 1/n]$

(b) $B_n = \{1/n\}$

(c) $C_n = [1/2 - 1/n, 1/2 + 1/n]$ for $n \geq 2$

13.4.17 In the metric space $M[a, b]$ of Examples 13.11 and 13.17 we saw that convergence of sequences is precisely uniform convergence. Is it true then in the subspace $\mathcal{P}[a, b]$ of all polynomials on $[a, b]$ with the same metric that a sequence $\{p_n\}$ of polynomials converges if and only if it converges uniformly?

13.4.18 Using the theory of real sequences as a guide, formulate a definition for what should be meant by a "Cauchy sequence" in a metric space. Prove that every convergent sequence in a metric space must be Cauchy but that the converse need not be true.

13.5 Sets in a Metric Space

To develop the notion of convergence we need to extend to an arbitrary metric space some of the concepts we discussed for $\mathbb{R}$ in Chapter 3 and for $\mathbb{R}^n$ in Chapter 11. As a start, you may wish to draw pictures to illustrate these notions in $\mathbb{R}^2$. But to get a good sense of these notions we must have many examples. See the exercises at the end of this section.

1. For $x_0 \in X$ and $r > 0$, the set
$$B(x_0, r) = \{x \in X : d(x_0, x) < r\}$$
is called the *open ball* with center x_0 and radius r.

2. The set
$$B[x_0, r] = \{x \in X : d(x_0, x) \leq r\}$$
is called the *closed ball* with center x_0 and radius r.

3. A set $G \subset X$ is called *open* if for each $x_0 \in G$ there exists $r > 0$ such that $B(x_0, r) \subset G$.

4. A set $F \subset X$ is called *closed* if its complement $X \setminus F$ is open.

5. For a nonempty set E the *diameter* of E is the number
$$\sup\{d(x, y) : x, y \in E\},$$
which may be infinite.

6. A nonempty set E is *bounded* if
$$\sup\{d(x, y) : x, y \in E\}$$
is finite (i.e., if the set E has finite diameter).

7. A *neighborhood* of x_0 is any open set G containing x_0.

8. If $G = B(x_0, \varepsilon)$ we call G the ε-*neighborhood* of x_0.

9. A point x_0 is called an *interior point* of a set A if x_0 has a neighborhood contained in A.

10. The *interior* of a set A consists of all interior points of A and is denoted by A^o or, occasionally, $\text{int}(A)$.

11. A point $x_0 \in X$ is a *limit point* or *point of accumulation* of a set A if every neighborhood of x_0 contains infinitely many points of A.

12. An *isolated point* of a set A is a point x_0 that has a neighborhood G for which $G \cap A = \{x_0\}$.

13. The *closure* $\overline{A}$ of a set A consists of all points that are either in A or limit points of A.

14. A *boundary point* of A is a point x_0 such that every neighborhood of x_0 contains at least one point of A and at least one point of the complement $X \setminus A$.

15. Let A and B be subsets of X. If $\overline{A} \supset B$ or, equivalently, if every open ball centered at a point of B contains a point of A, we say that A is *dense* in B. (Note that this does not require A to be a subset of B.) If $\overline{A} = X$, we simply say that A is *dense*.

16. A set A in a metric space X is said to be *nowhere dense* (in X) if every open ball $B(x, \varepsilon)$ contains another open ball $B(y, \delta)$ such that $B(y, \delta) \cap A = \emptyset$.

17. The *distance* between a point $x \in X$ and a nonempty set $A \subset X$ is defined as
$$\text{dist}(x, A) = \inf\{d(x, y) : y \in A\}.$$

You will have noticed that the definitions we gave of concepts such as *open set, closed set, limit point* and the like are entirely analogous to the corresponding definitions we had already given in $\mathbb{R}$ and in $\mathbb{R}^n$. It should be no surprise that many of the elementary relations among these concepts that hold in $\mathbb{R}$ carry over to arbitrary metric spaces. We highlight some of the more important ones in the exercises that follow this section.

We consider a few examples.

Example 13.18 Let X be any nonempty set and let (X, d) be the discrete space (Example 13.3). Let $x_0 \in X$ be any point in the space. You should verify that each of the following statements is true.

1. $B(x_0, 1) = \{x_0\}$
2. Every point is isolated in X
3. $B[x_0, 1] = X$
4. Every set is both open and closed
5. X is bounded
6. Every set containing $x_0 \in X$ is a neighborhood of x_0
7. Every ε-neighborhood of x_0 is either $\{x_0\}$ or X
8. For every set $A \subset X$, $A^o = \overline{A} = A$, and A has no boundary points
9. X has no accumulation points
10. The only dense subset of X is X itself

11. The closure of an open ball $B(x, \varepsilon)$ is not necessarily the closed ball $B[x, \varepsilon]$

If any of these statements seem to contradict your intuition about metric spaces, be sure to restudy the definitions. ◀

Example 13.19 Let K be the Cantor set viewed as a subspace of $[0, 1]$. Let $\{(a_k, b_k)\}$ be the sequence of intervals complementary to K and let S consist of the midpoints of those intervals. For our metric space we take $X = K \cup S$ furnished with the usual real metric.

1. Every point of S is isolated, and no point of K is isolated.

2. K is closed.

3. S is open.

4. $\overline{S} = X$ so S is dense.

5. Each subset of S is open.

Again, be sure to check each of these statements. ◀

Example 13.20 Consider the space $\mathcal{C}[a, b]$ furnished with the supremum metric (see Example 13.10). Let $f \in \mathcal{C}[a, b]$ and let $\varepsilon > 0$.

1. $B(f, \varepsilon)$ consists of all continuous functions g that satisfy
$$|f(t) - g(t)| < \varepsilon$$
for all $t \in [a, b]$.

2. g is a boundary point of $B(f, \varepsilon)$ if and only if $|f(t) - g(t)| \leq \varepsilon$ for all $t \in [a, b]$ and there exists t_0 such that $|f(t_0) - g(t_0)| = \varepsilon$.

3. Geometrically, $g \in B(f, \varepsilon)$ if and only if the graph of g lies strictly between the graphs of $f - \varepsilon$ and $f + \varepsilon$.

4. Similarly, g is a boundary point of $B(f, \varepsilon)$ if and only if the graph of g lies between the graphs of $f - \varepsilon$ and $f + \varepsilon$ and there exists t_0 such that $g(t_0) = f(t_0) + \varepsilon$ or $g(t_0) = f(t_0) - \varepsilon$.

See Figure 13.2. ◀

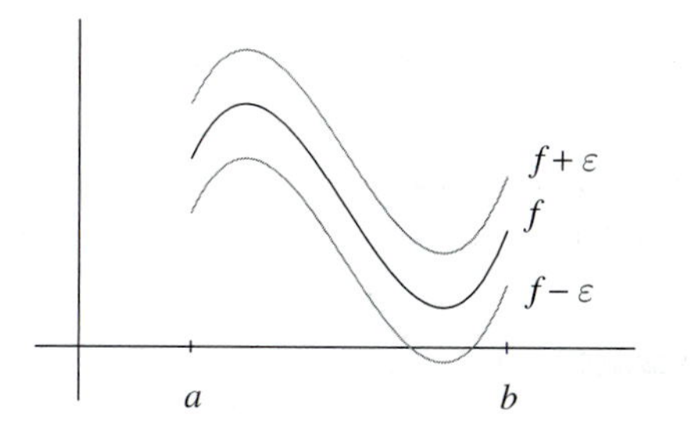

Figure 13.2. Ball centered at a function in $\mathcal{C}[a,b]$.

Comparison Between Two Metrics Suppose X is a set furnished with two different metrics d_1 and d_2. We then have two metric spaces, (X, d_1) and (X, d_2). In general, we expect that the concepts introduced in this section will be different. For example, a set might be bounded or closed with respect to one metric but not the other. In practice, though, there are often close connections between the properties.

For example, it might happen that the class $\mathcal{G}_1$ of sets open with respect to d_1 is the same as the class $\mathcal{G}_2$ of sets open with respect to d_2. In this case, the two spaces (X, d_1) and (X, d_2) will have exactly the same class of convergent sequences, and the two spaces are the same in a sense we shall make precise in Section 13.6.2. In general, however, the two spaces are different: The set X is the same, but the classes of open sets are not, and sequences that converge in one of the spaces need not converge in the other. Exercises 13.4.13 and 13.5.13 provide important illustrations of this phenomenon.

Exercises

13.5.1 Show that the definition of a point of accumulation is equivalent to the following:

(a) x_0 is a point of accumulation of A if for every $\varepsilon > 0$ the set
$$A \cap B(x_0, \varepsilon) \setminus \{x_0\}$$
is nonempty.

(b) x_0 is a point of accumulation of A if there is a sequence of points $x_n \in A$ so that $x_n \neq x_0$ and $x_n \to x_0$.

(c) x_0 is a point of accumulation of A if $x \in \overline{A}$ and x_0 is not an isolated point of A.

(d) x_0 is a point of accumulation of A if x_0 is not an interior point of $(X \setminus A) \cup \{x_0\}$.

13.5.2 Show that every point of a set is either isolated or a point of accumulation.

13.5.3 Let $S = \{1/k : k = 1, 2, 3, \dots\}$ and furnish S with the usual real metric. Answer the following questions about this metric space.

(a) Which points are isolated in S?

(b) Which sets are open and which sets are closed in S?

(c) Which sets have a nonempty boundary?

(d) Which sets are dense in S?

(e) Which sets are nowhere dense in S?

13.5.4 Let E be a closed set in a metric space (X, d) and let x be a point that is not in E. Show that

$$\inf\{d(x, y) : y \in E\} > 0.$$

Show that if E and F are disjoint closed sets in a metric space, then it is not necessarily true that

$$\inf\{d(x, y) : x \in E,\ y \in F\} > 0.$$

13.5.5 Compute the diameter of the set

$$\{\{x_1, x_2, \dots\} \in \ell_2 : |x_i| \leq 1\,,\ i = 1, 2, 3, \dots\}$$

as a subset of the metric space ℓ_2.

13.5.6 Prove the following elementary property of sequence convergence in a metric space: It is true that $x_k \to x$ if and only if for every $\varepsilon > 0$ there exists $N \in \mathbb{N}$ such that $x_k \in B(x, \varepsilon)$ for all $k \geq N$.

13.5.7 A sequence in a metric space is *bounded* if the range of the sequence is a bounded set. On the real line $\mathbb{R}$ every bounded sequence has a convergent subsequence. Is a similar statement true in any metric space?

13.5.8 Show, in a general metric space, that the open ball is open and that the closed ball is closed. Give an example of a metric space in which a closed ball $B[x, \varepsilon]$ is not necessarily the closure of the open ball $B(x, \varepsilon)$.

13.5.9 We chose to define a closed set as one whose complement is open. Show that the following are equivalent for a subset A of a metric space (X, d):

(a) $X \setminus A$ is open.

(b) A contains all its limit points.

(c) $A = \overline{A}$.

13.5.10 Let (X, d) be a metric space.

(a) Prove that X and $\emptyset$ are both open and closed.

(b) Prove that a finite union of closed sets is closed and a finite intersection of open sets is open.

(c) Prove that an arbitrary union of open sets is open and an arbitrary intersection of closed sets is closed.

13.5.11 Let X denote the set of points

$$\{0\} \cup \{1/k : k = 1, 2, 3, \dots\}$$

in $\mathbb{R}$ furnished with the usual real metric. Answer the following questions:

(a) Which points are isolated in X?

(b) Which sets are open and which sets are closed?

(c) Which sets are both open and closed?

(d) Is X bounded?

(e) Which sets have a nonempty boundary?

(f) Does X have any accumulation points?

(g) Describe all dense subsets of X.

(h) Is the closure of an open ball $B(x, \varepsilon)$ necessarily the closed ball $B[x, \varepsilon]$?

13.5.12 Which of the following subsets of $M[a, b]$ are closed?

(a) $\mathcal{C}[a, b]$

(b) $\mathcal{P}[a, b]$, the polynomials on $[a, b]$

(c) $\mathcal{R}[a, b]$, the Riemann integrable functions on $[a, b]$

(d) $\mathcal{P}_n[a, b]$, the polynomials of degree $\leq n$ on $[a, b]$

(e) $\triangle[a, b]$, the differentiable functions on $[a, b]$

(f) $\triangle'[a, b]$, the derivatives of differentiable functions with bounded derivatives on $[a, b]$

13.5.13 Consider the set $\mathcal{C}[0, 1]$ of continuous functions on $[0, 1]$ with the two metrics:

$$d_1(x, y) = \max_{0 \leq t \leq 1} |x(t) - y(t)|$$

$$d_2(x, y) = \int_0^1 |x(t) - y(y)|\, dt.$$

Let B_1 be an open ball in $(\mathcal{C}[0, 1], d_1)$ and let B_2 be an open ball in $(\mathcal{C}[0, 1], d_2)$.

(a) Is B_1 open in $(\mathcal{C}[0, 1], d_2)$?

(b) Is B_2 open in $(\mathcal{C}[0, 1], d_1)$?

13.5.14 Which of these are closed subsets of the metric space ℓ_∞ of bounded sequences (Example 13.9)?

(a) ℓ_2

(b) ℓ_1

(c) c, the set of all convergent sequences of real numbers

(d) c_0, the set of all those sequences that converge to 0

(e) p, the set of all bounded sequences of nonnegative real numbers

(f) r, the set of all bounded sequences of rational numbers

13.5.15 (The Hilbert Cube) Consider the following two subsets of Hilbert space ℓ_2:

(a) $H = \{(x_1, x_2, \dots) \in \ell_2 : |x_i| \leq 1/i\}$

(b) $G = \{(x_1, x_2, \dots) \in \ell_2 : |x_i| < 1/i\}$

H is called the Hilbert cube. Show that H is closed in ℓ_2. Is G open?

13.5.16 Which of the properties of closure and interior found in Exercises 4.3.6 and 4.3.7 are valid in every metric space? In particular, give an example of a metric space in which the examples sought in part (d) of these exercises don't exist.

13.5.17 Let $X = \mathbb{R}^2$. Sketch the unit spheres, that is, the set $\{x : d(x, 0) = 1\}$, for each of the following metrics, each defined for all $x = (x_1, x_2)$ and $y = (y_1, y_2)$ in $\mathbb{R}^2$.

$$d_1(x, y) = |x_1 - y_1| + |x_2 - y_2|$$
$$d_2(x, y) = \sqrt{(x_1 - y_1)^2 + (x_2 - y_2)^2}$$
$$d_\infty(x, y) = \max\{|x_1 - y_1|, |x_2 - y_2|\}$$

13.5.18 Let (X_1, d_1) and (X_2, d_2) be metric spaces and let $X_1 \times X_2$ be the product space as defined in Exercise 13.2.9.

(a) Show that every set of the form $A \times B$, where A is an open subset of (X_1, d_1) and B is an open subset of (X_2, d_2) is an open subset of $X_1 \times X_2$.

(b) Show that every open subset of $X_1 \times X_2$ can be expressed as a union

$$\bigcup_{i \in I} A_i \times B_i,$$

where each A_i is an open subset of (X_1, d_1) and B_i is an open subset of (X_2, d_2).

13.5.19 State and prove a version of the Bolzano-Weierstrass theorem that is valid in $\mathbb{R}^2$.

✂ **13.5.20** State and prove a version of the Heine-Borel theorem that is valid in $\mathbb{R}^2$.

13.5.21 A set A in a metric space X is called

(a) *nowhere dense* (in X) if every open ball $B(x, \varepsilon)$ contains another open ball $B(y, \delta)$ such that $B(y, \delta) \cap A = \emptyset$

(b) *dense in itself* if every point of A is a limit point of A

(c) *perfect* if it is closed and dense in itself

The following are are several subsets A of certain metric spaces (X, d). Which are nowhere dense? Which are dense in themselves? Which are perfect.

(a) $A = (0, 1)$ and $X = \mathbb{R}$

(b) $A = (0, 1)$ and $X = (0, 1)$

(c) $A = \mathcal{C}[a, b]$ and $X = M[a, b]$

(d) $A =$ constant functions and $X = \mathcal{C}[a, b]$

(e) $A = \{x \in c_0 : x_k = 0$ for all but finitely many values of $k\}$ and $X = c_0$ (see Exercise 13.5.14)

(f) $A = \{(x_1, x_2) \in \mathbb{R}^2 : x_1^2 + x_2^2 = 1\}$ and $X = \mathbb{R}^2$

(g) A is the collection of nonempty closed subsets of [0,1] with no more than k elements and $X = \mathcal{K}$ is the family of nonempty closed subsets of [0,1] furnished with the Hausdorff metric (Exercise 13.3.9)

(h) A is the union of the families of all sets in part (g) taken over $k = 1$, 2, 3, ... and $X = \mathcal{K}$ is the same as in part (g)

13.6 Functions

Our study of limits and continuity for functions $f : \mathbb{R} \to \mathbb{R}$ can be generalized by studying these same notions for functions mapping a metric space into another. Let us begin with some examples of functions from one metric space to another.

Example 13.21 We define a function $f : \mathbb{R}^2 \to \mathbb{R}$ by

$$f(x_1, x_2) = \frac{x_1 x_2}{x_1^2 + x_2^2} \qquad f(0, 0) = 0.$$

We would naturally be interested in the properties such a function would have when $\mathbb{R}^2$ and $\mathbb{R}$ are furnished with the usual euclidean metrics (see Example 13.5). Thus we consider this a function with domain one metric space and range in another metric space. ◀

Example 13.22 We can interpret the operation of integration as a function mapping the metric space $\mathcal{C}[a, b]$ into $\mathbb{R}$ by

$$T(f) = \int_a^b f(t)\, dt.$$

Thus integration can be considered a real-valued function on a metric space of functions. ◀

Example 13.23 We define a function S mapping the metric space $\mathcal{C}[a,b]$ into itself by

$$(S(f))(t) = \int_a^t f(s)\, ds.$$

Thus the notion of an indefinite integral is realized as an operation or function on the metric space $\mathcal{C}[a,b]$ with values in $\mathcal{C}[a,b]$. ◄

Example 13.24 The operation of differentiation can be interpreted as a function on an appropriate metric space. Let $\mathcal{C}^1[0,1]$ consist of those functions on [0,1] with continuous derivatives. We define a function $D : \mathcal{C}^1[0,1] \to \mathcal{C}[0,1]$ by

$$D(f) = f'.$$

If we use the sup metrics on these spaces, then we can interpret the operation of differentiation as a function from one metric space into another. ◄

Example 13.25 If f is a continuous function on the interval $[0,\pi]$, we write its Fourier sine series as

$$\sum_{k=1}^{\infty} b_k \sin kx,$$

where the coefficients are given by Fourier's formulas

$$b_k = \frac{2}{\pi}\int_0^{\pi} f(t) \sin kt\, dt.$$

With a slight shift in viewpoint this operation can be considered in the context of metric spaces. The input function f is transformed into a sequence of numbers $\{b_k\}$. Thus we can write it as $F(f) = \{b_k\}$ on the understanding that the terms of the sequence $\{b_k\}$ are given by the formulas.

But what sequence space is appropriate? One of the elementary inequalities of Fourier series shows which space to use. It is not difficult to prove that

$$\sum_{k=1}^{\infty} b_k^2 \leq \frac{2}{\pi}\int_0^{\pi} f^2(t)\, dt.$$

(This is called Bessel's inequality.) Evidently, we can interpret F as a mapping from $\mathcal{C}[0,\pi]$ into the sequence space ℓ_2. ◄

In each of these five examples we have introduced a function $f : X \to Y$ from one metric space (X,d) to another (Y,e). In Examples 13.22, 13.23, and 13.24 we follow a common practice of using uppercase letters (T, S and D) when the metric spaces are also vector spaces. We often use the terms *transformation* or *operator* in place of the term *function*.

Our study of functions between metric spaces begins with continuity. We then proceed to study two special kinds of continuous functions between metric spaces, isometries and homeomorphisms. Homeomorphisms are functions that are not merely continuous but whose inverses are continuous. Isometries are continuous functions that preserve distances.

13.6.1 Continuity

We wish to know, as in the case of functions from $\mathbb{R}$ to $\mathbb{R}$, whether these examples of functions (transformations) defined previously are "continuous." In defining continuity of functions between metric spaces we try to capture the following idea:

> If $T : X \to Y$, then T is continuous at $x_0 \in X$ provided that all points near x_0 are mapped to points near $T(x_0)$.

We make this precise in exactly the same way as we did for functions from subsets of $\mathbb{R}$ to $\mathbb{R}$ (in Section 5.1.2) and for functions from subsets of $\mathbb{R}^n$ to $\mathbb{R}^m$ (in Section 11.7).

Definition 13.26 Let (X, d) and (Y, e) be metric spaces and let $T : X \to Y$. We say that T is *continuous* at $x_0 \in X$ if for every sequence $\{x_k\}$ converging to $x_0 \in X$ the sequence $\{T(x_k)\}$ converges to $T(x_0) \in Y$. If T is continuous at every point of X, we say that T is *continuous.*

You can verify that the alternate characterizations of continuity that were valid for real functions, namely the ε-δ characterization and the neighborhood characterization, are also valid here. The proofs are the same. We list these characterizations for reference.

Theorem 13.27 *Let (X, d) and (Y, e) be metric spaces and let $T : X \to Y$.*

(i) T is continuous at x_0 if and only if for every neighborhood V of $T(x_0)$ there exists a neighborhood U of x_0 such that $T(U) \subset V$.

(ii) T is continuous at x_0 if and only if for every $\varepsilon > 0$ there exists $\delta > 0$ such that $e(T(x), T(x_0)) < \varepsilon$ whenever $d(x, x_0) < \delta$.

An important special case is the following. For continuity everywhere it is the way in which open sets are treated that is most significant. Note that it is not the image set $T(G)$ that is to be open when $G \subset X$ is open, but the preimage $T^{-1}(G)$ when $G \subset Y$ is open.

Theorem 13.28 *Let (X, d) and (Y, e) be metric spaces and let $T : X \to Y$. Then T is continuous at each point of X if and only if for each open set $G \subset Y$, the set*

$$T^{-1}(G) = \{x : T(x) \in G\}$$

is open in X.

You may wish to draw pictures in a familiar setting (such as $X = Y = \mathbb{R}^2$) to illustrate these characterizations.

Example 13.29 Let us look again at Example 13.21. Here

$$f(x_1, x_2) = \frac{x_1 x_2}{x_1^2 + x_2^2} \qquad f(0,0) = 0.$$

Let us check continuity at $(0,0)$. Since $f(0,0) = 0$ (by definition), f will be continuous at (0,0) if and only if

$$\lim_{in \to \infty} f(u_n, v_n) = 0$$

for every sequence of points $(u_n, v_n) \to (0,0)$. Any one example of a sequence for which this fails shows that f is not continuous there. For example, observe that $f(1/n, 1/n) \to 1/2$. [Perhaps we should redefine $f(0,0)$ to be $1/2$; in that case then again f is not continuous since $f(1/n, 0) \to 0$. Thus no value of $f(0,0)$ can make this function continuous there.]

At every other point it is easy to check that f is continuous. Take any sequence of points $(u_n, v_n) \to (a,b) \neq (0,0)$ and verify by the usual elementary sequence methods that $f(u_n, v_n) \to f(a,b)$. ◀

Example 13.30 Let us look again at Example 13.22. Here

$$T(f) = \int_a^b f(t)\, dt.$$

If $f_k \to f$ in X, then $f_k \to f$ uniformly on $[a,b]$. By Theorem 9.26

$$\int_a^b f_k(t)\, dt \to \int_a^b f(t)\, dt,$$

that is, $T(f_k) \to T(f)$ in $\mathbb{R}$. Thus T is continuous at f. Since this is true for all $f \in \mathcal{C}[a,b]$, T is continuous on $\mathcal{C}[a,b]$. ◀

Example 13.31 Let us look again at Example 13.23. We verify that S defined by

$$(S(f))(t) = \int_a^t f(s)\, ds$$

is continuous at every $f \in \mathcal{C}[a,b]$. Let $f_k \to f$ in that metric space. This means that

$$d(f_k, f) = \max_t |f_k(t) - f(t)| \to 0 \text{ as } k \to \infty$$

(that is, $f_k \to f$ uniformly on $[a, b]$). You should verify each step in these calculations:

$$\begin{aligned} d(S(f_k), S(f)) &= \max_t |(S(f_k))(t) - (S(f))(t)| \\ &= \max_t \left| \int_a^t (f_k(s) - f(s))\, ds \right| \\ &\le \max_t \int_a^t |f_k(s) - f(s)|\, ds = \int_a^b |f_k(s) - f(s)|\, ds \\ &\le (b-a) \max_t |f_k(t) - f(t)| = (b-a) d(f_k, f). \end{aligned}$$

Since $\lim_{k\to\infty} d(f_k, f) = 0$ by hypothesis, we conclude that

$$\lim_{k\to\infty} d(S(f)k), S(f)) = 0.$$

Thus $S(f_k) \to S(f)$, and S is continuous. ◀

Example 13.32 Let us look again at Example 13.24. We ask if the function $D : \mathcal{C}^1[0,1] \to \mathcal{C}[0,1]$ defined by $D(f) = f'$ is continuous. We must recall the metrics used. We use the sup metric on both spaces $\mathcal{C}^1[0,1]$ and $\mathcal{C}[0,1]$. Thus $f_k \to f$ if and only if $f_k \to f$ uniformly on $[0,1]$. A specific example suffices to show that D is not continuous. For each $k \in \mathbb{N}$, let $f_k(t) = t^k/k$. Then $f_k \to 0$ in $\mathcal{C}^1[0,1]$. In order for D to be continuous we must have $D(f_k) \to D(0) = 0$ in $\mathcal{C}[0,1]$. But

$$D(f_k)(t) = f_k'(t) = t^{k-1}$$

and this sequence does not converge in $\mathcal{C}[0,1]$. (Had we imposed a different metric on $\mathcal{C}^1[0,1]$ the answer might have been different.) ◀

Observe that Examples 13.22 and 13.23 involve operators defined by integrals. Such operators are often (but not always) continuous. Example 13.24 shows a differential operator. For such operators, continuity often fails. (But see Exercise 13.6.16 at the end of this section.)

Exercises

13.6.1 Let (X, d) be a discrete space.

(a) What functions $f : X \to \mathbb{R}$ are continuous everywhere?

(b) What functions $f : \mathbb{R} \to X$ are continuous everywhere?

13.6.2 Verify statements (i) and (ii) of Theorem 13.27 that provide characterizations of continuity.

13.6.3 Prove Theorem 13.28.

13.6.4 Prove that $T : X \to Y$ is continuous if and only if $T^{-1}(E)$ is closed for every closed set $E \subset Y$.

13.6.5 In Section 5.4.4 we defined continuity of a function $f : A \to \mathbb{R}$ when A is a subset of $\mathbb{R}$. Verify that the definition given there agrees with the definition given in this section when A is considered as a metric space with $d(x, y) = |x - y|$.

13.6.6 Verify each inequality and each equality in the calculation found in Example 13.31.

13.6.7 Let $f : \mathbb{R}^2 \to \mathbb{R}$ be defined by writing

$$f(x_1, x_2) = \frac{x_1^2 x_2}{x_1^4 + x_2^2} \quad f(0,0) = 0.$$

Show that $\lim_{x\to 0} f(x, mx) = 0$ for every $m \in \mathbb{R}$, but f is discontinuous at $(0, 0)$.

13.6.8 Let $f : \mathbb{R}^2 \to \mathbb{R}$. Suppose f is *separately continuous*, that is, $f(x_1, v)$ is a continuous function of v for every $x_1 \in \mathbb{R}$ and $f(u, x_2)$ is a continuous function of u for every $x_2 \in \mathbb{R}$. If the continuity with respect to one of the variables is uniform with respect to the other variable, then f is continuous. Make this statement precise and prove it.

13.6.9 Let (X, d) be a metric space. Prove that d is continuous on $X \times X$, where $X \times X$ is furnished with the product metric. See Exercise 13.2.9.

13.6.10 Let (X_1, d_1) and (X_2, d_2) be metric spaces and let $X_1 \times X_2$ be the product space as defined in Exercise 13.2.9. Show that the functions $f : X_1 \times X_2 \to X_1$ and $g : X_1 \times X_2 \to X_2$ defined by $f(x, y) = x$ and $g(x, y) = y$ are continuous.

13.6.11 Let (X, d) be a metric space and let A be a nonempty subset of X. Define $f : X \to \mathbb{R}$ by

$$f(x) = \text{dist}(x, A) = \inf\{d(x, y) : y \in A\}.$$

(a) Show that $|f(x) - f(y)| \leq d(x, y)$ for all x, $y \in X$.

(b) Show that f defines a continuous real-valued function on X.

(c) Show that $\{x \in X : f(x) = 0\} = \overline{A}$.

(d) Show that $\{x \in X : f(x) > 0\} = \text{int}(X \setminus A)$.

(e) Show that, unless X contains only a single point, there exists a continuous real-valued function defined on X that is not constant.

(f) If $E \subset X$ is closed and $x_0 \notin E$, show that there is a continuous real-valued function g on X so that $g(x_0) = 1$ and $g(x) = 0$ for all $x \in E$.

(g) If E and F are disjoint closed subsets of X, show that there is a continuous real-valued function g on X so that $g(x) = 1$ for all $x \in F$ and $g(x) = 0$ for all $x \in E$.

(h) If E and F are disjoint closed subsets of X, show that there are disjoint open sets G_1 and G_2 so that $E \subset G_1$ and $F \subset G_2$.

(i) In the special case where X is the real line with the usual metric and K denotes the Cantor ternary set, sketch the graph of the function $f(x) = \text{dist}(x, K)$.

(j) Give an example of a metric space, a point x_0, and a set $A \subset X$ so that $\text{dist}(x_0, A) = 1$ but so that $d(x, x_0) \neq 1$ for every $x \in \overline{A}$.

13.6.12 Let x and y be real-valued functions on $[0, 1]$. Define $f : [0, 1] \to \mathbb{R}^2$ by $f(t) = (x(t), y(t))$. If f is continuous, then f is called a *continuous curve.* Prove that f is a continuous curve if and only if both functions x and y are continuous.

13.6.13 Prove that the class of continuous real-valued functions on a metric space is closed under the arithmetic operations of addition, subtraction, and multiplication. (How about division?)

13.6.14 State precisely and prove a theorem that asserts under which conditions the composition $f \circ g$ of two continuous functions is continuous.

13.6.15 Prove that the function of Example 13.25 is continuous.

13.6.16 Let $\mathcal{C}^1[a, b]$ consist of the continuously differentiable functions on $[a, b]$. Define for $f, g \in \mathcal{C}^1[a, b]$

$$d(f, g) = \max_{a \leq t \leq b} |f(t) - g(t)| + \max_{a \leq t \leq b} |f'(t) - g'(t)|.$$

(a) Prove that d is a metric.

(b) Let $D : \mathcal{C}^1[a, b] \to \mathcal{C}[a, b]$ be defined by $D(f) = f'$. Prove that D is continuous. (Here, as usual, $\mathcal{C}[a, b]$ has the sup metric.)

13.6.17 Let $\mathcal{C}[0, 1]$ consist of the continuous functions on $[0, 1]$ and furnished with the metric

$$d(f, g) = \int_0^1 |f(t) - g(t)|\, dt.$$

Define $T : \mathcal{C}[0, 1] \to \mathbb{R}$ by

$$T(f) = \int_0^1 f(t)\, dt.$$

Is T continuous?

13.6.18 Extend the following concepts and results from functions of one variable to functions of two variables:

(a) Define *uniform continuity* for a function $f : \mathbb{R}^2 \to \mathbb{R}$.

(b) Prove that a continuous real-valued function f defined on a closed and bounded subset $K \subset \mathbb{R}^2$ is uniformly continuous on K.

(c) Prove that a continuous real-valued function f defined on a closed and bounded subset $K \subset \mathbb{R}^2$ is bounded on K.

(d) Prove that a continuous real-valued function f defined on a closed and bounded subset $K \subset \mathbb{R}^2$ achieves an absolute maximum on K.

13.6.19 Let $\{f_k\}$ be a sequence of real-valued functions on a metric space X. Define what it means for $\{f_k\}$ to approach a function f uniformly. Show that if each of the functions f_k is continuous and $f_k \to f$ uniformly on X, then f is continuous on X.

13.6.2 Homeomorphisms

A *homeomorphism* between metric spaces is a one-to-one onto mapping that is continuous and whose inverse is also continuous.

To motivate our discussion of homeomorphisms let us consider two problems that have been suggested in several of the exercises. First, suppose that a set X is furnished with two different metrics d_1 and d_2. While the spaces (X, d_1) and (X, d_2) may be different as metric spaces, it is still possible that they have closely related properties. When could we recognize that they have the same open sets, the same closed sets, the same convergent sequences, etc.? The following example discusses this problem in a concrete setting.

Example 13.33 Consider the three metrics d_1, d_2, and d_∞ on the plane $\mathbb{R}^2$ defined by

$$\begin{aligned} d_1(x, y) &= |x_1 - y_1| + |x_2 - y_2|, \\ d_2(x, y) &= \sqrt{(x_1 - y_1)^2 + (x_2 - y_2)^2}, \\ d_\infty(x, y) &= \max\{|x_1 - y_1|, |x_2 - y_2|\}, \end{aligned}$$

for points $x = (x_1, x_2)$, $y = (y_1, y_2)$. (We have seen these metrics in Example 13.4 and Exercise 13.2.6.)

While these are three different metrics on $\mathbb{R}^2$ we will see that that the open sets are the same under the three metrics and that convergent sequences are also identical. Why is this so?

An examination of the unit balls centered at the origin helps explain why. (Our analysis here repeats some of the discussion in Section 11.6.2 which you may have skipped.) Figure 11.4 on page 483 shows the three unit balls centered at the origin for these metrics. More generally, an open ball centered at x with radius r with respect to the metric d_∞ will be the inside of a square with sides parallel to the coordinate axes. Its center will be at x and its side length will be $2r$. For the metric d_2 the ball will be the inside of a circle of radius r, and for d_1 the ball will be the inside of a square of side length $r\sqrt{2}$ with sides parallel to the lines $x_2 = \pm x_1$.

Let us denote the balls in the three spaces

$$(\mathbb{R}^2, d_1), \ (\mathbb{R}^2, d_2), \ \text{and} \ (\mathbb{R}^2, d_\infty)$$

by $B_1(x, r)$, $B_2(x, r)$ and $B_\infty(x, r)$, respectively. It is easy to verify analytically that for every $x \in \mathbb{R}^2$ and $r > 0$

$$B_1(x,r) \subset B_2(x,r) \subset B_\infty(x,r). \tag{5}$$

Furthermore, for every $x \in \mathbb{R}^2$ and $r > 0$ there exists $s > 0$ such that

$$B_\infty(x,s) \subset B_1(x,r) \tag{6}$$

(Exercise 13.6.24). It follows that any ball centered at x with respect to one of the three metrics contains a ball centered at x with respect to either of the other two metrics. Thus any set that is open with respect to one of the metrics is also open with respect to the other two. ◀

This fact in our example, that the three different metrics give rise to the same family of open sets, has important consequences. If the open sets are the same, then the convergent sequences are the same. If the open sets are the same, then the continuous functions are the same. Thus any sequence $\{x_n\}$ converging to a point x_0 with respect to one of the metrics also converges to x_0 with respect to the other two. Further, a function f mapping $\mathbb{R}^2$ into a metric space (Y,e) will be continuous with respect to one of the metrics if and only if it is continuous with respect to to the others.

We could summarize our example by stating that the three metric spaces $(\mathbb{R}^2, d_1)$, $(\mathbb{R}^2, d_2)$, and $(\mathbb{R}^2, d_\infty)$ have the same "topological properties:" From the topological point of view they are indistinguishable. We need to make this notion precise. The key observation in our example concerned the open sets relative to each of the three metrics. The *topology* of a metric space is simply the family of open sets. Thus, loosely, a topological property is one that can be expressed in terms of the open sets and need not be expressed directly in terms of the metric defined on the space.

Before we state our definitions, consider the second problem. Suppose that two metric spaces (X,d) and (Y,e) are closely related in the sense that there is a one-to-one onto function $h : X \to Y$. Thus $Y = h(X)$ is a "copy" of X and each point $x \in X$ relates to a unique point $y = h(x) \in Y$. The problem is this: What additional properties should h have so that there is the same relation between the open sets of X and the open sets of Y, namely that G is an open subset of X precisely if $h(G)$ is an open subset of Y? The answer to this is given in the definition.

Definition 13.34 Let (X,d) and (Y,e) be metric spaces. A function

$$h : X \to Y$$

is called a *homeomorphism* if h meets the following conditions:

1. h is one-to-one,

2. h maps X onto Y,

3. h is continuous, and

4. h^{-1} is continuous.

Condition (4) is equivalent to h being an *open map*, that is, h maps open sets in X onto open sets in Y (Theorem 13.28). Two metric spaces are said to be *homeomorphic* or *topologically equivalent* if there is a homeomorphism mapping one space onto the other. A property that is preserved under homeomorphisms is called a *topological property.*

Some topological properties are listed in Exercise 13.6.21. Here are some examples of spaces that are topologically equivalent.

Example 13.35 The spaces $X = (-1, 1)$ and $Y = \mathbb{R}$, both furnished with the usual metrics, are topologically equivalent. [For an appropriate homeomorphism take $h(x) = \tan \pi x/2$.] ◀

Example 13.36 The space

$$X = \{(x_1, x_2) \in \mathbb{R}^2 : x_2 = 1/x_1,\ x_1 > 0\}$$

furnished with the euclidean metric and the interval $Y = (0, \infty)$ are topologically equivalent. [For a homeomorphism take $h(x_1, x_2) = x_1$.] ◀

Example 13.37 The three spaces $(\mathbb{R}^2, d_1)$, $(\mathbb{R}^2, d_2)$, and $(\mathbb{R}^2, d_\infty)$ of Example 13.33 are topologically equivalent. Between any pair take h as the identity map. Assertions 5 and 6 can be used to prove that h is a homeomorphism. Because the identity map is a homeomorphism we can say more: The open sets are the same for each of the three metrics. ◀

Note. In the last example the three metrics give rise to exactly the same families of open sets of $\mathbb{R}^2$. It is not always true, however, that if d and e are two metrics on a set X, and (X, d) is homeomorphic to (X, e), then d and e induce the same family of open sets. (See Exercise 13.6.30.) What *is* true is that if $h : (X, d) \to (X, e)$ is a homeomorphism, then a set $A \subset X$ is open in (X, d) if and only if the set $h(A)$ is open with respect to (X, e).

Sometimes it is not immediately clear whether two spaces are homeomorphic. One way to show that the spaces are *not* homeomorphic is to exhibit a topological property possessed by one of the spaces but not by the other.

Example 13.38 For instance, an open interval (a, b) in $\mathbb{R}$ cannot be homeomorphic to a closed interval $[c, d]$ (with respect to the usual metrics). The property that is not preserved is the Bolzano-Weierstrass property: Any sequence in $[c, d]$ must contain a convergent subsequence, but this is not true in (a, b). ◀

Example 13.39 $\mathbb{Q}$ cannot be homeomorphic to $\mathbb{R} \setminus \mathbb{Q}$. The property that is not preserved is countability: A homeomorphism cannot map a countable set to an uncountable one. ◀

We present some further examples of homeomorphic spaces in the exercises.

Exercises

13.6.20 Show that topologically equivalence is an equivalence relation, that is, prove the following:

(a) X is homeomorphic to itself.

(b) If X is homeomorphic to Y, then Y is homeomorphic to X.

(c) If X is homeomorphic to Y and Y is homeomorphic to Z, then X is homeomorphic to Z.

13.6.21 The following are several properties a set A in a metric space X might possess. Verify that each of these is a topological property; that is, if A has the property in (X, d), and (Y, e) is homeomorphic to (X, d) via the homeomorphism h, then $h(A)$ has the same property in (Y, e).

(a) A is open

(b) A is closed

(c) A is dense

(d) A is nowhere dense

(e) A is countable

13.6.22 The following are are several properties a set A in a metric space X might possess. Verify that each of these is a *not* a topological property. (Thus while these properties are defined in terms of the metric, they are not invariant under homeomorphisms.)

(a) A is bounded

(b) A has diameter equal to 1

(c) For any pair $x, y \in A$ there is an element $z \in A$ with $d(x, z) = d(y, z)$

(d) All points in A are equidistant from some point $z \in X$, that is, $d(x, z) = d(y, z)$ for all $x, y \in A$

13.6.23 In Example 13.35, $h^{-1}(\mathbb{R}) = (-1, 1)$ which is an open interval. Does this violate our claim that h^{-1} maps closed sets onto closed sets?

13.6.24 Verify the two assertions (5) and (6) of Example 13.33.

13.6.25 Show that if $h : X \to Y$ is a homeomorphism, then a sequence $\{x_k\}$ converges in X if and only if the sequence $\{h(x_k)\}$ converges in Y.

13.6.26 Show that $\mathbb{R}$ is homeomorphic to $(0, 1)$ but not to $[0, 1]$.

13.6.27 Show that $\mathbb{R}$ is homeomorphic to $(0,\infty)$ but not to $[0,\infty)$.

13.6.28 Let (X,d) be a metric space. Show that there is a metric e on X so that (X,d) and (X,e) are topologically equivalent and such that X has a finite diameter in (X,e).

13.6.29 Let (X,e) be a metric space and suppose that

$$\inf\{e(x,y) : x,y \in X,\ x \neq y\} > 0.$$

Show that (X,e) is topologically equivalent to (X,d), where d is the discrete metric on X. Is the converse true?

13.6.30 Let $X = (0,1) \cup (2,3)$. Define metrics d and e on X as follows. If $x_1 \neq x_2$, then

$$d(x_1,x_2) = \begin{cases} |x_1 - x_2|, & \text{if } x_1, x_2 \in (0,1) \\ 1, & \text{otherwise} \end{cases}$$

and

$$e(x_1,x_2) = \begin{cases} |x_1 - x_2|, & \text{if } x_1, x_2 \in (2,3) \\ 1, & \text{otherwise.} \end{cases}$$

Find a homeomorphism $h : (X,d) \to (X,e)$. Show the identity is not a homeomorphism between the spaces. Exhibit a set that is open with respect to d but not with respect to e and a sequence $\{x_k\}$ that converges with respect to d but not with respect to e.

13.6.31 A careless student states, "If d and e are metrics on a nonempty set X and the metric spaces (X,d) and (X,e) are homeomorphic, the two spaces have exactly the same open sets and the identity map is a homeomorphism between them. Indeed let $h : X \to X$ be a homeomorphism. Then h^{-1} is also a homeomorphism between the two spaces, so $h^{-1} \circ h$ is also a homeomorphism. But $h^{-1} \circ h$ is just the identity. It follows, too, that all the open sets are the same." In view of Exercise 13.6.30 this cannot be true. Where is the flaw in the argument?

13.6.32
(a) Let $X = (0,1)$, $Y = (2,3) \cup (4,5)$, both with the usual metrics. Prove that X and Y are not homeomorphic.

(b) Show that if X and Y are furnished with the discrete metric, they *are* homeomorphic.

13.6.33 Let $f : [a,b] \to \mathbb{R}$ be continuous. Show that the interval $[a,b]$ is homeomorphic to the graph of the function f, that is, to the set

$$\{(x,y) : x \in [a,b],\ y = f(x)\}$$

considered as a subset of $\mathbb{R}^2$.

13.6.34 For $x = (x_1,x_2)$, $y = (y_1,y_2)$ in $\mathbb{R}^2$ and $1 < p < \infty$ define

$$d_p(x,y) = (|x_1 - y_1|^p + |x_2 - y_2|^p)^{1/p}.$$

It can be shown that d_p is a metric on $\mathbb{R}^2$.

(a) Sketch the unit balls $B_p(0,1)$ centered at the origin for several values of p. (We already did this for $p=1$ and $p=2$ in Fig. 11.4.) Observe that $B_p(0,1)$ seems to approach $B_\infty(0,1)$.

(b) Prove analytically that for all $x, y \in \mathbb{R}^2$,

$$\lim_{p\to\infty} d_p(x,y) = d_\infty(x,y).$$

(c) Show that all the spaces $(\mathbb{R}^2, d_p)$ are topologically equivalent.

13.6.35 Show that the open interval $(0,1)$ is homeomorphic with the set

$$\left\{(x_1,x_2) \in \mathbb{R}^2 : x_2 = \sin\frac{1}{x_1},\ x_1 > 0\right\}$$

when $\mathbb{R}$ and $\mathbb{R}^2$ are furnished with the euclidean metrics.

13.6.36 Let (X,e) be a metric space. Show that (X,e) is topologically equivalent to (X,d), where d is the discrete metric if and only if any one of the following properties holds:

(a) Any intersection of a family of open sets is open.

(b) For any open set G the closure $\overline{G}$ is also open.

(c) Every point x in X is isolated.

13.6.37 A metric space X is called *connected* if it cannot be expressed as a disjoint union of two nonempty open sets.

(a) Prove that if $f : X \to Y$ is continuous and X is connected, then so is Y.

(b) Show that connectedness is a topological property,

(c) Characterize the connected subspaces of $\mathbb{R}$.

13.6.38 Let (X,d) be a connected metric space (see Exercise 13.6.37). Show that X either contains only one point or else uncountably many points.

13.6.39 Show that a metric space (X,d) is connected if and only if every continuous function $f : X \to \mathbb{R}$ has the intermediate value property.

13.6.40 Two points a and b in a metric space (X,d) can be *connected by a curve* if there is a continuous function $f : [0,1] \to X$ so that $f(0) = a$ and $f(1) = b$. Show that if every pair of points in X can be connected by a curve, then X is connected. Is the converse true?

13.6.41 Show $\mathbb{R}$ and $\mathbb{R}^2$ are not homeomorphic.

13.6.42 Show that the Cantor set K in $\mathbb{R}$ is homeomorphic to $K \times K$ in $\mathbb{R}^2$.

13.6.43 **(Cantor Space)** Denote by $2^{\mathbb{N}}$ the set of all sequences $x_0, x_1, x_2, \ldots$ of 0's and 1's furnished with the metric

$$d(x,y) = \sum_{k=0}^{\infty} \frac{|x_k - y_k|}{2^k}.$$

(a) Verify that $(2^{\mathbb{N}}, d)$ is a metric space.

(b) Show that if $x_k = y_k$ for all $k = 0, 1, 2, \ldots, n$, then $d(x, y) \leq 1/2^n$ and if $d(x, y) \leq 1/2^n$, then $x_k = y_k$ for all $k \leq n$.

(c) Show that $(2^{\mathbb{N}}, d)$ is homeomorphic to the Cantor set.

13.6.44 Let X and Y be closed and bounded subsets of $\mathbb{R}$ each having exactly one limit point.

(a) Prove that X and Y are homeomorphic.

(b) Part (a) establishes the fact that there exists a homeomorphism

$$h : X \to Y.$$

Is it necessarily true that there exists a homeomorphism $h : \mathbb{R} \to \mathbb{R}$ such that $h(X) = Y$?

(c) Is it necessarily true that there exists a homeomorphism $h : \mathbb{R}^2 \to \mathbb{R}^2$ that maps $X \times \{0\}$ to $Y \times \{0\}$?

13.6.45 Show that there is a homeomorphism h of the Cantor set such that $h(0) = 1/3$ and $h(1/3) = 0$. Does there exist a homeomorphism h of $\mathbb{R}$ onto $\mathbb{R}$ such that h maps the Cantor set onto itself for which $h(0) = 1/3$ and $h(1/3) = 0$?

13.6.3 Isometries

When two metric spaces (X, d) and (Y, e) are topologically equivalent they share certain properties—if one space has a property, so does the other. Topological properties of sets in X carry over to their images in Y. Examples of topological properties were given in Exercises 13.6.21 and 13.6.37. We shall study in Sections 13.7 and 13.12 two further topological properties, separability and compactness, that are important concepts in the theory of metric spaces.

But two topologically equivalent spaces can still have strikingly different properties. [For example, while $(0, 1)$ and $(0, \infty)$ are topologically equivalent, one is bounded and the other not.] What stronger notion of equivalence captures the idea that the spaces have identical metric space properties?

This stronger form of equivalence of metric spaces involves the concept of *isometry* or *congruence*. Recall that in elementary geometry we learn, for example, if the three sides of a triangle T_1 in a plane have the same lengths as the sides of a triangle T_2, then T_1 and T_2 are congruent. This means that T_1 can be *rigidly moved* onto the triangle T_2. We can make this notion of rigid motion precise in the general setting of metric spaces.

Definition 13.40 Let (X, d) and (Y, e) be metric spaces. A function h mapping X onto Y is called an *isometry* if

$$e(h(x), h(y)) = d(x, y)$$

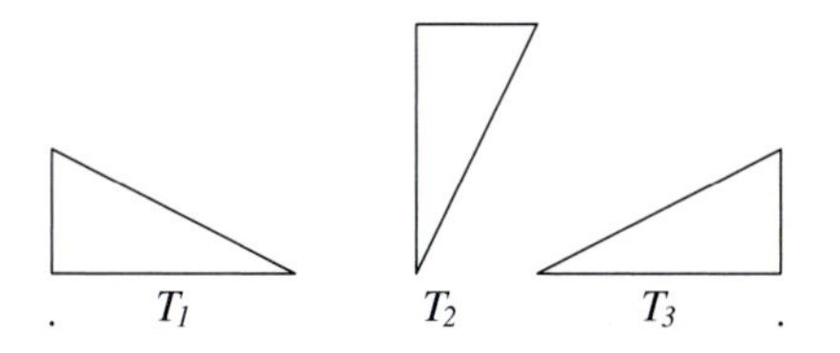

Figure 13.3. The triangles T_1, T_2 and T_3 are isometric.

for all $x, y \in X$. The two spaces are called *isometric* if there is an isometry of one onto the other. A property is called a *metric property* if it is preserved by isometries.

Observe that an isometry is just a homeomorphism that preserves distance. Thus two congruent triangles in the plane are isometric metric spaces when they are viewed as subspaces of the plane.

Note. We have to be a bit careful in our use of the term *rigid motion.* Consider the right triangles in Figure 13.3. Triangles T_1 and T_2 are congruent—the isometry h such that $h(T_1) = T_2$ can be a translation t followed by a rotation r: $h = r \circ t$. Thus T_1 is moved onto T_2 while staying in the plane. But there is no such motion (i.e., combination of translations and rotations) within $\mathbb{R}^2$ that maps T_1 onto T_3. Nonetheless, T_1 and T_3 are isometric. You may wish to show how, for example, we can construct an isometry between T_1 and T_3. (An isometry of $\mathbb{R}^3$ onto $\mathbb{R}^3$ can do the job, but this isn't necessary. All we need is an isometric mapping of T_1 onto T_3. The domain of that mapping need be only T_1, not some larger space.) Compare these remarks with Exercises 13.6.44 and 13.6.45.

Let's look at a few examples.

Example 13.41 Any two open intervals in $\mathbb{R}$ are homeomorphic. They are isometric if and only if they have the same length. ◀

Example 13.42 The two subspaces $\mathbb{N}$ and $\mathbb{N} \cup \{0\}$ of $\mathbb{R}$ are isometric. [Use the function $h(x) = x - 1$.] ◀

Example 13.43 Here is a more interesting example of spaces that are isometric. Recall the three metrics d_1, d_2, and d_∞ on $\mathbb{R}^2$ that we discussed at the beginning of Section 13.6.2. The spaces $(\mathbb{R}^2, d_1)$ and $(\mathbb{R}^2, d_\infty)$ are isometric. This might appear a bit surprising for the identity map is *not* an isometry. Instead take the function

$$h(x_1, x_2) = \left(\frac{x_1 + x_2}{2}, \frac{x_1 - x_2}{2} \right)$$

for $(x_1, x_2) \in \mathbb{R}^2$. We leave verification that h is an isometry as Exercise 13.6.50. ◀

Example 13.44 But $(\mathbb{R}^2, d_2)$ is *not* isometric to $(\mathbb{R}^2, d_\infty)$. To see this, observe that the d_∞ distance between any pair of the four points $(\pm\frac{1}{2}, \pm\frac{1}{2})$ is 1 and the images of these points under an isometry would have the same property with respect to d_2. This is impossible. (See Exercise 13.6.49.) ◀

Example 13.45 $\mathbb{R}$ and $\{(x_1, x_2) \in \mathbb{R}^2 : x_2 = 0\}$ are isometric. [Use the function $h(x) = (x, 0)$.] ◀

Embeddings Example 13.45 asserts that the real line and the x-axis in $\mathbb{R}^2$ are isometric spaces. Thus inside the space $\mathbb{R}^2$ is an isometric copy of the real line. We cannot say that the space $\mathbb{R}^2$ *contains* the real line, just that it contains an identical copy of the real line. In order to have some language with which to discuss such situations we introduce the notion of an embedding.

An *embedding* of a metric space X into another space Y is an isometric copy of the space X that is situated in Y. Thus we have $h(X) \subset Y$ with X and $h(X)$ isometric. Even though the two metric spaces are distinct and may contain formally no elements in common we can use the embedding notion to imagine that one is, for all practical purposes, a subspace of the other.

Exercises

13.6.46 Show that isometry is an equivalence relation, that is, prove the following:

(a) X is isometric to itself.

(b) If X is isometric to Y, then Y is isometric to X.

(c) If X is isometric to Y and Y is isometric to Z, then X is isometric to Z.

13.6.47 Is the following statement true? If X is isometric to a subspace of Y and Y is isometric to a subspace of X, then X and Y are isometric.

13.6.48 Prove that every isometry is also a homeomorphism but not conversely.

13.6.49 Show that a three-point discrete space can be embedded in $\mathbb{R}^2$ but that a four-point discrete space cannot.

13.6.50 Verify that the function

$$h(x_1, x_2) = \left(\frac{x_1 + x_2}{2}, \frac{x_1 - x_2}{2}\right)$$

is an isometry of $(\mathbb{R}^2, d_\infty)$ and $(\mathbb{R}^2, d_1)$. The Jacobian of this mapping is not 1. Does this violate anything you learned about Jacobians?

13.6.51 Show that $\mathbb{R}$ and $\mathbb{R}^2$ are not isometric (when they have the usual metrics). Find metrics e_1 and e_2 on $\mathbb{R}$ and $\mathbb{R}^2$ such that $(\mathbb{R}, e_1)$ is isometric to $(\mathbb{R}^2, e_2)$.

13.6.52 Let $X = (0, \infty)$ and $Y = \{(x_1, x_2) \in \mathbb{R}^2 : x_2 = 1/x_1, x_1 > 0\}$. Are X and Y isometric (using the usual metrics for $\mathbb{R}$ and $\mathbb{R}^2$)?

13.6.53 Let f be an increasing function on the interval $[0, 1]$. Let $d(x, y) = |x - y|$ be the ordinary metric on $X = [0, 1]$ and let e be the metric defined by

$$e(x, y) = |f(x) - f(y)|.$$

(a) Under what conditions on f are the spaces (X, d) and (X, e) topologically equivalent?

(b) Under what conditions on f are the spaces (X, d) and (X, e) isometric?

(c) Is there any choice of f so that (X, e) is topologically equivalent to X furnished with the discrete metric?

13.7 Separable Spaces

In our studies of elementary analysis in the earlier chapters we occasionally made use of the fact that the rational numbers formed a dense subset of $\mathbb{R}$. This was convenient for two reasons: First, the rational numbers are much easier to describe and handle than the real numbers, and second, they formed a countable subset. Generally, in metric space theory the existence of a countable dense subset within a metric space makes many arguments much simpler. This leads us to the following terminology.

Definition 13.46 A metric space (X, d) is said to be *separable* if it possesses a countable dense subset.

A separable space is "not too large" in the sense that some countable set can be used to approximate all members of the space. To show a space is separable we must show that there exists a countable dense subset.

Example 13.47 The space $\mathbb{R}$ of real numbers is separable. For a countable dense set take the rationals $\mathbb{Q}$. (There are many other countable dense sets in $\mathbb{R}$.) ◀

Example 13.48 The space $\mathbb{R} \setminus \mathbb{Q}$ of irrational numbers is separable. For a countable dense set take, for example, the set of all numbers of the form $m\sqrt{2}/n$, where m and n are integers. (Note that we cannot take $\mathbb{Q}$ this time since it is not a *subset* of $\mathbb{R} \setminus \mathbb{Q}$.) ◀

Example 13.49 The space $\mathbb{R}^n$ is separable. Take $\mathbb{Q}^n$ as a countable dense subset. ◀

Example 13.50 Let X be a nonempty countable set with any metric d. Then (X, d) is separable. For a countable dense set take X itself. ◀

Example 13.51 A famous theorem of Weierstrass states that every function that is continuous on the interval $[a, b]$ can be approximated uniformly by polynomials. (See Section 10.8.4.) Thus, given such a function f and an $\varepsilon > 0$ there is a polynomial p so that $|f(x) - p(x)| < \varepsilon$ for all $x \in [a, b]$. But any such polynomial p can itself be approximated by a polynomial with rational coefficients q merely by adjusting each of the coefficients, so that $|p(x) - q(x)| < \varepsilon$ for all $x \in [a, b]$. Putting these together we have $|f(x) - q(x)| < 2\varepsilon$ for all $x \in [a, b]$. Thus the class $\mathcal{P}_q$ of polynomials with rational coefficients is dense in $\mathcal{C}[a, b]$. Here, as usual, $\mathcal{C}[a, b]$ is furnished with the metric

$$d(f, g) = \max_{a \leq t \leq b} |f(t) - g(t)|.$$

Observe that $\mathcal{P}_q$ is countable, so $\mathcal{C}[a, b]$ possesses a countable dense subset and hence is a separable metric space. ◀

Example 13.52 The space c of convergent sequences of real numbers is a separable subspace of the space ℓ_∞ (which is itself not separable, as we shall argue later). To see that c is separable, let A_n denote the set of all sequences of the form

$$(r_1, r_2, r_3, \ldots, r_{n-1}, r_n, r_n, r_n, \ldots),$$

where each r_1, r_2, ... r_n is a rational number. Let $A = \bigcup_{n=1}^{\infty} A_n$. You can verify (Exercise 13.7.2) that A is a countable dense subset of c. ◀

Some Nonseparable Examples We can often show that a space is *not* separable by exhibiting an uncountable disjoint family of open sets. Since a dense set must intersect every nonempty open set, there could not exist a countable dense set.

Example 13.53 Any uncountable set X furnished with the discrete metric is not separable since each singleton set is open. No countable set can be dense. ◀

Example 13.54 Consider the space $M[a, b]$ of bounded functions on $[a, b]$ furnished with the sup metric (see Example 13.11). This space is larger than the separable subspace $\mathcal{C}[a, b]$. To see that this space is not separable, observe that if f and g are characteristic functions of distinct sets, then $d(f, g) = 1$. There are uncountably many distinct subsets of $[a, b]$ so the space $M[a, b]$ contains uncountably many disjoint open balls and is therefore nonseparable. ◀

Example 13.55 The space ℓ_∞ of bounded sequences is not separable. To see this, let

$$A = \{\{x_i\} : x_i = 0 \text{ or } x_i = 1\}.$$

This set is an uncountable subset of ℓ_∞. If $x, y \in A$ with $x \neq y$, then $d(x, y) = 1$. Thus the family

$$\{B(x, 1/2) : x \in A\}$$

is an uncountable disjoint family of open balls in ℓ_∞. Any dense set in ℓ_∞ must intersect each ball of this family and so must be uncountable. ◀

Separability Is a Topological Property Any space that is topologically equivalent to a separable metric space must be itself separable. We can obtain this from a slightly more general statement about how dense sets are preserved under continuous mappings.

Theorem 13.56 *Let (X, d) and (Y, e) be metric spaces and let $f : X \to Y$ be a continuous function. If D is dense in X, then the set $f(D)$ is dense in $f(X)$.*

Proof Let V be a nonempty open set in $f(X)$. Since f is continuous, the set $U = f^{-1}(V)$ is open in X. Since D is dense in X, there exists $x \in D \cap U$. Thus $f(x) \in V$. It follows that $f(D)$ is dense in $f(X)$. ■

Corollary 13.57 *If (X, d) is a separable metric space and (Y, e) is homeomorphic to (X, d), then (Y, e) is also separable. Thus, separability is a topological property of metric spaces.*

Exercises

13.7.1 Show that a metric space (X, d) is separable if and only if for every $\varepsilon > 0$ there is a countable set C_ε so that every point in the space is closer than ε to some point in C_ε.

13.7.2 Show that the set A in Example 13.52 is countable and dense in c.

13.7.3 Show that the space $\mathcal{K}$ of closed nonempty subsets of $[0, 1]$ with the Hausdorff metric (Exercise 13.3.9) is separable.

13.7.4 Show that the set of polygonal functions on $[a, b]$ with rational vertices is a countable dense subset of $\mathcal{C}[a, b]$.

13.7.5 Prove that a subspace of a separable space is separable.

13.7.6 Let c_0 denote the subspace of ℓ_∞ consisting of those sequences that converge to 0.

(a) Use Exercise 13.7.5 and Example 13.52 to show that c_0 is separable.

(b) Show that c_0 is separable by exhibiting a countable dense subset of c_0.

13.7.7 Let (X,d) be the space of Example 13.14. Show that this space is separable by showing any dense subset of $\mathcal{C}[a,b]$ is also dense in (X,d).

13.7.8 By a unit sphere we mean a set

$$\{x : d(x,x_0) = 1\}.$$

(a) Show that the unit sphere $\{x : d(x,0) = 1\}$ is separable in $\mathbb{R}^2$ with respect to any of the metrics d_1, d_2, d_∞.

(b) Give an example of a metric space and a unit sphere that is not separable.

13.7.9 Prove that a metric space X is separable if and only if there exists a countable collection $\mathcal{U}$ of open sets such that every open set in X can be expressed as a union of members of $\mathcal{U}$.

13.7.10 Prove that a product of two separable metric spaces, furnished with the product metric, is separable.

13.7.11 Let $X = \mathbb{R}$ and let d be the discrete metric on X. Determine which of the metric spaces ℓ_∞, $\mathcal{C}[0,1]$, or $M[0,1]$ contains an isometric copy of (X,d).

13.8 Complete Spaces

In our study of sequences in Chapter 2 we introduced the concept of a Cauchy sequence of real numbers and showed that every such sequence converges to some real number. In our study of uniform convergence in Chapter 9 we saw that a similar result is valid for a uniformly Cauchy sequence of functions (Theorem 9.12). This notion of a *Cauchy sequence* can be defined in any metric space.

Definition 13.58 Let (X,d) be a metric space and let $\{x_k\}$ be a sequence in X. This sequence is called a *Cauchy sequence* if for each $\varepsilon > 0$ there exists $N \in \mathbb{N}$ such that if $m \geq N$ and $n \geq N$, then $d(x_n, x_m) < \varepsilon$.

The statement in the definition is equivalent to the requirement that

$$\lim_{m,n\to\infty} d(x_m, x_n) = 0.$$

Example 13.59 In the metric space $M[a,b]$ a sequence $\{f_k\}$ is a Cauchy sequence if and only if the sequence of functions $\{f_k\}$ is uniformly Cauchy according to Definition 9.11. Thus Theorem 9.12 can be interpreted as stating that every Cauchy sequence in $M[a,b]$ converges in $M[a,b]$. ◀

It is not the case in a general metric space that all Cauchy sequences converge. This is so on the real line, and Example 13.59 shows that it is true in the space $M[a,b]$. A metric space that does have this property is said to be complete. We have already used this word to describe a certain property of the real numbers. In fact, completeness of $\mathbb{R}$, which we interpreted in

Chapter 1 by the least upper bound property, is exactly equivalent to the fact that Cauchy sequences converge.

Definition 13.60 A metric space (X, d) is *complete* if every Cauchy sequence in X converges (to an element of X).

We know that $\mathbb{R}$ and $M[a, b]$ are complete. Let's look at a few familiar examples of spaces that are not complete and observe why this is so.

Example 13.61 The space $\mathbb{Q}$, with the usual metric of $\mathbb{R}$, is not complete. Every Cauchy sequence in $\mathbb{Q}$ does converge to some real number, but not necessarily to a member of $\mathbb{Q}$. For example, the sequence $\{(1 + 1/n)^n\}$ is a Cauchy sequence of rational numbers and yet does not converge in $\mathbb{Q}$. ◀

Example 13.62 The space $\mathcal{P}[0, 1]$ of polynomials on $[0, 1]$, furnished with the sup metric is not complete. For example, the sequence $\{p_k\}$ with

$$p_k(t) = 1 + t + \frac{t^2}{2} + \cdots + \frac{t^k}{k!}$$

converges uniformly to the function e^t on [0,1]. This follows from our study of uniform convergence of power series in Section 10.3. Thus $\{p_k\}$ is a Cauchy sequence in the space $M[0, 1]$ (Exercise 13.8.1). It follows that it is a Cauchy sequence in the subspace $\mathcal{P}$ of $M[0, 1]$. But $\{p_k\}$ does not converge in $\mathcal{P}$, since the function e^t is not a polynomial. ◀

13.8.1 Completeness Proofs

How can we show that a given metric space (X, d) is complete? Sometimes certain theorems (such as Theorem 13.64, presented later in this section) can be applied to give a proof. But often we must simply use the definition and show directly that every Cauchy sequence in X converges in X. This can be achieved by applying the following steps to an arbitrary Cauchy sequence $\{x_k\}$ in X.

1. Find a natural "candidate" x_0 for the limit of the sequence.

2. Show that this candidate is in the space X.

3. Verify that $x_k \to x_0$.

It is important to realize that step (2) is essential. For example, as we observed in Example 13.61, the sequence $\{(1 + 1/n)^n\}$ is a Cauchy sequence in the space $\mathbb{Q}$ with a natural candidate for a limit, namely the number e, but e is *not* in $\mathbb{Q}$.

Example 13.63 To show how this process works, let us give a direct proof that $M[a,b]$ is complete using exactly these steps. (In Example 13.59 we obtained this as a consequence of Theorem 9.12; now we can obtain Theorem 9.12 instead as a consequence of the completeness of $M[a,b]$.)

Recall that the metric here is the sup metric

$$d(f,g) = \sup_{a \le t \le b} |f(t) - g(t)|,$$

and convergence reduces to what we called uniform convergence in Chapter 9. Let $\{f_k\}$ be a Cauchy sequence in $M[a,b]$.

Step 1. We wish to find a natural candidate for the limit. A bit of reflection on the meaning of uniform convergence suggests that we consider limits of the form $\lim_{k\to\infty} f_k(t)$ for each $t \in [a,b]$. We observe that for every $t \in [a,b]$ the sequence $\{f_k(t)\}$ is a Cauchy sequence of real numbers. This follows immediately from the inequality

$$|f_n(t) - f_m(t)| \le \sup_{a \le s \le b} |f_n(s) - f_m(s)| = d(f_n, f_m).$$

Since $\mathbb{R}$ is complete,

$$f_0(t) = \lim_{k\to\infty} f_k(t)$$

exists for each $t \in [a,b]$. This limit defines a function f_0 on $[a,b]$. The function f_0 is our candidate.

Step 2. We must verify that $f_0 \in M[a,b]$. To do this, we must show that f_0 is a bounded function. Observe that the fact that f_0 is by its definition the pointwise limit of a sequence of bounded functions does not in itself guarantee that f_0 is bounded. It is the fact that $\{f_k\}$ is a Cauchy sequence that guarantees that f_0 is bounded.

To see this, choose $N \in \mathbb{N}$ such that $d(f_m, f_n) < 1$ for all m, $n \ge N$. Then

$$|f_N(t) - f_m(t)| \le 1$$

for all $m \ge N$ and $t \in [a,b]$. Letting $m \to \infty$ in this inequality, we see that

$$|f_N(t) - f_0(t)| \le 1$$

for all $t \in [a,b]$. Since $f_N \in M[a,b]$, f_N is bounded, say $|f_N(t)| \le A$ for all $t \in [a,b]$. Then $|f_0(t)| \le A + 1$ for all $t \in [a,b]$, so f_0 is bounded, and therefore a member of $M[a,b]$.

Step 3. We must show that $f_k \to f_0$ in the space $M[a,b]$; that is, $f_k \to f_0$ uniformly on $[a,b]$. Again, it is not enough to know that $f_k \to f_0$ pointwise. We used pointwise convergence to get a candidate, knowing that if there is to be a limit, it must be the pointwise limit. Thus we still need to verify the convergence is uniform.

Let $\varepsilon > 0$. Since $\{f_k\}$ is a Cauchy sequence in $M[a,b]$, there exists $N \in \mathbb{N}$ such that if $n \geq N$, then

$$d(f_n, f_N) \leq \frac{\varepsilon}{2},$$

that is,

$$|f_N(t) - f_n(t)| < \frac{\varepsilon}{2} \text{ for all } t \in [a,b]. \tag{7}$$

Thus, for all $t \in [a,b]$,

$$|f_N(t) - f_0(t)| = \lim_{m\to\infty} |f_N(t) - f_m(t)| \leq \frac{\varepsilon}{2}. \tag{8}$$

It follows from (7) and (8) that, for $n \geq N$ and for $t \in [a,b]$,

$$|f_n(t) - f_0(t)| \leq |f_n(t) - f_N(t)| + |f_N(t) - f_0(t)| < \varepsilon.$$

This proves that $f_k \to f_0$ uniformly on $[a,b]$. ◄

13.8.2 Subspaces of a Complete Space

Suppose now we wish to prove that the space $\mathcal{C}[a,b]$ is complete. We could argue exactly as we did with $M[a,b]$, but it is easier to consider that $\mathcal{C}[a,b]$ is a subspace of the complete space $M[a,b]$. What property should a subspace have so that it too is complete? Theorem 13.64 supplies the answer and is a useful tool in checking for completeness of many spaces that we might encounter.

Theorem 13.64 *Let X be a complete metric space and let Y be a subspace of X. Then Y is complete if and only if Y is closed in X.*

Proof Suppose first that Y is closed and $\{y_k\}$ is a Cauchy sequence in Y. Since X is complete, $\{y_k\}$ converges to some point $x_0 \in X$. But Y is closed, so $x_0 \in Y$. Thus Y is complete.

Conversely, suppose that Y is complete. We show Y is closed. Let x_0 be a limit point of Y. Then there exists a sequence $\{y_k\}$ in Y such that $y_k \to x_0$. This sequence, being convergent in X, is a Cauchy sequence in X, and therefore in Y. Since Y is complete, the sequence $\{y_k\}$ converges to a point y_0 in Y. But limits are unique, so $y_0 = x_0$. Thus $x_0 \in Y$ and Y is closed. ■

13.8.3 Cantor Intersection Property

✂ *Enrichment*

Here is another criterion for completeness. You will recognize that this theorem is a direct generalization of the familiar Cantor intersection theorem in $\mathbb{R}$ (Section 4.5.2).

Theorem 13.65 *A metric space (X,d) is complete if and only if the intersection of every descending sequence of closed balls whose radii approach zero consists of a single point.*

We leave the proof as Exercise 13.8.8. See also Exercise 13.8.14 which shows that the requirement on the radii cannot be dropped. Thus, when the radii of the balls gets small, the intersection consists of a single point. But if the radii remain large, the intersection, instead of being large, might be empty!

You should also check the following useful version:

> *A metric space (X,d) is complete if and only if the intersection of every descending sequence of closed sets whose diameters approach zero consists of a single point.*

13.8.4 Completion of a Metric Space

✂ *Enrichment*

In all of our examples of metric spaces that were not complete we have simply chosen a subset of a complete metric space that was not closed. In that way our subset would have Cauchy sequences that do not converge in the subset. Examples 13.61 and13.62 were like this. It can be shown that this is, in a sense, the *only* way that a Cauchy sequence can fail to converge. It is always the case that an incomplete space resides within a larger complete metric space and Cauchy sequences in the former space that do not converge must, however, converge to a point in the larger space.

More precisely, if (X,d) is a metric space, there is always a complete metric space in which X can be isometrically *embedded.* Thus there exists a complete metric space (Y,e) and a function $h : X \to Y$ such that h maps X isometrically onto $h(X)$. Furthermore, this can be achieved in such a way that $(h(X),e)$ is dense in (Y,e). The space (Y,e) is called the *completion* of (X,d) and is *unique up to isometry.* This means that if (Y',e') is any other complete metric space into which (X,d) can be embedded as a dense subspace, then (Y',e') and (Y,e) are isometric

Let us state this formally as a theorem that we shall not prove.[1].

Theorem 13.66 *Let (X,d) be a metric space. Then there exists a complete metric space (Y,e) and a function $h : X \to Y$ such that h maps X isometrically onto $h(X)$, which has the property that $(h(X),e)$ is dense in (Y,e). The space (Y,e) with this property is unique up to an isometry and is called the* completion *of (X,d).*

---

[1]Exercise 13.15.16 can be made a basis for a proof. Another method, using Cauchy sequences, can be found in Bruckner, Bruckner, and Thomson, *Real Analysis,* Prentice Hall (1997), §9.6.7.

We can clearly consider the space $\mathbb{R}$ as the completion of the space $\mathbb{Q}$ of rational numbers. Note, however, that the goal of embedding $\mathbb{Q}$ as a dense subset of a larger complete metric space would be more ambitious than just this theorem might indicate: The algebraic and order structures would also need to be preserved, as indeed they are for $\mathbb{R}$.

Example 13.67 Since the space $\mathcal{P}[a,b]$ has been shown in Example 13.62 to be incomplete, what space might be used for its completion? A theorem of Weierstrass[2] implies that the completion of $\mathcal{P}[a,b]$ is $\mathcal{C}[a,b]$. ◀

Exercises

13.8.1 Show that a convergent sequence in a metric space must be a Cauchy sequence.

13.8.2 Show that every Cauchy sequence in a metric space is bounded.

13.8.3 Is the following statement true? A sequence $\{x_n\}$ in a metric space (X,d) is Cauchy if and only if $\lim_{n\to\infty} d(x_n, x_{n+1}) = 0$.

13.8.4 Let $\{x_n\}$ and $\{y_n\}$ be Cauchy sequences in a metric space (X,d). Show that $d(x_n, y_n)$ converges even if the sequences $\{x_n\}$ and $\{y_n\}$ themselves do not.

13.8.5 Let $\{x_n\}$ be a Cauchy sequence in a metric space (X,d). Show that there must be a subsequence $\{x_{n_k}\}$ with these two properties:

(a) $d(x_{n_{k+1}}, x_{n_k}) < 2^{-k}$ for $k = 0, 1, 2, \dots$.

(b) $B[x_{n_1}, 1] \supset B[x_{n_2}, 1/2] \supset \dots B[x_{n_k}, 1/2^{k-1}] \supset \dots$.

13.8.6 Prove that if any subsequence of a Cauchy sequence in a metric space converges, then the full sequence also converges.

13.8.7 Show that a metric space (X,d) is complete if and only if every sequence of points $\{x_n\}$ in X with the property that

$$\sum_{k=1}^{\infty} d(x_k, x_{k+1}) < \infty$$

converges.

13.8.8 Prove Theorem 13.65.

13.8.9 Let X be a nonempty set and let d be the discrete metric. Is (X,d) complete?

13.8.10 Obtain that each of the following spaces is complete by applying Theorem 13.64:

[2] See the material in Section 10.8.4; for a different proof see Bruckner, Bruckner, and Thomson, op. cit., §9.13.

(a) $C[a,b]$
(b) The Cantor set
(c) $\mathbb{N}$

13.8.11 Obtain that each of the following spaces is not complete by applying Theorem 13.64:

(a) The space $\mathcal{P}[0,1]$ of polynomials on $[0,1]$
(b) The set $\{1, 1/2, 1/3, 1/4, \dots\}$ with the usual real metric
(c) The open interval $(0,1)$

13.8.12 Show that euclidean n-dimensional space $\mathbb{R}^n$ is complete. (See Example 13.5.)

13.8.13 Show that the space $\mathcal{K}$ of Exercise 13.3.9 is complete.

13.8.14 **(Sierpinski's space)** Let $\mathbb{N}$ be furnished with the metric

$$d(m,n) = \begin{cases} \frac{1}{m+n} + 1, & \text{if } m \neq n \\ 0, & \text{otherwise.} \end{cases}$$

(a) Verify that d is a metric on $\mathbb{N}$.
(b) Show that every subset of $(\mathbb{N}, d)$ is open.
(c) Show that $(\mathbb{N}, d)$ is complete.
(d) Let $S_n = \left\{ m \in \mathbb{N} : d(m,n) \leq 1 + \dfrac{1}{2n} \right\}$.
Show that $S_n = \{n, n+1, n+2 \dots\}$.
(e) Show that $\{S_n\}$ is a descending sequence of closed balls whose intersection is empty.
(f) Reconcile part (e) with Theorem 13.65.

13.8.15 Verify that the spaces c (Exercise 13.5.14) and ℓ_∞ (Example 13.9) are complete. Is the subspace c_0 of c complete?

13.8.16 Let (X,d) and (Y,e) be metric spaces and let f be a continuous function mapping X onto Y.

(a) If X is separable, must Y be separable?
(b) If X is complete, must Y be complete?
(c) Is separability a topological property? Is completeness?
(d) Do the answers to (a) and (b) change if f is an isometry?

13.8.17 Let (X_1, d_1) and (X_2, d_2) be complete metric spaces. Is the product space $X_1 \times X_2$ also complete? (See Exercise 13.2.9 for the definition of the product metric.)

13.8.18 A metric space X is said to be *absolutely closed* if every isometric image of X into a space Y is closed in Y. Show that X is absolutely closed if and only if it is complete.

13.9 Contraction Maps

Let f be a continuous function mapping an interval $[a, b]$ into itself. Then f has a *fixed point*; that is, there exists a point $x \in [a, b]$ such that $f(x) = x$. To see this, observe that if a and b are not fixed points, then $f(a) > a$ and $f(b) < b$. The function $g(t) = f(t) - t$ satisfies $g(a) > 0 > g(b)$. Since g is continuous, the intermediate value theorem applies to show that there exists a point x such that $g(x) = 0$ (Theorem 5.52). From this it follows that $f(x) = x$.

A comparable statement is not valid for continuous functions mapping $\mathbb{R} \to \mathbb{R}$.

Example 13.68 The function $f(x) = x + 1$ is a function mapping $\mathbb{R}$ into itself and has no fixed points since $f(x) = x$ for no value of x. ◀

The existence of fixed points of mappings has proved to be of considerable importance in analysis. We could ask just what it takes about a space X that every continuous function $f : X \to X$ should have a fixed point. For example, a famous theorem of Luitzen Brouwer (1881–1966) asserts that any closed sphere in $\mathbb{R}^n$ would have this property. Instead we restrict our attention by considering not all continuous functions, but ones that are contractions in a sense that will be defined. For example, if a function $f : \mathbb{R} \to \mathbb{R}$ satisfies the Lipschitz condition

$$|f(x) - f(y)| \leq \alpha|x - y|$$

for all $x, y \in \mathbb{R}$, with $0 < \alpha < 1$, then f is "contractive" in the sense that any two points x and y move to points $f(x)$ and $f(y)$ that are closer together. This condition guarantees that f will have a fixed point.

Example 13.69 The functions $f(x) = x/2$ or $g(x) = \sin(x/2)$ map $\mathbb{R} \to \mathbb{R}$ and both satisfy such a Lipschitz condition with $\alpha = 1/2$. (Just check the derivatives and apply the mean value theorem.) In this case the fixed points of f and g are easy to find: Look for the point of intersection of the graph of f (or g) with the line $y = x$. ◀

This statement about the existence of fixed points for certain Lipschitz functions is a special case of an important property of complete metric spaces. The property, often called Banach's fixed point theorem, states that every *contraction map* of a complete metric space into itself has a unique fixed point. Let us give precise definitions for these terms.

Definition 13.70 Let (X, d) be a metric space and let $A : X \to X$. If there exists a number $\alpha \in (0, 1)$ such that

$$d(A(x), A(y)) \leq \alpha d(x, y) \text{ for all } x, y \in X,$$

we say A is a *contraction map*.

Definition 13.71 Let (X, d) be a metric space and let $A : X \to X$. A point $x \in X$ for which $A(x) = x$ is called a *fixed point.*

There are a few immediate properties of contraction maps that we should establish. The first shows that if a contraction map has a fixed point, then it has only one.

Lemma 13.72 *Let A be a contraction map that has a fixed point. Then that fixed point is unique.*

Proof To prove that there cannot be two fixed points for a contraction, observe that if $A(x) = x$ and $A(y) = y$, then

$$d(x, y) = d(A(x), A(y)) \leq \alpha d(x, y).$$

Since $\alpha < 1$, this implies that $d(x, y) = 0$ and $x = y$. ■

Another feature of the relation between contractions and fixed points is that if A is a contraction, then the sequence of iterates

$$x_0, A(x_0), A(A(x_0)), A(A(A(x_0))), \ldots$$

must converge to the fixed point or (if there is no fixed point) must diverge.

Lemma 13.73 *Let A be a contraction map, let x_0 be an element of the metric space and construct the sequence*

$$x_1 = A(x_0), \ x_2 = A(x_1), \ x_3 = A(x_2), \ \ldots.$$

Then

1. *If A has a fixed point x, then $x_n \to x$.*
2. *If $x_n \to x$, then x is the fixed point of A.*

Proof Let s, t be any elements of the metric space. Let A^n denote the composition of A with itself n times. Observe that, by the definition of a contraction, that

$$d(A(s), A(t)) \leq \alpha d(s, t),$$

that

$$d(A^2(s), A^2(t)) \leq \alpha d(A(s), A(t)) \leq \alpha^2 d(s, t)$$

and so, continuing in this way,

$$d(A^n(s), A^n(t)) \leq \alpha^n d(s, t) \tag{9}$$

for all n. Replace s by x_0 and t by x, where x is a fixed point of A so $x = A(x) = A^2(x) = \cdots = A^n(x)$ and obtain

$$d(A^n(x_0), x) \leq \alpha^n d(x_0, x) \to 0 \tag{10}$$

as $n \to \infty$. This proves the first assertion.

To prove the second assertion we first check that every contraction map is continuous (Exercise 13.9.5). Suppose that $x_n \to x$. Note that $x_{n+1} = A(x_n)$ and $x_{n+1} \to x$ and, by continuity, $A((x_n)) \to A(x)$, Thus $A(x) = x$ and x is a fixed point of A. ■

These two lemmas do not establish the existence of a fixed point for a contraction; indeed it is easy to give an example of a contraction with no fixed point. We shall see that a contraction map on a *complete* space must have a fixed point. But observe that a contraction map does more than "move points closer together:" It moves them together by an factor strictly smaller than 1. The following example shows a mapping that is nearly a contraction (on a complete space) but that has no fixed point.

Example 13.74 The function

$$f(x) = x + 1/x$$

maps the complete metric space $[2, \infty)$ into itself, and $f'(x) = 1 - x^{-2}$ satisfies $0 < |f'(x)| < 1$ for all $x \in [2, \infty)$. Thus if x and y are close together, $f(x)$ and $f(y)$ are even closer together. But this function has no fixed point because for all x,

$$x \neq x + 1/x.$$

Notice that it is not, however, a contraction map. The inequality

$$|x + 1/x - (y + 1/y)| \leq \alpha |x - y|$$

cannot hold for all x, $y \in [2, \infty)$ and any $\alpha < 1$ (although it does hold for $\alpha = 1$). [See Figure 13.4. Note that $A/B < 1$ but $\lim_{x_1, x_2 \to \infty} A/B = 1$ illustrating that the function f cannot be a contraction map on $[2, \infty)$.] ◄

Observe how, in Theorem 13.75, the fact that $\alpha < 1$ for a contraction map plays an essential role in the proof.

Theorem 13.75 (Banach) *A contraction map A defined on a complete metric space (X, d) has a unique fixed point.*

Proof We obtain the fixed point of A by starting at an arbitrary point x_0 in the space and iterating the function. Let $x_0 \in X$. Let $x_1 = A(x_0)$, $x_2 = A(x_1) = A^2(x_0)$, and, in general,

$$x_n = A(x_{n-1}) = A^n(x_0) \quad (n = 1, 2, \dots).$$

Here we are using the customary notation $A^{n+1}(x) = A(A^n(x))$.

We show that the sequence $\{x_n\}$ is a Cauchy sequence. Let $n \leq m$. Then, using the inequality (9), we obtain

$$d(x_n, x_m) = d(A^n(x_0), A^m(x_0))$$
$$= d(A^n(x_0), A^n(A^{m-n}(x_0))) \leq \alpha^n d(x_0, x_{m-n}).$$

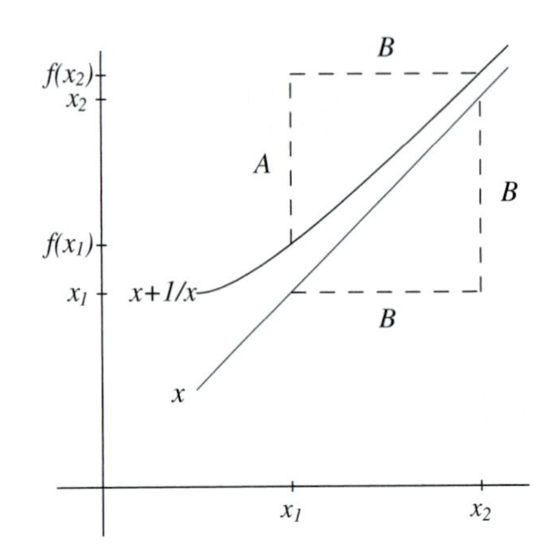

Figure 13.4. Graphs of $f(x) = x + 1/x$ and $g(x) = x$ with $A = d(f(x_1), f(x_2))$ and $B = d(g(x_1), g(x_2))$.

We can estimate the latter term, using the triangle inequality and the inequality (9),

$$\begin{aligned} d(x_0, x_{m-n}) &\leq d(x_0, x_1) + d(x_1, x_2) + \cdots + d(x_{m-n-1}, x_{m-n}) \\ &\leq d(x_0, x_1)[1 + \alpha + \alpha^2 + \cdots + \alpha^{m-n-1}] \\ &\leq \frac{d(x_0, x_1)}{1 - \alpha}. \end{aligned}$$

Let $\varepsilon > 0$. Choose N so that

$$\alpha^N d(x_0, x_1) < \varepsilon(1 - \alpha).$$

Then if $m \geq n \geq N$ we have from these inequalities that

$$d(x_n, x_m) \leq \frac{\alpha^n d(x_0, x_1)}{1 - \alpha} < \varepsilon$$

and we have established that the sequence is Cauchy. Since X is complete, there exists $x \in X$ such that $x_n \to x$. By Lemma 13.73 this point x is a fixed point of A. The uniqueness was given in Lemma 13.72. ■

Observe that the proof of Theorem 13.75 also provides a practical method for approximating the solution of an equation of the form $A(x) = x$. This method is often called the method of *successive approximations.*

We can choose x_0 to be *any* point in X. Then the sequence $\{A^n(x_0)\}$ converges to the unique solution of the equation $A(x) = x$. Indeed the inequality (10) shows that the convergence is as fast as some geometric progression $\{C\alpha^n\}$.

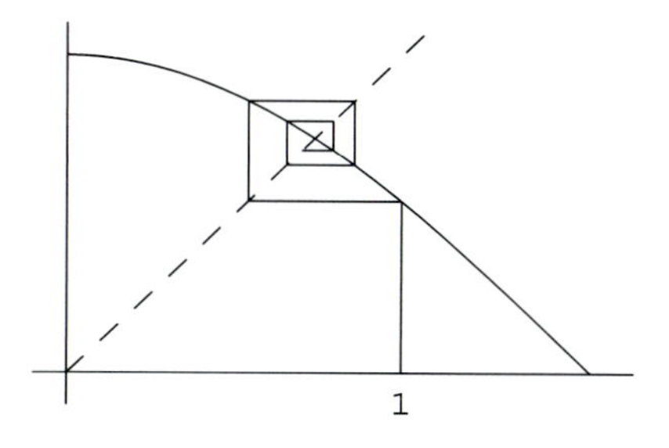

Figure 13.5. Approximate solution of $\cos x = x$. The "spiral" approaches the point of intersection of $y = \cos x$ and $y = x$.

Example 13.76 Let us solve the equation

$$\cos x = x.$$

Ordinary calculus techniques should convince you that there is a unique solution, but what practical method would offer an approximate answer? No algebraic methods or trigonometric identities will lead to a solution. Instead let us interpret this as the requirement to find the fixed point of the function

$$f(x) = \cos x.$$

On the interval $[1/2, 1]$ this is evidently a contraction. Thus if we start with any value $x_0 \in [1/2, 1]$ and follow the sequence

$$x_n = \cos x_{n-1}$$

for $n = 1, 2, 3, \ldots$ the proof of Theorem 13.75 shows that this sequence must converge to that fixed point. Indeed this process can be tried on any calculator: Start by entering the number 1 (say) and then repeatedly press the cosine key. You will see the numbers approach 0.73908513 quite quickly. It is instructive to see what is happening graphically. (See Fig. 13.5.) ◄

Example 13.77 Let $X = \mathcal{C}[0, 1/2]$ furnished, as usual, with the sup metric,

$$d(f_1, f_2) = \max_{x \in [0, \frac{1}{2}]} |f_1(x) - f_2(x)|.$$

Define a function A on X by

$$(A(f))(x) = g(x) = \int_0^x f(t)\, dt, \quad 0 \leq x \leq \frac{1}{2}.$$

Since g is continuous on $[0, 1/2]$, $g \in X$ and $A : X \to X$.

For $f_1, f_2 \in X$, let $g_1 = A(f_1)$ and $g_2 = A(f_2)$. Then

$$\begin{aligned} d(g_1, g_2) &= \max_{x\in[0,\frac{1}{2}]} |g_1(x) - g_2(x)| \\ &= \max_{x\in[0,\frac{1}{2}]} \left| \int_0^x [f_1(t) - f_2(t)]\, dt \right| \\ &\leq \max_{x\in[0,\frac{1}{2}]} \int_0^x |f_1(t) - f_2(t)|\, dt \\ &\leq \max_{x\in[0,\frac{1}{2}]} x \max_{t\in[0,\frac{1}{2}]} |f_1(t) - f_2(t)| \\ &\leq \frac{1}{2} d(f_1, f_2). \end{aligned}$$

Thus A is a contraction map.

From Theorem 13.75 we can conclude that there is a unique function $f \in \mathcal{C}[0, 1/2]$ such that $A(f) = f$, that is, such that

$$f(x) = \int_0^x f(t)\, dt \text{ for all } x \in [0, 1/2].$$

Since the zero function, $f(x) = 0$ for all $x \in [0, 1/2]$, satisfies this and the solution is unique, we see that the zero function is the only solution. Had we not observed this, we could have tried the method of successive approximations.

Starting with $f(t) = 1$, for example, we obtain the sequence

$$t, \frac{t^2}{2}, \frac{t^3}{6}, \dots, \frac{t^n}{n!}, \dots$$

which converges in the space $\mathcal{C}[0, 1/2]$ (that is, it converges uniformly) to the fixed point $f \equiv 0$. ◀

We shall see a variety of applications of Theorem 13.75 in the next section.

Exercises

13.9.1 Prove that if $f : \mathbb{R} \to \mathbb{R}$ is continuous and satisfies the Lipschitz condition

$$|f(x) - f(y)| \leq \alpha|x - y| \text{ for all } x, y \in \mathbb{R} \text{ with } 0 < \alpha < 1,$$

then f has a unique fixed point.

13.9.2 Let $f(x) = x + 1/x$. Choose a point $x_0 \in [2, \infty)$ and observe what happens to the sequence $\{x_k\} = \{f^k(x_0)\}$. Where would the proof of Theorem 13.75 break down if we try to apply it to this mapping, f?

13.9.3 Show that the function $f(x) = \cos x$ is a contraction mapping on $[1/2, 1]$ but is not a contraction map on $[0, \pi]$.

13.9.4 Show that the function $f(x) = \cos x$ is not a contraction mapping on $\mathbb{R}$ but that some iterate of f is a contraction map on $\mathbb{R}$. Can Theorem 13.75 be applied in this case?

13.9.5 Prove that a contraction map on a metric space is continuous.

13.9.6 Show that we cannot drop the requirement that X is complete in Theorem 13.75 without rendering the conclusion false.

13.9.7 A careless student claims the function e^x is a fixed point of the map in Example 13.77. Take this function as the initial approximation for a sequence of successive approximations. Obtain also the sequence of successive approximations when the initial function f_0 is $\sin x$.

13.9.8 Define a function $A : \mathcal{C}[0,1] \to \mathcal{C}[0,1]$ by

$$(A(f))(x) = g(x) = \int_0^x f(t)\, dt, \quad 0 \leq x \leq 1.$$

(a) Is A a contraction?

(b) Is A^2 a contraction?

(c) Does A have a unique fixed point?

13.9.9 Let A and B be contraction maps on a complete metric space (X, d) and suppose that

$$d(A(x), B(x)) < \varepsilon$$

for all $x \in X$. If a is the fixed point of A and b is the fixed point of B find an estimate for $d(a, b)$.

13.9.10 Let A be a contraction map on a complete metric space (X, d) such that $d(A(x), A(y)) \leq \alpha d(x, y)$ for all x, $y \in X$. Show that the "rate of convergence" of the sequence of iterates $\{A^n(x_0)\}$ in Theorem 13.75 to its limit x can be estimated by

$$d(A^n(x_0), x) \leq \frac{\alpha^n d(x_0, A(x_0))}{1 - \alpha}.$$

13.9.11 Let $\{A_n\}$ be a sequence of contraction maps on a complete metric space (X, d) such that

$$d(A_n(x), A_n(y)) \leq \alpha d(x, y) \tag{11}$$

for all x, $y \in X$ and $n \in \mathbb{N}$ and some $\alpha < 1$. Suppose that $\{A_n(x)\}$ converges to $A(x)$ for each $x \in X$. Show that A is a contraction mapping on X and that its fixed point can be computed as

$$a = \lim_{n \to \infty} a_n,$$

where a_n is the fixed point of the contraction map A_n.

13.9.12 In Hilbert space ℓ_2 consider the mappings $A_n : \ell_2 \to \ell_2$ defined so that if $(x_1, x_2, x_3, \dots) \in \ell_2$, then $A_n((x_1, x_2, x_3, \dots))$ is that element of ℓ_2 with all zero entries except for the number

$$\frac{1}{n} + \frac{n-1}{n} x_n$$

in the nth position. Show that A_n is a contraction mapping with a unique fixed point $a_n \in \ell_2$ but that the sequence $\{a_n\}$ does not converge in ℓ_2. Conclude from this example that the condition (11) in Exercise 13.9.11 cannot be dropped.

13.9.13 Let $\{A_n\}$ be a sequence of mappings of a complete metric space (X, d) into itself such that $\{A_n(x)\}$ converges uniformly to $A(x)$. Suppose further that A is a contraction mapping and that for each n there is at least one fixed point a_n of the mapping A_n. Show that the fixed point of A can be computed as

$$a = \lim_{n \to \infty} a_n.$$

13.10 Applications of Contraction Maps (I)

✂ *Enrichment*

A variety of problems in analysis involve the solving of some sort of equations. The equations could be ones involving numbers, or n-tuples of numbers, or sequences or functions or various other mathematical objects. Often such a problem can be cast in the following form: We observe that a certain operator arises naturally in connection with the problem, and that a solution to the problem is representable as a fixed point of the operator.

Let's look at a few examples. We will not consider the solutions until the next section. For now we only reinterpret the problem as a fixed point problem. We begin with a trivial example just to illustrate the approach.

Example 13.78 (A Simple Linear Equation) Consider the simple linear equation

$$ax = b, \quad (a \neq 0). \tag{12}$$

We can rewrite the equation as $x = (1-a)x + b$. Now consider the operator $A(x) = (1-a)x + b$. A solution to (12) is just a fixed point of the operator A. ◀

Example 13.79 (Systems of Linear Equations) Consider a system of linear equations

$$\begin{aligned} a_{11}x_1 + a_{12}x_2 + \cdots + a_{1n}x_n &= b_1 \\ a_{21}x_1 + a_{22}x_2 + \cdots + a_{2n}x_n &= b_2 \\ &\vdots \\ a_{n1}x_1 + a_{n2}x_2 + \cdots + a_{nn}x_n &= b_n \end{aligned} \tag{13}$$

We can rewrite this system in the form

$$\begin{aligned} x_1 &= c_{11}x_1 + c_{12}x_2 + \cdots + c_{1n}x_n + b_1 \\ x_1 &= c_{21}x_1 + c_{22}x_2 + \cdots + c_{2n}x_n + b_2 \\ &\vdots \\ x_n &= c_{n1}x_1 + c_{n2}x_2 + \cdots + c_{nn}x_n + b_n, \end{aligned} \tag{14}$$

where $c_{ij} = -a_{ij}$ if $j \neq i$, and $c_{ii} = 1 - a_{ii}$. We can then find solutions to the system (13) by solving the equivalent system (14).

It is now easy to view this problem in terms of a fixed point of an operator: For $x = (x_1, \ldots, x_n) \in \mathbb{R}^n$, let $y = A(x)$, where $y = (y_1, \ldots, y_n)$ with

$$y_i = \sum_{j=1}^{n} c_{ij}x_j + b_i.$$

Thus $A : \mathbb{R}^n \to \mathbb{R}^n$. A solution of (14) is just a fixed point of the operator A. ◀

Example 13.80 (Infinite Systems of Linear Equations) The preceding ideas can be applied to infinite systems of linear equations. In the late nineteenth century, a number of authors considered such systems arising, for example, in studies of algebraic equations and celestial mechanics. Curiously, the first person to encounter an infinite system of linear equations was Joseph Fourier (1768–1830) in his classic 1822 study. His methods were simple, but unjustified. After that, the subject received no further attention for another half century.

Suppose we have an infinite system of equations of the form

$$x_i = \sum_{j=1}^{\infty} c_{ij}x_j + b_i \qquad (i = 1, 2, 3 \ldots). \tag{15}$$

We seek a sequence $x = \{x_i\}$ that satisfies (15). As in Example 13.79, we can consider the operator A defined by $y = A(x)$, where $x = \{x_i\}$, $y = \{y_i\}$ and $y_i = \sum_{j=1}^{\infty} c_{ij}x_j + b_i$. As before, a solution to (15) could be viewed as a fixed point of the operator A, but here we must be a bit more careful. In Example 13.79, it was clear that the domain and range of the operator was $\mathbb{R}^n$. Here we have a number of sequence spaces that we might consider (e.g. ℓ_1, ℓ_2, ℓ_∞). (These spaces were described in Examples 13.7–13.9.) We continue our discussion of this example in the next section. ◀

Example 13.81 (Fredholm's Integral Equation) You may be familiar with certain "boundary value problems" from mathematical physics, such as the Dirichlet problem and the Neumann problem. Our next example involves an integral equation that together with its relatives plays a role in dealing with such problems.

Consider the equation

$$f(x) = \lambda \int_a^b K(x,y)f(y)\,dy + \phi(x), \tag{16}$$

where $\lambda \in \mathbb{R}$, ϕ is continuous on $[a,b]$, and K is continuous on $[a,b] \times [a,b]$. We seek a function $f \in \mathcal{C}[a,b]$ that satisfies (16). It is natural to consider the operator A defined on $\mathcal{C}[a,b]$ by $A(f) = g$, where

$$(A(f))(x) = g(x) = \lambda \int_a^b K(x,y)f(y)\,dy + \phi(x).$$

A fixed point of this operator provides a solution to (16). As in Example 13.80, we must be precise about the range of this operator. We continue our discussion in the next section. ◀

Example 13.82 (Differential Equations) Let D be an open set in $\mathbb{R}^2$ and let $f : D \to \mathbb{R}$. We wish to find a local solution to the differential equation

$$\frac{dy}{dx} = f(x,y)\,, \qquad y(x_0) = y_0.$$

For a specific and familiar type of example we could ask for the solution of

$$\frac{dy}{dx} = x^2 \sin xy + ye^{xy}$$

that "passes through" the point $(0,0)$.

Can we be sure such a solution exists? Naturally, some conditions on f must be imposed. For the moment, let's just see how to cast the problem in terms of operators.

We begin by reformulating the problem in terms of an integral equation. We seek a function ϕ such that

$$\phi'(x) = f(x,\phi(x)) \quad \text{and} \quad \phi(x_0) = y_0$$

on some open interval containing x_0. We can recast the problem as seeking ϕ such that

$$\phi(x) = y_0 + \int_{x_0}^x f(t,\phi(t))\,dt.$$

An appropriate operator A of the form

$$(A(\phi))(x) = y_0 + \int_{x_0}^x f(t,\phi(t))\,dt$$

would be natural for this problem. We must be careful to impose conditions on f that allow us to define A on an appropriate function space. We shall do this in the next section. ◀

We have presented several historically important examples of problems whose solutions involve fixed points of operators. In each case, appropriate restrictions on the class of sequences or functions to be considered allows us to use the Banach fixed point theorem (Theorem 13.75) to show that there is a fixed point—in fact a unique one—in the space under consideration. Each of the theorems we obtain in our next section addresses one of the examples mentioned in this section.

Exercises

13.10.1 What conditions on the values of a and b in Example 13.78 will force the function $A(x) = (1-a)x + b$ to be a contraction on $\mathbb{R}$? Examine the sequence of iterates in this case.

13.10.2 Apply the method of Example 13.79 to solve the system of two linear equations

$$\begin{aligned} a_{11}x_1 + a_{12}x_2 &= b_1 \\ a_{21}x_1 + a_{22}x_2 &= b_2. \end{aligned}$$

What sufficient conditions on the numbers a_{11}, a_{12}, a_{21}, a_{22}, b_1, and b_2 will force the function chosen by this method to be a contraction on $\mathbb{R}^2$?

13.10.3 Prove this special version of the intermediate value theorem: If f is a differentiable function on the interval $[a, b]$ with $f(a) < 0 < f(b)$ and $0 < c \leq f'(x) \leq C$, then there is a unique solution of the equation $f(x) = 0$ that can be obtained by iterating the function

$$F(x) = x - \frac{f(x)}{C}.$$

13.11 Applications of Contraction Maps (II)

Enrichment

In Section 13.10 we saw that solutions to various equations or systems of equations correspond to fixed points of operators associated with the equations. The fixed point theorem of Section 13.9 can sometimes be used to guarantee that there is a solution, in fact a unique one.

In this section we revisit each of the examples from Section 13.10. We obtain conditions under which the relevant operators are contraction maps. When they are, there will be unique solutions. And we can obtain the solutions by the method of successive approximations. Observe that in each case we must obtain a complete metric space on which the operator is a contraction. The conclusion of Theorem 13.75 will then apply in that space.

Example 13.83 (Example 13.78 Revisited.) A solution to the equation $ax = b$ is just a fixed point of the operator

$$A(x) = (1-a)x + b.$$

We are considering here the metric space $\mathbb{R}$ (with the usual metric). We find $A : \mathbb{R} \to \mathbb{R}$. In order to determine whether A is a contraction, we calculate for $x, y \in \mathbb{R}$,

$$\begin{aligned} d(A(x), A(y)) &= |A(x) - A(y)| = |(1-a)(x-y)| \\ &= |1-a||x-y| = |1-a|d(x,y). \end{aligned}$$

Thus A is a contraction if and only if $|1-a| < 1$. According to Theorem 13.75, the equation $ax = b$ will have a unique solution in $\mathbb{R}$ provided that $|1-a| < 1$, that is provided that $0 < a < 2$.

Students of elementary algebra know that the equation $ax = b$ has the unique solution $x = b/a$ provided that $a \neq 0$, no solution if $a = 0$ and $b \neq 0$, and every real number x as a solution if $a = b = 0$. The point here is that the application of the contraction mapping principle is available only under the restricted condition that $|1-a| < 1$. ◀

We presented this example partially as a warm-up and partially to emphasize that Theorem 13.75 provides a *sufficient* condition for a unique fixed point, not a *necessary* condition.

We continue with several examples of historically important problems to which the Banach fixed-point theorem can be applied. There are many other interesting applications. You can find one involving the space $\mathcal{K}$ of compact subsets of the plane furnished with the Hausdorff metric.[3] It pertains to the technique of "fractal image compression" that is useful for encoding and storing graphic images in computers. Another application, given as Exercise 13.11.3, provides a proof of an implicit function theorem.

13.11.1 Systems of Equations (Example 13.79 Revisited)

✂ *Enrichment*

We saw in Example 13.79 that the system of equations

$$\begin{aligned} a_{11}x_1 + a_{12}x_2 + \cdots + a_{1n} &= b_1 \\ a_{21}x_1 + a_{22}x_2 + \cdots + a_{2n} &= b_2 \\ &\vdots \\ a_{n1}x_1 + a_{n2}x_2 + \cdots + a_{nn} &= b_n \end{aligned} \tag{17}$$

can be solved by finding a fixed point of the operator $A : \mathbb{R}^n \to \mathbb{R}^n$ defined by $y = A(x)$, if

$$y_i = \sum_{j=1}^{n} c_{ij}x_j + b_i \quad \text{with} \quad x = (a_1, \ldots, x_n),\ y = (y_1, \ldots, y_n).$$

[3]There is material on the "collage theorem" in Bruckner, Bruckner, and Thomson, op. cit., p. 432.

In order for us to apply Theorem 13.75 we must specify the metric on $\mathbb{R}^n$—whether A is a contraction will depend on our choice of that metric. Let us choose the metric

$$d_\infty(x,y) = \max_{1\le i\le n} |x_i - y_i|.$$

(See Exercise 13.2.6.) In that case, let x, $x^* \in \mathbb{R}^n$ with $y = A(x)$, $y^* = A(x^*)$ and compute

$$\begin{aligned}
d(A(x), A(x^*)) &= d(y,y^*) = \max_i |y_i - y_i^*| \\
&= \max_i \left| \sum_i c_{ij}(x_j - x_j^*) \right| \\
&\le \max_i \sum_j |c_{ij}||x_j - x_j^*| \\
&\le (\max_i \sum_j |c_{ij}|)(\max_j |x_j - x_j^*|) \\
&\le \max_i \sum_j |c_{ij}| d(x,x^*).
\end{aligned}$$

Thus A will be a contraction map if there is a number α with

$$\sum_j |c_{ij}| \le \alpha < 1 \text{ for all } i = 1,\dots,n.$$

Observe that Theorem 13.75 guarantees a unique solution in the metric space $(\mathbb{R}^n, d_\infty)$ whenever there exists α such that

$$\sum_j |c_{ij}| \le \alpha < 1 \text{ for all } i = 1,\dots,n.$$

Exercise 13.11.1 provides a different condition involving column sums in place of row sums.

13.11.2 Infinite Systems (Example 13.80 revisited)

Enrichment

Here we are considering the infinite system of equations

$$x_i = \sum_{j=1}^{\infty} c_{ij}x_j + b_i \qquad (i = 1,2,3\dots). \tag{18}$$

This system leads to the operator $y = A(x)$, where

$$x = \{x_i\}, \quad y = \{y_i\} \quad \text{and} \quad y_i = \sum_{j=1}^{\infty} c_{ij}x_j + b_i. \tag{19}$$

If we wish to find conditions under which A is a contraction map, we must, as before, indicate which metric space we have under consideration. And here this decision is critical. Suppose, for example, we choose the space of bounded sequences known as m or as ℓ_∞ with the metric

$$d_\infty(x, y) = \sup_i(|x_i - y_i|).$$

(See Example 13.9.)

Since we wish A to map ℓ_∞ into itself, we impose the requirement that $\{b_i\} \in \ell_\infty$; that is, there exists $B < \infty$ such that

$$|b_i| < B \quad \text{for all} \quad i \in \mathbb{N}. \tag{20}$$

Our work with the previous example suggests the limitation

$$\sum_{j=1}^{\infty} |c_{ij}| \leq \alpha < 1 \quad \text{for all} \quad i \in \mathbb{N}. \tag{21}$$

Suppose then that the system (18) satisfies conditions (20) and (21) and that A is defined by (19). We wish to show that A is a contraction map on ℓ_∞.

We first verify that A maps ℓ_∞ into ℓ_∞. For $x = \{x_1, x_2, \dots\}$, an element of the space ℓ_∞, write $\|x\|_\infty = \sup_j |x_j|$. From (18), (20), and (21) we find that

$$|y_i| \leq \sum_{j=1}^{\infty} |c_{ij}|\|x\|_\infty + |b_i| \leq \alpha\|x\|_\infty + B. \tag{22}$$

Since (22) is valid for all $i \in \mathbb{N}$, we see that

$$\|A(x)\|_\infty = \|y\|_\infty = \sup_i |y_i| \leq \alpha \sup_j |x_j| + B,$$

so $A(x) \in \ell_\infty$. Thus A maps ℓ_∞ into ℓ_∞.

We next show that A is a contraction map. Let $x, x^* \in \ell_\infty$, $y = A(x)$ and $y^* = A(x^*)$. Then

$$y_i^* - y_i = \sum_{j=1}^{\infty} c_{ij}(x_j^* - x_j).$$

Using (21), we conclude that $|y_i^* - y_i| \leq \alpha\|x^* - x\|_\infty$, so

$$\|A(x^*) - A(x)\|_\infty = \sup_i |y_i^* - y_i| \leq \alpha\|x^* - x\|_\infty.$$

But this means that

$$d(A(x^*), A(x)) \leq \alpha d_\infty(x^*, x),$$

and we see that A is a contraction map.

We summarize this discussion as a theorem.

Theorem 13.84 *If the system of equations*

$$x_i = \sum_{j=1}^{\infty} c_{ij} x_j + b_i \ , \ i = 1, 2, 3 \ldots$$

satisfies the two conditions

1. *There exists $B < \infty$ such that $|b_i| < B$ for $i = 1, 2, \ldots,$ and*

2. *$\sum_{j=1}^{\infty} |c_{ij}| \leq \alpha < 1$ for $i = 1, 2, \ldots,$*

then this system has a unique solution in ℓ_∞.

Note. The conclusion of Theorem 13.84 guarantees a unique solution to the system of equations *in the space ℓ_∞*. Consider, for example, the system

$$x_1 = \frac{1}{2}x_2, \ x_2 = \frac{1}{2}x_3, \ x_3 = \frac{1}{2}x_4, \ldots \ .$$

For any $c \in \mathbb{R}$ the sequence $\{c, 2c, 4c, \ldots\}$ is a solution to this system. This does not contradict Theorem 13.84, however, since such a sequence is in ℓ_∞ if and only if $c = 0$.

13.11.3 Integral Equations (Example 13.81 revisited)

✂ *Enrichment*

Here we seek solutions to the integral equation

$$f(x) = \lambda \int_a^b K(x, y) f(y) \, dy + \phi(x) \tag{23}$$

The corresponding operator A is defined on $\mathcal{C}[a, b]$ by $A(f) = g$, where

$$(A(f))(x) = g(x) = \lambda \int_a^b K(x, y) f(y) \, dy + \phi(x).$$

We leave as Exercise 13.11.2 the fact that $A : \mathcal{C}[a, b] \to \mathcal{C}[a, b]$, that is, for each $f \in \mathcal{C}[a, b]$, $A(f) = g$ is also continuous on $[a, b]$.

We wish to find conditions under which A is a contraction map. Let $f_1, f_2 \in \mathcal{C}[a, b]$, let $g_1 = A(f_1)$, $g_2 = A(f_2)$, and let

$$M = \max\{|K(x, y)| : a \leq x \leq b, a \leq y \leq b\}.$$

(See Exercise 13.6.18.) Then

$$\begin{aligned} d(g_1, g_2) &= \max_{a \le x \le b} |g_1(x) - g_2(x)| \\ &\le |\lambda| M \max_{a \le x \le b} |f_1(x) - f_2(x)|(b-a) \\ &\le |\lambda| M(b-a) d(f_1, f_2). \end{aligned}$$

It follows that A is a contraction map provided that

$$|\lambda| < \frac{1}{M(b-a)}.$$

Thus, the method of successive approximations can be used to obtain the unique solution to (23) provided that $|\lambda|$ is not too big.

13.11.4 Picard's Theorem (Example 13.82 revisited)

Enrichment

Our aim is to prove the classical theorem of Picard, which is an application of contraction mappings to a problem in differential equations. We first need a definition. You will recall what was meant by a Lipschitz condition for a function $f : \mathbb{R} \to \mathbb{R}$. We now extend this meaning to allow for real-valued functions of two variables.

Definition 13.85 Let D be an open set in $\mathbb{R}^2$ and let $f : D \to \mathbb{R}$. We say that f satisfies a *Lipschitz condition in* y *on* D, with *Lipschitz constant* M if

$$|f(x, y_2) - f(x, y_1)| \le M|y_2 - y_1| \tag{24}$$

whenever (x, y_1) and (x, y_2) are in D.

Theorem 13.86 (Picard) *Let* f *be a continuous function on* D *and satisfy a Lipschitz condition in* y *on* D *with Lipschitz constant* M, *and let* $(x_0, y_0) \in D$. *Then there exists* $\delta > 0$ *such that the differential equation*

$$\frac{dy}{dx} = f(x, y) \tag{25}$$

has a unique solution $y = \phi(x)$ *in the interval* $[x_0 - \delta, x_0 + \delta]$ *for which* $\phi(x_0) = y_0$.

Proof As we indicated in Example 13.82, we formulate our problem in terms of the integral equation

$$\phi(x) = y_0 + \int_{x_0}^{x} f(t, \phi(t))\, dt, \tag{26}$$

which is to be valid on the interval $[x_0 - \delta, x_0 + \delta]$. Since D is open, there exists a closed sphere S centered at (x_0, y_0) and contained entirely inside the

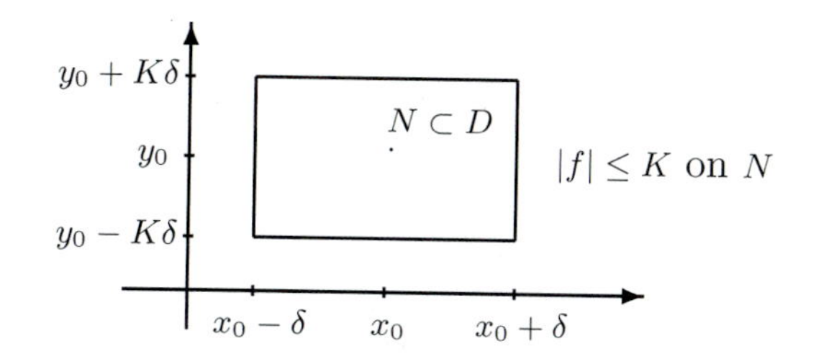

Figure 13.6. Choice of δ in the proof of Picard's theorem.

open set D. Since f is continuous on D, we may let K be the maximum of $|f(x, y)|$ for points (x, y) in the sphere S.

Now we use the Lipschitz constant M [defined as in the inequality (24)]. Choose $\delta > 0$ such that $\delta < 1/M$ and so that every point (x, y) with

$$|x - x_0| \leq \delta \quad \text{and} \quad |y - y_0| \leq K\delta$$

belongs to the sphere S. We then arrive at a rectangle

$$N = [x_0 - \delta, x_0 + \delta] \times [y_0 - K\delta, y_0 + K\delta]$$

that lies entirely inside the open set D (Fig. 13.6).

Consider now the map A defined so that

$$(A(\phi))(x) = \psi(x) = y_0 + \int_{x_0}^{x} f(t, \phi(t))\, dt$$

for $x_0 - \delta \leq x \leq x_0 + \delta$. Note that this may not defined as a map on the space $\mathcal{C}[x_0 - \delta, x_0 + \delta]$ since the values of the function $\phi(t)$ might not allow us to conclude that $(t, \phi(t))$ is in D so that $f(t, \phi(t))$ is defined. Thus we pass to a subspace.

Let $\mathcal{C}_1$ consist of those members of $\mathcal{C}[x_0 - \delta, x_0 + \delta]$ that satisfy $\phi(x_0) = y_0$ and for which

$$|\phi(x) - y_0| \leq K\delta \quad \text{for all } x \in [x_0 - \delta, x_0 + \delta].$$

Then $\mathcal{C}_1$ is a closed subspace of the space $\mathcal{C}[x_0 - \delta, x_0 + \delta]$ and is therefore complete by Theorem 13.64. We show that the operator A defined previously maps $\mathcal{C}_1$ into itself. Let $x \in [x_0 - \delta, x_0 + \delta]$ and suppose $\phi \in \mathcal{C}_1$. Then

$$\begin{aligned} |\psi(x) - y_0| &= \left| \int_{x_0}^{x} f(t, \phi(t))\, dt \right| \leq \int_{x_0}^{x} |f(t, \phi(t))|\, dt \\ &\leq K|x - x_0| \leq K\delta, \end{aligned}$$

so $\psi = A(\phi) \in \mathcal{C}_1$, and $A : \mathcal{C}_1 \to \mathcal{C}_1$.

We show that A is a contraction map on $\mathcal{C}_1$. To verify the contraction condition, let $\phi_1, \phi_2 \in \mathcal{C}_1$, let $\psi_1 = A(\phi_1)$ and let $\psi_2 = A(\phi_2)$. Then for all $x \in [x_0 - \delta, x_0 + \delta]$,

$$|\psi_1(x) - \psi_2(x)| \leq \int_{x_0}^{x} |f(t, \phi_1(t)) - f(t, \phi_2(t))|\, dt \qquad (27)$$
$$\leq M\delta \max_x |\phi_1(x) - \phi_2(x)|.$$

The last inequality is a consequence of the Lipschitz condition on f and the inequality $|x - x_0| \leq \delta$. Now (27) is valid for all x in the interval $[x_0 - \delta, x_0 + \delta]$, so

$$d(\psi_1, \psi_2) \leq M\delta d(\phi_1, \phi_2).$$

Since $M\delta < 1$, A is a contraction map, so the equation $\phi = A(\phi)$ has a unique solution in $\mathcal{C}_1$. In other words, the equation (26) and hence the equivalent equation (25) have unique local solutions. ■

Exercises

13.11.1 Show that the operator A of Example 13.79 revisited is a contraction on $(\mathbb{R}^n, d_1)$ if there exists α such that

$$\sum_{i=1}^{n} |c_{ij}| \leq \alpha < 1 \text{ for all } j = 1, \ldots, n.$$

13.11.2 Show that the operator A of Example 13.81 revisited maps $\mathcal{C}[a, b]$ into itself.

13.11.3 Use Theorem 13.75 to prove the following form of the implicit function theorem. For other versions of such theorems and different proofs, see Section 12.6.

> **Theorem** *Let $D = [a, b] \times \mathbb{R}$, and let $F : D \to \mathbb{R}$. Suppose that F is continuous on D and $\partial F/\partial y$ exists on D. If there exist positive real numbers α and β such that*
>
> $$\alpha \leq \frac{\partial F}{\partial y} \leq \beta$$
>
> *on D, then there exists a unique function $f \in \mathcal{C}[a, b]$ such that*
>
> $$F(x, f(x)) = 0 \text{ for all } x \in [a, b].$$
>
> *That is, the equation $F(x, y) = 0$ can be solved uniquely for y as a continuous function of x on $[a, b]$.*

13.12 Compactness

One of the goals of an abstract subject, such as the study of metric spaces, is to lift ideas and methods from our earlier studies into this abstract setting.

We have already seen how the Cauchy criterion has entered the general subject of metric spaces. On the real line the Cauchy criterion is a necessary and sufficient condition for a sequence to converge. In our new theory we have defined a space to be *complete* precisely when it has this same property. Thus we can lift our methods from the real line into metric space theory. We verify that some space that we are studying is complete and then we have available to us the theory of Cauchy sequences.

In an similar way we wish now to lift our compactness arguments from Section 4.5 into metric space theory. You will recall that the Bolzano-Weierstrass and Heine-Borel theorems offered a powerful tool in the study of properties of continuous functions. On the real line these tools could be used in any closed and bounded set. They are not, however, available in every metric space since closed and bounded sets do not necessarily have these properties. Instead we shall define a set to be compact *precisely when the Bolzano-Weierstrass and Heine-Borel theorems are available.*

In Section 13.12.1 we take the Bolzano-Weierstrass property as our definition of compactness and derive some consequences. In Section 13.12.2 we show that some of our theorems from the elementary analysis of real functions can be extended to continuous functions defined on compact sets in a metric space. In Section 13.12.3 we show that the Heine-Borel property and the Bolzano-Weierstrass property are equivalent in any metric space so that either could have been taken as a definition of compactness. If you are intending to go on to even more abstract levels of generality, we should warn that there, in the subject of topology, these two concepts are not equivalent and that compactness is usually defined using the Heine-Borel property only.

13.12.1 The Bolzano-Weierstrass Property

Every bounded sequence of real numbers has a convergent subsequence. This is false in a general metric space.

Example 13.87 Let X be an infinite set furnished with the discrete metric. Let $\{x_n\}$ be any sequence of distinct elements of X. Then $\{x_n\}$ is bounded; indeed the diameter of the whole space X is only 1 so every subset of X is bounded. But $\{x_n\}$ can have no convergent subsequence. ◀

Thus a special property of closed and bounded subsets of $\mathbb{R}$, namely the Bolzano-Weierstrass property, is false. This property asserts that if $K \subset \mathbb{R}$ is closed and bounded, then every sequence of points in K has a subsequence converging to a point in K. We have called closed and bounded subsets of $\mathbb{R}$ *compact*, but it was mainly this property of such sets that we needed. The natural thing to do in a general metric space is to recognize that "closed and bounded" no longer plays an important role but to turn the Bolzano-Weierstrass property itself into a definition of what it means to be compact.

Definition 13.88 A set K in a metric space is said to be *compact* provided that set has the Bolzano-Weierstrass property, namely that every sequence of points in K has a subsequence converging to a point in K.

If (X, d) is a metric space and the set X itself is compact (i.e., X has the Bolzano-Weierstrass property), then we say that the metric space is compact.

Let us begin by noting some properties that every compact set must have. The first has been left as Exercise 13.12.2.

Theorem 13.89 *Compactness is a topological property.*

Theorem 13.90 *Let K be a compact set in a metric space (X, d). Then*

1. *K is closed.*
2. *K is bounded.*
3. *K is complete.*
4. *K is separable.*

Proof If K has the Bolzano-Weierstrass property, then every sequence of points in K that converges to a point z must have that point $z \in K$. Consequently, K must be closed.

Further, K must be bounded. If not, then for every integer N and any fixed point x_0 there must be points $x_N \in K$ with $d(x_0, x_N) > N$. This constructs a sequence that has every subsequence unbounded and, hence, could have no convergent subsequence. Thus if K has the Bolzano-Weierstrass property, K cannot be unbounded.

Further, K must be complete. If $\{x_n\}$ is a Cauchy sequence in K, then, by the Bolzano-Weierstrass property, there is a subsequence $\{x_{n_k}\}$ convergent to a point $z \in K$. But any Cauchy sequence with a convergent subsequence is itself convergent (Exercise 13.8.6).

Finally, let us show that K is separable. We can work just in the metric space (K, d). Suppose that K is not separable. Then for some $\varepsilon > 0$ there does not exist a finite set of points $x_1, x_2, x_3, \ldots, x_p$ so that

$$K \subset \bigcup_{i=1}^{p} B(x_i, \varepsilon).$$

(This follows from Exercise 13.7.1.) Thus we can choose a sequence of points $x_1, x_2, x_3, \ldots$ in K with the property that

$$x_n \notin \bigcup_{i=1}^{n-1} B(x_i, \varepsilon).$$

Such a sequence can have no Cauchy subsequence since

$$d(x_N, x_m) > \varepsilon$$

for all $n > M$. Thus it has no convergent subsequence. Since this contradicts the Bolzano-Weierstrass property, the set K must be separable. ■

Note. In our proof we encountered a technical idea that is of independent interest. If K is compact, then we have shown that it is bounded in a strong sense: Given any radius ε we can cover K by a finite number of balls of that radius. More precisely, for every $\varepsilon > 0$ there exists $n \in \mathbb{N}$ and open balls

$$B(x_1, \varepsilon), \ldots, B(x_n, \varepsilon)$$

such that

$$K \subset \bigcup_{i=1}^{n} B(x_i, \varepsilon).$$

The set $\{x_1, x_2, \ldots, x_n\}$ is called an *ε-net* for K. It has the property that if $x \in K$ there exists i such that $d(x_i, x) < \varepsilon$. Clearly, this is a stronger condition than boundedness: Not merely is K contained in some large ball (i.e., is bounded) but K is contained in the union of a finite number of balls of any specified diameter. This will play a role in Section 13.12.4.

We might have hoped that the list of properties in Theorem 13.90, which are evidently necessary properties of compact sets, might also be sufficient. Theorem 13.89 gives a clue: Although compactness is a topological property, boundedness is not. A simple example also illustrates. Consider the real line $\mathbb{R}$ but furnished with a different metric from the usual, the metric of Exercise 13.2.1(c) or (d). Since that metric is equivalent to the usual metric, every closed subset of $\mathbb{R}$ has all the properties of Theorem 13.90; the only difference here is that under the new metric all sets are bounded. But the only closed subsets of $\mathbb{R}$ that are compact (under either metric) are the ones that are bounded in the usual metric.

Exercises

13.12.1 Show that every closed subset of a compact set is also compact.

13.12.2 Show that compactness is a topological property.

13.12.3 Show that every finite subset of a metric space is compact.

13.12.4 What subsets of the space (X, d), where d is the discrete metric, are compact?

13.12.5 Let $\{x_n\}$ be a convergent sequence in a metric space with limit z. Show that the set

$$\{z, x_1, x_2, x_3, \ldots\}$$

is compact.

13.12.6 Prove that a subset of $\mathbb{R}^n$ is compact if and only if it is closed and bounded.

13.12.7 Show that the Hilbert cube is a compact subset of ℓ_2.

13.12.8 Show that the unit sphere in $\mathcal{C}[a,b]$, that is, the set

$$\{f \in \mathcal{C}[a,b] : |f(x)| \leq 1\,, \quad x \in [a,b]\}$$

is not compact.

13.12.9 Show that if K is a compact subset of a metric space (X,d), then for any $x \in X$ there is a point $k \in K$ so that

$$d(k,x) = \inf\{d(x,y) : y \in K\}.$$

Show that if K is not compact, but merely closed, this would not necessarily be true. If K is complete but not compact, is this always true?

13.12.10 Show that if E and F are closed subsets of a metric space (X,d), at least one of which is compact, then there are points $e \in E$ and $f \in F$ so that

$$d(e,f) = \inf\{d(x,y) : x \in E,\ y \in F\}.$$

Show that if the sets E and F are not compact, but merely closed or complete, then this would not necessarily be true.

13.12.11 Show that the product of two compact metric spaces furnished with the product metric (Exercise 13.2.9) is also a compact metric space.

13.12.2 Continuous Functions on Compact Sets

Many of the standard theorems about continuous functions on compact subsets of $\mathbb{R}$ or $\mathbb{R}^n$ carry over to general metric spaces. If $f : [a,b] \to \mathbb{R}$ is continuous, then

1. f is uniformly continuous on $[a,b]$.

2. f is bounded.

3. f attains its maximum and minimum.

We ask now, what are the analogues of this for a continuous function

$$f : X \to Y$$

where X is a compact metric space?

Definition 13.91 If $f : (X,d) \to (Y,e)$ and for every $\varepsilon > 0$ there exists $\delta > 0$ such that $e(f(x), f(x')) < \varepsilon$ whenever $d(x,x') < \delta$, we say f is *uniformly continuous* on X.

Let us treat uniform continuity first. We prove, as for continuous functions defined on a compact subset of $\mathbb{R}$, that continuous functions on compact spaces are uniformly continuous. Since the proofs do not require new methods, they have been left as Exercises 13.12.17 and 13.12.18.

Theorem 13.92 *If X is compact and $f : X \to Y$ is continuous, then f is uniformly continuous.*

We now generalize the elementary theorem that asserts that a continuous real-valued function on a compact interval $[a, b]$ achieves absolute extrema on $[a, b]$. We can anticipate what form this will assume in a general metric space if we remember that for a continuous function $f : [a, b] \to \mathbb{R}$ the image set $f([a, b])$ must be itself a compact interval in $\mathbb{R}$.

Theorem 13.93 *If $f : X \to Y$ is continuous and X is compact, then the set $f(X)$ is compact in Y.*

Exercises

13.12.12 If $f : (X, d) \to (Y, e)$ and d is the discrete metric is f continuous? Is f uniformly continuous?

13.12.13 If $f : (X, d) \to (Y, e)$ and e is the discrete metric is f continuous? Is f uniformly continuous?

13.12.14 Show that every contraction mapping on a metric space is uniformly continuous.

13.12.15 Let (X, d) be a metric space and let A be a nonempty subset of X. Define $f : X \to \mathbb{R}$ by

$$f(x) = \mathrm{dist}(x, A) = \inf\{d(x, y) : y \in A\}.$$

In Exercise 13.6.11 we established that f is continuous. Is f uniformly continuous?

13.12.16 Show that if $f : X \to Y$ is uniformly continuous and $\{x_n\}$ is a Cauchy sequence in X, then $\{f(x_n)\}$ is a Cauchy sequence in Y. Show that this need not be true if f is merely continuous.

13.12.17 Prove that if X is compact and $f : X \to Y$ is continuous, then f is uniformly continuous.

13.12.18 Prove that if $f : X \to Y$ is continuous and X is compact, then the set $f(X)$ is compact in Y.

13.12.19 Let X and Y be metric spaces with X compact. Prove that a continuous, one-to-one mapping of X onto Y is necessarily a homeomorphism.

13.12.20 Let $f : X \to X$ be a continuous mapping from a compact space X into itself. Define the sequence of sets

$$X_1 = f(X),\ X_2 = f(X_1),\ \ldots,\ X_n = f(X_{n-1}).$$

Let $K = \bigcap_{i=1}^{\infty} X_i$. Show that K is nonempty, compact, and invariant under f in the sense that $f(K) = K$.

13.12.21 A function $f : X \to X$ on a metric space (X, d) is a *weak contraction* if

$$d(f(x), f(y)) < d(x, y) \quad x \neq y.$$

(a) Show that such a function f must be uniformly continuous.

(b) Show that such a function f need not be a contraction.

(c) Show that if X is compact, then a weak contraction f must have a unique fixed point.

(d) Show that if X is complete but not compact, then a weak contraction f need not have a fixed point, but if it does that fixed point is unique.

(e) How does the result in (c) compare to the Banach fixed point theorem?

13.12.22 Formulate and prove a generalization of Dini's theorem (Theorem 9.24) for sequences of continuous real-valued functions on a compact metric space.

13.12.3 The Heine-Borel Property

✂ *Enrichment*

We recall that on the real line a compact set can be characterized in several different ways. One of the most useful for the purposes of compactness arguments was the Heine-Borel property, that any covering of a closed and bounded set by a family of open intervals can be reduced to a finite subcovering.

This too is available in a general metric space. It is no longer true for all closed and bounded sets, but it is true for all compact sets as we have defined them in the preceding sections.

Let X be a metric space, and let $K \subset X$. A collection $\mathcal{U}$ of open sets is called an *open cover* of K if

$$K \subset \bigcup_{U \in \mathcal{U}} U.$$

Theorem 13.94 *The following conditions on a set K in a metric space X are equivalent.*

1. **(Bolzano-Weierstrass Property)** *K is compact.*

2. **(Heine-Borel Property)** *Every open cover of K can be reduced to a finite subcover.*

Proof Throughout we can assume that $X = K$. Let K satisfy (2), and let $\{x_n\}$ be a sequence in K. For each $N \in \mathbb{N}$, let $A_N = \{x_n : n \geq N\}$ and let

$U_N = X \setminus \overline{A}_n$. Note that $\{U_N\}$ forms an increasing sequence of open sets and that $U_N \neq K$ for any N. In particular, each of the sets U_N is open and no finite collection of the sets U_N covers K. Since K satisfies condition (2), this cannot be an open cover of K and so $\bigcup_{N=1}^{\infty} U_N \neq X$; that is,

$$\bigcap_{N=1}^{\infty} \overline{A}_N \neq \emptyset.$$

Let $x_0 \in \bigcap_{N=1}^{\infty} \overline{A}_N$. It follows directly from the definition of the sets $\overline{A}_N$ that x_0 is the limit of some subsequence of the sequence $\{x_n\}$. This completes the proof of (2) $\Rightarrow$ (1).

Now suppose that K satisfies condition (1). We need to recall from Theorem 13.90 that K is separable as this is needed in order to prove that K has the Heine-Borel property. Now let $\mathcal{U}$ be an open cover of X. It follows from Lindelöf's theorem (Exercise 13.12.30) that $\mathcal{U}$ can be reduced to a countable subcover $\{U_1, U_2, U_3, \dots\}$.

We now show that this subcover can be further reduced to a finite subcover. If this were not the case, then for each $N \in \mathbb{N}$ there exists

$$x_N \in X \setminus \bigcup_{i=1}^{N} U_i$$

since otherwise the collection $\{U_1, U_2, U_3, \dots, U_N\}$ would cover X. Since X has the Bolzano-Weierstrass property, the set $\{x_1, x_2, \dots\}$ has a limit point x_0. But $X = \bigcup_{i=1}^{\infty} U_i$, so there exists $j \in \mathbb{N}$ such that $x_0 \in U_j$. This implies that $x_i \in U_j$ for infinitely many $i \in \mathbb{N}$. This is impossible because our choice of the points x_N implies that $x_N \in X \setminus U_j$ when $N \geq j$. This contradiction implies that the collection $U_1, U_2, \dots$ can be reduced to a finite subcover, completing the proof of (1) $\Rightarrow$ (2). ■

Exercises

13.12.23 Show that the following subsets of $\mathbb{R}$ do not have the Heine-Borel property by constructing an open cover with no finite subcover.

(a) the set $\mathbb{N}$

(b) the set $(0, 1)$

(c) the set $\mathbb{Q} \cap [0, 1]$

13.12.24 Without appealing to Theorem 13.94, prove directly that a set with the Heine-Borel property is bounded.

13.12.25 Without appealing to Theorem 13.94, prove directly that a set with the Heine-Borel property is closed.

13.12.26 Without appealing to Theorem 13.94, prove directly that a set with the Heine-Borel property is separable.

13.12.27 Use the Heine-Borel property to describe completely what sets are compact in a discrete metric space.

13.12.28 Use the Heine-Borel property to prove that if X is compact and $f: X \to Y$ is continuous, then f is uniformly continuous.

13.12.29 Use the Heine-Borel property to prove that if $f: X \to Y$ is continuous and X is compact, then the set $f(X)$ is compact in Y.

13.12.30 **(Lindelöf's Theorem)** Prove that every open cover of a separable metric space has a countable subcover.

13.12.31 Show that to every open cover $\mathcal{C}$ of a compact set K in a metric space there corresponds a positive number L (called the *Lebesgue number* of the cover) such that if $x, y \in K$ and $d(x,y) < L$, then there is some member $U \in \mathcal{C}$ such that both x and y belong to U.

13.12.32 Use Exercise 13.12.31 to prove that if X is compact and $f: X \to Y$ is continuous, then f is uniformly continuous.

13.12.33 Show that the property of Exercise 13.12.31 is not equivalent to compactness in $\mathbb{R}$ (i.e., find a noncompact set $K \subset \mathbb{R}$ with this property).

13.12.34 Show that a metric space (X, d) is compact if and only if for every family $\mathcal{F}$ of closed subsets of X for which

$$\bigcap_{F \in \mathcal{F}} F = \emptyset$$

there must be a finite collection $F_1, F_2, \dots, F_m$ of sets in $\mathcal{F}$ so that

$$\bigcap_{i=1}^{m} F_i = \emptyset.$$

13.12.4 Total Boundedness

Enrichment

In $\mathbb{R}^n$ a set is compact if and only if it is both closed and bounded. We have seen already that this is false in a general metric space. Rather than simply abandon this idea we still can pursue the notion that for a set to be compact it should be sufficient that it is closed and "bounded" in some stronger sense.

The key idea has already appeared in the proof of Theorem 13.90. There we showed that if X is compact, then, for every $\varepsilon > 0$, there is an ε-net, that is, a finite set

$$\{x_1, x_2, \dots, x_n\} \subset X$$

such that the finite collection of balls $\{B(x_i, \varepsilon)\}$ covers X. When a space X has, for every $\varepsilon > 0$, an ε-net, we say that X is *totally bounded.* We express this formally.

Definition 13.95 Let X be a metric space. We say that a set $S \subset X$ is *totally bounded* if for every $\varepsilon > 0$ there is a finite set

$$\{x_1, x_2, \dots, x_n\} \subset X$$

such that

$$S \subset B(x_1, \varepsilon) \cup B(x_2, \varepsilon) \cup \cdots \cup B(x_n, \varepsilon).$$

We can also characterize total boundedness in terms of Cauchy sequences. Note that the language of this theorem is close to the language of the Bolzano-Weierstrass property.

Theorem 13.96 *A metric space X is totally bounded if and only if every sequence has a Cauchy subsequence.*

Proof Suppose that X is not totally bounded. Then for some $\varepsilon > 0$ there does not exist a finite set of points x_1, x_2, x_3, ... , x_p so that

$$K \subset \bigcup_{i=1}^{p} B(x_i, \varepsilon).$$

Thus we can choose a sequence of points x_1, x_2, x_3, ... in K with the property that

$$x_n \notin \bigcup_{i=1}^{n-1} B(x_i, \varepsilon).$$

Such a sequence can have no Cauchy subsequence since

$$d(x_N, x_m) > \varepsilon$$

for all $n > M$.

Conversely, if X is totally bounded, and $\{x_n\}$ is an arbitrary sequence, then we can construct a Cauchy subsequence as follows. X can be expressed as the union of a finite number of balls of radius 1. Our sequence $\{x_n\}$ must have a subsequence that is in one of these. That ball, in turn, can be subdivided using a finite number of balls of radius 1/2. By choosing an appropriate subsequence of that subsequence we can arrive at a subsequence that is in a ball of diameter 1/2. By continuing this process indefinitely and taking a sequence that is a subsequence of *all* of these we arrive at a final subsequence. We ask you, in Exercise 13.12.43, to give a precise description of this process and verify that the sequence constructed is a Cauchy subsequence of $\{x_n\}$. ■

We are now in a position to reinterpret our standard result that compact in $\mathbb{R}^n$ is equivalent to closed and bounded. Closed subsets of $\mathbb{R}^n$ are complete; bounded subsets of $\mathbb{R}^n$ are totally bounded. Thus we could equally say that compact subsets of $\mathbb{R}^n$ are those that are complete and totally bounded. Expressed this way we have a theorem that works in every metric space.

Theorem 13.97 *A metric space is compact if and only if it is complete and totally bounded.*

Proof Suppose that X is compact. Let $\{x_n\}$ be a Cauchy sequence in X. By the Bolzano-Weierstrass property $\{x_n\}$ has a convergent subsequence. But a Cauchy sequence with a convergent subsequence is itself convergent (Exercise 13.8.6). Thus X is complete.

In a similar way we see that X is totally bounded: If $\{x_n\}$ is an arbitrary sequence in X, then it has a convergent subsequence. That subsequence is Cauchy, and so it follows immediately from Theorem 13.94 that X is totally bounded.

Conversely, suppose that X is complete and totally bounded. If $\{x_n\}$ is an arbitrary sequence from X, then $\{x_n\}$ has a Cauchy subsequence, by Theorem 13.96. This subsequence converges, since X is complete. Thus X is compact by definition. ■

Exercises

13.12.35 Show that a set $S \subset X$ is *totally bounded* if and only if for every $\varepsilon > 0$ there is a finite set $\{x_1, x_2, \dots, x_n\} \subset S$ such that

$$S \subset B(x_1, \varepsilon) \cup B(x_2, \varepsilon) \cup \dots \cup B(x_n, \varepsilon).$$

13.12.36 The mathematician Herman Weyl is credited with joking that a "compact city is a city that can be guarded by a finite number of arbitrarily near-sighted policemen." Explain.

13.12.37 Show that every subset of a totally bounded set is totally bounded.

13.12.38 Show that the closure of a totally bounded set is totally bounded.

13.12.39 What sets are totally bounded in a discrete metric space?

13.12.40 Show that every totally bounded set is bounded but that the converse is not true.

13.12.41 Show directly that a set in $\mathbb{R}^n$ is totally bounded if and only if it is bounded.

13.12.42 Show that total boundedness is not invariant under continuous mappings. Is it a topological property? Is it invariant under uniformly continuous mappings? Is it invariant under isometries?

13.12.43 Supply all details needed to complete the proof of Theorem 13.96.

13.12.44 Show that closed balls in $\mathcal{C}[a, b]$, $M[a, b]$, and ℓ_∞ are not compact by using Theorem 13.97.

13.12.45 Prove that a totally bounded metric space must be separable.

13.12.46 Compare the following properties that a metric space might have.

(a) Every closed and bounded subset is compact.

(b) Every closed and bounded subset is totally bounded.

(c) Every bounded sequence has a convergent subsequence.

(d) Every bounded sequence has a Cauchy subsequence.

13.12.47 Show that a closed and bounded set E in ℓ_2 is compact if and only if

$$\lim_{N\to\infty}\sum_{i=N}^{\infty} x_i^2 = 0$$

uniformly for $x = (x_1, x_2, x_3, \dots) \in E$. State and prove the analogous version in the metric space ℓ_1.

13.12.5 Compact Sets in $\mathcal{C}[a, b]$

✂ *Enrichment*

In any metric space that is important to us it will be equally important to determine which sets are compact. Many theorems of existence and uniqueness in analysis can be obtained best from a compactness argument. Thus we shall need to know which sets in our space allow such arguments.

The space $\mathcal{C}[a, b]$ of continuous functions on an interval $[a, b]$ furnished with the supremum metric

$$d(f, g) = \max_{x\in X} |f(x) - g(x)|$$

is a complete separable metric space. Which subsets are compact?

While all compact sets must be closed and bounded, it is not true in this space that all closed and bounded sets are compact. Our purpose here is to obtain a useful characterization of the compact subsets of $\mathcal{C}[a, b]$. This characterization involves two properties that a family of functions on $[a, b]$ may or may not possess. For the first property, let us ask what characterizes the bounded subsets of $\mathcal{C}[a, b]$ since every compact set must also be bounded.

Definition 13.98 A family $\mathcal{F}$ of functions on an interval $[a, b]$ is said to be *uniformly bounded* on $[a, b]$ if there exists $M > 0$ such that $|f(x)| \leq M$ for all $x \in [a, b]$ and $f \in \mathcal{F}$.

This concept characterizes the bounded subsets of our metric space $\mathcal{C}[a, b]$. The proof is straightforward and is left to the exercises.

Lemma 13.99 *Let $\mathcal{F}$ be a family of continuous functions on $[a, b]$. Then $\mathcal{F}$ is uniformly bounded on $[a, b]$ if and only if $\mathcal{F}$ is a bounded subset of the metric space $\mathcal{C}[a, b]$.*

The other relevant notion concerns the uniformity of the continuity behavior of continuous functions in a compact subset of $\mathcal{C}[a, b]$. Let $f \in \mathcal{C}[a, b]$, let $x_0 \in X$, and let $\varepsilon > 0$. Then there exists $\delta > 0$ such that, if $|x - x_0| < \delta$, $|f(x) - f(x_0)| < \varepsilon$. The number δ depends on x_0, ε, and f and should perhaps be written $\delta = \delta(x_0, \varepsilon, f)$. Since any continuous function on $[a,b]$ is also uniformly continuous (Theorem 5.47), we know that δ can be chosen so as to be independent of x_0 for a given ε and f. If $\mathcal{F} \subset \mathcal{C}[a,b]$ and we can choose δ so as also to be independent of $f \in \mathcal{F}$, we say that $\mathcal{F}$ is an *equicontinuous* family. The concept is due to Giulio Ascoli (1843–1896).

Definition 13.100 A family $\mathcal{F}$ of functions on a metric space (X, d) is *equicontinuous* if for every $\varepsilon > 0$ there exists $\delta > 0$ such that, if $x, y \in X$ and $d(x,y) < \delta$, then $|f(x) - f(y)| < \varepsilon$ for all $f \in \mathcal{F}$.

For an easy example, note that a collection of functions that satisfies a uniform Lipschitz condition is equicontinuous.

Example 13.101 Let $M > 0$ and let

$$\mathcal{F} = \{f : [a,b] \to \mathbb{R} : |f(x) - f(y)| \le M|x-y| \text{ for all } x, y \in [a,b]\}.$$

Then $\mathcal{F}$ is an equicontinuous family of functions on $[a,b]$. It suffices to take $\delta = \varepsilon/M$. ◀

Our main theorem, usually attributed to both Ascoli and Cesare Arzelà (1847–1912), now uses the two concepts of uniform boundedness and equicontinuity to obtain a characterization of compactness in the space $\mathcal{C}[a,b]$.

Theorem 13.102 (Arzelà-Ascoli) *Let K be a closed subset of the metric space $\mathcal{C}[a,b]$. Then K is compact if and only if K is uniformly bounded and equicontinuous.*

Proof Note first that, since K is assumed to be a closed subset of a space that we already know is complete, K itself is complete. By Theorem 13.97 K is compact if and only if it is totally bounded. Thus we need prove only that K is uniformly bounded and equicontinuous if and only if K is totally bounded.

Suppose first that K is totally bounded in $\mathcal{C}[a,b]$. Then K is bounded in $\mathcal{C}[a,b]$ and is therefore a uniformly bounded family of functions. We show that K is equicontinuous. Let $\varepsilon > 0$, and let $f_1, f_2, \ldots, f_n$ be an $(\varepsilon/3)$-net in K. Let $f \in K$. There exists $j \le n$ such that

$$\max_{z \in [a,b]} |f(z) - f_j(z)| < \tfrac{1}{3}\varepsilon. \tag{28}$$

Then, for $x, y \in [a,b]$,

$$|f(x) - f(y)| \le |f(x) - f_j(x)| + |f_j(x) - f_j(y)| + |f_j(y) - f(y)|. \tag{29}$$

Each of the functions f_i is uniformly continuous. Thus, since there are only finitely many of them, there exists $\delta > 0$ such that

$$|x - y| < \delta,\ 1 \le i \le n \Rightarrow\ |f_i(x) - f_i(y)| < \varepsilon/3. \tag{30}$$

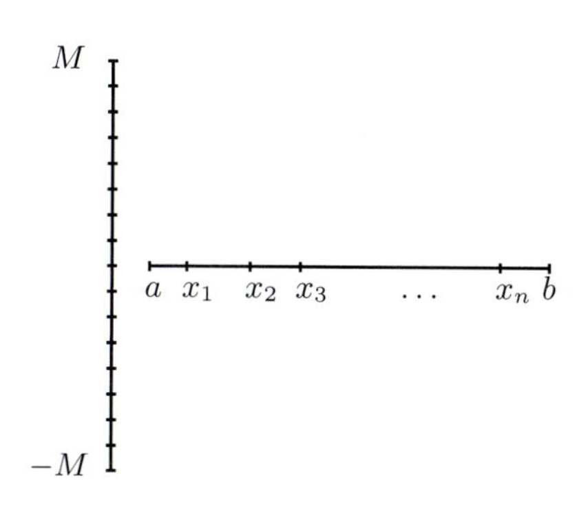

Figure 13.7. Partitioning the intervals $[-M, M]$ and $[a, b]$ in the proof of Theorem 13.102.

It now follows from (28), (29), and (30) that $|f(x) - f(y)| < \varepsilon$ for all x, $y \in [a, b]$ with $|x-y| < \delta$ and all $f \in K$. This shows that K is equicontinuous.

To prove the converse, suppose that K is uniformly bounded and equicontinuous. We show that K is totally bounded. Let $\varepsilon > 0$. We shall find an ε-net $\mathcal{F}$ in the space for K. This means $\mathcal{F}$ must be a finite collection of functions in $\mathcal{C}[a, b]$ such that every member of K is closer to some member of $\mathcal{F}$ than $\varepsilon > 0$. (Note that we do not need to select the members of $\mathcal{F}$ from K although it is possible to do so.)

Choose $M \in \mathbb{N}$ such that $|g(x)| \leq M$ for all $x \in X$ and $g \in K$. Let $\varepsilon > 0$. Since K is equicontinuous, there exists $\delta > 0$ such that

$$|x - y| < \delta, \ g \in K \ \Rightarrow \ |g(x) - g(y)| < \varepsilon/4. \tag{31}$$

Using this number δ, we subdivide $[a, b]$ by points $x_1, x_2, \ldots, x_n$ sufficiently close together so that every point in $[a, b]$ is within δ of one of these points. (Note that this would be a δ-net for the compact metric space $[a, b]$.)

Choose $m \in \mathbb{N}$ such that $1/m < \varepsilon/4$, and partition the interval $[-M, M]$ into $2Mm$ congruent intervals:

$$-M = y_0 < y_1 < \cdots < y_{2Mm} = M.$$

Consider now the grid of points in the rectangle $[a, b] \times [-M, M]$ whose first coordinates are from the points $x_1, x_2, \ldots, x_n$ and whose second coordinates are from the points $y_0, y_1, \ldots, y_{2Mm}$. Figure 13.7 illustrates the situation. Let $\mathcal{F}$ denote the set of all continuous functions on $[a, b]$ that are piecewise linear and whose corner points on the graph occur at points in the grid. There are clearly only finitely many such functions.

Let $g \in K$. There exists, because of the nature of the grid, at least one function $f \in \mathcal{F}$ so that

$$|g(x_j) - f(x_j)| < \varepsilon/4 \quad (j = 1, 2, \ldots, n). \tag{32}$$

Consider any point $x \in [a,b]$. There is some interval with $x \in [x_j, x_{j+1}]$ for some $1 \leq j < n$. We know that

$$|g(x_j) - g(x_{j+1})| < \varepsilon/4$$

because of (31). It follows also then, because of (32), that

$$|f(x_j) - f(x_{j+1})| < \varepsilon/2.$$

As f is linear on the interval $[x_j, x_{j+1}]$ we have as well the inequality

$$|f(x_j) - f(x)| < \varepsilon/2$$

for all $x \in [x_j, x_{j+1}]$. Together these inequalities show that

$$|g(x) - f(x)| \leq |g(x) - g(x_j)| + |g(x_j) - f(x_j)| + |f(x_j) - f(x)| < \varepsilon$$

which implies that

$$\max_{x \in [a,b]} |f(x) - g(x)| < \varepsilon.$$

We have shown that $\mathcal{F}$ is an ε-net, so K is totally bounded, as was to be proved. ■

Exercises

13.12.48 Prove that a family $\mathcal{F}$ of continuous functions on $[a,b]$ is uniformly bounded if and only if $\mathcal{F}$ is a bounded subset of the metric space $\mathcal{C}[a,b]$.

13.12.49 Is the set of functions $\{\sin kt : k = 1, 2, 3, \ldots\}$ equicontinuous on $[0, 2\pi]$?

13.12.50 Let $f : \mathbb{R} \to \mathbb{R}$ be uniformly continuous on $\mathbb{R}$ and let f_a denote the function $f_a(t) = f(t-a)$. Show that the family $\{f_a : a \in \mathbb{R}\}$ is equicontinuous on $\mathbb{R}$. Is this true if f is merely continuous?

13.12.51 Prove directly from the definition that the family of functions $f_n(x) = x^n$, $n = 1, 2, 3, \ldots$ is not equicontinuous on the interval $[0,1]$ but is equicontinuous on the interval $[0, 1/2]$.

13.12.52 Let A be a bounded subset of $\mathcal{C}[a,b]$. Prove that the set of all functions of the form

$$F(x) = \int_a^x f(t)\, dt$$

for $f \in A$ is an equicontinuous family.

13.12.53 Let K be continuous on $[a,b] \times [a,b]$ and ϕ continuous on $[a,b]$. The operator $A : \mathcal{C}[a,b] \to \mathcal{C}[a,b]$ is defined by

$$(A(f))(x) = \lambda \int_a^b K(x,y) f(y)\, dy + \phi(x)$$

(cf. Example 13.81). Show that the image under A of any bounded set in $\mathcal{C}[a,b]$ is an equicontinuous family.

13.12.54 Let σ be continuous and nondecreasing on $[0,\infty)$, with $\sigma(0) = 0$. A function $f \in \mathcal{C}[a,b]$ has *modulus of continuity* σ if

$$|f(x) - f(y)| \leq \sigma(|x-y|)$$

for all $x, y \in [a,b]$. Let $\boldsymbol{C}(\sigma)$ denote

$$\{f : \sigma \text{ is a modulus of continuity for } f\}.$$

(a) Show that every $f \in \mathcal{C}[a,b]$ has a modulus of continuity.

(b) Let σ be a modulus of continuity. Show that $\boldsymbol{C}(\sigma)$ is an equicontinuous family.

(c) Exhibit a modulus of continuity for the class of Lipschitz functions with constant M.

(d) Let σ be a modulus of continuity. Is it necessarily true that $\sigma \in \boldsymbol{C}(\sigma)$ on $[a,b]$? What if σ is concave down?

(e) Prove that the set

$$K = \left\{f \in \mathcal{C}[0,1] : |f(x) - f(y)| \leq \sqrt{|x-y|} \text{ and } f(0) = 0\right\}$$

is a compact subset of $\mathcal{C}[0,1]$. Is $\sqrt{x} \in K$? What about x^2?

13.12.55 Let $Lip_1[a,b]$ denote the family of functions f on $[a,b]$ that satisfy the Lipschitz condition

$$|f(x) - f(y)| \leq |x-y|$$

for all $x, y \in [a,b]$. We furnish this space with the usual supremum metric so that it is a subspace of $\mathcal{C}[a,b]$. Prove that a sequence of functions $\{f_n\}$ converges in $Lip_1[a,b]$ if and only if it converges pointwise.

13.12.56 Let $\{f_n\}$ be a sequence of real-valued functions defined on an interval $[a,b]$ and suppose that $\{f_n(x)\}$ is bounded for each $x \in [a,b]$.

(a) Show that there is a subsequence $\{f_{n_k}\}$ so that $\lim_{k\to\infty} f_{n_k}(r)$ exists for every rational number $r \in [a,b]$.

(b) Show that, if $\{f_n\}$ is equicontinuous, then in fact $\lim_{k\to\infty} f_{n_k}(x)$ exists for every $x \in [a,b]$ and the convergence is uniform.

(c) Use this to give another proof of the Arzelà-Ascoli theorem.

13.12.57 Use the Arzelà-Ascoli theorem to show that the set

$$\{f \in \mathcal{C}[0,1] : |f(x)| \leq 1\}$$

in $\mathcal{C}[0,1]$ cannot be covered by a sequence of compact sets. (This is also done, using the Baire category theorem, in Exercise 13.13.9.)

13.12.58 Prove the following more general version of the Arzelà-Ascoli theorem:

Let K be a closed subset of the metric space $C(X)$, where X is a compact metric space and $C(X)$ denotes the space of continuous real-valued functions on X furnished with the supremum metric. Then K is compact in $C(X)$ if and only if K is uniformly bounded and equicontinuous.

13.12.6 Peano's Theorem

✂ *Enrichment*

In Section 13.11.4 we saw how the contraction mapping principle can be used to prove an existence and uniqueness theorem for solutions to the differential equation $y' = f(x, y)$. The requirement that was needed in order to apply the contraction mapping principle involved a Lipschitz condition on the function f.

In order to obtain an existence theorem (this time without uniqueness) under weaker hypotheses we need a different approach. For this we go back to a simple idea of Euler from 1768. To solve numerically an initial value problem

$$y' = f(x, y) \quad y(x_0) = y_0$$

on an interval $[x_0, b]$, divide the interval into n equally spaced points (equal subdivision is not an essential feature, but is traditional in numerical methods)

$$x_0 < x_1 < x_2 < \cdots < x_n = b$$

and approximate a solution by a piecewise linear function. Write

$$y_1 = y_0 + f(x_0, y_0)(x_1 - x_0),$$

$$y_2 = y_1 + f(x_1, y_1)(x_2 - x_1),$$

and so on through to

$$y_n = y_{n-1} + f(x_{n-1}, y_{n-1})(x_n - x_{n-1}).$$

We can let $k_n(x)$ denote the function on $[x_0, b]$ that is continuous, passes through the points (x_0, y_0), (x_1, y_1), ... ,(x_n, y_n), and is linear in between. Note that $k_n'(x) = f(x_i, k(x_i))$ if x is in the ith interval (x_i, x_{i+1}) and that there is no derivative at the corner points.

In practice this method gives a reasonable approximation to solutions. We could make this the basis of an existence proof if we could show that the sequence $\{k_n\}$ converges uniformly to a function k and that the function k solves the initial value problem. Unfortunately, the hypothesis that we wish to use, the continuity of the function f, is too weak to allow a proof that that the sequence $\{k_n\}$ converges uniformly. But we can use a compactness argument to obtain a uniformly convergent subsequence. The key is to use the continuity of f to design an interval $[x_0, b]$ on which such a sequence

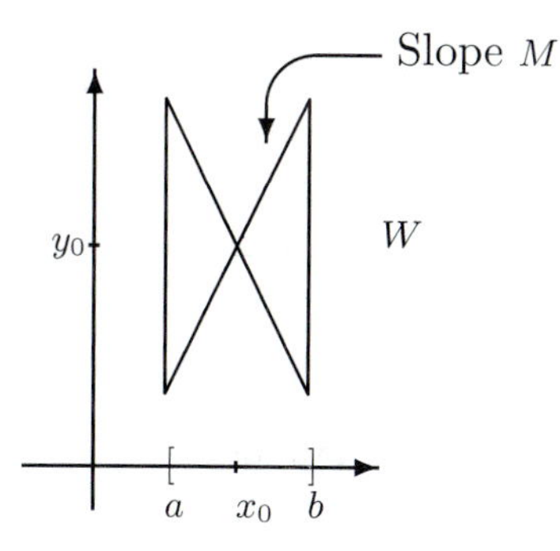

Figure 13.8. The set W and its projection to $[a, b]$ in the proof of Theorem 13.103.

of functions $\{k_n\}$ is bounded and equicontinuous. Then the Arzelà-Ascoli theorem supplies the subsequence.

This improvement of the classical existence theorems of Cauchy and Lipschitz was obtained by Peano in 1886. In 1890 he showed that this weakening of hypotheses occurs at the expense of uniqueness by supplying the example that we consider in Exercise 13.12.59.

Theorem 13.103 (Peano) *Let f be continuous on an open subset D of $\mathbb{R}^2$, and let $(x_0, y_0) \in D$. Then the differential equation*

$$y' = f(x, y)$$

has a local solution passing through the point (x_0, y_0).

Proof We are seeking an exact solution that is valid in some interval containing the point x_0. Thus we wish to find an interval $[a, b]$ containing x_0 and a differentiable function k defined on $[a, b]$ such that

$$k(x_0) = y_0 \quad \text{and} \quad k'(x) = f(x, k(x)) \quad \text{for all } x \in [a, b]. \tag{33}$$

Our strategy is essentially to construct a family K of approximate solutions through (x_0, y_0) on $[a, b]$. We then show that the set $\overline{K}$ is compact in $\mathcal{C}([a, b])$ and use compactness to show the existence of the function k as a point in $\overline{K}$.

We first select the interval $[a, b]$. Let R be a closed rectangle contained in D having sides parallel to the coordinate axes and having (x_0, y_0) as center. Let $M \geq 1$ be an upper bound for $|f|$ on R. Let

$$W = \{(x, y) \in R : |y - y_0| \leq M|x - x_0|\},$$

and let $[a, b]$ be the projection of W onto the x-axis, as in Figure 13.8.

We next obtain a family K of functions defined on $[x_0, b]$ that can be considered as approximate solutions to (33). Since W is compact in $\mathbb{R}^2$, f

is uniformly continuous on W. Thus, for every $\varepsilon > 0$, there exists $\delta \in (0,1)$ such that, if $(x,y) \in W$ and $(\overline{x},\overline{y}) \in W$ with $|x-\overline{x}| < \delta$ and $|y-\overline{y}| < \delta$, then

$$|f(\overline{x},\overline{y}) - f(x,y)| < \varepsilon.$$

Choose points $x_1, x_2, \ldots, x_n$ such that

$$x_0 < x_1 < x_2 < \cdots < x_n = b \text{ and } |x_i - x_{i-1}| < \delta/M$$

for all $i = 1, \ldots, n$. Define a function k_ε on $[x_0, b]$ as follows: $k_\varepsilon(x_0) = y_0$ and, on $[x_0, x_1]$, k_ε is linear with slope $f(x_0, y_0)$; on $[x_1, x_2]$, take k_ε to be linear with slope $f(x_1, k_\varepsilon(x_1))$; continuing in this way, we extend the definition of k_ε to all of $[x_0, b]$.

We have arrived at a function k_ε defined on $[x_0, b]$ whose graph is a polygonal arc through the point (x_0, y_0) and is contained in W. Since the slopes of the line segments composing the graph of k_ε are determined by values of the function f in W, we see that

$$|k_\varepsilon(x) - k_\varepsilon(\overline{x})| \le M|x - \overline{x}| \tag{34}$$

for all $x, \overline{x} \in [x_0, b]$. Now let $x \in [x_0, b]$, $x \ne x_i$, $i = 0, 1, \ldots, n$. Then there exists $j \in \{1, 2, \ldots, n\}$ such that $x_{j-1} < x < x_j$. Noting that

$$|x_j - x_{j-1}| < \delta/M$$

and using (34), we see that

$$|k_\varepsilon(x) - k_\varepsilon(x_{j-1})| \le M|x - x_{j-1}| < \delta.$$

This implies that

$$|f(x_{j-1}, k_\varepsilon(x_{j-1})) - f(x, k_\varepsilon(x))| < \varepsilon.$$

But

$$k_\varepsilon'(x) = f(x_{j-1}, k_\varepsilon(x_{j-1})),$$

so

$$|k_\varepsilon'(x) - f(x, k_\varepsilon(x))| < \varepsilon. \tag{35}$$

The inequality (35) is valid for all $x \in [x_0, b]$ except at points x in the finite set $\{x_0, \ldots, x_n\}$, at which k_ε need not be differentiable. By (35), we see that the functions k_ε are approximate solutions to (33).

We have constructed a family K of functions, one function corresponding to every $\varepsilon > 0$. The family K is uniformly bounded on $[x_0, b]$, since the graph of each of the functions k_ε is contained in W. It follows from (34) that K is an equicontinuous family. The Arzelà-Ascoli theorem now implies that $\overline{K}$ is compact in $C[x_0, b]$.

We can now complete the proof of the theorem. For all $x \in [x_0, b]$, we have

$$k_\varepsilon(x) = y_0 + \int_{x_0}^{x} k'_\varepsilon(t)\,dt \tag{36}$$
$$= y_0 + \int_{x_0}^{x} \left(f(t, k_\varepsilon(t)) + (k'_\varepsilon(t) - f(t, k_\varepsilon(t)))\right) dt.$$

The fact that k'_ε may fail to exist on the set $\{x_0, x_1, \ldots, x_n\}$ does not affect the integral.

Since $K \subset \overline{K}$ and $\overline{K}$ is compact the sequence $\{k_{(1/n)}\}$ contains a subsequence $\{k_{1/n_i)}\}$ that converges uniformly to some function $k \in \overline{K}$. Note that k must be continuous on $[x_0, b]$. Since f is uniformly continuous on W, the functions $f(t, k_{(1/n_i)}(t))$ converge uniformly to the function $f(t, k(t))$ on $[x_0, b]$. Noting (35), we now infer from (36) that

$$k(x) = y_0 + \int_{x_0}^{x} f(t, k(t))\,dt$$

for all $x \in [x_0, b]$. It follows that k is a solution to (33) on $[x_0, b]$.

In a similar manner, we obtain a solution $\overline{k}$ to (33) on $[a, x_0]$. The function y given by

$$y(x) = \begin{cases} k(x) & \text{for } x \in [x_0, b]; \\ \overline{k}(x) & \text{for } x \in [a, x_0], \end{cases}$$

satisfies (33) on all of $[a, b]$, as required. ■

Exercises

13.12.59 **(Peano's Example)** Show that the hypotheses given in Theorem 13.103 are not sufficient to guarantee uniqueness of solutions to the equation $y' = f(x, y)$ by taking, for example, the equation $y' = 3y^{2/3}$, $y(0) = 0$. Does this example conflict with the uniqueness assertion of Theorem 13.86?

13.13 Baire Category Theorem

In Section 6.4 we proved the Baire category theorem relative to the metric space $\mathbb{R}$. We are now in a position to establish this same theorem relative to any complete metric space. We introduce some of the basic ideas central to that theorem and prove the theorem in this section. Then, in Section 13.14, we provide some applications of this theorem. You may wish to review Sections 6.3 and 6.4 for a study of these ideas in the setting of the real line before proceeding to a development of the same ideas in an abstract metric space.

We begin by recalling the definition of a nowhere dense set in a metric space (as defined in Section 13.5 and used already in Exercises 13.5.21 and 13.6.21). After some examples of nowhere-dense sets we proceed to a proof of the Baire category theorem, which asserts that a complete space cannot be the union of any sequence of nowhere dense sets.

13.13.1 Nowhere Dense Sets

On the real line a set A is nowhere dense if it is dense in no interval; that would mean that, given any interval (a,b) there would be a subinterval $(c,d) \subset (a,b)$ that contains no points of A. In a metric space the role of the open intervals is played by open balls. Here is the formal definition.

Definition 13.104 Let (X,d) be a metric space. A set $A \subset X$ is called *nowhere dense* if, given any open ball $B(x,\varepsilon)$ in X, there exists an open ball $B(y,\delta) \subset B(x,\varepsilon)$ such that $A \cap B(y,\delta) = \emptyset$.

From this definition we see that A is nowhere dense provided that it fails to be dense in any open ball. It is easy to verify (Exercise 13.13.6) that A is nowhere dense if and only if $\overline{A}$ has empty interior. Thus a closed set is nowhere dense if and only if its complement is dense.

Let's look at some examples.

Example 13.105 Let K be the Cantor set in the interval $[0,1]$. Considered as a subset of $[0,1]$ it is clearly nowhere dense. [Be careful here: If instead the metric space is (K,d), where d is the usual metric, then K is not a nowhere dense set, it is dense.] ◀

Example 13.106 We shall show that the set $A = \mathcal{C}[a,b]$ of all continuous functions taken as a subset of $M[a,b]$, the space of bounded functions, is nowhere dense. (Again notice the fact that we consider this in the space $M[a,b]$, not in $\mathcal{C}[a,b]$.) Let $B(f,\varepsilon)$ be any open ball in $M[a,b]$. We shall find another open ball $B(g,\delta)$ contained in $B(f,\varepsilon)$ but containing no members of A, that is, this ball $B(g,\delta)$ must contain *no* continuous functions.

Now f, the center of $B(f,\varepsilon)$, may or may not be in A. We choose g as follows. If

$$\limsup_{t\to a} f(t) = \liminf_{t\to a} f(t), \tag{37}$$

then let

$$g(t) = \begin{cases} f(t) + \varepsilon/2, & \text{if } t \in [a,b] \cap \mathbb{Q} \\ f(t) - \varepsilon/2, & \text{if } t \in [a,b] \setminus \mathbb{Q}, \end{cases} \tag{38}$$

so

$$\limsup_{t\to a} g(t) - \liminf_{t\to a} g(t) = \varepsilon.$$

If, instead

$$\limsup_{t\to a} f(t) - \liminf_{t\to a} f(t) = \gamma > 0, \tag{39}$$

then let

$$g(t) = f(t) \quad \text{for all } t \in [a,b]. \tag{40}$$

We now obtain δ, for which $B(g,\delta)$ is the required ball. When (37) applies, take $\delta = \varepsilon/3$. When (39) applies, take $\delta = \gamma/3$. We must show

$$B(g,\delta) \subset B(f,\varepsilon) \tag{41}$$

and

$$B(g,\delta) \cap A = \emptyset. \tag{42}$$

To verify (41), we need only observe that

$$|g(t) - f(t)| = \frac{\varepsilon}{2} < \varepsilon$$

for all $t \in [a,b]$ when (38) applies and

$$|g(t) - f(t)| = 0 < \varepsilon$$

for all $t \in [a,b]$ when (40) applies.

To verify (42), let $h \in B(g,\delta)$. We show that $h \notin A$ by showing that

$$\limsup_{t\to a} h(t) - \liminf_{t\to a} h(t) > 0.$$

Now $h \in B(g,\delta)$, so $|g(t) - h(t)| < \delta$ for all $t \in [a,b]$. Thus

$$\limsup_{t\to a} h(t) \geq \limsup_{t\to a} g(t) - \delta \tag{43}$$

and

$$\liminf_{t\to a} h(t) \leq \liminf_{t\to a} g(t) + \delta. \tag{44}$$

Subtracting (44) from (43) we obtain

$$\limsup_{t\to a} h(t) - \liminf_{t\to a} h(t) \geq \limsup_{t\to a} g(t) - \liminf_{t\to a} g(t) - 2\delta.$$

When (38) applies, $2\delta = 2\varepsilon/3 < \varepsilon$, and when (40) applies, $2\delta = 2\gamma/3 < \gamma$. It now follows from (37) and (38) or (39) and (40) that

$$\limsup_{t\to a} h(t) - \liminf_{t\to a} h(t) > 0$$

so $h \notin A$.

◀

We observe that we could have presented a slightly simpler proof by showing that A is closed in $M[a,b]$ and that the complement of the set A is dense in $M[a,b]$. We leave this as Exercise 13.13.8. We provide such an argument in our next example.

Example 13.107 Recall the space ℓ_∞ of bounded sequences $\{x_i\}$ with metric

$$d_\infty(x,y) = \sup_i |x_i - y_i|.$$

(See Example 13.9.) We show that the subset c of convergent sequences is nowhere dense in ℓ_∞. To do this, we show that c is closed and its complement is dense.

To verify that c is closed, we show that its complement $\ell_\infty \setminus c$ is open. Let $x \in \ell_\infty \setminus c$. Then

$$L = \limsup_i x_i > \liminf_i x_i = l.$$

Take $\varepsilon = (L-l)/3$. If $d_\infty(x,y) < \varepsilon$, then

$$\limsup_i y_i \geq L - \frac{\varepsilon}{3} > l + \frac{\varepsilon}{3} \geq \liminf_i y_i,$$

so $y \in \ell_\infty \setminus c$. This proves that $\ell_\infty \setminus c$ is open: Each $x \in \ell_\infty \setminus c$ has a neighborhood contained in $\ell_\infty \setminus c$.

It remains to check that the complement of c is dense. Let $B(x,\varepsilon)$ be an open ball in ℓ_∞. We wish to show that $B(x,\varepsilon)$ contains points in $\ell_\infty \setminus c$. If $x \notin c$, there is nothing further to prove, so assume that $x \in c$. Let $L = \lim_i x_i$. Then there exists $N \in \mathbb{N}$ such that $|x_i - L| < \varepsilon/2$ if $i \geq N$. Consider the sequence $y = \{y_i\}$ such that

$$y_i = \begin{cases} x_i, & \text{if } i < N \\ L + \varepsilon/2, & \text{if } i \geq N,\ i \text{ odd} \\ L - \varepsilon/2, & \text{if } i \geq N,\ i \text{ even.} \end{cases}$$

Then $d_\infty(x,y) = \sup_i |x_i - y_i| < \varepsilon$, so $y \in B(x,\varepsilon)$. But

$$\limsup_i y_i = L + \frac{\varepsilon}{2} > L - \frac{\varepsilon}{2} = \liminf_i y_i,$$

so $y \notin c$. It is clear that $y \in \ell_\infty$. Thus $\ell_\infty \setminus c$ is dense in ℓ_∞.

We have shown c is closed and $\ell_\infty \setminus c$ is dense, from which it follows that c is nowhere dense in ℓ_∞. ◄

Exercises

13.13.1 Show that every subset of a nowhere dense set is also nowhere dense.

13.13.2 Show that the closure of every nowhere dense set is also nowhere dense.

13.13.3 Show that if $A_1, \dots, A_n$ are nowhere dense subsets of a metric space X, then $A_1 \cup \cdots \cup A_n$ is also nowhere dense in X.

13.13.4 Is it true in an arbitrary metric space that every finite set is nowhere dense?

13.13.5 Describe what property a metric space must have in order that every finite set is nowhere dense.

13.13.6 (a) Show that a set A in a metric space X is nowhere dense if and only if $\overline{A}$ has empty interior.

(b) Show that a closed set in a metric space X is nowhere dense if and only if its complement is dense in X.

13.13.7 Show that the Hilbert cube (Exercise 13.4.4) is a nowhere-dense subset of ℓ_2.

13.13.8 Show that $\mathcal{C}[a,b]$ is nowhere dense in $M[a,b]$ by verifying that $\mathcal{C}[a,b]$ is closed and its complement is dense.

13.13.9 Use the Baire category theorem to show that the set

$$\{f \in \mathcal{C}[0,1] : |f(x)| \leq 1\}$$

in $\mathcal{C}[0,1]$ cannot be covered by a sequence of compact sets. (This is also done, using the Arzelà-Ascoli theorem, in Exercise 13.12.57.)

13.13.10 Show that the function $f : \ell_2 \to \ell_2$ defined on Hilbert space by specifying

$$f(x_1, x_2, x_3, x_4, \dots) = (0, x_1, x_2, x_3, \dots)$$

for every $(x_1, x_2, x_3, x_4, \dots) \in \ell_2$ is an isometric map of ℓ_2 to a nowhere dense subset.

13.13.2 The Baire Category Theorem

On the real line the Baire category theorem asserts that any countable union of nowhere dense sets is small, so small in fact that its complement is dense. We might expect that this generalizes to a general metric space. All that needs to be added is the hypothesis that the space is complete.

Thus we have the statement of the Baire category theorem in a general metric space.

Theorem 13.108 (Baire) *Let (X, d) be a complete metric space, and let S be a countable union of nowhere dense sets in X. Then $X \setminus S$ is dense in X.*

Proof Let $S = \bigcup_{n=1}^{\infty} S_n$, where each of the sets S_n is nowhere dense, and let B_0 be a nonempty open ball in X. To show that $X \setminus S$ is dense in X we must prove that

$$(X \setminus S) \cap B_0 \neq \emptyset.$$

Choose, inductively, a nested sequence of balls

$$B_n = B_n(x_n, r_n)$$

with $r_n < 1/n$ such that

$$\overline{B}_{n+1} \subset B_n \setminus \overline{S}_{n+1}.$$

To see that this is possible, first note that

$$B_n \setminus \overline{S}_{n+1} \neq \emptyset,$$

since S_{n+1}, and therefore $\overline{S}_{n+1}$, is nowhere dense. Thus we can choose

$$x_{n+1} \in B_n \setminus \overline{S}_{n+1}.$$

Since $\overline{S}_{n+1}$ is closed,

$$\text{dist}(x_{n+1}, S_{n+1}) > 0,$$

so we can choose B_{n+1} as required. The sequence $\{x_n\}$ is a Cauchy sequence since, for $n, m \geq N$,

$$d(x_n, x_m) \leq d(x_n, x_N) + d(x_N, x_m) < 2/N.$$

Because X is complete, there exists $x \in X$ such that $x_n \to x$. But

$$x_{n+1} \in \overline{B}_{n+1}$$

for all n, so

$$x \in \bigcap_{n=1}^{\infty} \overline{B}_n \subset B_0 \cap (X \setminus S),$$

as was to be proved. ■

The following terminology is standard:

Definition 13.109 Let (X, d) be a metric space.

1. A set $A \subset X$ is called a *first category* set if A is a countable union of nowhere dense sets.

2. A set that is not of the first category is called a set of the *second category.*

3. The complement of a first-category set is called a *residual* set.

For complete metric spaces, first-category sets are the "small" sets and residual sets are the "large" sets in the sense of category. Second-category sets are merely "not small." Residual sets are large because they are dense and any intersection of a sequence of residual sets is still dense. First-category sets are small since (while they might be dense) their complement is always dense and any union of a sequence of first-category sets still has a dense complement.

For spaces that are not complete, a residual set can be empty. Note also that all these statements about sets in a space are relative to the space being considered. A set might be first category as a subset of some space but not so as a subset of some other space.

Example 13.110 The set of rational numbers $\mathbb{Q}$ considered as a subset of $\mathbb{R}$ is dense. It is also first category since each singleton set $\{q\}$ for $q \in \mathbb{Q}$ is a nowhere dense subset of $\mathbb{R}$ and

$$\mathbb{Q} = \bigcup_{q \in \mathbb{Q}} \{q\} \tag{45}$$

expresses $\mathbb{Q}$ as a countable union of nowhere dense sets. Note that this also means that the set of irrational numbers is residual. ◀

Example 13.111 The entire incomplete metric space $\mathbb{Q}$ with the usual real metric is of the first category. Equation (45) again serves to display this space as a countable union of sets nowhere dense in $\mathbb{Q}$. Thus the complement of $\mathbb{Q}$ in $\mathbb{Q}$, namely the empty set $\emptyset$, would be by our definition a residual set. (In an incomplete metric space a residual set need not be "large.") ◀

Example 13.112 Consider the subspace $\mathbb{N}$ of $\mathbb{R}$. As a subset of $\mathbb{R}$, $\mathbb{N}$ is of the first category, since $\{n\}$ is nowhere dense in $\mathbb{R}$ for each n and

$$\mathbb{N} = \bigcup_{n=1}^{\infty} \{n\} \tag{46}$$

expresses $\mathbb{N}$ as a countable union of nowhere dense sets. But considered as a space in itself, $\mathbb{N}$ is a complete metric space. By the Baire category theorem it cannot be expressed as a countable union of nowhere dense sets. But how can that be? Doesn't equation (46) express $\mathbb{N}$ as a countable union of nowhere dense sets? In fact, note that in the space $\mathbb{N}$ each set $\{n\}$ is dense in $B(n, \frac{1}{2})$. The only residual set in $\mathbb{N}$ is $\mathbb{N}$ itself and only the empty set $\emptyset$ is nowhere dense or first category. ◀

Exercises

13.13.11 Show that every subset of a first category set also first category.

13.13.12 The closure of every nowhere dense set is also nowhere dense. Is the closure of every first category set also first category?

13.13.13 In a metric space (X, d), where d is the discrete metric, determine which sets are nowhere dense, first category, or residual.

13.13.14 In a metric space, show that a countable union of first category sets is again first category and that a countable intersection of residual sets is again residual.

13.13.15 In a metric space, show that every dense set of type $\mathcal{G}_\delta$ is residual.

13.13.16 Let $\mathcal{P}$ denote the polynomials on $[a, b]$, and let $\mathcal{P}_n \subset \mathcal{P}$ denote the polynomials of degree at most n. Show that $\mathcal{P}_n$ is nowhere dense in $\mathcal{C}[a, b]$. Deduce that $\mathcal{P}$ is a first category subset of $\mathcal{C}[a, b]$.

13.13.17 Show that a complete metric space with no isolated points must be uncountable.

13.13.18 Let c_0 denote the set of all sequences of real numbers converging to zero, that is,

$$c_0 = \{x \in c : \lim_{i \to \infty} x_i = 0\}.$$

Prove that c_0 is nowhere dense in c.

13.13.19 Let c_Q denote the set of all sequences of real numbers converging to a rational number, that is,

$$c_Q = \{x \in c : \lim_{i \to \infty} x_i \in \mathbb{Q}\}.$$

Is c_Q nowhere dense in c? Is c_Q of the first category in c?

13.14 Applications of the Baire Category Theorem

✂ *Enrichment*

The Baire category theorem has numerous applications. Category arguments taken together with compactness arguments are among the most powerful and commonly used tools in mathematical analysis.

We have already encountered some simple applications of the Baire category theorem (Theorem 6.13, Exercises 6.4.7 and 13.13.17). The first two are theorems that show that certain "pointwise" conditions actually hold "uniformly" on some open set. The third shows that a complete metric space without isolated points cannot be countable.

In this section we emphasize the use of the Baire category theorem to prove the existence of objects that might be difficult to imagine or to construct. The method is to view these objects as members of an appropriate complete metric space, and then to show the objects with the required property form a residual subset of that space.

In this we are reminded of a argument of Cantor's. Do transcendental real numbers exist? The nontranscendental numbers (i.e., the algebraic numbers) form a countable set of real numbers. What remains is a large dense set of numbers, so that, indeed, transcendental real numbers do exist and they exist in abundance. (See Exercise 2.3.10 for the details.) Here we apply essentially the same reasoning to a complete metric space: The objects whose existence we seek to prove are those objects in the space that remain after a first category set is removed. By the Baire category theorem they must exist in abundance, forming a dense and residual set in the space.

13.14.1 Functions Whose Graphs "Cross No Lines"

We begin with an example in which the objects are continuous functions.

The continuous functions we encounter in elementary calculus have the property that there are many straight lines that cross the graph of the function. Let us make the notion of "crossing lines" precise. We mean simply that the line is above the graph on one side of a point and below on the other. The formal definition looks rather formidable but simply expresses this elementary idea.

Definition 13.113 Let $f : [a,b] \to \mathbb{R}$, and let $L : \mathbb{R} \to \mathbb{R}$ be a function whose graph is a straight line. We say L *crosses* f or (f *crosses* L) if there exists $x_0 \in [a,b]$ and $\delta > 0$ such that $f(x_0) = L(x_0)$ and either

(i) $L(x) \leq f(x)$ for all $x \in [x_0 - \delta, x_0] \cap [a,b]$ and
$L(x) \geq f(x)$ for all $x \in [x_0, x_o + \delta] \cap [a,b]$, or

(ii) $L(x) \geq f(x)$ for all $x \in [x_0 - \delta, x_0] \cap [a,b]$ and
$L(x) \leq f(x)$ for all $x \in [x_0, x_o + \delta] \cap [a,b]$.

Example 13.114 Consider the function $f(x) = \sin x$. At the point $x = 0$ the x-axis (i.e., the horizontal line $y = 0$), crosses f since in the interval $(-\pi, \pi)$ the line is above the graph of f on the left on $(-\pi, 0)$ and below on the right on $(0, \pi)$. Indeed every line through $(0,0)$ crosses f. At the point $(\pi/2, 1)$ every line through that point except the horizontal line $y = 1$ crosses f. Since the line $y = 1$ stays always above the graph of f, it does not cross. ◄

The situation in the example is typical of the case for functions familiar in calculus. Every line crosses the graph of such functions, except perhaps the tangent line may not.

Example 13.115 It is easy to give examples of continuous functions that "wiggle" so much near a specific point that it is impossible for any line to cross the graph at that point. The functions

$$f(x) = \sqrt{|x|}\sin(1/x), \quad f(0) = 0$$

and $g(x) = |f(x)|$ illustrate this at the point $(0,0)$. Note that at every other point on the graph of f or g there are many lines that cross (Fig. 13.9). ◄

Although it may be difficult to imagine a continuous function that crosses *no* lines, and more difficult to construct one, we shall see that "most" (in the sense of category) continuous functions have this property. In Sections 13.14.2 and 13.14.3 we shall use this to show that "most" continuous functions fail to be monotonic on any interval and fail to be differentiable at any point.

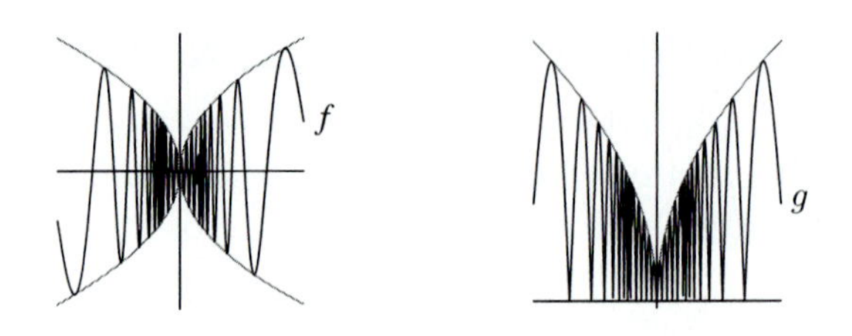

Figure 13.9. Graphs of $f(x) = \sqrt{|x|}\sin(1/x)$ and $g(x) = |f(x)|$. The asymptotes $y = \pm\sqrt{|x|}$ are also shown.

Before stating Theorem 13.116, we introduce a bit of notation that will be convenient for our proof. Let $f \in \mathcal{C}[a,b]$ and let $\gamma \in \mathbb{R}$. Define a function $f_{-\gamma}$ by

$$f_{-\gamma}(x) = f(x) - \gamma x.$$

Thus, $f_{-\gamma}$ is obtained from f by subtracting the linear function $L(x) = \gamma x$ from f. Observe that a line of slope γ crosses the graph of f at x_0 if and only if a corresponding horizontal line crosses the graph of $f_{-\gamma}$ at x_0.

Theorem 13.116 *The set*

$$Z = \{f \in \mathcal{C}[a,b] : f \text{ crosses no lines}\}$$

is a residual subset of $\mathcal{C}[a,b]$.

Proof We wish to express $X \setminus Z$ as a countable union of nowhere dense sets. Observe that if $f \in X \setminus Z$, there is at least one line that crosses f at some point x. In technical language, f (or $-f$) is in one of the sets A_n defined as follows.

For each $n \in \mathbb{N}$, let A_n denote those functions $f \in \mathcal{C}[a,b]$ for which there exists $\gamma \in [-n,n]$ and $x \in [a,b]$ such that

$$f_{-\gamma}(t) \leq f_{-\gamma}(x) \text{ when } t \in [a,b] \cap (x - 1/n, x)$$

and

$$f_{-\gamma}(t) \geq f_{-\gamma}(x) \text{ when } t \in [a,b] \cap (x, x + 1/n)\,.$$

In fact, if $f \in A_N$, then f is also in A_n for all $n > N$. Note that the number n plays two roles in this definition. For a function to be in A_n there must be *at least one* line whose slope is between $-n$ and $+n$ when it crosses the function. Furthermore, $1/n$ specifies the length of an interval in which that line must stay above or below the function (before crossing it again).

Let $A = \bigcup_{n=1}^{\infty} A_n$. We show that for each $n \in \mathbb{N}$, A_n is closed and that the complement of A_n is dense. It follows that each A_n is nowhere dense and hence that A is a first-category subset of $\mathcal{C}[a,b]$.

To verify that A_n is closed, let $\{f_k\}$ be a sequence of functions in A_n such that $f_k \to f$ in the space $\mathcal{C}[a,b]$. We must show $f \in A_n$. For each $k \in \mathbb{N}$, the function f_k is a member of A_n and so there exists $\gamma_k \in [-n,n]$ and $x_k \in [a,b]$ such that

$$f_{-\gamma_k}(t) \leq f_{-\gamma_k}(x_k) \text{ when } t \in [a,b] \cap (x_k - 1/n, x_k)$$

and

$$f_{-\gamma_k}(t) \geq f_{-\gamma_k}(x_k) \text{ when } t \in [a,b] \cap (x_k, x_k + 1/n)\,.$$

By the Bolzano-Weierstrass theorem (Theorem 2.40) applied to the sequence of points $\{x_k\}$ there exists a subsequence $\{x_{k_i}\}$ of $\{x_k\}$ that converges to some point x_0 in $[a,b]$. Applying the Bolzano-Weierstrass theorem again, we obtain a subsequence of that sequence, $\{x_{k_{i_j}}\}$ such that $\{\gamma_{k_{i_j}}\}$ converges to some $\gamma_0 \in [-n,n]$.

It is easy to verify (Exercise 13.14.1) that

$$f_{-\gamma_0}(t) \leq f_{-\gamma_0}(x_0) \text{ when } t \in [a,b] \cap (x_0 - 1/n, x_0)$$

and

$$f_{-\gamma_0}(t) \geq f_{-\gamma_0}(x_0) \text{ when } t \in [a,b] \cap (x_0, x_0 + 1/n)\,.$$

Thus $f \in A_n$. Since this is true for all convergent sequences chosen from A_n we conclude that A_n is closed in $\mathcal{C}[a,b]$.

To show that A_n is nowhere dense, we verify that every open ball in $\mathcal{C}[a,b]$ contains points that do not belong to A_n. Let $B(g,\varepsilon)$ be an open ball in $\mathcal{C}[a,b]$. It is easy to visualize (though tedious to verify analytically) that we can choose an appropriate sawtooth function f with many steep teeth such that $f \in B(g,\varepsilon) \setminus A_n$.

Intuitively, the line segments that make up the graph of f have such steep slopes and there are so many of these segments that no line whose slope is bounded by $-n$ and n can cross the graph as required for f to be in A_n. [See Fig. 13.10. The interval depicted has length $1/n$. The slopes of the "teeth" are so large that no line with slope between $+n$ and $-n$ can cross f only once in this interval. This means that $f \in B(g,\varepsilon) \setminus A_n$.]

Thus A_n is nowhere dense, and A is of the first category. Exactly the same arguments show that the set

$$B = \{f \in \mathcal{C}[a,b] : -f \in A\}$$

is first category. Consequently, $A \cup B$ is first category and it follows that

$$Z = \mathcal{C}[a,b] \setminus (A \cup B)$$

is residual. It can be checked that every member of Z crosses no lines. ∎

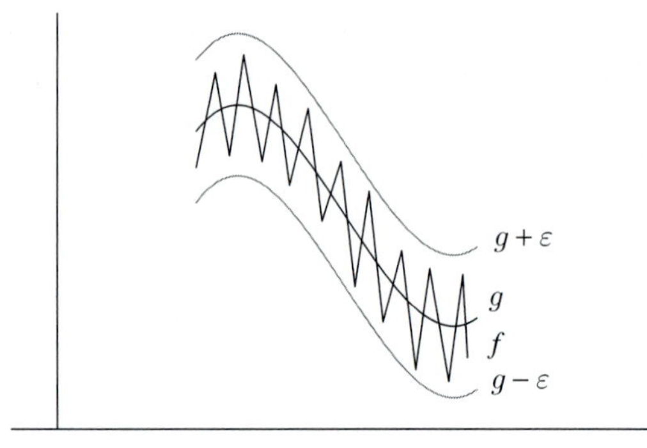

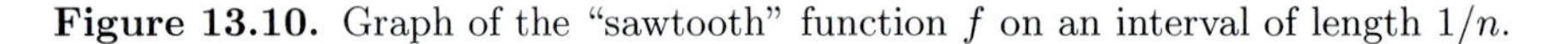

Figure 13.10. Graph of the "sawtooth" function f on an interval of length $1/n$.

Exercises

13.14.1 Verify that
$$f_{-\gamma_0}(t) \leq f_{-\gamma_0}(x_0)$$
when $t \in [a,b] \cap (x_0 - 1/n, x_0)$ and that
$$f_{-\gamma_0}(t) \geq f_{-\gamma_0}(x_0)$$
when $t \in [a,b] \cap (x_0, x_0 + 1/n)$ in the proof of Theorem 13.116.

13.14.2 Show that every $f \in Z$ (see the proof of Theorem 13.116) has $+\infty$ and $-\infty$ as derived numbers at every point.

13.14.2 Nowhere Monotonic Functions

All of the continuous functions that we encounter in a calculus class are monotonic or else they are piecewise monotonic: They are nondecreasing on some intervals and nonincreasing on remaining intervals. Indeed it is extremely hard to imagine a continuous function that behaves differently than this. Is it possible to prove the existence of continuous *nowhere monotonic* functions in the sense of the following definition?

Definition 13.117 A function f on an interval $[a,b]$ is said to be *nowhere monotonic* if there is no subinterval $[c,d] \subset [a,b]$ on which f is monotonic.

Theorem 13.118 *The class of continuous, nowhere monotonic functions forms a residual subset of* $C[a,b]$.

Proof When a continuous function f is monotonic on an interval $[c,d]$, it is clear that there are many lines that cross f. In fact, every horizontal line $y = k$ for k between $f(c)$ and $f(d)$ must cross f. Thus, any member of the class of functions Z discussed in the preceding section (Theorem 13.116) cannot be monotonic on any interval. Thus this theorem follows directly from Theorem 13.116 since the set of nowhere monotonic functions contains the residual set Z and hence must also be residual. ■

13.14.3 Continuous Nowhere Differentiable Functions

We have proved now, using the line crossing arguments together with Baire category arguments, the existence of continuous functions having geometric properties that are hard to visualize. Let us turn to an analytic property that is equally hard to visualize.

It is easy to construct a function that has no derivative anywhere: Simply take a function that is discontinuous everywhere. But can a continuous function exist that has no derivative anywhere? The function $f(x) = |x|$ illustrates how to arrange for at least one point of nondifferentiability. Any finite number of points can be just as easily handled. In the early nineteenth century a number of mathematicians were quite convinced that not much more could be said. Indeed some of them "proved" that a continuous function would have to have points of differentiability.

By the middle of the nineteenth century many mathematicians were aware of the existence of continuous functions that had no points of differentiability. Constructions of such functions involved summations of infinite series whose successive terms contributed increasingly to the nondifferentiability of their sum. One of the best known of such constructions was given by Weierstrass around 1875. Use of the Baire category theorem to prove the existence of such functions had to wait until 1931, at which time two Polish mathematicians Stefan Banach (1892–1945) and Stefan Mazurkiewicz (1888–1945), in separate papers in the journal *Studia Mathematica*, provided such proofs.

We call a function *nowhere differentiable* if it has a finite derivative at no point.

Theorem 13.119 *The class of continuous, nowhere differentiable functions forms a residual subset of $\mathcal{C}[a,b]$.*

Proof Observe that if a continuous function is differentiable at a point $x_0 \in (a,b)$, with $f'(x_0) = \gamma$, then any line whose slope is not γ and that passes through $(x_0, f(x_0))$ will cross f. This implies that each member of the class of functions Z discussed in Theorem 13.116 is differentiable at no point. Thus this theorem follows directly from Theorem 13.116 since the set of nowhere differentiable functions contains the residual set Z and hence must also be residual. ■

Note. In fact, a more careful analysis shows that for each $f \in Z$, $+\infty$ and $-\infty$

are derived numbers of f at every point. (See Section 7.8 for a discussion of Dini derivatives and derived numbers.) We leave this analysis as Exercise 13.14.2.

13.14.4 Cantor Sets

We turn now to an example in which the objects are closed sets. Recall that that the Cantor ternary or "middle-third set" was the first example of a nonempty, nowhere dense set of real numbers without isolated points. Such sets were extremely hard to visualize by the nineteenth-century mathematicians, and many mistakes were made before Cantor's construction was known. In honor of Cantor let us call such sets by his name.

Definition 13.120 A nonempty, nowhere dense compact set of real numbers without isolated points is said to be a *Cantor set.*

Suppose that we, like the nineteenth-century mathematicians, had not heard of Cantor sets, but, unlike nineteenth-century mathematicians, did know the Baire category theorem and did know that the space $\mathcal{K}$ of closed subsets of $[0,1]$ with the Hausdorff metric d is complete (Exercises 13.8.13 and 13.3.9). We might ask, "What do most nonempty closed subsets of $[0,1]$ look like?"

> **Question 1.** Do most nonempty closed subsets of $[0,1]$ contain interior points?

If a closed set has interior points, there exists a closed interval I with rational endpoints contained in that set. Thus every closed set that does contain interior points is in the family

$$\{K \in \mathcal{K} : K \supset I\}$$

for some such interval I. Take any $F \in \mathcal{K}$ and let $\varepsilon > 0$. By Exercise 13.7.3 (and its hint) there exists a finite set S of rational numbers such that $d(S,F) < \varepsilon$. We can choose S so that $S \cap I$ has at least two members. Let $S = \{s_1, \dots, s_n\}$ with s_1 and s_2 in I and no other points of S in between s_1 and s_2. Let δ be smaller than $|s_1 - s_2|/3$ and also sufficiently small that

$$B(S,\delta) \subset B(F,\varepsilon).$$

If $T \in B(S,\delta)$, then T cannot contain the point $m = (s_1+s_2)/2$ (Fig. 13.11). Thus T cannot contain I. Thus the ball $B(F,\varepsilon)$ contains the ball $B(S,\delta)$, no point of which is in the set

$$\{K \in \mathcal{K} : K \supset I\}.$$

It follows that this latter set is nowhere dense in $\mathcal{K}$ for any choice of interval I.

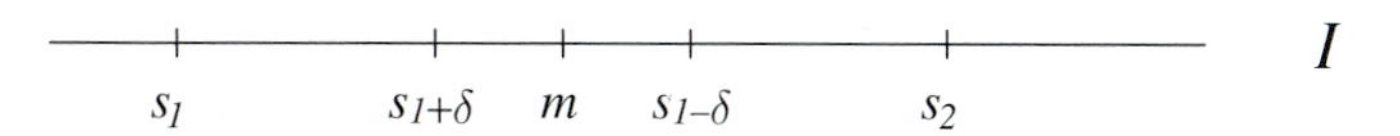

Figure 13.11. The interval I in the proof that most members of $\mathcal{K}$ are nowhere dense.

Now let $\{I_k\}$ be an enumeration of the closed subintervals of $[0,1]$ with rational endpoints. For each $k \in \mathbb{N}$ let

$$A_k = \{K \in \mathcal{K} : K \supset I_k\}.$$

We have shown that each of the sets A_k is nowhere dense in $\mathcal{K}$; thus

$$A = \bigcup_{k=1}^{\infty} A_k$$

is a first-category set. This shows that

$$\{K \in \mathcal{K} : K \text{ has nonempty interior}\}$$

is of the first category. Thus we have answered our first question and can assert that

$$\textit{Most members of } \mathcal{K} \textit{ are nowhere dense.} \tag{47}$$

We next ask about isolated points.

Question 2. Do most members of $\mathcal{K}$ have isolated points?

If $K \in \mathcal{K}$ has an isolated point x, then there exists $n \in \mathbb{N}$ such that

$$\operatorname{dist}(x, K \setminus \{x\}) \geq 1/n.$$

(See Exercise 13.6.11 for the definition of the function "dist.") Let

$$B_n = \{K \in \mathcal{K} : \exists\, x \in K \text{ for which } \operatorname{dist}(x, K \setminus \{x\}) \geq 1/n\}.$$

We verify easily (Exercise 13.14.3) that each of the sets B_n is nowhere dense in $\mathcal{K}$. Let $B = \bigcup_{n=1}^{\infty} B_n$. Then B is of the first category in $\mathcal{K}$. If K has any isolated points, then $K \in B$. Thus we have answered our second question and can assert that

$$\textit{Most members of } \mathcal{K} \textit{ have } \text{no} \textit{ isolated points.} \tag{48}$$

Combining (47) and (48), we see that we have proved the following theorem.

Theorem 13.121 *Let $\mathcal{K}$ be the metric space of nonempty closed subsets of $[0,1]$ with the Hausdorff metric, and let $\mathcal{S}$ consist of the Cantor sets in $[0,1]$. Then $\mathcal{S}$ is residual in $\mathcal{K}$.*

Now for nineteenth-century mathematicians (and for students of elementary calculus), closed sets in $[0, 1]$ were/are rather simple. They consist of intervals, isolated points, and limits of isolated points. With a little more effort, perhaps sets containing limits of limit points, or limits of limits of limit points could be imagined (Exercise 4.2.24). Our theorem shows that, in fact, *most* closed sets are Cantor sets, when "most" is interpreted in terms of category.

Exercises

13.14.3 Verify that the sets B_n that appear in the arguments leading to a proof of Theorem 13.121 are nowhere dense in $\mathcal{K}$.

13.14.4 Show that

$$Y = \{f \in M[a, b] : f \text{ is one-to-one}\}$$

is a residual subset of the space $M[a, b]$ of bounded functions on $[a, b]$.

13.14.5 Show that

$$Z = \{f \in M[a, b] : \text{ the range of } f \text{ is nowhere dense}\}$$

is a residual subset of the space $M[a, b]$ of bounded functions on $[a, b]$.

13.15 Challenging Problems for Chapter 13

13.15.1 Let X be a metric space and let $\mathcal{A}$ be a family of closed subsets of X such that if $A, B \in \mathcal{A}$, then $A \subset B$ or $B \subset A$. Let

$$E = \bigcup_{A \in \mathcal{A}} A.$$

Prove that E can be expressed as a countable union of closed subsets of X.

13.15.2 By a power of a map A we mean a composition with itself; thus

$$A^2(x) = A(A(x)),\ A^3(x) = A(A(A(x))),\ \ldots$$

are powers of A.

(a) Show that a map A defined on a complete metric space (X, d) for which one of the powers of A is a contraction has a unique fixed point.

(b) Show that the integral equation

$$f(x) = \lambda \int_a^x K(x, y) f(y)\, dy + \phi(x)$$

has a unique solution in $C[a, b]$. (Here λ, f, K, and ϕ are as in Example 13.81.)

13.15.3 Prove that $[0, 1]$ cannot be expressed as a countable disjoint union of closed sets (except in the obvious way as a union of one set!).

13.15.4 In a general metric space (X, d) take an arbitrary set $A \subset X$ and perform a sequence of complements or closures. How many distinct sets can arise in this way?

13.15.5 We know from elementary calculus that if a function $f : [0, 1] \to \mathbb{R}$ has an nth derivative that is identically zero, then f is a polynomial on $[0, 1]$. Prove that if f has derivatives of all orders on $[0, 1]$ and for each $x \in [0, 1]$ there exists $n(x) \in \mathbb{N}$ such that $f^{(n(x))}(x) = 0$, then f is a polynomial on $[0, 1]$.

13.15.6 Let (X, d) be a metric space and let $f : X \to X$ be an isometry, that is, so that $d(f(x), f(y)) = d(x, y)$ for all $x, y \in X$. Show that, in general, f need not be onto but must be if X is compact.

13.15.7 Suppose that every real-valued continuous function defined on a metric space X attains a maximum value. Show that X must be compact.

13.15.8 Suppose that every real-valued continuous function defined on a metric space X is uniformly continuous. Show that X must be complete but need not be compact.

13.15.9 Let (X, d) be a metric space. Show that (X, d) is compact if and only if for every equivalent metric ρ on X, the space (X, ρ) is complete.

13.15.10 (**Liouville numbers**) This example shows that a set can be small in the sense of measure, yet large in the sense of category. A real number z is called a *Liouville number* if z is irrational and has the property that for each n there exist integers p and $q > 1$ such that

$$\left| z - \frac{p}{q} \right| < \frac{1}{q^n}.$$

Prove the following statements about the set L of Liouville numbers [named after Joseph Liouville (1809–1882)].

(a) The set of Liouville numbers can be expressed as

$$L = (\mathbb{R} \setminus \mathbb{Q}) \cap \bigcap_{n=1}^{\infty} G_n,$$

where G_n are open sets defined as

$$G_n = \bigcup_{q=2}^{\infty} \bigcup_{p=-\infty}^{\infty} \left(\frac{p}{q} - \frac{1}{q^n}, \frac{p}{q} + \frac{1}{q^n} \right).$$

(b) L is a dense set of type $\mathcal{G}_\delta$, so L is large in the sense of category (see Exercise 13.13.15).

(c) L has measure zero and so is small in the sense of measure.

13.15.11 Use Exercise 13.14.4 to show that every bounded function on an interval $[a, b]$ is the sum of two bounded one-to-one functions on $[a, b]$.

13.15.12 Let $X = (0,1)$. Let $d(x,y) = |x-y|$ and let

$$e(x,y) = |x-y| + \left|\frac{1}{x} - \frac{1}{y}\right|.$$

(a) Prove that e is a metric on X.

(b) Prove that a sequence $\{x_k\}$ converges in (X,e) if and only if it converges in (X,d).

(c) Prove that a set $A \subset X$ is open in (X,e) if and only if it is open in (X,d).

(d) Prove that the identity map

$$h : (X,d) \to (X,e)$$

is a homeomorphism.

(e) Prove that (X,e) is complete.

(f) Consider a Cauchy sequence in (X,d) that does not converge. It can't converge in (X,e) either. So how can (X,e) be complete?

(g) Verify that the Baire category theorem holds in (X,d) even though the space is not complete.

13.15.13 The space (X,d) in Exercise 13.15.12 is not complete, but that exercise shows that it can be *remetrized* as (X,e) to be complete. This means that d and e are equivalent metrics and (X,d) and (X,e) are topologically equivalent (homeomorphic) metric spaces. Such a space is said to be *topologically complete.*

(a) Verify that the Baire category theorem holds in every topologically complete metric space.

(b) Give an example of a metric space that is not topologically complete.

13.15.14 **(Urysohn's Lemma)** Prove the following theorem, which is due to Pavel Urysohn (1898–1924).

> *Let A and B be disjoint nonempty closed subsets of a metric space (X,d). Then there exists a continuous function $g : X \to \mathbb{R}$ such that $g(x) = 0$ for all $x \in A$, $g(x) = 1$ for all $x \in B$ and $0 < g(x) < 1$ for all $x \in X \setminus (A \cup B)$.*

13.15.15 This problem requires a bit of knowledge about complex numbers and transcendental numbers. We exhibit two sets R and T in $\mathbb{R}^2$ such that R and T are congruent and each is congruent to their union $S = R \cup T$.

For each complex number z, let $t(z) = z + 1$, $r(z) = e^i z$. Thus t is just a right translation by 1 unit and r is a rotation by 1 radian. Let S consist of those points that can be obtained by a finite number of applications of t and r starting from the origin. Each member of S can be represented as a polynomial in e^i with positive integer coefficients. (For example, if we translate five times, then rotate twice, then translate once more, the resulting point can be represented as $5e^{2i}+1$.) Since e^i is transcendental, the representation is unique.

Let R consist of those points that have no constant term in their representation, and let $T = S \setminus R$. Prove that $t(S) = T$ and $r(S) = R$ so R, T and $S = R \cup T$ are pairwise isometric. Note that the isometries involved are isometries of $\mathbb{R}^2$ onto $\mathbb{R}^2$, not just isometries among the sets R, S, and T.

13.15.16 Let (X, d) be a metric space and let $M(X)$ denote the set of all bounded real-valued functions on X furnished with the sup metric. Choose a fixed $x_0 \in X$ and define a mapping $h : X \to M(X)$ by writing $h(x)$ for the function

$$(h(x))\,(y) = d(x, y) - d(y, x_0) \quad (y \in X).$$

(a) Show that each $h(x)$ is a bounded function on X for each $x \in X$.

(b) Show that h is an isometry of X to a subspace of $M(X)$.

(c) Show that *every* metric space is isometric to a subspace of some complete function space (i.e., a complete metric space of real-valued functions defined on a set and furnished with the sup metric).

APPENDIX A: BACKGROUND

A.1 Should I Read This Chapter?

This background chapter is not meant for the instructor but for the student. It is a mostly informal account of ideas that you need to survive an elementary course in analysis. The chapters in the text itself are more formal and contain actual mathematics. This chapter is *about* mathematics and should be an easier read.

You may skip around and select those topics that you feel you really need to read. For example, you may look through the section on notation (Section A.2) to be sure that you are familiar with the normal way of writing up many mathematical ideas, such as sets and functions.

The sections on proofs (Sections A.4, A.5, A.6, A.7, and A.8) should be read if you have never taken any courses that required an ability to write up a proof. For many students this course on real analysis is the first exposure to these ideas, and you may find these sections helpful.

A.2 Notation

If you are about to embark on a reading of the text without any further preliminaries, then there is some notation that we should review.

A.2.1 Set Notation

Sets are just collections of objects. In the beginning we are mostly interested in sets of real numbers. If the word "set" becomes too often repeated, you might find that words such as *collection*, *family*, or *class* are used. Thus a set of sets might become a family of sets. (We find such variations in ordinary language, such as flock of sheep, gaggle of geese, pride of lions.)

The statement $x \in A$ means that x is one of those numbers belonging to A. The statement $x \notin A$ means that x is *not* one of those numbers belonging to A. (The stroke through the symbol $\in$ here is a familiar device, even on road signs or no smoking signs.) Here are some familiar sets and notation.

(The Empty Set) $\emptyset$ to represent the set that contains no elements, the empty set.

(The Natural Numbers) $\mathbb{N}$ to represent the set of natural numbers (positive integers) 1, 2, 3, 4, etc.

(The Integers) $\mathbb{Z}$ to represent the set of integers (positive integers, negative integers, and zero).

(The Rational Numbers) $\mathbb{Q}$ to represent the set of rational numbers, that is, of all fractions m/n where m and n are integers (and $n \neq 0$).

(The Real Numbers) $\mathbb{R}$ to represent all the real numbers.

(Closed Intervals) $[a, b]$ to represent the set of all numbers between a and b, including a and b. We assume that $a < b$. This is called the closed interval with endpoints a and b. (Some authors allow the possibility that $a = b$, in which case $[a, b]$ must be interpreted as the set containing just the one point a. This would then be referred to as a *degenerate* interval. We have avoided this usage.)

(Open Intervals) (a, b) to represent the set of all numbers between a and b excluding a and b. This is called the open interval with endpoints a and b.

(Infinite Intervals) (a, ∞) to represent the set of all numbers strictly greater than a. The symbol ∞ is not interpreted as a number. [It might have been better for most students if the notation had been $(a, \rightarrow)$ since that conveys the same meaning and the beginning student would not have presumed that there is some infinite number called "$\rightarrow$" at the extreme right hand "end" of the real line.]

The other infinite intervals are

$$(-\infty, a), \ [a, \infty), \ (-\infty, a], \ \text{and} \ (-\infty, \infty) = \mathbb{R}.$$

(Sets as a List) $\{1, -3, \sqrt{7}, 9\}$ to represent the set containing precisely the four real numbers 1, -3, $\sqrt{7}$, and 9. This is a useful way of describing a set (when possible): Just list the elements that belong. Note that order does not matter in the world of sets, so the list can be given in any order that we wish.

(Set-Builder Notation) $\{x : x^2 + x < 0\}$ to represent the set of all numbers x satisfying the inequality $x^2 + x < 0$. It may take some time [see Exercise A.2.1], but if you are adept at inequalities and quadratic equations you can recognize that this set is exactly the open interval

$(-1, 0)$.) This is another useful way of describing a set (when possible): Just describe, by an equation or an inequality, the elements that belong. In general, if $C(x)$ is some kind of assertion about an object x, then $\{x : C(x)\}$ is the set of all objects x for which $C(x)$ happens to be true. Other formulations can be used. For example,

$$\{x \in A : C(x)\}$$

describes the set of elements x *that belong to the set* A and for which $C(x)$ is true. The example $\{1/n : n \in \mathbb{N}\}$ illustrates that a set can be obtained by performing computations on the members of another set.

Subsets, Unions, Intersection, and Differences The language of sets requires some special notation that is, doubtless, familiar. If you find you need some review, take the time to learn this notation well as it will be used in all of your subsequent mathematics courses.

1. $A \subset B$ (A is a subset of B) if every element of A is also an element of B.

2. $A \cap B$ (the intersection of A and B) is the set consisting of elements of both sets.

3. $A \cup B$ (the union of the sets A and B) is the set consisting of elements of either set.

4. $A \setminus B$ (the difference[1] of the sets A and B) is the set consisting of elements belonging to A but not to B.

In the text we will need also to form unions and intersections of large families of sets, not just of two sets. See the exercises for a development of such ideas.

De Morgan's Laws Many manipulations of sets require two or more operations to be performed together. The simplest cases that should perhaps be memorized are

$$A \setminus (B_1 \cup B_2) = (A \setminus B_1) \cap (A \setminus B_2)$$

and a symmetrical version

$$A \setminus (B_1 \cap B_2) = (A \setminus B_1) \cup (A \setminus B_2).$$

If you sketch some pictures these two rules become evident. There is nothing special that requires these "laws" to be restricted to two sets B_1 and B_2. Indeed any family of sets $\{B_i : i \in I\}$ taken over any indexing set I must obey the same laws:

$$A \setminus \left(\bigcup_{i \in I} B_i \right) = \bigcap_{i \in I} (A \setminus B_i)$$

[1]Don't use $A - B$ for set difference since it suggests subtraction, which is something else.

and

$$A \setminus \left(\bigcap_{i \in I} B_i \right) = \bigcup_{i \in I} (A \setminus B_i) .$$

Here $\bigcup_{i \in I} B_i$ is just the set formed by combining all the elements of the sets B_i into one big set (i.e., forming a large union). Similarly, $\bigcap_{i \in I} B_i$ is the set of points that are in all of the sets B_i, that is, their common intersection.

Augustus De Morgan (1806–1871), after whom these laws are named, had a respectable career as a Professor in London, although he is not remembered for any deep work. He was the originator in 1838 of the expression "mathematical induction" and the first to give a rigorous account of it. He has one interesting claim to fame, in addition to his "laws:" He was the tutor of Lady Ada Lovelace, who some say is the world's first computer programmer. A puzzle of his survives: He claims that he "was x years old in the year x^2."

Ordered Pairs Given two sets A and B, we often need to discuss pairs of objects (a, b) with $a \in A$ and $b \in B$. The first item of the pair is from the first set and the second item from the second. Since order matters here these are called *ordered pairs*. The set of all ordered pairs (a, b) with $a \in A$ and $b \in B$ is denoted

$$A \times B$$

and this set is called the *Cartesian product* of A and B.

Relations Often in mathematics we need to define a relation on a set S. Elements of S could be related by sharing some common feature or could be related by a fact of one being "larger" than another. For example, the statement $A \subset B$ is a relation on families of sets and $a < b$ a relation on a set of numbers. Fractions p/q and a/b are related if they define the same number; thus we could define a relation on the collection of all fractions by $p/q \sim a/b$ if $pb = qa$.

A relation R on a set S then would be some way of deciding whether the statement xRy (read as x is related to y) is true. If we look closely at the form of this we see it is completely described by constructing the set

$$R = \{(x, y) : \ x \text{ is related to } y\}$$

of ordered pairs. Thus a relation on a set is not a new concept: It is merely a collection of ordered pairs. Let R be any set of ordered pairs of elements of S. Then $(x, y) \in R$ and xRy and "x is related to y" can be given the same meaning. This reduces relations to ordered pairs. In practice we usually view the relation from whatever perspective is most intuitive. [For example, the order relation on the real line $x < y$ is technically the same as the set of ordered pairs $\{(x, y) : x < y\}$ but hardly anyone thinks about the relation this way.]

A.2.2 Function Notation

Analysis (indeed most of mathematics) is about functions. Do you recall that in elementary calculus courses you would often discuss some function such as $f(x) = x^2 + x + 1$ in the context of maxima and minima problems, or derivatives or integrals? The most important way of understanding a function in calculus was by means of the graph: For this function the graph is the set of all pairs $(x, x^2 + x + 1)$ for real numbers x, and often this graph was sketched as a set of points in two-dimensional space.

Definition of a Function What is a function really? Calculus students usually comprehend a formula $f(x) = x^2 + x + 1$ as defining a function, but begin to be confused when the term is used less concretely. For example, what is the distinction between the function $f(x) = x^2 + x + 1$ here and a statement such as $f(x^2 + x + 1)$?

Definition A.1 A function (or sometimes *map*) f from a set A into a set B is a rule that assigns a *value* $f(a) \in B$ to each element $a \in A$. The input set A is called the *domain of the function.* Note that f is the function, while $f(x)$ (which is not the function) is the value assigned by the function at the element $x \in A$. The set of all output values is written as

$$f(A) = \{b \in B : f(a) = b \text{ for some } a \in A\}$$

and is called the *range* of the function.

Thus the calculus example above really asserts that we are given a function named f, whose domain is the set of all real numbers, and which assigns to any number a the value $f(a) = a^2 + a + 1$. The range is not transparent from the definition and would need to be computed if it is required. (It is a simple exercise to determine that the range is the interval $[3/4, \infty)$.)

Mathematicians noted long ago that the graph of a function carried all the information needed to describe the function. Indeed, since the graph is just a set of ordered pairs $(x, f(x))$, the concept of a function can be explained entirely within the language of sets without any need to invent a new concept. Thus the function *is* the graph and the graph is a set. Thus you can expect to see the more formal version of this definition of a function given as follows.

Definition A.2 Let A and B be nonempty sets. A set f of ordered pairs (a, b) with $a \in A$ and $b \in B$ is called a *function* from A to B, written symbolically as

$$f : A \to B,$$

provided that to every $a \in A$ there is precisely one pair (a, b) in f.

The notation $(a, b) \in f$ is often used in advanced mathematics but is awkward in expressing ideas in calculus and analysis. Instead we use the familiar expression $f(a) = b$. Also, when we wish to think of a function as a graph we normally remind you by using the word "graph." Thus an analysis or calculus student would expect to see a question posed like this:

> Find a point on the graph of the function $f(x) = x^2 + x + 1$ where the tangent line is horizontal.

rather than the technically correct, but awkward looking

> Let f be the function
> $$f = \{(x, x^2 + x + 1) : x \in \mathbb{R}\}.$$
> Find a point in f where the tangent line is horizontal.

Domain of a Function The set of points A in the definition is called *the domain* of the function. It is an essential ingredient of the definition of any function. It should be considered incorrect to write

> Let the function f be defined by $f(x) = \sqrt{x}$.

Instead we should say

> Let the function f be defined with domain $[0, \infty)$ by $f(x) = \sqrt{x}$.

The first assertion is sloppy; it requires you to guess at the domain of the function. Calculus courses, however, often make this requirement, leaving it to you to figure out from a formula what domain should be assigned to the function. Often we, too, will require that you do this.

Range of a Function The set of points B in the definition is sometimes called the *range* or *co-domain* of the function. Most writers do not like the term "range" for this and prefer to use the term "range" for the set

$$f(A) = \{f(x) : x \in A\} \subset B$$

that consists of the actual output values of the function f, not some larger set that merely contains all these values.

One-To-One and Onto Function If to each element b in the range of f there is precisely one element a in the domain so that $f(a) = b$, then f is said to be *one-to-one* or *injective*. We sometimes say, about the range $f(A)$ of a function, that f maps A *onto* $f(A)$. If $f : A \to B$, then f would be said to be *onto* B if B is the range of f, that is, if for every $b \in B$ there is some $a \in A$ so that $f(a) = b$. A function that is onto is sometimes said to be *surjective*. A function that is both one-to-one and onto is sometimes said to be *bijective*.

Inverse of a Function Some functions allow an *inverse*. If $f : A \to B$ is a function, there is, sometimes, a function $f^{-1} : B \to A$ that is the reverse of f in the sense that

$$f^{-1}(f(a)) = a \text{ for every } a \in A$$

and

$$f(f^{-1}(b)) = b \text{ for every } b \in B.$$

Thus f carries a to $f(a)$ and f^{-1} carries $f(a)$ back to a while f^{-1} carries b to $f^{-1}(b)$ and f carries $f^{-1}(b)$ back to b. This can happen only if f is one-to-one and onto B. See the exercises for some practice on these concepts.

Characteristic Function of a Set Let $E \subset \mathbb{R}$. Then a convenient function for discussing properties of the set E is the function χ_E defined to be 1 on E and to be 0 at every other point. This is called the *characteristic function* of E or, sometimes, *indicator function.*

Composition of Functions Suppose that f and g are two functions. For some values of x it is possible that the application of the two functions one after another

$$f(g(x))$$

has a meaning. If so this new value is denoted $f \circ g(x)$ or $(f \circ g)(x)$ and the function is called the *composition* of f and g. The domain of $f \circ g$ is the set of all values of x for which $g(x)$ has a meaning and for which then also $f(g(x))$ has a meaning; that is, the domain of $f \circ g$ is

$$\{x : x \in \text{dom}(g) \text{ and } g(x) \in \text{dom}(f)\}.$$

Note that the order matters here so $f \circ g$ and $g \circ f$ have, usually, radically different meanings. This is likely one of the earliest appearances of an operation in elementary mathematics that is not commutative and that requires some care.

Exercises

A.2.1 This exercise introduces the idea of set equality. The identity $X = Y$ for sets means that they have identical elements. To prove such an assertion assume first that $x \in X$ is any element. Now show that $x \in Y$. Then assume that $y \in Y$ is any element. Now show that $y \in X$.

(a) Show that $A \cup B = B$ if and only if $A \subset B$.

(b) Show that $A \cap B = A$ if and only if $A \subset B$.

(c) Show that $(A \cup B) \cap C = (A \cap C) \cup (B \cap C)$.

(d) Show that $(A \cap B) \cup C = (A \cup C) \cap (B \cup C)$.

(e) Show that $(A \cup B) \setminus C = (A \setminus C) \cup (B \setminus C)$.

(f) Show that $(A \cap B) \setminus C = (A \setminus C) \cap (B \setminus C)$.

(g) Show that $\{x \in \mathbb{R} : x^2 + x < 0\} = (-1, 0)$.

A.2.2 This exercise introduces the notations $\bigcup_{n=1}^{N} A_i$ and $\bigcap_{n=1}^{N} A_i$ for the union and intersection of the sets A_1, A_2, ... A_N:

(a) Describe the sets

$$\bigcup_{n=1}^{N} (-1/n, 1/n) \text{ and } \bigcap_{n=1}^{N} (-1/n, 1/n).$$

(b) Describe the sets

$$\bigcup_{n=1}^{N} (-n, n) \text{ and } \bigcap_{n=1}^{N} (-n, n).$$

(c) Describe the sets

$$\bigcup_{n=1}^{N} [n, n+1] \text{ and } \bigcap_{n=1}^{N} [n, n+1].$$

A.2.3 This exercise introduces the notations $\bigcup_{n=1}^{\infty} A_i$ and $\bigcap_{n=1}^{\infty} A_i$ for the union and intersection of the sets A_1, A_2, ...

(a) Describe the sets

$$\bigcup_{n=1}^{\infty} (-1/n, 1/n) \text{ and } \bigcap_{n=1}^{\infty} (-1/n, 1/n).$$

(b) Describe the sets

$$\bigcup_{n=1}^{\infty} (-n, n) \text{ and } \bigcap_{n=1}^{\infty} (-n, n).$$

(c) Describe the sets

$$\bigcup_{n=1}^{\infty} [n, n+1] \text{ and } \bigcap_{n=1}^{\infty} [n, n+1].$$

A.2.4 Do you accept any of the following as an adequate definition of the function f? (The domain is not specified but it is assumed that you will try to find a domain that might work.)

(a) $f(x) = 1/\sqrt{1-x}$.

(b) $f(x) = x$ if x is rational and $f(x) = -x$ if x is irrational.

(c) $f(x) = 1$ if x contains a 9 in its decimal expansion and $f(x) = 0$ if not.

(d) $f(x) = 1$ if x contains a 7 in its decimal expansion and $f(x) = 0$ if not.

(e) $f(x) = 1$ if x is a prime number and $f(x) = 0$ if it is not.

A.2.5 This exercise promotes the use of the term *mapping* in the study of functions.

If $f : X \to Y$ and $E \subset X$, then

$$f(E) = \{y : f(x) = y \text{ for some } x \in E\} \subset Y$$

is called the *image* of E under f and we say f *maps* E to the set $f(E)$.

(a) Let $f : \mathbb{R} \to \mathbb{R}$. Give an example of sets A, B so that

$$f(A \cap B) \neq f(A) \cap f(B).$$

(b) Would $f(A \cup B) = f(A) \cup f(B)$ be true in general?

(c) Find a function $f : \mathbb{R} \to \mathbb{R}$ so that $f([0,1]) = \{1,2\}$.

A.2.6 This exercise concerns the notion of one-to-one function (i.e., injective function):

(a) Show that $f : \mathbb{R} \to \mathbb{R}$ is one-to-one if and only if

$$f(A \cap B) = f(A) \cap f(B)$$

for all sets A, B.

(b) Show that $f : \mathbb{R} \to \mathbb{R}$ is one-to-one if and only if $f(A) \cap f(B) = \emptyset$ for all sets A, B with $A \cap B = \emptyset$.

A.2.7 This exercise concerns the notion of preimage. If $f : X \to Y$ and $E \subset Y$, then

$$f^{-1}(E) = \{x : f(x) = y \text{ for some } y \in E\} \subset X$$

is called the preimage of E under f. [There may or may not be an inverse function here; $f^{-1}(E)$ has a meaning even if there is no inverse function.]

(a) Show that $f(f^{-1}(E)) \subset E$ for every set $E \subset \mathbb{R}$.

(b) Show that $f^{-1}(f(E)) \supset E$ for every set $E \subset \mathbb{R}$.

(c) Can you simplify $f^{-1}(A \cup B)$ and $f^{-1}(A \cap B)$?

(d) Show that $f : \mathbb{R} \to \mathbb{R}$ is one-to-one if and only if $f^{-1}(\{b\})$ contains at most a single point for any $b \in \mathbb{R}$.

(e) Show that $f : \mathbb{R} \to \mathbb{R}$ is onto, that is, the range of f is all of $\mathbb{R}$ if and only if $f(f^{-1}(E)) = E$ for every set $E \subset \mathbb{R}$.

A.2.8 This exercise concerns the notion of composition of functions:

(a) Give examples to show that $f \circ g$ and $g \circ f$ are distinct.

(b) Give an example in which $f \circ g$ and $g \circ f$ are not distinct.

(c) While composition is not commutative, is it associative, that is, is it true that

$$(f \circ g) \circ h = f \circ (g \circ h)?$$

(d) Give several examples of functions f for which $f \circ f = f$.

A.2.9 This exercise concerns the notion of onto function (i.e., surjective function): Which of the following functions map $[0,1]$ onto $[0,1]$?

(a) $f(x) = x$
(b) $f(x) = x^2$
(c) $f(x) = x^3$
(d) $f(x) = 2|x - \frac{1}{2}|$
(e) $f(x) = \sin \pi x$
(f) $f(x) = \sin x$

A.2.10 This exercise concerns the notion of one-to-one and onto function (i.e., bijective function):

(a) Which of the functions of Exercise A.2.9 is a bijection of $[0,1]$ to $[0,1]$?

(b) Is the function $f(x) = x^2$ a bijection of $[-1,1]$ to $[0,1]$?

(c) Find a linear bijection of $[0,1]$ onto the interval $[3,6]$.

(d) Find a bijection of $[0,1]$ onto the interval $[3,6]$ that is not linear.

(e) Find a bijection of $\mathbb{N}$ onto $\mathbb{Z}$.

A.2.11 This exercise concerns the notion of inverse functions: For each of the functions of Exercise A.2.9, select an interval $[a,b]$ on which that function has an inverse and find an explicit formula for the inverse function. Be sure to state the domain of the inverse function.

A.2.12 This exercise concerns the notion of an equivalence relation. A relation $x \sim y$ on a set S is said to be an equivalence relation if

(a) $x \sim x$ for all $x \in S$.
(b) $x \sim y$ implies that $y \sim x$.
(c) $x \sim y$ and $y \sim z$ imply that $x \sim z$.

(a) Show that the relation $p/q \sim a/b$ if $pb = qa$ defined in the text on the collection of fractions is an equivalence relation.

(b) Define a relation on the collection of fractions that satisfies two of the requirements of an equivalence relation but is not an equivalence relation.

(c) Define nontrivial equivalence relations on the sets $\mathbb{N}$ and $\mathbb{Z}$.

A.2.13 Set builder notation can be used to "describe" some curious sets. For example,

$$S_1 = \{S : S \text{ is a set}\}.$$

This has the peculiar property that $S_1 \in S_1$. (That is similar to joining a club where you find the club appearing on the membership list as a member of itself!) Worse yet is

$$S_2 = \{S : S \text{ is a set and } S \notin S\}.$$

This has the paradoxical property that if $S_2 \in S_2$, then $S_2 \notin S_2$, while if $S_2 \notin S_2$, then $S_2 \in S_2$. Any thoughts?

A.3 What Is Analysis?

The term "analysis" now covers large parts of mathematics. You almost need to be a professional mathematician to understand what it might mean.

For a course at this level, though, "real analysis" mostly refers to the subject matter that you have already learned in your calculus courses: limits, continuity, derivatives, integrals, sequences, and series. Calculus as a subject can be thought of as an eighteenth century development, analysis as a nineteenth-century creation. None of the ideas of calculus rested on very firm foundation, and the lack of foundations proved a barrier to further progress. There was much criticism by mathematicians and philosophers of the fundamental ideas of calculus (limits especially), and often when new and controversial methods were proposed (such as Fourier series) the mathematicians of the time could not agree on whether they were valid.

In the first decades of the nineteenth century the foundations of the subject were reworked, most notably by Cauchy (whose name will appear frequently in this text) and new and powerful methods developed. It is this that we are studying here.

We will look once again at notions of sequence limit, function limit, etc. that we have seen before in our calculus classes, but now from a more rigorous point of view. We want to know precisely what they mean and how to prove the validity of the techniques of the subject.

At first sight you might wonder about this. Are we just reviewing our calculus but now we do not get to skip over the details of proofs? If, however, you persist you will see that we are entering instead a new and different world. By looking closely at the details of why certain things work we gain a new insight. More than that we can do new things, things that could not have been imagined at a mere calculus level.

A.4 Why Proofs?

Can't we just do mathematics without proofs? Certainly there are many applications of mathematics carried on by people unable or unwilling to attempt proofs. But at the very heart and soul of mathematics is the proof, the careful argument that shows that a statement is true.

Compare this with the natural sciences. The advancement of knowledge in those subjects rests on the experiment. No scientist considers seriously whether students can skip over experimental work and just learn the result. At the core of all scientific discovery is the experimental method. It is too central to the discipline to be removed. It is the reason for the monumental success of the subject.

Mathematicians feel the same way about proofs. We can, with imagination and insight, make reasonable conjectures. But we can't be sure a conjecture is true until we prove it. The history of mathematics is filled with plausible (but false) statements made by mathematicians, even famous ones.

Proofs are an essential part of the subject. If you can master the art of reading and writing proofs, you enter properly into the subject. If not, you remain forever on the periphery looking in, a spectator able to learn some superficial facts about mathematics but unable to *do* mathematics.

What Is a Proof? Mathematicians are always prepared to define exactly what everything in their subject means. Certainly it is possible to define exactly what constitutes a proof. But that is best left to a course in logic.

For a course in analysis just understand that a proof is a short or long sequence of arguments meant to convince us that some statement is true. You will understand what a proof is after you have read some proofs and find that you do in fact follow the argument.

A proof is always intended for a specific audience. Proofs in this text are intended for readers who have some experience in calculus and good reasoning skills, but little experience in analysis. Proofs in more advanced texts would be much shorter and have less motivation. Proofs in professional research journals, intended for other professional mathematicians, can be terse and mysterious indeed.

Traditionally courses in analysis do not start with much of a discussion of proofs even though the students will be expected to produce proofs of their own, perhaps for the first time in their career. The best advice may be merely to jump in. Start studying the proofs in the text, the proofs given in lectures, the proofs attempted by your fellow students. Try to write them yourself. Read a proof, understand its main ideas, and then attempt to write the argument up in your own words.

How to Read a Proof While a proof may look like a short story, it is often much harder to read than one. Usually some of the computations will not seem clear and you will have to figure out how they were done. Some of the arguments (this is true and hence that is true) will not be immediate but will require some thinking. Many of the steps will appear completely strange, and it will seem that the proof is going off in a weird direction that is entirely mysterious. Basically you must unravel the proof. Find out what the main ideas are and the various steps of the proof.

One important piece of advice while reading a proof: Try to remember what it is that has to be proved. Before reading the proof decide what it is that must be proved exactly. Ask yourself, "What would I have to show to prove that?"

How to Write a Proof Practice! We learn to write proofs by writing proofs. Start by just copying nearly word for word a proof in a text that you find interesting. Vary the wording to use your own phrases. Write out the proof using more steps and more details than you found in the original. Try to find a different proof of the same statement and write out your new proof. Try to change the order of the argument if it is possible. If it is not possible you'll soon see why.

We all have learned the art of proof by imitation at first.

A.5 Indirect Proof

Many proofs in analysis are achieved as indirect proofs. This refers to a specific method.

The method argues as follows. I wish to prove a statement **P** is true. Either **P** is true or else **P** is false, not both. If I suppose **P** is false perhaps I can prove that then something entirely unbelievable must be true. Since that unbelievable something is not true, it follows that it cannot be the case that **P** is false. Therefore, **P** is true.

The method appears in the classical subject of rhetoric under the label *reductio ad absurdum* (I reduce to the absurd).

> Ladies and gentlemen my worthy opponent claims **P** but I claim the opposite, namely **Q** . Suppose his claim were valid. Then ... and then ... and that would mean .... But that's ridiculous so his claim is false and my claim must be true.

The pattern of all indirect proofs (also known as "proofs by contradiction") follows this structure: We wish to prove statement **P** is true. Suppose, in order to obtain a contradiction, that **P** is false. This would imply the following statements. (Statements follow.) But this is impossible. It follows that **P** is true as we were required to prove.

Here is a simple example. Suppose we wish to prove that

For all positive numbers x, the fraction $1/x$ is also positive.

An indirect proof would go like this.

Proof Suppose the statement is false. Then there is a positive number x and yet $1/x$ is not positive. This means

$$\frac{1}{x} \leq 0.$$

Since x is positive we can multiply both sides of the inequality by x and the inequality sign is preserved (this is a property of inequalities that we learned in elementary school and so we need not explain it). Thus

$$x \times \frac{1}{x} \leq x \times 0$$

or

$$1 \leq 0.$$

This is impossible. From this contradiction it follows that the statement must be true. ■

Indirect proofs are wonderfully useful and will be found throughout analysis. In some ways, however, they can be unsatisfying. After the statement "suppose not" the proof enters a fantasy world where all manipulations work toward producing a contradiction. None of the statements that you make along the way to this contradiction is necessarily of much interest because it is based on a false premise. In a direct proof, on the other hand, every statement you make is true and may be interesting on its own, not just as a tool to prove the theorem you are working on.

Also, indirect proofs reside inside a logical system where any statement not true is false and any statement not false is true. Some people have argued that we might wish to live in a mathematical world where, even though you have proved that something is not false, you have still not succeeded in proving that it is true.

Exercises

A.5.1 Show that $\sqrt{2}$ is irrational by giving an indirect proof.

A.5.2 Show that there are infinitely many prime numbers.

A.6 Contraposition

The most common mathematical assertions that we wish to prove can be written symbolically as

$$\mathbf{P} \Rightarrow \mathbf{Q},$$

which we read aloud as "Statement **P** implies statement **Q**." The real meaning attached to this is simply that if statement **P** is true, then statement **Q** is true.

A moment's reflection about the meaning shows that the two versions

If **P** is true, then **Q** must be true

and

If **Q** is false, then **P** must be false

are identical in meaning. These are called contrapositives of each other. Any statement

$$\mathbf{P} \Rightarrow \mathbf{Q}$$

has a contrapositive

$$\text{not } \mathbf{Q} \Rightarrow \text{not } \mathbf{P}.$$

To prove a statement it is sometimes better to prove the contrapositive.

Here is a simple example. Suppose that as calculus students we were required to prove that

> Suppose that $\int_0^1 f(x)\,dx \neq 0$. Then there must be a point $\xi \in [0,1]$ such that $f(\xi) \neq 0$.

At first sight it might seem hard to think of how we are going to find that point $\xi \in [0,1]$ from such little information. But let us instead prove the contrapositive. The contrapositive would say that if there is no point $\xi \in [0,1]$ such that $f(\xi) \neq 0$, then it would not be true that $\int_0^1 f(x)\,dx \neq 0$. Let's get rid of the double negatives. Restating this, now, we see that the contrapositive says that if $f(\xi) = 0$ for every $\xi \in [0,1]$, then $\int_0^1 f(x)\,dx = 0$. Even the C- students (none of whom are reading this book) would have now been able to proceed.

Exercises

A.6.1 Prove the following assertion by contraposition: If x is irrational, then $x+r$ is irrational for all rational numbers r.

A.7 Counterexamples

The polynomial

$$p(x) = x^2 + x + 17$$

has an interesting feature: It generates prime numbers for some time. For example, $p(1) = 19$, $p(2) = 23$, $p(3) = 29$, $p(4) = 37$ are all prime. More examples can be checked. After many more computations we would be tempted to make the claim

> For every integer $n = 1, 2, 3, \ldots$ the value $n^2 + n + 17$ is prime.

To prove that this is true (if indeed it is true) we would be required to show for any n, no matter what, that the value $n^2 + n + 17$ is prime. What would it take to disprove the statement, that is, to show that it is false?

All it would take is one instance where the statement fails. Only one! In fact there are many instances. It is enough to give one of them. Take $n = 17$ and observe that

$$17^2 + 17 + 17 = 17(17 + 1 + 1) = 17 \cdot 19,$$

which is certainly not prime. This one example is enough to prove that the statement is false. We refer to this as a *proof by counterexample.*

The Converse In analysis we shall often need to invent counterexamples. One frequent situation that occurs is the following. Suppose that we have just completed, successfully, the proof of a theorem expressed symbolically as

$$\mathbf{P} \Rightarrow \mathbf{Q}.$$

A natural question is whether the converse is also true. The *converse* is the opposite implication

$$\mathbf{Q} \Rightarrow \mathbf{P}.$$

Indeed once we have proved any theorem it is nearly routine to ask if the converse is true. Many converses are false, and a proof usually consists in looking for a counterexample.

For example, in calculus courses (and here too in analysis courses) it is shown that every differentiable function is continuous. Expressed as an implication it looks like this:

$$f \text{ is differentiable } \Rightarrow f \text{ is continuous}$$

and, hence, the converse statement is

$$f \text{ is continuous } \Rightarrow f \text{ is differentiable.}$$

Is the converse true? If it is then it, too, should be proved. If it is false, then a counterexample must be found. To prove it false we need supply just one function that is continuous and yet not differentiable. You may remember that the function $f(x) = |x|$ is continuous and yet not differentiable since at the point 0 there is no derivative.

Exercises

A.7.1 Disprove this statement: For any natural number n the equation

$$4x^2 + x - n = 0$$

has no rational root.

A.7.2 Every prime greater than two is odd. Is the converse true?

A.7.3 State both the converse and the contrapositive of the assertion "Every differentiable function is continuous." Is there a difference between them? Are they both true?

A.8 Induction

There is a convenient formula for the sum of the first n natural numbers:

$$1 + 2 + 3 + \ldots (n-1) + n = \frac{n(n+1)}{2}.$$

An easy direct proof of this would go as follows. Let S be the sum so that

$$S = 1 + 2 + 3 + \ldots (n-1) + n$$

or, expressed in the other order,

$$S = n + (n-1) + (n-2) + \cdots + 2 + 1.$$

Adding these two equations gives

$$2S = (n+1) + (n+1) + (n+1) + \cdots + (n+1) + (n+1)$$

and hence

$$2S = n(n+1)$$

or

$$S = \frac{n(n+1)}{2},$$

which is the formula we require.

Suppose instead that we had been unable to construct this proof. Lacking any better ideas we could just test it out for $n = 1$, $n = 2$, $n = 3$, ... for as long as we had the patience. Eventually we might run into a counterexample (proving the theorem is false) or have an inspiration as to why it is true. Indeed we find

$$1 = \frac{1(1+1)}{2}$$

$$1 + 2 = \frac{2(2+1)}{2}$$

$$1 + 2 + 3 = \frac{3(3+1)}{2}$$

and we could go on for some time. On a computer we could rapidly check for several million values, each time finding that the formula is valid.

Is this a proof? If a formula works this well for untold millions of values of n, how can we conceive that it is false? We would certainly have strong emotional reasons for believing the formula if we have checked it for this many different values, but this would not be a mathematical proof.

Instead, here is a proof that, at first sight, seems to be just a matter of checking many times. Suppose that the formula does fail for some value of n. Then there must be a first occurrence of the failure, say for some integer N. We know $N \neq 1$ (since we already checked that) and so the previous integer $N - 1$ does allow a valid formula. It is the next one N that fails. But if we can show that this never happens (i.e., there is never a situation with $N - 1$ valid and N invalid), then we will have proved our formula.

For example, if the formula

$$1 + 2 + 3 + \ldots M = \frac{M(M+1)}{2}$$

is valid, then

$$1 + 2 + 3 + \ldots M + (M+1) = \frac{M(M+1)}{2} + (M+1)$$

$$= \frac{M(M+1) + 2(M+1)}{2} = \frac{(M+1)(M+2)}{2},$$

which is indeed the correct formula for $n = M + 1$. Thus there never can be a situation in which the formula is correct at some stage and fails at the next stage. It follows that the formula is always true. This is a proof by induction.

This may be used to try to prove any statement about an integer n. Here are the steps:

Step 1 Verify the statement for $n = 1$.

Step 2 (The induction step) Show that whenever the statement is true for any positive integer m it is necessarily also true for the next integer $m + 1$.

Step 3 Claim that the formula holds for all n by the principle of induction.

In the exercises you are asked for induction proofs of various statements. You might try too to give direct (noninductive) proofs. Which method do you prefer?

Exercises

A.8.1 Prove by induction that for every $n = 1, 2, 3, \ldots,$

$$1^2 + 2^2 + 3^2 + \ldots n^2 = \frac{n(n+1)(2n+1)}{6}.$$

A.8.2 Compute for $n = 1$, 2, 3, 4 and 5 the value of

$$1 + 3 + 5 + \cdots + (2n - 1).$$

This should be enough values to suggest a correct formula. Verify it by induction.

A.8.3 Prove by induction for every $n = 1, 2, 3, \ldots$ that the number

$$7^n - 4^n$$

is divisible by 3.

A.8.4 Prove by induction that for every $n = 1, 2, 3, \ldots$

$$(1 + x)^n \geq 1 + nx$$

for any $x > 0$.

A.8.5 Prove by induction that for every $n = 1, 2, 3, \ldots$

$$1 + r + r^2 + \cdots + r^n = \frac{1 - r^{n+1}}{1 - r}$$

for any real number $r \neq 1$.

A.8.6 Prove by induction for every $n = 1, 2, 3, \ldots$ that

$$1^3 + 2^3 + 3^3 + \cdots + n^3 = (1 + 2 + 3 + \cdots + n)^2.$$

A.8.7 Prove by induction that for every $n = 1, 2, 3, \ldots$

$$\frac{d^n}{dx^n} e^{2x} = e^{2x + n \log 2}.$$

A.8.8 Show that the following two principles are equivalent (i.e., assuming the validity of either one of them, prove the other).

> **(Principle of Induction)** Let $S \subset \mathbb{N}$ such that $1 \in S$ and for all integers n if $n \in S$, then so also is $n + 1$. Then $S = \mathbb{N}$.

and

> **(Well Ordering of $\mathbb{N}$)** If $S \subset \mathbb{N}$ and $S \neq \emptyset$, then S has a first element (i.e., a minimal element).

well ordering of $\mathbb{N}$

A.8.9 Criticize the following "proof."

(Birds of a feather flock together) Any collection of n birds must be all of the same species.

Proof This is certainly true if $n = 1$. Suppose it is true for some value n. Take a collection of $n + 1$ birds. Remove one bird and keep him in your hand. The remaining birds are all of the same species. What about the one in your hand? Take a different one out and replace the one in your hand. Since he now is in a collection of n birds he must be the same species too. Thus all birds in the collection of $n + 1$ birds are of the same species. The statement is now proved by induction.

A.9 Quantifiers

In all of mathematics and certainly in all of analysis you will encounter two phrases used repeatedly:

> For all ... it is true that ...

and

> There exists a ... so that it is true that ...

For example, the formula

$$(x + 1)^2 = x^2 + 2x + 1$$

is true *for all* real numbers x. *There is a* real number x such that

$$x^2 + 2x + 1 = 0$$

(indeed $x = -1$).

It is extremely useful to have a symbolic way of writing this. It is universal for mathematicians of all languages to use the symbol $\forall$ to indicate "for all" or "for every" and to use $\exists$ to indicate "there exists." Originally these were chosen since it was easy enough for typesetters to turn the characters "A" and "E" around or upside down. These are called by the logicians *quantifiers* since they answer (vaguely) the question "how many?" For how many x is it true that

$$(x+1)^2 = x^2 + 2x + 1?$$

The answer is "For all real x." In symbols,

$$\forall x \in \mathbb{R},\ (x+1)^2 = x^2 + 2x + 1.$$

For how many x is it true that $x^2 + 2x + 1 = 0$? Not many, but there do exist numbers x for which this is true. In symbols,

$$\exists x \in \mathbb{R},\ x^2 + 2x + 1 = 0.$$

It is important to become familiar with statements involving one or more quantifiers whether symbolically expressed using $\forall$ and $\exists$ or merely using the phrases "for all" and "there exists." The exercises give some practice. You will certainly gain more familiarity by the time you are deeply into an analysis course in any case.

Negations of Quantified Statements Here is a tip that helps in forming negatives of assertions involving quantifiers. The two quantifiers $\forall$ and $\exists$ are complementary in a certain sense. The negation of the statement "All birds fly" would be (in conventional language) "Some bird does not fly." More formally, the negation of

For all birds b, b flies

would be

There exists a bird b, b does not fly.

In symbols let B be the set of all birds. Then the form here is

$\forall b \in B$ "statement about b" is true

and the negation of this is

$\exists b \in B$ "statement about b" is *not* true.

This allows a simple device for forming negatives. The negation of a statement with $\forall$ is a statement with $\exists$ replacing it, and the negation of a statement with $\exists$ is a statement with $\forall$ replacing it. For a complicated example, what is the negation of the statement

$\exists a \in A,\ \forall b \in B,\ \forall c \in C$
"statement about a, b and c" is true

even without assigning any meaning? It would be

$\forall a \in A,\ \exists b \in B,\ \exists c \in C,$
"statement about a, b and c" is not true.

Exercises

A.9.1 Let $\mathbb{R}$ be as usual the set of all real numbers. Express in words what these statements mean and determine whether they are true or not. Do not give proofs; just decide on the meaning and whether you think they are valid or not.

(a) $\forall x \in \mathbb{R}, x \geq 0$
(b) $\exists x \in \mathbb{R}, x \geq 0$
(c) $\forall x \in \mathbb{R}, x^2 \geq 0$
(d) $\forall x \in \mathbb{R}, \forall y \in \mathbb{R}, x + y = 1$
(e) $\forall x \in \mathbb{R}, \exists y \in \mathbb{R}, x + y = 1$
(f) $\exists x \in \mathbb{R}, \forall y \in \mathbb{R}, x + y = 1$
(g) $\exists x \in \mathbb{R}, \exists y \in \mathbb{R}, x + y = 1$

A.9.2 Form the negations of each of the statements in the preceding exercise. If you decided that a statement was true (false) before, you should naturally now agree that the negative is false (true).

A.9.3 Explain what must be done in order to prove an assertion of the following form:

(a) $\forall s \in S$ "statement about s" is true.
(b) $\exists s \in S$ "statement about s" is true.

Now explain what must be done in order to disprove such assertions.

A.9.4 In the preceding exercise suppose that $S = \emptyset$. Could either statement be true? Must either statement be true?

APPENDIX B: HINTS FOR SELECTED EXERCISES

1.3.3 Let F be the set of all numbers of the form $x + y\sqrt{2}$ where x, $y \in \mathbb{Q}$. Again to be sure that nine properties of a field hold it is enough to check, here, that $a + b$ and $a \cdot b$ are in F if both a and b are.

1.3.5 As a first step define what x^2 and $2x$ really mean. In fact, define 2. (It would be defined as $2 = 1+1$ since 1 and addition are defined in the field axioms.) Then multiply $(x + 1) \cdot (x + 1)$ using only the rules given here. Since your proof uses only the field axioms, it must be valid in any situation in which these axioms are true, not just for $\mathbb{R}$.

1.4.3 Suppose $a > 0$ and $b > 0$ and $a \neq b$. Establish that $\sqrt{a} \neq \sqrt{b}$. Establish that

$$(\sqrt{a} - \sqrt{b})^2 > 0.$$

Carry on. What have you proved? Now what if $a = b$?

1.6.4 You can use induction on the size of E, that is, prove for every positive integer n that if E has n elements, then

$$\sup E = \max E.$$

1.7.3 Suppose not, then the set

$$\{1/n : n = 1, 2, 3, \dots\}$$

has a positive lower bound, etc. You will have to use the existence of a greatest lower bound.

1.7.7 Not that easy to show. Rule out the possibilities $\alpha^2 < 2$ and $\alpha^2 > 2$ using the archimedean property to assist.

1.9.8 To find a number in (x, y), find a rational in $(x//\sqrt{2}, y//\sqrt{2})$. Conclude from this that the set of all (irrational) numbers of the form $\pm m\sqrt{2}/n$ is dense.

1.11.6 If $G = \{0\}$, then take $\alpha = 0$. If not, let $\alpha = \inf G \cap (0, \infty)$. Case 1: If $\alpha = 0$ show that G is dense. Case 2: If $\alpha > 0$ show that

$$G = \{n\alpha : n = 1, \pm 1, \pm 2, \pm 3, \dots\}.$$

For case 1 consider an interval (r, s) with $r < s$. We wish to find a member of G in that interval. To keep the argument simple just consider, for the moment, the situation in which $0 < r < s$. Choose $g \in G$ with $0 < g < s - r$. The set

$$M = \{n \in \mathbb{N} : ng \geq s\}$$

is nonempty (why?) and so there is a minimal element m in M (why?). Now check that $(m - 1)g$ is in G and inside the interval (r, s).

2.2.2 For the next term in the sequence some people might expect a 1. Most mathematicians would expect a 9.

2.2.3 Here is a formula that generates the first five terms of the sequence 0, 0, 0, 0, c, ...

$$f(n) = \frac{c(n-1)(n-2)(n-3)(n-4)}{4!}.$$

2.2.10 The formula is
$$f_n = \frac{1}{\sqrt{5}}\left\{\left(\frac{1+\sqrt{5}}{2}\right)^n - \left(\frac{1-\sqrt{5}}{2}\right)^n\right\}.$$
It can be verified by induction.

2.3.1 Find a function $f : (a,b) \to (0,1)$ one-to-one onto and consider the sequence $f(s_n)$, where $\{s_n\}$ is a sequence that is claimed to have all of (a,b) as its range.

2.3.4 We can consider that the elements of each of the sets S_i can be listed, say,
$$S_1 = \{x_{11}, x_{12}, x_{13}, \dots\}$$
$$S_2 = \{x_{21}, x_{22}, x_{23}, \dots\}$$
and so on. Now try to think of a way of listing all of these items, that is, making one big list that contains them all.

2.3.6 We need (i) every number has a decimal expansion; (ii) the decimal expansion is unique except in the case of expansions that terminate in a string of zeros or nines [e.g., $1/2 = 0.5000000\cdots = .49999999\dots$], thus if a and b are numbers such that in the nth decimal place one has a 5 (or a 6) and the other does not then either $a \neq b$, or perhaps one ends in a string of zeros and the other in a string of nines; and (iii) every string of 5's and 6's defines a real number with that decimal expansion.

2.3.10 Try to find a way of ranking the algebraic numbers in the same way that the rational numbers were ranked.

2.4.6 You will need the identity
$$1 + 2 + 3 + \cdots + n = n(n+1)/2.$$

2.4.7 You will need to find an identity for the sum of the squares similar to the identity $1 + 2 + 3 + \cdots + n = n(n+1)/2$.

2.5.6 To establish a correct converse, reword: If all $x_n > 0$ and $\frac{x_n}{x_n+1} \to 1$, then $x_n \to \infty$. Prove that this is true. The converse of the statement in the exercise is false (e.g., $x_n = 1/n$).

2.6.5 Use the same method as used in the proof of Theorem 2.11.

2.8.1 Give a counterexample. Perhaps find two sequences so that $s_n < 0 < t_n$ for all n and yet $\lim_{n\to\infty} s_n = \lim_{n\to\infty} t_n = 0$.

2.8.9 Take any number r strictly between 1 and that limit. Show that for some N, $s_{n+1} < rs_n$ if $n \geq N$. Deduce that
$$s_{N+2} < r^2 s_N$$
and
$$s_{N+3} < r^3 s_N.$$
Carry on.

2.8.10 Take any number r strictly between 1 and that limit. Show that for some N, $s_{n+1} > rs_n$ if $n \geq N$. Deduce that
$$s_{N+2} > r^2 s_N$$
and
$$s_{N+3} > r^3 s_N.$$
Carry on.

2.10.1 In terms of our theory of convergence this statement has no meaning since (as you should show) the sequence diverges. Even so, many great mathematicians, including Euler, would have accepted and used this formula. The fact that it is useful suggests that there are ways of interpreting such statements other than as convergence assertions.

2.11.13 If a sequence contains subsequences converging to every number in $(0,1)$ show that it also contains a subsequence converging to 0.

2.12.5 Consider the sequence
$$s_n = 1 + 1/2 + 1/3 + \dots 1/n.$$

2.12.10 Compare to
$$1 + 1 - \frac{1}{2} + \frac{1}{4} - \frac{1}{8} + \frac{1}{6} - \dots$$
which is the sum of a geometric progression.

2.13.15 Consider separately the cases where the sequence is bounded or not.

2.14.11 A sequence $\{x_n\}$ is periodic with period p if $x_{n+p} = x_n$ for all values of n and no smaller value of p will work. (Note that if $\{x_n\}$ is periodic with period p, then $x_n = x_{n+p} = x_{n+2p} = x_{n+3p} = \dots$.)

2.14.12 Clearly, no number larger than 1 or less than -1 could be such a limit. Show that in fact the interval $[-1,1]$ is the set of all such limit points. If $x \in [-1,1]$ there must be a number y so that $\cos y = x$ (why?). Now consider the set of numbers

$$G = \{n + 2m\pi : n, m \in \mathbb{Z}\}.$$

Using Exercise 1.11.6 or otherwise, show that this is dense. Hence there are pairs of integers n, m so that

$$|y - n + 2m\pi| < \varepsilon.$$

From this deduce that

$$|\cos y - \cos(n + 2m\pi)| < \varepsilon$$

and so $|x - \cos n| < \varepsilon$.

2.14.13 For (a) show that

$$|s_{n+1} - s_n| \leq \frac{1}{17}|s_n - s_{n-1}|$$

for all $n = 2, 3, 4, \ldots$. For (b) you will need to use the fact that the sum of geometric progressions is bounded, in fact that

$$1 + r + r^2 + \ldots r^n < (1-r)^{-1}$$

if $0 < r < 1$. Express for $m > n$,

$$|s_m - s_n| \leq |s_{n+1} - s_n|$$

$$+|s_{n+2} - s_{n+1}| + \cdots + |s_m - s_{m-1}|$$

and then use the contractive hypothesis. Note that

$$|s_4 - s_3| \leq r|s_3 - s_2| \leq r^2|s_2 - s_1|.$$

For (d) you might have to wait for the study of series in order to find an appropriate example of a convergent sequence that is not contractive.

2.14.15 This is from the 1947 Putnam Mathematical Competition.

2.14.16 This is from the 1949 Putnam Mathematical Competition.

2.14.17 This is from the 1950 Putnam Mathematical Competition.

2.14.18 This is from the 1953 Putnam Mathematical Competition.

2.14.19 Problem posed by A. Emerson in the *Amer. Math. Monthly*, **85** (1978), p. 496.

3.2.2 Define $\sum_{i \in I} a_i$ for I with zero or one elements. Suppose it is defined for I with n elements. Define it for I with $n+1$ elements and show well defined.

3.2.4 The answer is yes if I and J are disjoint. Otherwise the correct formula would be

$$\sum_{i \in I \cup J} a_i + \sum_{i \in I \cap J} a_i = \sum_{i \in I} a_i + \sum_{i \in J} a_i.$$

3.2.8 Try to interpret the "difference" $\Delta s_k = s_{k+1} - s_k = a_{k+1}$ as the analog of a derivative.

3.2.11 Use a telescoping sum method. Even if you cannot remember your trigonometric identities you can work backward to see which one is needed. Check the formula for values of θ with $\sin \theta/2 = 0$ and see that it can be interpreted by taking limits.

3.3.1 This is similar to the statement that convergent sequences have unique limits. Try to imitate that proof.

3.3.2 This is similar to the statement that convergent sequences are bounded. Try to imitate that proof.

3.3.3 This is similar to the statement that monotone, bounded sequences are convergent. Try to imitate that proof.

3.3.9 Compare with the sum

$$1 + \frac{1}{2} + \frac{1}{4} + \frac{1}{8} + \cdots = 2$$

given in the introduction to this chapter.

3.3.11 Here we are using, as elsewhere,

$$[X]^+ = \max\{X, 0\}$$

and

$$[X]^- = \max\{-X, 0\}$$

and note that

$$X = [X]^+ - [X]^- \quad \text{and} \quad |X| = [X]^+ + [X]^-.$$

3.3.12 Note that the index set is

$$I = \mathbb{N} \times \mathbb{N}.$$

Thus we can study unordered sums of double sequences $\{a_{ij}\}$ in the form

$$\sum_{(i,j) \in \mathbb{N} \times \mathbb{N}} a_{ij}.$$

3.4.10 Handle the case where each $a_k \geq 0$ separately from the general case.

3.4.15 Using properties of the log function, you can view this series as a telescoping one.

3.4.16 Consider that

$$\frac{1}{r-1} - \frac{1}{r+1} = \frac{2}{r^2-1}.$$

3.4.24 Establish the inequalities

$$\sum_{k=1}^{2^n-1} \frac{1}{k^p} \leq \sum_{k=1}^{\infty} \frac{2^{k-1}}{(2^{k-1})^p}$$

$$= \sum_{j=0}^{\infty} (2^{1-p})^j = \frac{2^{p-1}}{2^{p-1}-1}.$$

Conclude that the partial sums of the p-harmonic series for $p > 1$ are increasing and bounded. Explain now why the series must converge.

3.4.26 As a first step show that

$$\int_{2k\pi+\pi/4}^{2k\pi+3\pi/4} \frac{|\sin x|}{x}\,dx$$

$$\geq \frac{1}{\sqrt{2}} \int_{2k\pi+\pi/4}^{2k\pi+3\pi/4} \frac{1}{x}\,dx.$$

(Remember that in calculus an integral $\int_0^\infty$ is interpreted as $\lim_{X\to\infty} \int_0^X$.)

3.4.28 Establish that

$$\left| x - \sum_{i=1}^{n} \frac{k_i}{p^i} \right| \leq \frac{1}{p^n}.$$

3.5.5 Add up the terms containing p digits in the denominator. Note that our deletions leave only $8 \times 9^{p-1}$ of them. The total sum is bounded by

$$8(1/1 + 9/10 + 9^2/100 + \ldots) = 80.$$

3.5.8 Instead consider the series

$$\sum_{k=1}^{\infty} [a_k]^+ \text{ and } \sum_{k=1}^{\infty} [a_k]^-$$

where

$$[X]^+ = \max\{X, 0\}$$

and

$$[X]^- = \max\{-X, 0\}$$

and note that

$$X = [X]^+ - [X]^- \text{ and } |X| = [X]^+ + [X]^-.$$

3.5.15 Use the Cauchy-Schwarz inequality.

3.5.16 Use the Cauchy-Schwarz inequality.

3.6.3 The answer for (d) is $x < 1/e$.

3.6.5 Only one condition is sufficient to supply divergence. Give a proof for that one and counterexamples for the three others. Here is an idea that may help: Let $a_k = 0$ for all values of k except if $k = 2^m$ for some m in which case $a_k = 1/\sqrt{k}$. Note that $\limsup_{k\to\infty} \sqrt{k}a_k = 1$ in this case and that $\sum_{k=1}^{\infty} a_k$ will converge.

3.6.22 The exact value of γ, called *Euler's constant*, is not needed in the problem; it is approximately .5772156.

3.6.24 The integral test should occur to you while thinking of this problem. Start by checking that

$$\sum_{k=1}^{\infty} F'(k)$$

converges if and only if

$$\lim_{X\to\infty} F(X)$$

exists. Find similar statements for the other series.

3.7.13 Imitate the proof of the first part of Theorem 3.49 but arrange for the partial sums to go larger than α before inserting a term q_k. You must take the *first* opportunity to insert q_k when this occurs.

3.9.6 The name "Tauberian theorem" was coined by Hardy and Littlewood after a result of Alfred Tauber (1866–1942?). The date of his death is unknown; all that is certain is that he was sent by the Nazis to Theresienstadt concentration camp on June 28, 1942.

3.12.5 For (h) consider the series $\sum_{k=1}^{\infty} (s_{k+1} - s_k)/s_{k+1}$ where s_k is the sequence of partial sums of the series given.

3.12.6 For (b) use Abel's method and the computation in Exercise 3.2.11. Further treatment of some aspects of trigonometric series may be found in Section 10.8.

3.12.7 This is from the 1948 Putnam Mathematical Competition.

3.12.8 This is from the 1952 Putnam Mathematical Competition.

3.12.9 This is from the 1954 Putnam Mathematical Competition.

3.12.10 This is from the 1955 Putnam Mathematical Competition.

3.12.11 This is from the 1964 Putnam Mathematical Competition.

3.12.12 This is from the 1988 Putnam Mathematical Competition.

3.12.13 This is from the 1994 Putnam Mathematical Competition.

3.12.14 Problem posed by A. Torchinsky in *Amer. Math. Monthly*, **82** (1975), p. 936.

3.12.15 Problem posed by Jan Mycielski in *Amer. Math. Monthly*, **83** (1976), p. 284.

4.2.25 Let $\{q_n\}$ be an enumeration of the rationals. If x is isolated, then there is an open interval I_x containing x and containing no other point of the set. Pick the least integer n so that $q_n \in I_x$. This associates integers with the isolated points in a set.

4.3.1 Consider the set $\{1/n : n \in \mathbb{N}\}$.

4.3.23 The ternary expansion of a number $x \in [0,1]$ is given as

$$x = 0.a_1a_2a_3a_4 \cdots = \sum_{i=1}^{\infty} a_i/3^i$$

where the $a_i \in \{0,1,2\}$. (Thus this is merely the "base 3" version of a decimal expansion.) Observe that $1/3$ and $2/3$ can be expressed as $0.0222222\ldots$ and $0.200000\ldots$ in ternary. Observe that each number in the interval $(1/3, 2/3)$, that is the first stage component of G, must be written as $0.1a_2a_3a_4\ldots$ in ternary. How might this lead to a description of the points in G?

4.4.6 Consider the intersection of the family of *all* closed sets that contain the set E.

4.4.7 Consider the union of the family of *all* open sets that are contained in the set E.

4.5.1 Try this one: Define $f(x) = 0$ for x irrational and $f(x) = q$ if $x = p/q$ where p/q is a rational with p, q integers and with no common factors.

4.5.5 Take compact to mean closed and bounded. Show that a finite union or arbitrary intersection of compact sets is again compact. Check that an arbitrary union of compact sets need not be compact. Show that any closed subset of a compact set is compact. Show that any finite set is compact.

4.5.8 For a course in functions of one variable open covers can consist of intervals. In more general settings there may be nothing that corresponds to an "interval;" thus the more general covering by open sets is needed. Your task is just to look through the proof and spot where an "open interval" needs to be changed to an "open set."

4.5.9 Cousin's lemma offers the easiest proof, although any other compactness argument would work. Take the family of all intervals $[c,d]$ for which $f(c) < f(d)$ and check that the hypotheses of that lemma hold on any interval $[x,y]$.

4.5.18 Let $\mathcal{C} = \{V_\alpha : \alpha \in A\}$ be the open cover. Let N_1, N_2, ... be a listing of all open intervals with rational endpoints. For each $x \in E$ there is a $x \in V_\alpha$ and a k so that the interval N_k satisfies $x \in N_k \subset V_\alpha$. Call this choice $k(x)$. Thus

$$\mathcal{N} = \{N_{k(x)} : x \in E\}$$

is a countable open cover of E (but not the countable open cover that we want). But corresponding to each member of $\mathcal{N}$ is a member of $\mathcal{C}$ that contains it. Using that correspondence we construct the countable subcollection of $\mathcal{C}$ that forms a cover of E.

4.5.19 Lindelöff's theorem asserts that an open cover of any set of reals can be reduced to a countable subcover. The Heine-Borel theorem asserts that an open cover of any compact set of reals can be reduced to a finite subcover.

4.5.20 For (b)⇒(d) and for (c)⇒(d). Suppose that there is an open cover of A but no finite subcover. Step 1: You may assume that the open cover is just a sequence of open sets. (This is because of Exercise 4.5.18.) Step 2: You may assume that the open cover is an increasing sequence of open sets $G_1 \subset G_2 \subset G_3 \subset \ldots$ (just take the union of the first terms in the sequence you were given). Step 3: Now choose points x_i to be in $G_i \cap A$ but not in any previous G_j for $j < i$. Step 4: Now apply (b) [or (c)] to get a point $z \in A$ that is an accumulation point of the points x_i. This would have to be a point in some set G_N (since these cover A) but for $n > N$ none of the points x_n can belong to G_N.

4.6.5 This result may seem surprising at first since the Cantor set, at first sight, seems to contain only the endpoints of the open intervals that are removed at each stage, and that set of endpoints would be countable. (That view is mistaken; there are many more points.) Show that a point x in $[0, 1]$ belongs to the Cantor set if and only if it can be written as a ternary expansion $x = 0.c_1, c_2, c_3 \ldots$ (base 3) in such a way that only 0's and 2's occur. This is now a simple characterization of the Cantor set (in terms of string of 0's and 2's) and you should be able to come up with some argument as to why it is now uncountable.

4.6.9 You will need the Bolzano-Weierstrass theorem (Theorem 4.21). But this uncountable set E might be unbounded. How could we prove that an uncountable set would have to contain an infinite bounded subset? Consider

$$E = \bigcup_{n=1}^{\infty} E \cap [-n, n].$$

4.6.10 Select a rational number from each member of the family and use that to place them in an order.

4.7.11 For part (b) look ahead to part (c): Any such example must have A and B unbounded. For part (c) assume $\delta(A, B) = 0$. Then there must be points $x_n \in A$ and $y_n \in B$ with $|x_n - y_n| < 1/n$. As A is compact there is a convergent subsequence x_{n_k} converging to a point z in A. What is happening to y_{n_k}? (Be sure to use here the fact that B is closed.)

5.1.1 Model your answer after Example 5.2.

5.1.2 Consider the cases $a = 0$ and $a \neq 0$ separately. If it is easier for you, break into the three cases $a > 0$, $a < 0$, and $a = 0$.

5.1.3 Model your answer after Example 5.3.

5.1.4 Consider the cases $x_0 = 0$ and $x_0 \neq 0$ separately. Use the factoring trick in Example 5.3 and the device of restricting x to be close to x_0 by assuming that $|x - x_0| < 1$ at least.

5.1.8 Don't forget to exclude $x_0 < 0$ from your answer since it is not a point of accumulation of the domain of this function. Consider the cases $x_0 = 0$ and $x_0 > 0$ separately.

5.1.12 If $B \subset A$, then the existence of $\lim_{x \to x_0} g(x)$ can be deduced from the existence of $\lim_{x \to x_0} f(x)$. Can you find other conditions? If x_0 is a point of accumulation of $A \cap B$, then the equality of the two limits can be deduced, assuming that both exist.

5.1.16 Either find a single sequence

$$x_n \to 0$$

with $x_n \neq 0$ so that the limit

$$\lim_{n \to \infty} |x_n|/x_n$$

does not exist or else find two such sequences with different limits.

5.1.22 You could assume (i) that $L > 0$ or (ii) that $f(x) \geq 0$ for all x in its domain. Then convert to a statement about sequences.

5.1.28 At $x_0 \neq 0$ the two one-sided limits are equal. What are they? At $x_0 = 0$ they differ.

5.1.29 On one side the limit is zero and on the other the limit fails to exist. (Look ahead to Exercise 5.1.38, where you are asked to show that the limit is ∞ which means that the limit does not exist.) You may use the elementary inequality

$$0 < z < e^z$$

(which is valid for all $z > 0$) in your argument. Consider the sequences $1/n \to 0$ and $-1/n \to 0$.

5.1.30 Check the definition: There would be no distinction. The limit

$$\lim_{x \to 0-} \sqrt{x},$$

however, would be meaningless since 0 is not a point of accumulation of the domain of the square root function on the left.

5.1.34 Use the definitions in this section as a model. You will need a replacement for the "x_0 is a point of accumulation" of the domain condition. If you cannot think of anything better, then simply use the assumption that f is defined in some interval (a, ∞).

5.1.38 On one side at 0 the limit is zero and on the other the limit is ∞. See Exercise 5.1.29.

5.2.1 Model your proof after Theorem 2.8 for sequences.

5.2.3 If the theorem were false, then in every interval $(x_0 - 1/n, x_0 + 1/n)$ there would be a point x_n for which $|f(x_n)| > n$.

5.2.9 If x_0 is not a point of accumulation of

$$\text{dom}(f) \cap \text{dom}(g),$$

then the statement

$$\lim_{x \to x_0} f(x) + g(x) = L$$

does not have any meaning even though the two statements about $\lim_{x \to x_0} f(x)$ and $\lim_{x \to x_0} g(x)$ may have.

5.2.11 What exactly is the domain of the function $f(x)/g(x)$? Show that x_0 would be a point of accumulation of that domain provided that $g(x) \to C$ as $x \to x_0$ and $C \neq 0$.

5.2.28 It is enough to assume that $\lim_{x \to x_0} f(x)$ exists and to apply Theorem 5.25 with $F(x) = |x|$. Be sure to explain why this function F has the properties expressed in that theorem.

5.2.29 It is enough to assume that $\lim_{x \to x_0} f(x)$ exists and is positive and then apply Theorem 5.25 with $F(x) = \sqrt{x}$. Alternatively, assume that $f(x) \geq$ for all x in a neighborhood of x_0. Again be sure to explain why this function F has the properties expressed in that theorem.

5.2.32 Use the property of exponentials that $e^{a+b} = e^a e^b$ and the product rule for limits.

5.2.33 Use a trigonometric identity for $\sin(x - x_0 + x_0)$ and the sum and products rule for limits.

5.2.34 Take the function $H(x)$ of the text and consider instead $H(x) + x$.

5.2.36 This would be trivial if the sets A_i were disjoint. So it is the case where these are not disjoint that you need to address.

5.2.44 If x_0 is not in the Cantor set K, then it is in some open interval complementary to that set. Use that to prove the existence of the limit. If x_0 is in the Cantor set, then there must be sequences $x_n \to x_0$ and $y_n \to x_0$ with $x_n \in K$ and $y_n \not{i}nK$. Use that to prove the nonexistence of the limit.

5.3.5 Consider separately the cases $x_0 \in E$ and $x_0 \notin E$. Under what circumstances in the latter case would the lim sup be larger according to this revised definition?

5.4.15 One of the definitions treats isolated points in a special way. Note that each point in the domain of f is isolated.

5.4.17 You must arrange for $f(0)$ to be the limit of the sequence of values $f(2^{-n})$. No other condition is necessary.

5.4.19 At an isolated point x_0 of the domain the limit $\lim_{x \to x_0} f(x)$ has no meaning. But if x_0 is not an isolated point in the domain of f it must be a point of accumulation and then $\lim_{x \to x_0} f(x)$ is defined and it must be equal to $f(x_0)$.

5.4.20 For the converse consider the function $f(x) = \sqrt{x}$ on $[0, 1]$.

5.6.1 Let $a = \inf K$ and $b = \sup K$ and apply Cousin's lemma to the interval $[a, b]$ by taking the same collection nearly, namely $\mathcal{C}$ consist of all closed subintervals $[t, s]$ such that

$$|f(t') - f(s')| < \varepsilon/2$$

for all t', $s' \in K \cap [t, s]$. You will have to find a different choice of δ to make your argument work.

5.6.2 As usual in applications of Cousin's lemma, we should define first our collection of closed subintervals so as to have a desired property that can be extended to the whole interval $[a, b]$. Let $\varepsilon > 0$. Let $\mathcal{C}$ consist of all closed subintervals $[t, s]$ such that

$$|f(t') - f(s')| < \varepsilon/2$$

for all t', $s' \in [t, s]$. We check that $\mathcal{C}$ satisfies the hypotheses of Lemma 4.26.

For each $x \in [a, b]$ there exists $\delta(x) > 0$ such that if

$$t \in [a, b] \cap (x - \delta(x), x + \delta(x)),$$

then

$$|f(t) - f(x)| < \varepsilon/4.$$

It follows that if t' and s' are in the set

$$[a, b] \cap (x - \delta(x), x + \delta(x)),$$

then

$$|f(t') - f(s')| \leq |f(t') - f(x)| + |f(x) - f(s')| < \frac{\varepsilon}{4} + \frac{\varepsilon}{4} = \frac{\varepsilon}{2}.$$

Consequently, every interval $[t, s]$ inside

$$[a, b] \cap (x - \delta(x), x + \delta(x))$$

belongs to $\mathcal{C}$.

Thus Lemma 4.26 may be applied and there exists a partition

$$a = x_0 < x_1 < \cdots < x_n = b$$

such that if, for some $i = 1, \ldots, n$,

$$x_{i-1} \leq x, y \leq x_i,$$

then

$$|f(x) - f(y)| < \varepsilon/2.$$

Let

$$\delta = \min_{i=1,\ldots,n} |x_i - x_{i-1}|.$$

If $x < y$ and $|x - y| < \delta$, then either there exists i for which

$$x_{i-1} \leq x, y \leq x_i,$$

in which case

$$|f(x) - f(y) < \varepsilon/2,$$

or there exists i such that

$$x_{i-1} \leq x \leq x_i \leq y \leq x_{i+1},$$

in which case

$$|f(y) - f(x)| \leq |f(y) - f(x_i)| + |f(x_i) - f(x)| < \frac{\varepsilon}{2} + \frac{\varepsilon}{2} = \varepsilon.$$

Since this argument applies to any positive ε, we have proved that f is uniformly continuous on $[a, b]$.

5.6.5 If the set X has no points of accumulation this is possible. If the set X does have a point of accumulation, then it is possible to give an example of a function defined on X that is not uniformly continuous on X.

5.6.7 You need consider only two compact sets X_1, X_2. Since they are compact, there is a positive distance between them that you can use to help define your δ. For not closed consider $X_1 = (0, 1)$ and $X_2 = (1, 2)$ and define f appropriately. For not bounded use

$$X_1 = \{1, 2, 3, \ldots\}$$

and

$$X_2 = \{1, 2 + 1/2, 3 + 1/3, 4 + 1/4, \ldots$$

and define f appropriately.

5.6.9 For the converse consider the function $f(x) = \sqrt{x}$ on $[0, 1]$. By Theorem 5.47 we know that this function is uniformly continuous on $[0, 1]$.

5.6.10 Show that any function defined on a set X containing just one element is uniformly continuous. Then consider the sequence $X_i = \{x_i\}$, $i = 1, 2, \ldots, n$.

5.6.11 For the sequence of intervals you might choose $[1,2]$, $[2,3]$, $[3,4]$, ... (Why would you not be able to choose $[1/2,1]$, $[1/4,1/2]$, $[1/8,1/4]$, ...?)

5.6.12 This can be obtained merely by negating the formal statement that f is uniformly continuous on $[a,b]$.

5.6.13 Using the local continuity property, claim that there are open intervals I_x containing any point x so that

$$|f(y) - f(x)| < \varepsilon$$

for any $y \in I_x$. Now apply the Heine-Borel property to this open cover. Obtain uniform continuity from the finite subcover.

5.6.15 Let $\mathcal{C}$ be the collection of all closed intervals $I \subset [a,b]$ so that f is bounded on I. Use Cousin's lemma to find a partition of $[a,b]$ using intervals in $\mathcal{C}$.

5.6.16 Use an indirect proof. Show that if f is not bounded then there is a sequence $\{x_n\}$ of points in $[a,b]$ so that

$$|f(x_n)| > n$$

for all n. Now apply the Bolzano-Weierstrass property to obtain subsequences and get a contradiction.

5.6.17 Using the local continuity property, claim that there are open intervals I_x containing any point x so that

$$|f(y) - f(x)| < 1$$

for any $y \in I_x$. Now apply the Heine-Borel property to this open cover. Obtain boundedness of f from the finite subcover.

5.7.2 That is, prove that the image set

$$f(K) = \{f(x) : x \in K\}$$

is compact if K is compact and f is a continuous function defined at every point of K. Give a direct proof that uses the fact that a set is compact if and only if every sequence in the set has a subsequence convergent to a point in the set. Start with a sequence of points $\{y_n\}$ in $f(K)$, explain why there must be a sequence $\{x_n\}$ in K with $f(x_n) = y_n$ etc.

5.7.3 Let

$$M = \sup\{f(x) : a \leq x \leq b\}.$$

Explain why you can choose a sequence of points $\{x_n\}$ from $[a,b]$ so that

$$f(x_n) > M - 1/n.$$

Now apply the Bolzano-Weierstrass theorem and use the continuity of f.

5.7.5 If $f(x_0) = c > 0$, then there is an interval $[-N,N]$ so that $x_0 \in [N,N]$ and $|f(x)| < c/2$ for all $x > N$ and $x < -N$.

5.8.3 Suppose that the theorem is false and explain, then, why there should exist sequences $\{x_n\}$ and $\{y_n\}$ from $[a,b]$ so that $f(x_n) > c$, $f(y_n) < c$ and $|x_n - y_n| < 1/n$.

5.8.4 Suppose that the theorem is false and explain, then, why there should exist at each point $x \in [a,b]$ an open interval I_x centered at x so that either $f(t) > c$ for all $t \in I_x \cap [a,b]$ or else $f(t) < c$ for all $t \in I_x \cap [a,b]$.

5.8.5 You may take $c = 0$. Show that if $f(z) > 0$, then there is an interval $[z-\delta, z]$ on which f is positive. Show that if $f(z) < 0$, then there is an interval $[z, z+\delta]$ on which f is negative. Explain why each of these two cases is impossible.

5.8.6 The function must be onto. Hence there is a point x_1 with $f(x_1) = a$ and a point x_2 with $f(x_2) = b$. Now convince yourself that there is a point on the graph of the function that is also on the line $y = x$.

5.8.8 Condition (a) is the intermediate value property (IVP) according to Definition 5.27, while (b) can be interpreted as saying that connectedness is preserved by continuous functions. This latter interpretation requires a careful definition of connectedness in $\mathbb{R}$.

5.8.9 That is, prove that the image set $f([c,d])$ is a compact interval for any interval $[c,d]$ if f is a continuous function defined at every point of $[c,d]$. Apply Theorem 5.51 and Theorem 5.52.

5.9.13 You wish to show that (i) f is discontinuous at every point in C, indeed has a jump discontinuity at each such point; (ii) f is continuous at every point not in C; (iii) f is nondecreasing; (iv) f is increasing on any interval in which C is dense; and (v) f is constant on any interval containing no point of C.

The most direct and easiest proof that f is continuous at every point not in C would be to use "uniform convergence" but that is in a later chapter. Here you will have to use an ε-δ argument.

5.9.15 How large can the set of discontinuity points be?

5.9.16 The function f^{-1} is defined on the interval $J = [f(a), f(b)]$. Explain first why it exists (not all functions must have an inverse). Prove that it is increasing. Prove that it is continuous (using the fact that it is increasing).

5.10.1 The equation $f(x+y) = f(x)+f(y)$ is called a functional equation. You are told about this function only that it satisfies such a relationship and has a nice property at one point. Now you must show that this implies more. Show first that $f(0) = 0$ and that $f(x-y) = f(x) - f(y)$.

5.10.2 This continues Exercise 5.10.1. Show first that $f(r) = rf(1)$ for all $r = m/n$ rational. Then make use of the continuity of f that you had already established in the other exercise.

5.10.3 Show that either f is always zero or else $f(0) = 1$. Establish

$$f(x-y) = f(x)/f(y).$$

5.10.5 Consider the intersection

$$\overline{A} \cap \overline{B}.$$

5.10.15 You will need to use the fact that

$$\{x : \limsup_{x\to x_0^-} f(x) > \limsup_{x\to x_0^+} f(x)\}$$

is countable. See Exercise 5.10.8.

6.2.9 To make this true, assume that f is onto or else show that if E is dense then $f(E)$ is dense in the set (interval) $f(\mathbb{R})$.

6.3.1 If $q_1, q_2, q_3, \dots$ is an enumeration of the rationals, then each of the sets $\{q_i\}$, $i \in \mathbb{N}$, is nowhere dense, but

$$\bigcup_{i=1}^{n} \{q_i\} = \mathbb{Q}$$

is not nowhere dense. (Indeed it is dense.)

6.3.2 All of (a)–(e) and (h) are true. Find counterexamples for (f) and (g). The proofs that the others are true follow routinely from the definition.

6.4.1 Suppose that

$$A_n = \bigcup_{k=1}^{\infty} A_{nk}$$

with each of the sets A_{nk} nowhere dense. Then

$$\bigcup_{n=1}^{\infty}\bigcup_{k=1}^{\infty} A_{nk} = \bigcup_{n,k=1}^{\infty} A_{nk}$$

expresses that union as a first category set.

6.4.2 Let $\{B_n\}$ be a sequence of residual subsets of $\mathbb{R}$. Thus each of the sets B_n is the complement of a first category set A_n. For each n write

$$A_n = \bigcup_{k=1}^{\infty} A_{nk}$$

with each of the sets A_{nk} nowhere dense. Then

$$B_n = \mathbb{R} \setminus \bigcup_{k=1}^{\infty} A_{nk}.$$

Now use De Morgan's laws.

6.4.3 Suppose that X is residual, that is,

$$X = \mathbb{R} \setminus \bigcup_{n=1}^{\infty} Q_n$$

where each Q_n is nowhere dense. Show that for any interval $[a, b]$ there is a point in $X \cap [a, b]$ by constructing an appropriate descending sequence of closed subintervals of $[a, b]$.

6.4.4 Make sure your sets are dense but not both residual (e.g., $\mathbb{Q}$ and $\mathbb{R} \setminus \mathbb{Q}$).

6.4.5 This follows, with the correct interpretation, directly from the Baire category theorem.

6.4.7 Consider the sequence
$A_N = \{x \in [0,1] : |f_n(x)| \leq 1, \text{ all } n \geq N\}$.
Check that

$$\bigcup_{N=1}^{\infty} A_N = [0,1].$$

6.5.7 It is clear that there must be many irrational numbers in the Cantor ternary set, since that set is uncountable and the rationals are countable. Your job is to find just one.

6.5.10 Consider $G = (0,1) \setminus C$ where C is the Cantor ternary set.

6.6.7 Often to prove a set identity such as this the best way is to start with a point x that belongs to the set on the right and then show that point must be in the set on the left. After that is successful start with a point x that belongs to the set on the left. For example, if $f(x) > \alpha$, then

$$f(x) \geq \alpha + 1/m$$

for some integer m. But

$$f_n(x) \to f(x)$$

and so there must be an integer R so that $f_n(x) > \alpha + 1/m$ for all $n \geq R$, etc.

This exercise shows how unions and intersections of sequences of open and closed sets might arise in analysis. Note that the sets

$$\{x : f_n(x) \geq \alpha + 1/m\}$$

would be closed if the functions f_n are continuous. Thus it would follow that the set

$$\{x : f(x) > \alpha\}$$

must be of type $\mathcal{F}_\sigma$. This says something interesting about a function f that is the limit of a sequence of continuous functions $\{f_n\}$.

6.7.3 You need to recall Theorem 5.59, which asserts that monotone functions have left- and right-hand limits.

6.9.2 This is from the 1964 Putnam Mathematical Competition.

7.2.1 Write $x = x_0 + h$.

7.2.6 Write

$$f(x+h) - f(x-h)$$
$$= [f(x+h) - f(x)] + [f(x) - f(x-h)].$$

7.2.7 Use

$$1 - \cos x = 2\sin^2 x/2.$$

When you take the square root be sure to use the absolute value.

7.2.12 Just use the definition of the derivative. Give a counterexample with $f(0) = 0$ and $f'(0) > 0$ but so that f is not increasing in any interval containing 0.

7.2.13 Even for polynomials, $p(x)$ increasing does not imply that $p'(x) > 0$ for all x. For example, take $p(x) = x^3$. That has only one point where the derivative is not positive. Can you do any better?

7.2.14 Actually the assumptions are different. Here we assume $f'(x_0)$ does exist, whereas in the trapping principle we had to assume more inequalities to deduce that it exists.

7.2.15 Review Exercise 5.10.3 first.

7.2.16 Advanced (very advanced) methods would allow you to find a function continuous on $[0,1]$ that is differentiable at *no* point of that interval. For the purpose of this exercise just try to find one that is not differentiable at 1/2, 1/3, 1/4, ... (Novices constructing examples often feel they need to give a simple formula for functions. Here, for example, you can define the function on $[1/2, 1]$, then on $[1/4, 1/2]$, then on $[1/8, 1/4]$, and so on ... and then finally at 0.)

7.2.18 Find two examples of functions, one continuous and one discontinuous at 0, with an infinite derivative there.

7.2.19 Imitate the proof of Theorem 7.6. Find a counterexample to the question.

7.3.5 Use Theorem 7.7 (the product rule) and for the induction step consider

$$\frac{d}{dx}x^n = \frac{d}{dx}[x][x^{n-1}].$$

7.3.10 This formula is known as Leibniz's rule (which should indicate its age since Leibniz, one of the founders of the calculus, was born in 1646). It extends both Exercises 7.3.8 and 7.3.9. The formula is

$$(fg)^{(n)}(x_0)$$
$$= \sum_{k=0}^{n} \frac{n!}{k!(n-k)!} f^{(k)}(x_0) g^{(n-k)}(x_0).$$

7.3.11 Consider a sequence $x_n \to x_0$ with $x_n \neq x_0$ and $f(x_n) = f(x_0)$.

7.3.12 Let

$$f(x) = x^2 \sin x^{-1}$$

($f(0) = 0$) and take $x_0 = 0$. Utilize the fact that 0 is a limit point of the set $\{x : f(x) = 0\}$.

7.3.17 If $I(x)$ is the inverse function then $I(\sin x) = x$. The chain rule gives derivative as $I'(\sin x) = 1/\cos x$. This needs some work. Use

$$\cos x = \sqrt{1 - \sin^2 x}$$

and obtain

$$I'(\sin x) = \frac{1}{\sqrt{1 - \sin^2 x}}.$$

Now replace the $\sin x$ by some other variable. Caution: While doing this exercise make sure that you know how the arcsin function $\sin^{-1} x$ is actually defined. It is not the inverse of the function $\sin x$ since that function has no inverse.

7.3.19 Draw a good picture. The graph of $y = g(x)$ is the reflection in the line $y = x$ of the graph of $y = f(x)$. What is the slope of the reflected tangent line?

7.3.21 Use the idea in the example. If $f(x) = x^{1/m}$, then $[f(x)]^m = x$ and use the chain rule. If

$$F(x) = x^{n/m},$$

then

$$[F(x)]^m = x^n$$

and use the chain rule.

7.3.22 Once you know that

$$\frac{d}{dx}e^x = e^x$$

you can determine that

$$\frac{d}{dx}\ln x = 1/x$$

using inverse functions. Then consider

$$\frac{d}{dx}x^p = e^{(\ln p)x}.$$

7.3.23 The formula you should obtain is

$$a_k = \frac{p^{(k)}(0)}{k!}$$

for $k = 0, 1, 2, \ldots$.

7.3.24 If you succeed, then you have proved the binomial theorem using derivatives. Of course, you need to compute $p(0)$, $p'(0)$, $p''(0)$, $p'''(0)$, ... to do this.

7.5.6 Consider sets of the form

$$A_n = \{f(t) < f(x) :$$
$$\text{for } t \in (x - 1/n, x) \cup (x, x + 1/n)\},$$

and observe that

$$\bigcup_{n=1}^{\infty} A_n$$

is the set in question.

7.5.7 Modify the hint in Exercise 7.5.6.

7.6.3 Use Rolle's theorem to show that if x_1 and x_2 are distinct solutions of $p(x) = 0$, then between them is a solution of $p'(x) = 0$.

7.6.5 Use Rolle's theorem twice. See Exercise 7.6.7 for another variant on the same theme.

7.6.6 Since f is continuous we already know (look it up) that f maps $[a, b]$ to some closed bounded interval $[c, d]$. Use Rolle's theorem to show that there cannot be two values in $[a, b]$ mapping to the same point.

7.6.7 cf. Exercise 7.6.5.

7.6.8 First show directly from the definition that the Lipshitz condition will imply a bounded derivative. Then use the mean value theorem to get the converse, that is, apply the mean value theorem to f on the interval $[x, y]$ for any $a \leq x < y \leq b$.

7.6.9 Note that an increasing function f would allow only positive numbers in S.

7.6.12 Apply the mean value theorem to f on the interval $[x, x+a]$ to obtain a point ξ in $[x, x+a]$ with

$$f(x+a) - f(x) = af'(\xi).$$

7.6.13 Use the mean value theorem to compute

$$\lim_{x \to a+} \frac{f(x) - f(a)}{x - a}.$$

7.6.14 This is just a variant on Exercise 7.6.13. Show that under these assumptions f' is continuous at x_0.

7.6.15 Use the mean value theorem to relate

$$\sum_{i=1}^{\infty}(f(i+1) - f(i))$$

to

$$\sum_{i=1}^{\infty} f'(i).$$

Note that f is increasing and treat the former series as a telescoping series.

7.6.16 The proof of the mean value theorem was obtained by applying Rolle's theorem to the function

$$g(x) = f(x) - f(a) - \frac{f(b) - f(a)}{b - a}(x - a).$$

For this mean value theorem apply Rolle's theorem twice to a function of the form $h(x) = f(x) - f(a) - f'(a)(x-a) - \alpha(x-a)^2$ for an appropriate number α.

7.6.18 Write

$$f(x+h) + f(x-h) - 2f(x) =$$

$$[f(x+h) - f(x)] + [f(x-h) - f(x)]$$

and apply the mean value theorem to each term.

7.6.21 Let $\phi(x)$ be

$$\begin{vmatrix} f(a) & g(a) & h(a) \\ f(b) & g(b) & h(b) \\ f(x) & g(x) & h(x) \end{vmatrix}$$

and imitate the proof of Theorem 7.21.

7.7.1 Interpret as a monotonicity statement about the function

$$f(x) = (1 - x)e^x.$$

7.7.3 We do not assume differentiability at b. For example, this would apply to the function $f(x) = |x|$, with $b = 0$.

7.7.5 Interpret this as a monotonicity property for the function $F(x) = f(x)/x$. We need to show that F' is positive. Show that this is true if $f'(x) > f(x)/x$ for all x. But how can we show this? Apply the mean value theorem to f on the interval $[0, x]$ (and don't forget to use the hypothesis that f' is an increasing function).

7.7.6 If not, there is an interval $[a, b]$ with $f(a) = f(b) = 0$ and neither f nor g vanish on (a, b). Show that $f(x)/g(x)$ is monotone (increasing or decreasing) on $[a, b]$.

7.8.7 Let $\varepsilon > 0$ and consider $f(x) + \varepsilon x$.

7.8.9 Figure out a way to express $\mathbb{R}$ as a countable union of disjoint dense sets A_n and then let $f(x) = n$ for all $x \in A_n$. For an example subtract an appropriate linear function F from f such that $f - F$ is not an increasing function, and apply Theorem 7.30.

7.8.10 In connection with this exercise we should make this remark. If $A = \{a_k\}$ is any countable set, then the function defined by the series

$$\sum_{k=0}^{\infty} \frac{-|x - a_k|}{2^k}$$

has $D^+f(x) < D_-f(x)$ for all $x \in A$. This can be verified using the results in Chapter 9 on uniform convergence.

7.9.1 For the third part use the function $F(x) = x^2 \sin x^{-1}$, $F(0) = 0$ to show that there exists a differentiable function f such that $f'(x) = \cos x^{-1}$, $f(0) = 0$). Consider $g(x) = f(x) - x^3$ on an appropriate interval.

7.9.3 If either FG' or GF' were a derivative, so would the other be since

$$(FG)' = FG' + GF'.$$

In that case $FG' - GF'$ is also a derivative. But now show that this is impossible [because of (c)].

7.9.4 Use $fg' = (fg)' - f'g$. You need to know the fundamental theorem of calculus to continue.

7.9.5 If f' is continuous, then it is easy to check that E_α is closed. In the opposite direction suppose that every E_α is closed and f' is not continuous. Then show that there must be a number β and a sequence of points $\{x_n\}$ converging to a point z and yet $f'(x_n) \geq \beta$ and $f'(z) < \beta$. Apply the Darboux property of the derivative to show that this cannot happen if E_β is closed. Deduce that f' is continuous.

7.10.3 If f is convex on an interval I and g is convex *and also nondecreasing* on the interval $f(I)$, then you should be able to prove that $g \circ f$ is also convex. Show also that if the monotonicity assumption on g is dropped this might not be true.

7.10.5 Show that at every point of continuity of f'_+ the function is differentiable. How many discontinuities does the (nondecreasing) function f'_+ have?

7.10.10 Give an example of a convex function on the interval $(0, 1)$ that is not bounded above; that answers the first question. For the second question use Exercise 7.10.4 to show that f must be bounded below.

7.10.13 The methods of Chapter 9 would help here. There we learn in general how to check for the differentiability of functions defined by series. For now just use the definitions and compute carefully.

7.10.14 For (d) let

$$f(x) = \begin{cases} e^{-1/x^2}(\sin 1/x)^2, & \text{for } x > 0 \\ 0, & \text{for } x = 0, \\ -e^{-1/x^2}(\sin 1/x)^2, & \text{for } x < 0 \end{cases}.$$

The three definitions in the exercise are not equivalent even for infinitely differentiable functions. They are, however, equivalent for *analytic* functions; that is, functions represented by power series (a topic we cover in Chapter 10). Since the scope of elementary calculus is more or less limited to functions that are analytic on the intervals on which the functions are concave up or down, we might argue that on that level, the definition to take is the one that is simplest to develop. We should mention, however, that there are differentiable functions that are not concave-up or concave-down on any interval!

7.10.15 Order the terms so that

$$x_1 \leq x_2 \leq \cdots \leq x_n.$$

And write

$$p = \sum_{k=1}^{n} \alpha_k x_k.$$

Choose a number M between $f'_-(p)$ and $f'_+(p)$. Check that

$$x_1 \leq p \leq x_n.$$

Check that

$$f(x_k) \geq M(x_k - p) + f(p)$$

for $k = 1, 2, \ldots, n$. Now use these inequalities to obtain Jensen's inequality.

7.11.1 Use L'Hôpital's rule to find that $f(0)$ should be $\ln(3/2)$. Use the definition of the derivative and L'Hôpital's rule twice to compute

$$f'(0) = [(\ln 3)^2 - (\ln 2)^2]/2.$$

Exercise 7.6.13 shows that the technique in (c) part does in fact compute the derivative provided only that you can show that this limit exists.

7.11.2 Treat the cases $A > 0$ and $A < 0$ separately.

7.11.10 We must have $\lim_{x\to\infty} f'(x) = 0$ in this case. (Why?)

7.13.3 Consider the function
$H(x) = p(x) + p'(x) + p''(x) + \cdots + p^{(n)}(x)$
and note, in particular, the relation between H, H' and p.

7.13.7 Such functions are called *midpoint convex*. By the definition of convexity we need to show that if $x_1, x_2 \in I$ and $\alpha \in [0,1]$, then the inequality
$f(\alpha x_1 + (1-\alpha)x_2) \leq \alpha f(x_1) + (1-\alpha)f(x_2)$
is satisfied. Use the midpoint convexity condition to show that this is true whenever α is a fraction of the form $p/2^q$ for integers p and p. Now use continuity to show that it holds for all $\alpha \in [0,1]$. Without continuity this argument fails and, indeed, there exist discontinuous midpoint convex functions that fail to be convex. [For an extensive account of what is known about such conditions, see B. S. Thomson, *Symmetric Properties of Real Functions*, Marcel Dekker, (New York, 1994).]

7.13.8 If g does not vanish on (x_1, x_2), then Rolle's theorem applied to the quotient f/g provides a contradiction. Incidentally, Josef de Wronski (1778–1853), whose name was attached firmly to this concept in 1882 in a multivolume *History of Determinants*, was a rather curious figure whom you are unlikely to encounter in any other context. One biographer writes about him:

> For many years Wronski's work was dismissed as rubbish. However, a closer examination of the work in more recent times shows that, although some is wrong and he has an incredibly high opinion of himself and his ideas, there is also some mathematical insights of great depth and brilliance hidden within the papers.

7.13.9 Consider the function

$$H(x) = f(x) + cx^2 + ax + b$$

for $c > 0$ and various choices of lines $y = ax + b$ and make use of Exercise 7.10.14.

7.13.12 This is from the 1939 Putnam Mathematical Competition.

7.13.13 This is from the 1946 Putnam Mathematical Competition.

7.13.14 This is from the 1958 Putnam Mathematical Competition.

7.13.15 This is from the 1962 Putnam Mathematical Competition.

7.13.16 This is from the 1992 Putnam Mathematical Competition.

7.13.17 This is from the 1998 Putnam Mathematical Competition.

8.2.1 You will need to find a formula for

$$\sum_{k=1}^{n} k^3.$$

8.2.9 Be sure, first, to check that these associated points are legitimate. Show that each of these sums has the same value (think of telescoping sums!). What, then, would be the limit of the Riemann sums?

8.3.10 This is called the *Cauchy-Schwarz inequality* and is the analog for integrals of that inequality in Exercise 3.5.13. It can be proved the same way and does not involve any deep properties of integrals.

8.4.5 It would converge for all continuous functions.

8.5.2 Define

$$\int_{-\infty}^{\infty} f(x)\,dx$$

to be the sum of

$$\int_{-\infty}^{a} f(x)\,dx$$

and

$$\int_{a}^{\infty} f(x)\,dx.$$

Be sure to prove that this definition would not depend on the choice of a.

8.5.10 Compare with Exercise 3.4.26. Note, too, that it may seem to require special handling at the left-hand endpoint but it does not.

8.6.1 Note that this function is discontinuous everywhere and that

$$\omega f([c,d]) = 1$$

for every interval $[c,d]$.

8.6.3 The answer is no. It would be true if $|f| > c > 0$ everywhere. Equivalently, it is true if $1/f$ is bounded.

8.6.4 Step functions were defined in Section 5.2.6. If you sketch a picture of what the approximating sums look like, the step functions needed should be apparent.

8.6.6 The fact that the oscillation of a function f is smaller than η at each point of an interval $[c,d]$ is a local condition. Express it by using a $\delta(x)$ at each point. Now use a compactness argument (e.g., Heine-Borel) to get a uniform size that works.

8.7.1 Make ϕ' integrable and f continuous at each point $\phi(t)$ for $t \in [a,b]$.

8.7.9 For (a): What if F is discontinuous? For (b): Consider the Cantor function (Section 6.5.3). For (d): This is not easy! We will discuss this in Section 9.7.

8.9.1 The error is that the choice of δ depends on the point ξ considered and so is not a constant. This is an error you have doubtless made in other contexts: A local condition that holds for *each* point x is misinterpreted as holding uniformly for *all* x.

8.9.2 Consider the integral of a sum $f+g$, the integral formula $\int_a^b + \int_b^c = \int_a^c$, etc.x

8.10.1 This exercise develops the theory of the *Darboux integral*, which is equivalent to Riemann's integral but defined using infs and sups of "Darboux sums" rather than limits of Riemann sums. In preparation Exercise 8.2.17 should be consulted.

8.10.2 This is from the 1947 Putnam Mathematical Competition.

9.2.7 The statements that are defined by inequalities (e.g., bounded, convex) or by equalities (e.g., constant, linear) will not lead to an interchange of two limit operations, and you should expect that they are likely true.

9.2.8 As the footnote to the exercise explains, this was Luzin's unfortunate attempt as a young student to understand limits. The professor began by saying "What you say is nonsense." He gave him the example of the double sequence $m/(m+n)$ where the limits as $m \to \infty$ and $n \to \infty$ cannot be interchanged and continued by insisting that "permuting two passages to the limit *must not be done.*" He concluded with "Give it some thought; you won't get it immediately."

9.3.15 Use the Cauchy criterion for convergence of sequences of real numbers to obtain a candidate for the limit function f. Note that if $\{f_n\}$ is uniformly Cauchy on a set D, then for each $x \in D$, the sequence of real numbers $\{f_n(x)\}$ is a Cauchy sequence and hence convergent.

9.3.16

$$S_n(x) = \sum_{k=0}^{n} x^k = \frac{1-x^n}{1-x}.$$

9.4.10 For part (b) consider

$$F_n(x) = f_n(x) + Mx$$

and apply Exercise 9.4.6.

9.7.15 Suppose h' were integrable. Explain why

$$h(x) - h(a) = \int_a^x h'(t)\,dt$$

for all $x \in [a,b]$. Now by considering an appropriate Riemann sum, since $h' = 0$ on a dense set, we would have

$$h(x) - h(a) = 0$$

for all $x \in [a,b]$. That should be a contradiction.

9.8.1 What properties would F' have to have if the convergence were uniform?

9.9.2 You will need to use the Baire category theorem for the second part of this.

10.2.2 This follows immediately from the inequalities

$$\liminf_k \left|\frac{a_{k+1}}{a_k}\right| \leq \liminf \sqrt[k]{|a_k|}$$

$$\leq \limsup_k \sqrt[k]{|a_k|} \leq \limsup_k \left|\frac{a_{k+1}}{a_k}\right|$$

that we obtained in Exercise 2.13.16.

10.3.3 Write out the Cauchy criterion for uniform convergence on $(-r, r)$ and deduce that the Cauchy criterion for uniform convergence on $[-r, r]$ must also hold.

10.4.4

$$\int_0^1 \frac{1-e^{-sx}}{s}\,ds = \sum_{k=1}^{\infty}(-1)^{k-1}\frac{1}{k(k!)}x^k.$$

10.5.4 It is clear that $f^{(k)}$ exists for all $x \neq 0$. For $x = 0$ verify the following assertions:

1. $f^{(k)}(0)$ is of the form $R(x)e^{-1/x^2}$ for $x \neq 0$, where R is a rational function.
2. Show that
$$\lim_{x\to 0}\frac{1}{x^n}e^{-1/x^2} = 0$$
for all $n = 1, 2, \ldots$.
3. Conclude that
$$\lim_{x\to 0} f^{(k)}(x) = 0$$
for all $k = 1, 2, \ldots$.
4. Conclude that
$$f^{(k)}(0) = 0$$
for all k.

10.6.2 Just use Theorem 10.32.

10.7.1 Just use Theorem 10.33.

10.8.5 Easy, really. Just substitute

$$u = x + t$$

in the integral

$$\int_0^{\pi} f(x+t)D_n(t)\,dt$$

and expand the terms $\cos(ku - kx)$ using standard trigonometric identities.

10.8.12 First obtain a polynomial q so that

$$|f(x) - q(x)| < \varepsilon/2.$$

Then find a polynomial p with rational coefficients so that

$$|p(x) - q(x)| < \varepsilon/2.$$

10.8.13 Try $f(x) = e^x$.

10.8.14 Try $f(x) = 1/x$.

10.8.15 Show that f must be identically equal to zero. Use Theorem 10.37.

10.8.16 Define

$$G(t) = f(t/\pi)$$

for $t \in [0, \pi]$ and extend to $[-\pi, 0]$ by

$$G(-t) = -G(t).$$

Consider the Fourier series of G and show that it contains only sin terms (no cosine terms). Show that f must be identically equal to zero. Use Theorem 10.36.

11.1.5 See Definition 5.27 for the definition of the intermediate value property. The exercises in that section also provide a clue to the answer of this question.

11.1.6 For part (b) see Exercise 10.8.3.

11.2.7 Your answers to (b) and (c) illustrate a special feature of the euclidean norm. This norm relates to the dot product in certain useful ways. Our use of norms in this and the next chapter does not depend on this relationship, so we shall find it convenient to use $\|\cdot\|_1$ and $\|\cdot\|_\infty$ at certain times. See Chapter 13, Examples 13.7, 13.8, and 13.9 for more on this subject in a setting that involves infinite sequences in place of n-tuples. (A normed linear space is any vector space with a norm satisfying the conditions of Theorem 11.6. When that norm comes from a dot product via $\|\mathbf{x}\|^2 = \mathbf{x}\cdot\mathbf{x}$, the space is called a *Hilbert space* and enjoys many special properties.)

11.2.8 In fact, the only norm on $\mathbb{R}^n$ for which this identity is valid is the euclidean norm.

11.3.7 For a notion of connectedness that is different from this polygonal arc definition and that applies to sets that need not be open, see the discussion in Section 11.10.

11.3.8 The point of this problem is that since the open sets are exactly the same, so too will be all the other concepts whose definitions can be given entirely in terms of open sets. The same will be true when in later sections we consider convergence of sequences or limits, continuity, and differentiability of functions.

11.6.5 Use the ideas in the proof of Lemma 11.34.

11.7.11 Use Exercise 11.7.10.

11.7.12 Show that

$$\lim_{t\to 0} f(t,t^2) = 1.$$

11.7.13 For (a), look at the outline in conjunction with the example of Exercise 11.7.12. The hypothesis in (b) should involve an appropriate form of "uniform continuity in one of the variables with respect to the other variable." For (c) Example 11.32 of Section 11.6 and Exercise 11.7.12, illustrate that this does not imply that the double limit exists.

11.8.2 Section 13.12.3 establishes the equivalence of the Bolzano-Weierstrass and Heine-Borel properties in the more general setting of metric spaces.

11.9.4 For the counterexample you might use

$$\mathbf{f}(x) = (\cos x, \sin x)$$

for $0 \le x < 2\pi$.

11.10.6 Let

$$A_k = \{(0,0)\}\cup\{(1,0)\}\cup\{(x,y) : 0 < y < \frac{1}{k}\}.$$

12.2.6 Your definition should generalize the definition we gave for the case $n = 2$. The unit vector (u_1, u_2) will now have to be replaced by a unit vector $(u_1, u_2, u_3, \dots, u_n)$.

12.3.3 Let

$$G(u,v,w) = \int_u^v f(x,w)\,dx.$$

Use Leibniz's rule, the fundamental theorem of calculus and an appropriate chain rule to obtain

$$F'(y) = \int_{u(y)}^{v(y)} \frac{\partial f}{\partial y}(x,y)\,dx$$

$$+f(u,y)\frac{dv}{dx} - f(u,y)\frac{du}{dx}.$$

12.4.3 See Exercises 11.2.6 and 11.2.7 for further discussion of these equivalent norms for $\mathbb{R}^n$.

12.4.5 This means that f is differentiable at $\mathbf{x}$ with respect to to this definition if and only if it is differentiable at $\mathbf{x}$ with respect to Definition 12.16. Recall L is called *linear* if

$$L(\alpha\mathbf{x} + \beta\mathbf{y}) = \alpha L(\mathbf{x}) + \beta L(\mathbf{y})$$

for all α, $\beta \in \mathbb{R}$ and all $\mathbf{x}$, $\mathbf{y} \in \mathbb{R}^n$ and any linear function can be represented in the form

$$L(\mathbf{x}) = \sum_{i=1}^n a_i x_i$$

where

$$\mathbf{x} = (x_1, \dots, x_n).$$

Show that L must be as given in Definition 12.16, i.e., $a_i = f_i(\mathbf{x})$.

12.4.9 There are only two directions, $u_1 = 1$ and $u_1 = -1$. What are the two derivatives?

12.4.15 The picture illustrates a function that is mostly zero and has every directional derivative at $(0,0)$ zero but is not differentiable there.

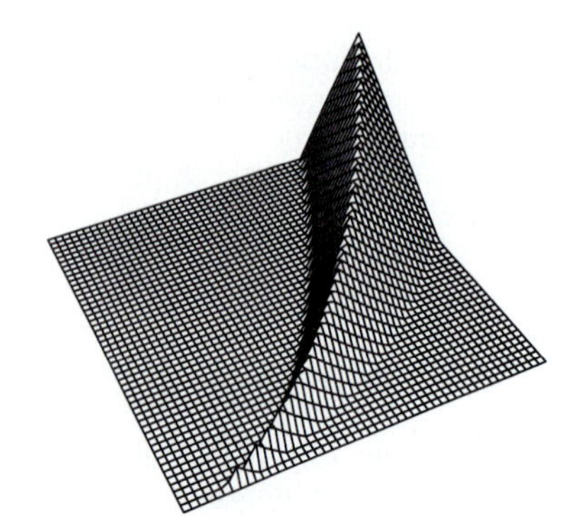

Construct f to be continuous and such that

1. $f(x,y) = 0$ unless $x > 0$ and
$$x^2 < y < 3x^2,$$
2. for each $x > 0$, $f(x, 2x^2) = x$, and
3. $0 \le f(x,y) \le x$ for all (x,y) with $x > 0$.

Then all directional derivatives vanish at $(0,0)$ but when $k = 2h^2$, we obtain
$$h = f_1(0,0)h + f_2(0,0)k + \varepsilon(h+k)$$
$$= \varepsilon(h + 2h^2)$$
so
$$\varepsilon = h/(h+2h^2) \to 1$$
as $h \to 0$. This example also shows that differentiability is not a necessary condition for formula (29) in the text to be valid.

12.4.16 Assume that all but one of the partials are continuous at $\mathbf{x} \in \mathbb{R}^n$ and the remaining partial is finite at $\mathbf{x}$.

12.5.8 For (a): Would you expect
$$\frac{\partial A}{\partial x}$$
to be independent of whether it is y or P that is held fixed?

12.5.15 Use the result of Exercise 12.5.14.

12.5.16 The equations that transform (z,r) to (ρ,ϕ) are the same as those from (x,y) to (r,θ). The result is
$$\frac{\partial^2 u}{\partial \rho^2} + \frac{1}{\rho}\frac{\partial^2 u}{\partial \phi^2}$$
$$+ \frac{1}{\rho^2 \sin^2\phi}\frac{\partial^2 u}{\partial \theta^2} + \frac{2}{\rho}\frac{\partial u}{\partial \rho} + \frac{\cos\phi}{\rho^2}\frac{\partial u}{\partial \phi}.$$

12.8.2 Calculate
$$\mathbf{A}(\mathbf{x}+\mathbf{h}) - \mathbf{A}(\mathbf{x}) - \mathbf{A}(\mathbf{h}).$$

12.8.10 Make explicit the domains of the functions, the differentiability assumed, and the concluding chain.

12.8.13 Solve the equation
$$\begin{pmatrix} \frac{\partial f}{\partial u} & \frac{\partial f}{\partial v} \\ \frac{\partial g}{\partial u} & \frac{\partial g}{\partial v} \end{pmatrix}\begin{pmatrix} a & b \\ c & d \end{pmatrix}$$
$$= \begin{pmatrix} 1 & 0 \\ 0 & 1 \end{pmatrix} \quad \text{for } a, b, c \text{ and } d.$$

13.1.2 The vector space axioms are taught in any elementary linear algebra course or text. We have used them in Theorem 11.1 to describe the vector space structure of $\mathbb{R}^n$. Note that, in this example, (d) is a special case of (b), but, in general, it is not true that scalar multiplication is a special case of multiplication of elements of the space. (Here, scalars can be viewed as constant functions.)

13.1.4 For example, define $f \preceq g$ if $g - f$ is both nonnegative and differentiable on $[0,1]$.

13.2.1 The first two are not metrics. The metrics in (d) and (e) are interesting since they are bounded, that is, $d(x,y) \le 1$ for all x, y, but nonetheless these metrics are closely related to the usual metric: Open sets, closed set, convergent sequences etc. are the same under these metrics. To check the triangle inequality in (e) check first that the function $f(t) = t/(1+t)$ is increasing on $[0,\infty)$. Then, use the fact that
$$f(|x-y|) \le f(|x-y| + |y-z|).$$
Finally, note that (f) is the discrete metric.

13.2.7 This idea was already used for the metric $d(x,y) = |x-y|$ in $X = \mathbb{R}$. See Exercise 13.2.1.

13.3.5 Check that all the properties of a metric hold except for condition (2).

13.3.6 While there is a subset relation here, there is no subspace relation because the metrics in the three spaces do not agree.

13.3.7 For this exercise we have merely to determine which of the sets given is a *subset* of $M(\mathbb{R})$. (Which of these classes might contain an unbounded function?)

13.4.1 Note that $d(x_n, x) < 1$ if and only if $x_n = x$.

13.4.2 Use the same ideas and methods as we used in Example 13.16.

13.4.3 The notation

$$x^{(n)} = (x_1^{(n)}, x_2^{(n)}, x_3^{(n)}, \dots)$$

might be a bit confusing at first. We have here a sequence of points $x^{(1)}$, $x^{(2)}$, $x^{(3)}$, $x^{(4)}$, ... in the space ℓ_2, each of which is, in turn, a sequence of real numbers. Only one direction in the exercise is true. Show that the condition is necessary but not sufficient. Find a counterexample with

$$x^{(n)} = (x_1^{(n)}, x_2^{(n)}, x_3^{(n)}, \dots)$$

converging coordinate-wise to zero but not convergent in the metric.

13.4.5 This is false. Construct a sequence of continuous functions with $d(f_n, 0) \to 0$ but not converging pointwise to the zero function.

13.4.6 The proofs of (a) and (b) can be obtained by copying fairly closely the corresponding proofs for real sequences. There is also an advantage here for us that we can use the theory of real sequences. Thus to prove (a) you can derive it directly from the inequality

$$d(x, y) \leq d(x, x_n) + d(y, x_n).$$

For further generalizations be careful: A general metric space has no addition, multiplication, or order. Even if a metric space *does* have some other structure, such as addition, it would still be necessary to assume some special properties of the metric on the space in order to obtain properties such as that

$$\lim_{n\to\infty} x_n + y_n = \lim_{n\to\infty} x_n + \lim_{n\to\infty} y_n.$$

13.4.9 Show that $t/(1+t) \leq t$ for all $t \geq 0$. From this deduce that convergence in d implies convergence in e_2. Show that

$$t/4 \leq t/(1+t)$$

for

$$0 \leq t \leq 1.$$

From this deduce that convergence in e_2 implies convergence in d.

13.4.10 Use the metrics from Exercises 13.2.7 and 13.2.8.

13.4.12 If (a) holds, then $x_n \to x$ in (X, d_2) implies that $x_n \to x$ in (X, d_1) but not necessarily conversely. A similar statement is true for (b). If (c) holds, then $x_n \to x$ in (X, d_1) if and only if $x_n \to x$ in (X, d_2).

13.4.14 For (a) part your answer will depend on $[a, b]$. For example, if $[a, b] = [0, 1/2]$, then the answer is that it does converge, while if $[a, b] = [0, 1]$, then it does not.

13.4.17 It is true that a sequence $\{p_n\}$ of polynomials converges in $\mathcal{P}[a, b]$ if and only if it converges uniformly *to a polynomial.*

13.4.18 For the converse show that $x_n = 1/n$ is a Cauchy sequence in the metric space $X = (0, 1)$ considered as a subspace of $\mathbb{R}$ with the usual metric.

13.5.4 For the counterexample, take the space as $\mathbb{R}^2$ and E and F are the graphs of $f(x) = e^x$ and $g(x) = -e^x$, respectively.

13.5.7 Consider the discrete space. (For unusual examples this is always worth a try first.)

13.5.12 This will require the application of some of the theorems of Chapter 9 since convergence in this space is exactly uniform convergence.

13.5.15 While it might seem plausible that G should be open, it is not. Show that G contains no ball $B(x_0, r)$ for

$$x_0 = (0, 0, 0, \dots)$$

and any $r > 0$.

13.5.19 An account can be found in Theorem 11.19.

13.5.20 A proof of the Heine-Borel theorem in $\mathbb{R}^2$ can be based on the Bolzano-Weierstrass theorem (Exercise 13.5.19). First show that from an open cover of any closed and bounded subset $K \subset \mathbb{R}^2$ you can extract a sequence $G_1, G_2, G_3, \ldots$ of open sets covering K. Then show that for some N the sets $G_1, G_2, G_3, \ldots G_N$ must cover K otherwise there is a sequence of points in K that has no convergent subsequence.

13.6.1 Show for the first question that all functions are continuous and for the other that only constant functions are continuous.

13.6.7 Calculate $f(x, x^2)$ and use this to find a sequence of points $(u_n, v_n) \to 0$ so that $f(u_n, v_n)$ does not converge to 0.

13.6.8 Write out the ε-δ statement that expresses the statement that "for each $x_2 \in \mathbb{R}$ the function $x \to f(x, x_2)$ is continuous." What would be the uniform version of that? For further discussion of these ideas, see Exercise 11.7.13.

13.6.11 For (g) use the function

$$\operatorname{dist}(x,E)[\operatorname{dist}(x,E) + \operatorname{dist}(x,F)]^{-1}.$$

Deduce (h) from (g). [For (h) do not be tempted to think that E and F must be a positive distance apart.]

13.6.12 Continuous curves don't always look like ones we've seen. In 1890 the Italian mathematician Giuseppe Peano (1858–1932) gave an example of a continuous curve that fills the unit square $[0,1] \times [0,1]$. Hausdorff, in 1914, stated that Peano's result was "one of the most remarkable results of set theory." Note that a continuous curve has to be continuous as a function from $[0,1]$ to $\mathbb{R}^2$ but does not have to be one-to-one (i.e., it can cross itself). It can be shown that a one-to-one continuous curve could not fill the unit square.

13.6.15 You can assume and use Bessel's inequality (given in the example). F. W. Bessel (1784–1846) is credited with this elementary and easy to prove inequality. His name is more famously attached to a special class of functions that have become an indispensable tool in applied mathematics, physics, and engineering. The interest in these functions arose in the treatment of the problem of the perturbation in the planetary system.

13.6.18 A closed and bounded subset $K \subset \mathbb{R}^2$ has the same properties that a closed and bounded set of real numbers has, namely the compactness properties of Section 4.5. (See Exercise 13.5.19 or Exercise 13.5.20.) Thus you can use similar proofs to those of Exercise 5.6.12 (Bolzano-Weierstrass property) or Exercise 5.6.13 (Heine-Borel property) to prove uniform continuity of continuous functions on closed and bounded sets. For (c) and (d) also imitate the proof for functions of one variable. (See also the discussion in Sections 11.8 and 11.9.)

13.6.26 See Example 13.36 for a homeomorphism of $\mathbb{R}$ and $(-1,1)$. A similar idea will show that $\mathbb{R}$ and $(0,1)$ are also homeomorphic. The Bolzano-Weierstrass theorem will help show that $\mathbb{R}$ and $[0,1]$ are not homeomorphic: Any sequence in $[0,1]$ would have to have a convergent subsequence.

13.6.27 If $h : [0,\infty) \to \mathbb{R}$ is a homeomorphism, show that h is either increasing or decreasing.

13.6.28 Compare with Exercise 13.2.7.

13.6.29 The answer is no for the converse. Show that the subspace

$$X = \{1, 1/2, 1/3, 1/4, \ldots\}$$

of $\mathbb{R}$ with the usual metric is topologically equivalent to a discrete space but that

$$\inf\{|x-y| : x, y \in X,\ x \neq y\} = 0.$$

13.6.37 See also the discussion of connected sets in $\mathbb{R}^n$ given in Section 11.10.

13.6.38 Use Exercise 13.6.11(d) to obtain, if X contains at least two points, a continuous real-valued function on X that is not constant. Explain why the range of this function cannot be finite or countable.

13.6.40 Even for subspaces of $\mathbb{R}^2$ the converse fails. Use the set

$$\{(x, \sin(1/x) : 0 < x < 1\}$$
$$\cup\{(0, x) : -1 \leq x \leq 1\}.$$

13.6.41 What set in $\mathbb{R}$ would map onto the x-axis in $\mathbb{R}^2$?

13.6.42 Consider the function

$$h : K \times K \to K$$

defined by

$$h(a, b) = .a_1b_1a_2b_2\ldots,$$

where $a = .a_1a_2\ldots$, $b = .b_1b_2\ldots$ are appropriate base 3 representations of $a, b \in K$.

13.6.43 For (c) let

$$h(\{a_1, a_2, a_3, \ldots\}) = .b_1b_2b_3\ldots,$$

where $b_k = 2a_k$.

13.6.44 For (b) contrast the case that the limit point is unilateral (or bilateral) in both X and Y with the case that one is a unilateral limit point and the other is a bilateral point.

13.6.47 Give an example of two subsets A and B of $\mathbb{R}$, each of which is isometric to a subset of the other but that are not themselves isometric. Use

$$A = \{2, 3, 4, \ldots\}$$

and

$$B = \{0\} \cup \{2, 3, 4, \ldots\}.$$

13.6.52 Note that $(0, 1) \subset X$ would map onto a set of diameter 1.

13.7.3 The family of finite sets of rational numbers in $[0, 1]$ forms a countable dense subset of $\mathcal{K}$.

13.7.4 By a polygonal function on $[a, b]$ we mean a continuous, piecewise linear function, the corners (the points at which the right- and left-hand derivatives are different) are called the vertices. Here both coordinates of any vertex are assumed to be rational numbers. Use uniform continuity to show that each function in $\mathcal{C}[a, b]$ can be approximated by such a function arbitrarily closely. Now show that the set of such functions is countable.

13.7.5 Here is a warning: The countable dense subset known to exist for the space might not be a subset of the subspace.

13.7.8 Several of the spaces we have considered are not separable and can be used for (b).

13.7.11 We are trying to find, if possible, an embedding of (X, d) into these spaces. You should be able to use the separability of $\mathcal{C}[0, 1]$ to determine whether this is possible. For the other two Examples 13.54 and 13.55 should assist in the construction of the embedding.

13.8.1 This follows easily from the triangle inequality.

13.8.2 The methods used to prove that convergent sequences of real numbers are bounded in Theorem 2.11 can be imitated here.

13.8.3 If $\{x_n\}$ is Cauchy, then show it is true that

$$\lim_{n\to\infty} d(x_n, x_{n+1}) = 0.$$

The converse is false. For example, the sequence $x_n = \sqrt{n}$ in $\mathbb{R}$ has the property that

$$\lim_{n\to\infty} d(x_n, x_{n+1}) = 0$$

but it is not Cauchy (nor even bounded).

13.8.4 Using the triangle inequality, show that the sequence $\{d(x_n, y_n)\}$ is a Cauchy sequence of real numbers.

13.8.5 For each positive $\varepsilon_k = 2^{-k}$, show that there is an integer n_k with

$$d(x_m, x_n) < \varepsilon_k$$

for m, $n \geq n_k$.

13.8.6 The methods used to prove this assertion for real numbers in Theorem 2.41 can be imitated here.

13.8.7 In one direction check that a sequence with the property that
$$\sum_{k=1}^{\infty} d(x_k, x_{k+1})$$
converges is Cauchy. In the other direction note that a Cauchy sequence need not have this property, but it must have a subsequence with that property. (To construct the subsequence Exercise 13.8.5 should help; then use Exercise 13.8.6.)

13.8.8 In one direction use the sequence of closed balls to select a Cauchy sequence. In the other direction, if $\{x_n\}$ is Cauchy you should be able to construct an appropriate sequence of closed balls with the help of Exercise 13.8.5.

13.8.13 (a) Show first that if $\{H_n\}$ is a decreasing sequence of nonempty closed sets in $[0,1]$ and $H = \bigcap_1^\infty H_n$, then $H_n \to H$ in $\mathcal{K}$. (b) Show that if $\{A_n\}$ is a Cauchy sequence in $\mathcal{K}$ and H_n denotes the closure of the set
$$\bigcup_{k=n}^{\infty} A_k,$$
then $\{H_n\}$ is a decreasing sequence of closed sets,
$$H = \bigcap_1^\infty H_n$$
is a nonempty closed set, and $H_n \to H$. (c) Finally, show $A_n \to H$.

13.8.16 It is important to realize that completeness is not a topological property. Consider $X = \mathbb{R}$ and $Y = (-1,1)$ and review Example 13.36.

13.8.18 Use Theorem 13.66.

13.9.8 A is not a contraction (in contrast to Example 13.77), but A^2 is. The unique fixed point is easy to find.

13.9.9 Show that
$$d(a,b) < \frac{\varepsilon}{1-\alpha},$$
where α has the property for all $x,\ y \in X$ either
$$d(A(x), A(y)) \le \alpha d(x,y)$$
or
$$d(B(x), B(y)) \le \alpha d(x,y).$$
Use the technique in the proof of Theorem 13.75 and observe that
$$d(a, B(a)) = d(A(a), B(a)) < \varepsilon.$$

13.9.13 Compare this with Exercises 13.9.11 and 13.9.12, where, contrary to this exercise, the mappings A_n are assumed to be contractions. Check the fact that
$$d(a_n, A(a_n) = d(A_n(a_n), A(a_n)) \to 0.$$
From this obtain that
$$|d(a_m, a_n) - d(A(a_m), A(a_n))| \to 0.$$
The fact that A is a contraction can be used to show that $\{a_n\}$ is Cauchy. Now show that the limit of that sequence is a fixed point of A and remember that A has only one fixed point.

13.10.1 The condition that is both necessary and sufficient is that $|1-a| < 1$. Display the iterates graphically using the scheme of Figure 13.5.

13.10.2 The condition depends on the metric chosen for $\mathbb{R}^2$. For this exercise you may use the usual euclidean metric.

13.10.3 Just show that F is a contraction on $[a,b]$ and that a fixed point of F is a zero of f. Theorem 13.75 implies that the sequence of iterates starting from any point of $[a,b]$ must converge to that fixed point.

13.11.1 Note that it is the d_1 metric here.

13.11.3 Let
$$(A(g))(x) = g(x) - cF(x, g(x)),$$
$c \in \mathbb{R}$, $c \ne 0$. Note that a fixed point of A solves the problem. Find c so that A becomes a contraction map. To do this, apply the mean value theorem to the expression
$$|(A(g))(x) - (A(f))(x)|$$
$$= |g(x) - f(x) - c[F(x,g(x)) - F(x,f(x))]|$$
and simplify the resulting expression when $c = 1/\beta$.

13.12.2 Show that if two spaces (X,d) and (Y,e) are topologically equivalent, then X is compact if and only if Y is compact.

13.12.4 Show that a set in this space is compact if and only if it is finite.

13.12.7 See Exercise 13.4.4.

13.12.9 For a counterexample when K is closed, but not compact you can use the space

$$X = \{-1, 1, 1/2, 1/3, 1/4, \dots\}.$$

with the usual real metric. For a counterexample when K is complete, but not compact, try using the space ℓ_∞.

13.12.21 Note that the hypothesis that f is a weak contraction is weaker than the hypothesis on the function in the Banach fixed point theorem. What other change in hypotheses are there? For (c) consider the function $h(x) = d(x, f(x))$ which is a real-valued continuous function on f. Explain why h has a minimum $z \in X$ and why $h(z) = 0$. That will supply you with a fixed point.

13.12.30 Let x_1, x_2, x_3, ... be a sequence that is dense in the metric space. Show that the set of open balls $\{B(x_i, 1/n)\}$ is countable. We can then consider the balls arranged into a sequence B_1, B_2, B_3, ... Now if $\mathcal{G}$ is a family of open sets covering the space, select out a sequence of open sets G_1, G_2, G_3, ... by choosing (when possible) any $G_k \in \mathcal{G}$ for which $B_k \subset G_k$. (Skip any k if there is no choice.) Why is this a countable subcover? If x is a point in the space, it belongs to some set $G \in \mathcal{G}$; explain why this means there is some stage k at which a set G_k will have been chosen with $x \in G_k$.

13.12.33 Kasahara showed in 1956 that a metric space has this "Lebesgue property" if and only if it is the union of a compact set and a discrete set.

13.12.34 Consider the family

$$\mathcal{G} = \{X \setminus F : F \in \mathcal{F}\}.$$

13.12.35 It takes a careful reading of this exercise to see that the stated condition is not merely the definition. S is totally bounded if you can always find an ε-net

$$\{x_1, x_2, \dots, x_n\} \subset X.$$

Here you must find an ε-net

$$\{x_1, x_2, \dots, x_n\} \subset S.$$

Find instead an $\varepsilon/2$-net contained in X and choose appropriate points in S to construct an ε-net

$$\{x_1, x_2, \dots, x_n\} \subset S.$$

13.12.41 Show this directly from the definitions, not using any theorem of this section.

13.12.43 Be careful to use the simplest notation. At each stage write $\{x_{nk}\}$ for the nth subsequence. Arrange that $\{x_{nk}\}$ is a subsequence of $\{x_{(n-1)k}\}$, that is in a ball of radius $1/n$. For the final subsequence that is itself a subsequence of every one of these, take

$$\{x_{11}, x_{22}, x_{33}, \dots\},$$

the diagonal sequence, and explain why this must be Cauchy.

13.12.45 The proof of Theorem 13.90 contains the argument needed.

13.12.46 Show that (a) and (c) are equivalent and that (b) and (d) are equivalent. What more can you say?

13.12.49 No.

13.12.55 Use the Arzelà-Ascoli theorem.

13.12.56 Start with an enumeration

$$\{r_1, r_2, r_3 \dots\}$$

of the rationals in $[a, b]$. Choose a subsequence $\{f_{1k}\}$ of $\{f_n\}$ so that $\{f_{1k}(r_1)\}$ converges (why is this possible?). Choose a subsequence $\{f_{2k}\}$ of $\{f_{1k}\}$ so that $\{f_{2k}(r_2)\}$ converges. Continue in this fashion and at the end select the diagonal sequence $\{f_{11}, f_{22}, f_{33}, \dots\}$. Verify that this works.

For part(b) write $\{g_k\}$ for the subsequence and let $\varepsilon > 0$ and choose δ according to the equicontinuity hypothesis. Choose M so that every point in $[a, b]$ is within δ of one of the points $\{r_1, r_2, r_3, \dots, r_M\}$. Choose N so that $|g_i(x) - g_j(x)| < \varepsilon/3$ if i, $j \geq N$ and $x \in \{r_1, r_2, r_3, \dots, r_M\}$. Now show that $\{g_k\}$ is uniformly Cauchy.

13.12.58 Just imitate the proof of the Arzelà-Ascoli theorem being careful not to use any special properties of $[a,b]$, which is now replaced by a compact metric space X.

13.12.59 Two solutions can be easily guessed. Show that

$$f(x,y) = 3y^{2/3}$$

does not satisfy the hypotheses of Theorem 13.86.

13.13.4 No. Give a counterexample.

13.13.5 The property needed is that the space is dense in itself. See Exercise 13.5.21.

13.14.4 Let $\mathcal{I}$ denote the family of open intervals in $[a,b]$ having rational endpoints. For I, J in $\mathcal{I}$, with $I \cap J = \emptyset$, let

$$A_{I,J} = \{f \in X : \exists\, x_1 \in I \text{ and } x_2 \in J \text{ such that } f(x_1) = f(x_2)\}.$$

Show that $A_{I,J}$ is nowhere dense.

13.15.1 If E is not a countable union of members of $\mathcal{A}$, then E is closed.

13.15.2 Hint for (b): Define

$$A : C[a,b] \to C[a,b]$$

in an appropriate manner and obtain $n \in \mathbb{N}$, $M > 0$ such that

$$|\lambda|^n M^n \frac{(b-a)^n}{n!} < 1.$$

Then apply part (a).

13.15.3 If $[0,1] = \bigcup E_k$, E_k closed and pairwise disjoint, then one of these sets contains an interval. Obtain a countable collection of closed intervals, each contained in one of the sets E_n, whose union is dense in $[0,1]$. Remove the interiors of these intervals and show that what remains is a Cantor set H. Apply the Baire category theorem to the set H and obtain a contradiction.

13.15.4 This problem is known as the Kuratowski fourteen set problem (which should give a hint as to the correct answer) as it was originally posed and solved by K. Kuratowski (1896-1980). See the article J. H. Fife, The Kuratowski Closure-Complement Problem, *Mathematics Magazine* **64**, No. 3, 180–182 (1991). Fife presents an easily readable proof, similar to, but not identical with Kuratowski's original 1922 version.

13.15.5 See the hint for Problem 13.15.3.

13.15.6 Let $X = \mathbb{N}$ with the discrete metric. Show that $f(n) = n+1$ is an isometry, but is not onto. If X is compact let x be any point in X and define the sequence $x_1 = f(x)$, $x_2 = f(x_1)$, ... By compactness, there is a convergent subsequence $\{x_{n_k}\}$. But note that

$$d(x, x_n) = d(x_m, x_{n+m})$$

for all n and m. Use this to show that

$$d(x, f(X)) = 0$$

and deduce that $f(X) = X$.

13.15.9 Since compactness is a topological property and every compact space is complete, one direction is easy. In the other direction, if X is not compact we need to construct an equivalent metric ρ on X so that (X,ρ) is not complete. (Assume that d is bounded, for if not there is an equivalent metric that is.) Use the fact that if X is not compact there must be a sequence

$$C_1 \supset C_2 \supset C_3 \supset \dots$$

of closed sets with an empty intersection. Define

$$\rho_i(x,y) = |d(x,C_i) - d(y,C_i)| + d(x,y)\min\{d(x,C_i), d(y,C_i)\}$$

and finally

$$\rho(x,y) = \sum_{i=1}^{\infty} 2^{-i}\rho_i(x,y).$$

It now remains to check that (i) d and ρ are equivalent, (ii) the sequence $\{C_i\}$ of closed sets in (X,ρ) is a descending sequence of closed sets with diameters approaching zero, and (iii) the fact that

$$\bigcap_{i=1}^{\infty} C_i = \emptyset$$

violates the Cantor intersection property so that the space cannot be complete.

13.15.11 For $h \in M[a,b]$, let
$H = \{h - g \in M[a,b] : g \text{ is one-to-one}\}$.
Show that H is residual in $M[a,b]$ and thus contains a one-to-one function f. Write $h = f + (h - f)$.

13.15.12 For (f): Show that a nonconvergent Cauchy sequence in (X, d) will not be a Cauchy sequence in (X, e).

13.15.14 Let

$$g(x) = \frac{\text{dist}(x, A)}{\text{dist}(x, A) + \text{dist}(x, B)}.$$

A.2.4 For (c) and (d): All numbers do not have a unique decimal expansion; for example, $1/2$ can be written as $0.5000000\ldots$ or as $0.499999999\ldots$. For (e): take the domain as the set $\mathbb{N}$. Are you troubled (some people might be) by the fact that nobody knows how to determine if x is a prime number when x is very large?

A.2.13 As a project, research the topic of Russell's paradox [named after Bertrand Russell (1872-1969], who discovered this in the early days of set theory and caused a crisis thereby].

A.5.1 Suppose not. Then $\sqrt{2}$ is rational. This means $\sqrt{2} = m/n$ where m and n are not both even. Square both sides to obtain $2n^2 = m^2$. Continue arguing until you can show that both m and n are even. That is your contradiction and the proof is complete.

A.5.2 Suppose not. Then it is possible to list all the primes

$$2, 3, 5, 7, 11, 13, \ldots P$$

where P is the last of the primes. Consider the number

$$1 + (2 \times 3 \times 5 \times 7 \times 11 \times \ldots P).$$

From this obtain your contradiction and the proof is complete. (To be completely accurate here we need to know the prime factorization theorem: Every number can be written as a product of primes.) This is a famous proof known in ancient Greece.

A.6.1 The contrapositive statement reads "if $x+r$ is not irrational for all rational numbers r, then x is not irrational." Translate this to "if $x + r$ is rational for some rational number r, then x is rational." Now this statement is easy enough to prove.

A.8.1 Check for $n = 1$. Assume that

$$1^2 + 2^2 + 3^2 + \ldots n^2 = \frac{n(n+1)(2n+1)}{6}$$

is true for some fixed value of n. Using this assumption (called the induction hypothesis in this kind of proof), try to find an expression for

$$1^2 + 2^2 + 3^2 + \ldots n^2 + (n+1)^2.$$

It should turn out to be exactly the correct formula for the sum of the first $n+1$ squares. Then claim the formula is now proved for all n by induction.

A.8.9 The induction step requires us to show that if the statement for n is true, then so is the statement for $n+1$. This step must be true if $n = 1$ and if $n = 2$ and if $n = 3$ $\ldots$, in short, for all n. Check the induction step for $n = 3$ and you will find that it does work; there is no flaw. Does it work for all n?

SUBJECT INDEX